ELEMENTARY
STRUCTURAL
ANALYSIS

**McGRAW-HILL
BOOK COMPANY**

New York
St. Louis
San Francisco
Auckland
Düsseldorf
Johannesburg
Kuala Lumpur
London
Mexico
Montreal
New Delhi
Panama
Paris
São Paulo
Singapore
Sydney
Tokyo
Toronto

CHARLES HEAD NORRIS, SC.D.

Dean and Professor Emeritus
College of Engineering
University of Washington
Honorary Member, American Society of Civil Engineers

JOHN BENSON WILBUR, SC.D.

Professor Emeritus of Engineering
Massachusetts Institute of Technology
Fellow, American Society of Civil Engineers

ŞENOL UTKU, SC.D.

Professor of Civil Engineering
Duke University
Member, American Society of Civil Engineers

Elementary Structural Analysis

THIRD EDITION

This book was set in Times New Roman.
The editors were B. J. Clark and Madelaine Eichberg;
the cover was designed by Anne Canevari Green;
the production supervisor was Leroy A. Young.
Fairfield Graphics was printer and binder.

Library of Congress Cataloging in Publication Data

Norris, Charles Head, date
    Elementary structural analysis.

    Previous editions are entered under J. B. Wilbur.
    Bibliography: p.
    1.   Structures, Theory of. I. Wilbur, John
Benson, date joint author. II. Utku, Senol,
joint author. III. Title. IV. Title: Structural
analysis.
TA645.W53   1976   624′.171   76-2407
ISBN 0-07-047256-4

**ELEMENTARY
STRUCTURAL
ANALYSIS**

7 8 9 0 F G R F G R 8 3 2 1

CONTENTS

PREFACE

Since this book was first published in 1948, there have been many very important developments in all aspects of structural engineering and construction. Improved fabrication and construction techniques have led to more efficient utilization of established structural materials such as steel, concrete, and wood. Much improved steels and specially designed concrete mixes are also now available and widely utilized by engineers and builders. The use of structural aluminum and plastics has expanded greatly, particularly for prefabricated elements used primarily in building construction. Postelastic behavior of structures is recognized, and structural design procedures have been modified to utilize the most useful characteristics of this behavior where appropriate. The use of electronic computers in structural analysis and design calculations has expanded tremendously in the last twenty-five years—most particularly during the last fifteen years of this period.

In view of the much expanded use of computers since 1960, when the second edition of this book was published, it is imperative in this new third edition to devote much more attention to the systematic arrangement of structural computations. Such well-organized computations are essential for the effective communications needed with a computer if we are to realize its full computational potentialities. In order to provide a treatment of systematic structural analysis having the breadth and depth desired, we have added a new and separate portion, Part II, consisting of five new chapters. To prepare Part II, the two senior authors enlisted, as a new coauthor, one of their finest former doctoral students, Professor Şenol Utku of Duke University.

In order to gain the space needed for this new addition to our third edition, we have deleted five chapters. One of these on graphic statics represents an area still having fine educational values for a student but covering computational techniques no longer widely used in structural engineering practice. Three chapters that previously provided an introduction to the model analysis of structures have been deleted because of the fine reservoir of technical literature developed in the last four decades to provide excellent coverage of the field of experimental stress analysis. The fifth chapter deleted was devoted to an introduction to matrix analysis of structures, but, in fact, this chapter has not been deleted but expanded into the new Part II. The material from the remaining fifteen chapters of the second edition has been rewritten and reorganized and is now presented in the third edition as a Part I of eleven chapters on basic structural theory and a Part III of two chapters providing an introduction to advanced structural analysis.

This new edition, like its predecessors, emphasizes fundamentals rather than professional practice and attempts to correlate the various procedures of structural analysis with the principles of engineering mechanics on which they are based. The scope of this book is restricted to the field of structural analysis and must be supplemented and complemented by books on structural design in order to obtain a well-balanced view of the entire field of structural engineering. Herein it is emphasized and reemphasized that *the function of a structural engineer is to design —not to analyze* and that analysis is a means to an end rather than the end itself. Today there are a number of truly excellent textbooks available covering various aspects of structural design, to which the reader is referred.

The authors wish to continue to acknowledge with appreciation the contributions of those who assisted in the preparation of the two previous editions: Prof. Donald R. F. Harleman, Prof. Myle J. Holley, Jr., the late Mrs. Grace M. Powers, and Mr. Harold D. Smith. The first listed author will be forever grateful and indebted to his most valued colleague, his wife, Martha M. Norris, who has provided all kinds of tangible and intangible assistance and loyal support in nursing the preparation of three editions of this book over a period of thirty years.

Both senior authors are also deeply grateful to all those responsible for their education in structural engineering, but most particularly to the late Prof. Charles Milton Spofford and to the late Prof. Charles Church More. They likewise acknowledge with deep appreciation the generous consideration, friendship, support, and understanding that they have received for many, many years from their M.I.T. structural engineering colleagues: Prof. John M. Biggs, Prof. Robert J. Hansen, Prof. Myle J. Holley, Jr., Prof. Eugene Mirabelli, and the late Profs. Walter M. Fife and John D. Mitsch.

CHARLES HEAD NORRIS
JOHN BENSON WILBUR
ŞENOL UTKU

This book is focused sharply on an elementary treatment of the analysis of structural behavior, with illustrations drawn almost exclusively from the field of civil engineering structures. Admittedly analysis is a means to an end—not the end itself—since *the primary objective of the structural engineer is to design, not to analyze.* Concentration on the study of analysis to the exclusion of the larger and more important problem of design does, however, allow the structural engineering student to develop his capacity for logical, rational, analytical thinking and to cultivate his appreciation, feeling, and intuition for structural behavior. Of course, he will never benefit fully from this educational experience in analytical thinking unless he correlates his analytical studies with a parallel study of structural design. By studying design, the student develops his judgment, his perception, his imagination and his creativity—in short his ability to synthesize.

The subject matter in this textbook is organized into three major subdivisions:

Part I: Basic Structural Theory contains eleven chapters devoted to the fundamental concepts and the basic definitions and analytical techniques that provide the foundation for the entire field of structural analysis.

Part II: Introduction to Systematic Structural Analysis consists of five chapters in which automatic computational procedures are developed for identifying and organizing input information for processing by digital computers through the various stages of the solution of some of the most basic problems encountered in the analysis of civil engineering structures.

Part III: Introduction to Advanced Structural Analysis includes two chapters; one contains a brief introduction to the principles of advanced structural mechanics, and the other is devoted to an introduction to methods of analyzing the plastic behavior of structures.

The presentations in Parts II and III build on the fundamentals presented in Part I. Considerations in Part I are based almost entirely on fundamental concepts previously introduced to a student in physics, mathematics, engineering mechanics, materials science, and strength of materials. It is presumed that a student has acquired previously a good sound understanding of mechanics (statics and dynamics) and that he is familiar with the differential and integral calculus and the basics of dealing with both ordinary and partial differential equations. Although vector notation is seldom used in Parts I and III, it is used extensively in Part II. It is assumed therefore that the reader is acquainted with vector notation and with the algebraic operations of addition and multiplication of vectors. If the reader's knowledge of vector notation is a bit rusty, he should review his old calculus and engineering mechanics textbooks or refer perhaps to one of the fine current mechanics texts that emphasizes both scalar and vector notations. Many readers will also be familiar with the essentials of matrix notation and matrix algebra. These essentials are, however, reviewed and summarized at the beginning of Chap. 12 in Part II.

Chapters 1 to 6 are devoted almost exclusively to a consideration of statically determinate structures. Here fundamental principles of static mechanics are reviewed and reemphasized as they apply to problems of structural analysis. Since this material is in the nature of a restatement and review, the student will progress through this portion quite rapidly, depending of course on how thoroughly he has assimilated these fundamentals in earlier courses.

Chapters 7 to 11 are devoted to the computation of deflections and the stress analysis of statically indeterminate structures. This will be substantially new material to most students, although they may have been introduced to these ideas earlier in mechanics and strength of materials courses. Although this is an elementary treatment, the reader will find it quite rigorous and comprehensive, so that, having mastered this subject matter, he should have no difficulty studying more advanced books on statically indeterminate structural analysis.

In Part II, systematic analysis and automatic computation are introduced in Chap. 12 by the consideration of techniques for describing structural problems to digital computers. In Chap. 13 the procedures for transforming physical data and various structural quantities from one coordinate system to another, or from one form to another, are discussed in considerable detail. In Chap. 14, stiffness and flexibility relations are developed for structural elements of discrete structures, i.e., structures composed of relatively long slender elements connected together at joints located at a certain number of discrete points. And in Chaps. 15

and 16 procedures are developed for the systematic analysis of discrete structures by the displacement method and by the force method, respectively.

Modern structural engineering has become increasingly involved not only with more complex structural forms but also with more refined design of both these complex forms and the standard conventional structural forms. It is therefore important for the structural engineer to become familiar with the analysis of plate and shell structures and with the analysis of buckling, dynamic response, and the stress distribution of various more complex structures and structural elements. Chapter 17 in Part III provides a very brief introduction to the principles of advanced structural mechanics involved in such analyses. That chapter and Chap. 18 are intended to serve as a transition from this elementary book to more advanced books in the areas of structural mechanics, structural analysis, and structural design. It is also hoped that these last two chapters will be useful to students who are proceeding to graduate studies or to those who are proceeding on their own to informal self-study in those advanced areas.

Up to Chap. 18, the emphasis is on the analysis of linear elastic structures. With the increased interest in structural design on the basis of ultimate strength, it becomes necessary to consider postelastic behavior. Chapter 18 is therefore devoted to the plastic behavior of structures. This treatment covers only the analysis of this behavior and not the methods and principles for the plastic design of structures.

Specific suggestions regarding certain features incorporated in this new edition may be helpful to the student reader:

1. The foreword which precedes the detailed technical presentation is designed to assist the reader in establishing a broad perspective of the entire field of structural engineering before he focuses his attention on the highly analytical subfield designated as *structural analysis. Please read this foreword carefully before starting Chap. 1.* Since it is expected that your appreciation of the foreword will grow with your mastery of structural analysis, *please reread the foreword as you develop your capability for analyzing and understanding structural behavior.*

2. In many of the chapters of Part I, there are numerous illustrative examples. These are intended to assist the reader to develop a full understanding of the application of the structural theories involved and also to assist him to develop good habits for organizing and carrying out his calculations. In addition, however, each of these examples has been carefully chosen to supplement and extend the ideas and concepts presented in the textual portions. To emphasize the supplementary ideas involved in these examples, discussions are appended after the calculations for most of these illustrations. *Please consider carefully the discussions appended to these illustrative examples.*

3. In Part II, the problems appended to the end of each section are designed both to serve some of the functions of illustrative examples and also to allow a student to test his understanding of the subject matter he has just studied. *Please do these problems so that you broaden and deepen your understanding of the textual material.*

4. Answers are supplied for approximately half the problems (the problem numbers of such problems are preceded by an asterisk). These answers are intended to help make your studying more efficient by enabling you to identify quickly whether you have made a careless computational error or perhaps a fundamental error in the solution of the problem. *Please do not use these answers as a crutch.* If you do so and do not start developing your capability to check yourself, you will not develop into a responsible professional engineer as soon as you should.

A good structural engineer continually makes approximate calculations to check himself. *Get in the habit of making an estimate of the answer to a problem before you start your detailed calculations.* You will find such mental exercise invaluable in developing your judgment and your intuition for structural behavior.

CHARLES HEAD NORRIS

JOHN BENSON WILBUR

SENOL UTKU

LIST OF SYMBOLS,
NOTATION, AND MEASURING UNITS

Symbols

Symbols are defined where they appear. The principal symbols are listed here for convenience. Note that some symbols have several definitions depending on the context in which they are being used.

A	cross-sectional area of member
C	carry-over factor in moment distribution method
D	flexural rigidity of plates and shells, $Eh^3/12(1 - v^2)$
DF	distribution factor in moment-distribution method
E	Young's modulus of elasticity (in tension or compression)
E_t	tangent modulus of elasticity
F	force in general; axial force or axial resisting force acting on a cross section of a member
$\mathscr{F}$	form factor of cross-sectional area, $\mathscr{L}/\mathscr{S}$
FEM	fixed-end moment of member used in slope-deflection method and in moment-distribution method
G	modulus of elasticity in shear

H	horizontal component of force; total shear acting on story of building frame; horizontal component of cable tension
I	moment of inertia of cross-sectional area
J	sidesway factor used in moment-distribution method
K	stiffness factor for flexural member, I/L for prismatic members
K^R	reduced stiffness factor for flexural member
K'	true stiffness factor for flexural member
K'^R	true reduced stiffness factor for flexural member
eff K	effective stiffness factor for flexural member
L	length of member; span length
M	moment of force (or couple); bending moment; resisting moment; end moment of member; bending or twisting stress couples in plate and shell theory
M_{yp}	resisting moment of cross section corresponding to stress condition where bending stress in extreme fiber has just reached the yield stress, $M_{yp} = \mathscr{S}\, \sigma_{yp}$
M_{pl}	resisting moment for a cross section which is fully plasticized in pure bending, $M_{pl} = \mathscr{Z}\, \sigma_{yp}$
N	membrane stress resultants used in plate and shell theory
P	external load
P_{cr}	critical load (for buckling)
P_{yp}	load which produces yield-point stress in most highly stressed fiber
P_{ult}	ultimate load (in plastic-behavior considerations, load at which excessive deflections impend)
π	total potential energy
π^*	total complementary potential energy
Q	imaginary external load used in application of method of virtual work; transverse shear-stress resultant used in plate and shell theory
R	reactive force; resultant
S	shear force (shear) in member; shear resisting force
$\mathscr{S}$	section modulus of cross-sectional area
T	cable tension
$\mathscr{U}$	strain energy
$\mathscr{U}^*$	complementary strain energy
$\mathscr{V}$	potential energy (or potential) of a force
$\mathscr{V}^*$	complementary potential energy (or potential) of a force
V	vertical component of force
W	weight of body
$\mathscr{W}_d$	internal virtual work of deformation
$\mathscr{W}_R$	external virtual work done by reactions
$\mathscr{W}_s$	external virtual work done by external force system (loads and reactions)
$\mathscr{W}_E$	external work done by external force system
W_m	elastic load (or weight) at joint m used in bar-chain method of computing truss deflections

X	redundant force (or couple) used in analysis of statically indeterminate structures
X, Y, Z	coordinate components of a force
X, Y, Z	coordinate components of body force used in theory of elasticity
$\overline{X}, \overline{Y}, \overline{Z}$	coordinate components of surface forces used in theory of elasticity
$\mathscr{Z}$	plastic modulus of cross-sectional area which is fully plasticized in pure bending
a	dimension of length; radius of curvature of shell
b	dimension of length; width of cross section; number of members in a frame or framework
c	dimension of length; distance from centroid to extreme point of cross section
d	dimension of length; depth of cross section; deflection of point on elastic curve from tangent at another point
e	lineal strain of a longitudinal element of member
e_0, e_1, etc.	lineal strain of longitudinal element of centroidal axis of member
g	acceleration of gravity
h	cable sag at mid-span of cable curve; story height of building; thickness of plate or shell
k	stiffness of a linear spring (force required to produce unit change of length of spring)
l	dimension of length
n	number of joints of a frame or framework
p	intensity per unit length of distributed loading on beam
q	intensity per unit area of distributed load applied to a plate or shell
r	number of independent reaction elements of a frame or framework; principal radius of curvature of shell; radius of gyration of cross-sectional area, $(I/A)^{1/2}$
s	dimension of axial length of member
t	uniform change of temperature; time
w	intensity of uniformly distributed loading
u, v, w	displacement components corresponding to x, y, and z coordinate directions, respectively
x, y, z	orthogonal coordinates
α	angle; rotational displacement; central angle subtended by line element of middle surface of shell
α_t	coefficient of thermal expansion
γ	shearing strain
Δ	deflection; increment of a quantity
δ	deflection; unit deflection used in superposition equation method for analyzing statically indeterminate structures
$\boldsymbol{\delta}$	variation symbol
ϵ	lineal strain
$\epsilon_x, \epsilon_y, \epsilon_z$	coordinate components of lineal strain

θ	angle; rotation of tangent elastic curve with respect to original direction of axis of member; sag ratio of cable, h/L
v	Poisson's ratio
σ	normal stress
σ_{yp}	yield point stress in tension
τ	shear stress; rotation of tangent to elastic curve with respect to chord of elastic curve
ϕ	angle; curvature of flexural member
χ	change in curvature or twist of middle surface of shell
ψ	angle; rotation of chord of an elastic curve with respect to original direction of axis of a member
ω	natural frequency of vibration

Notation

In this book both scalar notation and vector notation are used to describe vectorial quantities such as force, moment of a force, small displacements and rotations, and position vectors. With only a few exceptions, only scalar notation is used throughout Parts I and III. In Part II, however, vector notation is used almost exclusively.

Suitable combinations of light- or boldface type of roman or italic letters are used to distinguish scalar from vector quantities and to separate vector quantities from matrices. Thus, the notations used are as follows:

Lightface italics such as P, p designate scalars.
Boldface italics such as $\boldsymbol{P}$, $\boldsymbol{p}$ designate vectors.
Boldface roman letters such as $\mathbf{P}$, $\mathbf{p}$ designate matrices.

The specialized and unique symbols utilized for each of various methods of computation and analysis presented in this book are defined and described where each of these methods is first introduced.

Measuring Units

Both English and metric units are used in this new third edition. In Parts I and III, the English system of measuring units is used exclusively. In Part II, metric units are used wherever units are specified.

In the United States of America, the movement to replace the customary English system with the Système International d'Unites (SI) is under way. Obviously such an extensive conversion will not occur on a particular " SI Day." It will occur over a considerable transition period of time. During this transition, heavy users of measuring units must become more and more facile in converting from one system to the other. Eventually such users must truly become bilingual with these systems. The adequacy of his bilingualism will be apparent to a user when, one day, he finds he is *equally* capable of expressing his intuition for the " physical fitness of things" in one system or the other. Engineering students and teachers must join in working toward such a goal.

Table of Conversions

English unit	Abbreviation		Factor		SI unit	Abbreviation
Pound force	lb	×	4.448 N/lb	≈	newton	N
Slug mass	slug	×	14.59 kg/slug	≈	kilogram	kg
Second	sec	×	1.0	=	second	s
Foot	ft	×	0.3048 m/ft	≈	meter	m
Inch	in	×	0.0254 m/in	≈	meter	m
Statute mile	mi	×	1,609 m/mi	≈	meter	m

**ELEMENTARY
STRUCTURAL
ANALYSIS**

Foreword

Doubtless the planning, design, and construction of structures is essentially as old as man himself since nature provides on this planet an acceptable environment for man, but seldom are the detailed natural forms of this environment exactly suitable to his needs, convenience, and desires. Early man could probably find a natural cave or a hollow tree which would give him partial shelter from the elements, but it most likely would not be situated properly with respect to his source of food and water. So man started to supplement nature to meet his needs—and thereby built his first structure. Was his first structure a crude shelter, or may it have been a bridge made of a log beam or a stone slab or a vine?

Historical records tracing the evolution of structural engineering from its very early beginnings to the present are much poorer than those documenting the spiritual, philosophical, political, economic, and artistic development of man. On the other hand, archeological exploration has given us some limited physical evidence of the structures produced by early engineering even though the concepts and ideas behind their construction are lost and can only be inferred.

Man's basic needs for food and shelter of course are unchanging with time, but the standards which he accepts for the fulfillment of these needs continually change with the capabilities of his civilization. Man's ambitions and desires tend to exceed his achievements. Sometimes the differential between these two is so great that his ambitions drive him to try to accomplish his objective before he can really define what his objective is, or why he desires it, or what techniques or methods he should use to achieve it. The highlights of man's history are the record of his most daring efforts—sometimes the record of a sensational advance or breakthrough, other times the record of a catastrophic failure.

Down through the centuries, master builders and structural engineers have practiced their art assisted by, but not completely dependent upon, the mathematical and scientific knowledge available to them. Mathematics and science when available are convenient tools which the designer uses to determine the detailed

1

proportions and dimensions of his structures.   When theoretical knowledge has not yet been developed to satisfy his needs, the designer must rely on his experience, field and/or laboratory tests, or his intuition and feeling for structural behavior. The accomplishments of a successful builder are *really* bounded by *only* four limitations: the functional requirements and needs of his civilization; the properties of the materials available to him; the constructional techniques which are practical and economical for him to employ; and the creativity and imagination with which he applies his talents, knowledge, experience, and intuition.

The nineteenth and twentieth centuries have produced very impressive mathematical, scientific, material, and mechanical advances in the field of structural engineering.   So impressive is the store of knowledge which has been accumulated during this period that far too many structural engineers have been inclined to limit their designs to those which are encompassed by codes, standard practices, and available mathematical formulations.   In such a professional atmosphere, the truly outstanding efforts of modern master builders such as Maillart, Freyssinet, Torroja, and Nervi seem even more impressive in comparison with the work of their contemporaries.[1]   All these outstanding engineers have possessed creativity, imagination, and intuition to a very unusual degree, but it is noteworthy that all of them have likewise supplemented their mathematical and scientific knowledge generously with the intelligent and extensive use of not only their own and others' experience but also field and laboratory testing.

The careers of these outstanding men reveal the same lesson that can be drawn from such historical records as are available regarding the development of structural engineering and mechanics, namely, that the mathematical and scientific evaluation of a structural behavior or phenomena almost invariably antedates, rather than precedes, the initial application of such ideas in actual design and construction.

Historical Review.   The earliest construction forms of which we have any real knowledge are the lake dwellings at the bottom of some of the Swiss lakes.[2] These were very primitive houses erected on piles and date back to the Paleolithic Age.   There are earlier megalithic remains in Europe and Asia, but these show almost no evidences of a true structural art.   Archeological exploration and restoration have given us considerable evidence of Egyptian, Greek, and Roman con-

---

[1] For descriptions of work of these outstanding engineers see Eduardo Torroja, "Philosophy of Structures" (English version by J. J. Polivka and Milos Polivka), University of California Press, Berkeley, 1958; Eduardo Torroja, "The Structures of Eduardo Torroja," F. W. Dodge Corporation, New York, 1958; Pier Luigi Nervi, "Structures" (translated by Giuseppina and Mario Salvadori), F. W. Dodge Corporation, New York, 1956; Leonard Michaels, "Contemporary Structure in Architecture," Reinhold Publishing Corporation, New York, 1950.

[2] There are a number of excellent books tracing the history and development of structural mechanics and structural engineering, among which are Stephen P. Timoshenko, "History of Strength of Materials," McGraw-Hill Book Company, 1953; H. M. Westergaard, "Theory of Elasticity and Plasticity," Harvard University Press, Cambridge, Mass. (John Wiley & Sons, Inc., New York), 1952; David B. Steinman and Sara Ruth Watson, "Bridges and Their Builders," Dover Publications, Inc., New York, 1957.   See also R. S. Kirby, S. Withington, A. R. Darling, and F. G. Kilgour, "Engineering in History," McGraw-Hill Book Company, New York, 1956.

Figure F.1 Pont du Gard (*Bettman Archive*)

struction as well as that of the numerous ancient civilizations of Asia and Asia Minor.

Perhaps the structural forms used by early builders which are of greatest interest to structural engineers of today are the following: the post and lintel constructions of the classic Greek architecture, dating from about 500 B.C.; the Roman contributions to the development of the arch, the vault, the dome, the wood truss and concrete; the groined vault and vault rib developed by Romanesque builders between A.D. 500 and 1000; and the pointed arches and flying buttresses used by the Gothic builders in the magnificent cathedrals constructed in Europe during the Middle Ages between A.D. 1000 and 1500. Even in comparison with modern structures, a Roman aqueduct or bridge such as the famous Pont du Gard or the dome roof of the Santa Sophia cathedral in Istanbul (built about A.D. 540) is an impressive and inspiring sight. All the more is this so when one realizes the "state of the art" when such structures were designed and built.

No doubt the Egyptians and other ancient builders had formulated empirical rules based on their previous experience to guide them in planning a new structure, but there is no evidence that they had developed even the beginnings of a theory of structural behavior. The Greeks further advanced the art of building, but their contribution to structural theory was limited to that of certain of their philosophers such as Aristotle (384–322 B.C.) and Archimedes (287–212 B.C.), who laid the basis of structural mechanics by formulating certain fundamental principles of statics. The Romans were primarily builders and became very competent in developing and using certain structural forms. However, they apparently had little knowledge of the stress analysis of these forms. For example, they apparently did not know

Figure F.2   Santa Sophia dome   (*Bettman Archive*)

how to establish the proper shape of their arches and usually used semicircular short-span arches.

Most of the knowledge that the Greeks and Romans accumulated concerning structural engineering was lost during the Middle Ages and has been recovered only since the Renaissance. Leonardo da Vinci (1452–1519) was not only the leading artist of his time but was also a great scientist and engineer. From his notebooks, it is clearly evident that he was very perceptive and understood quite well certain fundamentals of the behavior of structural materials. This work of da Vinci appears to be the real beginning of the development of structural theory, but it was not widely publicized by being published in book form. As a result, Galileo (1564–1642) is properly acknowledged to be not only the founder of modern science but also the originator of the mechanics of materials. For, in his last publication, "Two New Sciences," completed in 1636 and published in 1638, he was the first to discuss the strength of certain structural members, including the failure of a cantilever beam.

Galileo investigated the failure of a beam cantilevered from a wall by rupture in the cross section of the beam adjacent to the wall. He considered the material to be rigid and assumed that, at failure, the compression was concentrated at the lower edge of the ruptured section and the tensile resistance was uniformly distributed over the depth of the section. Such a stress distribution would not be valid in the elastic range of a material following Hooke's law but could be true under some assumed law for a plastic range of a material. Whereas Galileo's analysis would not lead to a close absolute value of the strength of a cantilever beam, it is interesting to note that it did give him the correct ratio for the strengths of two beams of similar cross sections. Even more interesting to us today is the fact that Galileo and also da Vinci were interested in predicting *strengths* of members rather than *stresses* and *strains* developed in the members at loads less than the failure load. If all investigators over the last three hundred years had continued to have an interest in strength comparable to their interest in elastic behavior of structures, perhaps today the knowledge of structural mechanics would be better balanced between elastic behavior and strength of structures and structural members.

Subsequent investigators such as Hooke (1635–1703), Mariotte (1620–1684), Jacob Bernoulli (1654–1705), John Bernoulli (1667–1748), Daniel Bernoulli (1700–1782), Euler (1707–1783), Lagrange (1736–1813), Parent (1666–1716), and Coulomb (1736–1806) made important contributions to the development of this new field of knowledge. Of these, Coulomb's contributions were particularly outstanding, and his memoir written in 1773 was particularly significant though it did not attract the attention and exert the influence that it deserved. In this paper was presented for the first time a completely adequate elastic analysis for the flexure of beams. He also noted that, under certain conditions at rupture, the neutral axis could have shifted from its elastic position and thereby demonstrated an appreciation for plastic behavior.

It was not until 1826, when the first printed edition of Navier's (1785–1836) book on strength of materials was published, that some of the misconceptions corrected by Coulomb were widely publicized and the modern theory of mechanics of materials was established. Thus, Coulomb and Navier should be considered as two of the principal founders of this important field which is so basic to structural analysis. Whereas his predecessors in the eighteenth century had measured and calculated ultimate loads for structural members, Navier stated at the very beginning of his presentation that it is very important to know the limit up to which structures behave elastically, and he therefore concentrated on a consideration of the elastic behavior of structures. He suggested that formulas derived for elastic behavior should be applied to existing structures which were behaving satisfactorily in order to establish safe stresses which could be used in design. In retrospect, it is unfortunate that, in this very important and influential treatise, Navier had not also retained a proper interest in postelastic behavior and ultimate strengths of structures.

The remainder of the nineteenth century after the publication of Navier's book could be considered the "golden age" as far as the development of what is now considered to be the classical theory of structures is concerned. Some important contributors in this era were Lamé (1795–1870), Saint-Venant (1797–1886), Clapeyron (1799–1864), Clebsch (1833–1872), Rankine (1820–1872), Airy (1801–1892), Maxwell (1831–1879), Castigliano (1847–1884), Culmann (1821–1881), Mohr (1835–1918), Müller-Breslau (1851–1925), Engesser (1848–1931), Wöhler (1819–1914), A. Föppl (1854–1924), Jourawski (1821–1891), Bauschinger (1833–1893), von Tetmajer (1850–1905), and Jasinsky (1856–1899). These men and their contemporaries did a tremendous job of compiling, developing, and presenting the

Figure F.3   Saguenay River aluminum bridge   (*From the files of Engineering News-Record*)

theories of the mechanics of materials and of structural analysis into substantially the same form as we know them today.   Later in this book, as methods which were developed in this era are presented, the names of these men will be identified with some of their important contributions.   One of the classical methods discussed later, the slope-deflection method, was actually developed in the early *nineteen hundreds*, Bendixen, Maney, and Ostenfeld being the major contributors to this development.

Since 1900, the modern era of structural engineering has been characterized by the following major developments:

1. The publication of a number of excellent books[1] dealing with the theories of elasticity, plasticity, buckling, plates and shells, and vibrations

[1] Among the fundamental theoretical books which have had a very important influence on the development of structural mechanics, analysis, and design in America in the last fifty years are A. E. H. Love, "The Mathematical Theory of Elasticity," Cambridge University Press, London, 1892, 4th ed., 1927; A. Föppl and L. Föppl, "Drang und Zwang," R. Oldenburg-Verlag, Munich, 1920, 2d ed., 1928; H. Lorenz, "Technische Elastizitätslehre," R. Oldenburg-Verlag, Munich, 1913; A. Nádai, "Plasticity," McGraw-Hill Book Company, New York, 1931; S. P. Timoshenko, "Theory of Elasticity," McGraw-Hill Book Company, New York, 1934, 3d ed. with J. N. Goodier, 1970; S. P. Timoshenko, "Theory of Elastic Stability," McGraw-Hill Book Company,

2. The development of improved machines, instrumentation, and techniques for testing materials and structures and the increased use of experimental structural analysis in research

3. The development of the moment-distribution method and the related, but generalized, relaxation procedure

4. The renewed interest in ultimate strength and plastic behavior of structures and structural elements and the incorporation of these ideas into structural design procedures

5. The application of the probability theory and statistical methods to a re-evaluation of the probability of failure or the probability of inserviceability, and a comparison of these ideas with conventional factors of safety

6. The advent of the analog and digital electronic computers and the application of modern computational methods and procedures to structural analysis and design, these developments being accompanied since the late 1950s by a tremendous surge of publication of books and periodical literature dealing with matrix methods and methods of machine computation pertaining particularly to structural analysis and design

7. The improved understanding of reinforced-concrete behavior, the development of precast and prestressed concrete, and the application of these ideas to design

8. The shell, panel, and stressed-skin construction developed primarily by the aeronautical structural engineers and the increased incorporation of these ideas into civil engineering structural design of not only metal structures but also concrete structures

9. The development of improved structural materials—steels, aluminum, plastics, concretes, processed woods, laminates, sandwiches, ceramics, and fiber-reinforced composites

10. Last—but far from least—the continued and rapid development of new and improved construction equipment, and new methods and techniques pertaining to the fabrication, transportation, erection, and maintenance of structures

The most important contributions to structural theory in the last sixty years are no doubt the moment-distribution method, presented originally by the late Professor Hardy Cross, and the more general relaxation procedures, proposed by Professor R. V. Southwell. These analytical procedures contributed much to a better understanding of structural behavior by providing a means of converging (to the degree desired) on precise solutions for the stressed condition of frameworks,

---

New York, 1936; S. Timoshenko, "Theory of Plates and Shells," McGraw-Hill Book Company, New York, 1940, 2d ed. with S. Woinowsky-Krieger, 1960; R. V. Southwell, "An Introduction to the Theory of Elasticity for Engineers and Physicists," Oxford University Press, London, 1936; I. S. Sokolnikoff, "Mathematical Theory of Elasticity," 2d ed., McGraw-Hill Book Company, New York, 1956; F. Bleich, "Buckling Strength of Metal Structures," McGraw-Hill Book Company, New York, 1952; W. Flügge, "Statik und Dynamik der Schalen," Springer-Verlag, Berlin, 1934; K. Girkmann, "Flächentragwerke," Springer-Verlag, Berlin, 1946, 4th ed., 1956; W. Flügge, "Stresses in Shells," Springer-Verlag, Berlin, 1960.

frames, grids, or even panellike structures that could be approximated by equivalent jointed structural arrangements. Through the 1930s and 1940s, these successive approximation procedures overshadowed the classical methods, whose application often involved a prohibitive number of simultaneous equations.

Since the early 1950s, however, the accelerated development of electronic computers and their expanding application to structural engineering have had a tremendous impact on handling both the analytical and the design tasks encountered in professional practice. With the availability of digital computers and improved computing techniques, the practitioner is able to use classical structural theories whenever appropriate, even though large numbers of simultaneous equations may be involved. This modern computing power cannot be dramatized as readily as a new structural form, or a new development concerning structural materials, or even a new sophisticated structural theory. Nevertheless, computerization of design tasks is probably the most significant structural engineering development in the last quarter century.

Important as computers have been, they are just one of many recent structural developments that must be considered significant. Just in the last decade, a much improved balance has been achieved between strength criteria and working stress criteria in the structural design process. Great strides have been made in improving the quality and variety of structural steels, and analogous improvements have been achieved with concrete. Spectacular advances have been made in fabricating and constructing both steel and concrete structures. Of course, in steel the 100-story John Hancock Center and the 110-story Sears Tower in Chicago and the 110-story World Trade Center in New York are world-famous accomplishments, as are in concrete the 38-story Brunswick Building in Chicago, the 52-story One Shell Plaza Building in Houston, and the 50-story One Shell Square Tower in New Orleans. However, throughout the land, each one of us in his own locality has likewise seen dramatic new constructions in the last several decades, most of which have made great contributions to the well-being of the citizens of our communities.

The Structural Engineering Project. The structural engineer usually acts in a service capacity to the functional design engineer who normally provides the leadership in carrying out an engineering project. In the civil engineering field, he assists the transportation engineer, hydraulic engineer, or sanitary engineer by providing the structures needed to implement their projects. In building construction, he is one of the principal collaborators of the architect. In a similar way, he assists mechanical, chemical, or electrical engineers in designing the heavy machinery or facilities required for their projects. He may shift his entire activity into naval architecture or aeronautical engineering and become a specialist in the design of ship or airplane structures. Today, the structural engineer may be involved in providing special structures for launching or servicing space vehicles, or again he may use his knowledge in assisting with the structural design of the vehicle itself. Sometimes, of course, the structure itself may be the major feature of a project, and the structural engineer may provide the leadership for the undertaking, as, for example, in the case of a large bridge or dam, a wharf, or a large industrial facility.

Figure F.4 Mackinac Straits bridge, Michigan. Longest suspension bridge, anchorage to anchorage, 8,614 ft; third longest main span, 3,800 ft (*Photograph furnished and reproduced by courtesy of the designer*, the late *Dr. D. B. Steinman, Consulting Engineer, New York*)

A structural engineering project may be divided into three phases: planning, design, and construction.

The **planning phase** involves a consideration of the various requirements and factors which affect the general layout and dimensions of the structure and leads to the choice of one or perhaps several alternative types of structures, which offers the best general solution. The primary consideration is the function of the structure whether it is to enclose or house, to convey, or to support in space. Many secondary considerations are also involved, including aesthetic, sociological, legal, financial, economic, environmental, or resource-conservation factors. In addition, there are structural and constructional requirements and limitations which also may affect the structural type selected.

The **design phase** involves a detailed consideration of the alternative solutions evolved in the planning phase and leads to the most suitable proportions, dimensions, and details of the structural elements and connections for constructing each alternative structural arrangement being considered. Usually before the final design stage is reached the best solution has been identified, and final construction plans are prepared for this selection. Occasionally, the choice is dependent on economic and constructional features which cannot be accurately evaluated except by competitive bidding; so final bid plans have to be prepared for the competitive alternatives.

The **construction phase** involves procurement of materials, equipment, and personnel; shop fabrication of the members and subassemblies; transportation of them to the site; and the actual field construction and erection. During this phase, of course, some redesign may be required if unforseen foundation difficulties develop, or if specified materials cannot be procured, or for any number of other reasons.

Philosophy of Structural Design. As defined above, **structural design** involves determining the most suitable proportions of a structure and dimensioning the structural elements and details of which it is composed. This is the most highly technical and mathematical phase of a structural engineering project, but it cannot —or certainly should not—be conducted without being fully coordinated with the planning and construction phases of the project. The successful designer is at all times fully conscious of the various considerations which were involved in the preliminary planning for the structure and, likewise, of the various problems which may later be involved in its construction.

Specifically the structural design of any structure first involves the establishment of the loading and other design conditions which must be resisted by the structure and therefore must be considered in its design; then comes the analysis (or computation) of the internal gross forces (thrust, shears, bending moments, and twisting moments), stress intensities, strains, deflections, and reactions produced by the loads, temperature, shrinkage, creep, or other design conditions; finally comes the proportioning and selection of materials of the members and connections so as to resist adequately the effects produced by the design conditions. The criteria used to judge whether particular proportions will result in the desired behavior reflect accumulated knowledge (theory, field and model tests, and practical experience), intuition, and judgment. For most common civil engineering structures such as bridges, buildings, etc., the usual practice in the past has been to design on the basis of a comparison of allowable stress intensities with those produced by the service loadings and other design conditions. This traditional basis for design is called **elastic design** since the allowable stress intensities are chosen in accordance with the concept that the stress or strain corresponding to the yield point of the material should not be exceeded at the most highly stressed points of the structure. Of course, the selection of the allowable stresses may also be modified by a consideration of the possibility of failure due to fatigue, buckling, or brittle fracture or by consideration of the permissible deflections of the structure.

Depending on the type of structure and the conditions involved, the stress intensities computed in the analytical model of the actual structure for the assumed design conditions may or may not be in close agreement with the stress intensities produced in the actual structure by the actual conditions to which it is exposed. The degree of correspondence is not important, provided that the computed stress intensities can be interpreted in terms of previous experience. By the selection of the service conditions and the allowable stress intensities, a margin of safety against failure is provided. The selection of the magnitude of this margin depends on the degree of uncertainty regarding loading, analysis, design, materials, and construction and on the consequences of failure. For example, if an allowable tensile stress of 20,000 psi is selected for structural steel with a yield stress of 33,000 psi, the margin of safety (or factor of safety) provided against tensile yielding is 1.65, that is, 33,000/20,000.

The allowable-stress approach has an important disadvantage in that it does not provide a uniform overload capacity for all parts and all types of structures. As a result, there is today a rapidly growing tendency to base the design on the ultimate strength and the serviceability of the structure, with the older allowable-stress approach serving as an alternate basis for design. The newer approach currently goes under the name of **strength design** in reinforced-concrete-design literature and **plastic design** in steel-design literature. When proportioning is done on the strength basis, the anticipated service loading is first multiplied by a suitable **load factor** (greater than 1), the magnitude of which depends upon uncertainty of the loading, the possibility of its changing during the life of the structure, and, for a combination of loadings, the likelihood, frequency, and duration of the particular combination. In this approach for reinforced-concrete design, the theoretical capacity of a structural element is reduced by a capacity-reduction factor to provide for small adverse variations in material strengths, workmanship, and dimensions. The structure is then proportioned so that, depending on the governing conditions, the increased load would (1) cause a fatigue or a buckling or a brittle-fracture failure or (2) just produce yielding at one internal section (or simultaneous yielding at several sections) or (3) cause elastic-plastic displacement of the structure or (4) cause the entire structure to be on the point of collapse.

Proponents of this latter approach argue that it results in a more realistic design with a more accurately provided margin of strength over the anticipated service conditions. These improvements result from the fact that nonelastic and nonlinear effects which become significant in the vicinity of ultimate behavior of the structure may be accounted for.

In recent years, there has been a growing concern among many prominent engineers that not only is the term factor of safety improper and unrealistic, but worse still a structural design philosophy based on this concept leads in most cases to an unduly conservative and therefore uneconomical design, and in some cases to an unconservative design with too high a probability of failure. They argue that there is no such thing as certainty, either of failure or of safety of a structure but only a *probability* of failure or a *probability* of safety. They feel therefore that the variations of the load effects and the variations of the structural resistance should be studied in a statistical manner and the probability of survival or the probability of serviceability of a structure estimated.[1] It may not yet be practical to apply this approach to the design of each individual structure. However, it is believed to be practical to do so in framing design rules and regulations. It is highly desirable that building codes and specifications plainly state the factors and corresponding probabilities that they imply.

Structural Materials. Of course, the availability of suitable structural materials is one of the principal limitations on the accomplishments of an experienced structural engineer. Early builders depended almost exclusively on wood, stone, brick, and concrete. Although iron had been used by man at least since the building of the Egyptian pyramids, use of it as a structural material was limited because of the difficulties of smelting it in large quantities. With the industrial revolution, however, came both the need for iron as a structural material and the capability of smelting it in quantity.

John Smeaton, the English civil engineer, was the first to use cast iron extensively as a structural engineer in the mid-eighteenth century; but another Englishman, Abraham Darby, constructed the first cast-iron bridge in 1776–1779, the Coalbrookdale arch bridge with semicircular cast-iron arches. Cast-iron girders were unreliable, and a number of fatigue and brittle failures occurred under heavy moving loads. Malleable iron was developed as a more reliable material and was widely used after 1841. Iron-plate girders were very commonly used for short-span bridges. Since suspension bridges were too flexible for long-span railroad bridges, it was necessary to develop a more rigid type of structure, and the malleable-iron tubular design was evolved for the famous Britannia and Conway tubular bridges built in England just prior to 1850. At about this same time, the development of iron-truss construction started in the United States, England, and Europe. Incidentally, the first iron-chain suspension bridge had been built in England in 1741.

Whereas malleable iron was superior to cast iron, there were still too many

---

[1] A. L. Johnson, Strength, Safety and Economical Dimensions of Structures, *R. Inst. Technol., Div. Stat. Struct. Eng., Stockh. Bull.* 12, 1953; A. M. Freudenthal, The Safety of Structures, *Trans. ASCE*, vol. 112, p. 125, 1947; A. M. Freudenthal, Reflections on Standard Specifications for Structural Design, *Trans. ASCE*, vol. 113, p. 269, 1948; A. M. Freudenthal, Safety and Probability of Structural Failure, *Trans. ASCE*, vol. 121, p. 1337, 1956; O. G. Julian, Synopsis of First Progress Report of Committee on Factors of Safety, *J. Struct. Div., Proc. ASCE*, vol. 83, no. ST4, pap. 1316, July 1957; A. M. Freudenthal, J. M. Garrelts, and M. Shinozuka, The Safety of Structures, *Proc. ASCE, J. Struct. Div.*, vol. 92, no. ST1, February 1966; Structural Safety: A Literature Review, *Proc. ASCE*, vol. 98, no. ST4, pp. 845–884, April 1972.

structural failures and there was a need for a more reliable material. Steel was the answer to this demand. The invention of the Bessemer converter in 1856 and the subsequent development of the Siemens-Martin open-hearth process for making steel made it possible to produce structural steel at competitive prices and triggered the tremendous developments and accomplishments in the use of structural steel during the last hundred years.

Ordinary structural carbon steel is almost a perfect structural material. It can be manufactured reliably and economically and rolled into a large variety of suitable shapes and sizes. It can be fabricated and erected by a number of techniques and procedures without changing its physical properties appreciably. It has essentially the same strength in tension or compression. Up to over half of its strength, it behaves in a linearly elastic manner, and over this range it exhibits a high modulus of elasticity. At the end of this initial range of high stiffness, the steel will yield at constant stress for a strain of $1\frac{1}{2}$ to 2 per cent, thereby allowing stresses to equalize themselves at locations where loading or fabrication or erection has caused high stress concentrations. Following this yield range, the material regains some of its original stiffness—a small fraction, true, but enough to arrest sufficiently the rate of increase in deflection in many cases. Before actual fracture occurs in a tensile test, the specimen will have stretched 15 to 25 per cent of its original length. Refer to Fig. F.5 for the stress-strain diagrams of various materials.

The most serious disadvantage of steel is that it oxidizes easily and must be protected by paint or by some other suitable coating. When steel is used in an enclosure where a fire could occur, the steel members must be encased by a suitable fire-resistant enclosure such as masonry, concrete, vermiculite concrete, etc. Normally, steel members will not fail in a brittle manner unless an unfortunate combination of metallurgical composition, low temperature, and bi- or triaxial stresses exists.[1]

Structural aluminum is still not widely used in civil engineering structures, though its use is steadily increasing. By a proper selection of the aluminum alloy and its heat treatment, a wide variety of strength characteristics may be obtained. Some of the alloys exhibit stress-strain characteristics similar to structural steel, except that the modulus of elasticity for the initial linearly elastic portion is about 10,000,000 psi, or about one-third that of steel. Lightness and resistance to oxidation are of course two of the major advantages of aluminum. Because its properties are very sensitive to its heat treatment, care must be used when riveting or welding aluminum. In recent years, several techniques have been developed for prefabricating aluminum subassemblies which can be readily erected and bolted together in the field to form a number of beautiful and well-designed shell structures. This general procedure of prefabrication and field assembly by bolting seems to be the most promising way of utilizing structural aluminum.

---

[1] Brittle fracture of steel is a question which is not fully solved even though its mechanics are fairly well outlined. See M. E. Shank (ed.), "Control of Steel Construction to Avoid Brittle Failure," Plasticity Committee of Welding Research Council, New York, 1957; R. V. Phillips and S. M. Marynick, Brittle Fracture of Steel Pipeline Analyzed, *Civ. Eng. ASCE*, pp. 70–74, July 1972; S. T. Rolfe, Fracture Mechanics in Bridge Design, *Civ. Eng. ASCE*, pp. 37–41, August 1972.

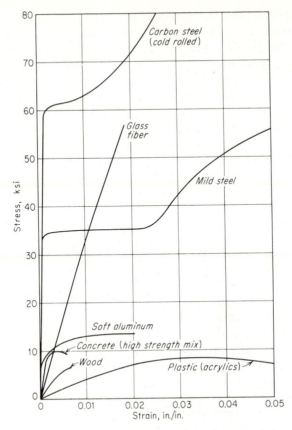

Figure F.5 Stress-strain diagrams for various materials

Reinforced and prestressed concrete share with structural steel the position of the most important civil engineering structural material. Natural cement concretes have been used for centuries. Modern concrete construction dates from the middle of the nineteenth century, though artificial portland cement was patented by Aspidin, an Englishman, about 1825. Although several builders and engineers experimented with the use of steel-reinforced concrete in the last half of the nine-teenth century, its dominant use as a building material dates from the early decades of this present century. The last forty years have seen the rapid and vigorous development of *prestressed*-concrete design and construction, founded largely on early work by Freyssinet in France and Magnel in Belgium.

Plain (unreinforced) concrete not only is a heterogeneous material but has one very serious defect as a structural material, namely, its very limited tensile strength, which is only of the order of one-tenth its compressive strength. Not only is tensile failure in concrete of a brittle type, but likewise compression failure occurs in a relatively brittle fashion without being preceded by the forewarning of large deformations. (Of course, in reinforced-concrete construction, ductile behavior can be obtained by proper selection and arrangement of the reinforcement.)

Unless proper care is used in the selection of aggregates and in the mixing and placing of concrete, frost action can cause serious damage to concrete masonry. Concrete creeps under long-term loading to a degree which must be considered carefully in selecting the design stress conditions. During the curing process and its early life, concrete shrinks a significant amount, which to a degree can be controlled by properly proportioning the mix and utilizing suitable construction techniques.

With all these potentially serious disadvantages, engineers have learned to design and build beautiful, durable, and economical reinforced-concrete structures for practically all kinds of structural requirements. This has been accomplished by careful selection of the design dimensions and the arrangement of the steel reinforcement, development of proper cements, selection of proper aggregates and mix proportions, careful control of mixing, placing, and curing techniques, and imaginative development of construction methods, equipment, and procedures.

The versatility of concrete, the wide availability of its component materials, the unique ease of shaping its form to meet the strength and functional requirements involved, together with the exciting potential of further improvements and developments of not only the new prestressed and precast concrete construction but also the conventional reinforced-concrete construction, combine to make concrete a strong competitor of other materials in a very large fraction of structures contemplated for the future. The use of precast-concrete elements which can be produced under carefully controlled plant conditions and then assembled in the field using prestressing techniques is bound to expand tremendously, particularly in this country with its high ratio of labor to material costs.

Wood, of course, was the only material used by early builders which had more or less equal tensile and compressive strength. Wood is characterized by its non-homogeneity due to its grain, knots, and other defects; by its anisotropic strength properties parallel and perpendicular to its grain; and by its poor durability due largely to its susceptibility to variations in humidity and moisture and to attack by bacteria, fungi, and insects. While its tensile and compressive strength is much less perpendicular than parallel to the grain, the most serious aspect of its anisotropic strength characteristics is its much smaller shear resistance parallel to the fibers than transverse to them. Wood also exhibits considerably lower strength under long-duration loadings. Wood deforms considerably under load, exhibiting essentially linear stress-strain characteristics at low stresses with a modulus of elasticity of about half that of concrete (and about a twentieth that of steel). Compression failures are not abrupt, but tension failures are—less so than concrete, however. The design of wood structures must allow properly for dimensional changes caused by temperature and moisture variations. The major design problem, however, is providing adequate, dependable, and practical connections.

In modern times, with the increased use of steel and reinforced-concrete construction, wood was relegated largely to accessory use during construction, to use in temporary and secondary structures, and to use for secondary members of permanent construction. Modern technology in the last forty-odd years has revitalized wood as a structural material, however, by developing vastly improved timber connectors, various treatments to increase the durability of wood, and laminated wood made of thin layers bonded together with synthetic glues using revolutionary gluing techniques. Plywood with essentially nondirectional strength properties is the most widely used laminated wood, but techniques have also been developed for building large laminated-wood members which for certain structures are competitive with concrete and steel.

Materials with future possibilities are the engineering plastics and the exotic metals and their alloys, such as beryllium, tungsten, tantalum, titanium, molybdenum, chromium, vanadium, and niobium. There are many different plastics available, and the mechanical properties exhibited by this group of materials vary over a wide range which encompasses the range of properties available among the more commonly used structural materials. Thus in many specific design applications it is possible to select a suitable plastic material for an alternative design. Experience with the use of plastics outdoors is limited. Generally speaking, however, plastics must be protected from the weather. This aspect of design is therefore a major consideration in the use of plastics for primary structural elements. One of the most promising potential uses of plastics is for panel and shell-type structures. Laminated or sandwich panels have been used in such structures with encouraging results which indicate an increased use in this type of construction in the future.

Another materials development with interesting possibilities is that of composites consisting of a matrix reinforced by fibers or fiberlike particles. Although glass-fiber-reinforced composites with a glass or plastic matrix have been used for years, they appear to have much broader possibilities for a large variety of secondary structural components. Fiber-reinforced concrete is another composite being actively studied and developed. Several experimental applications are being observed under service conditions. Experiments have been conducted with both steel and glass fibers, but most of the service experience has been with steel fibers.

Modes of Structural Failure and Inadequacy.[1] The inadequacy of a structure may be revealed in one of several ways. The most dramatic, even when no human casualties are involved, is the *collapse* of a structure. The collapse may be total and involving the entire structure, or localized and confined to one portion. Often collapse occurs during construction and is usually caused either by poor construction planning and scheduling or by inadequate support or bracing for the partially complete structure. Sometimes the collapse of a completed structure may result from a natural disaster, an explosion, a collision, or a severe fire; but often the failure may occur under the imposition of what appears to be a routine loading condition.

A less dramatic form of inadequacy occurs when a structure becomes *unserviceable* due to some structural deficiency. A structure becoming inadequate to fulfill its intended function may not be dramatic, but this situation may still result

---

[1] J. Feld, Structural Success or Failure?, *Proc. ASCE*, sep. 632, 1955; J. Feld, Lessons from Failures of Concrete Structures, *ACI Monogr.* 1,1964; Structural Failures: Modes, Causes, Responsibilities, *ASCE Res. Counc. Perform. Struct.*, 1973; F. A. Randall, Jr., Historical Notes on Structural Safety, *J. ACI*, pp. 669–679, October 1973.

in a very severe economic loss for its owner. For example, a structure may become unserviceable because of its stiffness and vibrational characteristics. In such a case, a structure may be so flexible that sensitive equipment and machinery supported by it cannot operate satisfactorily, so that unsightly cracks develop in plaster or other superficial finishes, or that many people using the structure are disturbed by its movement or by the visual appearance of its deflected shape. On the other hand, the structure may have vibrational characteristics that are poorly matched to the frequency of the dynamic loads acting on it. Such a circumstance could result in poor response characteristics for a building in an earthquake-prone zone or in the aerodynamic instability of a structure under a modest steady wind, as illustrated by the famous Tacoma Narrows Suspension Bridge failure in 1940.

Many failures other than the inadequacy of a main structural element may be the primary cause of the collapse of a structure. A faulty sprinkler system may allow a fire to get out of control, resulting in temperatures so high that deformation and loss of strength of steel truss members lead to the collapse of the entire roof system of a storage warehouse. Poor inspection may have allowed the application of much less than the specified fireproofing for the floorbeams above a laboratory in which an explosion caused a very intense and localized fire. Poor construction may have provided inadequate seals and paint protection for an important structural connection, resulting in corrosion and a major collapse triggered by corrosion fatigue in a key member leading into this joint. In another case, a design error led to a joint's being designed without taking into account some of the forces acting on it; the structure failed, partially assembled during construction. Another common design error is failure to provide adequate anchorage and bearing for steel beams framing into masonry walls. (*Joint design is terribly important*; perhaps all structural engineers and builders should repeat regularly to themselves a warning that paraphrases the old parable: "For want of a nail the shoe was lost; for want of a shoe the horse was lost; ...".) Just plain poor judgment in using an old structure or in converting an old structure to a new use may result in a collapse due to its being overloaded or subjected to a loading condition for which it was never designed. Finally structural failure or inadequacy may be caused by movement or settlement of poorly designed or constructed foundations.

Collapse of a structure may be triggered by any of the following modes of failure of one of its elements: plastic deformation or rupture of the material in a ductile manner, rupture of the material in a brittle manner, buckling, or fatigue of the material.

Seldom is the collapse of a structure triggered by rupture of the material in a ductile manner, since such rupture is preceded by such large strains that excessive deflections would become apparent. Under normal types of loading, there would therefore be opportunities to relieve the structure either by unloading it partially or by shoring it or strengthening it temporarily. Of course, in catastrophic situations—such as earthquakes, windstorms, floods, fires, and bomb blasts—there is no opportunity to relieve the structure, and collapse could be triggered by ductile rupture or excessive plastic deflection. Many collapsed structures show evidence of ductile rupture or excessive plastic deformation at a number of points, but these usually have developed during the collapsing process and did not trigger the failure.

Brittle fracture is much more likely to trigger the collapse of a structure than ductile fracture. Certain steels (and perhaps other alloyed metals), depending upon their metallurgical composition and manufacture and upon the minimum temperature of their environment, may be susceptible to this type of failure. This is particularly true if the member is designed and detailed so that at the location of potential failure a pronounced bi- or triaxial stress condition exists, thereby inhibiting plastic flow and permitting a cleavage type of fracture to occur. In reinforced-concrete structures, a brittle behavior can occur when portions are over-reinforced and the compressive strength of the concrete is reached before significant yielding occurs in the tensile steel. In such cases if the compression steel and stirrups are not detailed so as to prevent cracking off of large chunks of the compression zone, the member may fail in a precipitous brittle manner. Such behavior may also occur if the diagonal tension reinforcement is inadequate.

In civil engineering structures where the dominant loading is usually substantially static, the most common cause of collapse is a buckling failure. Buckling may occur locally in a manner that may or may not trigger collapse of the entire structure, such as in outstanding flanges, in flange or web plates of compression members, or in the web or compression flange of girders, etc. Compression members or beams may buckle as individual members, or structures as a whole may buckle. Occasionally in civil engineering structures, the buckling is an **elastic buckling**, wherein the portion recovers its original shape if unloaded. Such buckling resistance depends on the elastic flexural rigidity and the unsupported length (or some times the width) of the portion but not on the stress intensities involved. More often, however, because of the slenderness ratios commonly used, **plastic buckling** is involved, wherein the member or portions of it are stressed beyond the proportional limit of the material. Then, the stress intensities are a significant factor in the behavior, because the strain and flexural rigidity beyond the proportional limit depend upon the value of the stress. Sometimes, when elastic buckling occurs, the load on the portion is relieved as a result of the buckling displacements. In other cases, however, such relief does not occur, and under continued action of the load additional deflections are developed which may involve substantial plastic deformation and subsequent collapse of the structure.

Fatigue failures occur infrequently in conventional civil engineering structures, although unfortunately the frequency of such failures seems to be increasing. Such failures result in a brittle-type fracture, and sometimes it is difficult to classify the cause of the fracture properly. A case in point was the tragic collapse of the Point Pleasant Suspension Bridge in West Virginia Dec. 15, 1967. The Point Pleasant Bridge was an eye bar suspension bridge which had been in service more than fifty years when it collapsed with the loss of more than twenty lives. The failure was traced to a cleavage fracture in one of the eye bar heads caused by a very small crack radiating from a corrosion pit on the surface of the hole in the eye. Experts cannot agree on the mechanism by which the crack grew. Some believe stress corrosion; some corrosion fatigue. This issue was not resolved in the official final report.

When heavy dynamic loads or severe vibrations are involved, the possibility of fatigue failure must be considered. Elastic fatigue failures are of a brittle type, symptoms of which can be detected only by refined inspection techniques. So-called **plastic fatigue** can also occur, though it is rare. This is the phenomenon involved when one breaks a wire or nail by bending it into the plastic region, first in one direction and then the other. Obviously, it is seldom that a civil engineering structure is subjected to large reversible loadings which could cause this type of failure. Such possibilities should be recognized and considered, however.

The structural engineer must therefore design his structures so as to obtain a suitable improbability that they will collapse or become unserviceable. He must be prepared to analyze the behavior of the various structural forms which he uses so as to estimate the possibilities of collapse or poor performance under the loading and environment to which they are subjected. If the theory is inadequate for such predictions, he must make his evaluations and design decisions on the basis of his experience and model and/or field tests.

Structural Forms. The most important design decision made by a structural engineer is the selection of the most suitable structural form to satisfy the various requirements and objectives of a particular project. Most often, he is not able to identify immediately *the* best solution, and he must continue to consider several alternative structural types throughout the planning and design phases of the project until he is able to identify the best of the alternatives. The most suitable structural form is the one which satisfies the functional, economic, sociological, aesthetic, and other requirements to the highest degree and which may be economically and reliably built using the structural materials and construction methods which are the most appropriate of those which are available or can be devised.

Structural forms may be classified as being of one of the two following types according to the dominant stress condition developed under their most important design loads and conditions: **uniform-stress forms,** in which the stress is essentially uniform over the depth of a member or over the thickness of a panel e.g., cables, arches, truss members, membranes, shells; and **varying-stress forms,** in which the stress varies over the depth or thickness, usually from a maximum tensile stress at one surface to a maximum compressive stress at the other, e.g., beams, rigid frames, slabs, plates.

Obviously uniform-stress forms utilize the available strength of material more efficiently than the varying-stress forms, in which under elastic conditions the material between the outside faces is understressed except in regions where there are high shear stresses. Where functional requirements permit, therefore, the structural designer tries to use a uniform-stress form. Of course, sometimes the fabrication and/or construction difficulties involved in achieving the desired shape for the structure cost more than the material saved through its more efficient use, so that overall economy cannot be realized.

Steel-cable structures illustrate a very efficient structural form. Cables are almost perfectly flexible, i.e., they may be assumed to have no flexural strength and therefore *must* carry their loads by simple tension. As a result, they must adjust their shape to conform to the equilibrium polygon of the loads applied to them (see Chap. 6). This adjustment in shape is a significant geometrical change in shape, not simply the small deflections resulting from the stress-strain characteristics of deformable bodies. As shown by the solid-line position in Fig. F.6*a*, a cable has the shape of a catenary under its own dead weight. If a large load *P* (compared with its own weight) is now imposed on the cable, it shifts its shape markedly to the dashed-line position, which is substantially two straight line segments. On the other hand, as shown in Fig. F.6*b*, a load *P* small compared with the weight of the cable would require only a very small change in shape of the cable if it were applied. The principle is clear, however, from a consideration of the equilibrium of a small portion of the cable: the only two ways a cable can maintain simultaneously both vertical and horizontal equilibrium are (1) to change its tension and (2) to adjust its slope, i.e., its shape. In other words, *one* unknown (the cable tension) cannot in general satisfy *two* equations simultaneously—two unknowns (the cable tension and cable slope) are required to do so. If for func-

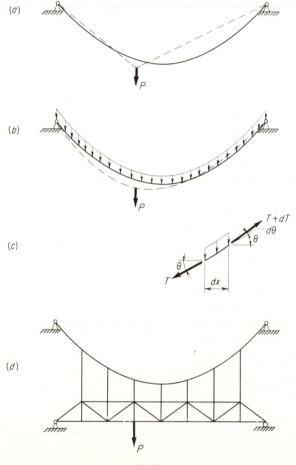

Figure F.6 Cable structures

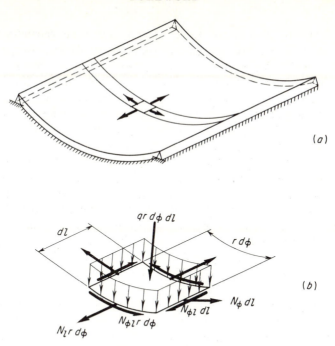

Figure F.7   Cylindrical membrane

tional reasons substantial changes in shape of cable structures cannot be tolerated, a stiffening truss, as illustrated in Fig. F.6d, must be connected to the cable to distribute large concentrated loadings.

A **flexible membrane** such as shown in Fig. F.7a at first glance seems to be the same as a large number of cables side by side, and it *would* be essentially this situation if the curved edges of this cylindrical membrane were unsupported. However, the generation of shear stresses between adjacent cable elements constitutes a very important difference between the cable and the membrane, which becomes apparent from a consideration of the equilibrium of a differential element of the membrane shown in Fig. F.7b.   In the case of the differential element of the membrane, there are *three* equilibrium conditions which must be satisfied—summation of forces in the longitudinal, the circumferential, and the radial directions must each equal zero—compared with only *two* conditions in the case of the cable. Note, however, that there are *three* forces N_l, N_ϕ, and $N_{\phi l}$ in the case of the membrane which by incremental changes in their values can satisfy these *three* equilibrium equations for the element *without* the necessity of changing its shape beyond the small strains or deformations associated with its stresses.   This characteristic of the three-dimensional membrane surface makes it essentially different structurally from the two-dimensional cable.   Such surfaces are true uniform-stress structural forms which are structurally stable and adequate, and capable of supporting *distributed* loadings with no changes of shape except the small deflections associated with the deformation of the material.   Of course, concentrated loadings on a flexible membrane—being discontinuities in the loading—can be

(*a*) Shape of arch axis selected to coincide with
shape of equilibrium polygon for applied loading

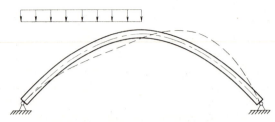

(*b*) Arch of same shape as (*a*) but subjected to
different loading, which causes arch axis to
bend into shape something like exaggeration
shown by dashed line

Figure F.8   Two-hinged arch

supported only by violent localized changes of shape of the membrane similar to those changes required of a cable.

An **arch,** such as shown in Fig. F.8*a*, is somewhat like an inverted cable provided that its shape conforms to the equilibrium polygon for the loading imposed on it. Even though the arch, unlike the flexible cable, possesses flexural stiffness and strength, it would carry such a loading essentially with a uniform stress condition over each cross section and no flexure.   Now, as shown in Fig. F.8*b*, if a completely different loading is applied to this arch, the arch tries to change its shape—just as a cable—to that of the equilibrium polygon for the new system of loading. Because of its flexural stiffness, however, it cannot change its shape without bending. As a result, therefore, it bends and changes its shape slightly—sufficiently, in fact, for a varying-stress condition to be developed in the arch whereby the arch carries the load partly by flexural stresses and partly by thrust.   That is, the arch thrust cannot in general satisfy both horizontal and vertical equilibrium without the assistance of bending stresses and the transverse shear stresses associated with them.   Of course, the one case where an arch does not bend materially is for the loading with an equilibrium polygon of the same shape as the arch.

From a design standpoint, the arch form may be an economical solution for cases where there is one predominant loading condition, the equilibrium polygon of which is used to define the shape of the arch. For this loading, the arch acts substantially as a uniform-stress structural form; for other loadings which may be relatively less important and infrequently applied, the arch acts satisfactorily, but less efficiently, as a varying-stress structural form.

The **shell** is to the membrane as the arch is to the cable. That is, a shell is a membrane which has flexural stiffness and strength, just as an arch is related in the same way to a cable. Most often, civil engineering shells are the inverts of shapes such as shown in Fig. F.7a so that they carry their loads primarily by compressive rather than tensile stresses and are often made of reinforced concrete as a result. A shell which was the invert of the membrane shown in Fig. F.7a would act distinctly differently from a series of arches side by side for exactly the same reason that a membrane would act differently from a series of cables, as discussed above.

Even though it may have flexural stiffness and strength, the shell possesses the basic structural stability and adequacy of its corresponding membrane shape as far as satisfying equilibrium requirements is concerned. That is, the *three* membrane forces available in the shell, N_l, N_ϕ, and $N_{\phi l}$, can satisfy the *three* equations of equilibrium without the assistance of bending stresses. As a result, as long as the loading is distributed, it carries the loading primarily by a membrane type of stress condition even though the shape of the shell may not correspond to the shape of the equilibrium surface for the loading involved. Because of the slight tendency of the membrane surface to deform and, hence, change shape, some slight secondary bending is induced. Only adjacent to the supports, however, where it is usually impossible to supply exactly the supporting forces required by the membrane stresses, do significant flexural stresses develop. Of course, if concentrated loadings or abruptly varying distributed loadings are applied to the shell, large localized flexural stresses are developed in the shell just as violent changes in shape occur in membranes under such conditions. Bending also occurs adjacent to abrupt changes in shell thickness or shape.

Another very important uniform-stress structural form is the **planar truss** (or **planar framework**). Of course, a single concentrated load P could be supported by the two-member arch form shown in Fig. F.9a, the axis of which conforms to the shape of the equilibrium polygon for such a load. If, however, the foundations were not suitable to provide the necessary horizontal reactions H, this resistance could be provided at both a and c by a tensile force H in the third member ac added as shown in Fig. F.9b to form, with the other two, a simple triangular truss. Additional concentrated loads at d and e could be supported as shown in Fig. F.9c by using two bars to connect each of these joints in turn to the framework already formed. It is quite apparent that such an arrangement permits the members to support a system of joint loads by uniform-stress conditions in each of the members —equivalent to tensile forces in some members and to compressive forces in others. If the load system to be supported does not lie in a plane but is three-dimensional in character, a suitable **space framework** can be arranged to connect the points of loading by starting with a tetrahedronal nucleus consisting of four joints and six bars, as shown by the heavy lines in Fig. F.9d. Three bars are used to connect each additional joint to the framework.

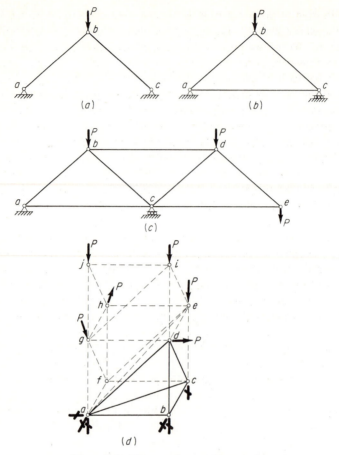

Figure F.9 Planar and space frameworks

In Fig. F.9, the encircled joints indicate frictionless pin joints for the planar truss (or planar framework) and ball joints for the space framework. Such joints are all that are needed for structural adequacy. Actually it is convenient and economical in modern truss fabrication and erection to use essentially rigid joint connections, using rivets, bolts, or welds. Doing so actually prevents the framework from carrying its loads by purely axial forces in its members. This is apparent when one considers that the members of an idealized pin- or ball-jointed framework change their lengths slightly in developing their axial forces. As a result, the angles between the members change slightly as there is displacement of the joints consistent with these changes of length in the members. If the members are rigidly connected together, however, at the joints, the angle changes cannot occur and slight bending of the members is necessary to accommodate the joint displacements. In a well-designed truss, the stresses associated with this flexure of the members are small compared with the uniform axial stresses and are therefore, in fact, referred to as **secondary stresses.**

It is interesting to note that in a sense rigid-jointed frameworks are very closely related to shell structures. Basically they want to belong to the uniform-stress family and try to act as pin- or ball-jointed frameworks, but because of their deformation they cannot do so exactly and thereby induce some nominal amount of flexure. Basically, shells want to act as membranes, but because of their flexural stiffness and the support arrangement along their boundaries they cannot deform as membranes and thereby induce flexural action which is usually of secondary importance except in the region adjacent to the edges.

It should also be noted that frameworks cannot be loaded directly on their members between joints without developing large flexural stresses in the members so loaded and in other nearby members connected to them. In other words, trusses prefer concentrated joint loadings, not distributed loadings; but conversely, shells and membranes prefer distributed loadings, not concentrated loadings.

Very often for reasons based on considerations of function, economy, or construction, varying-stress structural forms such as **beams, frames, plates,** and **slabs** are used, particularly in bridge and building design. The relative inefficiency of such forms in the utilization of the potential strength of all the material may be minimized by the selection of proper proportions for the cross sections of such members. For example, in the hypothetical case of pure bending shown in Fig. F.10*a*, all the material could be divided between the two flanges since, there

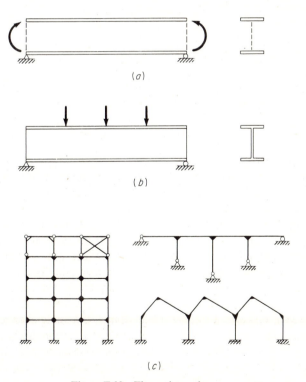

Figure F.10   Flexural members

being no shear, no web is required. In actual situations, of course, both bending and shear are involved and web material is needed as shown in Fig. F.10*b*. As a result, manufacturers of rolled-steel and aluminum beams have developed a number of series of standard I and WF (wide flange) beam sections which in many practical situations result in quite efficient use of material at least at the *most critical cross sections*. In recent years, considerable attention has been given to the fact that because of the inefficient use of material in flexural members, many beam and frame structures possess considerable load-carrying capacity beyond the loading which develops the yield stress at the most highly stressed point. Ductile structures made of structural steel or reinforced concrete may be subjected to further loading, which simply produces plastic deformation at locations where yield stresses have already been reached while mobilizing the full strength of less critical locations. Considerations and utilizations of such behavior are variously labeled strength design or plastic design, as mentioned previously.

Of course, many heavily loaded long-span beams exceed the capacity of the standard rolled-metal sections. In such cases, the required flexural strengths are obtained by designing special built-up sections out of plates, angles, and other standard shapes, these elements being connected together by rivets, bolts, or welds. Much more consideration could, and perhaps should, be given to the evolution of beam sections of reinforced concrete, wood, and other materials. Space limits the present discussion, and the reader is referred to standard textbooks on structural design for such information.

Sometimes beams are cantilevered or end-supported, as shown in Fig. F.10*a* and *b*. More often in modern construction, however, they are continuous over several intermediate supports or are connected to other members to form so-called **framed structures.** Everyone sees daily numerous examples of the almost infinite variety of configurations used in frame construction, some of which are indicated by the line diagrams in Fig. F.10*c*. Almost always these framed structures are three-dimensional assemblages of members, though often planar portions may be isolated, as depicted in the figure, for purposes of design and analysis. Sometimes the joints of these frames are detailed using rivets, bolts, and welds so as to be substantially *rigid* so that the ends of all members connected must not only translate but also rotate essentially identical amounts. Sometimes lighter connections are provided which are only *semirigid*. Rarely are the members actually pinned together at the joints, but sometimes the riveted or bolted joint details are so *flexible* that they transmit only negligible moments through the connection. In such cases, it may be necessary to add knee braces or diagonal bracing, as shown in the top story of the frame in Fig. F.10*c*, in order to stabilize the frame.

The members of framed structures are primarily flexural members subjected principally to bending moment and shear. In addition, the members are subjected to axial forces though these are not large except in members such as the vertical columns of these frames.

The last varying-stress form to be considered is the **slab** or **plate element.** The structural behavior of a plate may be visualized as being similar to that of two layers of beam strips, each layer consisting of strips which are cemented together side by side. The layers are then oriented so that the directions of the strips are

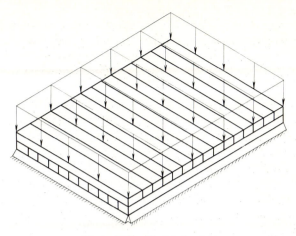

Figure F.11 Plate behavior

perpendicular as the layers are placed one on top of the other, as shown in Fig. F.11. If the edges of such a plate are supported and it is then subjected to a transverse (i.e., vertical) loading, both layers will collaborate in resisting the load, each layer bending and transmitting the load into the pair of edge supports at the ends of its beam strips. (It should be noted that for the visualized behavior to be correct, the thickness of each layer should be considered to be equal to the full thickness of the plate.) It is of interest to consider the behavior of a relatively long and narrow plate in terms of the beam-strip-layer analogy. In such a case, the layer with the long beam strips will be much less stiff than the layer with the short beam strips spanning in the narrow direction. As a result, most of the load is carried by the short beam strips, with practically no two-way plate action involved. In other words, to benefit significantly from two-way plate action, the length-width ratio of the plate should be 2 or less.

Analysis of Structural Behavior. One of the important assets of a structural engineer is his ability to analyze mathematically the behavior of a proposed structure for the specified design conditions. If he has available a theory which is correlated with and corroborated by laboratory and field experimentation, and by observed performance of existing structures, he then has a logical and rational basis for extrapolating his current knowledge and previous experience to new situations. Of course, when no theory is available, the successful engineer is prepared to use a combination of model and field experimentation, previous experience, judgment, and intuition to guide him in arriving at his design.

The structural engineer should be able to analyze[1] three basic aspects of structural behavior:

1. Stress, strain, and deflection characteristics under static or quasi-static loading or deformation conditions
2. Response and vibrational characteristics under dynamic loading conditions

---

[1] The analysis of the stress, strain, and deflection characteristics of structural behavior is referred to as **structural analysis.** The analysis of simply the internal force and stress condition in a structure is referred to as **stress analysis.**

3. Buckling characteristics usually under static loading conditions, but occasionally under dynamic loading conditions

In each of these respects, he must compare the limiting capabilities of the structure for each of the potential modes of failure which obtain, with the corresponding behavior of the structure for the design conditions.

Obviously, categorizing these aspects of structural behavior in this manner is somewhat arbitrary. However, analysis of aspects 2 and 3 requires specialized techniques beyond those used for 1; so it seems desirable to consider these aspects separately. Moreover, civil engineering structures must usually be designed so that there is sufficient improbability of brittle or ductile failure associated with the behavioral characteristics categorized under 1 and so that there is an *even greater* improbability of fatigue or unsuitable vibrational properties or of buckling failure associated with the behavioral characteristics categorized under 2 and 3, respectively.

Analysis of buckling and vibrational characteristics is beyond the scope of the present discussion, which will be focused on the analysis of the stress, strain, and deflection characteristics of a structure when it is subjected to static or quasi-static loading or deformation conditions. To solve such a problem completely it is necessary to obtain the unknown stresses and displacements throughout the structure so as to satisfy the governing equations and the known force and displacement conditions on the boundaries. These unknowns may be obtained by formulating a suitable number of independent equations which are obtained from the three following types of relations or conditions which govern the behavior of the structure:

1. Conditions for static equilibrium of the internal and external forces (and couples) acting on the structure
2. Relations between stress and strain of the material involved
3. Geometrical relations between strains and displacements of points on the structure

Of course, it is easy to combine relations 2 and 3 so as to eliminate the strains and obtain thereby direct relations between the stresses (or internal forces) and the displacements. Some analysts may therefore prefer to say that there are only two types of relations involved, namely:

1. Conditions for static equilibrium
2, 3. Relations between stresses (or internal forces) and displacements

It really makes no difference which point of view is preferred since in either case one of two basic approaches may be used to formulate the solution of the problem.

The two basic approaches are the **force**[1] **method** and the **displacement**[1] **method,** which may be described as follows. These approaches are described in a generalized fashion as they would apply to a **member structure.**[2]

---

[1] In these instances, the terms **force** and **displacement** are used in a generalized sense. Force may be interpreted to refer to either a force or a couple. Displacement may be interpreted to refer to either a lineal or a rotational displacement.

1. Force method

   a. First, the number of independent unknown forces (both external and internal) is established and compared with the number of independent equations of static equilibrium which may be written involving these unknowns. If the number of unknown forces is identical to the number of equilibrium equations, the problem is said to be **statically determinate** since the unknown forces can be determined directly from these equilibrium equations. If, however, the number of unknown forces exceeds the number of equilibrium equations, the problem is said to be **statically indeterminate** to a degree equal to this excess. In such a case, a number of the unknown forces equal to the degree of indeterminancy are designated as **redundant forces** and imagined to be removed as restraints on the actual structure to obtain a statically determinate (and stable) residual structure, called the **primary structure**. One equation may be written for each redundant force in which the displacement of its point of application on the primary structure is expressed in terms of the known forces and the unknown redundant forces. These equations may then be solved simultaneously for the unknown redundants.

   b. Once all the forces acting on the primary structure have been determined, the stresses can be computed throughout. Then the structural analysis can be easily completed by computing the strains from the stress-strain relations and the displacements using either the stress-displacement relations or the strain-displacement relations.

2. Displacement method

   a. First, the independent unknown displacement components involved in the structure are identified and considered as the basic unknowns involved in the problem. The internal forces in the structure may then be expressed in terms of these unknown displacements, using the stress-displacement relations. For each unknown displacement component, a corresponding equilibrium equation may be written in terms of known external forces and the

---

[2] For purposes of analysis, a structure may usually be classified as being either a member structure or a panel structure, though occasionally it is a mixture of these two types. A **member structure** consists of one or more members which are connected in some manner at their ends, and then the whole assemblage is supported in some fashion at certain discrete locations. Each member of such a structure is relatively long compared with its cross-sectional dimensions. The equilibrium or displacement conditions of a member structure may be defined by a certain number of algebraic equations involving the loads and reactions, the independent force components in the members, and/or the independent displacement components of the joints. In Part II, member structures are referred to as **discrete structures.** A **panel structure** consists of one or more panels which are connected together continuously along their edges or sizable portions thereof. The assemblage may be supported either at discrete points, or continuously over sizable areas, or along lines of considerable length. Each panel of such a structure may be either flat or curved, but of a thickness which is small compared with the length of its edges. The equilibrium or displacement conditions of a panel structure may be defined by a certain number of partial differential equations which involve continuous functions defining the internal stresses and the displacements throughout the structure. The solution of these equations must satisfy any prescribed condition regarding the external forces or displacements on the boundaries of the structure. In Part II, panel structures are referred to as **continuum structures.**

unknown internal forces which have been expressed in terms of the displacements. These equations, equal in number to the unknown displacements, may then be solved simultaneously for their values.

b. Once the displacements have been determined, the internal forces may be back-figured. Thus all the forces will have been computed except the unknown external forces, and these may easily be computed by using the remaining equilibrium equations which were not used initially in setting up the equations for the solution of the unknown displacement components.

Most of the methods of structural analysis that have evolved over the years are classified as force methods. The slope-deflection method and Castigliano's first theorem, to be discussed later in Chaps. 8 and 9, are both displacement methods, however.

Whether a force method or a displacement method is the more efficient method depends on several considerations. Up to the present time, force methods have been much more widely used in the field of civil engineering structures for a number of reasons. First it should be noted that in most cases a complete structural analysis of stresses and also displacements of all points of the structure has not been necessary for design purposes. A complete picture of the stresses has been needed, but only a few representative deflections have been computed. In addition, many civil engineering structures are substantially statically determinate; or if they are statically indeterminate, the degree of indeterminancy is considerably less than the number of unknown displacement components. There are of course certain structures such as many rigid frames where there are fewer unknown displacement components than the degree of statical indeterminancy.

There is one more very important truism which should be recognized regarding analysis of structural behavior. The majority of the actual structures which must be designed by the structural engineer are not susceptible to *precise* analysis. This is particularly true of structures involving complicated space frames and frameworks, slabs, shear walls, and shells. In such cases, an idealized and simplified analytical model of the actual structure is analyzed rather extensively to furnish a basis for designing the actual prototype. Where there is a backlog of experience with a particular type of structure, such simplifications are justified and provide a satisfactory basis for a new design when tempered by the judgment and intuition of an intelligent and experienced engineer.

When appraising the results of his analysis, the engineer must never forget that the results must satisfy the equilibrium requirements. To be "exact," the results must also satisfy the stress-displacement relations. A structure designed on the basis of an analysis which satisfies statics but only approximately satisfies the stress-strain relations will seldom be in serious trouble provided that the designer has been conservative with respect to satisfying buckling and vibrational requirements and has detailed the structure so that it is not susceptible to brittle failure. A structure so designed may deflect somewhat more and in a different manner than really desired. It may even crack or develop plastic deformations at lower loads than anticipated from the analysis. But, if an avenue is provided with sufficient strength to satisfy static equilibrium for the ultimate loading, given sufficient ductility and buckling resistance the structure will mobilize the strength provided before it collapses.

Challenges, Opportunities, and Responsibilities. Primarily, this foreword has been devoted to presenting a perspective of the entire field of structural engineering, with some attention being given to a brief summary of the historical background on which this branch of engineering is based. At this point, however, it is likewise important to acknowledge, at least briefly, some of the challenges that must be faced today in professional practice—not only by structural engineers but also by their fellow practitioners in all other fields of engineering and in all other professions. All of these professionals practice today in a highly dynamic world. In one sense, the basic problems and responsibilities they must face do not change greatly with time. But in another sense, however, rapidly expanding knowledge in each of their professional fields provides them with the capabilities to attack problems that they could not even consider handling in the past. Moreover, the changing needs of society create new problems and require more refined solutions to most all of the traditional problems. Perhaps the key characteristic of most of the more complex contemporary problems is that they are *interdisciplinary* problems. The solution of such a problem requires the combined talents and experiences of all members of the interdisciplinary professional group involved.

Modern society is faced with tremendous interdisciplinary problems involving technological, sociological, political, philosophical, legal, economic, and other factors. Currently, society is particularly concerned with world population growth, environmental pollution, energy shortages, and the waste and over-utilization of many of its natural resources, both replaceable and non-replaceable. Very gradually the public is becoming conscious of the likelihood that society must shift from an *economy of plenty* to an *economy of scarcity*, i.e., a shift from a "swidden" to a "sawah" approach.[1] This transition from one mode of operating society's activities to the diametrically opposite mode cannot be achieved without all of the principal professions counseling the public every step along the way.

In order to contribute such services, the combined engineering professions need to develop active counseling programs to replace the passive assistance the professions have been prone to offer society in the past. Most often when issues have developed, the engineering societies have offered their assistance to governmental and community leaders. But usually these leaders have had little appreciation or understanding of the potential value of the assistance being offered them. Even more unfortunately, the leaders seldom have had the technological vocabulary necessary to phrase their requests for assistance. Engineers therefore have sat in the corner wanting to be helpful, but usually no one called on them for help. As a result, communities and governments have often made their decisions without the benefit of badly needed inputs from engineering.

Perhaps what is needed is for the engineering professions to establish Better Engineering Bureaus in each major community. Such bureaus operating according

[1] Gerald Garvey, From "Swidden" to "Sawah": The Crisis in U.S. Resource Tradition, *M.I.T. Technol. Rev.*, pp. 24–31, March/April 1974.

to professional standards and in a nonpolitical and nonadvocatory manner could provide information services from which individuals and the public could obtain hard facts and documented evidence regarding issues. The bureaus acting as *amicus humani generis* could provide technological evaluations and pro and con arguments regarding legitimate alternate solutions available to society.

In addition to services to the public, the proposed bureaus could develop very valuable services to assist both public and private clients who wish to engage engineering services. In the end, such bureaus could very possibly prove themselves a tremendous boon both to the public and to the engineering profession.

In order to serve effectively and efficiently the ever increasing and more complex needs of his client and/or employer, the civil engineering practitioner must today (and even more so in the future) appreciate and utilize the new developments of science and engineering, which are evolving at a steadily increasing rate. To keep pace with these developments will require a practitioner to commit himself to a never-ending investment in his own continuing education.

Also, if practitioners are to practice in a truly professional manner, they must recognize that their planning and design efforts must blend the engineering and technological aspects with the socio-economic-political-environmental considerations involved. For most practitioners, this challenge will require them to invest in further continuing education aimed at broadening their backgrounds in these nontechnological areas so that they can communicate and work effectively with their fellow professionals from economics, political science, and other social sciences and humanities.

Part I
Basic Structural Theory

The fundamental concepts and the basic definitions and analytical techniques that provide the foundation for the entire field of structural analysis.

1

Introduction

1.1 Engineering Structures. The design of bridges, buildings, towers and other fixed structures is very important to the civil engineer. Such structures are composed of interconnected members and are supported in such a manner that they are capable of holding applied external forces in static equilibrium. A structure must also hold in equilibrium the gravity forces that are applied as a consequence of its own weight. A transmission tower, for example, is acted upon by its own weight, by wind and ice loads applied directly to the tower, and by the forces applied to the tower by the cables that it supports. The members of the tower must be so arranged and designed that they will hold these forces in static equilibrium and thus transfer their effects to the foundations of the tower.

There are many kinds of structures in addition to those mentioned above. Dams, piers, pavement slabs for airports and highways, penstocks, pipelines, standpipes, viaducts, and tanks are all typical **civil engineering** structures. Nor are structures of importance only to the *civil engineer.* The structural frame of an aircraft is important to the *aeronautical engineer;* the structure of a ship receives particular attention from the *naval architect;* the *chemical engineer* is concerned with the structural design of high-pressure vessels and other industrial equipment; the *mechanical engineer* must design machine parts and supports with due consideration of structural strength; and the *electrical engineer* is similarly concerned with electrical equipment and its housing.

The analysis of all these structures is based, however, on the same fundamental principles. In this book the illustrations used to demonstrate the application of these principles are drawn largely from civil engineering structures, but the methods of analysis described can be used for structures that are important in other branches of engineering.

Structural theory and the experimental evidence from laboratory and field tests provide an invaluable basis for structural analysis and design. However, mathematical theory and experimental evidence, no matter how extensive in depth and scope, can never be expected to provide a basis for developing *mathematically precise* design procedures.

First, in order to make a theoretical structural analysis mathematically possible, the geometry and configuration of the structure must be idealized and simplified to obtain an *analytical model* of the actual structure for computational purposes. As a result, the internal forces, stresses, strains, and displacements computed can only be approximations of those quantities for the actual structure.

Second, actual structures cannot be constructed exactly to dimension, nor can structural materials be manufactured precisely to specification. As a result, the resistance of a structure to imposed loads and deformations can be predicted only approximately. And third, actual structures are subjected to natural forces and service conditions that cannot be predicted exactly.

Although an engineer's experience and intuition are extremely valuable to him as he reaches his design decisions, they do not provide an adequate base for pushing much beyond the boundaries of previous engineering experience. Design engineers therefore value most highly structural theory, experimental evidence, and a deep knowledge of structural behavior as they extrapolate into new design areas.

1.2 Structural Design. A structure is designed to perform a certain function. To perform this function satisfactorily it must have sufficient strength and rigidity. Economy and good appearance are further objectives of major importance in structural design.

The complete design of a structure is likely to involve the following five stages:

1. Establishing the general layout to fit the functional requirements of the structure

2. Consideration of the several possible solutions that may satisfy the functional requirements

3. Preliminary structural design of the various possible solutions

4. Selection of the most satisfactory solution, considering an economic, functional, and esthetic comparison of the various possible solutions

5. Detailed structural design of the most satisfactory solution

Both the preliminary designs of stage 3 and the final detailed design of stage 5 may be divided into three broad phases, although in practice these three phases are usually interrelated. First, the loads acting on the structure must be determined. Next, the maximum stresses in the members and connections of the structure must be analyzed. Finally, the members and connections of the structure must be dimensioned, i.e., the make-up of each part of the structure must be determined.

That these three steps are interrelated may be seen from considerations such as the following: the weight of the structure itself is one of the loads that a structure must carry, and this weight is not definitely known until the structure is fully designed; in a statically indeterminate structure, the stresses depend on the elastic properties of the members, which are not known until the main members are designed. Thus, in a sense, the design of any structure proceeds by successive approximations. For example, it is necessary to assume the weights of members in order that they may be properly designed. After the structure is designed, the true weights may be computed; and unless the true weights correspond closely to those assumed, the process must be repeated.

In designing a structure, it is important to realize that each part must have sufficient strength to withstand the maximum stress to which it can be subjected. To compute such maximum stresses, it is necessary to know not only *what* loads may act but the exact *position* of these loads on the structure that will cause the stress under consideration to have its maximum value.

Thus, when a railroad locomotive crosses a bridge, a given portion of the bridge receives its maximum stress with the locomotive at a given position on the bridge. A second part of the structure may be subjected to its maximum stress with the locomotive in another position.

In this book, the emphasis is placed on structural analysis. But in order to discuss structural analysis satisfactorily, it is desirable to first give some attention to the loads acting on a structure and to the design of members and connections.

In the Foreword, it was noted that the traditional basis for structural design has been **elastic design**, in which stresses produced in the structure by service loadings and other design conditions are compared with allowable stresses for the material involved. However, it was also noted that there is a rapidly growing tendency in building codes and standard design specifications to base the design on the ultimate strength and the serviceability of the structure, with the older allowable-stress approach serving as an alternate basis for design. The newer approach goes under the name of **strength design** or **plastic design**, depending on whether reinforced concrete or structural steel is involved.

Some of the building codes and design specifications commonly encountered by civil engineers are the following:

AASHTO Standard Specifications for Highway Bridges, American Association of State Highway and Transportation Officials, Washington, D.C., 1973

ACI Building Code Requirements for Reinforced Concrete (ACI 318–71), American Concrete Institute, Detroit, 1971

AISC Specification for the Design, Fabrication, and Erection of Structural Steel for Buildings, American Institute of Steel Construction, New York, 1971

AISI Specification for the Design of Light Gage Cold-formed Steel Structural Members, American Iron and Steel Institute, New York, 1962

AREA Specifications for Steel Railway Bridges, American Railway Engineering Association, Chicago, 1965

ASTM Standards, American Society for Testing and Materials, Philadelphia, 1967

NBS American Standard Building Code, National Bureau of Standards, Washington, D.C.

UBC Uniform Building Code, International Conference of Building Officials (subsidiary of Pacific Coast Conference of Building Officials), Pasadena, Calif., 1967

1.3   Dead Loads. The dead load acting on a structure consists of the weight of the structure itself and of any other immovable loads that are constant in magnitude and permanently attached to the structure. Thus, for a highway bridge, the dead load consists of the main supporting trusses or girders, the floor beams and stringers of the floor system, the roadway slabs, the curbs, sidewalks, fences or railings, lampposts, and other miscellaneous equipment.

Since the dead load acting on a member must be assumed before the member is designed, one should design the members of a structure in such a sequence that to as great an extent as practicable the weight of each member being designed is a portion of the dead load carried by the next member to be designed. Thus, for a highway bridge, one would first design the road slab, then the stringers that carry the slab loads to the floor beams, then the floor beams that carry the stringer loads to the main girders or trusses, and finally the main girders or trusses.

In designing a member such as a floor slab, stresses due to dead loads are likely to be only a small percentage of the total stress in a member, so that even if dead loads are not very accurately estimated, the total stress can be predicted with fair accuracy and hence the first design can be quite satisfactory. For main trusses and girders, however, the dead loads constitute a greater portion of the total load to be carried, so that it is more important to make a reasonably accurate first estimate of dead weights. It should be emphasized, however, that the original dead-weight estimate is tentative. After a structure is designed, its actual dead weight should be accurately computed and the stress analysis and design revised as necessary. This is necessary for safety and desirable for economy.

1.4   Live Loads—General. As contrasted to dead loads, which remain fixed in both magnitude and location, it is usually necessary to consider live loads, i.e., loads that vary in position. It is sometimes convenient to classify live loads into movable loads and moving loads. Movable loads are those which may be moved from one position to another on a structure, such as the contents of a storage building. They are usually applied gradually and without impact. Moving loads are those which move under their own power, such as a railroad train or a series of trucks. They are usually applied rather rapidly and therefore exert an impact effect on the structure.

When live loads are involved, attention must be given to the placing of such loads on a structure so that the stress in the structural member or connection under consideration will have its maximum possible value. Thus, while we speak of dead stresses due to dead loads, we refer to maximum live stresses due to live loads.

1.5   Live Loads for Highway Bridges. The live load for highway bridges consists of the weight of the applied moving load of vehicles and pedestrians. Actually, the traffic over a highway bridge will consist of a multitude of different types of vehicles. It is designed, however, for a train of standard trucks, so chosen that the bridge will prove safe and economical in its actual performance. The live load for each lane of the roadway consists of a train of heavy trucks following each other closely. The weight and weight distribution of each truck vary with the specification under which one designs, but a typical example is afforded by the H-series trucks specified by the AASHTO.

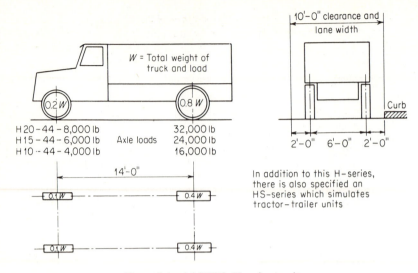

Figure 1.1 AASHTO H-series truck

These H-series trucks are illustrated in Fig. 1.1. They are designated H, followed by a number indicating the gross weight in tons for the standard truck. The choice as to which of the H-series trucks shall be used for the design of a given structure depends on circumstances such as the importance of the bridge and the expected traffic.

It is seen that the loading per lane of roadway consists of a series of concentrated wheel loads. The stress analysis involved in computing maximum live stresses due to a series of concentrated live loads may become rather complicated. Under some conditions it is permissible to substitute for purposes of stress analysis an equivalent loading consisting of a uniform load per foot of lane, plus a single concentrated load. Thus, for the H-20 loading, the equivalent live load consists of a uniform load of 640 lb per lin ft of lane plus a concentrated load of either 18,000 lb or 26,000 lb, depending on whether live moments or shears, respectively, are being computed. This equivalent live load is not exactly equivalent to the series of concentrated wheel loads, but it permits a simpler computation of maximum stresses that correspond closely enough to those which would be computed from the actual loads to be used for design purposes.

1.6 Live Loads for Railroad Bridges. The live load for railroad bridges consists of the locomotives and cars that cross them. The live load for each track is usually taken as that corresponding to two locomotives followed by a uniform load which represents the weight of the cars. For details of the design loadings refer to the AREA specifications.

1.7 Live Loads for Buildings. Live loads for buildings are usually considered as movable distributed loads of uniform intensity. The intensity of the floor loads to be used depends on the purpose for which the building is designed, as indicated in Table 1.1.

Table 1.1

	Minimum live load, lb per sq ft
Human occupancy:	
Private dwellings, apartment houses, etc.	40
Rooms of offices, schools, etc.	50
Aisles, corridors, lobbies, etc., of public buildings	100
Industrial or commercial occupancy:	
Storage purposes (general)	250
Manufacturing (light)	75
Printing plants	100
Wholesale stores (light merchandise)	100
Retail salesrooms (light merchandise)	75
Garages:	
All types of vehicles	100
Passenger cars only	80
Sidewalks 250 lb per sq ft or 8,000 lb concentrated, whichever gives the larger moment or shear	

1.8 Impact. Unless live load is applied gradually, the deformation of the structure to which the live load is applied is greater than it would be if the live load were considered as a static load. Since the deformation is greater, the stresses in the structure are higher. The increase in stress due to live load over and above the value that this stress would have if the live load were applied gradually is known as impact stress. Impact stresses are usually associated with moving live loads. For purposes of structural design, impact stresses are usually obtained by multiplying the live-load stresses by a fraction called the impact fraction, which is specified rather empirically. The determination of a wholly rational fraction for this purpose would be very complicated, since it depends on the time function with which the live load is applied, the portion of the structure over which the live load is applied, and the elastic and inertia properties of the structure itself.

For highway bridges, the impact fraction I is given in the specifications of the AASHTO by

$$I = \frac{50}{L + 125} \qquad \text{but not to exceed 0.300} \qquad (1.1)$$

in which L is the length in feet of the portion of the span loaded to produce the maximum stress in the member considered.

1.9 Snow and Ice Loads. Snow loads are often of importance, particularly in the design of roofs. Snow should be considered as a movable load, for it will not necessarily cover the entire roof and some of the members supporting the roof may receive maximum stresses with the snow covering only a portion of the roof. The density of snow, of course, will vary greatly, as will the fall of snow to be expected in different regions. In a given locality, the depth of snow that will gather on a given roof will depend on the slope of the roof and on the roughness of the roof surface. On flat roofs in areas subjected to heavy snowfalls, snow load may be as large as 60 or even 90 lb per sq ft. Whether or not snow and wind loads should be assumed to act simultaneously on a roof is problematical, since a high wind is likely to remove much of the snow.

Ice loads may also be of importance, as, for example, in designing a tower built up of relatively small members which have proportionately large areas on which ice may gather. Ice having a density equal approximately to that of water may build up to a thickness of 2 in. or more on such members. It may also build up to much greater thicknesses, but when it does, it is apt to contain snow or rime and hence have a lower density. When ice builds up on a member, it alters the shape and the projected area of the member. This should be considered in computing wind loads acting on members covered with ice.

1.10   Lateral Loads—General. The loadings previously discussed usually act vertically, although it is not necessary that live loads and their associated impact loads shall act in that direction. In addition, there are certain loads that are almost always applied horizontally, and these must often be considered in structural design. Such loads are called lateral loads. We shall now consider some of the more important kinds of lateral loads.

Wind loads, soil pressures, hydrostatic pressures, forces due to earthquakes, centrifugal forces, and longitudinal forces usually come under this classification.

1.11   Wind Loads. Wind loads are of importance, particularly in the design of large structures, such as tall buildings, radio towers, and long-span bridges, and for structures, such as mill buildings and hangars, having large open interiors and walls in which large openings may occur. The wind velocity that should be considered in the design of a structure depends on the geographical location and on the exposure of the structure. For most locations in the United States, a design to withstand a wind velocity of 100 mph is satisfactory.

Design loads for wind recommended in various specifications, like design loads for live load, should not be considered as attempts to represent actual wind loadings. Experience has shown that use of the recommended values generally results in a design of adequate strength and also of adequate rigidity.

The AASHTO specifies that the wind force on the superstructure of highway bridges shall be assumed as a movable horizontal load equal to 75 lb per sq ft for trusses and arches and 50 lb per sq ft for girders and beams. These forces are for a wind velocity of 100 mph. The exposed area is the sum of the areas of all members, including floor system and railing, as seen in elevation at 90° to the longitudinal axis of the structure.

The task Committee on Wind Forces of the Structural Division of the American Society of Civil Engineers (ASCE) compiled an excellent report on the wind force acting on various types of structures.[1] For tall buildings, the committee recommended the following approximate formula for the pressure p in pounds per square foot on exposed vertical surfaces normal to the wind:

$$p = 0.0033V^2 \tag{1.2}$$

[1] The final report of the committee, entitled Wind Forces on Structures, was published in the ASCE Transactions, vol. 5, no. 126, pt. 2, pp. 1124–1198, 1961.

in which V is the wind velocity in miles per hour. This pressure represents the combined net effect of the pressure on the windward side and the suction on the leeward side.

1.12 Soil Pressures. Load on retaining walls, on walls of buildings, and on other structures due to the pressure of soil must frequently be considered by the structural engineer. The lateral pressure caused by soil on a wall varies when the wall yields. After a small movement of the wall, the soil pressure reaches a minimum value known as the active pressure. If, on the other hand, the wall is forced into the backfill, the pressure between the wall and the backfill increases to a maximum value known as the passive pressure. Under usual conditions, the active pressure at any depth is about $\frac{1}{4}$ times the vertical pressure, and the passive pressure is about 4 times the vertical pressure. Since the lateral pressure would equal the vertical pressure if the material were a fluid, the approximate values of $\frac{1}{4}$ and

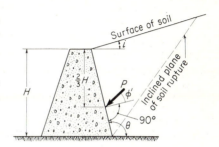

Figure 1.2 Retaining wall

4 are sometimes called **hydrostatic-pressure ratios** for the active and passive cases, respectively.

According to the above discussion, any wall that may yield without detrimental results may be designed on the basis of active pressure, although the pressure that will actually act on the wall will in general be somewhat above this value. Further, the distribution of pressure over such a wall may be assumed to be triangular, although this assumption is not strictly correct.

For a more complete treatment of soil pressures, the student is referred to books on soil mechanics.

1.13 Hydrostatic Pressures. Dams, tanks, etc., are subjected to hydrostatic loads that as a rule may be easily computed in accordance with the elementary principles of hydraulics. Hydrostatic loads should in general be considered as movable loads, inasmuch as critical stresses in a structure do not necessarily occur when the liquid involved is at its highest possible level. In some structures, the presence of certain hydrostatic pressures actually relieves stresses in a structure. Thus an underground tank might be more likely to collapse when empty than when full, or a tank built above the ground might undergo a critical-stress condition when it is only partly filled.

1.14 Earthquake Forces.[1] Important structures located in regions subject to severe earthquakes are often designed to resist earthquake effects. During an earthquake, structural damage may result from the fact that the foundation of the structure undergoes accelerations. Such accelerations are largely horizontal, and

---

[1] For a detailed discussion, refer to the Uniform Building Code and the Recommended Lateral Force Requirements of the Structural Engineers Association of California, San Francisco, 1967. Refer also to Robert L. Wiegel, "Earthquake Engineering," Prentice-Hall, Inc., Englewood Cliffs, N.J., 1970.

vertical components of acceleration are usually neglected.   In active earthquake zones, the maximum rate of horizontal acceleration of the foundations may reach values having a magnitude between 0.5 and 1.0 times g, the acceleration due to gravity; that is between 16 and 32 ft per sec^2.   If the structure is assumed to act as a rigid body it will accelerate horizontally at the same rate as its foundations. Hence each part of the structure will be acted upon by a horizontal inertia force equal to its mass multiplied by its horizontal acceleration, or, for example

$$\text{Lateral force} = \frac{\text{weight}}{g} \times 0.5g = \text{one-half of its weight}$$

Structures are sometimes designed to resist earthquakes on the foregoing basis, although it is quite approximate, inasmuch as the assumption that the entire structure accelerates as a rigid body is usually not particularly valid.

The horizontal acceleration of a structure such as a dam will produce not only horizontal inertia forces due to the mass of the dam but also hydrodynamic forces as the dam moves rapidly into the water that it retains.

1.15   Centrifugal Forces.   In designing a bridge on which the tracks or roadway are curved, vehicles crossing the structure exert centrifugal force that may be of sufficient magnitude to require consideration in design.   Such centrifugal forces are lateral loads and should be considered as moving loads.

1.16   Longitudinal Forces.   For a bridge, horizontal forces acting in the direction of the longitudinal axis of the structure, i.e., in the direction of the roadway, are called **longitudinal forces.**   Such forces are applied whenever the vehicles crossing the structure increase or decrease their speed.   Since they are inertia forces resulting from the acceleration or deceleration of vehicles, they act through the centers of gravities of the vehicles.   The magnitude of such forces is limited by the frictional forces that can be developed between the contact surfaces of the wheels of the vehicles applying these forces to the roadway or track and the surface of the roadway or track.

1.17   Thermal Forces.   Changes in temperature cause strains in the members of a structure and hence produce deformations in the structure as a whole.   If the changes in shape due to temperature encounter restraint, as is often the case in a statically indeterminate structure, stresses will be set up within the structure.   The forces set up in a structure as a result of temperature changes are often called **thermal forces.**   In addition to considering the forces set up by changes in temperature, it is important to take into consideration the expansion and contraction of a structure, particularly in connection with support details.

In a moderate climate, one should consider a variation in temperature of 0 to 120°F.   In cold climates, this range should be extended to from -30 to 120°F.

1.18   Make-up of Girders.   In a structure built of structural steel, rolled sections such as standard I beams or wide-flanged beams are commonly used to support loads across a span.   When rolled sections can be used, they are economical because they require less fabrication than built-up girders.   For longer spans and for heavier loadings, however, the bending moments and shears will be found to be too large to be carried safely by rolled sections, and built-up girders must be

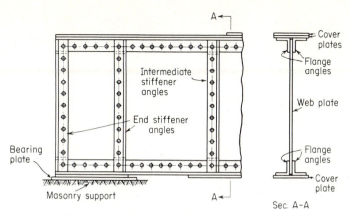

Figure 1.3 Plate girder components

used.   The most important parts of a typical built-up plate girder, as shown in Fig. 1.3, are the web plate, the top flange, which is composed of flange angles and cover plates, and the bottom flange, which is similarly composed.   Over the end bearing plates are vertical angles called **end stiffeners,** while at other points along the span it is usually necessary to have further vertical angles, which are called **intermediate stiffeners.**   The component parts of the girder shown in Fig. 1.3 are riveted together.   Welding is often used instead of riveting, in which case the details of the girder differ somewhat, but the essential component parts, namely, the web, top and bottom flange, and web stiffeners, must still be provided.

1.19   Make-up of Trusses.   Fabrication, shipping, and erection considerations usually limit the depth of built-up girders to about 10 ft.   When bending moments and shears are so large that they cannot be carried by a girder of that depth, it becomes necessary to employ a truss.   The layout of a typical truss is shown in Fig. 1.4a.   Members $L_0L_1, L_1L_2, \ldots, L_5L_6$ are called **bottom chords;** members $U_1U_2, U_2U_3, \ldots, U_4U_5$ are called **top chords;** members L_0U_1 and U_5L_6 are called **end posts** and are often included as top-chord members; members $U_1L_2, L_2U_3, \ldots, L_4U_5$ are called **diagonals;** members $U_1L_1, U_2L_2, \ldots, U_5L_5$ are called **verticals.**   Figure 1.4b shows a typical connection detail for the members of a riveted truss and the make-up of typical truss members.   The plates to which the members intersecting at a joint are connected are called **gusset plates.**   For long-span trusses, members are sometimes connected at a joint by a pin passing through the webs of the members themselves or through pin plates connected to these webs.   Trusses may be welded rather than riveted.

1.20   Make-up of Floor Systems.   For a plate-girder railroad bridge, if the girders are not too far apart, and if the track is located at the top of the girders, it is economical to have the ties rest directly on the cover plates of the top flanges of the girders.   Such bridges are called **deck structures.**   As the distance between girders becomes larger, such construction becomes uneconomical.   It is moreover obvious that it cannot be followed if the tracks are to be located below the tops of the girders.   Under either of the foregoing circumstances, it becomes necessary to

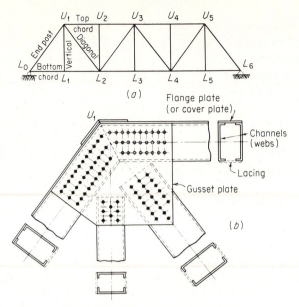

Figure 1.4 Riveted truss components

build up a floor system composed of stringers and floor beams, as shown in Fig. 1.5. If such a built-up floor system is at the top of the girders or trusses, the bridge is still called a deck structure; if it is at the bottom of the girders or trusses, the bridge is called a **through structure;** if it is at an intermediate elevation, it is called a **half-through structure.** Figures 1.5*a* and *b* illustrate a half-through single-track railroad bridge and show the make-up of floor systems for such bridges. The ties

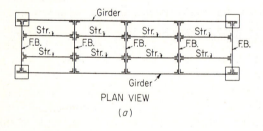

PLAN VIEW
(*a*)

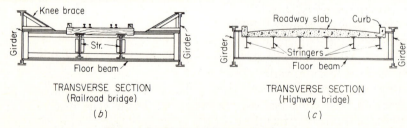

TRANSVERSE SECTION
(Railroad bridge)
(*b*)

TRANSVERSE SECTION
(Highway bridge)
(*c*)

Figure 1.5 Girder bridge cross sections

rest directly on members called **stringers,** which are parallel to the main girders. These stringers frame into members called floor beams, which are transverse to, and frame into, the main girders.   Thus a load applied to the rails is transferred by the ties to the stringers, which carry the load to the floor beam.   The floor beams carry the load to the girders, which in turn transfer the load to the foundations of the structure.

Figure 1.5c shows a transverse section through a typical highway bridge with a floor system.   Such a floor system is similar to that described for a railroad bridge, with the floor slab of the bridge resting on stringers that frame into floor beams, which in turn frame into the main girders.

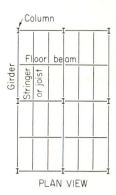

PLAN VIEW

Figure 1.6   Floor framing of building

Similar framing may be used for the floor of a building, as shown in Fig. 1.6.   In this case the floor slab would rest on the floor beams and girders as well as on the stringers, but loads applied to the floor slab directly over a stringer would be carried by the stringer to the floor beams, thence to the girders, thence to the columns, and finally to the foundations.

For bridges in which the main load-carrying elements are trusses rather than girders, floor systems are always used. The floor beams are located at the panel points (joints) of the loaded chord, so that the members of the truss will not be subjected to transverse loads and thus undergo primary bending.

1.21   Bracing Systems.   Trusses, girders, floor beams, stringers, columns, etc., which are designed primarily to carry vertical loads, are often referred to as the **main members** of a structure.   In addition to the main members, most structures require bracing systems, which serve a number of purposes, the most important being those of resisting lateral loads and preventing buckling.   The most important bracing systems of a typical through truss bridge are shown in Fig. 1.7.   The top-chord lateral system lies in the plane of the top chords and consists of a cross strut at each top-chord panel point and diagonals connecting the ends of these cross struts.   The bottom-chord lateral system lies in the plane of the bottom-chords and consists of the floor beams, which occur at each bottom-chord panel point and which serve as cross struts, and diagonals connecting the ends of the floor beams.   In the planes of the end posts is the so-called **portal bracing,** which stiffens the end posts laterally.

1.22   Allowable Stresses.   When a structure is being designed by the **elastic-design method,** the structural analyses are performed for the various design loading conditions to obtain the bending moments, the twisting moments, the shear forces, and the axial forces at various cross sections throughout the structure.   From this information, it is possible to compute the intensities of the normal and shear stresses on these cross sections.   In order that the structure shall perform satisfactorily, it is necessary that these stresses which have been developed by the design load shall not exceed certain allowable limits which have been specified for the

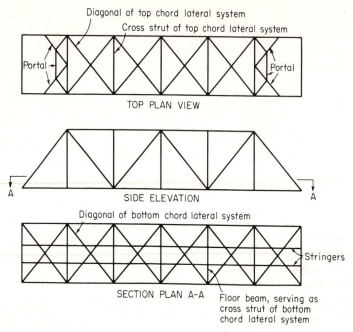

Figure 1.7 Lateral bracing systems of through truss

material being used. These limiting values which have been specified are called **allowable stresses.**[1]

In any practical problem there are always certain ambiguities regarding the physical dimensions which should be used in the calculations. The various assumptions which may be made in order to rationalize such difficulties can, of course, have a marked effect on the values computed. Specifications such as those of AASHTO, ACI, AISC, and AREA therefore not only state values for allowable stresses, but also define certain aspects of the basis on which they should be compared with the stresses developed by the design loads. For example, a transverse section through a structural member is likely to pass through rivets. Since a well-driven rivet is tight and substantially fills the rivet hole, it is customary to assume that in transferring compressive stresses the member is not weakened by rivet holes. The area of a transverse section, with no deduction made for rivet holes, is called the **gross area.** Even though the rivet is well driven, however, it is not customary to assume that tensile stresses can be transferred across the rivet hole. The area of a transverse section, reduced by the area of those rivet holes located where the section is in tension, is called the **net area.** In defining allowable stresses, it is necessary to define whether they are applicable to the net or gross area (or section).

[1] As used here, stress (or stress intensity) denotes force per unit area, such as pounds per square inch. Usage of this term is not uniform in structural engineering literature. Sometimes, for example, the total tensile or compressive force acting on the cross sections of a truss member is referred to as the stress in the member. Herein, the term **stress** denotes force per unit area, and the term **force** denotes the total of the stresses integrated over the area of a cross section.

1.23 Factor of Safety. In the Foreword, it is noted that the values of the allowable stresses used in the elastic-design method are chosen so as to provide an adequate *margin of safety* against exceeding the stress or strain corresponding to the yield of the material being used. Of course, the selection of the allowable stresses, design loads, and other design restrictions may also be modified by consideration of the possibility of failure due to fatigue, buckling, or brittle fracture, or by consideration of the permissible deflections. If for the design conditions a structure has been designed to utilize exactly the allowable tensile stress of 20,000 psi for structural steel with a specified yield stress of 33,000 psi, the margin of safety (or factor of safety) provided against tensile yielding is 1.65, that is, 33,000/20,000.

It should be obvious that in actuality this margin of safety cannot be defined this precisely. The actual loads on a structure, particularly those of certain types, cannot be predicted or defined with much precision. Certain simplifying assumptions are always made in the stress analysis, so that even for the assumed design loads the computed stresses are not precise. Even though care is used in manufacturing and fabricating the structural materials employed, some variations in the properties of these materials exist, even when the materials are new. With the passage of time, partial disintegration must be expected. There are always therefore certain variations in the estimates of the actual stress in the structure and of the stress at which it starts to yield.[1] Thus, ideally, the margin of safety should be studied in a statistical manner to ascertain if there is a satisfactorily small probability that the possible variation in the calculated design stress overlaps the possible variation in the yield stress.

Conventionally, the factor of safety is defined as above. That is, the factor of safety against yielding is the ratio of the yield stress of the material employed to the allowable design stress. Admittedly this is a rather crude way of handling this matter. It must be kept in mind, however, that the factors of safety implied in current standard specifications reflect years of experience during which design procedures have been correlated with the actual behavior of structures, usually successful behaviors but occasionally unsuccessful.[2]

The magnitudes of the load factors used in the plastic-design methods depend upon considerations similar to those noted above for the factor of safety.

---

[1] Due to such variations, the inflated value $F + \Delta F$ of the actual internal force F should not exceed the deflated value $R - \Delta R$ of the resistance of the material R. Thus,

$$F + \Delta F \leq R - \Delta R \qquad \text{or} \qquad F\left(1 + \frac{\Delta F}{F}\right) \leq R\left(1 - \frac{\Delta R}{R}\right)$$

The minimum factor of safety (FS) is then

$$\text{FS} = \frac{R}{F} = \frac{1 + \Delta F/F}{1 - \Delta R/R}$$

If each variation ΔF and ΔR is 25 per cent of F or R, respectively, the minimum factor of safety required is

$$\text{FS} = \frac{1 + 0.25}{1 - 0.25} = 1.67$$

[2] See discussion on p. 11 of Foreword.

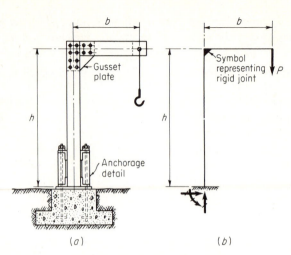

Figure 1.8 Idealized structure

1.24 Practical and Ideal Structures.   Rarely if ever does an actual structure correspond to the idealized structure that is considered in its analysis. The materials of which the structure is built do not have the exact properties assumed, nor do the dimensions correspond exactly with their theoretical values. Structural details such as lacing bars and gusset plates introduce effects that might make an analysis very complicated indeed; but because they have little actual effect, they are usually neglected in the analysis of stresses in main members. Because of the width of members, considerable difference may exist between clear spans and center-to-center spans which are ordinarily used in analyses. Support details may vary considerably from the idealized type assumed for purposes of analysis. A member may not actually be prismatic, and yet it may be assumed to be, to simplify computations.

The practical structure of Fig. 1.8*a*, for example, might be analyzed on the basis of the idealized structure of Fig. 1.8*b*, in which the footing has been assumed to be perfectly fixed, although it could not be so in nature; the extra column area of the anchorage detail and the material of the gusset plate have been ignored; the gusset plate has been assumed as a truly rigid connection, whereas it will actually permit some rotational yielding between the column and the horizontal girder; the effective height of the column has been taken as the distance from the top of the bedplate to the center line of the girder; and the effective span of the girder has been measured from the center line of the column to the center of the applied load.

It is necessary to idealize a structure in order to carry out a practical analysis. Experience and judgment are necessary in determining the idealized structure, i.e., the **analytical model** that should be used in a given case. In important structures, where doubt exists as to the most logical assumptions to be made in idealizing a structure, it is sometimes desirable to compute stresses on the basis of more than one possible analytical model, and to design the structure to resist the stresses corresponding to all the analyses.

2

Reactions

2.1 Definitions. Practically all structures are arranged in three-dimensional configurations of structural elements. However, under maximum design loading conditions, many of these configurations behave in such a manner that certain portions of the structure are substantially in a two-dimensional stress condition. As a result, the structural analyst is justified in isolating such portions and treating them as two-dimensional structures having the simple geometry associated with planar systems.

Since two-dimensional behavior is so common, the major portion of this book will be devoted to the analysis of so-called **planar structures**, i.e., structures that may be considered to lie in a plane that also contains the lines of action of all the forces acting on the structure. The analysis of three-dimensional, or space, structures involves no new fundamental principles beyond those required for planar structures, but the numerical computations are greatly complicated by the additional geometry introduced by the third dimension.

The discussion in this chapter will be limited to planar structures, and so all the force systems will be so-called **coplanar force systems**, i.e., systems consisting of several forces the lines of action of which all lie in one plane. Some of these systems have special characteristics, and it will be found convenient to classify them accordingly and to identify them by special names.

A **concurrent coplanar force system** is shown in Fig. 2.1a. This system consists of several forces the lines of action of which all intersect at a common point. A **parallel coplanar force system** consists of several forces the lines of action of which are all parallel, as shown in Fig. 2.1b A **general coplanar force system** consists of several forces the lines of action of which are in various directions and do not intersect in a common point, as shown in Fig. 2.1c. Another important system, a couple, is shown in Fig. 2.1d. A **couple** consists of two equal and opposite parallel forces that do not have a common line of action.

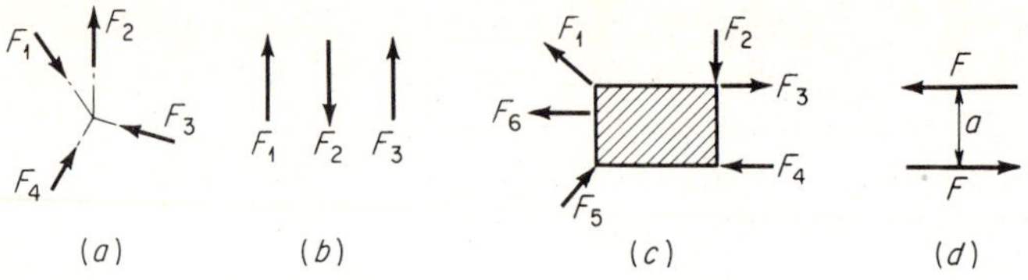

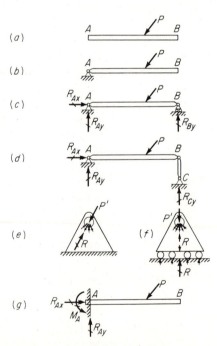

Figure 2.1   Coplanar force systems

Since, in a planar structure, the lines of action of all the forces lie in the plane of the structure, each of the forces F can be resolved into two components F_x and F_y, where the x and y reference axes may be taken in any direction as long as they do not coincide.   It is almost always desirable to select x and y axes that are mutually perpendicular, in which case F_x and F_y are called **rectangular components**.   Further, it is usually most convenient to take the x axis horizontally and the y axis vertically.

2.2   **General—Conventional Supports.**   Most structures are either partly or completely restrained so that they cannot move freely in space.   Such restrictions on the free motion of a body are called **restraints** and are supplied by supports that connect the structure to some stationary body.   For example, consider a planar structure such as the bar AB shown in Fig. 2.2.   If this bar were a free body and were acted upon by a force P, it would move freely in space with some combined translatory and rotational motion.   If, however, a restraint were introduced in the form of a hinge that connected the bar to some stationary body at point A, then the motion of the bar would be partly restricted and could consist only of a rotational movement about the hinge.   During such a rotation, point B would move along an arc with point A as the center.   Instantaneously, point B could be considered to move normal to the line AB, or, in this case, vertically.   If, therefore, another restraint were introduced that would not allow point B to move instantaneously in a vertical direction, the rotation about the hinge at point A would be prevented and thus the free motion of the bar would be completely restricted.   It is evident that this type of restraint would be supplied by the supports shown at B in either c or d of Fig. 2.2.

The supports at A and B, in restricting the free motion of the bar, are called upon to resist the action that the force P imposes upon them through the bar.   The

Figure 2.2   Support reactions

resistances they thus develop to counteract the action of the bar upon them are called **reactions**. The effect of the supports may therefore be replaced by the reactions that they supply to the structure.

In future discussions, it will be necessary to deal continually with the reactions that different types of supports supply, and it therefore will be convenient to use a conventional symbol for describing these different types. A **hinge support**, as shown in Fig. 2.2e, is represented by the symbol ⟋⟍ . In such a support, if it is assumed that the pin of the hinge is frictionless in the pinhole, the contact pressures between the pin and its hole must remain normal to the circular contact surface and must therefore be directed through the center of the pin. The reaction R that the support supplies to the structure must completely counteract the action of the force P', and therefore R and P' must be collinear and numerically equal but must act opposite to one another. It is therefore evident that a hinge support supplies a reactive force the line of action of which is known to pass through the center of the hinge pin but the magnitude and direction of which are unknown.[1] These two unknown elements of such a reaction could also be represented by the unknown magnitudes of its horizontal and vertical components, R_x and R_y, respectively, both acting through the center of the hinge pin.

A **roller support**, as shown in Fig. 2.2f, is represented by either the symbol ⟋⟍ or ⟋⟍ . In the same manner as above, it may be reasoned that the reactive force of a roller support must be directed through the center of the pin. In addition, however, if the rollers are frictionless, they can transmit only a pressure which is normal to the surface on which they roll. Hence, a roller support supplies a resultant reactive force which acts normal to the surface on which the rollers roll and is directed through the center of the hinge pin. It is therefore evident that a roller support supplies a reactive force which is applied at a known point and acts in a known direction but the magnitude of which is unknown. Roller supports are usually detailed so that they can supply a reaction acting either away from or toward the supporting surface.

Consideration of the **link support** BC shown in Fig. 2.2d shows that for small movements it effectively reproduces the action of the roller support shown at B in Fig. 2.2c. By the same approach as used above, it may be reasoned that if the pins at the ends of this link are frictionless, the force transmitted by the link must act through the centers of the pins at each end. Therefore, a link support also supplies a reaction of a known direction and a known point of application but of an unknown magnitude. A link support is denoted by the symbol ⟋⟍ .

One other common type of support used for planar structures is that shown at A in Fig. 2.2g and called a **fixed support**. Such a support encases the member so that both translation and rotation of the end of the member are prevented. A fixed support therefore supplies a reaction, the magnitude, point of application, and direction of which are all unknown. These three unknown elements may also

---

[1] According to this terminology, the **direction of a force** is intended to define the slope of its line of action, while the **magnitude** indicates not only its numerical size but also the sense in which the force acts along this line of action, i.e., whether toward or away from a body.

be considered to be a force, which acts through a specific point but has an unknown magnitude and direction, and a couple of unknown magnitude. For example, the three unknown elements could be selected as a couple and a horizontal and a vertical force, the two latter acting through the center of gravity of the end cross section. A fixed support is designated by the symbol ⊢ .

The external forces acting on a body therefore consist of two distinct types—the applied loads P (shown thus →) and the reactions R (shown thus ↦). A reactive force at support a will be designated as R_a, while its x and y components will be called R_{ax} and R_{ay}. A reactive couple at support a will be called M_a.

2.3 Equations of Static Equilibrium—Planar Structures. If the supports of a planar structure are considered to be replaced by the reactions they supply, the structure will be acted upon by a general coplanar force system consisting of the known applied loads and the unknown reactions. In general, the resultant effect of a general coplanar force system could be either a resultant force, acting at some point and in some direction in the plane, or a resultant couple.

A body that is initially at rest and remains at rest when acted upon by a system of forces is said to be in a **state of static equilibrium**. *For such a state to exist, it is neccessary that the combined resultant effect of the system of forces shall be neither a force nor a couple; otherwise, there will be a tendency for motion of the body.* If the combined resultant effect of a general system of forces acting on a planar structure is not to be equivalent to a resultant force, the algebraic sum of all the F_x components must be equal to zero and the algebraic sum of all the F_y components must be equal to zero. If the combined resultant effect is not to be equivalent to a couple, the algebraic sum of the moments M_z of all the forces about any axis parallel to the z reference axis and normal to the plane of the structure must also be equal to zero. The three following conditions must therefore be fulfilled simultaneously by the loads and reactions of a planar structure for the structure to remain in a state of static equilibrium:

$$\sum F_x = 0 \qquad \sum F_y = 0 \qquad \sum M_z = 0 \tag{2.1}$$

These equations are called the **equations of static equilibrium** of a planar structure subjected to a general system of forces.

In the special case where a planar structure is acted upon by a concurrent system of loads and reactions, it is impossible for the resultant effect of the system to be a couple, for the lines of action of all the forces of a concurrent system intersect in a common point. Therefore, for the structure to remain in static equilibrium in such a case, it is necessary only that the two following conditions be satisfied:

$$\sum F_x = 0 \qquad \sum F_y = 0 \tag{2.2}$$

These equations are the equations of static equilibrium for the special case of a planar structure subjected to a concurrent system of forces.

Quite often in the discussions that follow, structural members will be referred to as being **rigid bodies**. In an exact sense, a rigid body is one in which there is *no* relative movement between any two particles of the body. Of course, any structural element is never absolutely rigid since it is made of materials that deform slightly under the loads imposed on them. However, such deformation is so slight that the changes of dimension, the shifting of the lines of action of forces, etc., may usually be neglected during the investigation of the condition of equilibrium of the body. Thus, in most problems, in applying the equations of static equilibrium to structural elements, it will be assumed that they are rigid bodies for all practical purposes and hence that the geometry after application of the loads is essentially the same as before.

Vector Notation. Most statics problems can be handled easily by a direct scalar-notation approach, which also makes it easy to observe and appreciate the concepts and the structural behavior involved. However, dynamics problems, complex derivations, and the systematic organization of computations are all facilitated by utilizing vector notation with its formalized rules and conventions. With one or two minor exceptions, scalar notation is used throughout Parts I and III. In Part II, however, vector notation is frequently used.

The essentials of vector notation and vector algebra are summarized here for ready reference (see Fig. 2.3).

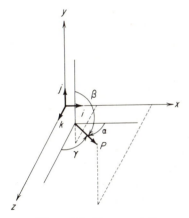

Figure 2.3   Vector notation

Let P be any vector (force, moment, or position vector); let P_x, P_y, and P_z be the x, y, and z components, respectively, of this vector; let α, β, and γ be the angles the direction of this vector makes with the x, y, and z axis, respectively; and let l, m, and n be the direction cosines corresponding, respectively, to α, β, and γ. Let i, j, and k be unit vectors on the direction of the positive x, y, and z axes, respectively; and let p be a unit vector in the direction of P.

Then typically, the following relationships exist between vectors and their corresponding scalar values:

$$P = |P|\, p \quad \text{or} \quad P = Pp$$

and

$$\begin{aligned} P &= P_x + P_y + P_z \\ &= P_x i + P_y j + P_z k \\ &= Pl\, i + Pm\, j + Pn\, k \end{aligned} \tag{2.3}$$

Specifically, for a force, moment, or position vector

$$F = F_x i + F_y j + F_z k = Fl\, i + Fm\, j + Fn\, k$$
$$M = M_x i + M_y j + M_z k = Ml\, i + Mm\, j + Mn\, k$$
$$r = r_x i + r_y j + r_z k = rl\, i + rm\, j + rn\, k$$

Vector Algebra. A few of the essential operations of vector algebra for dealing mathematically with vectors are the following:

1. Addition or subtraction of vectors is commutative:

$$A + B = B + A$$
$$A - B = A + (-B) = -B + A$$

2. Addition or subtraction of vectors is associative:

$$A + (B + C) = (A + B) + C$$
$$A - (B + C) = -(B + C) + A$$

3. Multiplication of a vector by a scalar is commutative, associative, and distributive:

$$mA = Am$$
$$m(nA) = n(mA) = mnA$$
$$(m + n)A = mA + nA$$
$$m(A + B) = mA + mB$$

4. The **scalar,** or **dot, product** of two vectors is a scalar quantity,

$$A \cdot B = |A| |B| \cos \theta \tag{2.4}$$

where θ is the smaller angle between vectors A and B placed tail to tail.

Note that scalar products are commutative and distributive:

$$A \cdot B = B \cdot A$$
$$A \cdot (B + D) = A \cdot B + A \cdot D$$

Note that for orthogonal unit vectors the following scalar-product relations exist:

$$i \cdot i = |i| |i| = 1 \qquad j \cdot j = 1 \qquad k \cdot k = 1$$
$$i \cdot j = j \cdot i = 0 \qquad i \cdot k = k \cdot i = 0 \qquad j \cdot k = k \cdot j = 0$$

5. The **vector,** or **cross, product** of two vectors is a vector quantity,

$$A \times B = C \tag{2.5}$$

$$\text{where} \quad |C| = |A| |B| \sin \theta \tag{2.5a}$$

where θ is the smaller angle between the two vectors placed tail to tail and C is a vector acting normal to the plane containing A and B and with its positive sense defined by the advancement of a right-handed screw revolving from A to B through the angle θ.

Note that vector products are distributive but noncommutative:

$$A \times B = -B \times A$$
$$A \times (B + D) = A \times B + A \times D$$

Note that for orthogonal unit vectors the following cross-product relations exist:

$$i \times i = 0 \qquad j \times j = 0 \qquad k \times k = 0$$
$$i \times j = k = -j \times i \qquad i \times k = -j = -k \times i$$

and
$$j \times k = i = -k \times j$$

Equilibrium Conditions Expressed in Vector Notation. For equilibrium of a three-dimensional system of forces F and couples C, the following two vectorial equations must be satisfied:

$$\sum F = 0 \tag{2.6}$$
$$\sum r \times F + \sum C = 0 \tag{2.7}$$

When the vectors are expressed in terms of their scalar equivalents, these equations become

$$\sum (F_x i + F_y j + F_z k) = 0$$
$$\sum [(xi + yj + zk) \times (F_x i + F_y j + F_z k)] + \sum (C_x i + C_y j + C_z k) = 0$$

When the indicated operations are performed, the equations expand to the following:

$$i \sum F_x + j \sum F_y + k \sum F_z = 0$$

$$i \left[\sum (F_z y - F_y z) + \sum C_x \right] + j \left[\sum (F_x z - F_z x) + \sum C_y \right] + k \left[\sum (F_y x - F_x y) + \sum C_z \right] = 0$$

Since the unit vectors i, j, and k are each equal to unity, the only way that these equations can be satisfied is for the coefficients of the unit vectors to be equal to zero, or

$$\sum F_x = 0 \qquad \sum F_y = 0 \qquad \sum F_z = 0 \qquad\qquad (2.8)$$

and

$$\sum (F_z y - F_y z) + \sum C_x = \sum M_x = 0$$

$$\sum (F_x z - F_z x) + \sum C_y = \sum M_y = 0$$

$$\sum (F_y x - F_x y) + \sum C_z = \sum M_z = 0$$

$$(2.9)$$

Equations (2.8) and (2.9) are immediately recognized as the six *scalar* equations of static equilibrium that must be satisfied if a three-dimensional force system is to be in equilibrium.[1]

2.4 Equations of Condition. Many structures consist of simply one rigid body—a truss, frame, or beam—restrained in space by a certain number of supports. Sometimes, however, the structure may be built up out of several rigid bodies partly connected together in some manner, the whole assemblage then being mounted on a certain number of supports. In either type of structure, the force system consisting of the loads and reactions must satisfy the equations of static equilibrium if the structure is to remain at rest. In the latter type of structure, however, the details of the method of construction used to connect the separate bodies together may enforce further restrictions on the force system acting on the structure. The separate parts may be connected together by hinges, links, or rollers in some way, and in each case these details can transmit only a certain type of force from one part of the structure to the other.

Figure 2.4 illustrates the latter type of structure. It is composed of two rigid members *ab* and *bc* connected together by a frictionless hinge at point *b* and supported by hinge supports at points *a* and *c*. Since a frictionless hinge cannot transmit a couple, its insertion imposes the condition that the action of one portion of the structure on the other portion connected by the hinge can consist only of a force acting through the center of the hinge pin. Therefore, the algebraic sum of the moments of the loads and reactions applied to any *one* such portion of the structure taken about an axis through the center of the hinge pin must be equal to zero. Such conditions introduced by the method of construction (other than the manner in which the supports are detailed) result in so-called **equations of construction** or **condition.**

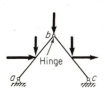

Figure 2.4 Three-hinged arch

2.5 Static Stability and Instability—Statically Determinate and Indeterminate Structures Not Involving Equations of Condition. Consider first a planar structure which is acted upon by a general system of loads and into which no equations of construction have been introduced. If the supports are replaced by the reactions

---

[1] For a planar structure loaded in its plane *xy* by a planar load system, note that Eqs. (2.8) and (2.9) reduce to Eqs. (2.1).

that they supply to the structure, the structure will be acted upon by a general system of forces consisting of the known loads and the unknown reactions.   If the structure is in static equilibrium under these forces, the three equations of static equilibrium may be written in terms of the known loads and the unknown elements defining the reactions.   The simultaneous solution of these three equations will in certain cases determine the magnitude of the unknown reaction elements.   Whether or not these three equations are sufficient for the complete determination of the reactions, they must be satisfied for the structure to be in static equilibrium and therefore they form a partial basis for the solution to obtain the reactions of any structure that is in static equilibrium.

If there are fewer than three unknown independent reaction elements, there are not enough unknowns to satisfy the three equations of static equilibrium simultaneously. Fewer than three unknown reaction elements are therefore insufficient to keep a planar structure in equilibrium when it is acted upon by a general system of loads. Under such a condition, a structure is said to be **statically unstable.**

Under certain special conditions, a planar structure having fewer than three unknown independent reaction elements may be in static equilibrium.   Of course, if the system of applied loads acting on the structure is in equilibrium itself, no reactions are required; also, if the loads and reactions have certain mutual charac- teristics, fewer than three reaction elements may be sufficient for equilibrium.   For example, considering the bar shown in Fig. 2.2b, if the resultant effect of the applied loads is a force whose line of action goes through the center of the hinge pin at point A, the forces acting on the structure are concurrent and the horizontal and vertical components of the reaction at A will be capable of maintaining static equilibrium.   Moreover, if the bar were supported at both points A and B by rollers on horizontal surfaces, the reactions supplied by such supports could main- tain in a state of static equilibrium any system of applied loads of which the resultant effect is either a couple or a vertical force.   Such structures, although stable under special types of loading but unstable under the general case of loading, are said to be in a state of **unstable equilibrium** and are still classed as unstable structures.

Since three unknowns can be obtained from the solution of three independent simultaneous equations, the reactions of a stable planar structure having exactly three unknown reaction elements may be obtained from the simultaneous solution of the three equations of static equilibrium.   In such a case, the reactions of the structure are said to be **statically determinate.**   However, if there are more than three un- known independent reaction elements supplied to a stable planar structure, the three equations of static equilibrium are not sufficient to determine the unknown reactions.   This is evident since all but three of the unknowns could be assigned arbitrary values and then the remaining three determined from the simultaneous solution of the three equations of static equilibrium.   In such cases, there are an infinite number of related sets of values for the unknown reactions that could satisfy the conditions of static equilibrium.   The correct values for the reactions cannot therefore be determined simply from these three equations but must also satisfy certain deformation conditions of the structure, as will be discussed later in this book.   *If the unknown reaction elements cannot be determined simply by the* *equations of static equilibrium, the reactions of the structure are said to be* **statically indeterminate.** *The structure is then said to be indeterminate to a degree equal to the number by which the unknowns exceed the available equations of statics.*

From the above discussion it may be concluded that at least three independent reaction elements are *necessary* to satisfy the conditions of static equilibrium for a planar structure acted upon by any general system of loads. It may easily be demonstrated, however, that three or more elements are not always *sufficient*, and therefore a planar structure having three or more independent reaction elements may still be unstable. This is the reason why a *stable* planar structure was specified in the discussion of the previous paragraph.

The question of sufficiency of the reactions for stability may be discussed by an extension of the approach used in the early part of Art. 2.2. For example, the bars shown in Figs. 2.2c and d are stable when supported by the horizontal and vertical components of the reaction at A and the vertical reaction at B. These structures are equivalent to that shown in Fig. 2.5, where the hinge support has been replaced

Figure 2.5 Stable structure

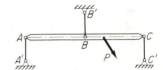

Figure 2.6 Geometrically unstable structure

by two links attached to the hinge pin at point A. These two links may be in any directions as long as they are not collinear. This structure would be stable as long as the line of action of the link at B did not also pass through the center of the hinge pin at A. If this line of action did pass through A, the structure would be only partly restrained and therefore unstable under the general system of loads because there would be nothing to prevent the instantaneous rotation of the structure about the hinge at point A. *In such cases, where there are nominally sufficient reaction elements but the geometrical arrangement is such that the structure is unstable, the structure is said to be* **geometrically unstable**.

One other case where three independent reaction elements are not sufficient for stability should be discussed. Consider a bar supported by three parallel links as shown in Fig. 2.6. It is apparent that there is no restraint to prevent a small translation of the structure normal to the direction of the links. Hence, if the resultant effect of the applied loads has a component in this direction, motion will be produced and such a structure must be classed as geometrically unstable. These and similar considerations lead to the following conclusion: *If the reactions are equivalent to those supplied by a system of three or more link supports that are either concurrent or parallel, they are not sufficient to maintain static equilibrium of a planar structure subjected to a general system of loads even if there are three or more unknown reaction elements.* In other words, *the stability of a structure is determined not only by the number of reaction elements but also by their arrangement.*

It should be noted that unstable structures having three or more independent reaction elements usually could also be classed as statically indeterminate. Consider the structure shown in Fig. 2.6. While it is unstable and starts to translate horizontally under the load P, it is not completely unrestrained. Instantaneously the bar translates horizontally, and the three links rotate about points A', B', and C', respectively. After a finite rotation of the links, points A, B, and C will have moved vertically as well as horizontally. A vertical movement of these points can be accomplished only by making the bar AC bend and by stretching the links. The final equilibrium position is determined not only by the geometry of the structure but also by its elastic deformation due to the stresses developed in the links and the bar. In this final position, the links will have moved through such finite rotations that the algebraic sum of the horizontal components of the internal forces in the links is equal to the horizontal component of the load. Hence, the analysis of this structure in its final equilibrium position involves not only the equations of static equilibrium but also its deformation properties, and the structure may therefore be classed as statically indeterminate. The structure shown in Fig. 2.5 also falls into this same category if the line of action of the link at B passes through the hinge at A.

If the structure shown in Fig. 2.6 were acted upon by a system of vertical loads, there would be no tendency for it to move horizontally, i.e., the structure would be in unstable equilibrium. In such a case, note that the reactions would also be statically indeterminate.

2.6 Stability and Determinancy of Structures Involving Equations of Condition. So far the discussion has been restricted to structures in which no special construction conditions have been introduced. If, however, such conditions are introduced into a planar structure, a like number of equations of condition are added to the three equations of statics. It is then necessary for the loads and reactions acting on the structure to satisfy simultaneously both the equations of static equilibrium and the equations of condition. If the number of unknown reaction elements is fewer than the total number of equations, the structure is statically unstable or possibly in unstable equilibrium for certain special conditions of loading. If the number of reaction elements is more than the total number of equations, the structure is classed as statically indeterminate. If, however, there are the same number of reaction elements as there are equations, the structure is statically determinate unless the reactions are so located and arranged that geometrical instability is possible.

Such a condition of instability may be simply illustrated by a consideration of a structure of the type shown in Fig. 2.4. The structure shown there is, of course, stable; but if the hinge at b lay on the line joining points a and c, there would be no restraint to the instantaneous rotation of members ab and bc about points a and c, respectively. After a certain finite rotation of both members, tensile forces will be developed in members ab and bc that in view of the new slopes of the members will have vertical components keeping the load P in equilibrium. The computation of the equilibrium position of joint b and the resulting tensile forces that are developed involves a consideration of the deformation properties of the structure. Therefore, this structure is not only unstable in its original position but also statically indeterminate.

Geometrical instability similar to that just illustrated is most likely to occur whenever equations of construction are introduced into an originally stable structure. It is therefore evident that care must be used in introducing construction hinges, etc., so that geometrical instability is not produced. Such a condition will always be apparent, for solution of the combined equations of static equilibrium and equations of condition will yield inconsistent, infinite, or indeterminate values for the unknown reaction elements.[1]

2.7 Free-body Sketches. In the previous discussion, it is shown that the unknown reaction elements of a statically determinate and stable structure may be computed from the simultaneous solution of the equations of static equilibrium, augmented, in certain cases, by equations of condition. All the equations involve some or all of the forces acting on the structure, including both the applied loads and the reactions supplied by the supports. To assist in formulating these equations, it is desirable to draw **free-body sketches** of either the entire structure or some portions of it. The importance of drawing an adequate number of these free-body sketches cannot be overemphasized to the student. Such sketches are the basis of the successful stress analysis of structures. The student should be admonished that *it is impossible to draw too many free-body sketches* and that time spent in so doing is never wasted.

A free-body sketch of the entire structure is drawn by isolating the structure from its supports and showing it acted upon by the applied loads and all the possible reaction components that the supports may supply to the structure. Such a sketch is illustrated in Fig. 2.8*b*. In this manner, any portion of the structure can be isolated by passing any desired section through the structure and a free-body sketch drawn showing this portion acted upon by the applied loads and reactions, together with any forces that may act on the faces of the members cut by the isolating section. Any force the magnitude of which is unknown may be assumed to act in either sense along its line of action. The assumed sense is used in writing any equation involving such a force. When the magnitude of such a force is determined from the solution, if the sign is positive, the force is then known to act in the assumed sense; if negative, in the opposite sense.

Sometimes it becomes desirable to isolate several portions of the structure and to draw free-body sketches of these portions. In such cases, it is necessary to show the internal forces acting on the internal faces that have been exposed by the isolating section. If free-body sketches are drawn for two adjacent portions of the structure and the internal forces have been assumed to act in certain senses on an internal face of one portion, the corresponding forces must be assumed to act with the same numerical values but in *opposite* senses on the matching face of the adjacent portion.[2] This is evident since the action and reaction of one body on

---

[1] For a discussion of this subject, see W. M. Fife and J. B. Wilbur, "Theory of Statically Indeterminate Structures," art. 8, p. 9, McGraw-Hill Book Company, New York, 1937.

[2] Note that this convention is required when the internal forces are represented as *scalars*, as is done throughout Part I of this book. If, however, the interacting forces are represented as *vectors*, the vectors on one face are represented as the negative of the corresponding vectors on the matching face. **This is a very important difference between using scalars and vectors.**

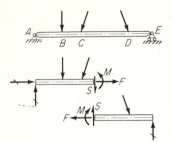

Figure 2.7 Free-body sketches

another must be numerically equal but opposite in sense. Such free-body sketches are shown in Fig. 2.7. If this practice is not followed, equations of static equilibrium written using two such free-body sketches will not be consistent with one another and it will be impossible to obtain a correct solution from them. It is of course obvious that any particular reaction which is assumed to act in a certain sense on one sketch must be shown in the same sense on all other sketches in which it appears.

2.8 Computation of Reactions. If no equations of condition have been introduced into a statically determinate structure, the computation of the reactions involves only a straightforward application of the equations of static equilibrium. Such computations may be illustrated by considering the simple end-supported beam shown in Fig. 2.8a. The unknown reaction elements may be taken as the vertical and horizontal components of the reaction at A and the vertical component of the reaction of D. These may be assumed to act as shown in the free-body sketch, Fig. 2.8b.[1]

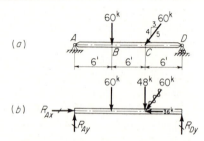

Figure 2.8 Computation of reactions

To obtain these three unknowns there are available the three equations of static equilibrium $\sum F_x = 0$, $\sum F_y = 0$, and $\sum M_z = 0$, and therefore such a structure is statically determinate. It is possible to write the following three equations of static equilibrium and solve them simultaneously for the three unknowns R_{Ax}, R_{Ay}, and R_{Dy}:

$$\sum F_x = 0, \overset{+}{\rightarrow} \qquad R_{Ax} - 36 = 0$$

$$\sum F_y = 0, \uparrow+ \qquad R_{Ay} + R_{Dy} - 60 - 48 = 0$$

$$\sum M_C = 0, +\circlearrowleft \qquad 12R_{Ay} - 6R_{Dy} - (60)(6) = 0$$

While such a solution is always possible, it is not very ingenious and it is inefficient, particularly in a complicated structure. Consider the advantages of proceeding as follows. By taking the summation of moments about an axis through point A, the only unknown entering the equation will be R_{Dy}, and a direct solution for it will be possible:

$$\sum M_A = 0, +\circlearrowleft \qquad (60)(6) + (48)(12) - (R_{Dy})(18) = 0$$

$$\therefore R_{Dy} = {}^{936}\!/_{18} = 52 \text{ kips } \uparrow$$

[1] To facilitate computations, loads are usually given in units of kips, 1 kip being equal to 1,000 lb.

In the same manner, R_{Ay} may be found directly from

$$\sum M_D = 0, +\circlearrowleft \qquad (R_{Ay})(18) - (60)(12) - (48)(6) = 0$$
$$\therefore R_{Ay} = {}^{1,008}\!/_{18} = 56 \text{ kips} \uparrow$$

Of course, these values should satisfy the equation $\sum F_y = 0$, and the following check is obtained:

$$\sum F_y = 0, \uparrow+ \qquad 56 + 52 - 60 - 48 = 0 \qquad \therefore \text{ O.K.}$$

From $\sum F_x = 0$, R_{Ax} is obtained directly,

$$\sum F_x = 0, \xrightarrow{+} \qquad R_{Ax} - 36 = 0 \qquad \therefore R_{Ax} = 36 \text{ kips} \rightarrow$$

Thus, by ingenuity in applying the equations of static equilibrium, the solution for the reactions is facilitated.

To enlarge on this discussion, *it should be noted that the three conventional equations of static equilibrium $\sum F_x = 0$, $\sum F_y = 0$, and $\sum M_z = 0$ may always be replaced by the independent moment equations $\sum M_A = 0$, $\sum M_B = 0$, and $\sum M_C = 0$, when points A, B, and C are three points which do not lie on a straight line.* This can be verified in the following manner. If a system of forces satisfies any one of these moment equations, such as $\sum M_A = 0$, then the resultant effect of the system cannot be a couple but may be a resultant force acting through point A. If the system also satisfies the equation $\sum M_B = 0$, then the resultant effect of the system may still be a resultant force; but, if so, such a force can act only along the line joining points A and B. If in addition, however, the system also satisfies the equation $\sum M_C = 0$ (where C does not lie on a line going through A and B), this eliminates the possibility of the resultant force existing and acting along the line AB. Therefore, if the system satisfies all three moment equations, its resultant effect can be neither a couple nor a resultant force and the system must be in a condition of static equilibrium.[1]

This principle may often be used to advantage in writing equations of static equilibrium, for it is often possible to select a moment axis so that only one unknown reaction element is involved in the summation of moments about that axis, thus leading to a direct solution for that one unknown. The student should study the illustrative examples in Art. 2.11 in conjunction with this article, since methods

---

[1] **Important Questions:**

1. If two moment equations were obtained, first from $\sum M_A = 0$ and then from $\sum M_B = 0$, would a third independent equation be obtained by summing up the force components perpendicular to the line AB and setting this sum equal to zero? Why? Would summing up the force components in any direction other than parallel to AB result in a satisfactory independent equation?

2. It has been shown that it is often desirable to replace a force equilibrium equation with a moment equilibrium equation in the case of a planar structure subjected to a general system of forces. Is the following statement valid in the case of a planar structure subjected to a concurrent force system?

The two conventional equations of equilibrium $\sum F_x = 0$ and $\sum F_y = 0$ may be replaced by the two moment equations $\sum M_A = 0$ and $\sum M_B = 0$ if the point of concurrency does not lie on the line through points A and B.

of arranging the computations and applying the above general approach will be discussed with a view to facilitating the numerical computations.

2.9 Computation of Reactions—Structures Involving Equations of Condition. Whenever special conditions of construction have been introduced into a structure, the evaluation of the reactions involves both the equations of static equilibrium and the equations of condition. Even when such structures are statically determinate, the computation of the reactions is more difficult and requires more ingenuity than cases where no equations of condition have been introduced.

To illustrate the method of attack in such cases, consider the structure shown in Fig. 2.9. This structure is of a type already discussed and may be shown to be statically determinate and stable. There are four unknown reaction elements, but in addition to the three equations of static equilibrium there is the equation of condition introduced by the frictionless hinge at point b. As discussed previously (Art. 2.4), such a hinge is incapable of transmitting a couple either from the portion ab to the portion bc or vice versa. It is therefore necessary that the algebraic sum of the moments about an axis through point b of all the forces acting on *either*

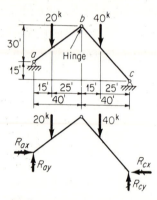

Figure 2.9 Three-hinged arch

the portion ab or the portion bc add up to zero; in equation form,

$$\overset{ab}{\sum} M_b = 0 \qquad \text{or} \qquad \overset{bc}{\sum} M_b = 0$$

At first glance, a student might infer from this that such a hinge really introduces two independent equations of condition. This is not so. There is only *one* independent equation introduced, as may easily be shown by the following reasoning. One of the equations of static equilibrium requires that for the entire structure the summation of the moments about any axis of all the forces shall be equal to zero, and hence, with b as an axis, $\sum M_b = 0$. If one then writes the equation of condition that, for the portion ab, $\overset{ab}{\sum} M_b = 0$, it immediately implies that the algebraic sum of the moments about b of all the remaining forces acting on the remainder of the structure bc must add up to zero. Therefore, $\overset{bc}{\sum} M_b = 0$ is not an independent relation: it is simply equal to the equation $\sum M_b = 0$ minus the equation $\overset{ab}{\sum} M_b = 0$. The student should constantly keep these ideas in mind when setting up the equations of condition of the structure. He should never be fooled into thinking that he has more independent equations than he actually has. In this case any two of these three equations, $\sum M_b = 0$, $\overset{ab}{\sum} M_b = 0$, $\overset{bc}{\sum} M_b = 0$, can be used independently, but the remaining one is not an independent relation.

To proceed now with the solution of this example, the reactions can easily be obtained if ingenuity is used in setting up the equations of statics:

$$\sum M_a = 0, \; \tilde{\curvearrowright} \quad (20)(15) + (40)(55) - 80R_{cy} + 15R_{cx} = 0$$
$$\therefore \; R_{cy} = 31.25 + \tfrac{3}{16} R_{cx} \tag{a}$$

$$\overset{bc}{\sum} M_b = 0, \; \tilde{\curvearrowright} \quad (40)(15) - 40R_{cy} + 45R_{cx} = 0$$
$$\therefore \; R_{cx} = \tfrac{8}{9}R_{cy} - 13.33^{\dagger} \tag{b}$$

Substituting for R_{cy} from Eq. (a) into Eq. (b) gives

$$R_{cx} = (\tfrac{8}{9})(31.25 + \tfrac{3}{16} R_{cx}) - 13.33 \qquad \therefore \; R_{cx} = 17.33 \text{ kips} \leftrightarrow$$

And then substituting back in (a) leads to

$$R_{cy} = 34.5 \text{ kips} \uparrow$$

In a similar manner,

$$\sum M_c = 0, \; \tilde{\curvearrowright} \quad 80R_{ay} + 15R_{ax} - (20)(65) - (40)(25) = 0$$
$$\therefore \; R_{ay} = 28.75 - \tfrac{3}{16}R_{ax} \tag{c}$$

$$\overset{ab}{\sum} M_b = 0, \; \tilde{\curvearrowright} \quad 40\,R_{ay} - 30R_{ax} - (20)(25) = 0$$
$$\therefore \; R_{ax} = \tfrac{4}{3} R_{ay} - 16.67 \tag{d}$$

Then substituting from Eq. (c) into Eq. (d) gives

$$R_{ax} = (\tfrac{4}{3})(28.75 - \tfrac{3}{16}R_{ax}) - 16.67 \qquad \therefore \; R_{ax} = 17.33 \text{ kips} \rightarrow$$

And substituting back in (c) gives

$$R_{ay} = 25.5 \text{ kips} \uparrow$$

Using equations $\sum F_x = 0$ and $\sum F_y = 0$ for checks on the solution leads to

$$\sum F_x = 0, \; \tilde{\curvearrowright} \quad 17.33 - 17.33 = 0 \qquad \therefore \text{ O.K.}$$
$$\sum F_y = 0, \; \uparrow+ \quad 25.5 - 20 - 40 + 34.5 = 0 \qquad \therefore \text{ O.K.}$$

If the supports of a and c of this structure had been on the same elevation, it is evident that the solution of the reactions would have been much easier, for in that case a direct solution for the vertical components of the reactions would be obtained from the equations $\sum M_a = 0$ and $\sum M_c = 0$. From a consideration of the illustrative examples in Art. 2.11, it will be apparent that the solution of the reactions of complicated structures involving special conditions of construction can be expedited in some cases by isolating internal portions as free bodies and applying the equations of static equilibrium to those portions. It should again be emphasized to the student, however, that such a technique does not add any new independent equations besides the three equations of static equilibrium for the entire structure plus the equations of condition resulting from the special conditions of

---

† Numbers such as 13.333 . . . will often be written 13.3̇, the dot over the last digit indicating that it may be repeated indefinitely.

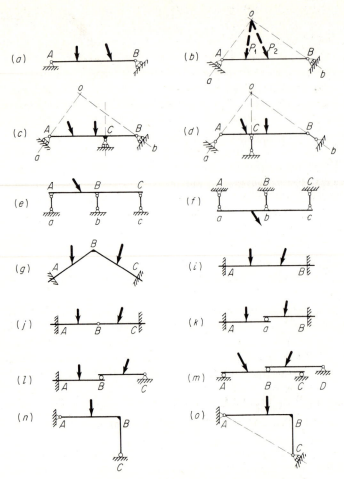

Figure 2.10   Examples for classification

construction.   The equations used may be in a different form, but they are not new, independent ones.

2.10 Examples for Classification.   In this article, examples are discussed illustrating the methods of determining whether a structure is stable or unstable and whether it is statically determinate or indeterminate with respect to its reactions. Note that all beams are represented by straight lines coinciding with their centroidal axes and that their depth is not shown in the sketches of Fig. 2.10.   This will be common practice in the remainder of the book whenever the depth does not essentially affect the solution of the problem.

Consider the beam shown in Fig. 2.10a.   The unknown independent reaction elements are the magnitude and direction of the reaction at A and the magnitude of the reaction at B or a total of three.   These elements may also be considered as the magnitude of the horizontal and vertical components of the reaction at A and of either the horizontal or vertical component at *B*. Note that if the point of application and the direction of a reaction are known, the one unknown element of this reaction may be considered as the magnitude of the resultant reaction itself or the magnitude of either its vertical or horizontal component. Since the reaction supplied by the support at *B* does not pass through point *A*, this structure is stable. The three unknown reaction elements can therefore be found from the three available equations of static equilibrium, and the structure is statically determinate.

Consideration of the beam in Fig. 2.10*b* shows that there are only two unknown reaction elements, the magnitude of the reactions at *A* and *B*. These two unknown elements may also be considered as either the horizontal or the vertical component of each reaction. Since the three equations of static equilibrium cannot be satisfied simultaneously by these two independent unknown reaction elements, this structure is statically unstable under a general system of applied loads. The lines of action of the two reactions intersect at point *O*. If the resultant effect of the applied loads is a resultant force whose line of action also passes through point *O*, the two reactions will be capable of keeping such a special system of loads in equilibrium. While the structure is still classed as unstable, it will be in a state of unstable equilibrium under this loading and its reactions can be determined by the equations of static equilibrium.

If a roller support is added at point *C*, as shown in Fig. 2.10*c*, the reactions of the structure will be equivalent to three links whose lines of action are neither concurrent nor parallel. The structure will therefore be stable and also statically determinate, for the three unknown reaction elements—the three unknown magnitudes— can be found from the three equations of static equilibrium. On the other hand, if the link support is applied at point *C* in Fig. 2.10*d*, the structure will be geometrically unstable since the supports will consist of three links whose lines of action all intersect at point *O*. The structure is not completely unrestrained, of course, but will rotate instantaneously about point *O* through some finite angle until equilibrium is attained. Since the new position assumed by the structure is a function of the deformations of the structure, the reactions acting on the structure in its displaced position are considered to be statically indeterminate.

The structure shown in Fig. 2.10*g* is obviously stable and has a total of six unknown reaction elements—a horizontal and vertical component and a couple at each support. Since there are only three equations of static equilibrium available, the structure is indeterminate to the third degree. The insertion of the hinge in the structure shown in Fig. 2.10*j* introduces one condition equation and therefore makes the structure indeterminate to the second degree. The insertion of the roller in Fig. 2.10*k* makes it possible to transmit only a vertical force from one part of the structure to the other. This in effect introduces two equations of condition, one that the sum of the moments about *a* of the forces acting on *either* portion shall be equal to zero and the other that the sum of the horizontal components of the forces acting on *either* portion shall be equal to zero. As a result, this structure is indeterminate to the first degree only.

In the figures from *p* on, all the truss portions should be considered as rigid bodies whose bar forces are statically determinate once the reactions have been calculated.

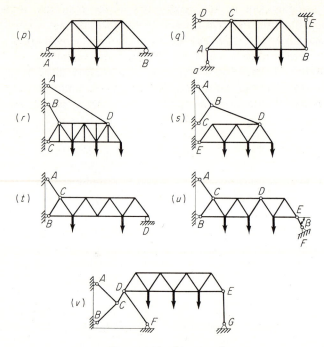

Figure 2.10 (*Continued*)

The arrangement of the bars of a truss necessary for stability, etc., is discussed in detail in Chap. 4.   For the purposes of the present discussion, the trussed portions may be considered as solid bodies.   The structures in Figs. 2.10*p* and *q* are easily seen to be stable and statically determinate under any general system of loads. In Fig. 2.10*r* the structure has four unknown reaction elements—the magnitude of the reactions at the link supports *A* and *B* and both the magnitude and direction of the reaction of the hinge support at *C*.   To solve for these four unknowns there are only three equations of static equilibrium.   The structure is therefore stable and statically indeterminate to the first degree.

For structures similar to that shown in Fig. 2.10*s*, it is often better to use a different approach to determine their stability or statical determinacy.   In such structures, the counting of the available equations may be awkward and obscure. Suppose instead that the structure is broken up into separate parts and the solution for the forces connecting each part to the rest of the structure is considered.   If in this case an isolating section is passed through the link *BD* and the hinge support at *E* is replaced by the horizontal and vertical reaction components that it supplies, the truss portion will be isolated as a free body acted upon by the applied loads and three unknown forces—the link force and the two components of the reaction at *E*. If this free body is to be in equilibrium, these three unknowns may be determined so as to satisfy the three equations of static equilibrium.   Taking $\sum M_E = 0$ on the free body will yield an equation involving the link force as the only unknown.

Knowing the force in the link *BD*, we can find the forces in links *AB* and *BC* by $\sum F_x = 0$ and $\sum F_y = 0$ applied to hinge *B* isolated as a free body. The reactions of this structure can therefore be found by the equations of static equilibrium, and the structure is thus said to be stable and statically determinate.

2.11 Illustrative Examples of Computation of Reactions. The student should study the following examples carefully. All the structures are statically determinate, but the student should investigate them independently for practice.

Example 2.1 *Determine the reactions of the beam ab.*

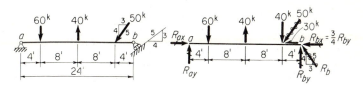

Method A:

$$\sum M_a = 0, \curvearrowleft + \qquad (60)(4) - (40)(12) + (40)(20) - (R_{by})(24) = 0 \qquad \therefore R_{by} = {}^{560}\!/_{24} = 23.\dot{3}^k \uparrow$$

$$R_{bx} = \tfrac{3}{4} R_{by} = \underline{17.5^k} \leftleftarrows$$

$$\sum M_b = 0, \curvearrowleft + \qquad (R_{ay})(24) - (60)(20) + (40)(12) - (40)(4) = 0 \qquad \therefore R_{ay} = {}^{880}\!/_{24} = \underline{36.\dot{6}^k} \uparrow$$

$$\sum F_x = 0, \xrightarrow{+} \qquad R_{ax} - 30 - 17.5 = 0 \qquad \therefore R_{ax} = \underline{47.5^k} \rightrightarrows$$

Check: $\qquad \sum F_y = 0, \uparrow + \qquad 36.\dot{6} - 60 + 40 - 40 + 23.3 = 0 \qquad \qquad 0 = 0 \qquad \therefore O.K.$

Method B:

$$\sum M_b = 0, \curvearrowleft +$$

	+	−
$-(60)(20) =$		1,200
$+(40)(12) = $	480	
$-(40)(4) \ =$		160
	$+ \ \overline{480}$	$- \ \overline{1,360}$
	$+$	480

$$R_{ay} = \underline{36.\dot{6}^k} \uparrow = - \ \frac{880}{24}$$

$$R_{ax} = \underline{47.5^k} \rightrightarrows$$

$$\sum M_a = 0, \curvearrowleft +$$

	+	−
$+(60)(4) \ =$	240	
$-(40)(12) =$		480
$+(40)(20) =$	800	
	$+ \ \overline{1,040}$	$- \ \overline{480}$
	$-$	480

$$R_{by} = + \ \frac{560}{24} = 23.\dot{3}^k \uparrow$$

$$R_{bx} = = (\tfrac{3}{4})(23.3) = \underline{17.5^k} \leftleftarrows$$

Discussion:

The two methods of solving for the reactions are fundamentally the same and differ only in the details of the arrangement of the computations. Method A is probably the best way to organize the computations for an unusual or complicated structure. However, the systematic arrangement of method B will be found most useful for the simple or conventional types of beams and trusses.

Note that it is usually extremely convenient to replace inclined forces by their horizontal and vertical components and to use these components instead of the forces in writing the equations of static equilibrium.[1]

---

[1] If both a force and its components are shown on a free-body sketch, a wavy line should be drawn along the shank of the force arrow, thereby indicating that the force has been replaced by its components.

The analyst should clearly indicate his results by underlining them and also by indicating the units and directions of the forces. Remember that if an answer comes out plus, the force was assumed acting in the correct direction on the free-body sketch; if the answer comes out minus, the force acts opposite to the direction assumed.

Note that in this problem two moment equations and one force equation were used in the solution. The check equation $\sum F_y = 0$ gives a check on the vertical components of the reactions but does not check the horizontal components. If $\sum M = 0$ were written about any axis that did not lie on a line through a and b, a check would be obtained on the values of the horizontal reactions.

Example 2.2   *Determine the reactions of this structure.*

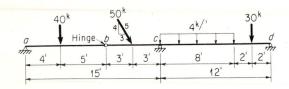

Solution:

Isolate ab:

$\sum M_b = 0, \curvearrowright +$
$9R_{ay} - (40)(5) = 0$
$\therefore R_{ay} = 22.\dot{2}^k \uparrow$

$\sum M_a = 0, \curvearrowright +$
$(40)(4) - (S)(9) = 0$
$S = 17.\dot{7}^k \uparrow$

$\sum F_x = 0$
$\therefore R_{ax} = 30^k \leftarrow\!\!\leftarrow$

Isolate bd:

$\sum M_d = 0, \curvearrowright +$
$-(17.\dot{7})(18) = \quad -320$
$-(40)(15) = \quad -600$
$-[(4)(8)](8) = \quad -256$
$-(30)(2) = \quad -60$
$R_{\cdot y} = 103^k \uparrow = \dfrac{-1{,}236}{12}$

$\sum F_x = 0$
$\therefore F = 30^k \leftarrow$

$\sum M_c = 0, \curvearrowright +$

	$+$	$-$
$-(17.\dot{7})(6) =$		$106.\dot{6}$
$-(40)(3) =$		120
$+[(4)(8)](4) =$	128	
$+(30)(10) =$	300	
	$+428$	$-226.\dot{6}$
	$-226.\dot{6}$	

$R_{dy} = \dfrac{+201.\dot{3}}{12} = 16.\dot{7}^k \uparrow$

Check: $\sum F_y = 0, \uparrow +,$ *for entire structure*
$$22.\dot{2} - 40 - 40 + 103 - 32 - 30 + 16.\dot{7} = 0$$
$$142 - 142 = 0 \qquad \therefore O.K.$$

Discussion:

When isolating the two portions ab and bd from one another, imagine that the hinge pin is removed. The hinge may transmit a force acting in any direction through the center of the hinge pin. If the horizontal and vertical components of this force are designated by F and S and assumed to act as shown on the portion ab, they must act in the opposite senses on the portion bd. Note that an independent check could be obtained on both the horizontal and the vertical reactions by taking $\sum M = 0$ about some axis which does not lie on the line through a and d.

Example 2.3   *Determine the reactions of this structure.*

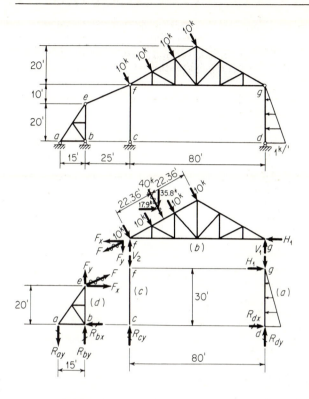

Free-body sketch *a*:

$$\sum M_d = 0, \curvearrowright +   (H_1)(30) - \frac{(1)(30)}{2}(10) = 0   H_1 = 5^k \rightarrow\!\!\rightarrow$$

$$\sum M_g = 0, \curvearrowright +   \frac{(1)(30)}{2}(20) - (R_{dx})(30) = 0   R_{dx} = 10^k \rightarrow\!\!\rightarrow$$

Free-body sketch *b*:

$$\sum M_f = 0, \curvearrowright +   (40)(22.36) - (V_1)(80) = 0   V_1 = 11.18^k \uparrow   \therefore R_{dy} = 11.18^k \updownarrow$$

$$\sum F_x = 0, \rightarrow +   17.9 - 5 - F_x = 0   F_x = 12.9 \leftarrow   \therefore F_y = (\tfrac{2}{5})(12.9) = 5.16^k \downarrow$$

$$\sum M_g = 0, \curvearrowright +   (17.9)(10) - (35.8)(60) - (5.16)(80) + (V_2)(80) = 0   V_2 = 29.77 \uparrow$$

$$\therefore R_{cy} = 29.77^k \updownarrow$$

Check:

$$\sum F_y = 0, \uparrow +   -5.16 + 29.77 - 35.8 + 11.18 = 0$$

$$\therefore -0.01 = 0   \therefore O.K.$$

Free-body sketch *d*:

$$\sum M_b = 0, \curvearrowright +   (12.9)(20) - (R_{ay})(15) = 0   \therefore R_{ay} = 17.2^k \updownarrow$$

$$\sum M_a = 0, \curvearrowright +   (12.9)(20) - (5.16)(15) - (R_{by})(15) = 0   \therefore R_{by} = 12.04^k \updownarrow$$

$$\sum F_x = 0   R_{bx} = 12.9^k \leftarrow\!\!\leftarrow$$

Check on structure as a whole:

$$\sum F_x = 0, \xrightarrow{+} \qquad -12.9 + 17.9 + 10 - \frac{(1)(30)}{2} = 0 \qquad \therefore 0 = 0 \qquad O.K.$$

$$\sum F_y = 0, \uparrow + \qquad -17.2 + 12.04 + 29.77 - 35.8 + 11.18 = 0 \qquad -0.01 = 0 \qquad O.K.$$

$$\sum M_a = 0, \curvearrowright +$$

$$-(12.04)(15) - (29.77)(40) - (11.18)(120) + (17.9)(40) + (35.8)(60) - \frac{(1)(30)}{2}(10) = 0$$

$$-2{,}863 + 2{,}864 = 0 \qquad O.K.$$

Discussion:

In problems of this type, the structure is broken up into its separate structural elements, and the free-body sketches are drawn for each element. The interacting forces between elements may be assumed to act in either sense, but they must act in opposite senses on two adjacent elements. For example, if force F is assumed to act down to the left on sketch b, it must act up to the right on sketch d. Since the structure as a whole is in equilibrium, each of its elements must be in equilibrium. The equations of static equilibrium for each element must be satisfied, and they therefore form a basis for the solution of the unknown reactions and unknown interacting forces.

Note that when all the unknown reactions have been obtained, the results can be checked by applying the equations of static equilibrium to the structure as a whole to see whether or not they are satisfied.

2.12 Superposition of Effects. At this point, a brief discussion of the **principle of superposition** is appropriate. This principle is used continually in structural computations; e.g., in Example 2.1, it would have been easy to compute the reactions due to each one of the three loads acting separately and then superimpose these separate effects to obtain the total reactions produced by the three loads acting simultaneously.

Such a procedure is usually permissible. However, there are two important cases in which the principle of superposition is *not* valid: (1) when the geometry of the structure changes an essential amount during the application of the loads; (2) when the strains in the structure are not directly proportional to the corresponding stresses, even though the effect of change in geometry can be neglected. The latter case occurs whenever the material of the structure is stressed beyond the elastic limit or when it does not follow Hooke's law through any portion of its stress-strain curve.

In Art. 2.3 it was pointed out that usually the deformations of a structure are so slight that it is permissible to consider the structure as a rigid body in applying the equations of static equilibrium and therefore to neglect the effect of slight changes in geometry on the lever arms of forces, the inclinations of members of the structure, etc. However, consider the structure shown in Fig. 2.11. In this case, all three hinges lie along the same straight line in the unloaded structure. It will be found necessary to consider the alteration of the geometry caused by the deformation

Figure 2.11   Nonlinear structure

of the structure, for the lever arms, slopes of the members, etc., are changed by an important amount. As a result, it will be found that the forces and deflections in the structure are not directly proportional to the load P even though the material of the structure may follow Hooke's law. This is therefore an illustration of the first case noted above where the principle of superposition is not valid. In this structure, the effects of a load $2P$ are not twice the effects of a load of P, nor are the effects of a load equal to $P_1 + P_2$ equal to the algebraic sum of the effects of P_1 and of P_2 acting separately. From applied mechanics, a more important case can be recalled where superposition is not valid—the case of a slender strut acted upon by both axial and transverse loads. There the stresses, moments, deflections, etc., due to an axial load of $P_1 + P_2$ are not equal to the algebraic sum of the values caused by P_1 and P_2 acting separately. Fortunately, most cases of this type where superposition is not valid are easily recognized.

It was mentioned above that superposition is not valid in cases where although the effect of change in geometry can be neglected, the material of the structure does not follow Hooke's law. If such structures are also statically determinate, then quantities that can be found by statically determinate stress analysis (such as reactions, shears, and bending moments) can be superimposed but stress intensities and deflections cannot be superimposed. For example, in the case of an end-supported cast-iron beam, the reactions and the shear force and bending moment on the tranverse cross sections are statically determinate quantities and can be superimposed. However, the stress intensities and deflections produced by the bending moment due to a load $2P$ are not equal to twice those due to load P, and hence such quantities cannot be superimposed. If the reactions of a cast-iron beam are statically indeterminate, none of these quantities may be superimposed, since the stress analysis is then a function of the deformation of the structure.

2.13 Problems for Solution

*Problem 2.1 Classify the structures shown in Figs. 2.10e, f, i, l to o, and t to v as stable or unstable and statically determinate or indeterminate. Discuss and state reasons for your answer.

*Problem 2.2 Determine the reactions of the structures of Fig. 2.12.

---

* An asterisk in front of a problem number indicates that answer to this problem appears in the Answer section at the end of the book.

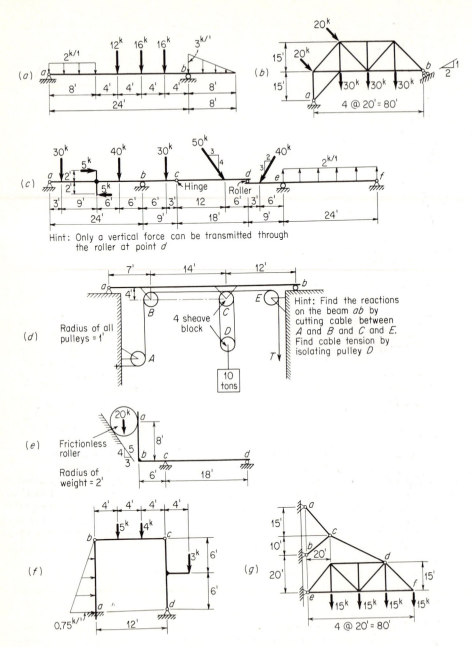

Hint: Only a vertical force can be transmitted through the roller at point d

Hint: Find the reactions on the beam ab by cutting cable between A and B and C and E. Find cable tension by isolating pulley D

Figure 2.12   Problem 2.2

3

Flexural Members—Shear and Bending Moment

3.1 General. The ultimate aim of all stress analysis is to determine the adequacy of a structure to carry the loads for which it is being designed. The criteria for determining this involve a comparison of the stresses developed by the applied loads with certain allowable stress levels for the structural material being used. The stresses acting on any cross section can be studied by passing an imaginary section that cuts through the structure along this cross section and isolates *any* convenient portion of the structure as a free body. If all the other forces acting on this isolated portion have already been determined, the required resultant effects of the stresses acting on the cross section being investigated can easily be computed by the equations of static equilibrium.

One of the commonest structural elements to be investigated in this manner is a **beam,** i.e., a member that is subjected to bending or flexure by loads acting transversely to its centroidal axis or sometimes by loads acting both transversely and parallel to this axis. The following discussions are limited to straight beams, i.e., beams in which the axis joining the centroids of the cross sections (centroidal axis) is a straight line, and to beams the cross sections of which are of such a shape that the shear center[1] and the centroid are coincident. It is also assumed that all the loads and reactions lie in a single plane which also contains the centroidal axis of the flexural member and a principal axis of every cross section. If these conditions are satisfied, the beam will simply bend in the plane of loading without twisting.

3.2 Determination of Stresses in Beams. Suppose that in determining the adequacy of the statically determinate beam in Fig. 3.1 it is necessary to compute the stresses on a transverse cross section *mn*. The reactions necessary for static equilibrium can easily be computed and are shown in free-body sketch *a*. The portions of the beam to the left and right of cross section *mn* may be imagined to be isolated from one another by cutting the structure in two along this section. Free-body sketches *b* and *c* can then be drawn showing all the forces acting on these two portions of the beam.

---

[1] For a discussion of the location of the shear center of a beam cross section, see standard textbooks on mechanics of materials such as S. H. Crandall, N. C. Dahl, and T. J. Lardner, "An Introduction to the Mechanics of Solids," 2d ed., p. 470, McGraw-Hill Book Company, New York, 1972.

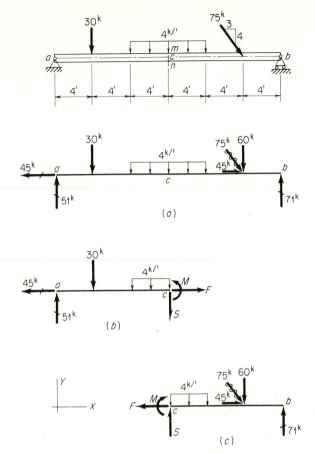

Figure 3.1 Stresses in beams

When one considers the external forces acting on either of the portions b or c, it is immediately apparent that the portions are not in static equilibrium under the external forces acting alone. If the beam as a whole is in equilibrium, however, then each and every portion of it must be in equilibrium. It is therefore necessary that there shall be internal forces or stresses distributed over the internal faces which have been exposed by the imaginary cut. These stresses must be of such a magnitude that their resultant effect balances that of the external forces acting on the isolated portion and therefore maintains the portion in a state of static equilibrium.

The stresses acting on the exposed internal faces may be broken down into two components, one component acting normal to the face and called the **normal stresses** and the other acting parallel to the face and called the **shear stresses.** In the free-body sketches b and c, these stresses have been replaced by their resultant effect as represented by the forces S and F, acting through the centroid of the cross section, and the couple M. Note that the resultant effects S, F, and M of the stresses acting on the portion in sketch b are shown to be numerically equal but opposite in sense to the corresponding effects shown in sketch c. That this must be so is apparent from the following considerations.

Consider the free-body sketch a showing the entire beam acted upon by all the external loads and reactions. Suppose that the resultant of all the external forces applied to the beam on the left of section mn is computed in magnitude and position. Suppose that similar computations are made for the resultant of the remaining external forces applied to the right of *mn*. Now if the beam as a whole is in equilibrium under *all* the forces acting on it, it must be apparent that the resultant of those forces applied to the left of *mn* must be collinear, numerically equal, but opposite in sense to the resultant of the remaining forces applied to the right of *mn*. It therefore follows that the resultant of the external forces in sketch *b* must be numerically equal but opposite in sense to the resultant of the external forces in sketch *c*, and hence the resultant effects of the stresses in sketches *b* and *c* must likewise bear the same relation.

It is convenient to assign names to *F*, *S*, and *M*, the resultant effects of the stresses acting on a cross section of a member. The axial force *F* acts through the centroid of the cross section and will be called the **axial resisting force.** The transverse force *S* will be called the **shear resisting force,** and the couple *M* will be called the **resisting moment.**

In order to satisfy the three equations of static equilibrium for either the portion of the beam in sketch *b* or the portion in sketch *c*, the magnitudes of *F*, *S*, and *M* must be such as to counteract the resultant of the external forces acting on the portion considered. Which portion is used makes no difference theoretically. The portion having the fewer external forces is ordinarily used to simplify the computations. Static equilibrium will be maintained by values computed in the following manner. The axial resisting force *F* and the shear resisting force *S* must be equal and opposite to the axial and transverse components, respectively, of the resultant of the external forces acting on the portion of the beam under consideration. Upon taking moments about an axis through the centroid of the cross section, i.e., the point of intersection of the resisting forces *F* and *S*, it is then apparent that the resisting moment *M* must be equal in magnitude but opposite in direction to the moment of the resultant of the external forces acting on this portion.

Once the axial resisting force, the shear resisting force, and the resisting moment have been determined at any section, the intensities of the normal and shear stresses at any point on the cross section can be computed by using well-known equations given in standard textbooks on the strength of materials.

3.3 Shear and Bending Moment Defined; Sign Convention. From the previous discussion, it is evident that in order to determine the magnitudes of the axial resisting force, the shear resisting force, and the resisting moment acting on a cross section of a beam, it is advisable to compute first the magnitude and position of the resultant of the external forces acting on the portion of the beam on either side of that cross section. It is usually convenient to represent this resultant by its axial component, its transverse component, and its moment about an axis through the centroid of the cross section under consideration. These three elements are statically equivalent to this resultant and are respectively called the **axial force, shear force,** and **bending moment.** The definition of these three terms may therefore be summarized as follows:

Axial force F   **The axial force at any transverse cross section of a straight beam is the algebraic sum of the components acting parallel to the axis of the beam of all the loads and reactions applied to the portion of the beam on** *either* **side of that cross section.**

Shear force (shear) S   **The shear force at any transverse cross section of a straight beam is the algebraic sum of the components acting transverse to the axis of the beam of all the loads and reactions applied to the portion of the beam on** *either* **side of the cross section.**

Bending moment M   **The bending moment at any transverse cross section of a straight beam is the algebraic sum of the moments, taken about an axis passing through the centroid of the cross section, of all the loads and reactions applied to the portion of the beam on** *either* **side of the cross section. The axis about which the moments are taken is, of course, normal to the plane of loading.**

While it is not the purpose of the authors to encourage students to memorize structural principles, formulas, etc., these definitions recur constantly and are so fundamental to structural engineering that students should study and understand them so thoroughly and completely that they are indelibly impressed on their minds.

With these definitions introduced, this discussion can now be summarized by saying that the axial resisting force acting on a cross section will be equal but opposite to the axial force at that section; the shear resisting force will be equal and opposite to the shear force (or shear); and the resisting moment will be equal and opposite to the bending moment.

In subsequent calculations, the following sign convention will be used to designate the directions of axial force, shear, and bending moment at any transverse cross section of a beam. The convention is that ordinarily used in structural engineering. It is clear and simple to employ and will be referred to as the **beam convention.** As shown in Fig. 3.2, axial force is plus when it tends to pull two portions of a member apart, i.e., when it tends to produce a tensile stress on the cross section. Shear force is plus when it tends to push the left portion upward with respect to the right. Bending moment is plus when it tends to produce tension in the lower fibers of the beam and compression in the upper fibers, i.e., to bend the beam concave upward. Many beams are horizontal, and this convention can be applied without confusion. When a member is not horizontal, however, either side may be selected as the "lower side" and the beam convention applied to correspond.

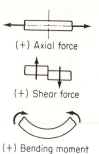

(+) Axial force

(+) Shear force

(+) Bending moment

Figure 3.2 Beam convention

3.4 Method of Computation of Shear and Bending Moment. The procedure for computing the axial force, shear force, and bending moment at any section of a beam is straightforward and can easily be explained by the illustrative example shown in Fig. 3.3. In this example, it is desired to compute the axial force, shear, and bending moment at the cross sections at points b, c, and d. The computation of the axial force is simple and needs no explanation in this case, as is true of this computation with respect to most beams. If the couple applied by the bracket at point c is assumed to be applied along the cross section at point c, there will be an abrupt change in the bending moment at this point and it is necessary to compute the bending moment, first on a cross section an infinitesimal distance to the left of c and then on a cross section an infinitesimal distance to the right of c.

It will be recalled that the shear and bending moment at any cross section can be computed by considering *all* the external loads and reactions applied to the portion of the beam on *either* side of the cross section under consideration. *Either* portion may be used, but the computations can usually be performed more efficiently by using that involving the smaller number of forces. One portion having been selected, *only the loads and reactions acting on that portion* are included in the summation of the force components or moments.

The left-hand portion of the beam was chosen for computing the shear and bending moment at section b and at the sections to the left and right of point c. Free-body sketches b, c, and d, respectively, show the portion used in each case. To illustrate the advantage of using one portion instead of the other, the shear and bending moment at section d are computed by using first the left- and then the right-hand portion, as shown in free-body sketches e and f, respectively. Note the simplicity of the computations for sketch f as compared with sketch e.

To explain typical computations, consider those for the shear and bending moment on the cross section just to the left of point c, as shown in sketch c. The shear force at this section is the algebraic sum of the 57-kip reaction, the uniformly distributed load totaling 16 kips, and the concentrated load of 30 kips, where the reaction causes positive shear (tends to push left portion up) and the two loads cause negative shear (tend to push the left portion down). Hence, the shear force is

$$S = +57 - 16 - 30 = +11 \text{ kips}$$

having a resultant positive effect tending to push the left portion up with respect to the right. For equilibrium, the shear resisting force obviously must be 11 kips acting down on the cross section in sketch c. In the same manner, the bending moment is the algebraic sum of the moments about the centroid of the cross section of the same three forces. Since the reaction tends to produce tension in the lower fibers at the section and the two loads tension in the upper fibers, the bending moment is

$$M = (57)(12) - (16)(10) - (30)(4) = +404 \text{ kip-ft}$$

having a resultant positive tendency to produce tension in the lower fibers. Again for equilibrium,

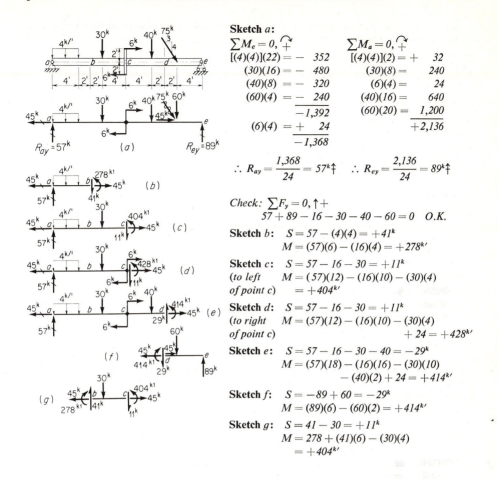

Sketch a:

$$\sum M_e = 0, \curvearrowright +$$

$$
\begin{aligned}
[(4)(4)](22) &= -\ 352\\
(30)(16) &= -\ 480\\
(40)(8) &= -\ 320\\
(60)(4) &= -\ 240\\
\hline
&\quad -1{,}392\\
(6)(4) &= +\ \ 24\\
\hline
&\quad -1{,}368
\end{aligned}
$$

$$\sum M_a = 0, \curvearrowright +$$

$$
\begin{aligned}
[(4)(4)](2) &= +\ \ \ \ 32\\
(30)(8) &= \ \ \ \ 240\\
(6)(4) &= \ \ \ \ \ \ 24\\
(40)(16) &= \ \ \ \ 640\\
(60)(20) &= \ 1{,}200\\
\hline
&\quad +2{,}136
\end{aligned}
$$

$$\therefore R_{ay} = \frac{1,368}{24} = 57^k\!\Updownarrow \qquad \therefore R_{ey} = \frac{2,136}{24} = 89^k\!\Updownarrow$$

Check: $\sum F_y = 0, \uparrow +$

$$57 + 89 - 16 - 30 - 40 - 60 = 0 \quad O.K.$$

Sketch b: $\quad S = 57 - (4)(4) = +41^k$

$\qquad\qquad\quad M = (57)(6) - (16)(4) = +278^{k'}$

Sketch c: $\quad S = 57 - 16 - 30 = +11^k$

(to left $\quad M = (57)(12) - (16)(10) - (30)(4)$

of point c) $\quad = +404^{k'}$

Sketch d: $\quad S = 57 - 16 - 30 = +11^k$

(to right $\quad M = (57)(12) - (16)(10) - (30)(4)$

of point c) $\qquad\qquad\qquad\qquad + 24 = +428^{k'}$

Sketch e: $\quad S = 57 - 16 - 30 - 40 = -29^k$

$\qquad\qquad M = (57)(18) - (16)(16) - (30)(10)$

$\qquad\qquad\qquad\qquad - (40)(2) + 24 = +414^{k'}$

Sketch f: $\quad S = -89 + 60 = -29^k$

$\qquad\qquad M = (89)(6) - (60)(2) = +414^{k'}$

Sketch g: $\quad S = 41 - 30 = +11^k$

$\qquad\qquad M = 278 + (41)(6) - (30)(4)$

$\qquad\qquad\quad = +404^{k'}$

Figure 3.3 Computation of axial force, shear, and bending moment

it is necessary for the resisting moment to be 404 kip-ft acting counterclockwise on the cross section in sketch *c*. The remaining computations are self-explanatory.

It is important to note that, if the axial force, shear, and bending moment are known at one cross section, similar items can be computed at any other cross section by using these known quantities rather than working with all the external forces on the entire portion of the beam on either side of the new cross section. For example, the axial force, shear, and bending moment at *c* could be computed with the quantities already computed at *b*. This is apparent since the shear, axial force, and bending moment at *b* are statically equivalent to the resultant of the external forces applied to the left of *b*. Hence, the contribution of these forces to the resultant effect of all the forces to the left of *c* may be evaluated by using their statical equivalent rather than the forces themselves. The advantage of this procedure increases with the number of external forces applied to the left of point *b*. Such computations are illustrated under sketch *g* of Fig. 3.3, and the forces acting on the isolated portion *bc* are shown in this sketch.

3.5 Shear and Bending-Moment Diagrams. When a beam is being analyzed or designed for a stationary system of loads, it is helpful to have diagrams available from which the value of the shear and bending moment at any cross section can readily be obtained. Such diagrams can be constructed by drawing a base line corresponding in length to the axis of the beam and then plotting ordinates at points along this base line which indicate the value of the shear or bending moment at that cross section of the beam. Plus values of shear or bending moment are plotted as upward ordinates from the base line and minus values as downward ordinates. Diagrams drawn connecting the ends of all such ordinates along the base line are called **shear** and **bending-moment diagrams.** In Fig. 3.4, shear and bending-moment diagrams are shown for the beam in Fig. 3.3.

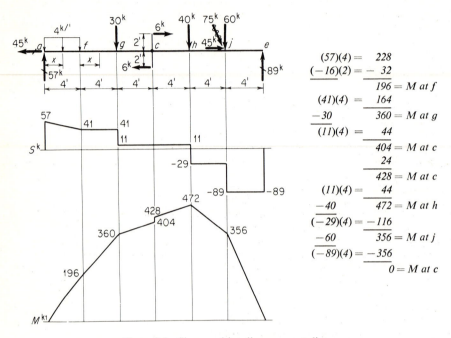

$$(57)(4) = \quad 228$$
$$(-16)(2) = -\ 32$$
$$\overline{\qquad 196} = M \ at \ f$$
$$(41)(4) = \quad 164$$
$$-30 \qquad\qquad \overline{360} = M \ at \ g$$
$$(11)(4) = \qquad 44$$
$$\overline{\qquad 404} = M \ at \ c$$
$$24$$
$$\overline{\qquad 428} = M \ at \ c$$
$$(11)(4) = \qquad 44$$
$$-40 \qquad\qquad \overline{472} = M \ at \ h$$
$$(-29)(4) = -116$$
$$-60 \qquad\qquad \overline{356} = M \ at \ j$$
$$(-89)(4) = -356$$
$$\overline{\qquad 0} = M \ at \ c$$

Figure 3.4   Shear and bending-moment diagrams

The construction of these diagrams is quite straightforward but needs some explanation. The shear on a cross section an infinitesimal distance to the right of point a is $+57$ kips, and therefore the shear diagram rises abruptly from zero to $+57$ at this point. In the portion af, the shear on any cross section a distance x from point a is

$$S = 57 - 4x$$

which indicates that the shear diagram in this portion is a straight line decreasing from an ordinate of $+57$ at a to $+41$ at point f. Since no additional external loads are applied between points f and g, the shear remains $+41$ on any cross section throughout this interval and the shear diagram is a horizontal line as shown. An infinitesimal distance to the left of point g the shear is $+41$, but at an infinitesimal distance to the right of this point the 30-kip load has caused the shear to be further reduced to $+11$. Therefore, at point g, there is an abrupt change in the shear diagram from $+41$ to $+11$. In this same manner, the remainder of the shear diagram can easily be verified. It should be noted that, in effect, a concentrated load is assumed to be applied at a point and hence at such a point the ordinate to the shear diagram changes abruptly by an amount equal to the load. Physically, it is impossible for a load to be applied at a point without developing an infinite contact pressure, and it is therefore necessary for such loads to be distributed over a small area. However, for such computations as for shear and bending moment, such inconsistencies are ignored, and it is considered mathematically possible for concentrated loads to be applied at a point.

In the portion af, the bending moment at a cross section a distance x from point a is $M = 57x - 2x^2$. Therefore, the bending-moment diagram starts at 0 at point a and increases along a curved line to an ordinate of $+196$ kip-ft at point f. In the portion fg, the bending moment at any point a distance x from point f is $M = 196 + 41x$. Hence, the bending-moment diagram in this portion is a straight line increasing from an ordinate of 196 at f to 360 at g. Likewise, in the portion gc, the bending-moment diagram is a straight line increasing to a value of 404 at a cross section an infinitesimal distance to the left of point c. However, at a cross section an infinitesimal distance to the right of point c, the bending moment has increased by 24 to 428. Assuming the external couple of 24 kip-ft to be applied exactly on the cross section at point c, there will be an abrupt change in the bending-moment curve similar to the abrupt changes in the shear diagram already discussed. In an analogous manner, the remainder of the bending-moment diagram may easily be verified. The computations for the controlling ordinates of the diagram are shown in Fig. 3.4.

3.6 Relations between Load, Shear, and Bending Moment. In cases where a beam is subjected to transverse loads, the construction of the shear and bending-moment diagrams can be facilitated by recognizing certain relationships that exist between load, shear, and bending moment. For example, consider the beam shown in Fig. 3.5. Suppose that the shear S and the bending moment M have been computed at the cross section for any point m. Point m is located by the distance x, which is measured from point a, being positive when measured to the right from that point. Suppose that the shear and bending moment are now computed at the cross section at point n, a differential distance dx to the right of point m. Assuming that a uniformly distributed *upward* load of an intensity p per unit length of beam has been applied to the beam between m and n, the shear and bending moment will have increased by differential amounts to values of $S + dS$ and $M + dM$, respectively.

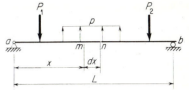

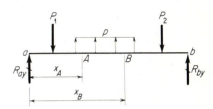

Figure 3.5 Relations between load, shear, and bending moment

The new values of shear and bending moment at point n can be computed by using the values already computed at point m, as is discussed in Art. 3.4. Thus,

$$S + dS = S + p\,dx \qquad (a)$$

$$M + dM = M + S\,dx + p\,dx\,\frac{dx}{2} \qquad (b)$$

Therefore, from (a) it is evident that

$$\frac{dS}{dx} = p \qquad (c)$$

and, differential quantities of the second order being neglected, from Eq. (b) it can be found that

$$\frac{dM}{dx} = S \qquad (d)$$

It should be particularly noted that in addition to the usual beam convention being used for shear and bending moment, upward loads have been considered as positive and x has been assumed to increase from left to right.

The relationships stated mathematically in Eqs. (c) and (d) are tremendously helpful in constructing shear and bending-moment curves.   Consider first Eq. (c).   It states that the rate of change of shear at any point is equal to the intensity of load applied to the beam at that point, i.e., that the slope of the shear curve at any point is equal to the intensity of the load applied to the beam at that point.   The change in shear dS between two cross sections a differential distance dx apart is

$$dS = \frac{dS}{dx} dx = p \, dx$$

Therefore, the difference in shear at two cross sections A and B is

$$S_B - S_A = \int_{x_A}^{x_B} p \, dx \quad \text{or} \quad S_B = S_A + \int_{x_A}^{x_B} p \, dx$$

Thus, the difference in the ordinates of the shear diagram at points A and B is equal to the total load applied to the beam between these two points.

According to Eq. (c) and the sign convention used in its derivation, if the load is upward, or positive, at a point on the beam, the shear is changing at a positive rate at this point.   This means that if the shear is computed on a cross section just to the right of this point, i.e., at a slightly greater distance x from the left support, it will tend to be *more* positive, or algebraically larger, than it was at the first point.   Of course, if the load is downward, or negative, at a point, just the reverse will be true.   If we think of this interpretation in terms of slope of the shear diagram and use the ordinary calculus convention for slope of a curve, if dS/dx is plus, the diagram slopes upward to the right, since positive values of S are plotted upward and x increases from left to right.   If dS/dx is minus, the shear diagram slopes downward to the right.

To apply these ideas, if a uniformly distributed load is applied to a portion of a beam, p will be constant and therefore the shear will change at a constant rate and the shear diagram will be a straight sloping line in such a portion.   However, if the load is distributed but its intensity varies continuously, the shear diagram will be a curved line whose slope changes continuously to correspond.   If no load is applied to a beam between two points, the rate of change of shear will be zero, i.e., the shear will remain constant and the shear diagram will be straight and parallel to the base line in this portion.   At a point where a concentrated load is applied to a beam, the intensity of load will be infinite, and therefore the slope of the shear diagram will be infinite, or vertical.   At such a point, there will be a discontinuity in the shear diagram, and the difference in ordinates from one side of the load to the other will be equal to the concentrated load.   These ideas conform to the discussion of the previous article.

Equation (d) may be interpreted in the same manner.   It states that the rate of change of bending moment at any point is equal to the shear at that point in the beam, i.e., the slope of the bending-moment diagram at any point is equal to the ordinate of the shear diagram at that point. The change in bending moment dM between two cross sections a differential distance dx apart is

$$dM = \frac{dM}{dx} dx = S \, dx$$

Therefore, the difference in the bending moment at two cross sections A and B is

$$\int_{M_A}^{M_B} dM = M_B - M_A = \int_{x_A}^{x_B} S \, dx \quad \therefore M_B = M_A + \int_{x_A}^{x_B} S \, dx$$

or the difference in the ordinates of the bending-moment diagram at points A and B is equal to the area under the shear diagram between the two points.

From Eq. (d) it is evident that, if the shear is positive at a point in a beam, the rate of change of bending moment is also positive at this point.   This means that if the bending moment is computed on a cross section just to the right of this point, i.e., at a slightly greater distance x

from the left support, it will tend to be *more* positive, or algebraically larger, than it was at the first point.   If the shear is negative, just the reverse will be true.   In terms of slope of the bending-moment diagram, it may be said that if dM/dx is positive (or negative), the slope of the bending-moment diagram at this point is upward (or downward) to the right, since positive values of M are plotted upward and x increases from left to right.

If the shear is constant in a portion of the beam, the bending-moment diagram will be a straight line in this portion.   However, if the shear varies in any manner within a portion, the bending-moment diagram will be a curved line.   At a point where a concentrated load is applied, there is an abrupt change in the ordinate of the shear diagram and, therefore, an abrupt change in the slope of the bending-moment diagram at such a point.   At a point where the shear diagram goes through zero and the ordinates to the left of the point are positive and those to the right negative, the slope of the bending-moment diagram will change from positive at the left of the point to negative at the right of the point.   Therefore, the ordinate of the bending-moment diagram will be a maximum at such a point.   If, on the other hand, the shear diagram goes through zero in the reverse manner, the ordinate at that point on the bending-moment diagram will be a minimum.

3.7   Construction of Shear and Bending-Moment Diagrams.   The ideas of Art. 3.6 can be utilized most efficiently in constructing shear and bending-moment diagrams for beams subjected to transverse loads if the following procedure is adopted.   After computing the reactions of a beam, first plot a load diagram.   The load diagram is a diagram the ordinates of which show the intensity of the distributed load applied to the beam at any point.   In addition, all concentrated loads should be indicated.   Upward, or positive, loads should be drawn above the base line and negative loads below.   Then the shear and bending-moment diagrams can be constructed in turn, proceeding from left to right across the beam and establishing the *shape* of the diagrams by using the following principles, which are summarized from the above discussion:

1. The slope of the shear diagram at any point is equal to the intensity of the distributed load at that point.

2. Abrupt changes in the ordinates of the shear diagram occur at points of application of concentrated loads.

3. The slope of the bending-moment diagram at any point is equal to the ordinate of the shear diagram at that point.

4. At points where concentrated loads are applied, there are abrupt changes in the ordinates of the shear diagram and, hence, abrupt changes in the slopes of the bending-moment diagram.

It is usually necessary to compute the numerical values of the ordinates of the shear and bending-moment diagrams only at the points where the shapes of the diagrams change or at points where the maximum or minimum values occur.   Such values can usually be computed most easily by direct computation, as in Fig. 3.3.   Such computations can be checked using the following principles if the value of one ordinate of a diagram is known:

5. The difference in the ordinates of the shear diagram between any two points is equal to the total load applied to the beam between these two points, i.e., the area under the load diagram between these two points plus any concentrated loads applied within this portion.

6. The difference in the ordinates of the bending-moment diagram between any two points is equal to the area under the shear diagram between these two points.

This method of constructing shear and bending-moment diagrams will be illustrated in the examples that follow.

Although all these relations and this discussion apply specifically to the case of a beam loaded by transverse loads, it should not be inferred that they are useless in analyzing a beam subjected to a more general condition of loading.   The method of handling such cases will be discussed in the examples that follow.   For cases of loading involving anything more complicated than transverse loads, it will be seen that a load diagram loses its utility and becomes impractical.   While some of the above relations may be used to advantage, it will be found that in most of the more complicated cases of loading they must be revised.   To illustrate, in Example 3.4 it will be seen that abrupt changes in the ordinates of the moment diagram occur at points where external couples are applied to the beam. Therefore, the difference in the ordinates of the bending-moment diagram between any two points will be equal to the area under the shear diagram between these two points plus or minus the sum of any external couples applied to the beam within this portion. However, the student will find that the experience gained in drawing the shear and bending-moment diagrams for the simpler cases of transverse loadings will enable him to proceed to the more complicated problems with very little difficulty.

3.8 Illustrative Examples—Statically Determinate Beams). The following examples illustrate the construction of shear and bending-moment diagrams for statically determinate beams, utilizing the ideas and principles discussed above.

Example 3.1

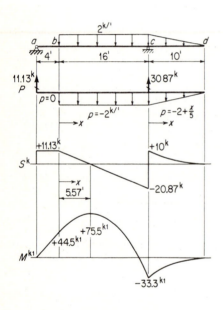

$$\sum M_c = 0, +$$
$$-(2)(16)(8) = -256$$
$$+(2)(10\tfrac{1}{2})(10\tfrac{1}{3}) = +33.3$$
$$\overline{-222.6}$$

$$\therefore R_{ay} = 11.13^k \uparrow$$

$$\sum M_a = 0, +$$
$$(2)(16)(12) = +384$$
$$(2)(10\tfrac{1}{2})(23.3) = +233.3$$
$$\overline{+617.3}$$

$$\therefore R_{cy} = 30.87^k \uparrow$$

$$\sum F_y = 0, \uparrow +,$$
$$11.13 + 30.87 - 32 - 10 = 0 \qquad \therefore O.K.$$

Shear
$$S_c(left) = 11.13 - (2)(16) = -20.87^k$$

Location of point of S = 0 between b and c,
$$S_x = 11.13 - 2x = 0 \qquad \therefore x = 5.57'$$

Bending moment:
$$M_b = +(11.13)(4) = +44.52^{k'}$$

$$M_c = -\frac{(2)(10)}{2}\frac{10}{3} = -33.3^{k'}$$

Between b and c,

$$M_{max} = (11.13)(9.57) - \frac{(2)(5.57)^2}{2} = +75.48^{k'}$$

Discussion:
In establishing the shape of the shear and bending-moment diagrams, follow the ideas of Art. 3.7. Starting at the left end of the shear diagram, the diagram rises abruptly to a value of +11.13. From a to b, since p = 0, the shear diagram is horizontal. From b to c, since p = −2, the shear diagram slopes downward to the right at a constant slope to a value of −20.87 just to the left of c. At c, the reaction causes an abrupt increase in the ordinate of the shear diagram to +10 just to the right of point c. From c to d, since p = −2 + x/5, the shear diagram slopes downward to the right with a slope that varies linearly from −2 at c to 0 at d.

In the same way, the bending-moment diagram starts at 0 at a and progresses from a to b with a constant positive slope (upward to the right). To the right of b, the slope decreases linearly from a slope of $+11.13$ at b to zero at the point of maximum moment, and further to a slope of -20.87 at c. There is an abrupt change at c to a value of $+10$ just to the right of c. Between c and d, the slope decreases from $+10$ to 0 at d.

The numerical value of the controlling ordinates of the shear and bending moment can most easily be computed by direct computation in the manner discussed in Art. 3.4.

Example 3.2

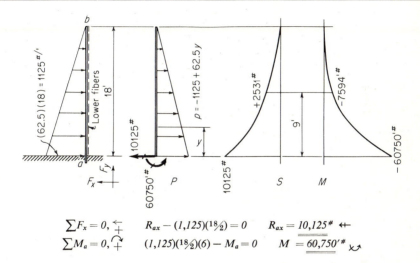

$$\sum F_x = 0, \; \underset{+}{\leftarrow} \qquad R_{ax} - (1,125)(^{18}\!/_2) = 0 \qquad R_{ax} = 10,125^{\#} \; \leftarrow\!\leftarrow$$

$$\sum M_a = 0, \; \underset{+}{\curvearrowright} \qquad (1,125)(^{18}\!/_2)(6) - M_a = 0 \qquad M = 60,750'^{\#} \; \underset{\nearrow}{\curvearrowright}$$

Shear and bending moment:

At $y = 9'$:

$$p = -1,125 + (62.5)(9) = -1,125 + 562.5 = -562.5^{\#/'}$$
$$S = +(562.5)(^9\!/_2) = +2,531.3^{\#}$$
$$M = -(562.5)(^9\!/_2)(3) = -7,593.8'^{\#}$$

Discussion:

After the reactions have been determined, the load, shear, and bending-moment diagrams can be drawn, the fibers on the right-hand side of the cantilever being considered as the "lower fibers" in applying the beam convention. Then, the uniformly varying load would be considered downward or negative in plotting the load diagram.

The shear diagram rises abruptly at a to a value of $+10,125$. Progressing toward b, the shear diagram starts downward with a negative slope of 1,125 but gradually flattens out to zero slope as well as a zero ordinate at b.

On the other hand, the bending-moment diagram starts from an ordinate of $-60,750$ at a with a positive slope of $+10,125$. Proceeding toward b, the magnitude of slope remains positive but steadily decreases until at b both the slope and the ordinate of the bending-moment diagrams are zero.

The shape of these diagrams having been established, an ordinate of either the shear or the bending-moment diagram at any intermediate point such as $y = 9$ can most easily be computed directly by considering the portion of the beam between that point and b.

Example 3.3

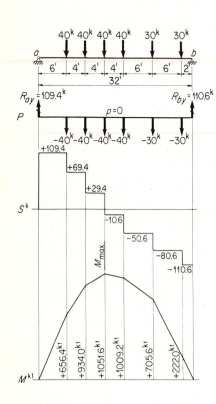

$$\sum M_b = 0, \curvearrowleft +$$

$$
\begin{array}{rr}
(30)(\ 2) = & 60 \\
(30)(\ 8) = & 240 \\
(40)(14) = & 560 \\
(40)(18) = & 720 \\
(40)(22) = & 880 \\
(40)(26) = & 1{,}040 \\
\hline
220 \qquad & 3{,}500 \\
\end{array}
$$

$$\therefore R_{ay} = 109.37^k \ddagger$$

$$\sum M_a = 0, \curvearrowright +$$

$$
\begin{array}{rr}
(30)(30) = & 900 \\
(30)(24) = & 720 \\
(40)(18) = & 720 \\
(40)(14) = & 560 \\
(40)(10) = & 400 \\
(40)(\ 6) = & 240 \\
\hline
& 3{,}540 \\
\end{array}
$$

$$\therefore R_{by} = 110.63^k \ddagger$$

$$\sum F_y = 0, \uparrow + \quad 109.37 + 110.63 - 220 = 0$$
$$\therefore O.K.$$

$$
\begin{array}{lll}
S = & (109.4)(6) = & 656.4 \\
& -40 & \overline{656.4} = M_6 \\
S = & (69.4)(4) = & +277.6 \\
& -40 & \overline{934.0} = M_{10} \\
S = & (29.4)(4) = & +117.6 \\
& -40 & \overline{1{,}051.6} = M_{14} \\
S = & (-10.6)(4) = & -42.4 \\
& -40 & \overline{1{,}009.2} = M_{18} \\
S = & (-50.6)(6) = & -303.6 \\
& -30 & \overline{705.6} = M_{24} \\
S = & (-80.6)(6) = & -483.6 \\
& -30 & \overline{222.0} = M_{30} \\
S = & (-110.6)(2) = & -221.2 \\
& +110.6 & \cancel{0.8} = M_{32} \\
\hline
& 0 & 0 \\
& \therefore O.K. & \therefore O.K. \\
\end{array}
$$

Note: M_{max} *occurs where shear passes through zero (14 ft from a).*

Discussion:

In computing ordinates to the shear and bending-moment diagrams for concentrated load systems, it is convenient to arrange the computations in this manner, the ordinates being computed successively from left to right by means of principles 5 and 6 stated near the end of Art. 3.7. Note that a check of the computations is obtained if both diagrams come back to zero at point b.

Example 3.4

All pulleys have a 2-ft diameter.

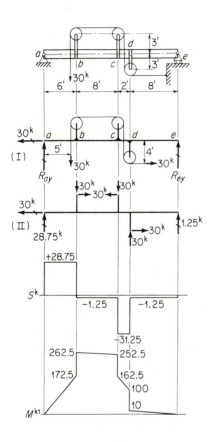

$$\sum M_a = 0, \curvearrowright +$$
$$(5)(30) - (4)(30) - (R_{ey})(24) = 0$$
$$\therefore R_{ey} = 1.25^k$$

$$\sum M_e = 0, \curvearrowright +$$
$$(R_{ay})(24) - (30)(19) - (30)(4) = 0$$
$$\therefore R_{ay} = 28.75^k$$

$$\sum F_y = 0, \uparrow + \quad 28.75 + 1.25 - 30 = 0$$
$$\therefore O.K.$$

Bending Moments:

At b (*just to left*),
$$M_{b_L} = + (28.75)(6) = +172.5^{k'}$$

At b (*just to right*),
$$M_{b_R} = (28.75)(6) + (30)(3) = +262.5^{k'}$$

At c (*just to left*),
$$M_{c_L} = 262.5 - (1.25)(8) = +252.5^{k'}$$

At c (*just to right*), *isolate element of beam between two cross sections—one just to left, one just to right of point c. This element will have a differential length, say, of $0 + ft.$*

$$\sum M_{c_R} = 0, \curvearrowright +$$
$$+252.5 - (30)(3) - (1.25)(0+)$$
$$- M_{c_R} = 0$$
$$M_{c_R} = 252.5 - 90 - 0 = +162.5^{k'}$$

At d (*just to left*),
$$M_{d_L} = (1.25)(8) + (30)(3) = +100^{k'}$$

At d (*just to right*)
$$M_{d_R} = (1.25)(8) = +10^{k'}$$

Discussion:

As is evident in Chap. 2, the computation of the reactions of such structures can be carried out without first computing the forces that the individual pulleys apply to the beam. However, for the construction of shear and bending-moment diagrams and the computation of internal stresses in the beam, it is necessary to compute the detailed manner in which the pulleys apply the loads. The reactions from sketch I and the pulley loads having been computed, sketch II can be drawn, showing the precise manner in which the structure is loaded.

As explained in the latter part of Art. 3.7, it is unwise to attempt to draw a load diagram in such problems. Instead, the load diagram is replaced by a free-body sketch such as sketch II. The student may now proceed to draw the shear bending-moment diagrams, using the fundamental definitions and methods of computation and, of course, utilizing the experience gained in simple problems such as Examples 3.1 to 3.3. All the above calculations may be followed without difficulty.

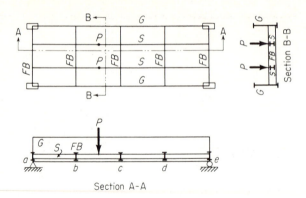

Section A-A

Figure 3.6   Girder bridge

3.9 Illustrative Example—Girders with Floor Beams. In all the previous examples, the loads have been applied directly to the beam itself. Quite often, however, loads are applied indirectly through a floor system that is supported by girders. A typical construction of this kind is shown diagrammatically in Figs. 1.5 and 3.6. In such a structure, the loads P are applied to the longitudinal members S, which are called **stringers**. These are supported by the transverse members FB, called **floor beams**. The floor beams are in turn supported by the girders G. Therefore, no matter whether the loads are applied to the stringers as a uniformly distributed load or as some system of concentrated loads, their effect on the girder is that of concentrated loads applied by the floor beams at points a, b, c, etc.

To illustrate the construction of the shear and bending-moment diagrams for girders loaded in this manner, consider Example 3.5. For simplicity, it will be assumed in this example that the loads are applied to stringers supported on the top flange of the girder as shown. The stringers and girder will be assumed to lie in the same plane. It will further be assumed that the stringers are supported as simple end-supported beams, with a hinge support at one end and a roller at the other. As a first step, it is necessary to obtain the stringer reactions and from them to determine the concentrated forces acting on the girder. From this point on, the construction of the shear and bending-moment diagrams for the girder proceeds as for any beam acted upon by a system of concentrated loads.

The student should study the following questions concerning this type of structure: How do the shear and bending-moment diagrams differ in the two cases, i.e., with and without the stringers? Is there any notable similarity? If a uniformly distributed load is applied to the structure, how will the shear and bending-moment diagrams compare with and without stringers? For concentrated loads applied to the stringers directly at the floor-beam locations, does it make any difference, as far as the floor-beam forces are concerned, whether we think of the loads as being applied to the stringer on the left or right of the floor-beam location, or whether we divide such loads between these two stringers? Note at b and d of Example 3.5 such loads are assumed to be applied to the stringers on the left. If the stringers are not supported as simple end-supported beams by the girder, will the answers to the previous questions be altered? Problems at the end of the chapter will emphasize some of these points.

Example 3.5

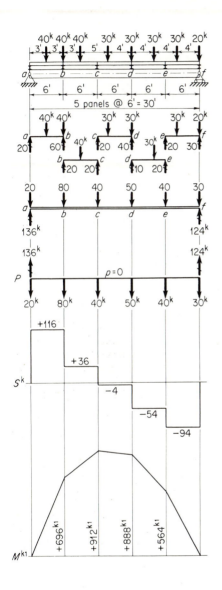

Considering isolated girder acted upon by stringer reactions,

$$\sum M_a = 0, \curvearrowright + \qquad\qquad \sum M_f = 0, \curvearrowright +$$

$(80)(1P) = 80P$	$(80)(4P) = 320P$
$(40)(2P) = 80P$	$(40)(3P) = 120P$
$(50)(3P) = 150P$	$(50)(2P) = 100P$
$(40)(4P) = 160P$	$(40)(1P) = 40P$
$(5P)(R_{fy}) = \overline{470P}$	$(5P)(R_{ay}) = \overline{580P}$
Net $R_{fy} = \underset{=}{94^k} \uparrow$	Net $R_{ay} = \underset{=}{116^k} \uparrow$
30^k	20^k
Gross $R_{fy} = 124^k \uparrow$	Gross $R_{ay} = 136^k \uparrow$

$$\sum F_y = 0, \uparrow +$$
$$124 + 136 - (3)(40) - (4)(30)$$
$$- 20 = 0$$
$$260 - 120 - 120 - 20 = 0 \quad \therefore \text{ O.K.}$$

These reactions may also be checked by using the applied loads directly.

Shear and bending moment on girder:

$$
\begin{aligned}
\text{Gross } R_{ay} = &\quad 136^k \\
&\quad \underline{-20} \qquad\qquad 0 = M_a \\
S_{a-b} = &\ (116^k)(6) = +696^{k'} \\
&\qquad\qquad\quad \overline{+696^{k'}} = M_b \\
&\quad \underline{-80} \\
S_{b-c} = &\ (36)(6) = +216 \\
&\qquad\qquad\quad \overline{+912} = M_c \\
&\quad \underline{-40} \\
S_{c-d} = &\ (-4)(6) = -24 \\
&\qquad\qquad\quad \overline{+888} = M_d \\
&\quad \underline{-50} \\
S_{d-e} = &\ (-54)(6) = -324 \\
&\qquad\qquad\quad \overline{+564} = M_e \\
&\quad \underline{-40} \\
S_{e-f} = &\ (-94)(6) = -564 \\
&\qquad\qquad\qquad\quad 0 = M_f \\
&\quad \underline{-30} \\
&\quad -124 \\
\text{Gross } R_{fy} = &+124 \\
&\quad \overline{\ 0\ }
\end{aligned}
$$

Discussion:

Note the terms *gross* and *net* reactions. The gross reaction is the total force supplied by the support and includes any load applied to the beam immediately over the support point. The net reaction is the reaction at a support due to all loads except the one applied right at this support. Note that only the net reaction enters the computations for shear and bending moment.

3.10 Illustrative Example—Statically Indeterminate Beams. From the discussion in Chap. 2 it will be recalled that the stress analysis of indeterminate structures involves the satisfaction of not only the equations of static equilibrium but also certain conditions of deformation. The analysis of such structures is discussed in detail later in this book. In these later chapters, it will be seen that after some of the unknown stress components, e.g., reactions, shears, and moments, have been found so as to satisfy the conditions of deformation, the remaining unknowns can be found so as to satisfy the equations of static equilibrium. That is, the remaining portion of the problem is statically determinate and can be handled by means of the techniques explained in Chaps. 2 and 3 for statically determinate structures.

Once the reactions of a statically indeterminate beam have been determined, the shear and bending moment can be computed at any desired cross section in the same manner as that used for statically determinate beams. The same principles may also be followed in constructing shear and bending-moment diagrams.

Example 3.6 *This statically indeterminate continuous beam is not only acted upon by the loads shown but is also subjected to certain support movements. The following bending moments have been computed by using methods discussed later for analysis of statically indeterminate structures:*

$$M_a = -85.17^{k'} \qquad M_b = -60.05^{k'} \qquad M_c = 0$$

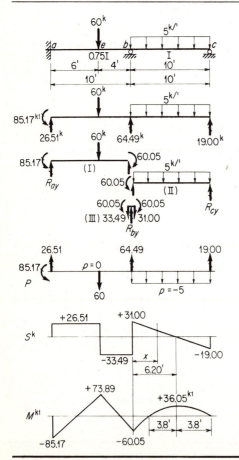

Sketch I:
$$\sum M_b = 0, \; \curvearrowright +$$
$$(R_{ay})(10) + 60.05 - 85.17 - (60)(4) = 0$$
$$\therefore R_{ay} = 26.51^k \uparrow$$
$$\therefore S_{bL} = 26.51 - 60 = -33.49^k$$

Sketch II:
$$\sum M_b = 0, \; \curvearrowright +$$
$$(5)(10)(5) - 60.05 - (R_{cy})(10) = 0$$
$$\therefore R_{cy} = 19.00^k \uparrow$$
$$\therefore S_{bR} = -19.00 + 50 = +31.00^k$$

Sketch III:
$$\sum F_y = 0$$
$$R_{by} = 33.49 + 31.00 = 64.49^k \uparrow$$

Bending moments:
$$M_e = -85.17 + (26.51)(6) = +73.89^{k'}$$

Between b and c, find M_{max},
$$S_x = 31.00 - 5x$$
When $S_x = 0 = 31 - 5x,$ $\qquad \therefore x = 6.20'$
$$M_{max} = -60.05 + (31.00)\left(\frac{6.20}{2}\right) = +36.05^{k'}$$

3.11   Problems for Solution

Problem 3.1   Draw the shear and bending-moment diagrams for the conditions of loading of a simple end-supported beam as shown in Fig. 3.7.

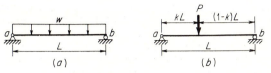

Figure 3.7   Problem 3.1

Suggestion: What is the maximum bending moment in each case?   If, in part *b*, *k* equals 0.5, what is the maximum bending moment?

*Problem 3.2   Draw the load, shear, and bending-moment diagrams for the beam of Fig. 3.8.

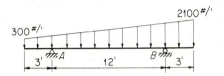

Figure 3.8   Problem 3.2

*Problem 3.3   Draw the shear and bending-moment diagrams for the beam of Fig. 3.9.

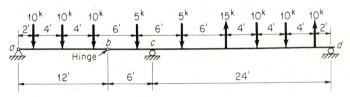

Figure 3.9   Problem 3.3

Problem 3.4   Draw the shear and bending-moment diagrams for the beams shown in Probs. 2.2*a* and *c*.

*Problem 3.5   Draw the shear and bending-moment diagrams for beam *AB* of Fig. 3.10.

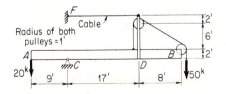

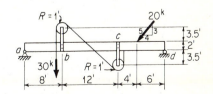

Figure 3.10   Problem 3.5   **Figure 3.11**   Problem 3.6

*Problem 3.6   Draw the shear and bending-moment diagrams for the beam of Fig. 3.11.

Problem 3.7   Draw the shear and bending-moment diagrams for beam *ab* of Prob. 2.2*d* and for beam *bcd* of Prob. 2.2*e*.

Problem 3.8 Draw the shear and bending-moment diagrams for members *ab* and *bc* of the structure of Fig. 3.12.

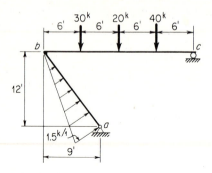

Figure 3.12   Problem 3.8

*Problem 3.9 Draw the shear and bending-moment diagrams for girder *ab* of Fig. 3.13.

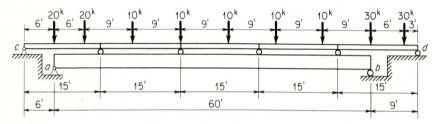

Figure 3.13   Problem 3.9

Problem 3.10 Draw the shear and bending-moment diagrams for girder *ab* of Fig. 3.14.

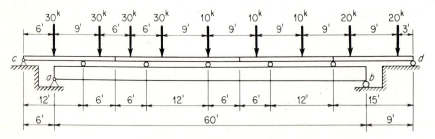

Figure 3.14   Problem 3.10

4

Plane and Space Frameworks

4.1 Introduction. Although most engineering structures are three-dimensional it is often permissible to break a three-dimensional structure down into component planar structures and to analyze each planar structure for loads lying in its plane. Consider, for example, a typical through-parallel-chord truss highway bridge. Such a structure is three-dimensional, but it can be broken down into six component structures, each of which is planar, the two main vertical trusses, the top-chord lateral system, the bottom-chord lateral system, and the two end portals. Often a given member must be considered as part of more than one component planar structure; a bottom chord of a vertical truss, for example, is also likely to be a chord of the bottom-chord lateral system. This introduces no difficulty, since for such members the stresses can be computed for each component planar structure in which it participates and these two contributions superimposed to give the total stress in the member.

In some three-dimensional structures, however, the stresses are so interrelated between members not lying in a plane that the analysis cannot be carried out on the basis of component planar structures. For such structures, a special consideration of the analysis of three-dimensional structures is necessary.

Structures that may fall into this classification include towers, guyed masts, derricks, framing for domes, and framing for aircraft, to mention only a few. Such structures may be either statically determinate or statically indeterminate.

In this chapter, the general theory of the stress analysis of both two- and three-dimensional frameworks is discussed. Consideration is also given to the manner in which the members (bars) of plane and space frameworks must be arranged in order to obtain stable structures. In Chap. 6 the stress analysis of some of the more important types of bridge and roof trusses is considered in detail under design loading conditions.

Consideration in this chapter is focused on statically determinate frameworks. In Chaps. 8 and 9 methods of computing deflections and of analyzing statically indeterminate structures are discussed in detail. These methods are applicable in principle to both plane and space frameworks, although almost all the illustrative examples in Chaps. 8 and 9 involve only two-dimensional structures.

4.2 General Definitions. A **plane framework** (usually called a **truss**) may be defined as a structure composed of a number of bars, all lying in one plane and hinged together at their ends in such a manner as to form a rigid configuration. For the purposes of the discussion in this chapter, it will be assumed that the following conditions exist: (1) The members are connected together at their ends by frictionless pin joints. (2) Loads and reactions are applied to the truss only at the joints. (3) The centroidal axis of each member is straight, coincides with the line connecting the joint centers at each end of the member, and lies in a plane that also contains the lines of action of all the loads and reactions. Of course, it is physically impossible for all these conditions to be satisfied exactly in an **actual truss,** and therefore a truss in which those idealized conditions are assumed to exist is called an **ideal truss.**

Any member of an ideal truss can be isolated as a free body by disconnecting it from the joints at each end. Since all external loads and reactions are applied to the truss only at the joints and no loads are applied between the ends of the members themselves, the isolated member would be acted upon by only two forces, one at each end, each such force representing the action on the member by the joint at that end. Since all the pin joints are assumed to be frictionless, each of these two forces must be directed through the center of its corresponding pin joint. For these two forces to satisfy the three conditions of static equilibrium for the isolated member, $\Sigma F_x = 0$, $\Sigma F_y = 0$, and $\Sigma M = 0$, it is apparent that the two forces must both act along the line joining the joint centers at each end of the member and must be numerically equal but opposite in sense. Since the centroidal axes of the members of an ideal truss are straight and coincide with the line connecting the joints at each end of the member, every transverse cross section of a member will be subjected to the same axial force but to no bending moment or shear force. *The stress analysis of an ideal truss is essentially completed, therefore, when the axial forces have been determined for all the members of the truss, since the normal stress intensities on the cross sections of the members can then easily be computed.*

Three-dimensional structures composed of a number of bars connected together by ball-and-socket joints so as to form a rigid configuration are called **space frameworks.** Such structures are discussed in detail in Arts. 4.15 to 4.21.

Trusses and space frameworks are structures which are arranged specifically to carry loads applied at their joints. As a result of these joint loads, the members of such frameworks are subjected primarily to axial forces and to little or no shear and bending. If transverse loads are applied to the members between joints, one of the requirements for truss behavior is violated and significant shear and bending effects are produced. If the joints are substantially frictionless pin joints or ball joints, these flexural effects are confined to the members which are subjected to the between-joint loads. (Problem 4.6 illustrates such a case.) If, however, the joints are substantially rigid, the structure acts as if it were a rigid frame (see Art. 4.14).

In such a case, the flexural effects in the transversely loaded members are transmitted through the joints to other members so that significant shear and bending can be produced in all members of the structure. Methods of analyzing statically indeterminate rigid frames will be discussed in Chaps. 7 and 9.

4.3 Ideal vs. Actual Trusses. While it is true that the ideal truss is hypothetical and can never exist physically, the stress analysis of an actual truss based on the assumption that it acts as an ideal truss usually furnishes a satisfactory solution for the axial forces in the members of the actual truss. The axial forces in the members, or bars, of a truss will be referred to as the **bar forces**. The stress intensities due to the bar forces computed on the basis that the truss acts as an ideal truss are referred to as **primary stress intensities**.

The pins of an actual pin-jointed truss are never really frictionless; moreover, most modern trusses are made with riveted or welded joints so that there can be no essential change in the angles between the members meeting at a joint.[1] As a result, even when the external loads are applied at the joint centers, the action of the joints on the ends of a member may consist of both an axial and transverse force and a couple. The transverse cross sections of a member may be subjected, therefore, to an axial force, a shear force, and a bending moment. In addition, the dead weight of the members themselves must necessarily be distributed along the members and therefore contributes to further bending of the members. If good detailing practice is followed and care is taken to see that the centroidal axes of the members coincide with the lines connecting the joint centers, additional bending of the members due to possible eccentricities of this type may be eliminated or minimized.

All these departures from the conditions required for an ideal truss not only may develop a bending of the members of an actual truss but also may cause bar forces that are somewhat different from those in an ideal truss. The differences between the stress intensities in the members of an actual truss and the primary stress intensities computed for the corresponding ideal truss are called **secondary stress intensities.** It may be demonstrated, however, that in the case of the usual truss, where it is detailed so that the centroidal axes of the members meet at the joint centers and where the members are relatively slender, the secondary stress intensities are small in comparison with the primary stress intensities.[2] The primary stress intensities computed on the basis that the truss acts as an ideal truss are therefore usually satisfactory for practical design purposes.

In subsequent discussions, the term *truss* will be used to denote a framework that either is actually an ideal pin-jointed truss or may be assumed to act as if it were an ideal truss.

4.4 Arrangement of Members of a Truss. In Art. 4.2, it was stated that the members of a truss must be hinged together in such a manner as to form a rigid configuration. The term *rigid* as used in this instance has the same significance as when used previously in Art. 2.3, that is, a configuration is said to be rigid if there

[1] See also first three paragraphs of Art. 4.14.

[2] See J. I. Parcel and R. B. B. Moorman, "Analysis of Statically Indeterminate Structures," chap. 9, John Wiley & Sons, Inc., New York, 1955.

is no relative movement between any of its particles beyond that caused by the small elastic deformations of the members of the configuration. In this sense, a rigid configuration can be obtained by arranging the truss members in many different ways. When the bars have been satisfactorily arranged in one of these ways, the entire truss can be supported in some manner and used to carry loads just like a beam.

Suppose that it is necessary to form a truss with pin joints at points *a*, *b*, *c*, and *d*. If this is attempted by pin connecting four bars together as shown in Fig. 4.1*a*, the resulting framework will not be rigid and may collapse in the manner shown due to the loads *P*, until, in this case, joints *a*, *d*, and *c* are lying along a straight line. After a little thought, it is apparent that any attempt like this to connect four or more joints together with a like number of bars hinged at their ends will result in a framework which will collapse under all but a few special conditions of loading. If, however, points *a* and *b* are first connected by a bar *ab*, then two other bars of lengths *ad* and *bd* can be hinged at *a* and *b*, respectively. If the *d* ends of these bars are then hinged together at point *d*, a rigid triangle will be formed connecting joints *a*, *b*, and *d*. Bars of lengths *dc* and *bc* can then be connected to the pin joints at *d* and *b*, respectively. The *c* ends of these bars can then be made to coincide at point *c*

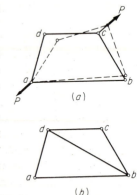

Figure 4.1 Frameworks: (*a*) nonrigid; (*b*) rigid

and pinned together at this point, thus rigidly connecting joint *c* to the triangle *abd* and resulting in a rigid framework of five bars with joints at *a*, *b*, *c*, and *d*. As alternative arrangements, point *c* can be connected to joints *a* and *b* by bars *ac* and *bc* or to joints *a* and *d* by bars *ac* and *dc*. Several other alternative arrangements can be used by first forming a triangle with joints *a*, *b*, and *c*, or with *a*, *d*, and *c*, or with *b*, *c*, and *d*. Any of these arrangements will result in a rigid framework capable of withstanding any system of joint loads without collapsing, as long as none of the members buckles or is strained beyond the yield point of the material.

In this same manner, any number of pin joints can be connected together with bars to form a rigid framework. The procedure is first to select three joints that do not lie along a straight line. These three points can then be connected by three bars pinned together to form a triangle. Each of the other joints can then be connected in turn, two bars being used to connect it to any two suitable joints on the framework already constructed. Of course, the new joint and the two joints to which it is connected should never lie along the same straight line. Each of the trusses shown in Fig. 4.2 has been formed in this manner by starting with a rigid triangle *abc* and using two additional bars to connect each of the other joints in alphabetical order.

Trusses the bars of which have been arranged in this manner are called **simple trusses,** for this is the simplest and commonest type of bar arrangement encountered in practice.

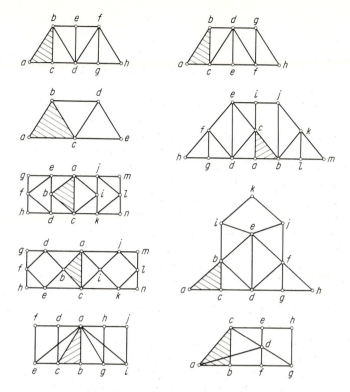

Figure 4.2 Simple trusses

In all truss diagrams like those shown in Fig. 4.2, the members will be represented by single lines and the pin joints connecting them by small circles. Sometimes bars may cross each other but may be arranged in such a manner that they are not connected together by a joint at their point of intersection.

When the members have been arranged to form a simple truss, the entire framework can then be supported in the same manner as a beam. In order to approach the conditions of an ideal truss, the supports should be detailed so that the reactions are applied at the joints of the truss. When the discussion in Art. 2.5 is recalled, it is apparent that if the supports of the truss are arranged so that they are equivalent to three-link supports neither parallel nor concurrent, the structure is stable and its reactions are statically determinate under a general condition of loading. Illustrations of a simple truss supported in a stable and statically determinate manner are shown in Fig. 4.3.

Sometimes it is desirable to connect two or more simple trusses together to form one rigid framework. In such cases, the composite framework built up in this manner is called a **compound truss.** One simple truss can be rigidly connected to another simple truss if the two trusses are connected together at certain points by three links neither parallel nor concurrent or by the equivalent of this type of connection. Hence, two trusses connected in this manner will form a composite framework that is completely rigid. Additional simple trusses can be connected in a similar manner to the framework already assembled to form a more elaborate compound truss.

　　Several examples of compound trusses are shown in Fig. 4.4. In all these cases the simple trusses that have been connected together are shown crosshatched. In trusses *a*, *e*, and *f*, the simple trusses have been connected together by bars 1, 2, and 3. In cases *b* and *c*, the trusses have been hinged together at one common joint, thus requiring only one additional bar to form a rigid composite framework. In case *d*, the additional bar connecting trusses *A* and *B* together has been replaced by the simple truss *C*.

　　The members having been arranged to form a compound truss, the entire framework can be supported in the same manner as a simple truss.

　　4.5 Notation and Sign Convention Used in Truss Stress Analysis. Before the stress analysis of trusses is discussed, it is necessary to establish a definite notation and sign convention for designating the bar forces in the members of a truss.

　　The various members of a truss will be designated by the names of the joints at each end of the

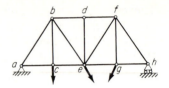

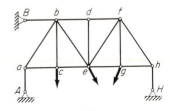

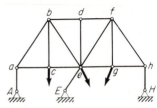

Figure 4.3　Stable and statically determinate simple trusses

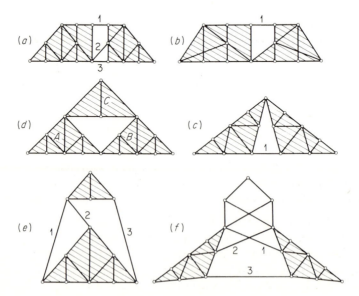

Figure 4.4　Compound trusses

member. The letter F will be used to denote the bar force in a member. Thus, subscripts being used to denote the bar, F_{ah} denotes the bar force in member ah. The values of the bar forces in the members of a truss are often tabulated or written alongside the various members on a line diagram of the truss. For this purpose, it is convenient to have a *definite convention for designating the character of stress in a member, i.e., whether the internal stress is tension or compression. The most convenient convention is to use a plus (+) sign to designate a tension and a minus (−) sign to designate a compression.* Thus, +10 means a tension of 10, and −10 a compression of 10. A plus sign is used to designate tension because such a bar force causes an elongation, or an **increase** in length, of the bar. Thus, a plus force causes a plus change in length. On the other hand, a compression, or minus, bar force causes a decrease, or minus change in the length of a bar.

In the stress analysis of trusses, it is often convenient to work with two rectangular components of a bar force rather than the bar force itself. For this purpose, two orthogonal directions x and y are selected (usually horizontal and vertical, respectively), and the two corresponding components in bar ah are designated as X_{ah} and Y_{ah}. It is particularly important for the student to be completely familiar with the various relationships between a force and its two rectangular components. These relationships are so important in truss analysis that a student must have complete facility with them. For this reason, some of them are reviewed at this time. When it is realized that a bar force acts along the axis of a member, the following statements are self-evident:

1. *The horizontal (or vertical) component of a bar force is equal to the bar force multiplied by the ratio of the horizontal (or vertical) projection of a member to its axial length.*

2. *The bar force of a member is equal to its horizontal (or vertical) component multiplied by the ratio of the axial length of the member to its horizontal (or vertical) projection.*

3. *The horizontal component of a bar force is equal to the vertical component multiplied by the ratio of the horizontal to the vertical projection of the axial length, and vice versa.*

The following principle is also useful and important in dealing with bar-force computations:

4. *For purposes of writing equilibrium equations for a particular isolated body, any force may be replaced by its rectangular components as long as the components are both assumed to act at some one convenient point along the line of action of the force.*

4.6 Theory of Stress Analysis of Trusses. To determine the adequacy of a truss to withstand a given condition of loading, it is first necessary to compute the bar forces developed in the members of the truss to resist the prescribed loading. The fundamental approach to studying the internal stresses in any body is the same whether it is a beam, a truss, or some other type of structure. In the case of a truss, this approach consists in passing an imaginary section that cuts through certain of the bars and isolates some convenient portion of the truss as a free body. Acting on the internal cross sections exposed by the isolating section will be internal stresses. In the case of a member of an ideal truss, the resultant of these stresses is simply an axial force referred to as the bar force in the member.

If the truss as a whole is in static equilibrium, any isolated portion of it must likewise be in equilibrium. Any particular isolated portion of a truss will be acted upon by a system of forces which may consist of certain external forces and the bar forces acting on the exposed faces of those members which have been cut by the isolating section. It is often possible to isolate portions of a truss so that each portion is acted upon by a limited number of unknown bar forces, which may then be determined so as to satisfy the equations of static equilibrium for that portion.

This procedure may be explained quite easily by considering a specific example such as the simple truss shown in Fig. 4.5. This truss is supported in such a manner that the reactions are statically determinate and are easily computed, as shown in the figure. Proceeding now with the determination of the bar forces in this truss, suppose an imaginary section is passed around joint a, cutting through bars ah and ab and thus completely isolating joint a from the rest of the truss, as shown in free-body sketch a of this figure.

Such an isolated joint will be a body acted upon by a concurrent system of forces since the bar forces in the cut bars of an ideal truss and the external forces are all forces the lines of action of which are directed through the center of the isolated joint. The resultant of a concurrent system of forces cannot be a couple, and hence the force system will be in equilibrium if $\Sigma F_x = 0$ and $\Sigma F_y = 0$. Therefore, if there are only two unknown bar forces acting on a given isolated joint and these do not have the same line of action,[1] the two conditions for static equilibrium will yield two independent equations that can be solved simultaneously for the two unknown forces. If there are more than two unknown bar forces, the values of all the unknowns cannot be determined immediately from these two equations alone.

In this particular case, however, the isolated joint a is acted upon by the known reaction and only two unknown bar forces F_{ah} and F_{ab}. The slopes of the members being known, the horizontal and vertical components of the two unknown bar forces may be expressed in terms of the bar force as shown in sketch a. The two equations of static equilibrium can then be written, assuming that both F_{ah} and F_{ab} are tensions, as

$$\Sigma F_y = 0, +\uparrow \qquad 58 + \tfrac{4}{5}F_{ah} = 0 \qquad\qquad (a)$$

$$\Sigma F_x = 0, \overset{\rightarrow}{+} \qquad \tfrac{3}{5}F_{ah} + F_{ab} = 0 \qquad\qquad (b)$$

Then, from Eq. (a),

$$F_{ah} = -72.5 \text{ kips} \qquad \text{(compression)}$$

and hence, from Eq. (b),

$$F_{ab} = -\tfrac{3}{5}F_{ah} = -(\tfrac{3}{5})(-72.5) = +43.5 \text{ kips} \qquad \text{(tension)}$$

---

[1] If the two unknown bar forces have the same line of action, will the two equations be consistent and independent?

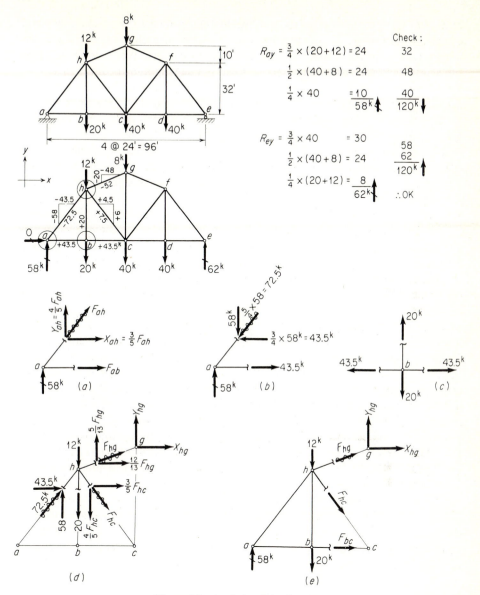

$$R_{ay} = \tfrac{3}{4} \times (20+12) = 24$$
$$\tfrac{1}{2} \times (40+8) = 24$$
$$\tfrac{1}{4} \times 40 = \underline{10}$$
$$58^k$$

$$R_{ey} = \tfrac{3}{4} \times 40 = 30$$
$$\tfrac{1}{2} \times (40+8) = 24$$
$$\tfrac{1}{4} \times (20+12) = \underline{8}$$
$$62^k$$

Check:
32
48
$$\underline{40}$$
$$120^k$$

58
62
$$\underline{120^k}$$
∴ OK

Figure 4.5   Analysis of simple truss

Then the components of F_{ah} are

$$X_{ah} = (\tfrac{3}{5})(-72.5) = -43.5 \text{ kips}$$
$$Y_{ah} = (\tfrac{4}{5})(-72.5) = -58 \text{ kips}$$

Thus, the minus sign indicates that F_{ah} is opposite to the assumed sense (a compression), while the plus sign indicates that F_{ab} is in the assumed sense (a tension).

The sign of the results, therefore, automatically conforms to the sign convention that has been adopted to indicate the character of stress. These results may now be recorded on the line diagram of the truss as -72.5 and $+43.5$, the signs indicating the proper character of stress. The components of the force in ah may be recorded conveniently as shown on the line diagram.

Such conformity in the signs of the results will always be obtained if in drawing the free-body sketches and setting up the equations of statics the sense of the unknown bar forces is assumed to be tension. If this is done, then a plus sign for an answer indicates that the assumed sense is correct and therefore tension, while a minus sign indicates that the assumed sense is incorrect and therefore compression. Thus the signs of the results will automatically conform with the established sign convention.

The procedure just used may be applied in principle to solve for the unknown bar forces at any isolated joint that is acted upon by only two unknown bar forces. In this particular truss, the remaining unknown bar forces could be computed easily by isolating the remaining joints one after the other, always selecting the next joint to be isolated so that the forces in all but two (or fewer) of the cut bars have previously been computed. Of course, it is also necessary for these two unknown bar forces to have different lines of action. This technique of passing a section so as to isolate a single joint of a truss is called the **method of joints.**

Sometimes it is more expedient to pass a section that isolates a portion containing several joints of a truss. This latter technique of passing the cutting section is called the **method of sections.** An isolated portion consisting of several joints of a truss will be a body acted upon by a nonconcurrent system of forces that may consist of certain external forces and the bar forces in those bars cut by the isolating section. For equilibrium of such a portion, the three equations $\Sigma F_x = 0$, $\Sigma F_y = 0$, and $\Sigma M = 0$ must be satisfied by the forces acting on this part of the truss. Therefore, if there are only three unknown bar forces acting on this part and these three bars are neither parallel nor concurrent, the values of the three unknown forces can be obtained from these three equilibrium equations.

A typical application of the method of sections is shown in free-body sketch e of Fig. 4.5. In this case, the cutting section is passed through bars hg, hc, and bc, thus isolating the portion of the truss to the left of this section. The unknown forces in these three cut bars can now be determined by solving the three equations of equilibrium for the isolated portion. In previous discussions of the computations of reactions, it was demonstrated that it is often possible to expedite the solution in such cases of nonconcurrent force systems by using ingenuity in writing the equations of statics. For example, to solve for F_{hg}, take moments about point c, the point of intersection of F_{hc} and F_{bc}, and resolve F_{hg} into its horizontal and vertical components at point g. Then only X_{hg} enters the moment equation, and

$$\Sigma M_c = 0, \quad \overset{\curvearrowright}{+} \qquad (X_{hg})(42) + (58)(48) - (32)(24) = 0$$

whence $X_{hg} = -48$ and by proportion $Y_{hg} = -20$ and $F_{hg} = -52$. In a similar manner,

$$\Sigma M_h = 0, \quad \overset{\curvearrowright}{+} \qquad (58)(24) - (F_{bc})(32) = 0$$

whence $F_{bc} = +43.5$. Then either $\Sigma F_x = 0$ or $\Sigma F_y = 0$ may be used to obtain the horizontal or vertical component, respectively, of F_{hc}:

$$\Sigma F_x = 0, \; \overset{\rightarrow}{+} \qquad X_{hc} + 43.5 - 48 = 0$$

whence $X_{hc} = +4.5$ and by proportion $Y_{hg} = +6$ and $F_{hg} = +7.5$. Of course, any three independent equations of statics could be written and solved for these three unknown forces. However, if ingenuity is not used, the three equations may all contain all three unknowns and have to be solved simultaneously, whereas, as just shown, it is possible to write three equations, each of which contains only one unknown.

4.7 Application of Method of Joints and Method of Sections. In the previous article, the equations involved in applying both the method of joints and the method of sections were set up in a rather formal manner. However, it is often unnecessary to do this. For example, consider joint a, which was used in illustrating the method of joints in the previous article. Consider now this isolated joint as shown in free-body sketch b of Fig. 4.5. By inspection, it is obvious that for $\Sigma F_y = 0$ to be satisfied, the vertical component in bar ah must push downward with a force of 58 kips in order to balance the reaction. Then, by proportion, the horizontal component and the bar force itself in this bar are equal to 43.5 and a compression of 72.5, respectively, acting in the directions shown. The horizontal component in ah being known, it is now apparent that, for $\Sigma F_x = 0$ to be satisfied, the force in ab must be a tension of 43.5 acting to the right to balance the horizontal component in ah acting to the left.

Since the bar force in ab is known, it is a simple matter to find the forces in bc and bh by passing a section that isolates joint b as shown in free-body sketch c of Fig. 4.5. Again, this simple case can easily be solved in an informal manner to obtain the two unknown bar forces F_{bc} and F_{bh}. To satisfy $\Sigma F_x = 0$, it is apparent that F_{bc} must be a tension of 43.5, and to satisfy $\Sigma F_y = 0$, F_{bh} must be a tension of 20.

If joint h is isolated in a similar manner, as shown in free-body sketch d, the isolating section will cut through four bars, two in which the bar forces are known and two in which they are unknown. Again these two unknowns can be found from the equilibrium conditions $\Sigma F_x = 0$ and $\Sigma F_y = 0$ for the isolated joint. Assuming the unknown forces to be tension, the two equations may be written

$$\Sigma F_x = 0, \; \overset{\rightarrow}{+} \qquad \tfrac{12}{13}F_{hg} + \tfrac{3}{5}F_{hc} + 43.5 = 0$$
$$\Sigma F_y = 0, +\uparrow \qquad \tfrac{5}{13}F_{hg} - \tfrac{4}{5}F_{hc} - 12 - 20 + 58 = 0$$

In this case, unfortunately, both equations contain both unknowns, and it is necessary to solve the equations simultaneously for these two values. Of course, the two unknowns can be obtained quite easily in this manner, but consider the advantages of proceeding as follows.

In the discussion of computation of reactions in Chap. 2, it will be recalled that it was often advantageous to replace either or both $\Sigma F_x = 0$ and $\Sigma F_y = 0$ by one or two moment equations. A similar technique is likewise desirable in the present case of the isolated joint h. Suppose that in free-body sketch d the positions of joints a, c, and g are located in space as shown. Then, $\Sigma M_c = 0$ could be used instead of the equation $\Sigma F_y = 0$ or $\Sigma F_x = 0$. Taking moments about point c not only eliminates F_{hc} from the equation but also makes it possible to simplify the computation of the moments of the forces in bars ah and hg. These two bar forces can now be resolved into their vertical and horizontal components at joints a and g, respectively, and then only the vertical component of F_{ah} and the horizontal component of F_{hg} will enter the moment equation and the lever arms of both these components are easily obtained. In this way,

$$\Sigma M_c = 0, \, \overset{\curvearrowright}{+} \qquad (X_{hg})(42) + (58)(48) - (32)(24) = 0 \qquad \therefore X_{hg} = -48$$

and by proportion $Y_{hg} = -20$ and $F_{hg} = -52$.

F_{hg} and its two components being known, it is an easy matter to use either $\Sigma F_x = 0$ or $\Sigma F_y = 0$ and obtain, respectively, either the horizontal or vertical components of the force in hc directly. For example, since $X_{hg} = -48$, the horizontal component in hc must act to the right with 4.5 kips so as to balance the 43.5 kips in ah and make $\Sigma F_x = 0$. This means that this bar is in tension and by proportion the bar force and vertical component are $+7.5$ and $+6.0$, respectively. All these computations can now be checked by seeing whether the results satisfy $\Sigma F_y = 0$.

In the following illustrative examples, additional techniques and tricks will be used to expedite the application of the method of joints and the method of sections. In the first example all of the free-body sketches will be shown in detail. To train the student to visualize free-body sketches when possible, such sketches will be omitted in the later examples. When desirable, the numerical computations will be carried out in an informal manner. If the student finds it difficult to follow these short cuts, he should draw the necessary free-body sketches and set up the equilibrium equations in a fundamental manner. The student should recognize that it is desirable to develop a facility for visualizing free-body sketches and solving equilibrium equations in an informal manner; but he should also recognize that even the expert has to go back to fundamentals—draw sketches and write equations—whenever he is confused or faced with a difficult problem.

The student should also note carefully the technique of drawing free-body sketches of isolated portions of a truss. Any bar force that is known in magnitude[1] from previous computations should be shown acting with this known magnitude on any sketch that is drawn subsequently. For example, in drawing free-body sketch d of Fig. 4.5, the forces in bars ah and bh have previously been computed and recorded on the line diagram of the truss as -72.5 and $+20$, respectively. Hence, the bar forces acting on the stub ends of these two bars should be shown pushing the stub end of ah into joint h and pulling the stub end of bh out of this joint. Thus the sense of these known bar forces having been indicated by arrows, the forces should be labeled with their numerical value only, i.e., simply by 72.5 and 20, not by -72.5 and $+20$. As suggested in the previous article, the free-body sketch is completed by showing the unknown bar forces as being tensions.

When the value of a bar force is recorded on the line diagram of the truss, it will be found helpful also to draw arrows at each end of the member, indicating the direction in which the force in the member acts *on the joint*. This procedure will be followed in recording the bar forces in the remaining illustrations of this chapter.

---

[1] Note that magnitude has previously been defined as including the sense in which a force acts.

Example 4.1   *Compute the bar forces in members Cc, CD, cd, cD, and DE of this truss due to the loads shown.*

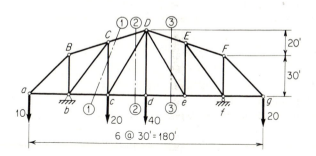

$$\Sigma M_b = 0, \; \widehat{+}$$
$$
\begin{aligned}
(20)(1) &= 20 \\
(40)(2) &= 80 \\
(20)(5) &= \underline{100} \\
& 200 \\
-(10)(1) &= \underline{-10} \\
& 190
\end{aligned}
$$
$$R_{fy} = 47.5 \uparrow$$

$$\Sigma M_f = 0, \; \widehat{+}$$
$$
\begin{aligned}
-(40)(2) &= -80 \\
-(20)(3) &= -60 \\
-(10)(5) &= \underline{-50} \\
& -190 \\
(20)(1) &= \underline{20} \\
& -170
\end{aligned}
$$
$$R_{by} = 42.5 \uparrow$$

$$\Sigma F_y = 0, \; \uparrow +$$
$$47.5 + 42.5 - 10 - 20$$
$$- 40 - 20 = 0$$

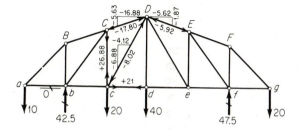

$$\Sigma M_m = 0, \; \widehat{+}$$
$$
\begin{aligned}
(10)(2) &= +20 \\
-(42.5)(3) &= \underline{-127.5} \\
& -107.5 \; \curvearrowleft
\end{aligned}
$$

$$\therefore F_{Cc} = + \frac{107.5}{4} = +26.88$$

$$\Sigma M_c = 0, \; \widehat{+}$$
$$
\begin{aligned}
(42.5)(1) &= +42.5 \\
-(10)(2) &= \underline{-20} \\
& +22.5 \; \curvearrowright
\end{aligned}
$$

$$\therefore X_{CD} = \frac{-(22.5)(30)}{40}$$

$$= -16.88$$

$$\therefore F_{CD} = -(16.88)(^{31 \cdot 63}\!/_{30})$$

$$= -17.80$$

Bar *Cc*   Section ① – ①
Bar *CD*

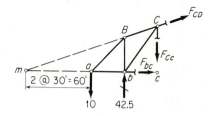

Bar *cd* Section ②–②
Bar *cD*

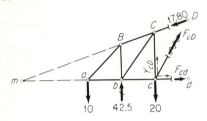

$$\Sigma M_m = 0, \; \overset{\curvearrowright}{+}$$
$$(10)(2) = \; +20$$
$$(20)(4) = \; \underline{\quad 80}$$
$$+100$$
$$-(42.5)(3) = \; \underline{-127.5}$$
$$-27.5 \; \curvearrowleft$$

$$\therefore \; Y_{cD} = \frac{-27.5}{4} = -6.88$$

$$\therefore \; F_{cD} = -6.88 \times \frac{58.3}{50}$$
$$= -8.02$$

$$\Sigma M_D = 0, \; \overset{\curvearrowright}{+}$$
$$-(20)(1) = \; -20$$
$$-(10)(3) = \; \underline{-30}$$
$$-50$$
$$(42.5)(2) = \; \underline{+85}$$
$$+35 \; \curvearrowright$$
$$\therefore \; F_{cd} = (+35)(^{30}\!/_{50}) = +21$$

$$\Sigma M_e = 0, \; \overset{\curvearrowright}{+}$$
$$-(47.5)(1) = \; -47.5$$
$$(20)(2) = \; \underline{+40}$$
$$-7.5 \; \curvearrowleft$$

Bar *DE* Section ③–③

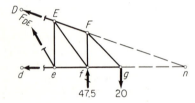

$$\therefore \; X_{DE} = (-7.5)(^{30}\!/_{40}) = -5.62$$

$$\therefore \; F_{DE} = (-5.62)(^{31.6}\!/_{30})$$
$$= -5.92$$

Discussion:

Note that after the force in *Cc* has been found, the vertical component of the force in *cD* can easily be found by isolating joint *c*. Note also that the force in *cD* being known, the vertical component in *CD* can be found from $\Sigma F_y = 0$ rather than $\Sigma M_c = 0$ for the free-body sketch for section 2-2. The force in *CD* being known, the force in *cd* can be obtained from $\Sigma F_x = 0$ applied to this free-body sketch.

The force in member *DE* can be obtained by isolating the portion of the truss to either the right or the left of section 3-3. The portion to the right was chosen because it has fewer external forces acting on it.

In all these computations, the moments of the vertical forces are computed in terms of panel lengths. When necessary, the panel length of 30 ft is substituted at the end of the computation. The trick simplifies the numerical work in such computations.

Example 4.2   *Determine the bar forces in members dg, eg, gh, and hm.*

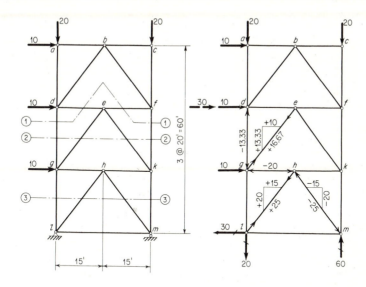

Consider the equilibrium of the structure as a whole to obtain the reactions. From $\Sigma M_l = 0$, obtain R_{my}; and from $\Sigma M_m = 0$, obtain R_{ly}. Use $\Sigma F_y = 0$ as a check.

Section 1-1: *Isolate the structure above this section and apply $\Sigma M_f = 0$ to obtain F_{dg}.*

Section 2-2: *Isolate the structure above this section and, with F_{dg} known, use $\Sigma M_k = 0$ to obtain Y_{eg}. Then isolate joint g and, with X_{eg} known, use $\Sigma F_x = 0$ to obtain F_{gh}.*

Next isolate joint h and note from $\Sigma F_y = 0$ that in the absence of any applied vertical load, the vertical components in bars lh and hm must be equal and opposite. Thus if $Y_{lh} = A$, then $Y_{hm} = -A$, and $X_{lh} = \frac{3}{4}A$ and $X_{hm} = -\frac{3}{4}A$.

Section 3-3: *Isolate the structure above this section and apply $\Sigma F_x = 0$. Since the horizontal components of lh and mh have been expressed in terms of A, it is easy to find A to be equal to $+20$ from this equation, thereby defining completely the bar forces in lh and mh.*

Discussion:

If the forces in all the bars are required, the method of joints can be applied, the joints being isolated successively in the following order: a, c, b, d, f, e, g, k, h, l, m. This is probably the most efficient method of finding all the bar forces in this particular truss. If only a few particular bar forces are required, the computation can be carried out as illustrated in this example.

Example 4.3 *Compute the bar forces in members bc, BC, aC, and bC.*

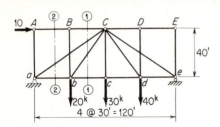

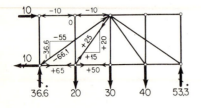

Bar bc: *Portion to left of 1-1,* $\Sigma M_C = 0$, $\curvearrowright +$,

$$(36.\dot{6})(2p) = \quad 73.\dot{3}p$$
$$(20)(1p) = -20.0p$$
$$\overline{}$$
$$F_{bc} = \frac{53.\dot{3}p + (10)(40)}{40} = +50$$

Bar BC: *Isolate joint A, then B.* $F_{BC} = -10$

Bar aC: *Portion to left of 2-2,* $\Sigma F_y = 0$
$$\therefore Y_{aC} = -36.\dot{6}$$

$$\therefore F_{aC} = \frac{\sqrt{3^2 + 2^2}}{2}(-36.\dot{6}) = -66.1$$

Bar bC: *Portion to left of 1-1, with force in aC known, from* $\Sigma F_y = 0$ $Y_{bC} = +20$

or, from joint B, $F_{Bb} = 0$; *then, from joint b,*
$\Sigma F_y = 0$ $\therefore Y_{bC} = +20$

Example 4.4 *Compute the bar forces in members cd, BC, bm, nf, nF, and md.*

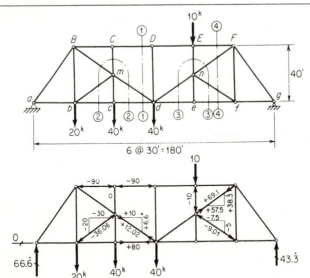

Bar cd: *Portion to left of 1-1, from* $\Sigma M_B = 0$, $F_{cd} = +80$

Bar BC: *Same portion,* $\Sigma M_d = 0$, $F_{CD} = -90$
Then, isolate joint C, $\Sigma F_x = 0$, $F_{BC} = -90$

Bar bm: *Isolating joint C shows* $F_{Cm} = 0$. *Hence, considering portion isolated by section 2-2,* $\Sigma M_d = 0$, $\therefore Y_{bm} = -20$

Bar nf: *Isolating joint E shows* $F_{En} = -10$. *Hence, considering portion isolated by section 3-3,* $\Sigma M_d = 0$, $\therefore Y_{nf} = -5$

Bar nF: *Portion to right of 4-4,* $\Sigma F_y = 0$, $\therefore Y_{nF} = +38.\dot{3}$

Bar md: *Portion to right of 1-1,* $\Sigma F_y = 0$, $\therefore Y_{md} = +6.\dot{6}$

Example 4.5   *Determine the bar forces in all members of this truss.*

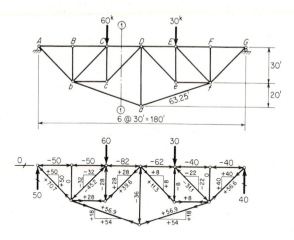

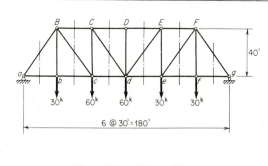

Discussion:

In the solution of this problem, it is possible to start out and apply the method of joints successfully at joints A, B, G, and F.   However, attempting to apply this same procedure at any of the remaining joints proves impossible since there are more than two unknown bar forces at each of these joints. It is therefore desirable to resort to the method of sections.   In this problem, the bar force in member bd is found by considering the portion to the left of section 1-1. The method of joints can then be applied at joint b and then successively at each of the remaining joints.

It should be noted that this is a compound truss.   In such trusses it is usually impossible to solve for all the bar forces by using simply the method of joints.   As illustrated in this problem, it is usually necessary to make at least one application of the method of sections.

Example 4.6   *Determine the bar forces in all members of this truss.*

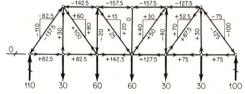

Discussion:

In the solution of this example, it is possible to start at one end of the truss and work through to the other end, using only the method of joints to compute the bar forces. This is a simple truss, and it is always possible to find all the bar forces in such a truss in this manner once the reactions have been computed.

It also should be noted that the vertical components of the bar forces in the diagonals can also be easily obtained by using $\Sigma F_y = 0$ on the portions either to the right or to the left of the indicated vertical sections through the panels. The force in the verticals can then be obtained from $\Sigma F_y = 0$ at the joints, and the force in the chords from $\Sigma F_x = 0$ at the joints, working across from end to end of the truss.

4.8 Discussion of Method of Joints and Method of Sections. The examples in the previous article illustrate that isolations of portions of the truss using both the method of joints and the method of sections must be employed in the stress analysis of trusses. Experience in such computations will teach the student how to combine these two methods most effectively   It is the purpose of this particular article to summarize and clarify the important points concerning them.

In the previous discussion of the method of joints, it was pointed out that this procedure enables one to determine immediately all the unknown bar forces acting on an isolated joint, provided there are not more than two unknown forces and they have different lines of action. It sometimes happens that there is only one unknown bar force acting on an isolated joint. In such a case, one of the two available equilibrium equations may be used to solve for the one unknown force, and the other may be used as a check that must be satisfied by all the forces acting on this joint. If there are more than two unknown forces acting on an isolated joint, it is usually impossible to obtain an immediate solution for any of the unknowns from the two equations of equilibrium that are available at that joint. In such cases, it is necessary to isolate additional joints and write two additional equations for each joint. In this manner, it is sometimes possible to obtain n independent equations, involving n unknown bar forces. Then these n equations can be solved simultaneously for the n unknowns.

There is one important case where there are more than two unknown bar forces acting on an isolated joint but so arranged that it is possible to obtain the value of one of them immediately. If all the unknown forces except one have the same line of action, the force in that one particular bar can be determined immediately. Such a case is shown in sketch a of Fig. 4.6. If the x axis is taken as being parallel to the line bac and the y axis perpendicular to this line, the y component of the unknown force F_{ad} can be determined immediately from the equation $\Sigma F_y = 0$. No immediate solution can be obtained from $\Sigma F_x = 0$ at such a joint since this equation involves both the unknowns F_{ab} and F_{ac}. A special case of this type is shown in sketch b of Fig. 4.6, where the joint is acted upon by only the three unknown bar forces. If F_{ab} and F_{ac} have the same line of action, it is apparent that the only remaining force at the joint, F_{ad} must be zero. It is also of interest to consider the case shown in sketch c of Fig. 4.6. In this case, the joint is acted upon by only two forces, which do not have the same line of action; therefore, in order to satisfy $\Sigma F_x = 0$ and $\Sigma F_y = 0$ for such a joint, both F_{ab} and F_{ac} must be equal to zero.

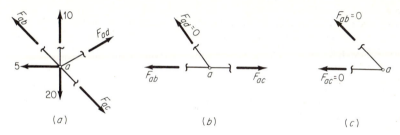

Figure 4.6   Special cases

It is also interesting to note that once the reactions have been determined on a *simple* truss, all the bar forces can be determined by using only the method of joints and never resorting to the method of sections. It is apparent that this is so, since there are only two unknown bar forces acting on the joint that was located *last* in arranging the layout of the bars of the truss. After these two bar forces have been determined by isolating this joint, it will be found that the joints at the far ends of these two members are acted upon by only two unknown bar forces. Thus, by isolating the joints in the reverse order from that in which they were established in laying out the truss, the method of joints can be used to determine all the bar forces. This explains why it is possible to solve Example 4.6 in this manner. It should be noted, however, that in many cases of simple trusses the calculation of all the bar forces can be expedited by combining the use of both the method of sections and the method of joints, as illustrated by some of the examples in Art. 4.12.

When applying the method of sections, if the isolated portion of the truss is acted upon by three unknown bar forces that are neither parallel nor concurrent, all three unknown forces can be determined from the three equilibrium equations available for the isolated portion. It is presumed, of course, that the reactions acting on any such portion have been determined previously. Of course, if there are only one or two unknown bar forces, they can be determined by using a like number of the available equations. The remainder of the equations that must be satisfied by the system of forces acting on the isolated portion can then be used simply as a check on the calculations up to this point.

It is sometimes possible to find some of the unknown forces by the method of sections even when there are more than three unknowns acting on the isolated portion. For example, suppose that the lines of action of all but one of the unknown forces intersect at a point a. Then the stress in this one remaining bar can be determined from the equation $\Sigma M_a = 0$, that is, by summing up the moments of the forces about point a. Another similar case would be one where all the unknown bar forces except one are parallel. The stress in this remaining bar can then be determined by summing up all the force components that are perpendicular to the direction of the other unknown bar forces. In each of the above cases, the remaining two equilibrium equations for the isolated portion will involve more unknowns than there are equations, and hence no immediate solution for these remaining unknowns is possible.

In applying either the method of joints or the method of sections, it is important to realize that it makes no difference how many bars have been cut in which the bar forces are *known*. Only the number of unknown bar forces is important.

In the latter half of the nineteenth century, graphical methods of stress analysis were widely used, with outstanding analysts such as Culmann, Maxwell, Mohr, and Müller-Breslau being leaders in developing these computational techniques. Graphical procedures for applying the method of joints in truss analysis were particularly useful, especially for trusses having unusual geometrical configurations. In earlier editions of this text, a 20-page chapter was devoted to a condensed treatment of some of the more useful graphical methods. Nowadays, of course, computerized solutions are available to handle truss analyses and other similar problems that were cumbersome and tedious to solve by hand calculations, whether graphical or algebraic. Since the study of graphical methods helps a person develop his skills for visualizing and representing physical and spatial relationships, it was with great reluctance that we finally decided to delete the chapter on graphic statics in order to contribute space for the treatment of systematic structural analysis in Part II.

4.9 Statical Stability and Statical Determinacy of Truss Structures. Up to this point the emphasis has been placed on the methods of computing the bar forces in trusses. For this purpose, all the examples have been statically determinate and stable. With these ideas as a background, it is now possible to discuss the question of statical stability and determinacy of trusses from a general standpoint.

In discussing the arrangement of the members of a simple truss, it was shown that a rigid truss is formed by using three bars to connect three joints together in the form of a triangle and then using two bars to connect each additional joint to the framework already constructed. Thus, to form a rigid simple truss of n joints, it is necessary to use the three bars of the original triangle plus two additional bars for each of the remaining $n - 3$ joints. If b denotes the total number of bars required, then

$$b = 3 + 2(n - 3) = 2n - 3$$

This is the minimum number of bars that can be used to form a rigid simple truss. To use more is unneccessary, and to use fewer results in a nonrigid or unstable truss. If a simple truss having n joints and $2n - 3$ bars is supported in a manner that is equivalent to three links that are neither parallel nor concurrent, the structure is stable under a general condition of loading and the reactions are statically determinate. In the previous article, it was pointed out that once the reactions are found, all the bar forces of a simple truss can be computed by the method of joints.

It can therefore be concluded that a stable simple truss having three independent reaction elements and $2n - 3$ bars is statically determinate with respect to both reactions and bar forces. If there are more than three reaction elements, the structure is statically indeterminate with respect to its reactions; if there are more than $2n - 3$ bars but only three reaction elements, it is indeterminate with respect to the bar forces; and if there is an excess of both bars and reaction elements, the structure is indeterminate with respect to both reactions and bar forces.

The same discussion and conclusions apply equally well to a compound truss. Suppose a compound truss is formed by connecting two simple trusses together by means of three additional bars that are neither parallel nor concurrent.   If the two simple trusses have n_1 and n_2 joints, respectively, the total number of bars b in the compound truss is

$$b = (2n_1 - 3) + (2n_2 - 3) + 3 = 2(n_1 + n_2) - 3$$

or if n denotes the total number of joints in the compound truss, i.e., if $n = n_1 + n_2$, then

$$b = 2n - 3$$

Thus, the minimum number of bars that can be used to form a rigid compound truss is the same as in the case of a simple truss.   If the remainder of the discussion were carried out in a similar manner for a compound truss, it would be found that the conclusions of the previous paragraph apply equally well to both simple and compound trusses.

It is desirable to discuss this question of determinacy and stability from a more general standpoint.   Suppose a truss structure has r independent reaction elements, b bars, and n joints.   If the truss as a whole is in equilibrium, every isolated portion must likewise be in equilibrium.   To isolate an entire bar or some portion of it would produce no new information since the equilibrium conditions of the bars were considered during the establishing of the definition of the term bar force. However, it is possible to isolate each of the n joints in turn and to write for each of these joints two new and independent equations of static equilibrium, $\Sigma F_x = 0$ and $\Sigma F_y = 0$.   In this manner, a total of $2n$ independent equations would be obtained, involving as unknowns the r reaction elements and the b bar forces, a total of $r + b$ unknowns.   These $2n$ equations must be satisfied simultaneously by the $r + b$ unknowns.   By comparing the number of unknowns with the number of independent equations, it is possible to decide whether a truss structure is unstable, statically determinate, or indeterminate.   The reasoning involved is similar to that used in Art. 2.5.   *If $r + b$ is less than $2n$, there are not enough unknowns available to satisfy the $2n$ equations simultaneously and therefore the truss structure is said to be statically unstable.   If $r + b$ is equal to $2n$, the unknowns can be obtained from the simultaneous solution of the $2n$ equations and therefore the structure is said to be statically determinate.   If $r + b$ is greater than $2n$, there are too many unknowns to be determined from these $2n$ equations alone and therefore the structure is said to be statically indeterminate.   The criterion establishes the combined degree of indeterminancy with respect to both reactions and bar forces, provided that the structure is neither statically nor geometrically unstable.*   It is apparent that these conclusions agree with the foregoing discussion of simple and compound trusses.

At first glance, it might seem that the total number of independent equations of static equilibrium in a truss structure should include not only the $2n$ equations noted in the previous paragraph but also the three equations $\Sigma F_x = 0$, $\Sigma F_y = 0$, and $\Sigma M = 0$ applied to the entire structure as a

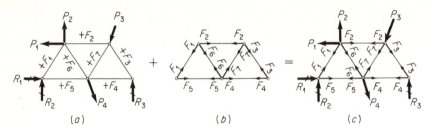

Figure 4.7   Equilibrium considerations

free body.   However, the following demonstration will prove that this offhand opinion is not so and that there are only $2n$ independent equations.   Consider any truss as a free body acted upon by its reactions and applied loads such as the truss shown in Fig. 4.7a.   Suppose that the system of forces shown in Fig. 4.7b is superimposed on the system in Fig. 4.7a; the combination of these two loading systems will then be as shown in Fig. 4.7c.   The load system in b is a special system consisting of several pairs of equal and opposite forces, one pair for each member of the truss. For any particular member, both forces of its pair act along the member, one acting on the joint at one end of the member and the other on the joint at the other end.   Each of the forces is numerically equal to *the bar force in that member produced by the forces in system a* and acts in a sense that is the same as the action of this bar force on the joint.   Each pair of forces is in equilibrium, of course, and hence all the pairs acting together form a system that is in equilibrium.

Considering now the combined system in c, it is found that the forces acting at each joint of the truss are the same forces that would be acting on that joint if, under the loading of a, it were isolated by itself as a free body.   If, however, the applied forces, reactions, and bar forces satisfy simultaneously the $2n$ equations of statics obtained by isolating the n joints and writing $\Sigma F_x = 0$ and $\Sigma F_y = 0$ for each joint, the forces acting on *each* joint in c form a concurrent system of forces that are in equilibrium.   Since at each joint the forces are in equilibrium, the combined system in c of all joints is in equilibrium and satisfies the equations $\Sigma F_x = 0$, $\Sigma F_y = 0$, and $\Sigma M = 0$ for the entire truss.   Since the combined system in c is in equilibrium and the portion of this system shown in sketch b is also in equilibrium by itself, then the remaining portion of the system as shown in sketch a must be in equilibrium and therefore must satisfy the equations $\Sigma F_x = 0$, $\Sigma F_y = 0$, and $\Sigma M = 0$ for the entire truss.   It may be concluded, therefore, that, if the reactions, bar forces, and applied loads satisfy the $2n$ equations of equilibrium obtained by isolating the joints of the truss, then the reactions and applied loads will automatically satisfy the three equations of equilibrium for the truss as a whole and that thus there are only $2n$ independent equations of static equilibrium involved in a truss.

It should be noted that comparing the count of the unknowns and the independent equations establishes a criterion which is necessary but not always sufficient to decide whether a truss is stable or not.   If $b + r$ is less than $2n$, this comparison is sufficient for deciding that the truss is statically unstable.   If, however, $b + r$ is equal to or greater than $2n$, it does not automatically follow that the truss is stable.   This statement can be verified by considering the examples shown in Fig. 4.8.   In all four of these cases, the structures are unstable, whereas the count by itself indicates that a and b are statically determinate and c and d are indeterminate to the first degree. a and d are unstable under a general condition of loading because in each case the reactions are equivalent to parallel links. b and c are unstable, not because of the arrangement of the reactions, but because of the arrangement of the bars. In b, for example, the reactions are statically determinate, but the truss is unstable

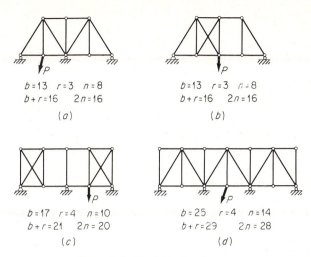

$b=13 \quad r=3 \quad n=8$
$b+r=16 \quad 2n=16$

(a)

$b=13 \quad r=3 \quad n=8$
$b+r=16 \quad 2n=16$

(b)

$b=17 \quad r=4 \quad n=10$
$b+r=21 \quad 2n=20$

(c)

$b=25 \quad r=4 \quad n=14$
$b+r=29 \quad 2n=28$

(d)

Figure 4.8   Unstable structures

and would collapse because there is nothing to carry the shear in the second panel from the right end.

These and other considerations lead to the conclusion that, even though the count indicates that the truss structure is either statically determinate or indeterminate, for it to be stable also, it is necessary that the following conditions shall likewise be satisfied: (1) *The reactions must be equivalent to three or more links that are neither parallel nor concurrent.* (2) *The bars of the truss must also be arranged in an adequate manner.*   It is sometimes difficult to determine whether or not the arrangement of the bars is adequate.   In such cases, if the arrangement is inadequate, it will become apparent; for when a stress analysis is attempted, it will yield results that are inconsistent, infinite, or indeterminate.

4.10   Examples Illustrating the Determination of Stability and Determinancy.   It is easy to investigate the stability and determinacy of a truss structure that is formed by supporting in some manner a truss that is in itself a rigid body.   The truss itself may be merely a simple or compound truss or, in some cases, a simple or compound truss modified by adding more than the necessary number of bars. In either case, the bars, reactions, and joints can be counted and the criteria of the last article applied to decide whether the structure is unstable, statically determinate, or statically indeterminate.   This count, of course, enables one to classify the structure with respect to both bar forces and reactions.   If the count shows that the structure is statically determinate or indeterminate, the question of stability must still be decided, for the count by itself is not sufficient to prove that the structure is stable.

It is also easy to classify simple or compound trusses with respect to their reactions only.   If there are fewer than three independent reaction elements, the structure is statically unstable under a general condition of loading regardless of how the bars of the truss are arranged.   If there are three or more independent reaction elements

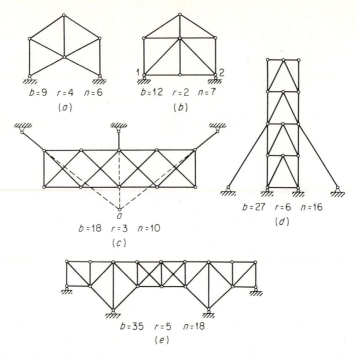

Figure 4.9   Examples for classification

and they are arranged so as to be equivalent to three or more links that are neither parallel nor concurrent, the structure is stable with respect to its reactions.   For a stable structure, if there are exactly three reaction elements, these elements are statically determinate; if there are more than three reaction elements, the structure is statically indeterminate with respect to its reactions alone to a degree that is equal to the number of reaction elements in excess of three.   Structures in this general category are shown in Fig. 4.9.

The count of the bars, joints, and reaction elements is shown in each of the sketches in Fig. 4.9.   Considering only the reactions, structure a is stable and statically indeterminate to the first degree.   Since $b + r = 13$ and $2n = 12$, it is also indeterminate to the first degree, considering both reactions and bar forces. The count of structure b indicates that it is statically determinate since $b + r$ and $2n$ are both equal to 14.   A consideration of the reactions, however, discloses that this structure is actually unstable.   Likewise, the count indicates that structure c is statically indeterminate to the first degree, but consideration of the reactions shows that it is unstable.   Both the count and the consideration of the reactions indicate that structure d is indeterminate to the first degree.   With respect to the reactions only, structure e is indeterminate to the second degree, but a count of both bars and reactions discloses that it is actually indeterminate to the fourth degree.

There is another important type of truss structure that is built up out of more than just one rigid truss.   In this type, the structure is composed of several rigid

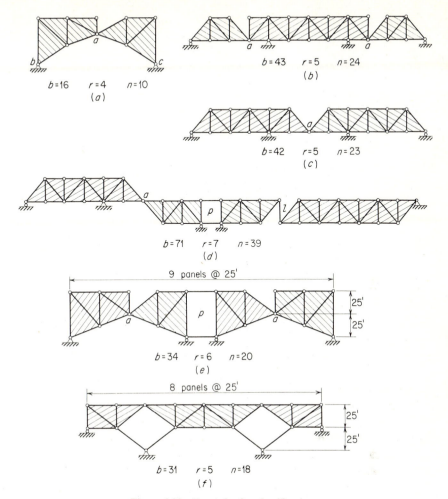

Figure 4.10 Examples for classification

trusses connected together in some manner and then the whole assemblage mounted on a certain number of supports. In such cases, the supports are usually arranged so as to provide more than three independent reaction elements. The connections between the several rigid trusses are not completely rigid, however, so that certain equations of condition (or construction) are introduced to reduce the degree of indeterminancy or perhaps even to make the reactions statically determinate. This type of structure is the hardest to analyze from a stability or determinancy standpoint. However, some of the most important trussed structures, e.g., cantilever and three-hinged arch bridges, belong in this category, and therefore it is important for the student to master the methods of investigating this type. Structures of this general type are illustrated in Fig. 4.10.

The stability and determinancy of structures of the type shown in Fig. 4.10 can be investigated by comparing the count of the bars and reaction elements with the count of the joints.   With this criterion, it will be concluded that structures *a*, *b*, *d*, *e*, and *f* are statically determinate and structure *c* is indeterminate to the first degree.   In structures of this type, it is also important to consider whether or not the structure is statically determinate with respect to its reactions alone.   This can be done by comparing the count of the unknown reaction elements with the number of available equations in the same manner as in Arts. 2.5 and 2.6.   In these cases, the available equations include the three equations of static equilibrium for the structure as a whole plus any equations of condition which may be introduced by the manner in which the several rigid trusses are connected together.

If two trusses are hinged together at a common joint, such as the joints marked *a* in Figs. 4.10*a* to *e*, one equation of condition is introduced, namely, that the bending moment about that point must be zero since the hinge cannot transmit a couple from one truss to the other.   If two trusses are connected together by a link or roller, such as the link marked *l* in structure *d*, two equations of condition are introduced since then both the direction and point of application of the interacting force are known.   This means, therefore, (1) that the bending moment about *either* end of the link must be zero or, in other words, (2) that the interacting force between the two trusses cannot have a component perpendicular to the link. If two trusses are connected together by two parallel bars, as is done in panels *p* of structures *d* and *e*, one equation of condition is introduced, namely, that the interaction between the two trusses cannot involve a force perpendicular to the two bars.   In the case of structures *d* and *e*, this means that the shear acting on panel *p* must be zero.

From this discussion, it is apparent that one equation has been introduced in structure *a*, two in *b*, one in *c*, four in *d*, and three in *e*.   It will be concluded, therefore, that with respect to reactions only structures *a*, *b*, *d*, and *e* are statically determinate and structure *c* is indeterminate to the first degree.   Structure *f* is a special type of structure, called a **Wichert**[1] **truss** in this country, that can be counted only by considering the bars, joints, and reactions.

There is no obvious instability in any of the structures of Fig. 4.10.   If, however, one attempts to compute the reactions and bar forces for either structures *e* or *f*, the results will be inconsistent, infinite, or indeterminate; therefore, these structures are actually unstable.   In both cases, by changing only the geometry of the structure, it is possible to make the structure stable.   Structures *e* and *f* are therefore said to be geometrically unstable.   This type of instability may arise whenever equations of condition are introduced by the arrangement of the structure.   Sometimes the instability is obvious, but usually it does not become apparent until one attempts to compute the reactions, etc.[2]

4.11   Conventional Types of Bridge and Roof Trusses.   The members of a truss can be arranged in an almost unlimited number of ways, but the vast majority of trusses encountered in bridge or building work belong to one of the common types shown in Figs. 4.11 and 4.12.   Since they are encountered so frequently, the student should be familiar with the names of these conventional types.

---

[1] D. B. Steinman, "The Wichert Truss," D. Van Nostrand Company, Inc., New York, 1932.

[2] For a more complete discussion see W. M. Fife and J. B. Wilbur, "Theory of Statically Indeterminate Structures," McGraw-Hill Book Company, New York, 1937.

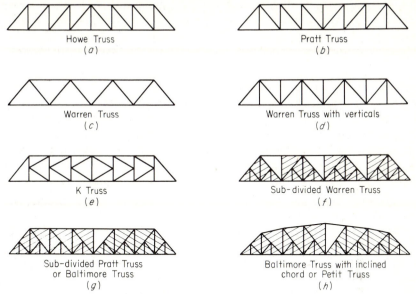

Figure 4.11   Conventional bridge trusses

Trusses a, b, c, d, and e of Fig 4.11 are simple trusses, while the remaining trusses are compound trusses built up out of the simple trusses (shaded). In order to achieve economical design of single-span steel-truss bridges, it is essential for the ratio of depth of truss to length of span to be between $\frac{1}{5}$ and $\frac{1}{10}$, for the diagonals to slope at approximately $45°$ to the horizontal, and for the panel lengths not to exceed 30 to 40 ft. Trusses a, b, c, and d can meet these requirements if the span is not too long. For long-span bridges, however, it becomes necessary to use one of the subdivided types such as f, g, or h.

All the roof trusses shown in Fig. 4.12 are simple trusses, with the exception of the Fink truss, which is a compound truss.

4.12. Illustrative Examples of Stress Analysis of Determinate Trusses. The following examples illustrate the application of the previous discussions to the stress analysis of several conventional types of trusses. The analysis of such trusses is discussed further in Chap. 6.

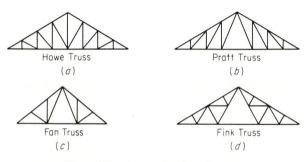

Figure 4.12   Conventional roof trusses

Example 4.7   *Determine the bar forces in all members of this Pratt truss with a curved top chord.*

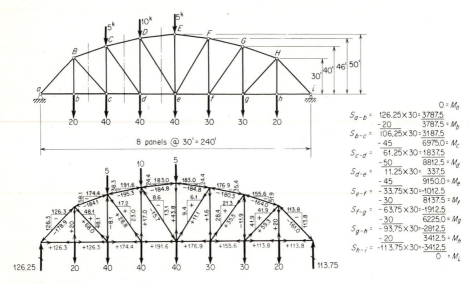

$$0 = M_a$$

$$S_{a-b} = \frac{126.25 \times 30 = 3787.5}{-20}$$
$$3787.5 = M_b$$

$$S_{b-c} = \frac{106.25 \times 30 = 3187.5}{-45}$$
$$6975.0 = M_c$$

$$S_{c-d} = \frac{61.25 \times 30 = 1837.5}{-50}$$
$$8812.5 = M_d$$

$$S_{d-e} = \frac{11.25 \times 30 = 337.5}{-45}$$
$$9150.0 = M_e$$

$$S_{e-f} = \frac{-33.75 \times 30 = -1012.5}{-30}$$
$$8137.5 = M_f$$

$$S_{f-g} = \frac{-63.75 \times 30 = -1912.5}{-30}$$
$$6225.0 = M_g$$

$$S_{g-h} = \frac{-93.75 \times 30 = -2812.5}{-20}$$
$$3412.5 = M_h$$

$$S_{h-i} = \frac{-113.75 \times 30 = -3412.5}{0} = M_i$$

Discussion:

This Pratt truss is a simple truss and can therefore be analyzed by using simply the method of joints. This procedure, however, is not particularly efficient in a case where the two chords are not parallel. Probably the best procedure is first to find the horizontal components in the members of the curved chord. These can be computed by passing a vertical section through a panel and taking moments about the appropriate joint on the bottom chord. These computations are facilitated if the bending moments are known at the various joints along the bottom chord.

The bending moments at the bottom-chord joints and the shears in the panels can conveniently be computed as shown. In this case, where all the loads and reactions are vertical, the bending moment about any top-chord joint is the same as that about the bottom-chord joint directly under it. Of course, if there are horizontal loads, this relationship is not necessarily true.

The horizontal components in the top chord being known, the remainder of the stress analysis can be accomplished by the method of joints. It should also be noted that it is easy to compute the vertical components in the diagonals once the shears in the panels and the vertical components in the top chords are known.

Example 4.8   *Determine the bar forces in all members of this Fink roof truss.*

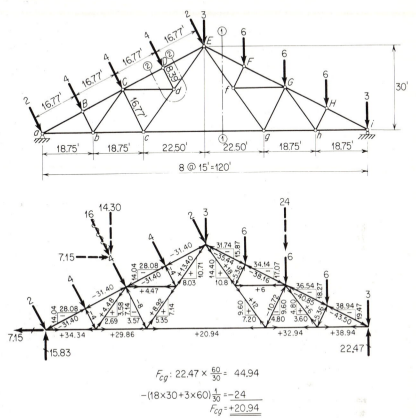

$$F_{cg}: 22.47 \times \frac{60}{30} = 44.94$$

$$-(18 \times 30 + 3 \times 60)\frac{1}{30} = \underline{-24}$$

$$F_{cg} = \underline{+20.94}$$

Discussion:

Although this Fink truss is a compound truss, it can be analyzed by using only the method of joints. For example, after the reactions have been computed, the method of joints can be applied successively at joints *i*, *H*, and *h*. Since there are more than two unknown bar forces at each of joints *g* and *G*, it is not possible to consider these joints as the next step in the analysis; however, one can determine the force in bar *Ff* by isolating joint *F* and the force in bar *fG* by isolating joint *f*. Then the stress analysis can be completed, still by means of joints. This procedure is possible only because joints *E*, *F*, and *G* and likewise joints *E*, *f*, and *g* lie on straight lines.

However, it is usually preferable to proceed as follows. After applying the method of joints at joints *i*, *H*, and *h*, the force in bar *cg* can be obtained by isolating the portion to the right of section 1-1 and taking moments about point *E*; the rest of the analysis can then be completed by the method of joints.

Note also that the stress analysis can be expedited by finding the forces in bars *Cd*, *Cb*, *Gf*, and *Gh*, using sections similar to section 2-2 and taking moments about points such as point *E* in the case of section 2-2.

The main point to remember is to consider the application of both the method of joints and the method of sections and combine the two approaches in such a way as to expedite the calculations.

Note that the geometry of the truss is rather complicated. This is often true of roof trusses. In such cases it is often easier to accomplish the stress analysis by graphical methods.

Example 4.9   *Determine the forces in all members of this Howe truss.*

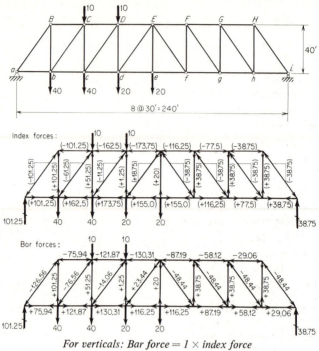

For verticals: Bar force = 1 × index force
For diagonals: Bar force = $^{50}/_{40}$ × index force
For chords: Bar force = $^{30}/_{40}$ × index force

Discussion:

After the reactions have been computed, it is possible to compute the vertical components in all the web members by working from one end of the truss to the other, using either the method of joints or the method of sections. Then it is possible to compute the horizontal components in the diagonals and apply the method of joints to obtain the chord forces. In this case, however, the computation of the chord forces depends only on the horizontal components in the various diagonals. Since all the diagonals have the same slope, the ratio between the vertical and horizontal components is the same for all; it is in this case 40 : 30. In applying $\Sigma F_x = 0$ at the various joints to obtain the chord forces, it is therefore permissible to use the values of vertical components temporarily in place of the horizontal components in the diagonals. In this manner, the values obtained for the chord forces are not equal to the true bar forces in the chords, but the ratio between these values and the true values is constant and equal to the ratio between the vertical and horizontal components in the diagonals.

These values of the chord forces which must be multiplied by a constant factor to obtain the true bar force are called **index forces** *for the chords. Likewise, the vertical components in the web members may be referred to as index forces for the webs. These index forces are easily written down, as shown by the numbers in parentheses in the first force diagram. Then, the true bar forces can be obtained by multiplying the index forces by certain factors, as indicated.*

The use of index forces is helpful in analyzing parallel-chord trusses that have equal panels and are acted upon by transverse loads. In other cases, the index-force method becomes involved and is usually inferior to the other methods already discussed.

4.13 Exceptional Cases. Occasionally trusses cannot be classified as either simple or compound. Such a truss is shown in Fig. 4.13a. In these cases, it is usually difficult to tell by inspection whether the truss is rigid or not and whether it is statically determinate or indeterminate. In this particular case, a count of the structure shows that there are nine bars and six joints, which indicates that the structure is statically determinate. Whether or not the truss is stable is not apparent, but one way of finding out is to attempt a stress analysis and discover whether the results are consistent.

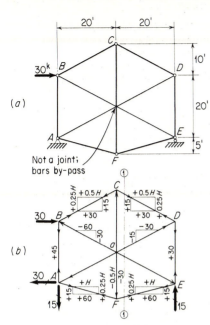

(a)

Not a joint;
bars by-pass

(b)

Figure 4.13   Complex truss

After computing the reactions, it is found that there is no joint at which there are only two unknown bar forces. Applying the method of joints will therefore not yield immediate solutions for the bar forces as in the case of simple trusses. It is also found that the method of sections likewise will not yield an immediate solution for any of the bar forces. Of course, it is possible to set up and solve nine simultaneous equations involving the nine unknown bar forces by using nine of the twelve equations that result from applying the method of joints to the six joints of the structure. If the three reactions have already been computed, the three remaining equations can be used for checking the results obtained for the nine bar forces.

Setting up nine equations in this manner is a poor way to solve this problem, however. Several other approaches are much superior, one of which is to proceed as follows. After computing the reactions, assume that the horizontal component of the bar force in member FE is a tension of H. Then, from joint F, the horizontal component in FA must also be $+H$, and the bar force in FC is $-0.5H$. By isolating joint C, it is then found that the horizontal and vertical components in both bars BC and CD are $+0.5H$ and $+0.25H$, respectively. The forces in these five bars having been found in terms of H, it is now possible to pass section 1-1 through the truss, thus isolating the portion to the right of this section. When the moments about point a are summed, an equation is obtained involving H as the only unknown:

$$\overset{R}{\sum} M_a = 0, \overset{\curvearrowright}{+} \qquad 15H - (20)(0.5H) - (15)(20) = 0$$

whence $H = +60$. With H known, all the other bar forces can be found by the method of joints as shown in Fig. 4.13b. Since in this manner it is possible to make a consistent stress analysis of this truss under any condition of loading, it can be concluded that it is statically determinate and stable.

Trusses of this type, which cannot be classified as either simple or compound, may be called **complex trusses.** Professor S. Timoshenko uses this terminology.[1] In his excellent discussion of complex trusses, Timoshenko describes a general method of analysis of complex trusses called **Henneberg's method.**[2]

While the student should know how to recognize a complex truss and something about investigating its stability and stress analysis, he will not encounter this type often enough to warrant

[1] S. Timoshenko and D. H. Young, "Engineering Mechanics," vol. I, "Statics," McGraw-Hill Book Company, New York, 1956.

[2] This method was developed by L. Henneberg in his "Statik der starren Systeme," Darmstadt, 1886.

devoting more space to the subject here.   If additional information is required, the student is referred to Timoshenko's book.[1]   Several problems at the end of this chapter emphasize the fact that complex trusses may often be arranged so as to be geometrically unstable.   Cases of this type are not always obvious and may not become apparent until the stress analysis is attempted and found to lead to inconsistent results.

4.14   Two-Dimensional Rigid Frames.   Before closing this treatment on two-dimensional structures, it is important to call to the attention of the student the difference between an ideal truss and a so-called **rigid frame**.   The members of a rigid frame are usually connected by moment-resisting (rigid) joints instead of being hinged together as in an ideal truss.   **Thus, a rigid frame may be defined as a structure composed of a number of members all lying in one plane and connected so as to form a rigid configuration by joints, some or all of which are moment-resisting (rigid) instead of hinged.**

A moment-resisting joint is capable of transmitting both a force and a couple from one member to the other members connected by the joint.   Such a joint can be formed by riveting or welding all the members to gusset plates.   The detail of such joints is such that the angles between the ends of the various members at a joint remain essentially unchanged as the frame deforms under load.   For this reason, moment-resisting joints are usually referred to as rigid joints.

By a strict interpretation of these definitions, a modern truss with riveted or welded joints should actually be classified as a rigid frame.   However, since a satisfactory stress analysis can usually be obtained by assuming that such a truss acts as if it were pin-jointed, such structures are called trusses.   The term rigid frame is reserved to designate structures of the type shown in all of Fig. 4.14 except sketch *b*.   When rigid frames are represented by line diagrams, as in this figure, moment-resisting joints are designated by indicating little fillets between the members meeting at a joint.   Any pin joints are represented in the usual manner.

The stability and determinacy of rigid frames can be investigated by methods similar to those used for trusses.   For this purpose, a criterion may be established comparing the number of unknowns with the number of independent equations of static equilibrium available for their solution.

The total number of independent unknowns is equal to the sum of the number of unknown reaction elements plus the number of independent unknown internal-force components in the members.   In a frame with rigid joints, the action of a joint on a member may consist of a couple as well as a force.   Likewise, this force may have both axial and transverse components.   As a result, the cross sections of a member may be subjected to an axial force, shear, and bending moment.   However, if the axial force, shear, and bending moment are known at one end of a member, similar quantities can be found for any other cross section of the member.   There are therefore only three independent internal-force components for each member of the frame.   If the number of reaction elements is r and the number of members is b, the total number of independent unknowns in a rigid frame is equal to $3b + r$.

[1] Timoshenko and Young, *op. cit.*

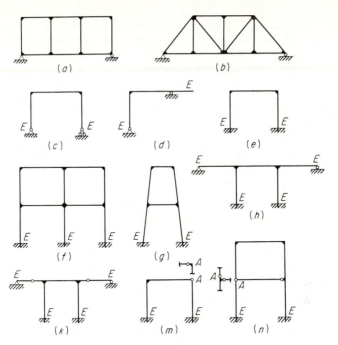

Figure 4.14   Rigid frames

If a rigid joint is isolated as a free body, it will be acted upon by a system of forces and couples.   For equilibrium of such a joint, this system, therefore, must satisfy three equations of static equilibrium, $\Sigma F_x = 0$, $\Sigma F_y = 0$, and $\Sigma M = 0$.   If the entire frame is in equilibrium, each of its joints must be in equilibrium. If there are n rigid joints in the frame, each of these joints can be isolated as a free body and a total of $3n$ equations of static equilibrium obtained. As in the discussion of trusses, it can be shown that the three equations of equilibrium of the entire structure are not independent of these equations, and therefore it can be concluded that there are only $3n$ equations of static equilibrium for the entire rigid frame.

Occasionally hinges or some other special conditions of construction are introduced into the structure.   *If in this manner s special equations of condition are introduced, the total number of equations available for the solution of the unknowns will equal $3n + s$.   The criterion for stability and determinancy of the rigid frame is obtained by comparing the number of unknowns, $3b + r$, with the number of independent equations, $3n + s$.   As before, it can therefore be concluded that*

If $3n + s > 3b + r$, the frame is unstable.
If $3n + s = 3b + r$, the frame is statically determinate.
If $3n + s < 3b + r$, the frame is statically indeterminate.

If the criterion indicates that the frame is statically determinate or indeterminate, it should be remembered from the similar discussion in Art. 4.9 that the count alone does not prove absolutely that the structure is stable: it may be statically or geometrically unstable.

This criterion establishes the combined degree of indeterminancy with respect to both reactions and internal-force components. The degree of indeterminancy with respect to reactions only can be established in the same manner as discussed in Art. 4.10 for truss structures and also in Arts. 2.5 and 2.10.

Table 4.1 shows the application of the above criterion to the frames in Fig. 4.14.

Table 4.1

Frame	n	s	b	r	$3n + s$	$3b + r$	Classification
a	8	0	10	3	24	33	Indeterminate, nineth degree
b	8	0	13	3	24	42	Indeterminate, eighteenth degree
c	4	0	3	3	12	12	Determinate
d	4	0	3	3	12	12	Determinate
e	4	0	3	6	12	15	Indeterminate, third degree
f	9	0	10	9	27	39	Indeterminate, twelfth degree
g	6	0	6	6	18	24	Indeterminate, sixth degree
h	6	0	5	9	18	24	Indeterminate, sixth degree
k	6	2	5	9	20	24	Indeterminate, fourth degree
m	4	1	3	6	13	15	Indeterminate, second degree
n	6	3	6	6	21	24	Indeterminate, third degree

In applying this criterion, any extremity of the frame, e.g., those marked E in Fig. 4.14, should be counted as a rigid joint even though just one member is connected to it. Sometimes the count of the s special equations of condition is rather difficult to make. It is quite obvious in structure k of Fig. 4.14 that the insertion of the two hinges has introduced two equations of condition. The insertion of the hinge joint A in structure m introduces one equation of condition, but the insertion of a similar joint A as shown in structure n introduces two equations of condition. In each case, the validity of these counts is more apparent if one considers the auxiliary sketches of the joints shown in each case. The auxiliary sketches show how the same structural action can be obtained at these joints by insertion of hinges in the ends of the members meeting at the joint. These alternative arrangements require one hinge for structure m and two hinges for the joint A in structure n. To generalize, it may be stated that the number of special condition equations introduced by the insertion of a hinge joint in a rigid frame is equal to the number of bars meeting at that joint minus 1. If the s special equations are counted in this manner, the criterion yields the correct results.

After reading this last paragraph, the reader will no doubt appreciate the truth of the statement that it is almost impossible to count some structures properly without first knowing the answer. Because of the difficulties encountered in counting some structures, the authors feel that, while criteria such as the above are sometimes very useful, *the stress analyst should rely on a more fundamental approach to determine the degree of indeterminancy of an indeterminate structure. The most fundamental approach is to remove supports and/or to cut members until the structure has been reduced to a statically determinate and stable structure. The number of restraints that must be removed to accomplish this result is equal to the degree of in-determinacy of the actual structure.*

4.15   Statical Conditions for Three-dimensional Frameworks.   In this treatment of three-dimensional structures, analysis will be made with reference to three coordinate axes. *ox* and *oy* will be used, as in planar structures, i.e., with *ox* horizontal and *oy* vertical; the third axis *oz* is horizontal and perpendicular to the plane *xoy*.

It should be pointed out that the basic approach to the analysis of three-dimensional structures is the same as for planar structures.   Any equation of statics can be applied to the structure as a whole or to any portion of the structure.   There are, however, more equations of statics available, since forces can be summed up along a new coordinate axis and moments can be taken about two new coordinate axes.

It is usually assumed that the members of a three-dimensional framework are connected with universal joints in such a manner that the members carry axial force only.[1]   Hence there is only one independent component of bar force for each member of the framework.   Although each member can have three components of bar force, one parallel to each of the three coordinate axes, the relations between these three components can be computed from the projections of the member.

At a point of support of a space framework, it is possible to have three independent components of force reactions, although the structure may be designed at a point of support so that one or more of these reaction components equal zero. In Fig. 4.15a, the hinge support shown, which is actually a universal joint, can develop three independent reactions, R_{ox}, R_{oy}, and R_{oz}.   Suppose that the hinge is replaced with a roller, as shown in Fig. 4.15b.   Since this resists horizontal movement in the *z* direction only, the reactions R_{oy} and R_{oz} can be developed, but R_{ox} equals zero.   If the roller is replaced by a spherical ball, there can be no horizontal reaction whatever and only the vertical reaction R_{oy} can be developed.   This condition is shown in Fig. 4.15c.   Sometimes it may be convenient to represent these three typical supports by an equivalent number of *links*. For example, the **universal-joint support** shown in Fig. 4.15a is equivalent to three concurrent but nonplanar links such as shown isometrically in Fig. 4.15d; the **roller support** in Fig. 4.15b, to the two concurrent links in the plane *yoz* as shown in Fig. 4.15e; and the **ball support** in Fig. 4.15c, to the one link in the *oy* direction as shown in Fig. 4.15f.

To represent these three types of support in plan view, heavy dotted lines will be drawn along the lines of action where horizontal reactions can exist.   This is illustrated in Fig. 4.16, where (*a*) represents a hinge-type support in which horizontal reactions can be developed in both the *x* and *z* direction; (*b*) represents a roller-type support, with the roller so placed that a horizontal reaction can be developed in the *z* direction only; (*c*) represents a roller-type support, with the roller so placed that a horizontal reaction can be developed in the *x* direction only;

---

[1] Such a universal (or ball) joint will be represented by small open circles in the line-diagram representation of space frameworks.

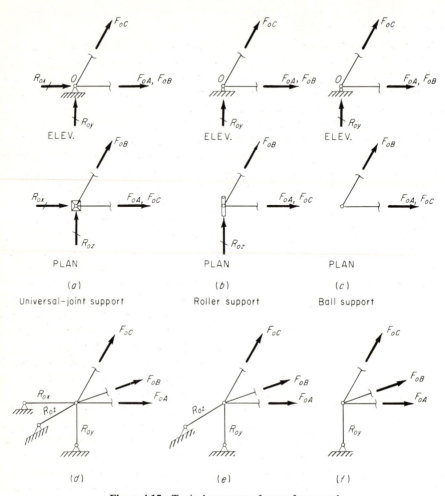

Figure 4.15 Typical supports of space frameworks

and (*d*) represents a ball-type support, in which there can be no horizontal reaction.

Thus the total number of independent unknown stress elements present in the analysis of a three-dimensional framework equals the number of bars plus the number of independent reaction components (or links), there being one, two, or three of the latter at each point of support, depending on the type of construction used at the reaction point.

For a three-dimensional framework, six independent equations of statics may be written regarding the equilibrium of the external loads and reactions acting on the entire structure. If *ox*, *oy*, and *oz* represent the three coordinate axes, these equations are $\Sigma F_x = 0$, $\Sigma F_y = 0$, $\Sigma F_z = 0$, $\Sigma M_x = 0$, $\Sigma M_y = 0$, and $\Sigma M_z = 0$. ΣM_x denotes the sum of the moments about the *ox* axis of all the forces acting on the structure, etc.

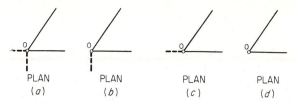

Figure 4.16   Indication of horizontal reaction components

It may therefore be concluded that a necessary (although not sufficient) condition for statical determination of a three-dimensional framework with respect to its outer forces is that the total number of independent reaction components shall equal six.

If we now consider *both the internal and external forces, three independent equations of statics* can be written *for each joint,* namely, $\Sigma F_x = 0$, $\Sigma F_y = 0$, and $\Sigma F_z = 0$. Equations of statics applied to the structure as a whole will not furnish further independent equations. *It may therefore be concluded that a necessary (although not sufficient) condition for statical determination of a three-dimensional framework with respect to both inner and outer forces is that the total number of bars plus the total number of independent reaction components shall equal 3 times the number of joints.*

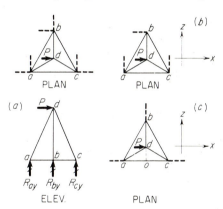

Figure 4.17   Arrangement of reactions for stability

The application of these principles can be illustrated by considering Fig. 4.17*a*. Considering first only the external forces, if the horizontal reactions are arranged as shown in the plan view, there is a total of 9 independent reaction components, so that the structure is statically indeterminate to the 9 − 6 = third degree. If rollers are substituted for hinges, so that the horizontal reactions act as shown in Fig. 4.17*b* the number of independent reactions is 6 and the structure is statically determinate. Suppose, however, that the rollers are placed so that the horizontal reactions have the directions indicated in Fig. 4.17*c*. Then, while the numerical count indicates that this structure is statically determinate with respect to its outer forces, it is actually unstable. The *z* reaction at *b*, for example, will apparently have two values, depending on whether it is determined by applying $\Sigma F_z = 0$ to the entire structure (in which case it equals zero) or whether it is computed by applying the equation $\Sigma M_y = 0$ about a vertical axis through *a* (in which case it must have a value). This shows that the numerical count, while a necessary condition for statical determination, is not a sufficient criterion. *The reactions must be placed so that they can resist translation along and rotation about each of the three coordinate axes if a three-dimensional structure is to be stable.* The reactions shown in Fig. 4.17*c* all pass through point *o*; they cannot resist rotation about a vertical axis passing through that point.

Considering now both internal and external forces, in Fig. 4.17*a*, there are present 15 independent unknowns, 6 bar forces and 9 reaction components. There are 4 joints and hence (4)(3) = 12 independent equations of statics. Therefore the structure is statically indeterminate to the $15 - 12 =$ third degree. With the horizontal reactions arranged as shown in Fig. 4.17*b*, there are only 12 independent unknowns, 6 bar forces and 6 reactions. There are still 12 independent equations of statics. Hence the structure is statically determinate with respect to both inner and outer forces.

4.16 Determination of Reactions and Bar Forces of Three-dimensional Frameworks. If a three-dimensional framework is statically determinate with respect to its outer forces, and if it is supported at only three points, its reactions can be readily determined by applying the equations of statics to the structure as a whole. When there are more than three points of support, it is usually necessary to determine some or all of the bar forces before the reactions can be evaluated. In Example 4.10 consideration will be given to a structure where the reactions can be determined directly. In Example 4.11 a case will be considered where reactions cannot be computed without first computing the bar forces.

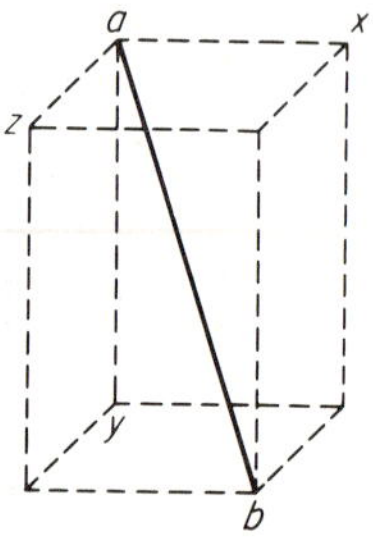

Figure 4.18
Projections of bar

As with planar structures, the particular equations of statics used and the order in which they are applied may be varied in accordance with the ingenuity of the analyst.

A bar of a three-dimensional framework may have projections on each of three coordinate axes. This is illustrated in Fig. 4.18, where the bar *ab* has the projections *ax*, *ay*, and *az* in the directions of the *ox*, *oy*, and *oz* axes, respectively. In terms of these projected lengths, the length of the bar *ab* is given by

$$ab = [(ax)^2 + (ay)^2 + (az)^2]^{1/2}$$

Since the force F_{ab} is axial, the components of F_{ab} parallel to the coordinate axes are

$$X_{ab} = F_{ab}\frac{ax}{ab} \qquad Y_{ab} = F_{ab}\frac{ay}{ab} \qquad Z_{ab} = F_{ab}\frac{az}{ab}$$

By combining these relations, it is easy to express one component of force in terms of either of the other force components. For example, $X_{ab} = Y_{ab} \, ax/ay$, etc.

At any joint where the converging bars do not lie in a plane, three equations of statics are available for bar-force determination. Hence, if not more than three bars with unknown forces meet at such a joint, these bar forces can be determined. This general method of procedure, which is the method of joints expanded to three dimensions, can be illustrated by its application to joint *d* of the structure in Example 4.10. Note also that this application in this particular example is given solely for the purpose of illustrating a procedure of general importance. The solution at joint *d* could be substantially simplified by the application of Theorem 1, Art. 4.17.

The reader will also be interested in a useful procedure originally introduced in

English by Southwell[1] for computing bar forces. This procedure amounts to setting up the equations of equilibrium for the joints of a framework in terms of the direction cosines of the bars. At joint a the bar force F_{ab} in bar ab may be expressed as $F_{ab} = f_{ab}L_{ab}$, where f_{ab} is called the **tension (or force) coefficient.** Then, when setting up the force-equilibrium equations at a joint, the contribution of the component of F_{ab} to equilibrium in the x direction will simply be f_{ab} multiplied by the x projection of ab, etc. A convenient rote is involved when bar-force calculations are set up in terms of tension coefficients and bar projections, and the computations can readily be tabulated.

Example 4.10   *Determine first the reactions and then the bar forces for this space framework.*

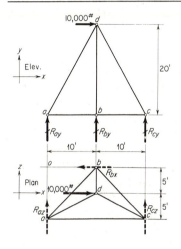

Member	Length	Projections		
		x	y	z
ad	22.9	10	20	5
bd	20.6	0	20	5
cd	22.9	10	20	5
ab	14.14	10	0	10
bc	14.14	10	0	10
ac	20.0	20	0	0

If moments are taken about a horizontal axis passing through any two points of supports, the vertical reaction at the third point of support will be the only unknown in the resulting equation:

$$\Sigma M_x = 0 \text{ about } ac \qquad \therefore R_{by} = 0$$

$$\Sigma M_z = 0 \text{ about } ao \qquad (10,000)(20) - (R_{cy})(20) = 0 \qquad \therefore R_{cy} = +10,000^{\#} \quad \therefore \uparrow$$

Then, $\Sigma F_y = 0$ *for the entire structure* $\qquad \therefore R_{ay} = -10,000^{\#} \quad \therefore \downarrow$

If moments are taken about a vertical axis passing through the intersection of the lines of action of any two horizontal reactions, the third horizontal reaction will be the only unknown involved.

$$\Sigma M_y = 0, \text{ axis through } 0, \; -10,000(5) - R_{cz}(20) = 0 \qquad R_{cz} = -2,500^{\#} \quad \therefore \downarrow$$

Then $\qquad \Sigma F_z = 0$ *for entire structure* $\qquad \therefore R_{az} = 2,500^{\#} \quad \therefore \uparrow$

and $\qquad \Sigma F_x = 0$ *for entire structure* $\qquad \therefore R_{bx} = 10,000^{\#} \quad \therefore \leftarrow$

where arrows indicate directions in the plan view.

---

[1] R. V. Southwell, "Theory of Elasticity," Oxford University Press, London, 1936.

Consider now the computation of the bar forces. At joint d, the three unknown bar forces can be obtained from the simultaneous solution of the following three force-equilibrium equations for the isolated joint, in which all bars are assumed to be in tension:

$$\Sigma F_x = 0 \qquad 10{,}000 - {}^{10}\!/_{22.9}F_{ad} + {}^{10}\!/_{22.9}F_{cd} = 0$$

$$\Sigma F_y = 0 \qquad -{}^{20}\!/_{22.9}F_{ad} - {}^{20}\!/_{20.6}F_{bd} - {}^{20}\!/_{22.9}F_{cd} = 0$$

$$\Sigma F_z = 0 \qquad +{}^{5}\!/_{22.9}F_{ad} - {}^{5}\!/_{20.6}F_{bd} + {}^{5}\!/_{22.9}F_{cd} = 0$$

Simultaneous solution of these equations leads to

$$F_{ad} = +11{,}450^{\#} \qquad F_{bd} = 0 \qquad F_{cd} = -11{,}450^{\#}$$

Actually in this problem, once the vertical reactions have been determined, the bar forces in ad, bd, and cd can be computed more easily from $\Sigma F_y = 0$ applied in turn to the isolated joints a, b, and c, respectively. It is important, however, to understand the general approach based on writing three simultaneous equations for each joint, since in more complicated structures it may be the only procedure that can be used.

The bar forces in the horizontal bars can easily be determined once the horizontal reactions and the bar forces in ad, bd, and cd have been found. For example, at joint a,

$$\Sigma F_z = 0 \qquad 2{,}500 + (11{,}450)({}^{5}\!/_{22.9}) + (F_{ab})({}^{10}\!/_{14.14}) = 0 \qquad \therefore F_{ab} = -7{,}070^{\#}$$

Now from $\Sigma F_x = 0$ at joint a, F_{ac} can be computed. F_{bc} can be obtained from $\Sigma F_z = 0$ at joint c.

Example 4.11 *Compute the bar forces and reactions of this structure.*

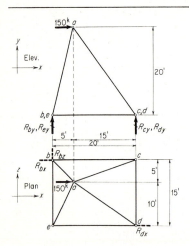

Bar	Projection			Length
	x	y	z	
ab	5	20	5	21.2
ac	15	20	5	25.5
ad	15	20	10	26.9
ae	5	20	10	22.9
bc	20	0	0	20.0
cd	0	0	15	15.0
de	20	0	0	20.0
eb	0	0	15	15.0

This structure is statically determinate with respect to both inner and outer forces. Whereas the three horizontal reactions can be determined by a consideration of the structure as a whole, the four vertical reactions cannot be computed without first computing some of the bar forces. If the forces in ab, ac, ad, and ae had been computed, the four vertical reactions could be determined from the vertical equilibrium equations for joints b, c, d, and e.

Unfortunately at joint a, initially the four unknown bar forces cannot be determined from the three joint-equilibrium equations available. Even though the horizontal reactions have been computed, there still remain four unknowns at each of the four base joints b, c, d, and e. Of course, three force equilibrium equations could be written at each of the five joints, and the simultaneous solution of these fifteen equations could be carried out, but this might be laborious. Under these conditions, a convenient alternate approach is to adopt one of the bar forces as a temporary unknown and proceed as follows:

At joint c: $\Sigma F_z = 0$ $F_{cd} + \frac{5}{25.5} F_{ac} = 0$ $\therefore F_{ac} = -5.10 F_{cd}$

At joint d: $\Sigma F_z = 0$ $F_{cd} + \frac{10}{26.9} F_{ad} = 0$ $\therefore F_{ad} = -2.69 F_{cd}$

At joint c: $\Sigma F_y = 0$ $R_{cy} + \frac{20}{25.5} F_{ac} = 0$ $\therefore R_{cy} = -0.784 F_{ac}$
$$= +4.00 F_{cd}$$

At joint d: $\Sigma F_y = 0$ $R_{dy} + \frac{20}{26.9} F_{ad} = 0$ $\therefore R_{dy} = -0.744 F_{ad}$
$$= +2.00 F_{cd}$$

Now with R_{cy} and R_{dy} expressed in terms of F_{cd}, taking moments about be for the entire structure yields

$$\Sigma M_z = 0 \qquad (150)(20) - (R_{cy})(20) - (R_{dy})(20) = 0$$

whence $3{,}000 - (4.00 F_{cd})(20) - (2.00 F_{cd})(20) = 0$ $\therefore F_{cd} = \underline{+25^k}.$

Then, of course, R_{ay}, R_{dy}, F_{ac}, and F_{ad} can quickly be evaluated. With these known, the remainder of the analysis proceeds without difficulty.

4.17   Special Theorems for Three-dimensional Frameworks. While three-dimensional frameworks can be analyzed by the methods that have been presented, the following theorems are of importance because application of them often results in an appreciable saving in computations:

Theorem 1: *If all the bars meeting at a joint, with the exception of one bar n, lie in a plane, the component normal to that plane of the force in bar n is equal to the component normal to that plane of any external load or loads applied at that joint.*

That this theorem is correct can be seen from a consideration of the static equilibrium of the joint, summing up all the forces normal to the plane that contains all bars except *n*. In the structure of Example 4.10, for example, suppose that this theorem is applied to joint *d*. Bars *ad* and *dc* lie in the plane *adc*; the component of bar force in bar *bd*, normal to plane *adc*, must equal the component of the applied load normal to the same plane. For this particular case, the applied load also lies in the plane *adc* and hence has no component normal to that plane. We can conclude, then, that the bar force in bar *bd* is zero. Recognition of this fact would simplify the analysis of joint *d* carried out in Example 4.10 by means of three simultaneous equations, since only two equations would be necessary.

On the basis of Theorem 1, two corollary theorems may be stated:

Theorem 2: *If all the bars meeting at a joint, with the exception of one bar n, lie in a plane, and if no external load is applied at this joint, the force in bar n is zero.*

Theorem 3: *If all but two bars at a joint have no bar force and these two are not collinear, and if no external load acts at that joint, the bar force in each of these two bars is zero.*

4.18   Application of Special Theorems—Schwedler Dome. The importance of these three theorems in the analysis of three-dimensional frameworks may be illustrated by considering the Schwedler dome shown in Fig. 4.19, acted upon by a vertical load P applied at joint A. The bars shown by dotted lines in the plan view have no bar force, as may be concluded from the foregoing theorems, applied as follows.

At joint F, bars EF, KF, and LF all lie in a plane, but AF does not. Since no load is applied at joint F, the force in bar AF is zero, in accordance with Theorem 2. A similar consideration of joints E, D, C, and B leads to the conclusion that bars EF, DE, CD, and BC, respectively, all have zero bar force.

Considering joint F again, since the bar forces in FA and FE are zero, bars KF and LF comprise two bars meeting at a joint where no load is applied. Hence the bar forces in KF and LF are zero, on the basis of Theorem 3. A similar consideration of joints E, D, and C leads to the conclusion that bars KE and JE, JD and ID, and IC and HC, respectively, all have zero force.

Now, considering joint K, since KE and KF have zero force, bar KL is a single bar lying outside the plane of bars JK, PK, and QK. Hence bar KL has zero force. In a similar manner, a consideration of joints J and I shows that bars JK and IJ have zero force.

Considering joint K again, since bars JK, EK, FK, and LK have zero force, bars QK and PK comprise two bars meeting at a joint where no external load is applied. The forces in these two bars are therefore zero. A similar consideration of joint J shows that bars PJ and OJ also have zero force.

Hence, when this framed dome is acted upon by a vertical load at A, only the bars shown by solid lines in the plan view of Fig. 4.19 carry force. To complete the analysis of this structure, the method of joints can now be applied successively to joints A, B, L, G, H and I, and the bar forces in all the bars except those of the base ring thus determined. The vertical reactions can then be determined by applying $\Sigma F_y = 0$ at each point of support.

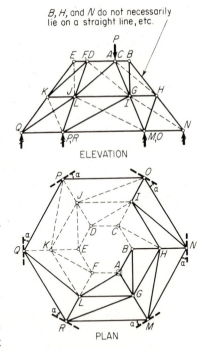

Figure 4.19   Schwedler dome

To determine the horizontal reactions and the bar forces in the base-ring bars, all of which lie in a plane, it is necessary to take the force in one of the base-ring bars as a temporary unknown. Suppose F_{RM} is taken as the temporary unknown. At joint R, the application of $\Sigma F_x = 0$ and $\Sigma F_z = 0$ permits one to express F_{QR} and the horizontal reaction at R in terms of F_{RM}. (Note that joint R is also acted upon by the x and z components of bar force in bars LR and GR.) Proceeding clockwise around the base ring and successively writing similar equations at joints Q, P, O, and N enable one to express the bar forces in all the base-ring bars (and all the horizontal reactions except that at M) in terms of F_{RM}. If the equations $\Sigma F_x = 0$ and $\Sigma F_z = 0$ are now applied at joint M, the values of F_{RM} and the horizontal reaction at M can be obtained. Since all the other bar forces and horizontal reactions have been previously expressed in terms of F_{RM}, their values can now be determined.

It is possible for a Schwedler dome to be geometrically unstable, even though the statical count holds, if the angles between the horizontal reactions and the base-ring bars have certain values.[1]

4.19 Towers.   Unless the legs of a framed tower have a constant batter throughout their length, the structure should be analyzed on the basis of three-dimensional considerations. Figure 4.20a shows the side elevation of a tower of triangular cross section, in which the batter of the legs is not constant; Fig. 4.20b shows the arrangement of horizontal reactions. This structure is statically determinate, as can be verified, and it can be analyzed panel by panel, beginning with the top panel and working down. Whenever, as in this case, adjacent legs of the top panel lie in a plane, the bar force in any of the top ring bars can be computed most easily by utilizing the three theorems of Art. 4.17. For example, F_{ab} can be computed by noting that its component normal to the face $acfd$ must balance the component of the external load at joint a, which is also normal to this plane. Once the top ring bar forces are known, the forces in the legs and diagonals of the top panel can be computed. The forces in these bars together with any external loads at joints d, e, and f comprise the loading on the second panel which can now be analyzed similarly, etc.

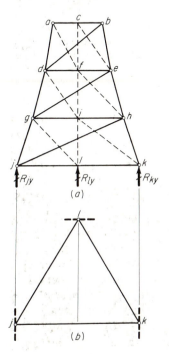

Figure 4.20   Triangular tower

If however, adjacent legs of a panel do not lie in a plane, a more general approach must be used to obtain the ring bar forces. Starting as before with the top panel, the bar force in any top ring bar may be taken as a temporary unknown. If F_{ab} is chosen for this purpose, application of the method of joints at joint b permits one to express F_{bc} in terms of F_{ab}. Similar treatment at c gives F_{ca} in terms of F_{ab}. Finally, F_{ab} can be evaluated by applying the method of joints at joint a.

It is customary to have secondary bracing members in each horizontal plane where the batter of the tower legs changes and often at panel points where no change in batter occurs. In a tower with a rectangular cross section, for example, this bracing may consist of horizontal diagonals connecting diagonally opposite panel points. The presence of such members may make a tower statically indeterminate. It is common practice, however, to assume for the purpose of analyzing the main members of the tower that the bar forces in these secondary bracing

[1] C. M. Spofford, "Theory of Structures," 4th ed., chap. 16, Space Frameworks, McGraw-Hill Book Company, New York, 1939.

members are zero, thus permitting such analyses to be carried out by the principles of statics.

4.20   Tower with Straight Legs.   If the batter of the tower legs is constant over the entire height, the tower can be analyzed on the basis of component planar trusses.   Such a structure is shown in Fig. 4.21.   A load P, applied at any joint, can be resolved into three components: C_1, parallel to the tower leg; C_2, horizontal and lying in the plane of one adjacent face of the tower; and C_3, horizontal and lying in the plane of the other adjacent face of the tower.

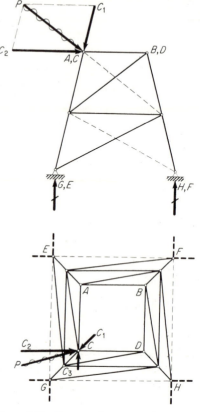

It is easy to show by the theorems of Art. 4.17 that C_1 causes bar forces in the bars of leg GC only, C_2 causes forces in the bars of tower side $CDGH$ only, and C_3 causes forces in the bars of tower side $ACEG$ only.

Thus the bar forces due to each of the components C_1, C_2, and C_3 can be obtained by carrying out a separate planar analysis, and the total force in any bar due to load P can be obtained by superposition of the effects of its three components.   Since each panel load can be handled in the forgoing manner, this constitutes a general procedure for analysis.

If all the faces of such a tower are identical, influence data can be prepared, giving the bar forces in each bar of one of the faces due to (1) a unit horizontal load applied successively at

Figure 4.21   Rectangular tower with straight legs

each joint of that face and (2) a unit load parallel to the tower leg applied successively at each joint of that face.   Such influence data, prepared for one face of the tower, will be applicable to all faces of the tower.   By resolving panel loads into components as previously outlined and by using these influence data, forces in any member, due to any condition of external loading, can be obtained by superposition.

4.21   General Theory for Three-dimensional Frameworks.   Many space frameworks used in conventional structures can be analyzed by the application of the methods and procedures discussed above.   In order to handle with facility some of the problems regarding the more unconventional frameworks, however, it is necessary to study the statical stability and statical determinancy of space frameworks in a more rigorous manner,[1] similar to that used for trusses (or planar frameworks). A brief outline of such a treatment follows.

Arrangements of members of simple space framework. A group of points which do not lie in a plane could be connected together by a rigid and noncollapsible space framework by constructing this framework in the following manner. (See Art. 4.4 for definitions of rigid and noncollapsible.) First form a tetrahedronal nucleus by using six bars to connect together four points which do not lie in a plane, as shown by the solid lines connecting joints 1, 2, 3, and 4 in Fig. 4.22. Joint 5 could then be rigidly connected to this nucleus by bars from joints 1, 3, and 4, as indicated by the dashed lines. Joints 6 to 12 could be connected in sequence as shown, using *three bars which do not lie in a plane* to connect each new joint to the framework previously formed.

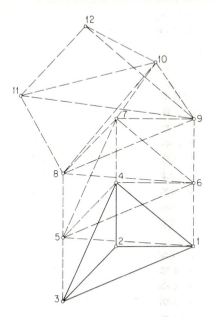

Using the same terminology as for trusses, a space framework formed in the manner just described is called a **simple space framework**. The number of bars b required to form such a framework of n joints is obviously $3(n - 4) + 6$, or $b = 3n - 6$.

Figure 4.22   Simple space framework

Formation of compound space frameworks. Several simple space frameworks may be connected together to form a so-called **compound space framework**, as illustrated in Fig. 4.23, by using *six bars the axes of which do not intersect a single straight line.* In Fig. 4.23, the two simple space frameworks are shaded, and the six connecting bars are shown by the dashed lines. Note that the total number of bars used in such a case is $(3n_1 - 6) + (3n_2 - 6) + 6$, or $b = 3(n_1 + n_2) - 6$, or $b = 3n - 6$, where n_1 and n_2 represent the number of joints in the individual simple space frameworks and n represents the total number of joints in the entire structure.

Complex space frameworks. Many practical space frameworks cannot be classified as either simple or compound and are therefore categorized as **complex space frameworks**. An illustration of such a framework is shown in Fig. 4.24. Note that this framework would be a simple space framework if a member 1-4 were substituted for member 2-6. As before, the number of bars required to construct a rigid and noncollapsible complex space framework is given by $b = 3n - 6$.[2]

Supports for space frameworks. If a rigid noncollapsible space framework is supported such that the *reaction components are equivalent to six links the lines of*

---

[1] For an excellent discussion of statically determinate space frameworks, the reader is referred to S. P. Timoshenko and D. H. Young, "Theory of Structures," 2d ed., McGraw-Hill Book Company, New York, 1965.

[2] As emphasized in Art. 4.9, such a criterion is a *necessary but not sufficient* condition that the framework be rigid and noncollapsible. Though the number of bars may be adequate, if they are not properly arranged, the space framework may be *geometrically unstable.*

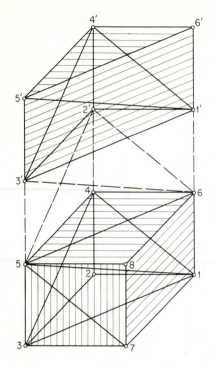

Figure 4.23 Compound space framework

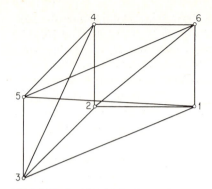

Figure 4.24 Complex space framework

action of which do not intersect a single straight line, the structure will be *stable* and these six reaction components are *statically determinate* from the six equations of statics noted in Art. 4.15 and applied to the structure as a whole. Such an arrangement is shown in Fig. 4.25 by the links shown by the heavy black lines and numbered 1 to 6.

Obviously if link 6 had a line of action coinciding with line *aa*, the structure would be supported in an unstable fashion since there would be no reaction capable of resisting rotation of the structure about an axis coinciding with axis of link 3. This would likewise be true if link 6 were replaced by the vertical link 6′. Note that in either of these situations the axis of the six links involved intersect a vertical line coinciding with link 3 since links 4, 5, and 6′ which are parallel to link 3,

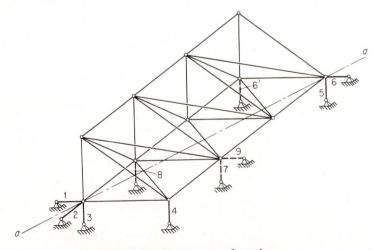

Figure 4.25 Arrangement of reactions

intersect this vertical line at infinity.   Of course if additional links such as 7, 8, and 9 were added to the original set, links 1 to 6, the structure would be stable but the reactions would be statically indeterminate, in this case to the third degree.

In many practical space frameworks, there are more than six reaction components provided.   Often in such cases, if there are r reaction components, $r - 6$ bars have been omitted from the $3n - 6$ required for a rigid and noncollapsible framework, and, as a result,

$$b = (3n - 6) - (r - 6)$$

or $b + r = 3n$.   In other words, the omission of the $r - 6$ bars is equivalent to introducing $r - 6$ equations of condition (or construction).   When this is done properly so that geometric instability is avoided, the structure is stable and statically determinate.   Sometimes the reactions can be determined directly from the combination of the equations of condition and the equations of statics for the structure as a whole.   Often, however, the reactions cannot be obtained without first determining some of the bar forces as illustrated in Example 4.11.

These and similar considerations lead to the following conclusions regarding the stability of a space framework structure and its combined statical determinancy with respect to both reaction components and bar forces.   For a space framework, a total of $3n$ independent equations of statics are available by isolating each joint and writing the three equations $\Sigma F_x = 0$, $\Sigma F_y = 0$, and $\Sigma F_z = 0$. Involved among these equations are a total of $b + r$ unknowns, that is, b bar forces and r reaction components.   Comparison of the number of unknowns with the number of equations indicates that

If $b + r < 3n$, the structure is *unstable.*
If $b + r = 3n$, the structure is *statically determinate.*
If $b + r > 3n$, the structure is *statically indeterminate.*

If this criterion indicates that the frame is statically determinate or indeterminate, it should be remembered that the count alone does not prove it is also stable: it may still be statically or geometrically unstable if the bars or reaction components are improperly arranged.

Stress analysis of statically determinate space frameworks.   The stress analysis of many space frameworks can be carried out by using a combination of the method of joints and the method of sections as discussed previously for trusses (or planar frameworks).   The additional geometry of the three-dimensional structure impedes the rate of progress with such computations very considerably, of course.

Consider, for example, the case of a *simple* space framework which is supported in a statically determinate fashion.   In such a case, the reactions can be determined from a consideration of the equilibrium of the entire structure.   Once the reactions are known, the definition of the external forces acting on the structure is complete. Then one can isolate the last joint established in the pseudo construction of the simple space framework (such as joint 12 in Fig. 4.22).   At this joint, there are only three unknown bar forces, which can then be determined from the three equilibrium equations for this joint.   If one works backward through the structure considering the joints in the reverse order—11, 10, 9, etc.—there will be only three unknown bar forces at each joint.   Thus, the entire solution for the bar forces can be completed simply by successive application of the method of joints.

In the case of a *compound* space framework, even after the reactions have been determined, one can proceed just so far using the method of joints before finding that a joint cannot be isolated where there are only three unknown bar forces.   At such a point in the solution, the method of sections must be applied to obtain some of the bar forces.   In the case of *complex* space frameworks, even after the reactions have been determined, one finally reaches a point in the solution where neither the method of joints nor the method of sections will lead to a direct solution. In such a case, one must resort to a procedure such as described in Art. 4.13 for planar frameworks or to Henneberg's method (see reference cited in Art. 4.13).

In recent years, some extremely complicated space frameworks have been involved in radar transmitters and receivers.   Such structures involve vast numbers of joints and bars and sometimes cumbersome geometry.   Often, deflection limitations and vibrational characteristics rather than strength requirements control the design. Modern digital computers have been used extensively in the analysis of such structures, utilizing displacement-matrix methods such as described in Chap. 15.

4.22   Problems for Solution

* Problem 4.1   Classify the truss structures of Fig. 4.26 first as being simple, compound, or

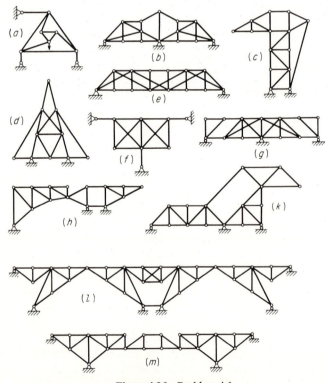

Figure 4.26   Problem 4.1

complex trusses and then as being statically determinate or indeterminate, stable or unstable.   If
the structure is indeterminate, state the degree of indeterminancy both with respect to reactions
and bar forces and with respect to reactions only.   If the structure is unstable, state the reason
for the instability.

*Problem 4.2   Compute the bar forces in the lettered bars of the trusses of Fig. 4.27 due to the
loads shown.

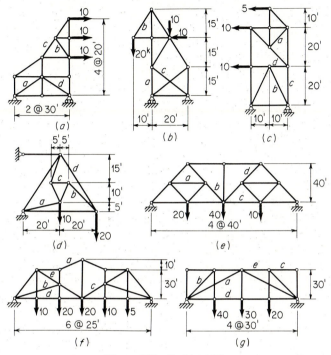

Figure 4.27   Problem 4.2

Problem 4.3   Compute the reactions for each of the structures shown in Fig. 4.28.

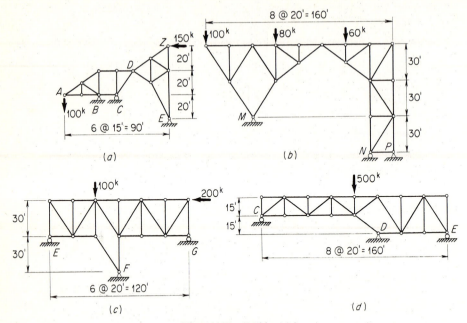

Figure 4.28   Problem 4.3

Problem 4.4   Compute all the bar forces in the trusses of Fig. 4.29 due to the loads shown.

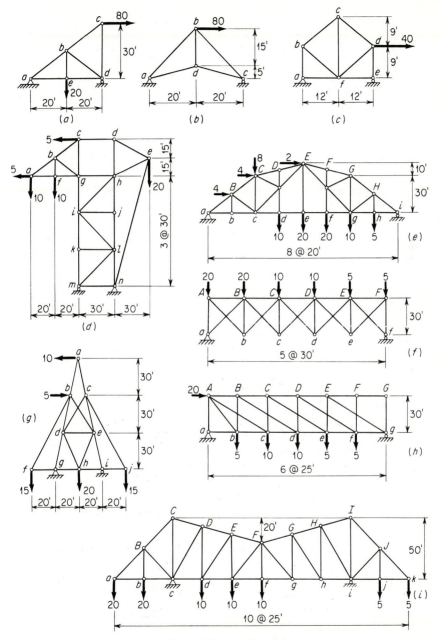

Figure 4.29   Problem 4.4

*Problem 4.5 Compute the bar forces in the structures of Fig. 4.30. (*Hint:* Remember that structures of this type may be geometrically unstable.)

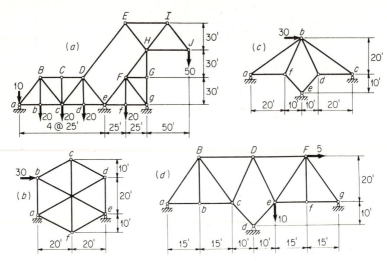

Figure 4.30 Problem 4.5

*Problem 4.6 Compute the bar forces in the structures of Fig. 4.31. Also draw the shear and bending-moment diagrams for those members in which such stress conditions exist.

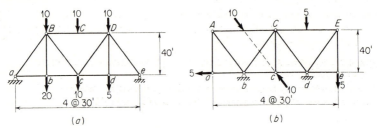

Figure 4.31 Problem 4.6

Problem 4.7 Compute the bar forces in the bars noted on each of the structures shown in Fig. 4.32.

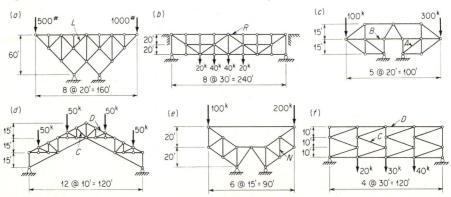

Figure 4.32 Problem 4.7

*Problem 4.8   Find the reactions on the structure of Fig. 4.33 due to the load of 1,000 lb acting as shown.

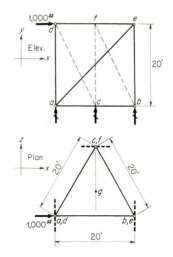

Figure 4.33   Problems 4.8 and 4.9, 4.10

*Problem 4.9   Find the bar forces in the structure of Fig. 4.33 due to the load of 1,000 lb acting as shown.

*Problem 4.10   Find the reactions and bar forces of the structure of Fig. 4.33 if the load of 1,000 lb is applied at joint *d* but with a direction such that it passes through point *g*, which lies at the center of the equilateral triangle *def*.

*Problem 4.11   (*a*) Show that the structure of Fig. 4.34 is statically indeterminate to the first degree.

(*b*) Find the reactions on the structure of Fig. 4.34 and the bar forces in all the bars, for the load of 10,000 lb acting as shown, assuming that bar *ef* is in compression and has a force equal to half the applied load.

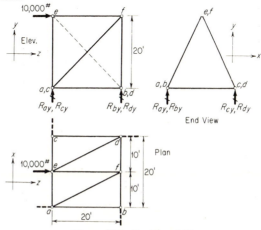

Figure 4.34   Problem 4.11

Theorem 2:

The value of a function due to the action of a single concentrated live load equals the product of the magnitude of the load and the ordinate to the influence line for that function, measured at the point of application of the load.[1]

This follows from the principle of superposition.   Further, the total value of a function due to more than one concentrated load can be obtained by superimposing the separate effects of each concentrated load, as determined by Theorem 2.

To illustrate the application of these two theorems, suppose that a concentrated live load of 10,000 lb is applied to the beam of Example 5.1.   If the influence line (*b*) of Example 5.1 is used, the maximum positive shear that this load can cause at *D* occurs with the load just to the right of *D* and equals $(10,000)(+\frac{7}{10}) = +7,000$ lb.   The maximum negative shear at the same section occurs with the load just to the left of *D* and equals $(10,000)(-\frac{3}{10}) = -3,000$ lb.   From influence line (*d*) of Example 5.1, the maximum positive moment at *D* occurs with the load at *D* and equals

$$(10,000)(+2\frac{1}{10}) = +21,000 \text{ ft-lb}$$

From the definition of an influence line, the following theorem, dealing with uniformly distributed live loads, is apparent:

Theorem 3:

To obtain the maximum value of a function due to a uniformly distributed live load, the load should be placed over all those portions of the structure for which

---

[1] In the above definition of an influence line, it is suggested that the function involved be computed for a 1-lb load.   As a result, the ordinates of typical influence lines would have the following units: reactive forces, pounds; shear and axial forces, pounds; bending moments, foot-pounds; etc.   On this basis, to compute the effect of some other concentrated load the influence-line ordinate involved should be multiplied by the dimensionless magnitude of this load, i.e., by simply the *number* of 1-lb loads which it represents.   For example, from Fig. 5.2*b*, the bending moment at *C* produced by a load of 750 lb at *D* may be found to be

$$(2.5 \text{ ft-lb})(750) = 1,875 \text{ ft-lb}$$

Some structural engineers prefer to define an influence line as showing functions caused by a dimensionless load of unity.   On such a basis, the product just noted would be obtained by multiplying by a number of 750 lb, or

$$(2.5 \text{ ft})(750 \text{ lb}) = 1,875 \text{ ft-lb}$$

Still other structural engineers prefer to define the ordinates of an influence line as being the ratio of the function involved to the dimensional 1-lb load causing it and therefore say that the ordinates in Fig. 5.2*b* have units of foot-pounds per pound.   On this basis, the product noted would again be obtained by multiplying by a dimensional number of 750 lb, or

$$(2.5 \text{ ft-lb per lb})(750 \text{ lb}) = 1,875 \text{ ft-lb}$$

This latter approach has some merit although it will not be used here.   According to this usage, the ordinates of influence lines are referred to as **influence coefficients.**

the ordinates to the influence line for that function have the sign of the character of the function desired.

To compute, from the influence line, the actual value of the function due to a uniformly distributed live load, the following theorem should be used:

Theorem 4:

The value of a function due to a uniformly distributed live load is equal to the product of the intensity of the loading and the net area under that portion of the influence line, for the function under consideration, which corresponds to the portion of the structure loaded.

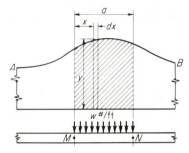

Figure 5.3　Portion of influence line

That the foregoing theorem is correct can be seen from the following. Let AB be the influence line for a given function F, for a portion of a structure, as shown in Fig. 5.3, that is subjected to a uniformly distributed load of w lb per ft, applied continuously to the structure between two points M and N. That portion of the uniform load applied in distance dx may be treated as a concentrated load equal to $w\,dx$. By Theorem 2, the value of the function F due to this differential load is given by $dF = w\,dx\,y$. The total value of F due to the load between M and N is obtained by integrating dF between $x = 0$ and $x = a$, or $F = \int_0^a wy\,dx = w\int_0^a y\,dx = w$ multiplied by the area under that portion of the influence line which corresponds to the portion of the structure loaded.

To illustrate the application of Theorems 3 and 4, suppose that a uniform live load of 1,000 lb per ft is applied to the beam of Example 5.1. In accordance with influence line (*b*) for this beam, to obtain the maximum positive shear at D the uniform load should extend from C to A and from D to B of the beam. The value of this maximum positive shear at D is given by

$$(1,000)[(\tfrac{1}{2})(5)(+\tfrac{1}{2}) + (\tfrac{1}{2})(7)(+\tfrac{7}{10})] = +3,700 \text{ lb}$$

For maximum negative shear at D, the structure should be loaded from A to D, leading to a resultant shear at D equal to

$$(1,000)[(\tfrac{1}{2})(3)(-\tfrac{3}{10})] = -450 \text{ lb}$$

For maximum positive moment at D, refer to influence line (*d*) of Example 5.1. The structure should be loaded from A to B; the resultant moment equals

$$(1.000)[(\tfrac{1}{2})(10)(+\tfrac{21}{10})] = +10,500 \text{ ft-lb}$$

For maximum values of functions due to a concentrated live load and uniformly distributed live load acting simultaneously, the maximum function due to each acting separately should be computed by the methods already given and the results superimposed. For example, to obtain the maximum negative moment at A, in the beam of Example 5.1, due to a uniform load of 1,000 lb per ft and a single concentrated load of 10,000 lb, it is found by reference to influence line (c) of Example 5.1 that the uniform load should extend from C to A and the concentrated load should be placed at C. The maximum negative moment at A is then given by

$$(1,000)[(\tfrac{1}{2})(5)(-5)] + (10,000)(-5) = -62,500 \text{ ft-lb}$$

Suppose that a uniform load of 1,000 lb per ft extends over the entire length of the beam of Example 5.1. Functions are then computed on the basis of the algebraic sum of the component areas that constitute the entire influence line. From influence line (d) of Example 5.1, the resultant moment at D would, for example, be given by

$$(1,000)[(\tfrac{1}{2})(5)(-\tfrac{7}{2}) + (\tfrac{1}{2})(10)(+\tfrac{21}{10})] = +1,750 \text{ ft-lb}$$

5.5 Illustrative Examples—Influence Lines for Beams and Girders. Examples 5.1 to 5.3 illustrate the construction of typical influence lines for beams and girders.

For statically determinate structures, influence lines for all functions except deflections turn out to be composed of straight-line segments. For such structures, ordinates of the influence line for a particular function may be computed by one of the three following methods:

1. A unit load is placed successively at a number of the positions it could occupy in a transit across the structure; for each such position of the load, the particular function is computed; plotting the computed values of the function traces out the desired influence line. This is a straightforward but laborious procedure.

2. The position of the unit load is located by the distance x from the end or some other key point along the structure, and the particular function of interest is then computed in terms of x. This expression for the function can likewise be interpreted as the equation defining the shape and position of the influence line. As soon as the unit load moves past a support point or another key location along the structure, the expression for the function changes. As the load moves along into the next portion of the structure, its position will be defined by an x distance measured from a new convenient origin and a new expression for the function must be computed. With experience, one soon acquires the ability to move quickly across the structure, constructing the influence line by this procedure.

3. With more experience, one finally acquires the ability to spot the points where discontinuities or changes in slope occur in an influence line. Computation of the particular function for a unit load at these key locations gives the corresponding key ordinates, which can then be connected by straight-line segments to obtain the complete influence line.

In Chap. 10 construction of influence lines for statically indeterminate structures will be discussed. Such influence lines are curved lines or, at best, a series of chords of a curved line. As a result, their construction involves much more computation and proceeds more slowly than for the statically determinate cases now being considered.

Example 5.1 *For the beam shown, construct the influence lines for (a) the shear at the cross section to the left of A, (b) the shear at D, (c) the bending moment at A, and (d) the bending moment at D.*

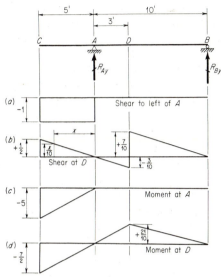

Discussion:

Sketch *a: When a unit load is applied at any location to the left of point A, the shear on this section is negative and equal to the unit load. For positions of the load between A and B, there is no shear on the cross section of interest.*

Sketch *b: For positions of the unit load to the left of D, the shear at D can be computed from the reaction at B. For the unit load at C, R_{By} is 0.5 down, and S_D is therefore +0.5. As the unit load moves from C to A, R_{By}, and therefore S_D, decreases linearly to zero. S_D continues to change linearly from 0 to −0.3 as the load moves from A to D. As is often preferable, when the unit load moves from B to D, it is easier to work with the portion of the structure on the side away from the load because there are fewer forces to consider. In this fashion, it is easy to see as the load moves from B to D, R_{Ay}, and therefore S_D, increases linearly from 0 to +0.7. Of course, as the unit load passes from just to the right to just to the left of D, S_D changes abruptly from +0.7 to −0.3.*

Sketch *c: By proceeding in much the same fashion as in (a), it is easy to construct the influence line for M_A.*

Sketch *d: Construction of the influence line for M_D proceeds much as in part (b).*

Example 5.2 *For the girder shown, construct the influence lines for (a) shear in panel BC and (b) moment at E.*

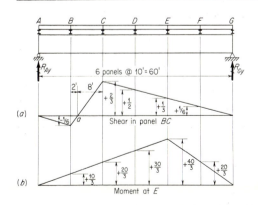

Discussion:

Refer to Art. 3.9 and Example 3.5 regarding girders with floor systems supported by floor beams as used in this example. In such structures, live loads may be considered as traveling along the stringers, which can transmit forces into the girder only through the floor beams at the panel points. As a result, the live-load shear remains constant at any cross section within a given panel of the girder. In the case of this girder, the stringers act as end-supported beams spanning between adjacent floor beams. Therefore, as a unit load travels from one panel point to another, the reactions of the stringer in the loaded panel, which are also the forces applied to the girder by the floor beams, vary linearly with the position of the load. Hence, any stress function in the girder (such as shear in a given panel or moment at a panel point) varies linearly also. Girder influence lines are therefore straight lines between panel points for girders with end-supported stringers.

Sketch a: *When the load is to the right of C, there are no floor beam forces at A or B and S_{BC} is equal to R_{Ay}. As the load travels from G to C, R_{Ay}, and therefore S_{BC}, increases linearly from 0 to $\frac{1}{6}$. When the load is to the left of B, S_{BC} is most easily computed by isolating the portion to the right and noting that S_{BC} is equal to R_{Gy}, which increases linearly from 0 to $\frac{1}{6}$ as the load travels from A to B.*

Sketch b: *The bending moment at E can most easily be computed by isolating the portion away from the load and noting that M_E is equal to the reaction times the distance from the support to E. As the load travels from A to E, M_E therefore increases linearly from 0 to $(\frac{2}{3})(20) = 40\frac{}{3}$.*

Example 5.3   *Note the unusual stringer arrangement for this girder. Construct the influence line for the bending moment in the girder at E.*

Discussion:

In the girder in Example 5.2 with conventional end-supported stringers, panel points were the key points where influence lines might change slope or have discontinuities. For unusual stringer arrangements, as in the present example, discontinuities in influence lines may also occur at key points other than panel points, for instance at C-D.

When the load is to the left of C, M_E is most easily computed from the portion of the girder to the right, being equal to simply $10R_{Fy}$. Thus, as the unit load travels from A all the way to C, M_E

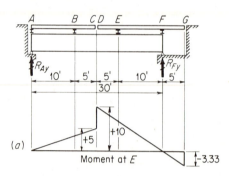

(a) Moment at E

increases linearly from 0 to $(10)(0.5)$. When the load is to the right of D, M_E can be computed from the isolated girder to the left of E and is equal simply to $20R_{Ay}$ since for such positions of the load there are no floor-beam forces at A or B. As the load travels from D to G, M_E changes linearly from 10 to $-3.\dot{3}$.

5.6   Interpretation of Influence Lines for Girders with Floor Systems.

The four theorems of Art 5.4 dealing with the use of influence lines are perfectly general and are applicable to influence lines for girders with floor systems. Suppose that live loads consisting of a uniform load of 1,000 lb per ft and a single concentrated load of 10,000 lb are applied to the structure of Example 5.2. To obtain the maximum live shear in panel BC, refer to the influence line of Example 5.2a. It is first necessary to locate point a at which this influence line crosses the base line. Such a point is called a **neutral point**, since a load applied at this point has no effect on the function under consideration. This point can be located by similar triangles; its distance from B will be found to be 2 ft. The maximum positive live shear in panel BC occurs when the uniform load extends from the neutral point to G and when the concentrated load is at C; it is equal to

$$(1,000)[(\tfrac{1}{2})(+\tfrac{2}{3})(48)] + (10,000)(+\tfrac{2}{3}) = 22,667 \text{ lb}$$

Maximum negative live shear in this panel occurs when the uniform load extends from A to the neutral point and the concentrated load is at B; it has a value equal to

$$(1,000)[(\tfrac{1}{2})(-\tfrac{1}{6})(12)] + (10,000)(-\tfrac{1}{6}) = -2,667 \text{ lb}$$

In Example 5.2b the maximum positive live moment at panel point E, due to the same live load, occurs with the uniform load extending over the entire span and with the concentrated load at E. Its value is equal to

$$(1,000)[(\tfrac{1}{2})(+\tfrac{40}{3})(60)] + (10,000)(+\tfrac{40}{3}) = +533,333 \text{ ft-lb}$$

The foregoing method of computing maximum live shears and moments, based on locating neutral points and using exact areas under influence lines is *exact*. The following *approximate* method is of importance, since it often involves less computation and is well suited to efficient organization of computations for complicated structures. In the approximate method, it is assumed that for uniform live load there is acting at each panel point either a full panel load or no panel load whatever, depending on whether the ordinate to the influence line indicates that a load at that panel point increases or decreases the value of the function for which a maximum value is desired.

A full panel load is the maximum possible load that can be applied to a girder by a floor beam. It can occur only when the stringers adjacent to the panel are fully loaded, and it is equal (for panels of equal length) to wl, where w is the intensity of the uniform load and l is the length of the panel.

Consider again the structure of Example 5.2 acted upon by live loads consisting of a uniform load of 1,000 lb per ft and a single concentrated load of 10,000 lb. For the uniform live load, the full panel load equals $(1,000)(10) = 10,000$ lb. To compute the maximum positive live shear in panel BC by the approximate method, this full panel load is placed at C, D, E, and F, since the influence line of Example 5.2a has positive ordinates at these panel points. No panel load will be placed at B, where the ordinate to the influence line is negative. As in the exact method, the concentrated load will be placed at C. The resultant maximum positive live shear in panel BC is equal to

$$(10,000)(\tfrac{2}{3} + \tfrac{1}{2} + \tfrac{1}{3} + \tfrac{1}{6}) + (10,000)(\tfrac{2}{3}) = 23,333 \text{ lb}$$

The corresponding value was 22,667 lb by the exact method, so that the result by the approximate method is seen to be slightly on the safe side, i.e., slightly larger than the exact value. The approximate method assumes a full panel load acting at C, which could not occur without completely loading stringer BC; loading stringer BC would cause a floor-beam reaction at B equal to half a full panel load, which by itself would cause negative shear in panel BC. Because the negative shear due to the partial panel load applied at B is neglected in the approximate method, the resultant positive shear computed is necessarily on the safe side. The approximate method of computing maximum values of functions never gives smaller values than the exact method.

To find the maximum positive live moment at E for the same structure and loading by the approximate method, refer to the influence line of Example 5.2b. For the uniform load, a full panel load of 10,000 lb is applied at all intermediate panel points, since all the corresponding ordinates to the influence line are positive.

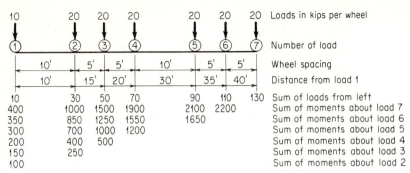

Figure 5.4 Moment chart

The concentrated load is placed at *E*. The maximum positive live moment is given by

$$(10,000)(+\tfrac{10}{3} + \tfrac{20}{3} + \tfrac{30}{3} + \tfrac{40}{3} + \tfrac{20}{3}) + (10,000)(+\tfrac{40}{3}) = +533,333 \text{ ft-lb}$$

This is the same value as that obtained by the exact method.

5.7 Series of Concentrated Live Loads—Use of Moment Chart. The methods of using the influence line as previously presented apply to uniformly distributed live loads and to single concentrated live loads. They cannot, however, be used directly when the live load consists of a series of concentrated loads of given magnitude and spacing, such as are actually applied by the wheels of a locomotive or of a series of trucks. When there is more than one concentrated load, it is not possible, in general, to tell by inspection which of the concentrated loads should be placed at the maximum ordinate of the influence line in order to make the given function a maximum.

The method that should be followed for such a live load is essentially one of trial. In order to expedite the various trial solutions, it is desirable to organize such an analysis carefully, so as to minimize computations. For a series of concentrated loads, a moment chart, like that shown in Fig. 5.4, can be used to advantage. This particular moment chart is computed for the seven concentrated loads spaced as shown. The chart is practically self-explanatory. The numbers in the six bottom rows can be explained by a single illustration. The number 1,900 under load 4 and in the horizontal line labeled Sum of moments about load 7 represents the moment about load 7 of loads 1 to 4; thus

$$(10)(40) + (20)(30) + (20)(25) + (20)(20) = 1,900$$

To illustrate the use of the moment chart, suppose that it is desired to compute the moment at load 3 in the beam of Fig. 5.5 due to the loading of Fig. 5.4 located as shown in Fig. 5.5. The moment about *B* of the applied loads is equal to 1,650 (the moment of loads 1 to 5 about load 6) plus 110 (the sum of the loads from 1 to 6) multiplied by 2 (the distance from load 6 to point *B*), whence 1,650 + (110)(2) = 1,870 kip-ft. Dividing this moment by the span of the beam, gives R_{Ay} equal 1,870/50 = +37.4 kips. Hence the moment at load 3, working from the

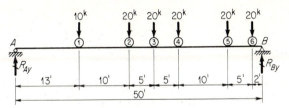

Figure 5.5 Use of moment chart

forces to the left, is given by $(+37.4)(28) - 250 = +798$ kip-ft. It is to be noted that the moment of 250 kip-ft, which was subtracted, is the moment of loads 1 and 2 about load 3.

As a second example of the use of the moment chart, the shear in panel BC of the girder of Fig. 5.6 will be computed for the loads shown acting, these loads being a part of the loading of Fig. 5.4. For the loads so placed, load 1 is not on the span. The girder reaction at A is given by

$$R_{Ay} = \frac{(1,650 - 350) + (110 - 10)(3)}{36} = +44.5 \text{ kips}$$

The sum of the floor-beam reactions at A and B is equal to

$$20 + (\tfrac{5}{9})(20) = 31.1$$

Hence the shear in panel BC equals $+44.5 - 31.1 = +13.4$ kips.

In using the moment chart it is usually convenient to reproduce it to scale on cardboard and place it in the proper position on a drawing of the structure to be analyzed, which is drawn to the same scale.

5.8 Series of Concentrated Live Loads—Computation of Maximum Moment. The computation of maximum moment at a given section in a girder will be illustrated by computing the maximum moment at C of the girder of Fig. 5.7a due to the live loading corresponding to the moment chart of Fig. 5.4. The influence line for the moment at C is first constructed, as shown in Fig. 5.7b. The maximum moment at C will occur when one of the concentrated loads is at C; the first part of the problem consists in finding out which load should be at C in order to cause this maximum moment.

Before attempting this trial solution, the slope of each portion of the influence line, going from right to left, is first computed. For example, the portion of the influence line for the moment at C, which runs from F to C, has a slope of $+\tfrac{12}{30} = +\tfrac{2}{5}$.

Using the moment chart of Fig. 5.4, place load 1 at C. This causes a certain

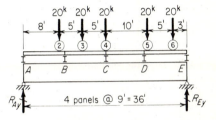

Figure 5.6 Use of moment chart

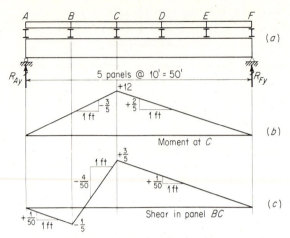

Figure 5.7 Girder influence lines

moment at C, which, however, will not be computed at this stage of the analysis. Instead, the entire system of loads will be moved to the left until load 2 is at C, and computations will be carried out to determine whether the moment at C has been increased or decreased by this change in the position of the loads. To see whether the moment has become larger or smaller, it is convenient to divide the loads under consideration into three groups: (1) those loads which were on the structure before the loads were moved and which remain on the structure after the loads are moved; (2) those loads which were on the structure before the loads were moved but which have passed off the structure after the loads are moved; (3) those loads which were not on the structure before the loads were moved but which are on the structure after the loads are moved. For convenience we shall refer to these three load groups as load group 1, load group 2, and load group 3, respectively.

The following computations determine whether this load move has increased or decreased the moment at C. It should be noted that if a load P moves a distance d, and if the slope of the influence line is m, the corresponding change in moment equals Pdm.

Load 1 at section; move up load 2	Increase in moment	Decrease in moment
Load group 1: loads 1 to 5	$(80)(10)(+\tfrac{2}{5}) = +320$	$(10)(10)(-\tfrac{3}{5}) = -60$
Load group 2: none	0	0
Load group 3: load 6	$(20)(5)(+\tfrac{2}{5}) = +40$	0
All loads combined	$+360$	-60

The net change in moment is $+360 - 60 = +300$ kip-ft, so that a larger moment at C occurs with load 2 at the section, i.e., at C, than with load 1. However, it may be that a still larger moment occurs with load 3 at the section. The loads will now be moved to the left until load 3 is at C, and computations will be made to determine whether this new movement of loads has increased or decreased the moment at C.

Load 2 at section; move up load 3	Increase in moment	Decrease in moment
Load group 1: all loads	$(100)(5)(+\frac{2}{5}) = +200$	$(30)(5)(-\frac{3}{5}) = -90$
Load group 2: none	0	0
Load group 3: none	0	0
All loads combined	$+200$	-90

Since 200 is greater than 90, the moment has again increased. We shall now find out whether there will be a still further increase if load 4 is moved to the section.

Load 3 at section; move up load 4	Increase in moment	Decrease in moment
Load group 1: all loads	$(80)(5)(+\frac{2}{5}) = +160$	$(50)(5)(-\frac{3}{5}) = -150$
Load group 2: none	0	0
Load group 3: none	0	0
All loads combined	$+160$	-150

Again, the moment has increased. We shall now move up load 5.

Load 4 at section; move up load 5	Increase in moment	Decrease in moment
Load group 1: loads 2 to 7	$(60)(10)(+\frac{2}{5}) = +240$	$(60)(10)(-\frac{3}{5}) = -360$
Load group 2: load 1	0	0
Load group 3: none	0	0
All loads combined	$+240$	-360

Note that although load 1 was considered as on the structure with load 4 at the section, it caused no moment at C, so that no change occurred when it passed off the structure. Since 240 is less than 360, moving up load 5 caused a decrease in the moment at C. Hence the maximum moment at C occurs with load 4 at C. With some experience in moving up loads, one might have foreseen that the maximum moment at C would not occur with load 1 at the section and that it probably would not occur with load 2 at the section. This would have eliminated a portion of the foregoing computations.

With the position of loads causing maximum moment at C known, the value of this moment can now be computed, either from the ordinates of the influence line directly or by using the moment chart. By the latter procedure,

$$R_{Ay} = \frac{2,200 + (130)(10)}{50} = +70 \text{ kips}$$

The moment of the floor-beam reactions at A and B about C is equal to the moment of loads 1, 2, and 3 about load 4, which is 500 kip-ft. Hence the maximum positive live moment at C equals

$$(+70)(20) - 500 = +900 \text{ kip-ft}$$

5.9   Absolute Maximum Live Shear. The methods which have been given for the computation of maximum shear due to live loads assume that the section or panel in which the shear is to be computed is known. It is often desirable to compute the absolute maximum live shear in a member, i.e., the maximum live shear that can occur at any section in the member. For a simple end-supported beam or girder, absolute maximum live shear will occur at a section immediately adjacent to one of the end reactions. If the beam or girder is not a simple end-supported member, the absolute maximum live shear will occur on one side of one of the reactions. The true value of absolute maximum live shear can be determined only by computing the maximum live shear at each such section.

5.10   Absolute Maximum Live Moment. Similarly, the methods which have been given for the computation of maximum moment due to live loads assume that the section at which maximum live moment is to be computed is known. It is often necessary to compute the absolute maximum live moment for a beam or girder. For a simple end-supported beam, this occurs at mid-span for either a uniform live load or a single concentrated live load. For a simple end-supported girder with a floor system, absolute maximum live moment occurs at the panel point nearest the center of the span. For a girder wholly

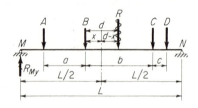

Figure 5.8   Position for absolute maximum moment

or partly cantilevered, absolute maximum live moment is likely to occur at a reaction. If the section where absolute maximum live moment occurs cannot be definitely identified by inspection, it is necessary to compare maximum moments computed for various sections where absolute maximum live moment is likely to occur.

A special case of importance consists in determining the absolute maximum live moment due to the action of a series of concentrated live loads on an end-supported beam, as shown in Fig. 5.8. The moment curve for a series of concentrated loads is a series of straight lines intersecting at the positions of the loads, so that absolute maximum live moment must occur directly beneath one of the loads. Two questions must be answered: (1) Under which load does absolute maximum live moment occur? (2) What is the position of this load when absolute maximum live moment occurs?

The answer to the first question must often be determined by trial, but the second question is subject to direct analysis. Assume that in Fig. 5.8 the absolute maximum live moment will occur under load B. Let the distance from the center of the span to load B be denoted by x and the distance from load B to the resultant R of all the loads A, B, C, and D be denoted by d. We wish to determine the value of x that will make the moment at load B a maximum. The value of R_{My} may be determined by taking moments about N and considering the resultant force R rather than the actual loads A, B, C, and D. Thus

$$R_{My} = \frac{R(L/2 + x - d)}{L} = \frac{R}{2} + \frac{Rx}{L} - \frac{Rd}{L}$$

Denoting by M_B the moment under load B gives

$$M_B = R_{My}\left(\frac{L}{2} - x\right) - Aa = \left(\frac{R}{2} + \frac{Rx}{L} - \frac{Rd}{L}\right)\left(\frac{L}{2} - x\right) - Aa$$

$$= \frac{RL}{4} - \frac{Rd}{2} - \frac{Rx^2}{L} + \frac{Rxd}{L} - Aa$$

For a maximum value of M_B,

$$\frac{dM_B}{dx} = -\frac{2Rx}{L} + \frac{Rd}{L} = 0$$

whence $x = d/2$.

We may therefore conclude that *the maximum moment directly beneath one of a series of concentrated live loads that are applied to a simple end-supported beam occurs when the center of the span is halfway between that particular load and the resultant of all the loads on the span.*

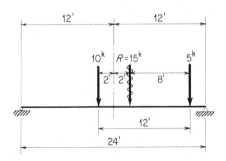

Figure 5.9   Absolute maximum moment for two loads

If there are only two concentrated loads to consider, the absolute maximum live moment will occur under the heavier of the two loads. Such a case is illustrated in Fig. 5.9, where the distance from the 10-kip load to the resultant R of the two loads equals $[(5)(12)]/15 = 4$ ft. For absolute maximum moment, the 10-kip load is placed 2 ft from the center of the span, and thus the resultant R is 2 ft on the other side of the span center. At this point one should check to see whether or not both loads are on the span. If not, the absolute maximum moment occurs at mid-span when the heavy load is at the center of the span. For this case, both loads are on the span. The absolute maximum live moment occurs directly beneath the 10-kip load and is given by

$$M = \frac{(15)(12 - 2)^2}{24} = +62.5 \text{ kip-ft}$$

If there are more than two concentrated loads, it may not be possible to tell by inspection under which load the absolute maximum live moment will occur. It will usually occur under a large load near the center of the group of loads. The maximum moment that can occur under each of the loads can be determined by the foregoing method, and the largest of these moments will be the absolute maximum live moment.

5.11   Influence Lines for Trusses—General.   Influence lines can be constructed for the bar forces in truss members and are important in determining the location of live loads leading to maximum bar forces in truss members as well as for computing the actual values of these maximum bar forces.   The same general procedure as that used for constructing influence lines for beams and girders is applicable to trusses.   It is always possible to compute the ordinate to the influence line for a unit load at each panel point of the truss.   Usually the stringers act as end-supported beams between the floor beams, so that the influence line is a straight line between panel points.   As with beams and girders, it is often possible to reduce the amount of computation by recognizing that the influence line is a straight line for several successive panels.

Once the influence line has been constructed for the force in a given truss member, the interpretation of the curve with respect to loading criteria and stress analysis is identical with that for beams and girders.

Influence lines for trusses are drawn to correspond to a unit load traveling across the *loaded chord*, i.e., the chord containing the panel points at which the live load is applied to the truss by the floor beams that support the floor system.

5.12   Illustrative Examples—Influence Lines for Trusses.   The first of following examples demonstrates that influence-line ordinates can be computed directly and easily for bar forces of Pratt, Howe, and Warren trusses.   More complicated cases, e.g., the K truss with a curved top chord in Example 5.5, may require the construction of several other influence lines before it is possible to construct the influence line actually desired.

Example 5.4   *For the through Pratt truss shown, construct the influence lines for the bar forces in members L_2L_3, U_2L_2, and U_1L_1.*

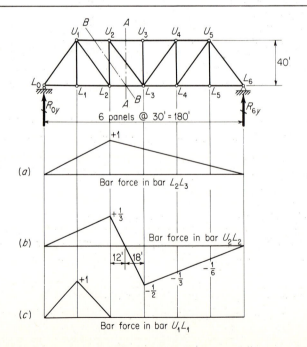

Discussion:

Sketch a: *Ordinates of the influence line for the bar force in $L_2 L_3$ are most easily computed by considering the unloaded portion of the truss isolated by section A-A. Taking moments about point U_2 involves only the reaction and $F_{L_2L_3}$. As the load moves from L_0 to L_2, R_{6y} increases linearly from 0 to $\frac{1}{3}$, and therefore $F_{L_2L_3}$ increases linearly from 0 to $(\frac{1}{3})(^{120}\!\!/\!_{40}) = +1$. As the load moves from L_6 to L_2, similar calculations for the portion to the left of A-A show that $F_{L_2L_3}$ again increases linearly from 0 to $+1$.*

Sketch b: *The bar force in $U_2 L_2$ can be computed by applying $\Sigma F_y = 0$ to the unloaded portion of the truss isolated by section B-B. In this fashion as the unit load travels from L_0 to L_2, vertical equilibrium of the right portion shows that $F_{U_2L_2}$ must balance R_{6y} and therefore its influence line ordinates increase from 0 to $+\frac{1}{3}$. As the unit load travels from L_6 to L_3, similar calculations for the portion to the left of B-B show that $F_{U_2L_2}$ changes from 0 to $-\frac{1}{2}$. For end-supported stringers between L_2 and L_3, it is easy to confirm that the influence line is a straight line from $+\frac{1}{3}$ at L_2 to $-\frac{1}{2}$ at L_3.*

Sketch c: *Influence lines for chord members and for most web members extend over the entire length of the truss, as just computed for $L_2 L_3$ and $U_2 L_2$. Such members are called **primary** truss members. Verticals $U_1 L_1$ and $U_5 L_5$ are called **secondary** truss members because they are stressed only when the unit load is in certain panels of the span. The method of joints at L_1 shows that $F_{L_1U_1}$ is different from zero only when there is a floor-beam force applied at L_1, that is, only when the unit load is in the first and second panels. In a similar fashion, consideration of the equilibrium of joint L_5 leads to the conclusion that the ordinates of the influence line for $F_{L_5U_5}$ are different from zero only when the unit load is in the fifth and sixth panels.*

Example 5.5 *For the K truss shown, construct the influence line for the bar force in member $U_2 M_3$. (Hint: Note that this truss is actually internally statically indeterminate. However, the indeterminancy is confined to the two center panels, and the members in the three panels at each end are statically determinate.)*

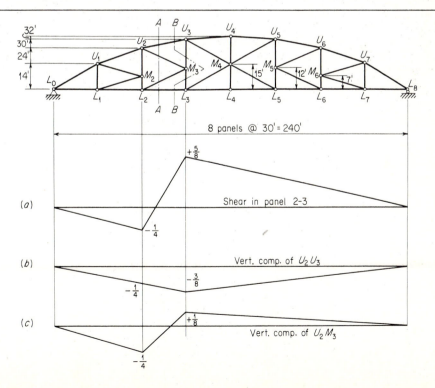

8 panels @ 30' = 240'

(a)   Shear in panel 2-3

(b)   Vert. comp. of $U_2 U_3$

(c)   Vert. comp. of $U_2 M_3$

Discussion:

Consideration of the horizontal equilibrium of M_3 shows, in the absence of any horizontal loads at M_3, that the horizontal components of the bar forces in $U_2 M_3$ and $L_2 M_3$ are always equal in magnitude but opposite in character. Since the slopes of these two bars are the same, the vertical components of these bar forces are likewise equal in magnitude but opposite in character; thus they act in the same direction when holding in equilibrium the vertical forces applied to the portion of the structure on one side or the other of section A-A, one of which vertical forces is the vertical component in the top chord $U_2 U_3$.

If the influence line for the shear in panel 2-3 is first drawn as in (a), the influence line for the vertical component in $U_2 U_3$ can be combined with the shear influence line to obtain the influence line for the vertical component in $U_2 M_3$ or $L_2 M_3$. Note that the bar force in $U_2 U_3$ can easily be determined by summing moments about L_3 of the forces acting on one side or the other of section B-B. Thus, the influence line for the vertical component in $U_2 U_3$ is a triangle with its apex at L_3, where the ordinate is equal to $-(\frac{5}{8})(\frac{90}{30})(\frac{6}{30}) = -\frac{3}{8}$. With the vertical component in $U_2 U_3$ known from (b), it is now easy to consider the vertical equilibrium of the portion one side or the other of section A-A to determine the ordinates of the influence line for the vertical component of $U_2 M_3$ shown in (c).

5.13 Influence Tables. It is often advantageous to express influence data in the form of influence tables instead of curves. The influence table (Table 5.1) refers to the truss of Example 5.4. It gives the force in each bar of the truss due to a unit load at each panel point. Forces in bars $L_2 L_3$, $U_2 L_2$, and $U_1 L_1$ were taken directly from Example 5.4. Forces for other bars may be checked by the student.

In utilizing an influence table to compute maximum live bar forces by the approximate method, it is convenient to prepare a second table that is a summary of the influence table.

Table 5.1 Influence Table for Truss of Example 5.4

Bar	Bar force due to unit load at:						
	L_0	L_1	L_2	L_3	L_4	L_5	L_6
$L_0 L_1$	0.000	+0.625	+0.500	+0.375	+0.250	+0.125	0.000
$L_1 L_2$	0.000	+0.625	+0.500	+0.375	+0.250	+0.125	0.000
$L_2 L_3$	0.000	+0.500	+1.000	+0.750	+0.500	+0.250	0.000
$L_0 U_1$	0.000	−1.041	−0.833	−0.625	−0.417	−0.208	0.000
$U_1 U_2$	0.000	−0.500	−1.000	−0.750	−0.500	−0.250	0.000
$U_2 U_3$	0.000	−0.375	−0.750	−1.125	−0.750	−0.375	0.000
$U_1 L_2$	0.000	−0.208	+0.833	+0.625	+0.417	+0.208	0.000
$U_2 L_3$	0.000	−0.208	−0.417	+0.625	+0.417	+0.208	0.000
$U_1 L_1$	0.000	+1.000	0.000	0.000	0.000	0.000	0.000
$U_2 L_2$	0.000	+0.167	+0.333	−0.500	−0.333	−0.167	0.000
$U_3 L_3$	0.000	0.000	0.000	0.000	0.000	0.000	0.000

In the summary of the influence table, the sum of the plus and the negative ordinates and the sum of all the ordinates are listed for the influence line for each bar.   Also listed are the maximum ordinates, positive and negative, and the loaded lengths of the influence lines, positive and negative.   If the dead load is represented by an equivalent uniform load and the live load by the combination of a uniform load and a single concentrated load placed so as to produce a maximum effect, the bar forces due to dead and live load can easily be computed using the data in the summary of the influence table.

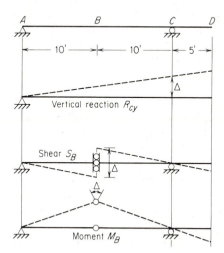

Figure 5.10 Influence lines obtained by Müller-Breslau's principle

5.14 Alternative Approach for Determination of Influence Lines.   An alternative approach to the construction of influence lines may be made by introducing an imaginary deformation into the truss member or girder section under consideration.   This method is of more interest than use in connection with statically determinate structures, but it is of importance in constructing influence lines for statically indeterminate structures, both by analytical and experimental procedures.   This alternate method is based on Müller-Breslau's principle, discussed in Art. 10.3.

The principle is restated here to illustrate its application to several simple statically determinate structures as shown in Figs. 5.10 and 5.11.   *The shape of the influence line for any stress element (such as bar force, shear, bending moment, or reaction) is similar to the deflection curve of the structure obtained by removing the restraint corresponding to that element and introducing in its place a corresponding deformation into the residual structure that remains.*

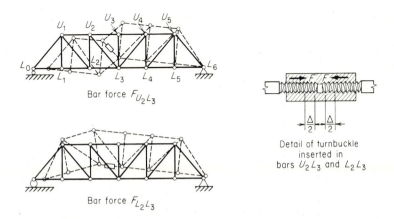

Figure 5.11 Bar-force influence lines obtained by Müller-Breslau's principle

5.15 Problems for Solution

*Problem 5.1 Referring to Fig. 5.12, construct the influence lines for (*a*) shear at *a*; (*b*) moment at *a*; (*c*) reaction at *b*.

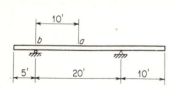

Figure 5.12 Problems 5.1 and 5.2

Figure 5.13 Problems 5.3 and 5.4

*Problem 5.2 Referring to Fig. 5.12 and using a live load consisting of a uniform load of 500 lb per ft and a concentrated load of 5,000 lb, compute (*a*) the maximum upward reaction at *b*; (*b*) the maximum positive and negative moments at *a*; (*c*) the maximum positive and negative shears at a section just to the right of the support at *b*; (*d*) If the dead load is 1,000 lb per ft, compute, from the influence line, the maximum moment at *a* due to dead plus live load.

Problem 5.3 For the structure of Fig. 5.13, construct influence lines for (*a*) shear in panel *AB*; (*b*) moment at panel point *C*.

Problem 5.4 Compute, by both the exact and the approximate methods, the maximum live shear in panel *AB* and the maximum live moment at panel point *C* of the structure of Fig. 5.13 due to a uniform live load of 1,200 lb per ft and a single concentrated live load of 18,000 lb.

*Problem 5.5 For the structure of Fig. 5.14, construct influence lines for (*a*) shear in panel *DE*; (*b*) moment at panel point *E*.

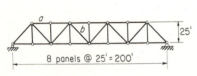

Figure 5.14 Problems 5.5 and 5.6

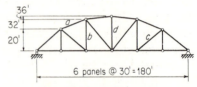

Figure 5.15 Problems 5.7 and 5.8

Problem 5.6 Compute, by the approximate method, the maximum live shear in panel *DE* and the maximum live moment at panel point *E* of Fig. 5.14, due to the live load of Prob. 5.4.

Problem 5.7 Construct a moment chart for the loading of Fig. 5.15.

Problem 5.8 Using the moment chart for the loading of Fig. 5.15 and the structure of Fig. 5.13, determine (*a*) the left girder reaction when load 3 is at *C*; (*b*) the left girder reaction when load 5 is at *C*; (*c*) the shear in panel *AB* when load 3 is at *E*; (*d*) the moment at panel point *D* when load 2 is at *D*.

*Problem 5.9 Compute the maximum live bar force in bars *a* and *b* of the truss of Fig. 5.16 due to a uniform live load of 750 lb per ft. Consider both tension and compression.

Figure 5.16 Problem 5.9

Figure 5.17 Problem 5.10

*Problem 5.10 The top-chord panel points of the truss of Fig. 5.17 lie on a parabola. Draw influence lines for (*a*) horizontal component of the force in bar *a*; (*b*) force in bar *b*; (*c*) force in bar *c*; (*d*) force in bar *d*.

6

Bridge and Roof Trusses and
Long-Span Structures

6.1 Introduction. Loadings for trusses were discussed in Chap. 1; stress analysis for trusses was treated in Chap. 4; influence lines for trusses and the determination of maximum live bar forces were considered in Chap. 5.   In this chapter, we shall bring together the ideas contained in the earlier chapters and apply these concepts to the general analysis of a typical roof truss and a typical bridge truss. Finally, consideration will be given to movable bridges and long-span structures, including cantilevers, arches, and suspension bridges.

The trusses considered will be analyzed as planar structures, but it should be understood that actually they are portions of three-dimensional frameworks.   To see why this procedure is permissible, refer to the deck bridge of Fig. 6.1, in which only the lateral loads $P_1, P_2, \ldots, P_5$ will be assumed as acting, these loads lying in the plane of the top-chord lateral system.   Suppose that a horizontal plane is passed through the structure at some elevation between the top and bottom chords. Considering the isolated part of the structure above this plane, it will be seen that the horizontal components of the bar forces in the sway-bracing diagonals at the ends of the structure must hold the top-chord lateral system, acted upon by the lateral loads, in equilibrium.[1]   These horizontal components provide the end reactions on the top-chord lateral system, which can therefore be analyzed as a planar truss.   The vertical components of the bar forces in the sway-bracing diagonals must be held in equilibrium by the bar forces in the end verticals of the main vertical trusses.   Thus the sway bracing in each end acts as a planar truss that transfers a reaction on the top-chord lateral system to the foundation.

---

[1] This statement is not completely correct, because under an unsymmetrical system of lateral loads, the horizontal components of the bar forces in the diagonals of the main trusses would likewise assist in furnishing the reactions to the top-chord lateral system.

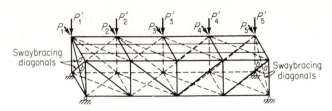

Figure 6.1 Truss bridge

Next, consider the same structure, with only the vertical loads P_1', P_2', ..., P_5' assumed to be acting, these loads lying in the plane of the rear vertical truss. Suppose that a vertical plane, parallel to the main vertical trusses, is passed through the structure between these trusses. The sway-bracing diagonals at each end will be the only members cut that are capable of carrying vertical components. If the diagonals *a* are capable of carrying compression, they will be stressed and this will affect the bar forces in the end verticals of the rear vertical truss. This effect is of secondary importance, however, so that it is permissible and on the safe side to analyze the rear vertical truss as a planar truss acted upon by the loads P_1', P_2', ..., P_5'.

If additional sway bracing is introduced at intermediate panel points, the situation becomes more complicated. As the panel points of the loaded vertical truss deflect vertically, the sway bracing at each intermediate panel point acts in a manner such that the corresponding panel point on the unloaded vertical truss must also undergo a certain amount of vertical movement. This produces what are called **participating stresses** in the unloaded vertical truss. However, if both vertical trusses have the same loading, there would, because of symmetry, be no participating stresses of this type. In an actual bridge, the vertical loads on the two main vertical trusses will not always be the same, but they are usually so nearly the same for the maximum loading conditions which control the design that the bar forces in the diagonals of sway bracing at intermediate panel points may be assumed as zero. Each main vertical truss can then be analyzed separately as a planar structure.

6.2 General Analysis of a Roof Truss. The general analysis of a roof truss includes not only the computation of bar forces in each member due to the various types of load that must be supported by the truss but the combination, for each truss member, of the bar forces due to each type of loading, in a manner such that the maximum bar force that can result from the combined effects of the different types of loading is obtained. To illustrate this procedure, consider the wall-supported roof truss of Fig. 6.2, which will be taken as an intermediate truss of a series of trusses spaced at 20 ft center to center. It will be assumed that the framing of the roof is such that loads from the roof are applied to the truss at the top-chord panel points only. The roofing, including the purlins that support the roofing between trusses, will be assumed to weigh 4.3 lb per sq ft of roof area; the truss itself will be assumed to weigh 75 lb per horizontal foot, with the weight equally divided between the top- and bottom-chord panel points. The snow load will be taken as 20 lb per sq ft of projected horizontal area of the roof. The ice load will be taken as 10 lb per sq ft of projected horizontal area of the roof. Wind loads shall be in accordance with the recommendations of the ASCE report[1] discussed in Art. 1.11 and shall be based on a maximum wind velocity of 100 mph. The analysis will be limited to the following combinations of loads: (1) dead plus snow over the entire roof, (2) dead plus wind plus snow on the leeward side,[2] (3) dead plus ice over the entire roof plus wind. It is to be noted that the wind can blow from either the right or the left.

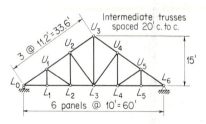

Figure 6.2   Roof truss

The dead bar forces will first be computed. For the bottom chord, the dead panel loads are given by $(7\frac{1}{2})(10) = 375$ lb; for the top chord, the dead panel loads are given by $375 + (4.3)(11.2)(20) = 1,345$ lb. These panel loads, together with the dead bar forces they produce, are shown in Fig. 6.3. Since the dead bar forces are symmetrical about the center line of the truss, only half the truss is shown. The computations leading to the determination of the dead bar forces are omitted, since they involve no new procedures.

Loading condition 1 calls for snow over the entire roof; loading condition 2 calls for snow on the leeward side only. We shall first compute the truss bar forces for snow on the leeward side only; because the truss is symmetrical, we can then determine the bar forces due to snow over the entire roof by the principle of super position, thus avoiding a complete second stress analysis. The snow panel load is given by

$$(20)(10)(20) = 4,000 \text{ lb}$$

---

[1] Wind Forces on Structures, *Trans. ASCE*, vol. 5, no. 126, part. 2, pp. 1150–1152, 1961.

[2] This combination of loads is considered in the present discussion because it is one frequently investigated. It corresponds to a reasonable condition if wind loads are computed by a formula like that by Duchemin (see below), since the snow might be likely to be blown from the roof by the pressure on the windward slope but remain on the leeward slope where there is neither pressure nor suction. When the recommendations of the ASCE report are followed, however, this particular load combination does not appear to be so likely to occur, since one may encounter, as in this particular problem, suction on both slopes, with the greater suction on the leeward slope. With this condition of wind load, the snow, if it remained on the roof at all, would appear more likely to remain on the windward side. However, it is always possible that snow, having fallen previous to the action of the wind, would have crusted over in a manner such as to act as assumed for loading condition 2.

According to Duchemin, P_n, the intensity of normal pressure on a given surface, is given by

$$P_n = P \frac{2 \sin \alpha}{1 + \sin^2 \alpha}$$

in which P is the intensity of pressure on a vertical surface and α is the angle made by the given surface with the horizontal. Such a formula leads to wind pressure on the windward slope of a gable roof and makes no attempt to account for suction on the leeward slope.

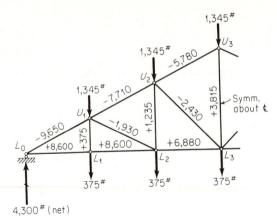

Figure 6.3 Bar forces due to dead load

With the snow on the leeward side only, the snow panel load at U_3 equals only half this value.   These panel loads, together with the bar forces they produce, are shown in Fig. 6.4.   Since this is an unsymmetrical loading condition, bar forces will be computed for the entire truss.

To obtain the bar forces due to full snow load, we may use the principle of superposition, as indicated by Fig. 6.5, since the bar forces due to loading a (as already computed in Fig. 6.4), superimposed upon the bar forces due to loading b (which can be obtained from Fig. 6.4 by symmetry), equal the bar forces due to loading c, which corresponds to full snow load.   For example, the bar force in bar L_0U_1 due to full snow load equals

$$-6,720 - 15,680 = -22,400$$

for bar U_2L_3, it equals $0 - 5,660 = -5,660$; etc.   Since this is a symmetrical loading condition, only half the truss is shown.   The bar forces in Fig. 6.6 could, of course, have been obtained by a separate stress analysis for the full snow-load condition.

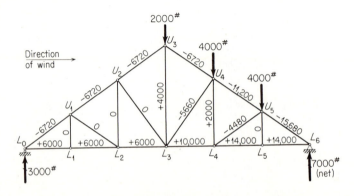

Figure 6.4 Bar forces due to snow on leeward side

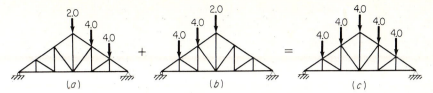

Figure 6.5   Superposition of loadings

Loading condition 3 calls for ice over the entire roof.   Since the intensity of the ice load is half the intensity of the snow load, the bar forces due to full ice load can be obtained from Fig. 6.6 by direct proportion, equaling half the bar forces due to full snow load.   Bar forces due to full ice load are shown in Fig. 6.7.

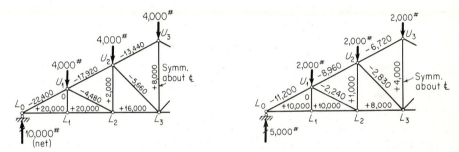

Figure 6.6   Bar forces due to full snow load   **Figure 6.7**   Bar forces due to full ice load

According to the ASCE report cited, q, the dynamic air pressure in pounds per square foot, can be computed from the following expression using the specified wind velocity of 100 mph:

$$q = 0.002558V^2 = (0.002558)(100)^2 = 25.6 \text{ lb per sq ft}$$

For the symmetrical gabled roof, $\tan \alpha = 15/30 = 0.5$, $\alpha = 26.6°$, whence by the expression $p = (0.07\alpha - 2.1)q$ from the ASCE report, the windward slope is

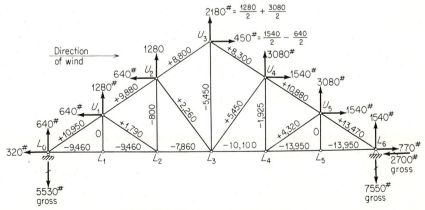

Figure 6.8   Bar forces due to wind load

Table 6.1　Bar-Force Table for Roof Truss of Fig. 6.2[†]
(All bar forces in pounds)

Bar	D	S	S_L	I	W	(1) $D+S$	(2) $D+W+S_L$	(3) $D+W+I$	Maximum bar force	Load combinations
L_0L_1	+8,600	+20,000	+ 6,000 / +14,000	+10,000	− 9,460 / −11,250	+28,600	+ 5,140 / +11,350	+9,140 / +7,350	+28,600	1
L_1L_2	+8,600	+20,000	+ 6,000 / +14,000	+10,000	− 9,460 / −11,250	+28,600	+ 5,140 / +11,350	+9,140 / +7,350	+28,600	1
L_2L_3	+6,880	+16,000	+ 6,000 / +10,000	+ 8,000	− 7,860 / − 7,400	+22,880	+ 5,020 / + 9,480	+7,020 / +7,480	+22,880	1
L_0U_1	−9,650	−22,400	− 6,720 / −15,680	−11,200	+10,950 / +13,470	−32,050	− 5,420 / −11,860	−9,900 / −7,380	−32,050	1
U_1U_2	−7,710	−17,920	− 6,720 / −11,200	− 8,960	+ 9,880 / +10,880	−25,630	− 4,550 / − 8,030	−6,790 / −5,790	−25,630	1
U_2U_3	−5,780	−13,440	− 6,720 / − 6,720	− 6,720	+ 8,800 / + 8,300	−19,220	− 3,700 / − 4,200	−3,700 / −4,200	−19,220	1
U_1L_2	−1,930	− 4,480	0 / − 4,480	− 2,240	+ 1,790 / + 4,320	− 6,410	− 140 / − 2,090	−2,380 / + 150	+ 150 / − 6,410	3 / 1
U_2L_3	−2,430	− 5,660	0 / − 5,660	− 2,830	+ 2,260 / + 5,450	− 8,090	− 170 / − 2,640	−3,000 / + 190	+ 190 / − 8,090	3 / 1
U_1L_1	+ 375	0	0 / 0	0	0 / 0	+ 375	+ 375	+ 375	+ 375	1,2,3
U_2L_2	+1,235	+ 2,000	0 / + 2,000	+ 1,000	− 800 / − 1,925	+ 3,235	+ 435 / + 1,310	+1,435 / + 310	+ 3,235	1
U_3L_3	+3,815	+ 8,000	+ 4,000	+ 4,000	− 5,450	+11,815	+ 2,365	+2,365	+11,815	1

† D = dead bar force, S = bar force due to snow load on both sides, S_L = bar force due to snow on leeward side only, I = bar force due to ice on both sides, W = bar force due to wind.

subjected to a suction of

$$p = [(0.07)(26.6) − 2.10](25.6) = 6.4 \text{ lb per sq ft}$$

The suction on the leeward slope equals $(0.6)(25.6) = 15.4$ lb per sq ft. Thus, for the windward side, the wind panel load equals

$$(6.4)(11.2)(20) = 1,435 \text{ lb}$$

For purposes of analysis it is more convenient to treat the vertical and horizontal components of the wind panel load, which equal 1,280 lb upward and 640 lb upwind, respectively. For the leeward side, the wind panel load equals $(15.4)(11.2)(20) = $ 3,450 lb. This panel load has an upward vertical component of 3,080 lb and a downwind horizontal component of 1,540 lb. Figure 6.8 shows the truss acted upon by the vertical and horizontal components of the wind panel loads and the bar forces for that condition of loading. The half panel loads at L_0 and L_6 were included in the analysis since the horizontal component of load applied at L_0 causes stresses in the truss. Since the wind loading is unsymmetrical, bar forces are computed for the entire truss.

In Table 6.1, we shall list the bar forces as given in Figs. 6.3, 6.4, and 6.6 to 6.8. Then we shall combine these bar forces to give the total bar force in each member corresponding to the three specified loading combinations. Finally we shall choose, for each member, the maximum bar force of each character which occurs in any of the three loading combinations, these maximum bar forces being those which would control the design of the members. We shall list in the table only the bars in half the truss; but, in entering bar forces for the columns headed S_L and W, we shall enter first the bar force in the member itself, due to the loading under consideration, and second the bar force in the symmetrically placed member, due to the same loading.[1] This second entry covers the case for the wind direction reversed, a condition for which we must provide. Computations for bars L_6L_5, L_5L_4, and L_4L_3 produce more extreme max/min values under loading combination (3) than shown for L_0L_1, L_1L_2, and L_2L_3, although such values would not control the design of the lower chords.

6.3   General Analysis of a Bridge Truss. The general analysis of a bridge truss consists in computing the bar force in each member due to each type of loading and the combination, for each truss member, of these forces into the maximum total bar force that will control design. To illustrate this procedure, we shall consider the Warren type of highway truss of Fig. 6.9, which will be analyzed

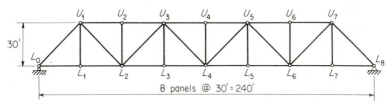

Figure 6.9 Highway bridge truss

for dead, live, and impact loads. It will be assumed that the truss, including details and secondary bracing, weighs 0.500 kip per ft, with this dead weight divided equally between the top- and bottom-chord panel points. The weight per foot of that portion of the floor system carried by the truss will be taken as 0.800 kip, acting at the bottom-chord panel points. For live load, an equivalent live-load system will be used, consisting of a uniform load of 0.650 kip per ft and a single concentrated load of 20.0 kips. Impact will be computed in accordance with Eq. (1.1). Maximum live bar forces are to be computed by the approximate

---

[1] This procedure was not followed for the lower-chord members. Why?

method. For each member of the left half of the truss, the maximum total bar force of each character is to be obtained.

We shall first compute the ordinates to the influence line for each member of the left half of the truss, considering the effect of a unit load at each bottom-chord panel point of the entire truss. These ordinates are summarized in the influence table (Table 6.2). Since no unusual conditions are encountered in the computations of these values, the details of the computations will be omitted.

A summary of the influence table (Table 6.2) will next be prepared (Table 6.3).

Table 6.2   Influence Table for Truss of Fig. 6.9

Bar	Bar force due to unit load at								
	L_0	L_1	L_2	L_3	L_4	L_5	L_6	L_7	L_8
$L_0 L_2$†	0	+0.875	+0.750	+0.625	+0.500	+0.375	+0.250	+0.125	0
$L_2 L_4$	0	+0.625	+1.250	+1.875	+1.500	+1.125	+0.750	+0.375	0
$L_0 U_1$	0	−1.238	−1.060	−0.885	−0.708	−0.530	−0.354	−0.177	0
$U_1 U_3$	0	−0.750	−1.500	−1.250	−1.000	−0.750	−0.500	−0.250	0
$U_3 U_4$	0	−0.500	−1.000	−1.500	−2.000	−1.500	−1.000	−0.500	0
$U_1 L_2$	0	−0.177	+1.060	+0.885	+0.708	+0.530	+0.354	+0.177	0
$L_2 U_3$	0	+0.177	+0.354	−0.885	−0.708	−0.530	−0.354	−0.177	0
$U_3 L_4$	0	−0.177	−0.354	−0.530	+0.708	+0.530	+0.354	+0.177	0
$U_1 L_1$	0	+1.000	0	0	0	0	0	0	0
$U_2 L_2$	0	0	0	0	0	0	0	0	0
$U_3 L_3$	0	0	0	+1.000	0	0	0	0	0
$U_4 L_4$	0	0	0	0	0	0	0	0	0

† By the method of joints, the force in $L_0 L_1$ is equal to the force in $L_1 L_2$ as long as only vertical loads are applied at joint L_1. Hence to save space in this and the following tables, these two members will be treated as a single member.

Table 6.3   Summary of Influence Table for Truss of Fig. 6.9

Bar	Sum of ordinates			Maximum ordinates		Loaded length, ft	
	Positive	Negative	All	Positive	Negative	Tension	Compression
$L_0 L_2$	+3.500	0	+3.500	+0.875	0	240	0
$L_2 L_4$	+7.500	0	+7.500	+1.875	0	240	0
$L_0 U_1$	0	−4.952	−4.952	0	−1.238	0	240
$U_1 U_3$	0	−6.000	−6.000	0	−1.500	0	240
$U_3 U_4$	0	−8.000	−8.000	0	−2.000	0	240
$U_1 L_2$	+3.714	−0.177	+3.537	+1.060	−0.177	206	34
$L_2 U_3$	+0.531	−2.654	−2.123	+0.354	−0.885	69	171
$U_3 L_4$	+1.769	−1.061	+0.708	+0.708	−0.530	137	103
$U_1 L_1$	+1.000	0	+1.000	+1.000	0	60	0
$U_2 L_2$	0	0	0	0	0	0	0
$U_3 L_3$	+1.000	0	+1.000	+1.000	0	60	0
$U_4 L_4$	0	0	0	0	0	0	0

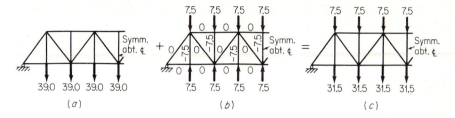

Figure 6.10   Computation of dead bar forces

Preparatory to the construction of a bar-force table for the members under consideration, we shall compute the panel loads for dead and live load. For dead load, the top-chord panel load equals $(30)(0.250) = 7.5$ kips; the bottom-chord panel load equals $7.5 + (30)(0.800) = 31.5$ kips. For uniform live load, the panel load equals $(30)(0.650) = 19.5$ kips; the concentrated live load, which is already expressed as a panel load, equals 20.0 kips.

In the bar-force table (Table 6.4), the dead bar forces are computed by the principle of superposition, as shown in Fig. 6.10. The bar forces due to the loading of Fig. 6.10*a* are first computed. This loading assumes that the total dead load is applied at the bottom chord, so that the dead panel load for bottom-chord panel points equals $31.5 + 7.5 = 39.0$. Since the influence table is based on the application of loads to the bottom-chord panel points, the bar forces for this condition of loading are obtained by multiplying, for each member, the sum of all the ordinates to the influence line by the total dead panel load of 39.0. The bar forces due to the loading of Fig. 6.10*b* are next computed, the value 7.5 being that of the top-chord dead panel load. This stress analysis is exceedingly simple and results, as shown in the figure, in a bar force of -7.5 in all verticals and zero force in all other members. If the loadings of Fig. 6.10*a* and *b* are superimposed, the loading of Fig. 6.10*c*, which corresponds to the actual dead loading, is obtained. Hence the dead bar forces can be obtained by superimposing the bar forces computed for the loadings of Figs. 6.10*a* and *b*. To summarize, the dead bar forces for a given member can be computed as follows: (1) Multiply the *total* dead panel load by the sum of all the ordinates to the influence line. (2) To the bar force from step 1, subtract, *for the verticals only*, the magnitude of the top-chord dead panel load.

Had the dead load not been uniformly distributed, the bar forces corresponding to the loading of Fig. 6.10*a* would require recourse to the influence table, with bar forces obtained by summing the cross products of the individual panel loads and corresponding ordinates from the influence table. It is, of course, possible, and sometimes desirable, to obtain dead bar forces by a separate analysis for dead loads, without making use of influence data.

Maximum tension and compression due to uniform live load are entered in the next column. These values are obtained by multiplying the uniform live panel load (19.5) by the sum of the positive and negative ordinates, respectively, to the influence line. Maximum tension and compression due to concentrated live load are entered in the following column. These values are obtained by multiplying the concentrated live panel load (20.0) by the maximum positive and negative ordinates, respectively, to the influence line. Total live bar force of each character is then obtained by summing the effects of the uniform and concentrated live loads.

Impact fractions, computed in accordance with Eq. (1.1), are then entered; these impact fractions multiplied by the *total* live bar forces give the impact bar forces, which are then added to the total live bar forces, leading to the total bar forces due to live load and impact. Total bar forces due to dead load plus live

Table 6.4   Bar-Force Table for Truss of Fig. 6.9
(All bar-forces in kips)

Bar	Dead bar force D	Live bar force L			Impact fraction	Impact bar force I	$L + I$	Total $= D + L + I$
		Uniform	Concen-trated	Total				
$L_0 L_2$	$+136.5$ —	$+ 68.3$ —	$+17.5$ —	$+ 85.8$ —	0.137 —	$+11.8$ —	$+ 97.6$ —	$+234.1$ —
$L_2 L_4$	$+292.5$ —	$+146.2$ —	$+37.5$ —	$+183.7$ —	0.137 —	$+25.2$ —	$+208.9$ —	$+501.4$ —
$L_0 U_1$	— -193.0	— $- 96.5$	— -24.7	— -121.2	— 0.137	— -16.6	— -137.8	— -330.8
$U_1 U_3$	— -234.0	— -117.0	— -30.0	— -147.0	— 0.137	— -20.2	— -167.2	— -401.2
$U_3 U_4$	— -312.0	— -156.0	— -40.0	— -196.0	— 0.137	— -26.8	— -222.8	— -534.8
$U_1 L_2$	$+138.0$ —	$+ 72.5$ $- 3.5$	$+21.2$ $- 3.5$	$+ 93.7$ $- 7.0$	0.151 0.300	$+14.2$ $- 2.1$	$+107.9$ $- 9.1$	$+245.9$ —
$L_2 U_3$	— $- 83.0$	$+ 10.4$ $- 51.8$	$+ 7.1$ -17.7	$+ 17.5$ $- 69.5$	0.258 0.169	$+ 4.5$ -11.8	$+ 22.0$ $- 81.3$	— -164.3
$U_3 L_4$	$+ 27.6$ —	$+ 34.5$ $- 20.7$	$+14.2$ -10.6	$+ 48.7$ $- 31.3$	0.191 0.219	$+ 9.3$ $- 6.9$	$+ 58.0$ $- 38.2$	$+ 85.6$ $- 10.6$
$U_1 L_1$	$+ 31.5$ —	$+ 19.5$ —	$+20.0$ —	$+ 39.5$ —	0.270 —	$+10.7$ —	$+ 50.2$ —	$+ 81.7$ —
$U_2 L_2$	— $- 7.5$	— —	— —	— —	— —	— —	— —	— $- 7.5$
$U_3 L_3$	$+ 31.5$ —	$+ 19.5$ —	$+20.0$ —	$+ 39.5$ —	0.270 —	$+10.7$ —	$+ 50.2$ —	$+ 81.7$ —
$U_4 L_4$	— $- 7.5$	— —	— —	— —	— —	— —	— —	— $- 7.5$

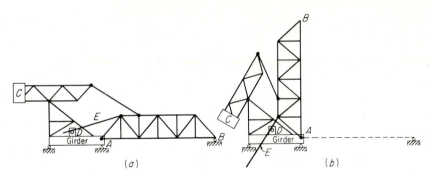

Figure 6.11　Bascule bridge

load plus impact are entered in the final column. It is to be noted that live and impact bar forces of both characters are carried forward in the bar-force table until it is definitely ascertained that no stress reversal can occur. In the actual design of a truss, bar forces due to other causes, such as wind loads, would be considered, such forces being combined with dead, live, and impact bar forces in additional columns in the bar-force table. In this example, bar forces due to wind have been omitted because the analysis of lateral systems and portals has been reserved for discussion in Chap. 7.

6.4　Movable Bridges—General. When the topography of a bridge site is such that it is desirable to have the roadway close to the surface of the body of water crossed by the bridge, the vertical underclearance requirements of the navigation passing beneath the bridge may require a movable bridge. A moveable bridge is one that can be moved to permit the passage of navigation. The three most important types of movable bridges are (1) bascule bridges, (2) vertical-lift bridges, and (3) horizontal-swing bridges. The type to be used depends largely upon the horizontal and vertical clearance requirements of the navigation. Whether a low-level movable bridge or a high-level fixed bridge should be used in a given site can usually be determined only by a careful economic study.

6.5　Bascule Bridges. A bascule bridge may prove economical where horizontal navigation requirements do not necessitate too long a span and where a high vertical clearance is required. A typical bascule bridge is shown in Fig. 6.11. Motive power drives a pinion at D, which engages the rack E, thus opening or closing the span. The required motive power is reduced by the action of the counterweight C.

The dead-load bar forces in a bascule span change as the bridge is opened or closed, and it is possible that the dead bar forces in certain members during such an operation may exceed the total bar forces with the bridge closed and subjected to traffic.

To find the maximum dead bar forces that occur while the span is being raised or lowered,[1] let F_H be the dead bar force in any member with the span horizontal and

[1] See O. E. Hovey, "Movable Bridges," vol. I, p. 219, John Wiley & Sons, Inc., New York, 1926.

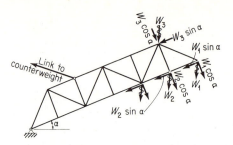

Figure 6.12 Bascule bridge, partially open

F_V be the dead bar force in the same member after the span is vertical, i.e., after having rotated through 90° from its closed position, both these values being easily computed by the usual methods of analysis. With the bridge partly opened and the bottom chord making an angle α with the horizontal, as shown in Fig. 6.12, each dead panel load can be resolved into two components, one perpendicular and one parallel to the bottom chord. The components of dead load that are perpendicular to the bottom chord will cause bar forces equal to $F_H \cos \alpha$, while the components of dead load that are parallel to the bottom chord will cause bar forces equal to $F_V \sin \alpha$. Hence, for any angle α, the total dead bar force F_D in any member is given by

$$F_D = F_V \sin \alpha + F_H \cos \alpha \qquad (a)$$

Placing the derivative of F_D with respect to α equal to zero gives

$$\frac{dF_D}{d\alpha} = F_V \cos \alpha - F_H \sin \alpha = 0$$

whence $\tan \alpha = F_V/F_H$. Substituting this value of α into Eq. (a), we have

$$\max F_D = F_V \frac{F_V}{\sqrt{F_V^2 + F_H^2}} + F_H \frac{F_H}{\sqrt{F_V^2 + F_H^2}} = \sqrt{F_V^2 + F_H^2} \qquad (6.1)$$

With the bridge closed, the dead-load reaction at the free end will be zero, since the counterweight holds the dead loads in equilibrium, but live loads produce reactions at each end, in the same manner as for an end-supported span.

6.6 Vertical-Lift Bridges. When the horizontal-clearance requirement is greater than the vertical-clearance requirement for navigation, a vertical-lift bridge is likely to prove economical. A typical vertical-lift bridge is shown in Fig. 6.13. The span AB is raised or lowered vertically by cables running over sheaves at D that are supported at the tower tops. The motive power required is reduced by the counterweights C. These counterweights are usually designed to balance the entire dead load of the movable span, so that the dead-load reactions are taken by the cables. Live loads on the movable span produce reactions on the piers at A and B, however.

6.7 Horizontal-Swing Bridges. Horizontal-swing bridges give unlimited vertical clearance, but the center pier constitutes an obstruction to traffic. A large horizontal area is required for this type of movable bridge. Horizontal-swing

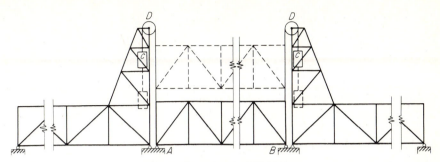

Figure 6.13   Vertical-lift bridge

bridges may be of either the center-bearing type, as shown in Fig. 6.14*a*, or the rim-bearing type, as shown in Fig. 6.14*b*.   In either case, the bridge is opened by swinging it horizontally about the vertical center line.   When the bridge is open, the two spans cantilever from the center pier and are statically determinate.   When the bridge is closed, the trusses are continuous and hence statically indeterminate. Stress analysis for the closed condition depends on principles discussed in the portion of this book dealing with statically indeterminate structures.

When a swing bridge is closed, the dead reactions developed at the outer ends of the structure depend on the design.   If these ends just touched their supports, any live load on one span would cause uplift at the far end of the other span.   This condition is usually avoided by lifting the ends a slight amount when they are closed.   The desired dead-load reaction can be computed so that it will exceed the maximum live-load reaction of opposite character.

6.8   Skew Bridges.   If the abutments are not perpendicular to the longitudinal axis of a bridge, the bridge is said to be **skewed**.   Figure 6.15*b* shows the plan view of such a structure.   In order to keep the connections from becoming complicated, the floor beams are usually kept perpendicular to the main trusses.   This is likely to make the main trusses unsymmetrical.   The two end posts of a truss should have the same slope so that the end portals will lie in a plane.   This may lead to an inclined hanger as shown by bar *a*, Fig. 6.15*a*.

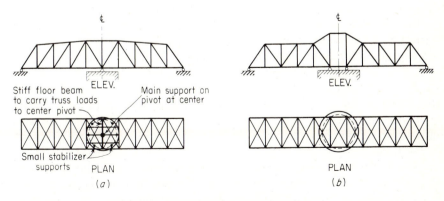

Figure 6.14   Horizontal-swing bridges

When analyzing bar forces in the members of a truss of a skewed bridge, one may proceed by the same general approach followed for a bridge that is perpendicular to the abutments. However, in dealing with live loads, one should take into consideration the irregularity of the floor system. This will alter the details of the computation of live panel loads, which may not be the same for all panel points.

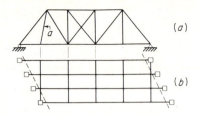

Figure 6.15   Skew bridge

6.9   Long-Span Structures. As the span of a structure becomes larger, the bending moments to which the structure is subjected increase rapidly if simple end-supported structures are used. Even if the load per foot to be carried by the structure did not increase with the span, the moment due to distributed loads would vary with the square of the span. Actually, the dead weight increases with the span, so that the bending moment increases at a rate greater than the square of the span. Since the chord stresses in trusses depend on bending moments to be carried by the truss, these considerations are of importance in the design of trusses as well as in the design of beams and girders.

For an economical structure, it is desirable, for the case of long spans, to adopt some means of construction that will reduce the bending moments to values less than would occur for simple end-supported structures. There are a number of methods by which this can be accomplished. In the remainder of this chapter, some of these methods will be illustrated by considering the analysis of several types of long-span structures.

6.10   Cantilever Structures—General. In a cantilever structure, bending moments are reduced by shortening the span in which positive bending occurs, by supporting an end-supported beam, of a length shorter than the total span, on cantilevered arms that act in negative bending. The structure of Fig. 6.16 shows cantilever construction and is statically determinate. The maximum moment in beam BC equals

$$\frac{1}{8}w\left(\frac{2L}{3}\right)^2 = \frac{wL^2}{18}$$

The maximum moment in the cantilever arm AB occurs at A and equals

$$-\frac{w}{2}\left(\frac{2L}{3}\right)\left(\frac{L}{6}\right) - w\left(\frac{L}{6}\right)\left(\frac{L}{12}\right) = \frac{-5wL^2}{72}$$

Had a simple span of length L been used, the maximum moment would have been $+wL^2/8$. Hence, in this particular case, the reduction in maximum moment resulting from cantilever construction is from $wL^2/8$ to $5wL^2/72$, or about 45 per cent. It should be pointed out that maximum moment is not the only criterion by which the relative merit of alternative types of construction should be judged, but it is an important factor in determining the desirability of using a particular type of structure for a given span. Moments and shears along the entire length of the

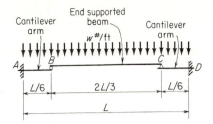

Figure 6.16 Cantilever construction

structure, computed for all types of loads including live loads, must be considered in the actual design.

To obtain the fixity of supports at A and D for the cantilevers AB and CD of Fig. 6.16 would be difficult under many conditions. The same principle of cantilever construction is present, however, in the structure of Fig. 6.17, where this difficulty has been eliminated. The moments in this structure, between A and D, are identical with the moments in the structure of Fig. 6.16. The moments at A and D, however, are resisted by the flanking spans AE and DF. The moment at A for example, is held in equilibrium by the reaction at E and the load applied between E and A.

6.11   Statical Condition of Cantilever Structures. For a cantilever structure to be statically determinate with respect to its outer forces, there must be as many independent equations available for the determination of these outer forces as there are independent reactions for the structure. Except for a simple cantilever beam, there are always more than three independent reactions for a cantilever structure, but there are only three independent equations of statics that can be applied to the entire structure. To make a cantilever structure statically determinate, it is therefore necessary to introduce certain construction features that make it possible to apply the equations of statics to certain portions of the structure, thus obtaining additional independent equations of condition. Some of these construction features are inherent to cantilever construction; others must be specially provided.

Such construction features are illustrated by the structure of Fig. 6.18. The hinge at a constitutes one such construction feature, since $\sum M_a = 0$ may be applied to those forces acting on that portion of the structure lying on *either* side of joint a. The hinge at b is similar in its effect, since it permits one to apply $\sum M_b = 0$ to those forces acting on that portion of the structure lying on either side of joint b.

The hinge at c permits the application of $\sum M_c = 0$ to those forces acting on that portion of the structure lying on either side of joint c. Instead of utilizing the hinge at c in this manner, the following alternative interpretation is usually advan-

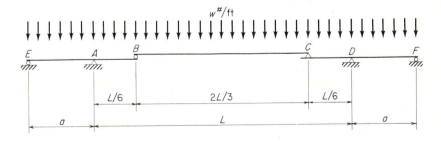

Figure 6.17 Typical cantilever girder bridge

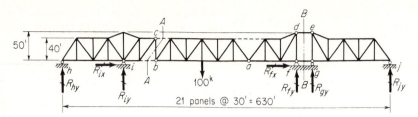

Figure 6.18   Cantilever truss bridge

tageous. Because there are hinges at both b and c, the hanger bc carries axial force only and hence has no component of force perpendicular to bc; that is, bc is a link. One may therefore conclude that if a section is taken through bc, the sum of the forces perpendicular to bc which act on either side of the section equals zero. For the case under consideration, this means that the equation $\sum F_x = 0$ may be applied to those forces acting on that portion of the structure lying on one side of section A-A.

The omission of the diagonal in panel $defg$ constitutes another construction feature that permits an additional independent equation of statics with respect to external forces, since, as a result, no shear can be carried by this panel. Hence the equation $\sum F_y = 0$ may be applied to the forces acting on one side of section B-B.

There are, therefore, the following seven independent equations available for determining the reactions of this structure:

(1) $\sum M = 0$ for all forces acting on structure

(2) $\sum F_x = 0$ for all forces acting on structure

(3) $\sum F_y = 0$ for all forces acting on structure

(4) $\sum M_a = 0$ for all forces acting on one side of hinge a

(5) $\sum M_b = 0$ for all forces acting on one side of hinge b

(6) $\sum F_x = 0$ for all forces acting on one side of section A-A

(7) $\sum F_y = 0$ for all forces acting on one side of section B-B

There are also seven independent reactions, R_{hy}, R_{iy}, R_{ix}, R_{fy}, R_{fx}, R_{gy}, and R_{jy}. Therefore this structure is statically determinate with respect to its outer forces. That the structure is also statically determinate with respect to both outer and inner forces can be verified from the fact that the total number of bars (75) plus the total number of reactions (7) equals twice the number of joints [(2)(41) = 82].

The available equations can often be applied in different sequences; but if the structure is stable, the same results will be obtained, regardless of the sequence followed.

Once the reactions are known, the bar forces can be computed by the usual methods of analysis for statically determinate trusses. Since the analysis can be carried out for a unit load at any panel point, the construction of influence lines for reactions or bar forces involves no special difficulties, although it is often advantageous first to construct influence lines for reactions or bar forces other than the

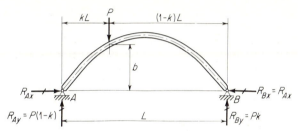

Figure 6.19   Two-hinged arch

one under consideration and use the data thus obtained in constructing the influence line actually desired.

6.12   Arches—General.   Another method of reducing maximum moments in long-span structures consists in adopting a structural layout in which applied vertical loads produce horizontal reactions acting so that the moments due to these horizontal reactions tend to reduce the moments that would otherwise exist.   Figure 6.19 shows an arch, which is a structure that develops horizontal *thrust* reactions under the action of vertical loads.   This particular arch is of the two-hinged type. The vertical reactions on this structure can be determined by statics by taking moments about an end hinge of all the forces acting on the arch; for the load P acting as shown, they have the values given on the figure.   The relation between the horizontal reactions R_{Ax} and R_{Bx} can be determined by statics ($\sum F_x = 0$), but the actual values of these reactions can be obtained only on the basis of an elastic analysis, since with four independent reactions this two-hinged arch is statically indeterminate to the first degree.

If the hinge at one end were replaced with a roller, as shown in Fig. 6.20, the structure would be, not an arch, but a statically determinate curved beam and the bending moment at the point of application of the load would equal $P(1 - k)kL$. For the two-hinged arch of Fig. 6.19, however, this moment is reduced by the moment $R_{Ax} b$.

An arch can be made statically determinate by building in a third hinge at some internal point, such as at the crown, in addition to the end hinges.   Such a structure is shown in Fig. 6.21 and is called a **three-hinged arch.**   This structure has four independent reactions.   Three equations of statics can be applied to the structure as a whole, and one equation of condition can be obtained by taking moments about

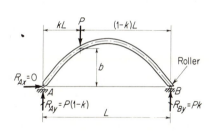

Figure 6.20   Curved beam

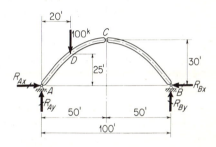

Figure 6.21   Three-hinged arch

the hinge at C of the forces acting on *either* side of the hinge. Thus, taking moments about A of all the forces acting on the structure leads to

$$R_{By} = +(100)(^{20}\!/_{100}) = +20 \text{ kips}$$

Similarly, taking moments about hinge B gives $R_{Ay} = +(100)(^{80}\!/_{100}) = +80$ kips. Now, taking moments about the hinge C of the forces acting on that portion of the structure to the right of the hinge gives

$$(+R_{Bx})(30) - (20)(50) = 0 \qquad R_{Bx} = +33.\overset{.}{3} \text{ kips}$$

Applying $\sum F_x = 0$ to the entire structure leads to $R_{Ax} = +33.\overset{.}{3}$ kips.

The moment at the point of application of the load is given by

$$M_D = (+80)(20) - (33.3)(25) = +767 \text{ kip-ft}$$

For a simple end-supported beam of the same span and loading, the moment at the load would equal $(+80)(20) = +1{,}600$ kip-ft. Hence the arch construction has reduced this moment by 52 per cent. The arch ribs, however, must carry compression that is not present in the end-supported beam. For example, the compression at the crown C in the arch of Fig. 6.21, where the rib is horizontal, equals the horizontal reaction and therefore has a value of -33.3 kips for the load considered. It is usually more economical, however, to carry loads by uniform axial stress than by bending stresses, although if the axial stress is compression, one must provide stability against buckling.

6.13 Analysis of Three-Hinged Trussed Arch. The arch ribs AC and BC of the structure of Fig. 6.21 may be replaced by trusses, as shown in Fig. 6.22. Since there are four reactions on this structure, it would be statically indeterminate were it not for the hinge at e, which is effective since the bar EF, shown dashed, is connected at its ends in such a manner that it can carry no axial force.

The reactions for this structure can be computed as follows. Taking moments about a of all the forces acting on the structure, we have

$$(+100)(30) + (200)(2)(30) + (300)(6)(30) - (R_{iy})(8)(30) = 0$$
$$R_{iy} = +287.5 \text{ kips}$$

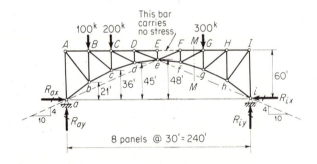

Figure 6.22 Three-hinged trussed arch

Taking $\sum F_y = 0$, for all forces acting on the structure we have

$$+ R_{ay} - 100 - 200 - 300 + 287.5 = 0 \qquad R_{ay} = +312.5 \text{ kips}$$

To obtain R_{ix}, apply $\sum M = 0$ about the hinge at e, considering the forces acting on the part of the structure to the right of the hinge:

$$(+300)(2)(30) + (R_{ix})(48) - (287.5)(4)(30) = 0 \qquad R_{ix} = +344 \text{ kips}$$

Since $\sum F_x = 0$ for the entire structure, R_{ax} also equals $+344$ kips. R_{ax} might have been computed by taking moments about the hinge at e of the forces to the left of the hinge, leading to

$$(+312.5)(4)(30) - (R_{ax})(48) - (100)(3)(30) - (200)(2)(30) = 0$$
$$R_{ax} = +344 \text{ kips}$$

In computing bar forces, the effect of the horizontal reactions must not be overlooked. To compute the force in FG, for example, taking moments about f of the forces to the right of section M-M gives

$$(+300)(30) + (344)(45) - (287.5)(90) - (F_{FG})(15) = 0 \qquad F_{FG} = -93 \text{ kips}$$

If there is no load between the center hinge and one end of the truss, the resultant reaction at that end of the truss must be directed so that it passes through the center hinge, because the moment about the hinge of the forces acting on that side of the hinge must equal zero. Thus, if we consider the action of a unit vertical load at B on the structure of Fig. 6.22, $R_{iy} = +\frac{1}{8}$; since the resultant reaction at i lies along the dashed line drawn through i and e, we conclude immediately that

$$R_{ix} = (+\frac{1}{8})(10\frac{1}{4}) = +\frac{5}{16}$$

This fact is often convenient in analysis, particularly in constructing influence lines.

If the three-hinged arch of Fig. 6.22 were subjected to equal vertical panel loads at each top-chord panel point (or at each bottom-chord panel point), the following stress condition would result: (1) The force in each top chord would be zero. (2) The force in each diagonal would be zero. (3) The force in each vertical would equal the top-chord panel load. (4) The horizontal component of force in each bottom chord would be the same and equal to the horizontal reactions. These facts may be verified by the student and would be useful, for example, in computing dead bar forces for a uniformly distributed dead load. Such conditions exist because the bottom-chord panel points of this structure lie on a parabola. If a funicular polygon were drawn for the loads under consideration so that it passed through the three hinges, the polygon would coincide with the location of the bottom chords.

6.14 Influence Lines for Three-Hinged Trussed Arch. Influence lines for a three-hinged trussed arch can be constructed by considering successive positions of the unit load, but a method similar to the following will often prove advantageous. We shall construct the influence line for bar FG of the structure of Fig. 6.22. We shall first construct the influence line for that portion of the force in bar FG due to the unit load and the vertical reactions only. Since these reactions have the same values they would have for an end-supported beam, the influence line for this portion of the force is a triangle with its maximum value occurring when the load is directly over the center of moments f, where the ordinate equals $(-\frac{5}{8})(\frac{90}{15}) = -3.75$, as shown in Fig. 6.23$a$. We shall next construct an influence line for $R_{ax} = R_{ix}$. As the unit load travels from A to E, R_{iy}, and hence R_{ix} (which equals $\frac{5}{2}R_{iy}$), increases linearly. With the load at E, R_{ix} equals $(\frac{5}{2})(+\frac{1}{2}) = +1.25$. Hence the influence line for R_{ix} is a straight line from zero at A to $+1.25$ at E. By similar reasoning, considering the reactions at a, the influence line for $R_{ax} = R_{ix}$ is a straight line from $+1.25$ at E to zero at I. The influence line for the magnitude of this horizontal reaction is shown in Fig. 6.23b. The force in bar FG due to the horizontal reactions equals

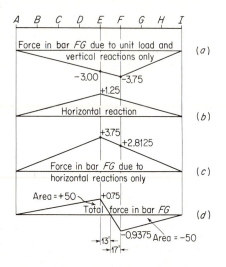

(a)

(b)

(c)

(d)

Figure 6.23   Influence lines for arch in Fig. 6.22

$$(+R_{ix})(\tfrac{45}{15}) = +3R_{ix}$$

Hence, the influence line for this portion of the force in bar FG is a triangle with its apex at E, where the ordinate equals $(+3)(+1.25) = +3.75$, as shown in Fig. 6.23c. The influence line for the total force in bar FG is now obtained by superimposing the influence lines of Fig. 6.23a and c, leading to the influence line of Fig. 6.23d.

It will be noted that the net area under this influence line is zero, as should be the case, since a uniform load extending over the entire structure would cause no force in the top chords.

6.15   Three-Hinged Trussed Arches with Supports at Different Elevations.   The points of support of a three-hinged trussed arch may be at different elevations, as shown in Fig. 6.24. The vertical reactions will then differ from the values they would have for an end-supported truss, since when moments are taken about one point of support, the horizontal reaction at the far end enters into the equation. The reactions can, however, be obtained by statics. Referring to Fig. 6.24a and taking moments about a of all the forces acting on the structure gives

$$(+100)(60) - (R_{ix})(20) - (R_{iy})(240) = 0$$

whence $R_{ix} = +300 - 12R_{iy}$.

Now, taking moments about the hinge at e of the forces acting to the right of the hinge, we have

$$(+R_{ix})(38) - (R_{iy})(120) = 0$$

$$R_{iy} = +\tfrac{38}{120}R_{ix} = (+\tfrac{38}{120})(+300 - 12R_{iy})$$

whence $R_{iy} = +19.8$ kips and $R_{ix} = +300 - (12)(19.8) = +62.5$ kips.

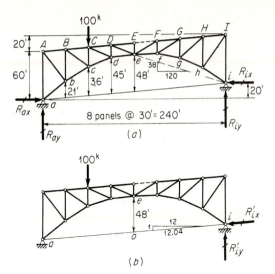

Figure 6.24 Supports at different elevations

The foregoing solution required the solution of two simultaneous equations, which might have been avoided by taking the reactions as shown in Fig. 6.24*b*, where R'_{ix} is taken as acting along a line passing through the two points of support and R'_{iy}, while a vertical reaction, differs from R_{iy}, since R'_{ix} has a vertical component. Taking moments about *a* gives

$$(+100)(60) - (R'_{iy})(240) = 0 \qquad R'_{iy} = +25.0$$

Taking moments about *e* and considering R'_{ix} to act at point *o* gives

$$(+\,{}^{12.00}\!/_{12.04}R'_{ix})(48) - (25.0)(120) = 0$$

whence $R'_{ix} = +62.7$ kips. Hence the actual horizontal reaction at *i* is given by

$$R_{ix} = (+62.7)({}^{12.00}\!/_{12.04}) = +62.5 \text{ kips}$$

while the actual vertical reaction at *i* is given by

$$R_{iy} = +25.0 - (\tfrac{1}{12.04})(62.7) = +19.8 \text{ kips}$$

For the particular case under consideration, since there are no loads applied to the right of the hinge at *e*, we might have concluded immediately that the resultant reaction at *i* passed through *e*, whence

$$R_{ix} = {}^{120}\!/_{38}R_{iy} = +3.16R_{iy}$$

Then, taking moments about *a* gives

$$(+100)(60) - (3.16R_{iy})(20) - (R_{iy})(240) = 0$$

whence $R_{iy} = +19.8$ kips and $R_{ix} = (+3.16)(+19.8) = +62.5$ kips.

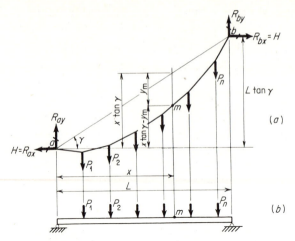

Figure 6.25 Derivation of cable theorem

6.16 Cables. Cables are used in many important types of engineering structures. They form the main load-carrying elements for suspension bridges and
cable-car systems. They are used extensively for permanent guys on structures
such as derricks and radio towers. They are also used for temporary guys during
erection. Although exact cable analyses may require mathematical procedures
beyond the scope of this book, a knowledge of certain fundamental relationships
for cables is important in structural engineering.

When a cable supports a load that is uniform per unit length of the cable itself,
such as its own weight, it takes the form of a catenary; but unless the sag of the
cable is large in proportion to its length, the shape taken may often be assumed to
be parabolic, the analysis being thus greatly simplified.

6.17 General Cable Theorem. Consider the general case of a cable supported
at two points a and b, which are not necessarily at the same elevation, and acted
upon by any system of vertical loads $P_1, P_2, \ldots, P_n$ as shown in Fig. 6.25a. The
cable is assumed to be perfectly flexible, so that the resisting moment at any point
on the cable must be zero and the stress condition in the cable can only be an
axial tension. Since all the loads are vertical, the horizontal component of cable
tension, which will be denoted by H, has the same value at any point on the cable
and the horizontal reactions are each equal to H. Let

$\sum M_b$ = sum of moments about b of all loads $P_1, P_2, \ldots, P_n$
$\sum M_m$ = sum of moments about any point m on cable of those of loads $P_1, P_2, \ldots,$
$\quad\quad P_n$ that act on cable to left of m

Taking moments about b of all the forces acting on the cable gives

$$+ H(L \tan \gamma) + R_{ay} L - \sum M_b = 0$$

from which
$$R_{ay} = \frac{\sum M_b}{L} - H \tan \gamma \tag{a}$$

Taking moments about m of those forces acting on that portion of the cable to the left of m gives

$$+H(x \tan \gamma - y_m) + R_{ay}x - \sum M_m = 0$$

Substituting R_{ay} from Eq. (a) and simplifying gives

$$Hy_m = \frac{x}{L}\sum M_b - \sum M_m \qquad (b)$$

In interpreting Eq. (b), it should be noted that y_m is the vertical distance from the cable at point m to the cable chord ab, which joins the points of cable support. The right side of Eq. (b) may be seen to equal the bending moment that would occur at point m (see Fig. 6.25b) if the loads $P_1, P_2, \ldots, P_n$ were applied to an end-supported beam of span L and m were a point on this imaginary beam located at distance x from the left support.

From Eq. (b), we may therefore state the general cable theorem.

General cable theorem:

At any point on a cable acted upon by vertical loads, the product of the horizontal component of cable tension and the vertical distance from that point to the cable chord equals the bending moment which would occur at that section if the loads carried by the cable were acting on an end-supported beam of the same span as that of the cable.

It is to be emphasized that this theorem is applicable to any set of vertical loads and holds true whether the cable chord is horizontal or inclined.

6.18   Application of General Cable Theorem.   Suppose that the loading on a cable is defined and that the distance from the cable to the cable chord is known at one point, as is the case in Fig. 6.26.   Neglecting the weight of the cable itself, the bending moment at point d on an imaginary beam of equal span is equal to

$$(2{,}330)(20) - (1{,}000)(10) = 36{,}600 \text{ ft-lb}$$

Hence, by the general cable theorem, $10H = 36{,}600$, or $H = 3{,}660$ lb.   To determine the distance of any other point such as c from the cable chord, the general cable theorem is applied at section c, leading to $3{,}660y_c = (2{,}330)(10)$, from which $y_c = 6.38$ ft.   The segment of the cable between a and c lies along a straight line, since the weight of the cable has been neglected, and has a length equal to

$$\sqrt{(10)^2 + (6.38)^2} = 11.85 \text{ ft}$$

Since the horizontal component of cable tension equals 3,660 lb, the actual cable tension between a and c equals $(3{,}660)$ $(11.85/10) = 4{,}340$ lb.   The left vertical

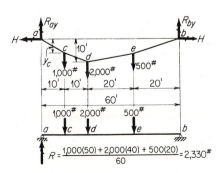

Figure 6.26   Application of cable theorem

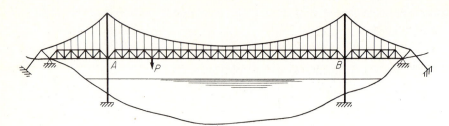

Figure 6.27   Suspension bridge

reaction on the cable equals the vertical component of cable tension in segment ac and is given by $(3,660)(6.38/10) = 2,330$ lb.   For this particular case, this value equals the left vertical reaction on the imaginary end-supported beam.   Had the cable chord been inclined, however, these two vertical reactions would have had different values.

6.19   Suspension Bridges.   An important method of reducing bending moments in long-span structures consists in providing partial support at points along the span by means of a system of cables, as in a suspension bridge.   Referring to Fig. 6.27, a suspension bridge is usually so erected that all the dead load is carried by the cable.   When live load is applied to such a structure, tension in the hangers transfers a large portion of the live load to the cable.   Hence, the stiffening truss AB is subjected to no dead moments, and the live moments it must carry are substantially reduced.   For long-span structures, this is of particular importance, since so much of the load is carried by the cable in tension, which is a highly efficient manner of carrying loads.

A suspension bridge like that in Fig. 6.27 is statically indeterminate.   By the introduction of certain features of construction, it can however, be made statically determinate.   The analysis of statically determinate suspension bridges is treated in Art. 6.20.

6.20   Statically Determinate Suspension Bridges.   A suspension bridge is usually constructed so that the dead loads are carried entirely by the cables.   A large portion of the dead load comes from the roadway and is uniform.   It is commonly assumed that the entire dead load is uniform per horizontal foot.   On the basis of this assumption, the cables are parabolic under dead load only.   When live load is applied, with partial loadings so as to give maximum forces in members, the cables tend to change their shape.   In order to prevent local changes of slope in the roadway due to live load from being too large, the floor beams of the floor system are usually framed into stiffening trusses, which in turn are supported by hangers running to the cables.   These stiffening trusses distribute the live load to the various hangers, in such a manner that even under live loads the cable may be assumed to remain essentially parabolic.   As long as the cable remains parabolic, it must be acted upon by a load that is uniform per horizontal foot.   Since the hangers are equally spaced, this is equivalent to stating that the hanger tensions in a given span must be equal.   In the **elastic theory** of suspension bridges, all the

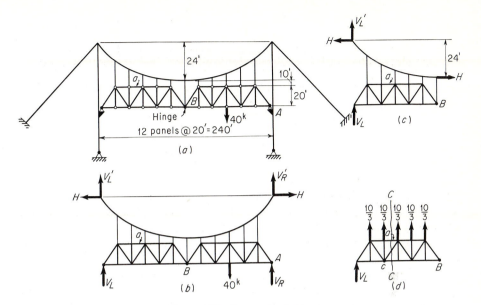

Figure 6.28 Suspension bridge

hangers in a given span are assumed to have equal tensions. That this assumption is not strictly correct can be shown by the more accurate and more complicated **deflection theory** of suspension bridges. However, unless a suspension bridge is long and flexible, the elastic theory may lead to results that are not greatly in error.

If the stiffening trusses of a suspension bridge are arranged and supported as shown in Fig. 6.28a, with a hinge at some intermediate point in the main-span stiffening truss, the structure is statically determinate provided that it is assumed that all the hangers in a given span have equal tensions. The application of the various relations for cables to a statically determinate suspension bridge will be illustrated by an analysis of the structure shown in that figure.

Suppose this bridge is subjected to a live load of 40 kips acting as shown in the figure. Consider the equilibrium of all the forces acting on that portion of the structure shown in Fig. 6.28b. The horizontal components of cable reaction at each end of the cable are equal and have the same line of action. Taking moments about point A gives

$$(V_L + V_L')(240) - (40)(60) = 0 \qquad V_L + V_L' = +10 \text{ (upward)}$$

Now consider the equilibrium of all the forces acting on that portion of the structure shown in Fig. 6.28c, taking moments of these forces about the hinge at B:

$$(V_L + V_L' = 10)(120) + (H)(30) - (H)(54) = 0 \qquad H = 50.0 \text{ kips}$$

The maximum cable tension in the main span occurs at the ends of the cable and equals $(50.0)(1 + {}^{16}\!/_{100})^{1/2} = 53.9$ kips.[1]

Let X equal the tension in each hanger. The equivalent uniform load of the hangers equals $X/20$ kips per ft. To evaluate X, use the relation $H = wL^2/8h$, where $w = X/20$:

$$50.0 = \frac{X}{20}\frac{(240)^2}{(8)(24)} \qquad X = +\frac{10}{3}\text{ kips}$$

Once the hanger tensions have been determined, the forces in the bars of the stiffening truss are readily evaluated. For example, to find the force in bar a, first find V_L by taking moments about B of the forces acting on the portion of the structure shown in Fig. 6.28d.

$$+ 120V_L + ({}^{10}\!/_3)(20 + 40 + 60 + 80 + 100) = 0 \qquad V_L = -{}^{25}\!/_3\text{ (down)}$$

Now taking moments about c of the forces acting on the portion of the structure to the left of section C-C (Fig. 6.28d),

$$(-{}^{25}\!/_3)(40) + ({}^{10}\!/_3)(20) + (F_a)(20) = 0 \qquad F_a = +13.33\text{ kips}$$

For this particular structure, the side spans are not suspended. The cables of the side spans act as guys to the towers. The horizontal component of cable tension is the same for side and center spans, as can be seen by taking moments about the hinge at a tower base of the forces acting on a tower. V_L is assumed to act on the center line of the tower.

6.21   Problems for Solution

Problem 6.1   Figure 6.29 shows a bascule span acted upon by its dead panel loads in kips.

(a) Compute the dead bar forces, including the tension in the link to the counter-weight, with the span in the closed position as shown.

(b) Compute the maximum dead bar force that will occur as the span is raised through an angle of 90°, for all members except $L_0 U_1$.

(c) Is the method of analysis used in part b applicable to the determination of the maximum dead bar force in $L_0 U_1$?

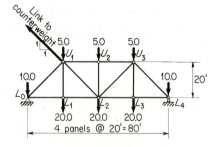

Figure 6.29   Problem 6.1

[1] For a uniformly loaded cable hanging in the shape of a parabola, whose sag at mid-span is h measured with respect to the chord of the cable (which slopes upward as in Fig. 6.25 with an angle γ with respect to the horizontal) and whose sag ratio h/L is called θ, the maximum cable tension occurs at one end:

For $x = 0$: $T_{\max} = H(1 + 16\theta^2 + \tan^2\gamma - 8\,\theta\tan\gamma)^{1/2}$
For $x = L$: $T_{\max} = H(1 + 16\theta^2 + \tan^2\gamma + 8\,\theta\tan\gamma)^{1/2}$

If the chord is horizontal, $\tan\gamma = 0$, whence at either end $T_{\max} = H(1 + 16\theta^2)^{1/2}$.

Problem 6.2   Figure 6.30 shows a horizontal-swing bridge truss acted upon by its dead panel loads in kips.   Dashed-line members carry tension only.

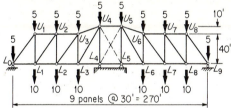

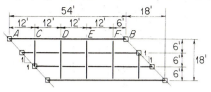

Figure 6.30   Problem 6.2

(a) Compute the dead bar forces when the bridge is swung into its open position.

(b) When the bridge is closed, each end of the structure is raised 1 in. above the elevation it has when the bridge is open. If a force of 10 kips applied upward at L_0 when the bridge is open would raise L_0 by 1 in. and lower L_9 by $\frac{1}{4}$ in., compute the dead reaction at L_0 and L_9 with the bridge in the closed position.

(c) Compute the dead bar forces in the truss with the bridge in the closed position, based on the dead reactions corresponding to part b.

Problem 6.3   For the single-track railroad skew bridge for which the plan view is shown in Fig. 6.31, the weight of the track and ties is 500 lb per ft of track; the weight of each stringer, including details, is 125 lb per ft; the weight of each floor beam, including details, is 175 lb per ft.

(a) Compute those portions of the dead panel loads acting on girder AB that are applied by the floor beams to the girder at panel points C, D, E, and F.

(b) If the track is subjected to a uniform live load of 5,000 lb per ft, extending over the entire structure, what are the live panel loads acting on the girder AB that are applied by the floor beams to the girder at panel points C, D, E, and F?

Figure 6.31   Problem 6.3

*Problem 6.4   Referring to the cantilever bridge of Fig. 6.18, construct an influence line for (a) the force in the hanger bc; (b) the vertical reaction at i; (c) the bar force in the top chord de. (d) Compute the maximum force in bar de due to the following loading: dead load, 2,000 lb per ft; uniform live load, 1,000 lb per ft; concentrated live load, 10,000 lb.

*Problem 6.5   Construct an influence line for the force in bar bc of the structure of Fig. 6.24.

*Problem 6.6   What is the maximum force in bar Bc of the structure of Fig. 6.24 due to the following loading: dead load, 1,000 lb per ft; uniform live load, 500 lb per ft; concentrated live load, 5,000 lb?

7

Approximate Analysis of Statically Indeterminate Structures

7.1 Introduction. From a broad viewpoint, the analysis of every structure is approximate, for it is necessary to make certain assumptions in order to carry out the analysis. For example, in computing the stresses in a pin-connected truss, it is assumed that the pins are frictionless, so that the truss members carry axial force only. It is, of course, impossible to build a pin connection that is frictionless, and as a result the stress analysis of a pin-connected truss is approximate. It may therefore be said that there is no such thing as an "exact" analysis.

However, if proper judgment is exercised in making the assumptions upon which the analysis of a given structure is based, the resultant errors will be small. A stress analysis based on the usual assumptions that underlie structural theory is often called "exact," although it may be seen that, strictly speaking, this term is not used correctly. It is, however, a convenient term to use, for it is desirable to distinguish between analyses based on the usual assumptions which are relatively exact and analyses based on further assumptions which introduce further errors and which are therefore frankly approximate.

When one speaks of an approximate analysis for a given structural problem, one does not necessarily refer to any particular set of assumptions and resultant approximations. The particular approximate method to be used under any given circumstances will depend on the time available for the analysis and the degree of accuracy considered necessary.

For structural types that occur commonly in structural analysis, one may take advantage of approximate methods of analysis worked out by others and investigated as to their accuracy so that they can be used with a fair degree of confidence. The approximate methods described in engineering literature do not, however, cover all cases. A good stress analyst should be familiar enough with the action of statically indeterminate structures to be able to set up his own assumptions when he encounters circumstances not covered by the literature.

In this chapter, a number of approximate solutions for common types of statically indeterminate structures are given.   A knowledge of these methods is of importance but of perhaps greater importance is the fact that the procedures here outlined will serve as a basis for making intelligent assumptions that will permit simplified approximate analyses of other types of statically indeterminate structures.

7.2   Importance of Approximate Methods in Analyzing Statically Indeterminate Structures.   The analysis of a statically *determinate* structure does not depend on the elastic properties of its members.   Because of this, relatively simple "exact" stress analyses can be carried out for such structures.

In a statically *indeterminate* structure, however, stress analysis depends on the elastic properties of members.   These elastic properties include modulus of elasticity, cross-sectional area, cross-sectional moment of inertia, and length of member.   That this is so can be visualized by reference to Fig. 7.1.   Suppose that the stiffness of beam AB is made very small in comparison with the stiffness of beam CD. This can be accomplished by making $E_1 I_1 / L_1^3$ very small in comparison with $E_2 I_2 / L_2^3$. Then beam CD will carry a greater portion of load P than beam AB.   Suppose further that the tension tie EF that connects the two beams is given a very small stiffness by making $E_3 A / h$ small.   This might be accomplished by constructing the tie EF of rubber, which has a very low value of E.   Under this condition the beam CD would carry still more of the load P.

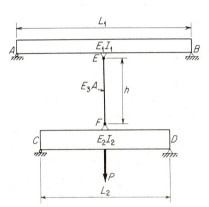

Figure 7.1   Effect of elastic properties on stress analysis

When the design of a statically indeterminate structure is first begun, the areas and moments of inertia of its members are not known.   It is therefore necessary to carry out an approximate analysis for stresses in the structure, so as to obtain some idea on the required sizes of these members.   Once these tentative sizes have been assigned, an elastic analysis can be carried out.   In general, the first elastic analysis will show that the actual stresses in the structure are not satisfactory, and it is only by successive designs that a satisfactory final result can be obtained. Approximate analyses of statically indeterminate structures are therefore important in preliminary design stages.

If the quantities determining the stiffness of members in a statically indeterminate structure are known, a so-called "exact" analysis may be carried out that will yield results of the same order of accuracy as can be obtained for statically determinate structures.   In actual practice, however, the following factors may prevent an exact analysis:

1. The configuration and complexity of the structure may be such that exact methods of analysis are either not available or impractical to apply.

2. The time required to carry out a statically indeterminate analysis may be so great that an exact solution must be abandoned.

In some cases, the need of meeting time schedules may be the controlling factor. In other instances, economic considerations may make it desirable to use an approximate method of analysis. It may be less expensive to use more material, as a result of basing design on approximate stresses and on a higher apparent factor of safety with respect to the computed stresses, than to save material by basing design on exact stresses and a lower apparent factor of safety. This attitude may sometimes be properly taken in designing relatively unimportant structures or secondary portions of important structures. The analyst will also be influenced in this connection by his judgment as to the magnitude of the errors likely to be introduced by the approximate method he proposes to use.

7.3 Number of Assumptions Required. It has previously been pointed out that for an analysis of a structure to be possible on the basis of the equations of statics only, there must be available as many independent equations of statics as there are independent components of force in the structure. If there are n more independent components of force than there are independent equations of statics, the structure is statically indeterminate to the nth degree. It will then be necessary to make n independent assumptions, each of which supplies an independent equation or relation, in order that an approximate solution can be worked out on the basis of statics only.

If fewer than n assumptions are made, a solution based only on statics will not be possible. If more than n assumptions are made, the assumptions will not in general be consistent with each other and the application of the equations of statics will lead to inconsistent results, depending on which equations are used and the order in which they are used. Therefore, the first step in the approximate analysis of a statically indeterminate structure is to find the degree to which the structure is indeterminate and hence the number of assumptions to be made.

7.4 Parallel-Chord Trusses with Two Diagonals in Each Panel. Trusses of this type occur frequently in structural engineering, e.g., in the top- and bottom-chord lateral systems of a bridge, as described in Art. 1.21. The approximate analysis of such a truss will be illustrated by considering the truss of Fig. 7.2, in which it is assumed that all members are capable of carrying either compression or tension. It should first be noted that this truss is statically indeterminate to the sixth degree; that this is true can readily be seen from the fact that if one diagonal were removed from each panel, the remaining members would form a statically determinate truss. It is therefore necessary to make six independent assumptions about stress conditions. *It will be assumed that in each panel the shear is equally divided between the two diagonals;* since there are six panels, this amounts to six independent assumptions. It is now easy to complete the analysis of the structure, by using only relations of statics. The solution is shown in Fig. 7.2, the method of index forces having been employed. The shear in each panel is first computed from the external forces. As an illustration, the shear in panel 1-2 is -39 kips. This shear is equally divided between L_1L_2' and $L_1'L_2$, so that the index forces in these two bars

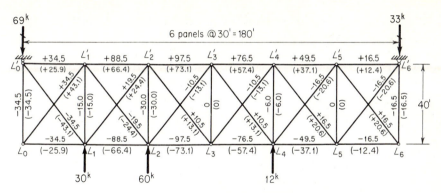

Figure 7.2   Truss with double diagonal system

are $+19.5$ and -19.5, respectively. Index forces for all diagonals are determined in a similar manner.[1] From these index forces, the index forces in all other members can be determined in the usual manner. In Fig. 7.2 the actual bar forces are given in parentheses for each member.

In trusses of this type, the diagonals are often designed as tension members only, by making their slenderness ratios (unsupported length divided by radius of gyration) large, since then, when a diagonal is subjected to compression, it will buckle slightly and carry only a negligible load. When the diagonals are designed in this manner, the total shear in each panel is carried in tension by a single diagonal. A consideration of the total shear on each panel enables one to tell which diagonal has a tendency to buckle and therefore carries no load. In Fig. 7.2, if it is assumed that the diagonals were designed so that they could carry tension only, all the diagonals that carried compression in the previous analysis would now have zero force. This, in effect, makes the truss statically determinate, and hence no difficulty is encountered in completing the analysis.

7.5 Portals. Portal structures, similar to the end portals of the bridge described in Art. 1.21, have as their primary purpose the transfer of horizontal loads applied at their top to their foundations. Clearance requirements usually lead to the use of statically indeterminate structural layouts for portals, and approximate solutions are often used in their analyses. Consider the portal shown in Fig. 7.3a, all the members of which are capable of carrying bending and shear as well as axial force. The legs are hinged at their base and rigidly connected to the cross girder at the top. This structure is statically indeterminate to the first degree; hence, one assumption must be made. Solutions of this type of structure, based on elastic considerations, show that the total horizontal shear on the portal will be divided almost equally between the two legs; it will therefore be assumed that the horizontal reactions for the two legs are equal to each other and therefore equal to $P/2$.[2] The remainder of the analysis can now be carried out by statics. The

---

[1] See Example 4 9.

[2] This assumption is reasonable if the frame is symmetrical and the bending stiffnesses of the legs are equal. What would be a reasonable assumption if the stiffnesses of the legs are not equal?

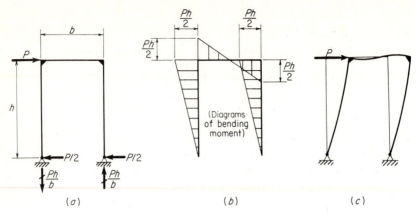

Figure 7.3 Portal frame

vertical reaction on the right leg can be obtained by taking moments about the hinge at the base of the left leg. The vertical reaction on the left leg can then be found by applying $\Sigma F_y = 0$ to the entire structure. Once the reactions are known, the diagrams of bending moment and shear are easily computed, leading to values for bending moment as given in Fig. 7.3b. It is well to visualize the deformed shape of the portal under the action of the applied load. This is shown, to an exaggerated scale, in Fig. 7.3c.

Consider now a portal similar in some ways to that of Fig. 7.3a but with the bases of the legs fixed, as shown in Fig. 7.4a. This structure is statically indeterminate to the third degree, so that three assumptions must be made. As when the legs were hinged at their base, it will again be assumed that the horizontal reactions for the two legs are equal and hence equal to $P/2$. Figure 7.4c shows the deformed shape of the portal under the action of the applied load. It will be noted that near the center of each leg there is a point of reversal of curvature. These are points of inflection, where the bending moment is changing sign and hence has zero value. It will therefore be assumed that there is a point of inflection at the center of each

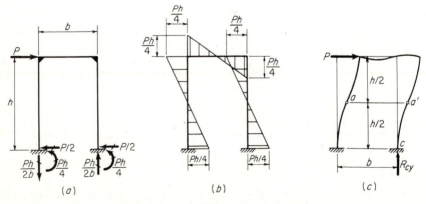

Figure 7.4 Portal frame with fixed legs

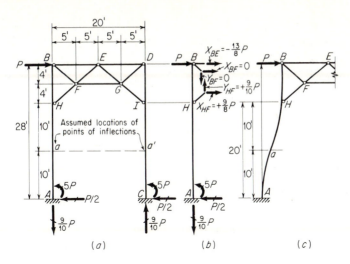

Figure 7.5 Bridge portal

leg; this is structurally equivalent to assuming that hinges exist at points a and a', as shown in Fig. 7.4c.   The vertical reactions on this portal equal the axial forces in the portal legs and can be determined by successively taking moments about a and a' of all the forces acting on that portion of the structure above a and a'.   For example, taking moments about a gives

$$+P\frac{h}{2} - R_{cy}b = 0 \qquad R_{cy} = +\frac{Ph}{2b}$$

The moment reaction at the base of each leg equals the shear at the point of inflection in the leg multiplied by the distance from the point of inflection to the base of the leg and therefore equals $(P/2)(h/2) = Ph/4$.   Once the reactions are known, the diagrams of shear and bending moment for the members of the portal are easily determined by statics.   The diagrams of bending moment for this structure and loading are given in Fig. 7.4b.

Portals for bridges are often arranged in a manner similar to that shown in **Fig. 7.5a**.   In such a portal, the legs AB and CD are continuous from A to B and C to D, respectively, and are designed to be capable of carrying bending moment and shear as well as axial force.   The other members that constitute the truss at the top of the portal are considered as pin-connected and carrying axial force only.   Such a structure is statically indeterminate to the third degree; the following three assumptions will be made:

1. The horizontal reactions are equal.
2. A point of inflection occurs midway between the base A of the leg AB and the end H of the knee brace for leg AB.
3. A point of inflection occurs midway between the base C of the leg CD and the end I of the knee brace for leg CD.

The horizontal reactions therefore each equal $P/2$.   The moment at the base of each leg equals the shear in the leg multiplied by the distance from the point of inflection to the base of the leg and has a value of $(P/2)(10) = 5P$. Vertical reactions can be obtained by successively taking moments about the points of inflection a and a' of the forces acting on that portion of the structure above a and a'. These equations show the vertical reactions each equal to $9P/10$ and acting in the directions shown in the figure.

To find the forces in the bars connected to the legs, one may proceed as follows. Considering the leg AB as a free body and taking moments about B of the forces acting on the leg lead to

$$+\frac{P}{2}(28) - 5P - (X_{HF})(8) = 0 \qquad X_{HF} = +\frac{9P}{8}$$

$$\therefore Y_{HF} = +\frac{9P}{8}\frac{4}{5} = +\frac{9P}{10}$$

To find the force in bar BF, apply $\Sigma F_y = 0$ to all the forces acting on leg AB:

$$+Y_{BF} - \frac{9P}{10} + \frac{9P}{10} = 0 \qquad \therefore Y_{BF} = 0 \qquad \therefore X_{BF} = 0$$

To find the force in bar BE, apply $\Sigma F_x = 0$ to all the forces acting on leg AB:

$$+X_{BE} + 0 + \frac{9P}{8} + P - \frac{P}{2} = 0 \qquad \therefore X_{BE} = -\frac{13P}{8}$$

With all the forces acting on leg AB known, as shown in Fig. 7.5b, the axial force on any section of the leg can be computed, and diagrams of shear and moment for the leg can be constructed. Leg CD can be analyzed in a similar manner. The remaining bar forces can then be computed without difficulty.

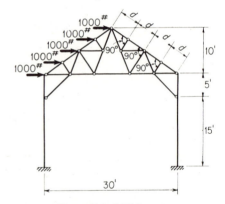

Figure 7.6   Mill bent

7.6 Mill Bents. Bents of mill buildings are often constructed as shown in Fig. 7.6. An approximate analysis of stresses in such a bent when it is acted on by lateral loads can be carried out on the basis of assumptions identical with those made for the portal of Fig. 7.5, namely, the horizontal reactions at the bases of the legs are equal, and a point of inflection occurs in each leg at an elevation of 7.5 ft above the base of the leg. The application of the equations of statics to carry out the analysis after these assumptions have been made follows the same general procedure as that employed for the portal.

7.7 Towers with Straight Legs.[1] If the legs of a statically determinate tower have a constant batter throughout their length, each face of the tower lies in a plane. For such a tower, stress analysis for lateral loads can be carried out by resolving all lateral loads into components lying in the planes of the faces of the tower that are adjacent to the joints where the lateral loads are applied. Each face of the tower can then be analyzed as a planar truss, acted upon by forces lying in the plane of that truss. Such an analysis leads directly to the forces in the web members of each face, while the forces in the legs are obtained by superimposing the bar forces resulting from the analyses of the adjacent faces.

The planar truss of each face, however, may be statically indeterminate. For such a tower, stress analysis may still be carried out on the basis of planar trusses, but it is necessary to make certain assumptions if the analysis is to be based on statics only. In Fig. 7.7a, let it be required to determine the bar forces in the diagonals of panel $abcd$. Passing section M-M through this

[1] Refer to Arts. 4.19 and 4.20.

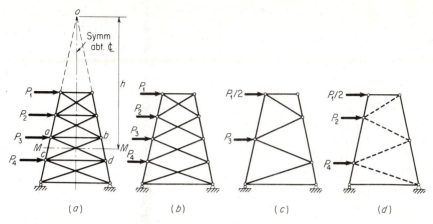

Figure 7.7 Tower truss

panel, at the elevation of the intersection of the diagonals, take moments about o, which is at the intersection of the extended legs, of all the forces acting on that portion of the truss above section *M-M*. Since the legs extended pass through the origin of the moments, the moments of forces P_1, P_2, and P_3 are held in equilibrium by the moments of the horizontal components of the forces in the diagonals ad and bc, which act with the lever arm h.

If the diagonals can carry tension only, diagonal ad will have zero bar force, so that only one unknown appears in the foregoing equation, and the horizontal component of force in bar bc is obtained directly. If, however, the diagonals can carry compression as well as tension, it may be assumed that the horizontal components of force in the two diagonals are numerically equal but opposite in sign. This reduces the number of independent unknowns in the foregoing equation to one, and the horizontal components of the diagonal bar forces can thus be determined.

If the horizontal bars at the intermediate panel points are omitted, as shown in Fig. 7.7b, the tower truss may be considered as being composed of two component trusses as shown in Fig. 7.7c and d and each truss can be analyzed by statically determinate procedures for the loads assigned to it. Resultant forces in the actual truss can then be determined by superimposing the bar forces from the two component trusses.

7.8 Stresses in Building Frames Due to Vertical Loads. A building frame consists primarily of girders, which carry vertical loads to columns, and of the columns themselves. While such a frame might be built like that shown in Fig. 7.8a, which is statically determinate, it would have little resistance against horizontal forces, such as wind loads, which it must also carry. It is therefore actually built as shown in Fig. 7.8b, in which the girders are rigidly connected to the columns so that all the members can carry bending moment, shear, and axial force. Such a frame is called a rigid frame; it is also referred to as a building bent. Because of the rigid construction, a building frame is highly indeterminate. The degree to which it is indeterminate can be investigated by an examination of Fig. 7.8c. Suppose that each girder is cut near mid-span, as shown. The resulting structure will be statically determinate, since each column, together with its girder stubs, acts as a cantilever. To arrive at this condition, however, it is necessary to remove the bending moment, shear, and axial force in each girder where it is cut. If n is the number of girders in the bent, it is necessary to remove $3n$ redundants to make the bent statically determinate; hence the bent is indeterminate to the $3n$th degree.

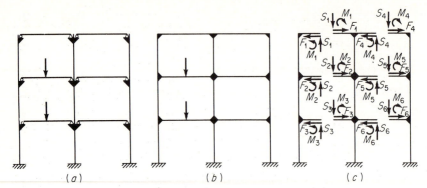

Figure 7.8 Building bent

The bent of Fig. 7.8b is therefore statically indeterminate to the 18th degree. A bent with 100 stories and 8 stacks of columns would include 700 girders and be statically indeterminate to the 2,100th degree. It is therefore highly advantageous to have approximate methods available for the analysis of such structures.

Since a building frame of the type just considered is statically indeterminate to a degree equal to three times the number of girders, it will be necessary to make three stress assumptions for each girder in the bent if an analysis is to be carried out on the basis of statics only. In Fig. 7.9a, if a girder is subjected to a load of w lb per ft, extending over the entire span, both of the joints A and B will rotate as shown in Fig. 7.9b, since while they are partly restrained against rotation, the restraint is not complete. Had the supports at A and B been completely fixed against rotation, as shown in Fig. 7.9c, it can easily be shown from a consideration of bending moments in a fixed-end beam that the points of inflection would be located at a distance of $0.21L$ from each end. If the supports at A and B were hinged, as shown in Fig. 7.9d, the points of zero moment would be at the end of the beam. For the actual case of partial fixity, the points of inflection may be assumed to lie somewhere between the two extremes of $0.21L$ and $0.00L$ from the ends of the beam. If they are assumed to be located at one-tenth of the span length from each end joint, a reasonable approximation has been made.

Solutions of building bents based on elastic action show that under vertical loads the axial force in the girders is usually very small.

The following three assumptions will therefore be made for *each girder*, in analyzing a building bent acted upon by vertical loads:

1. The axial force in the girder is zero.

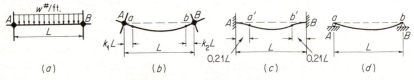

Figure 7.9 Girder for building bent

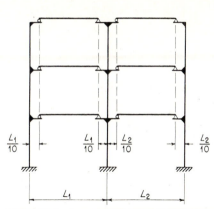

Figure 7.10   Assumptions regarding girders

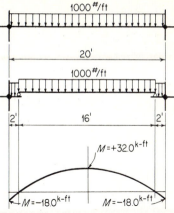

Figure 7.11   Girder bending moments

2. A point of inflection occurs at the one-tenth point measured along the span from the left support.

3. A point of inflection occurs at the one-tenth point measured along the span from the right support.

This is equivalent to assuming that the bent acts structurally like the statically determinate bent of Fig. 7.10. Girders can then be analyzed by statics, as will be illustrated by considering Fig. 7.11. The maximum positive moment occurs at span center and is given by

$$M = (+\tfrac{1}{8})(1.0)(16)^2 = +32.0 \text{ kip-ft}$$

The maximum negative moment occurs at either end of the span and is given by

$$M = (-8.0)(2) - (1.0)(2)(1) = -18.0 \text{ kip-ft}$$

The maximum shear occurs at each end of the span and is given by

$$S = (8.0)(1) + (2.0)(1) = 10 \text{ kips}$$

Since the end shears acting on the girders are equal to the vertical forces applied to the columns by the girders, the axial forces in the columns are easily found by summing up the girder shears from the top of a column down to the column section under consideration.

To produce maximum compression in a column, bays on both sides of the column are loaded. For interior columns, moments applied at a given floor by the two girders oppose each other; hence the column moments are small and are often neglected in design. For exterior columns, however, girders apply moments to only one side of the columns; hence the column moments are larger and must be considered in design. In computing column moments, however, assumption 1 is invalid. Girder moments at a given floor level should be divided between the columns above and below in proportion to their stiffnesses.

7.9   Stresses in Building Frames Due to Lateral Loads—General. In Art. 7.8, it was pointed out that approximate methods of analyzing bents are of importance because such structures are highly indeterminate. It was shown that the degree of statical indetermination for a bent like that of Fig. 7.12a equals three times the number of girders in the bent. The number of assumptions that must be made to permit an analysis by statics depends on the structure itself so that for this bent one must make three times as many assumptions as there are girders, regardless of the

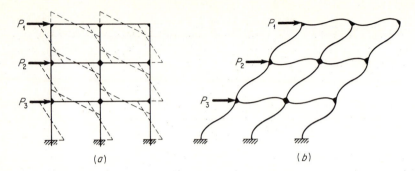

Figure 7.12 Building frame subjected to lateral loading

type of loading considered.   The assumptions made in analyzing building bents acted upon by vertical loads will not, however, be suitable for lateral-load analysis, for the structural action of bents is entirely different when lateral loads are considered.   This can be seen by a consideration of Fig. 7.12b, which illustrates, to an exaggerated scale, the shape that a building frame takes under the action of lateral loads.   It will be noted that while points of inflection occur in the members, they do not occur in the same manner as under the action of vertical loads.   Actually, when a building bent is acted upon by lateral loads, there will be, as shown in Fig. 7.12b, a point of inflection near the center of each girder and each column. The assumption that points of inflection occur at the mid-points of all members is therefore a reasonable one and is often among those made to carry out by statics an approximate analysis of building bents under lateral loads.   The moment curves for the structure of Fig. 7.12 are of the type shown by the dotted lines of Fig. 7.12a.

In this treatment, three approximate methods for analyzing building frames acted upon by lateral loads will be given.   These methods are (1) the portal method, (2) the cantilever method, and (3) the factor method.

In order that the relative accuracies of these methods can be considered, all three methods are applied to the same bent.   This bent and its loading are shown in Fig. 7.13.   It is possible to make a so-called "exact" solution for this bent and loading, using, for example, the slope-deflection method of analyzing statically indeterminate rigid frames, as treated in Chap. 9.   Such a solution is of interest for the comparison of the results of the approximate methods and has led to the values of end moments in foot-pounds in the girders and columns of the bent of Fig. 7.13 that are shown by the numbers at the ends of the members.

For the slope-deflection solution and also for the factor-method solution, it is necessary to know the relative stiffness of each member.   Relative stiffness is denoted by K and for a given member is obtained by dividing its cross-sectional moment of inertia by its length.   Values of K are shown for each member of the bent of Fig. 7.13.

7.10   The Portal Method.   In the portal method, the following assumptions are made:

1. There is a point of inflection at the center of each girder.

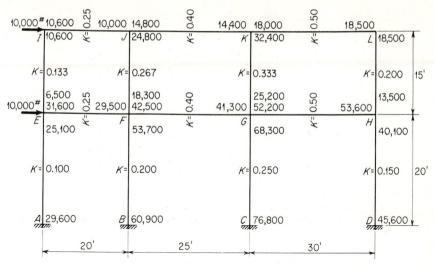

Figure 7.13   "Exact" solution for end moments caused by lateral loading

2. There is a point of inflection at the center of each column.

3. The total horizontal shear on each story is divided between the columns of that story so that each interior column carries twice as much shear as each exterior column.

This last assumption is arrived at by considering each story to be made up of a series of portals, as shown in Fig. 7.14. Thus, while an exterior column corresponds to a single portal leg, an interior column corresponds to two portal legs, so that it becomes reasonable to assume interior columns to carry twice the shear of exterior columns.

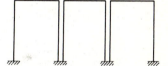

Figure 7.14   Series of portals equivalent to building frame

If there are m columns in a story, assumption 3 is equivalent to making $m - 1$ assumptions per story, regarding column-shear relations.

With reference to the bent of Fig. 7.13, application of the portal methods results in making the following number of assumptions:

Inflection points in girders	$(2)(3) =$	6
Inflection points in columns	$(4)(2) =$	8
Column shear relations	$(2)(3) =$	6
Total		20

Since there are six girders in this bent, the structure is indeterminate to the eighteenth degree. Hence, the portal method makes more assumptions than necessary. However, it so happens that the additional assumptions are consistent with the necessary assumptions, and no inconsistency of stresses, as computed by statics, results.

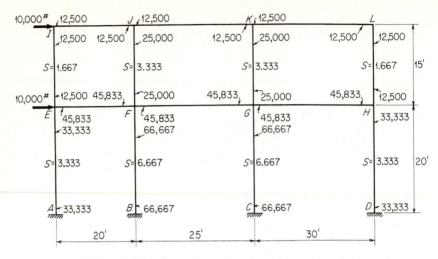

Figure 7.15 Solution for end moments by portal method

To illustrate the application of the portal method, it will now be applied to the bent of Fig. 7.13. The following discussion refers to Fig. 7.15, where the results of the portal-method analysis are given.

Column Shears. In accordance with assumption 3, let $x =$ shear in each exterior column of a given story; then $2x =$ shear in each interior column of the same story. For the first story,

$$x + 2x + 2x + x = 6x = 10{,}000 + 10{,}000 = 20{,}000$$
$$x = 3{,}333 \qquad 2x = 6{,}667$$

For the second story, $6x = 10{,}000$; $x = 1{,}667$; $2x = 3{,}333$.

Column Moments. In accordance with assumption 2, the moment at the center of each column is zero. Hence, each end moment for a given column equals the shear on that column multiplied by half the length of that column. For example, M_{AE}, the moment at the A end of column AE, equals $(3{,}333)(10) = 33{,}333$ ft-lb; $M_{FJ} = (3{,}333)(7.5) = 25{,}000$ ft-lb; etc.

Girder Moments. Reference to Fig. 7.12b, which shows the type of deformation occurring in a building frame acted upon by lateral forces, indicates that girder and column moments act in opposite directions on a joint. This fact is further clarified in Fig. 7.16a, from which the following equation may be written: $M_{c1} + M_{c2} = M_{g1} + M_{g2}$. Hence we conclude that for any joint the sum of the column end moments equals the sum of the girder end moments. This relation can

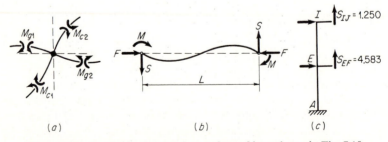

| | | |
| (a) | (b) | (c) |

Figure 7.16 Free-body sketches for portions of bent shown in Fig. 7.15

be used to determine girder end moments, since the column end moments have already been evaluated. At joint E, for example,

$$M_{EF} = 33,333 + 12,500 = 45,833 \text{ ft-lb}$$

Since by assumption 1 there is a point of inflection at the center of girder EF, M_{FE} also equals 45,833 ft-lb. Equating girder moments to column moments at joint F gives $M_{FG} + 45,833 = 66,667 + 25,000$, whence M_{FG} also equals 45,833 ft-lb. Continuing across the girders of the first floor in this manner, we find that all the end moments in the girders of the first floor equal 45,833 ft-lb. Girder end moments in the roof can be determined in a similar manner; each will be found to equal 12,500 ft-lb.

Girder Shears. In Fig. 7.16b, if $\Sigma M = 0$ is written for the forces acting on a girder, taking moments about one end of the girder gives $SL = 2M$, whence $S = 2M/L$. Hence the shear in girder EF is given by

$$S_{EF} = \frac{(2)(45,833)}{20} = 4,583 \text{ lb} \qquad S_{IJ} = \frac{(2)(12,500)}{20} = 1,250 \text{ lb} \qquad \text{etc.}$$

Column Axial Forces. In Fig. 7.16c, the axial forces in the columns can be obtained by summing up, from the top of the column, the shears applied to the column by the girders. Thus

$$F_{EI} = +1,250 \text{ lb} \qquad F_{AE} = +1,250 + 4,583 = +5,833 \text{ lb} \qquad \text{etc.}$$

Girder axial forces, while not usually important in design, can be obtained in a similar manner, by summing up, from one end of the girder, the shears applied to the girder by the columns; one would, of course, include the effects of the lateral loads themselves in such a summation.

7.11 The Cantilever Method. In the cantilever method, the following assumptions are made:

1. There is a point of inflection at the center of each girder.
2. There is a point of inflection at the center of each column.
3. The intensity of axial stress in each column of a story is proportional to the horizontal distance of that column from the center of gravity of all the columns of the story under consideration.

This last assumption is arrived at by considering that the column axial stress intensities can be obtained by a method analogous to that used for determining the distribution of normal stress intensities on a transverse section of a cantilever beam.

If there are m columns in a story, assumption 3 is equivalent to making $m - 1$ assumptions regarding column axial-stress relations, for each story. Hence, as with the portal method, the cantilever method makes more assumptions than are necessary, but again the additional assumptions prove to be consistent with the necessary assumptions.

To illustrate the application of the cantilever method, it will now be applied to the building frame of Fig. 7.13. The following discussion refers to Fig. 7.17b, where the results of the cantilever-method analysis are given.

Column Axial Forces. Assuming that all columns have the same cross-sectional area, the center of gravity of the columns in each story is found by the following equation:

$$x = \frac{20 + 45 + 75}{4} = 35.0 \text{ ft from } AEI$$

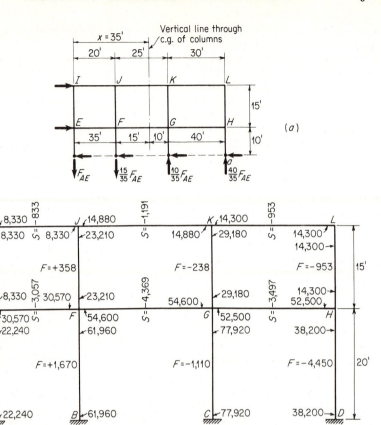

Figure 7.17 Solution for end moments by cantilever method

For the first story, refer to Fig. 7.17*a*. If the axial force in *AE* is denoted by $+F_{AE}$, then, by assumption 3, the axial forces in *BF*, *CG*, and *DH* will be $+\tfrac{15}{35} F_{AE}$, $-\tfrac{10}{35} F_{AE}$, and $-\tfrac{40}{35} F_{AE}$, respectively. Taking moments about *a*, the point of inflection in column *DH*, of all the forces acting on that part of the bent lying above the horizontal plane passing through the points of inflection of the columns of the first story gives

$$(+10{,}000)(25) + (10{,}000)(10) - (F_{AE})(75) - (\tfrac{15}{35} F_{AE})(55) + (\tfrac{10}{35} F_{AE})(30) = 0$$

whence $F_{AE} = +3{,}890$; $F_{BF} = (+\tfrac{15}{35})(+3{,}890) = +1{,}670$; etc.

For the second story, the column axial forces are in the same ratio to each other as they are in the first story. They would be evaluated in a similar manner, taking moments about the point of inflection in *HL* of all the forces acting on that portion of the bent lying above a horizontal plane passing through the points of inflection of the columns of the second story. Column axial forces are shown at the center of each column of Fig. 7.17*b*.

Girder Shears.   The girder shears can be obtained from the column axial forces at the various joints. For example, at joint *E*,

$$S_{EF} = +833 - 3{,}890 = -3{,}057$$

at joint F, $S_{FG} = -3,057 + 358 - 1,670 = -4,369$; etc. Girder shears are shown at the center of each girder of Fig. 7.17b.

Girder Moments. Since the moment at the center of each girder is zero, the moment at each end of a given girder equals the shear in that girder multiplied by half of the length of that girder. For example,

$$M_{EF} = (3,057)(10) = 30,570 \text{ ft-lb} \qquad M_{KJ} = (1,191)(12.5) = 14,880 \text{ ft-lb}$$

etc.

Column Moments. Column moments are determined by beginning at the top of each column stack and working progressively toward its base, as shown in the following illustration. At joint J, the column moment equals the sum of the girder moments, whence

$$M_{JF} = 8,330 + 14,880 = 23,210 \text{ ft-lb}$$

Since there is a point of inflection at the center of FJ, M_{FJ} also equals 23,210 ft-lb. At joint F, $M_{FB} + 23,210 = 30,570 + 54,600$, whence $M_{FB} = 61,960$ ft-lb. M_{BF} also equals 61,960 ft-lb, since a point of inflection is assumed midway between B and F.

7.12 The Factor Method.[1] The factor method of analyzing building frames acted upon by lateral loads is more accurate than either the portal or the cantilever method. Whereas the portal and cantilever methods depend on certain stress assumptions that make possible a stress analysis based on the equations of statics, the factor method depends on certain assumptions regarding the elastic action of the structure which make possible an approximate slope-deflection analysis of the bent. Although it is based upon the slope-deflection method of analysis, it is possible to formulate a relatively simple set of rules by which the method can be applied without knowledge of the elastic principles upon which it is based.

Before applying the factor method, it is necessary to compute the value of $K = I/L$ for each girder and each column. It is not necessary to use absolute values of K, since the stresses depend upon the relative stiffness of the members of the bent. It is, however, necessary for the K values for the various members to be in the correct ratio to each other.

The factor method is applied by carrying out the following six steps:

1. For each joint, compute the girder factor g by the following relation: $g = \Sigma K_c/\Sigma K$, where ΣK_c denotes the sum of the K values for the columns meeting at that joint and ΣK denotes the sum of the K values for all the members of that joint. Write each value of g thus obtained at the near end of each girder meeting at the joint where it is computed.

2. For each joint compute the column factor c by the following relation: $c = 1 - g$, where g is the girder factor for that joint as computed in step 1. Write each value of c thus obtained at the near end of each column meeting at the joint where it is computed. For the fixed column bases of the first story, take $c = 1$.

3. From steps 1 and 2, there is a number at each end of each member of the bent. To each of these numbers, add half of the number at the other end of the member.

4. Multiply each sum obtained from step 3 by the K value for the member in which the sum occurs. For columns, call this product the column moment factor C; for girders, call this product the girder moment factor G.

---

[1] J. B. Wilbur, A New Method for Analyzing Stresses Due to Lateral Loads on Building Frames, *Boston Soc. Civ. Eng.*, vol. 21, no. 1, January 1934.

5. The column moment factors C from step 4 are actually the approximate relative values for column end moments for the story in which they occur. The sum of the column end moments in a given story can be shown by statics to equal the total horizontal shear on that story multiplied by the story height. Hence, the column moment factors C can be converted into column end moments, by direct proportion, for each story.

6. The girder moment factors G from step 4 are actually approximate relative values for girder end moments for each joint. The sum of the girder end moments at each joint is equal, by statics, to the sum of the column end moments at that joint, which can be obtained from step 5. Hence, the girder moment factors G can be converted into girder end moments, by direct proportion, for each joint.

The factor method will now be illustrated by applying it to the bent of Fig. 7.13. The following discussion refers to Fig. 7.18, where computations for the factor method and the results obtained by it are shown. Values of K are written on the various members as part of the given data. For each story, on the right side of the figure, values of H, the total horizontal shear on the story, and Hh, the product of H and the story height h, are first worked out. Illustrative details of the solution follow:

Step 1: Computation of Girder Factors:

For joint E:
$$g_E = \frac{0.133 + 0.100}{0.133 + 0.100 + 0.250} = 0.482$$

This number is written at the left end of girder EF.

For joint F:
$$g_F = \frac{0.267 + 0.200}{0.267 + 0.200 + 0.250 + 0.400} = 0.418$$

This number is written at the left end of girder FG and at the right end of girder EF.

For joint I:
$$g_I = \frac{0.133}{0.133 + 0.250} = 0.347$$

This number is written at the left end of girder IJ.

Girder factors for all other joints are computed in a similar manner and written at the near end of each girder meeting at the joint where the girder factor is computed.

Step 2: Computation of Column Factors:

For joint E: $\qquad\qquad c_E = 1 - g_E = 1.000 - 0.482 = 0.518$

This number is written at the top of column AE and at the bottom of column EI.

For joint J: $\qquad\qquad c_J = 1.000 - 0.291 = 0.709$

This number is written at the top of column FJ.

For joint A: $\qquad\qquad c_A = 1.000$

since this is a fixed column base of the first story. This number is written at the bottom of column AE.

Column factors for all other joints are computed by similar procedures and written at the near end of each column meeting at the joint where the column factor is computed.

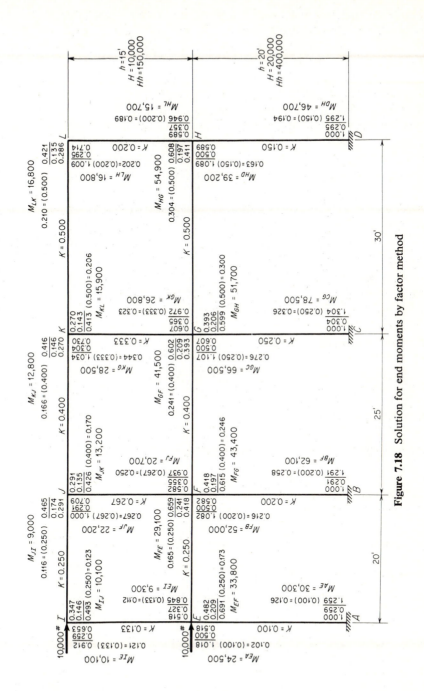

Figure 7.18 Solution for end moments by factor method

209

Step 3: Increasing the Number at Each End of Each Member by Half of the Number at the Other End of That Member:

For joint A: Member AE: $1.000 + (0.5)(0.518) = 1.259$
For joint E: Member EI: $0.518 + (0.5)(0.653) = 0.845$
                        Member EF: $0.482 + (0.5)(0.418) = 0.691$
                        Member EA: $0.518 + (0.5)(1.000) = 1.018$

Similar computations for all joints are made directly on Fig. 7.18.

Step 4: Computation of Column Moment Factors and Girder Moment Factors:

For joint A: Member AE: $C_{AE} = (1.259)(0.100) = 0.126$
For joint E: Member EI: $C_{EI} = (0.845)(0.133) = 0.112$
                        Member EF: $G_{EF} = (0.691)(0.250) = 0.173$
                        Member EA: $C_{EA} = (1.018)(0.100) = 0.102$

Similar computations for all joints are made directly on Fig. 7.18.

Step 5: Determination of Column Moments: Since the column moment factors are relative values of column end moments for each story of the bent, this is another way of saying that

$$M_{AE} = \mathscr{A}C_{AE} \qquad M_{EA} = \mathscr{A}C_{EA} \qquad M_{BF} = \mathscr{A}C_{BF} \qquad \text{etc.}$$

where M_{AE}, M_{EA}, M_{BF}, etc., are the actual moments at the ends of the columns and $\mathscr{A}$ has the same value for all the columns of a given story. The sum of the end moments can therefore be expressed

$$\Sigma \text{ column end moments} = \mathscr{A}(C_{AE} + C_{EA} + C_{BF} + C_{FB} + C_{CG} + C_{GC} + C_{DH} + C_{HD})$$
$$= \mathscr{A}\Sigma C \text{ for story} \qquad (a)$$

Consider the static equilibrium of all the forces acting on all the columns of a given story. Refer to Fig. 7.19, and take moments about the base of the right-hand column at point a:

$$(S_1 + S_2 + S_3 + S_4)h = M_1 + M_2 + M_3 + M_4 + M_5 + M_6 + M_7 + M_8$$

The sum $S_1 + S_2 + S_3 + S_4$ equals H, the total horizontal shear on the story. The sum $M_1 + M_2 + \cdots + M_8$ equals the sum of the column end moments for the story. Hence

$$\Sigma \text{ column end moments} = Hh \qquad (b)$$

From Eqs. (a) and (b),

$$\mathscr{A} = \frac{Hh}{\Sigma C \text{ for story}} \qquad (c)$$

For each story, $\mathscr{A}$ can be determined by Eq. (c). The end moment for each column of that story may then be obtained by multiplying the respective column moment factor by $\mathscr{A}$.

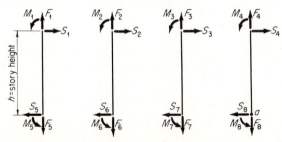

Figure 7.19 Isolated columns of a given story

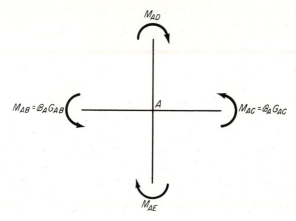

Figure 7.20   Isolated joint

This procedure is illustrated by its application to the first story of the bent of Fig. 7.18:

$$\mathscr{A}_1 = \frac{400,000}{0.126 + 0.102 + 0.258 + 0.216 + 0.326 + 0.276 + 0.194 + 0.163} = 241,000$$

$$M_{AE} = (0.126)(241,000) = 30,300 \text{ ft-lb}$$
$$M_{EA} = (0.102)(241,000) = 24,500 \text{ ft-lb}$$
$$M_{BF} = (0.258)(241,000) = 62,100 \text{ ft-lb}$$

Moments in the other column ends for the first story are similarly obtained by using $\mathscr{A}_1 =$ 241,000. Moments in the column ends of the second story are obtained from $\mathscr{A}_2$, which is computed by applying Eq. (c) to the second story, leading to $\mathscr{A}_2 = 83,000$.

Step 6: *Determination of Girder Moments*: Since the girder moment factors are relative values of girder end moments for a given joint, this is another way of saying, with reference to Fig. 7.20, that $M_{AB} = \mathscr{B}_A G_{AB}$ and $M_{AC} = \mathscr{B}_A G_{AC}$, where $\mathscr{B}_A$ has the same value in each of the foregoing relations. Moreover, since at any joint the sum of the girder moments equals the sum of the column moments, $\mathscr{B}_A$ can be evaluated from the relation

$$\mathscr{B}_A G_{AB} + \mathscr{B}_A G_{AC} = M_{AE} + M_{AD}$$

whence, at any joint N,

$$\mathscr{B}_N = \frac{\text{sum of column moments at joint } N}{\text{sum of girder moment factors at joint } N} \qquad (d)$$

For each joint, $\mathscr{B}_N$ can be evaluated by Eq. (d). The end moment for each girder at that joint can then be obtained by multiplying the respective girder moment factor by $\mathscr{B}_N$. This procedure is illustrated by its application to joint F of the bent of Fig. 7.18.

$$\mathscr{B}_F = \frac{52,000 + 20,700}{0.165 + 0.246} = 176,500$$

$$M_{FE} = (0.165)(176,500) = 29,100 \text{ ft-lb}$$
$$M_{FG} = (0.246)(176,500) = 43,400 \text{ ft-lb}$$

It will be noted that the application of this procedure for step 6 to exterior joints of a bent results in the discovery that girder end moments at these joints equal the sum of the column end moments, as they should by statics. Thus girder end moments at exterior columns can be obtained directly from column end moments by statics, and the computation of girder moment factors at these joints is not necessary.

The shears and axial forces in the columns and girders can be computed by the equations of statics, once the end moments are known.

7.13   Problems for Solution

*Problem 7.1   Compute the maximum bar force of each character in the following bars of the truss of Fig. 7.2: (a) L_3L_4; (b) $L_3'L_4$, due to a uniform live load of 500 lb per ft.   (Diagonals can carry compression.)

Problem 7.2   The portal of Fig. 7.3a is acted upon by a uniformly distributed wind load of 200 lb per ft along the entire length of the left column.   If $h = 40$ ft and $b = 30$ ft, construct the moment diagrams for all the members of the portal.

Problem 7.3   Solve Prob. 7.2 using the portal of Fig. 7.4a.

Problem 7.4   Solve Prob. 7.2 using the portal of Fig. 7.5a with dimensions as shown on the figure.

*Problem 7.5   For the mill bent of Fig. 7.6,

(a) Draw the shear and moment diagrams for the left supporting column.

(b) What are the forces applied to the roof truss by the columns and knee braces?

Problem 7.6   A building bent has three equal bays of 20 ft each and three stories of 12 ft each. The columns of the first story are fixed at their bases.   For each girder the dead load is 500 lb per ft, and the live load is 300 lb per ft.   Determine (a) the maximum positive girder moment, (b) the maximum negative girder moment, (c) the maximum girder shear, (d) the maximum exterior column compression, (e) the maximum interior column compression, (f) the maximum exterior column moment, and (g) the maximum interior column moment occurring in the bent.

*Problem 7.7   The building bent of Prob. 7.6 is acted upon by a horizontal force of 5,000 lb applied at each girder elevation on the left exterior columns.   Determine the bending moment at each end of each member by the portal method.

*Problem 7.8   Solve Prob. 7.7 by the cantilever method.

*Problem 7.9   Solve Prob. 7.7 by the factor method if the cross-sectional moment of inertia of each girder is three times as great as the cross-sectional moment of inertia of each column.

Problem 7.10   Use the cantilever method to determine the shears and end moments in the girders of the building frame shown in Fig. 7.21.   Areas of columns are shown at the top of the figure.

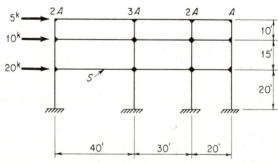

Figure 7.21   Problem 7.10

8

Deflections of Structures

8.1 Introduction. Engineering structures are constructed from materials that deform slightly when subjected to stress or a change in temperature.   As a result of this deformation, points on the structure undergo certain movements called **deflections.**   Provided that the elastic limit of the material is not exceeded, this deformation and the resulting deflection disappear when the stress is removed and the temperature returns to its original value.   This type of deformation or deflection is called **elastic** and may be caused either by loads acting on the structure or by a change in temperature.

Sometimes, the deflection of the structure is the result of settlement of the supports, play in pin joints, shrinkage of concrete, or some other such cause.   In such cases, the cause of the deflection remains in action permanently, and therefore the resulting deflections never disappear.   This type of deflection may be called **nonelastic** to distinguish it from the elastic type mentioned above.   For either type, it should be noted that deformation and deflection may occur with or without stress in the structure.   This will be discussed in detail later.

The structural engineer often finds it necessary to compute deflections.   For example, in the erection of cantilever or continuous bridges or in designing the lifting devices for swing bridges, it is imperative to compute the deflection of various points on the structure.   In certain cases, computations must be made to see that the deflection of a structure does not exceed certain specified limits.   For example, the deflection of the floor of a building must be limited to minimize the cracking of plaster ceilings, and the deflection of a shaft must be limited to ensure the proper functioning of its bearings.   Deflections must also be computed in order to analyze the vibration and dynamic response characteristics of a structure.   Perhaps the most important reason for the structural engineer's interest in deflection computations is that the stress analysis of statically indeterminate structures is based largely on an evaluation of their deflection under load.

Numerous methods have been presented in the literature for computing deflections. Of the various methods the following are considered the most fundamental and useful and will therefore be discussed in this chapter:

1. Methods to compute one particular deflection component at a time
   a. Method of virtual work (applicable to any type of structure)
   b. Castigliano's second theorem (applicable to any type of structure)
2. Methods to compute several deflection components simultaneously
   a. Williot-Mohr method (applicable to trusses only)
   b. Bar-chain method (applicable to trusses only)
   c. Moment-area method (elastic-load method or conjugate-beam method) (applicable to beams and frames)

8.2 Nature of the Deflection Problem. The computation of the deflection of a structure is essentially a problem in geometry or trigonometry. Of course, it is first necessary to define the deformation of the particles or elements of the structure, but once this is done the deflections can be computed by using geometrical or trigonometrical principles.

This is particularly evident in the case of a simple truss, which is usually composed of triangles. The configuration of these triangles can be determined if the lengths of their three sides are known. Thus, if the lengths of the members before and after deformation are known, the position of the joints before and after can be calculated by trigonometry. From the difference of the two positions of any joint, its deflection can then readily be determined. This procedure, though simple in theory, is a laborious one and therefore not suitable for practical application.

In the case of a truss, it is also possible to solve the deflection problem graphically simply by superimposing the layouts of the deformed and undeformed truss. This procedure is obvious and simple in theory, but in order to achieve any accuracy such a large scale would have to be used that it would be physically impossible for a draftsman to make the drawing.

The so-called **method of rotation** is another means of computing truss deflections that is simple in theory but impractical in application. This method gives us some useful ideas concerning the kinematics of truss deflection, however. In applying this method, the deflection of any joint of a simple truss due to a change in length of any one member can be determined by investigating the resulting rotation of one portion of the truss with respect to the other, the latter being assumed as fixed in position. By considering the effect of each member separately and summing up the results, the total deflection of any joint due to the change in length of all members can be determined.

To illustrate this procedure, consider first the effect of a change in length of the upper chord $U_2 U_3$ of the truss of Fig. 8.1. Since all deformations are small, it is permissible to assume that the rotations of members are so small that for a rotation α

$$\alpha = \sin \alpha = \tan \alpha$$

It is further permissible to consider that the arc along which a point actually travels as a body is rotated through a small angle coincides with its tangent for all practical purposes. To consider the effect of the change in length of $U_2 U_3$ on the deflections, first remove the pin at joint U_3 and then allow this change in length, ΔL, to take place. If the left-hand portion of the truss is considered to be held fixed in position, it then is necessary to rotate member $U_2 U_3$ about U_2 and the

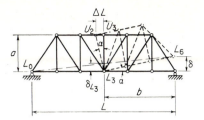

Figure 8.1 Truss deflection caused by change in length of a chord bar

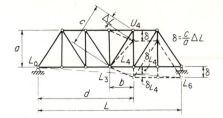

Figure 8.2 Truss deflection caused by change in length of a diagonal bar

dotted portion of the truss about L_3 until the points U_3 coincide again. During these rotations, point U_3 on member $U_2 U_3$ may be considered to move vertically, and on the dotted portion horizontally. In such a case, the intersection of these two paths will fall along the original position of $U_2 U_3$, and the final position of U_3 is located as shown. Then

$$\delta = \alpha b = \frac{\Delta L}{a} b = \frac{b}{a} \Delta L$$

Now if the supports do not settle, joint L_6 must not actually move vertically; so the entire truss must be rotated clockwise about L_0 until L_6 is back on its support. The dashed line joining L_0 and L_6 will therefore correspond to the line of zero deflection, and the downward vertical deflection of joint L_3 will be, by proportion,

$$\delta_{L_3} = \frac{L-b}{L} \delta = \frac{L-b}{L} \frac{b}{a} \Delta L$$

In a similar manner, the deflection caused by the change in length of a typical diagonal can be evaluated by proceeding as indicated in Fig. 8.2. These considerations demonstrate the impracticality of this method, but the ideas involved are directly applicable to the Williot-Mohr method, which is discussed in Art. 8.12.

While the various methods discussed above are impractical, it is important to recognize their existence, for it gives us confidence to know that the deflection problem can be solved using simple everyday ideas. Further, it is now obvious that some refinement must be introduced in theory to reduce the labor in the practical solution of such problems.

8.3 Bernoulli's Principle of Virtual Work for Rigid Bodies. Perhaps the most general, direct, and foolproof method for computing the deflections of structures is the **method of virtual work.** This method is based on an application of an alternate form of the **principle of virtual displacements,** which was originally formulated by John Bernoulli in 1717. This alternate form of these ideas will be referred to in this book as **Bernoulli's principle of virtual work for rigid bodies.** This alternate principle can be developed from the following considerations.

Consider a truly rigid body which is in static equilibrium under a system of forces Q. In this sense, a rigid body is intended to mean an undeformable body in which there can be no relative movement of any of its particles. Suppose first that, as shown in Fig. 8.3, this rigid body is translated without rotation a small amount by some other cause which is separate from, and independent of, the Q-force system. Upon selecting an origin o and two coordinate reference axes x and y, this translation may be defined by δ_o, the actual translation of the origin o, or by

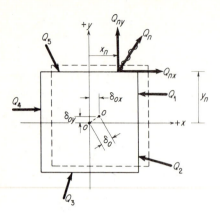

Figure 8.3 Translation of rigid body

the two components δ_{ox} and δ_{oy} in the x and y directions, both assumed to be plus when in the sense shown. Since the body is rigid, every point on the body will be translated through exactly the same distance as point o.

All the Q forces can be resolved into x and y components, designated as Q_{nx} and Q_{ny} for any particular force Q_n and assumed to be plus when in the same sense as the plus sense of the corresponding coordinates. Since these Q forces are in static equilibrium, the following equations are satisfied by the components of these forces:

$$\sum Q_{nx} = 0 \qquad \sum Q_{ny} = 0$$
$$\sum (Q_{nx} y_n - Q_{ny} x_n) = 0 \qquad (a)$$

Consider now the work $\mathscr{W}_Q$ done by only these Q forces as they "ride along" when the rigid body is translated a small amount δ_o by some *other* cause. Since this translation is small, all the Q forces may be assumed to maintain the same position and direction relative to the body and to each other and hence to remain in equilibrium during the translation. Then, we can write

$$\mathscr{W}_Q = \sum (Q_{nx} \delta_{ox} + Q_{ny} \delta_{oy}) = \delta_{ox} \sum Q_{nx} + \delta_{oy} \sum Q_{ny} = 0$$

and, therefore, in view of Eqs. (a), the total work done by the Q forces in such a case is equal to zero.

In a similar manner, we may consider the work done by the Q forces during a small rotation α_o of the rigid body about point o. During a small angular rotation, a point may be assumed to move along the normal to the radius drawn from the center of rotation to that point, i.e., along the tangent rather than the arc. Hence, the x and y components of the displacement of any point n can be computed to be as shown in Fig. 8.4. Since the angular rotation is small, the Q forces again may be assumed to remain in equilibrium. Then, we can write

$$\mathscr{W}_Q = \sum (Q_{nx} \alpha_o y_n - Q_{ny} \alpha_o x_n) = \alpha_o \sum (Q_{nx} y_n - Q_{ny} x_n) = 0$$

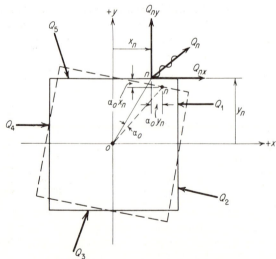

Figure 8.4 Rotation of rigid body

and, therefore, in view of Eqs. (*a*), the total work done by the *Q* forces during the rotation of a rigid body is also equal to zero.

After a little thought, it is evident that *any* small displacement of a rigid body can be broken down into a translation of a given point on the body plus a rotation of the rigid body about that point. Since, in either translation or rotation, the work done by the *Q* system (which is a system in equilibrium) has been shown to equal zero, the following principle is obviously true in the general case where a rigid body may be given *any* type of small displacement:

Bernoulli's principle of virtual work for rigid bodies:

If a system of virtual forces Q acting on a rigid body is in equilibrium and remains in equilibrium as the body is given any small displacement, the virtual work done by the Q-force system is equal to zero.

In this statement, the term **virtual** has been used to describe the *Q*-force system to emphasize that this system is *separate from*, and *independent of*, the action producing the displacement. The work done by the virtual *Q*-force system as it rides along during an imposed displacement is called **virtual work** in order to distinguish that effort from the **real work** done by the forces and other actions producing the displacement. Note that the real work is of no consequence when one applies Bernoulli's principle since this principle deals with a relationship involving *only* the *virtual work* done by the *virtual-force system*.

8.4   Principle of Virtual Work for Deformable Bodies. Bernoulli's principle of virtual work for rigid bodies can now be used to develop the basis for the method of virtual work for computing the real deflections of structures. This method is applicable to any type of structure—beam, truss, or frame, planar or space frameworks. For simplicity, however, consider any planar structure such as that shown in Fig. 8.5. Suppose that this structure is in static equilibrium under the external loads and reactions of the virtual *Q*-force system shown.

Since the body as a whole is in equilibrium, any particular particle such as the crosshatched one may be isolated and will also be in equilibrium under the internal

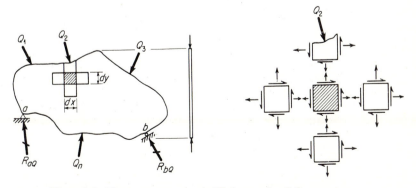

Figure 8.5   Planar structure in equilibrium under *Q*-force system

virtual Q stresses developed by the external virtual Q forces. If this particle and its adjoining particles are isolated, it will be acted upon by internal Q stresses on any internal boundaries with adjacent particles but by external Q forces on any external boundaries. On the adjacent internal boundaries of any two adjoining particles, the internal stresses will be numerically equal but act in opposite directions.

Now suppose that the body is subjected to a small change in shape caused by some source *other* than the virtual Q-force system. Owing to this change in shape, any particle such as the crosshatched one might be deformed as well as translated and rotated as a rigid particle. Hence, the boundaries of such a particle maybe displaced, and therefore the virtual Q stresses acting on such boundaries would move and hence do virtual work. Let the virtual work done by the Q stresses on the boundaries of the differential particle be designated by $d\mathcal{W}_s$. Part of this virtual work will be done because of the movements of the boundaries of the particle caused by the deformation of the particle itself; this part will be called $d\mathcal{W}_d$. The remaining part of $d\mathcal{W}_s$ will be the virtual work done by the Q stresses during the remaining part of the displacement of the boundaries and will be equal to $d\mathcal{W}_s - d\mathcal{W}_d$. However, this remaining displacement is caused by the translation and rotation of the particle as a *rigid body*, and according to the Bernoulli's principle of virtual work for rigid bodies, the virtual work done in such a case is equal to zero. Hence,

$$d\mathcal{W}_s - d\mathcal{W}_d = 0 \qquad \text{or} \qquad d\mathcal{W}_s = d\mathcal{W}_d$$

If the virtual work done by the Q stresses on all particles of the body is now added up, this equation becomes

$$\mathcal{W}_s = \mathcal{W}_d \tag{8.1}$$

To evaluate first $\mathcal{W}_s$, we recognize that this term represents the total virtual work done by the virtual Q stresses and forces on all the boundaries of all particles. However, for every *internal* boundary of a particle, there is an adjoining particle whose adjacent boundary is actually the same line on the body as a whole, and therefore these adjacent boundaries are displaced exactly the same amount. Since the forces acting on the two adjacent internal boundaries are numerically equal but opposite in direction, the total virtual work done on the pair of adjoining internal boundaries is zero. Hence, since all the internal boundaries occur in pairs of adjoining boundaries, there is no net virtual work done by the forces on all the *internal* boundaries. $\mathcal{W}_s$, therefore, consists only of the work done by the *external* Q forces on the *external* boundaries of the particles. Equation (8.1) may therefore be interpreted in the following manner:

Principle of virtual work for deformable bodies:

If a deformable body is in equilibrium under a virtual Q-force system and remains in equilibrium while it is subjected to a small and compatible deformation, then the external virtual work done by the external Q forces is equal to the internal virtual work of deformation done by the internal Q stresses.

This principle of virtual work for deformable bodies is the specific basis of the method of virtual work. Before such computations can be made, suitable expressions must be developed so that the external virtual work and internal virtual work of deformation may be evaluated. In addition, certain tricks must be used in selecting a suitable Q system so that the desired deflection components can be computed. All this will be explained in the following articles.

It is important to emphasize the assumptions and limitations of this development in order to appreciate the flexibility and generality of the method of virtual work.

1. The only requirement of the external Q forces and the internal Q stresses is that they shall form a system of forces which are in equilibrium and remain in equilibrium throughout the deformation produced by an action that is *separate from*, and *independent of*, the virtual Q-force system. This requirement will not be satisfied if the deformation has varied the geometry of the structure appreciably.

2. The virtual-work relations that have been derived are independent of the cause or type of deformation; they are true whether the deformation is due to loads, temperature, errors in lengths of members, or other causes or whether the material follows Hooke's law or not. Further, the principle that has been derived involves only the virtual work done by the virtual Q-force system; it involves in no way the real work done by the active-force system or effects producing the displacements of the structure.

3. In addition to being *small* so that the geometry of the structure is not altered appreciably, the deformations must be **compatible**; i.e., the elements of the structure must deform so as to fit together after deformation and to satisfy the conditions of restraint at the supports.

8.5 Fundamental Expressions Used in Method of Virtual Work. It is easy to evaluate $\mathscr{W}_s$, the external virtual work done by a system of virtual Q forces acting on any structure. Let δ denote the displacement of the point of application of a Q force during the actual deformation imposed on the structure. This displacement δ is to be measured along the same direction as the line of action of its corresponding Q force. The virtual work done by the force Q_1 will then be Q_1 times δ_1, and therefore the total external virtual work done by all the Q forces, including both loads and reactions will be

$$\mathscr{W}_s = Q_1\delta_1 + Q_2\delta_2 + Q_3\delta_3 + \cdots$$

This expression can be represented as

$$\mathscr{W}_s = \sum Q\delta \tag{8.2}$$

which simply means that the products of Q times δ must be evaluated for each Q force and summed up for all the Q forces, including both loads and reactions. Note that assuming the work to be positive in this formula implies that δ should be considered plus when in the same sense as its corresponding Q force.

It is also relatively easy to evaluate $\mathscr{W}_d$, the internal virtual work of deformation. In this instance, however, an appropriate expression for $\mathscr{W}_d$ must be developed

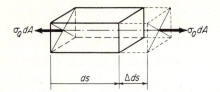

Figure 8.6 Axial change of length

for each different type of deformation, i.e., for axial change in length, shear deformation, bending deformation, etc. Consider first the simple case of a differential element having an undeformed length of ds and a cross-sectional area of dA, as shown in Fig. 8.6. Assume that the Q stress intensities σ_Q are uniformly distributed over the cross section dA so that the axial force is $\sigma_Q\,dA$. If the actual imposed deformation of this element is simply a uniform axial strain e, the axial change in length Δds will be equal to $e\,ds$. The internal virtual work of deformation $d\mathcal{W}_d$ done by the Q stresses in this case will simply be

$$d\mathcal{W}_d = \sigma_Q\,dA\,\Delta ds = \sigma_Q\,dA\,e\,ds = \sigma_Q\,e\,dA\,ds \tag{8.3}$$

This expression can now be used to evaluate the internal virtual work of deformation for a beam or a member of a truss or frame.

Consider the case of a member deformed by a two-dimensional system of P loads or by a change in temperature. Suppose that the centroidal axis of the member is straight and that all the cross sections of the member have axes of symmetry lying in the plane of the P loads. The resultant of all the internal P stresses, then, will also lie in this same plane. The axial force F_P at any cross section will produce an axial strain that is uniform for all the fibers at this section. Suppose that the strain produced by the change in temperature is also uniform at this cross section. Let e_o denote the uniform axial strain at any particular cross section produced by these two effects. The P-load system may also produce a shear and bending moment at this cross section. Because of these two effects, a longitudinal element such as that of Fig. 8.7 will be subjected to shear and normal stresses as shown. More detailed analysis reveals that unless a member is very deep in comparison with its length, the effect of the shear stress is small in comparison with the deflection caused by the elongation and contraction of these elements under the longitudinal normal stresses. In this book, therefore, the effect of shearing stresses will be neglected.

Neglecting the effect of shear deformation, it is then easy to evaluate the virtual work of deformation done simply as a result of the elongation and contraction of

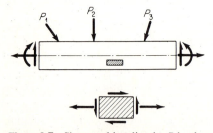

Figure 8.7 Shear and bending by P loads

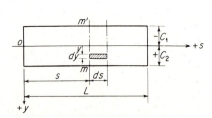

Figure 8.8 Coordinate axes

the longitudinal elements. Suppose that the virtual Q-force system lies in the same plane as the P forces of Fig. 8.7 and exerts some axial force, shear, and bending moment on the cross sections of the member. We now want to develop an expression for the internal virtual work of deformation done by the resulting virtual Q stresses as the member is subjected to an imposed deformation of the type described in the previous paragraph. We select the orthogonal axes at o, s, y, and z so that the s axis coincides with the centroidal axis of the member and the z axis is normal to the plane of the paper. If we now consider a longitudinal element located at the position (s,y), in this member, as shown in Fig. 8.8, it is apparent that under the conditions described this element is precisely similar to the one shown in Fig. 8.6. The virtual work of deformation for a member can therefore be evaluated by integrating the expression for dW_d given in Eq. (8.3) for all the differential elements ds of the entire member.

Assuming that the normal stresses can be found by elementary beam theory, let

M_P, M_Q = bending moment on section mm' due to P loads (which cause deformation) or to Q loads, respectively

F_P, F_Q = axial force on section mm' due to P loads or to Q loads, respectively

σ_P = normal stress at point (s,y) due to P loads = $F_P/A + M_P y/I$

σ_Q = normal stress at point (s,y) due to Q loads = $F_Q/A + M_Q y/I$

e, e_o = axial strain of a longitudinal element at a point (s,y) due to P loads and temperature change t or at a point on the centroidal axis, respectively

I = moment of inertia of cross section mm' about oz axis

A = cross-sectional area of cross section mm'

E = modulus of elasticity of material

α_t = coefficient of thermal expansion of the material

b = width of cross section mm' at fiber y

Considering now the longitudinal element of the member at point (s,y), the length of which is ds, the width b, and the depth dy, we have

$$e = \alpha_t t + \frac{1}{E}\left(\frac{F_P}{A} + \frac{M_P y}{I}\right) = e_o + \frac{M_P y}{EI} \qquad \text{where } e_o = \alpha_t t + \frac{F_P}{EA}$$

The total virtual work of deformation for the entire member is therefore

$$W_d = \int_0^L \int_{-C_1}^{+C_2} \left(\frac{F_Q}{A} + \frac{M_Q y}{I}\right)\left(e_o + \frac{M_P y}{EI}\right)(b \, dy) \, ds$$

With

$$\int_{-C_1}^{C_2} b \, dy = A \qquad \int_{-C_1}^{C_2} yb \, dy = 0 \qquad \int_{-C_1}^{C_2} y^2 b \, dy = I$$

this expression simplifies to

$$\mathscr{W}_d = \int_0^L F_Q e_o \, ds + \int_0^L \frac{M_Q M_P}{EI} \, ds \tag{8.4}\dagger$$

This expression can now be applied and further simplified for the common types of members encountered in planar structures, such as beams and members of trusses and frames.

8.6 Deflection of Trusses, Using Method of Virtual Work. An expression for the principle of virtual work as applied specifically to trusses may be obtained by substituting in Eq. (8.1) from Eqs. (8.2) and (8.4). Consider first the case of an ideal pin-jointed truss where both the deforming P loads and the virtual Q loads are applied only at the joints of the truss. In such a case, the individual members will be subjected only to axial bar forces F_P with no shear or bending moment involved, and the second term in Eq. (8.4) will disappear. Furthermore, the bar force F_Q will be constant throughout the length L of a given member, and since

$$\int_0^L e_o \, ds = \text{axial change in length of member} = \Delta L$$

the virtual work of deformation for one particular truss member becomes

$$\mathscr{W}_d = F_Q \int_0^L e_o \, ds = F_Q \, \Delta L$$

When such products for all the members of the truss are summed, the internal virtual work of deformation for the entire truss may be represented as

$$\mathscr{W}_d = \sum F_Q \, \Delta L$$

and therefore the principle of virtual work as applied to an ideal pin-jointed truss becomes

virtual force system

$$\sum Q\delta = \sum F_Q \, \Delta L \tag{8.5}$$

deformations caused by actual force system

Suitable expressions for ΔL can easily be developed, depending on whether the imposed change in length is produced by the P loads, by a change in temperature,

---

† In a similar fashion, the following third and fourth terms could be added to Eq. (8.4), giving the contributions of the deformations due to bending shears S_P and twisting moments T_P to the total

$$C_s \int_0^L \frac{S_Q S_P}{AG} \, ds + \int_0^L \frac{T_Q T_P}{K_T G} \, ds$$

where S_Q, T_Q = shear and twisting moment on section mm' due to Q loads
      C_s = shape factor varying with shape of cross section
      K_T = torsional constant for cross section (equals polar moment of inertia for circular cross section)
      G = shear modulus of material

in addition to the notation introduced previously.

or by some other cause.   For a prismatic member having a constant cross-sectional area A and a constant modulus of elasticity E, *if the deformation is due to joint loads P on the truss,*

$$\Delta L = \frac{F_P L}{AE} \tag{8.5a}$$

if the deformation is due to a uniform change in temperature t,

$$\Delta L = \alpha_t t L \tag{8.5b}$$

and if the deformation is caused by both these effects acting simultaneously,

$$\Delta L = \frac{F_P L}{AE} + \alpha_t t L \tag{8.5c}$$

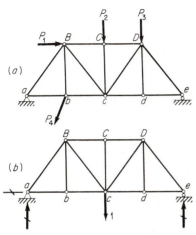

Figure 8.9   Application of method of virtual work to pin-jointed truss

Equation (8.5) is the basis for the method of virtual work for computing the deflection of ideal pin-jointed trusses. Suppose, for example, that we wish to compute the vertical component of the deflection of joint c caused by the P loads shown in Fig. 8.9a.   Suppose that we select as the virtual Q-load system a unit vertical load at joint c together with its reactions.   If we imagine that we first apply this Q system to the structure, then when we apply the actual deforming loads P, the Q loads will be given a ride and will do a certain amount of external virtual work.   According to the principle of virtual work, the internal Q stresses will do an equal amount of internal virtual work as the members change length owing to the F_P stresses. Applying Eq. (8.5) gives

$$(1)(\delta_c^1) + \mathscr{W}_R = \sum F_Q \frac{F_P L}{AE}$$

where $\mathscr{W}_R$ represents the virtual work done by the Q reactions if the support points move and could be evaluated numerically if such movements were known. If the supports are unyielding, $\mathscr{W}_R = 0$ and

$$(1)(\delta_c^1) = \sum F_Q F_P \frac{L}{AE}$$

The bar forces F_Q and F_P due to the Q- and P-load systems, respectively, can easily be computed.   These data combined with the given values of L, A, and E give us enough information to evaluate the right-hand side of the above equation and therefore to solve for the unknown value of δ_c.

Figure 8.10 shows how to select suitable Q systems for use in the computation of other deflection components that may be required.   Note that the trick is simply

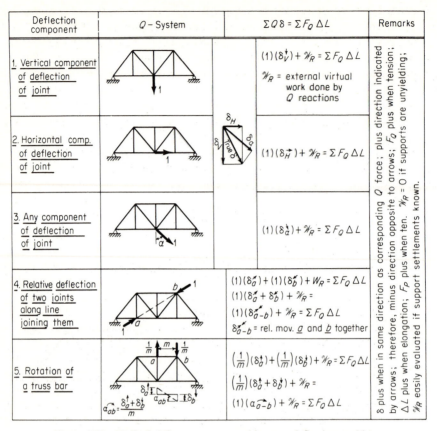

Figure 8.10 Typical Q-force systems used in truss-deflection problems

to select the virtual Q-force system in such a way that the desired deflection is the only unknown δ appearing on the left-hand side of the equation. Some students worry about the deflection produced by the Q system. We do not care what this is. We want to find the deflection produced by the given cause of deformation. The virtual Q system is purely a system that rides along and does virtual work, thereby enabling us to compute the desired deflection. Note also that in these virtual-work computations we never have to worry about or compute the *real work* done by the P loads as they deflect the structure.

The following illustrative examples show how to organize these computations in certain typical problems. In using the principle of virtual work, it is particularly important to note the sign convention involved. In setting up the formulas for external and internal virtual work done by the Q system, it was assumed that the work was positive. This implies first that δ is to be considered positive when it is in the same direction as its corresponding Q force. Furthermore, this implies that both F_Q and ΔL are to be considered positive when in the same sense. If F_Q is considered plus when tension, then ΔL is plus when an elongation and therefore F_P is plus when tension and t plus when an increase in temperature.

After studying the following examples, it will be evident that there are two principal sources of difficulties—units and signs. The novice will probably have less difficulty with units if force units are assigned to the Q forces and stresses, although some authorities treat these forces as dimensionless. Ordinarily it is advisable to use the same length units throughout a problem; however, in some cases it is desirable to mix such units in order to obtain more convenient numbers. For instance, in these examples A and E are used in inch units, while L is used in feet. What is done in this respect is largely a matter of personal preference, but everyone should follow one rule: Be sure the units are consistent. As for signs, no difficulty should be encountered if one is careful and follows the convention noted above. Be sure to check the signs of all products, however. Note also that F_Q and F_P are *bar forces*, not the horizontal or vertical components.

Example 8.1   (a) *Compute the vertical component of deflection of joint c due to the 100^{kip} load shown. $E = 30 \times 10^3$ kips per sq in.*

(b) *Compute the vertical component of deflection of joint c, due to a decrease of temperature of $50°F$ in the bottom chord only. $\alpha_t = 1/150,000$ per $°F$.*

(a) $$\sum Q\delta = \sum F_Q \, \Delta L = \sum F_Q \, F_P \frac{L}{AE}$$

$$(1^k)(\delta_c^{\downarrow}) = \frac{1}{E} \sum F_Q \, F_P \frac{L}{A}$$

$$= \frac{+325.01^{k^2 \prime/in^2}}{(30 \times 10^3)^{k/in^2}}$$

$$\therefore \delta_c = +0.01083 \, ft \qquad \therefore \, down$$

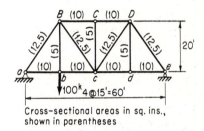

Cross-sectional areas in sq. ins., shown in parentheses

(b) $$\sum Q\delta = \sum F_Q \, \Delta L = \sum F_Q \, \alpha_t t L$$

$$(1^k)(\delta_c^{\downarrow}) = \alpha_t \sum F_Q t L$$

$$= \left(\frac{1}{150,000} \, per \, °F \right) (-1,125^{k°F\prime})$$

$$\therefore \delta_c = -0.0075 \, ft \qquad \therefore \, up$$

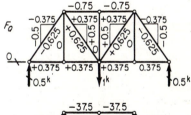

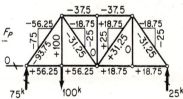

Bar	L	A	$\dfrac{L}{A}$	F_Q	F_P	$F_Q F_P \dfrac{L}{A}$	t	$F_Q t L$
Units	$'$	$''^2$	$'/''^2$	k	k	$k^2{}'/''^2$	$°F$	$k°F'$
ab	15	10	1.5	+0.375	+ 56.25	+ 31.64	−50	− 281.25
bc	15	10	1.5	+0.375	+ 56.25	+ 31.64	−50	− 281.25
cd	15	10	1.5	+0.375	+ 18.75	+ 10.55	−50	− 281.25
de	15	10	1.5	+0.375	+ 18.75	+ 10.55	−50	− 281.25
BC	15	10	1.5	−0.75	− 37.5	+ 42.19	0	0
CD	15	10	1.5	−0.75	− 37.5	+ 42.19	0	0
aB	25	12.5	2	−0.625	− 93.75	+117.19	0	0
Bc	25	12.5	2	+0.625	− 31.25	− 39.06	0	0
cD	25	12.5	2	+0.625	+ 31.25	+ 39.06	0	0
De	25	12.5	2	−0.625	− 31.25	+ 39.06	0	0
bB	20	5	4	0	+100	0	0	0
cC	20	5	4	0	0	0	0	0
dD	20	5	4	0	0	0	0	0
Σ						+325.01		−1,125

Example 8.2   *Compute the horizontal component of the deflection of joint E due to the load shown.* $E = 30 \times 10^3$ *kips per sq in.*

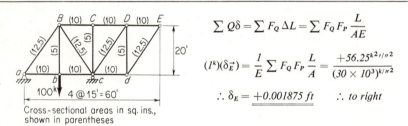

$$\sum Q\delta = \sum F_Q \,\Delta L = \sum F_Q F_P \frac{L}{AE}$$

$$(1^k)(\delta_{\vec{E}}) = \frac{1}{E} \sum F_Q F_P \frac{L}{A} = \frac{+56.25^{k^2{}'/''^2}}{(30 \times 10^3)^{k/''^2}}$$

$$\therefore \delta_E = +0.001875 \; ft \qquad \therefore \; to \; right$$

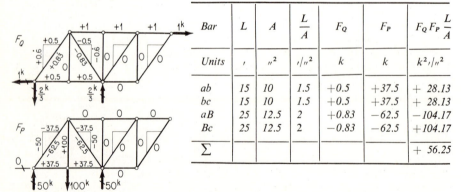

Bar	L	A	$\dfrac{L}{A}$	F_Q	F_P	$F_Q F_P \dfrac{L}{A}$
Units	$'$	$''^2$	$'/''^2$	k	k	$k^2{}'/''^2$
ab	15	10	1.5	+0.5	+37.5	+ 28.13
bc	15	10	1.5	+0.5	+37.5	+ 28.13
aB	25	12.5	2	+0.83	−62.5	− 104.17
Bc	25	12.5	2	−0.83	−62.5	+104.17
Σ						+ 56.25

Discussion:

Note that any bar in which either F_Q or F_P is zero may be omitted from the tabulation since the product of $F_Q F_P(L/A)$ would be zero for such a bar.

Example 8.3 *For the truss in Example 8.2, compute the horizontal component of the deflection of joint E due to the following movements of the supports:*

$$At\ a,\ horizontal = 0.5''\ to\ left$$
$$At\ a,\ vertical = 0.75''\ down$$
$$At\ c,\ vertical = 0.25''\ down$$

Use the stress analysis for the Q system from the previous problem. In this example, the deflection is caused simply by support movements. There are no changes in length of the members, that is, $\Delta L = 0$ for all members.

$$\sum Q\delta = \sum F_Q\,\Delta L = 0$$
$$(1^k)(\delta\vec{E}) + (1^k)(0.5'') + (\tfrac{2}{3}^k)(0.75'') - (\tfrac{2}{3}^k)(0.25'') = 0$$
$$\therefore\ \delta_E = -0.5 - 0.5 + 0.167 = \underline{\underline{-0.833\ in.}}\qquad \therefore\ to\ left$$

Discussion:

In evaluating the external work done by the Q reactions, be careful to include the proper sign for the work. Any particular reaction does plus or minus virtual work depending on whether its point of application moves in the same or the opposite sense as the reaction, respectively.

Deflections due to support settlement can also be evaluated simply by a consideration of the kinematics of the problem. In sketch a, the solid line shows the outline of the truss in its original position. The truss may then be translated as a rigid body until the support at a is in its final position; the translated position is shown by the dashed line. The truss must then be rotated counterclockwise about a until the support at c is in its proper position. The final position of the truss is shown by the dotted line. In these sketches, the movements have been exaggerated tremendously to clarify the mechanics of the problem. The horizontal movement of point E can therefore be computed as

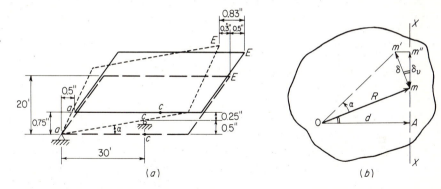

Due to translation	0.5″ to left
Due to rotation about a	$\dfrac{0.5''}{30'}\,(20') = 0.3''\ to\ left$

Total horizontal movement of point E $0.5'' + 0.3'' = 0.83''$ to left

The computation of the movement during rotation requires some explanation, and it likewise illustrates the application of a useful theorem. Consider the movement of a point m as a rigid body is rotated about some center O through some small angle α. In sketch b, this angle has been exaggerated tremendously so that the geometry is clearer. Actually, the angle α is so small that the angle (in radians), its sine, and its tangent are all essentially equal to each other. This means that it is legitimate to consider that m moves to its rotated position m' along the tangent shown rather than along the actual arc. Suppose that it is desired to obtain the component of the displacement *mm' along some given direction through m such as XX. Drop a perpendicular OA from point O to this direction XX. From the sketch it is apparent that triangles mm'm" and OmA are similar. Then*

$$\frac{\delta_v}{\delta} = \frac{d}{R} \qquad \delta_v = \frac{\delta}{R}d = \alpha d$$

since $\alpha \approx \delta/R$. Therefore, the following theorem may be stated:

If a rigid body is rotated about some center O through some small angle α, the component of the displacement of a point m along some direction XX through that point is equal to the angle α times the perpendicular distance from O to the line XX.

Applied to the above truss,

$$\alpha = \frac{0.5''}{30'}$$

Therefore, the horizontal movement of E during rotation about O is

$$\frac{0.5''}{30'}(20') = 0.3''$$

Example 8.4 *Compute the relative deflections of joints b and D along the line joining them due to the following causes:*

(a) The loads shown. $E = 30 \times 10^3$ kips per sq in.

(b) An increase in temperature of $80°F$ in the top chord; a decrease of $20°F$ in the bottom chord. $\alpha_t = 1/150,000$ per $°F$.

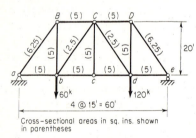

Cross-sectional areas in sq. ins. shown in parentheses

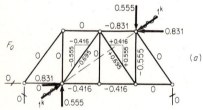

(a)

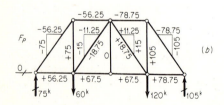

(b)

$(a) \quad \sum Q\delta = \sum F_Q \, \Delta L = \sum F_Q F_P \frac{L}{AE}$

$(1^k)(\delta_b^{\nearrow}) + (1^k)(\delta_D^{\nwarrow}) = \frac{1}{E} \sum F_Q F_P \frac{L}{A}$

$(1^k)(\delta_{b-D}^{\nwarrow}) = \frac{55.0^{k^2 \prime/\prime\prime^2}}{(30 \times 10^3)^{k/\prime\prime^2}}$

$\therefore \delta_{b-D}^{\nwarrow} = +0.00183 \, ft \qquad \therefore \, together$

$(b) \quad \sum Q\delta = \sum F_Q \, \Delta L = \sum F_Q \alpha_t \, tL = \alpha_t \sum F_Q \, tL$

$(1^k)(\delta_{b-D}^{\nwarrow}) = (6.6 \times 10^{-6})^{/°F}(-747.2^{k°F\prime})$

$\therefore \delta_{b-D}^{\nwarrow} = -0.00499 \, ft \qquad \therefore \, apart$

Bar	L	A	$\dfrac{L}{A}$	F_Q	F_P	$F_Q F_P \dfrac{L}{A}$	t	$F_Q tL$
Units	$'$	$''^2$	$'/''^2$	k	k	$k^2'/''^2$	$°F$	$k°F'$
bc	15	5	3	−0.416	+ 67.5	− 84.5	−20	+125
cd	15	5	3	−0.416	+ 67.5	− 84.5	−20	+125
CD	15	5	3	−0.831	− 78.75	+197.0	+80	−997.2
bC	25	2.5	10	−0.695	− 18.75	+130	0	0
Cd	25	2.5	10	+0.695	+ 18.75	+130	0	0
dD	20	5	4	−0.555	+105	−233	0	0
Σ						+ 55.0		−747.2

Discussion:

　Many students are confused about temperature or settlement deflection problems dealing with statically determinate trusses. They "feel" that stresses must be developed in the members in such cases. However, no reactions can be developed on a statically determinate truss unless there are loads acting on the structure. This can be proved by applying equations of statics. If there are no reactions or external loads, then there can be no internal bar forces. From a physical standpoint, the deformation of such trusses that is the result of support settlements or changes in length due to temperature may take place without encountering resistance, and therefore no reactions and bar forces can be developed.

　All the trusses discussed so far have been ideal pin-connected trusses acted upon by Q or P loads, which were always applied at the joints. If a pin-connected truss is deformed by P loads some of which are applied to certain members between the joints, such members are subjected to M_P bending moments. If, however, only joint deflections are desired, the virtual Q system consists simply of certain joint loads, which cause no M_Q bending moments. The second term on the right side of Eq. (8.4) therefore disappears, and the principle of virtual work for such a case is expressed the same as Eq. (8.5). The same thing is likewise true in the cases where in order to compute the deflection desired it is necessary to apply Q loads between the joints but the P loads causing deformation are applied only at the joints so that there are no M_P bending moments.

　The case of a pin-connected truss having both P and Q loads applied between the joints is handled like that of a rigid-frame structure, which is discussed in the next article. The case of riveted trusses is also discussed there.

　8.7 Deflection of Beams and Frames, Using Method of Virtual Work. An expression for the principle of virtual work as applied to beams or frames can likewise be obtained by substituting in Eq. (8.1) from Eqs. (8.2) and (8.4). Consider first the case of a beam deformed by transverse loads. In such a case, if the reactions have no axial components, the cross sections of the beam are subjected only to shear and bending moment with no axial force. The first term of Eq. (8.4) will disappear, and the principle of virtual work as applied to this case will be simply

$$\sum Q\delta = \int M_Q M_P \frac{ds}{EI} \tag{8.6}$$

Remember that the effect of shearing deformation has been neglected in the development of Eq. (8.4). Such a case is shown in Fig. 8.11. To find the vertical, horizontal, or, in fact, any component of the deflection of a point on such a beam, a unit load is applied in the appropriate direction. This unit load together with its reactions constitutes the Q system, which rides along during the deformation of the beam. The solution of a beam-deflection problem is basically similar to the solution of a truss-deflection problem but differs in the detail of evaluating the right-hand side of Eq. (8.6).

Before the integration of the right-hand side can be accomplished, both M_Q and M_P must first be expressed as functions of s. It is usually necessary to separate the integration for the entire beam into the sum of several integrals, one for each of several portions of the beam. The integration must be broken at points where there is a change in the functions representing M_Q, M_P, or I in terms of s. The integration process can often be simplified by selecting different origins for the measurement of s for these various portions of the beam. The technique of organizing such computations will be illustrated in the following examples.

Note particularly the sign convention used for the various terms in Eq. (8.6). Any suitable convention may be used for M_Q and M_P as long as the same convention is used for both. Usually, the ordinary beam convention is the most satisfactory. δ is, of course, plus when in the same sense as its corresponding Q force.

Often it is necessary to find the change in slope of some cross section of a beam. To do this, select as the virtual Q system a distributed load, such as that shown in Fig. 8.12, and its reactions. This load is distributed across the cross section so as to be equivalent to a unit couple. Let the intensity of this load at a distance y from the centroidal axis be q_y. Considering only the effect of bending deformation, a cross section that was plane before bending remains plane and normal to the elastic curve after bending. If a cross section is rotated through a small angle α, a point at distance y from the centroidal axis would move αy. Then the external virtual work done by the distributed q load during the rotation of the cross section caused by the P system would be

$$\int (q_y b\, dy)(\alpha y) = \alpha \int q_y by\, dy = (1)(\alpha)$$

since the moment of the q load about the centroidal axis $= \int q_y by\, dy = 1$. Having recognized that the external virtual work done by such a distributed q load is simply its resultant unit couple times α, we may henceforth consider that we apply

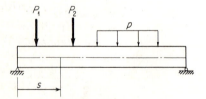

Figure 8.11   Deflection of beams

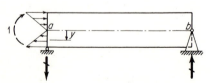

Figure 8.12 Q-system used to obtain change of slope of cross section

a unit couple to the cross section and not trouble about showing the distributed load in detail. Applying Eq. (8.6) in this case therefore yields

$$(1)(\alpha_a^{\curvearrowright}) + \mathcal{W}_R = \int M_Q M_P \frac{ds}{EI}$$

From this point on, the rest of the solution is similar to that for a vertical deflection.

Example 8.5 *Compute the vertical deflection of a, due to the load shown.*

$$E = 30 \times 10^3 \text{ kips per sq. in.} \qquad I = 200 \text{ in.}^4$$

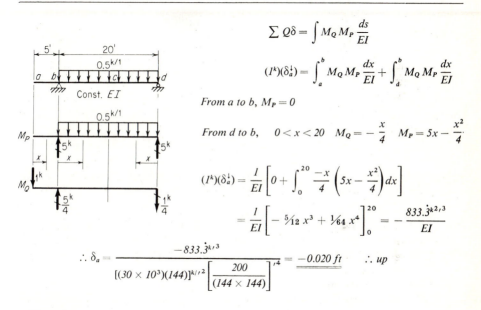

$$\sum Q\delta = \int M_Q M_P \frac{ds}{EI}$$

$$(1^k)(\delta_a^{\downarrow}) = \int_a^b M_Q M_P \frac{dx}{EI} + \int_d^b M_Q M_P \frac{dx}{EJ}$$

From a to b, $M_P = 0$

From d to b, $\quad 0 < x < 20 \quad M_Q = -\frac{x}{4} \quad M_P = 5x - \frac{x^2}{4}$

$$(1^k)(\delta_a^{\downarrow}) = \frac{1}{EI}\left[0 + \int_0^{20} \frac{-x}{4}\left(5x - \frac{x^2}{4}\right)dx\right]$$

$$= \frac{1}{EI}\left[-\tfrac{5}{12}x^3 + \tfrac{1}{64}x^4\right]_0^{20} = -\frac{833.\dot{3}^{k^2/3}}{EI}$$

$$\therefore \delta_a = \frac{-833.\dot{3}^{k/3}}{[(30 \times 10^3)(144)]^{k//2}\left[\dfrac{200}{(144 \times 144)}\right]^{/4}} = \underline{-0.020 \text{ ft}} \qquad \therefore \text{ up}$$

Discussion:

The origin for measuring x in any given portion may be selected at will, but note that the same origin must be used for x in the expressions for both M_Q and M_P for a given portion. An origin should be selected that minimizes the number of terms in the expressions for M_Q and M_P and also reduces the labor in substituting the limits of integration.

Example 8.6 *Compute the change in slope of the cross section at point a caused by the load shown.* $E = 30 \times 10^3$ kips per sq. in. $I_1 = 150$ in.^{4} $I_2 = 200$ in.^{4}

$$\sum Q\delta = \int M_Q M_P \frac{ds}{EI}$$

$$(1^{k\prime})(\alpha_a^{\curvearrowright}) = \int_a^b M_Q M_P \frac{dx}{EI}$$

$$+ \int_b^c + \int_e^d + \int_d^c$$

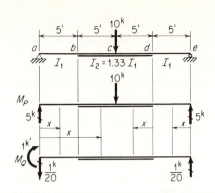

From a to b, $0 < x < 5$ I_1 $M_Q = 1 - \dfrac{x}{20}$ $M_P = 5x$

From b to c, $5 < x < 10$ $1.33 I_1$ $M_Q = 1 - \dfrac{x}{20}$ $M_P = 5x$

From e to d, $0 < x < 5$ I_1 $M_Q = \dfrac{x}{20}$ $M_P = 5x$

From d to c, $0 < x < 5$ $1.33 I_1$ $M_Q = \dfrac{1}{4} + \dfrac{x}{20}$ $M_P = 25 + 5x$

$$(1^{k\prime})(\alpha_a^{\curvearrowright}) = \frac{1}{EI_1}\left[\int_0^5 \left(1 - \frac{x}{20}\right)(5x)\,dx + \int_5^{10}\left(1 - \frac{x}{20}\right)(5x)\frac{dx}{1.33}\right.$$

$$\left. + \int_0^5 \left(\frac{x}{20}\right)(5x)\,dx + \int_0^5\left(\frac{1}{4} + \frac{x}{20}\right)(25 + 5x)\frac{dx}{1.33}\right] = \frac{1}{EI_1}\left\{\left[\frac{5x^2}{2} - \frac{x^3}{12}\right]_0^5\right.$$

$$+ \frac{1}{1.33}\left[\frac{5x^2}{2} - \frac{x^3}{12}\right]_5^{10} + \left[\frac{x^3}{12}\right]_0^5 + \frac{1}{1.33}\left[\frac{25x}{4} + \tfrac{5}{4}x^2 + \frac{x^3}{12}\right]_0^5\right\}$$

$$= \frac{1}{EI_1}\left\{(\tfrac{5}{2})(25) + (\tfrac{1}{1.33})\left[\tfrac{5}{2}(100 - 25) - \frac{1{,}000 - 125}{12}\right]\right.$$

$$\left. + (\tfrac{1}{1.33})\left[(\tfrac{25}{4})(5) + \tfrac{5}{4}(25) + \frac{125}{12}\right]\right\}$$

$$= \frac{1}{EI_1}[62.5 + (\tfrac{1}{1.33})(187.5 - 62.5 + 62.5)] = \frac{1}{EI}(62.5 + 140.6) = \frac{203.1^{\,k2,3}}{EI_1}$$

$$\therefore \alpha_a = \frac{203.1^{\,k\prime 2}}{[(30 \times 10^3)(144)]^{\,k\prime\prime 2}\left[\dfrac{150}{(144)(144)}\right]^{\prime 4}} = \underline{\underline{+0.0065\ radian}} \qquad \therefore clockwise$$

Discussion:

    The selection of the origins for measuring x is not necessarily the best in this solution but was intended to illustrate several of the possible ways of handling the problem. Whenever cancellations are made before integration or before substitution of limits, be very sure that the cancellation is legitimate; check to see that both the terms and the limits are the same.

Consider now the more general case of a beam or rigid frame where the cross sections of the members are subjected to an axial force as well as shear and bending moment. Several cases of this type are shown in Fig. 8.13. The deflection of such structures may also be caused by temperature changes as well as by the axial forces and bending moments developed by the P loads. As a result, both terms in Eq. (8.4) must be considered in evaluating the internal virtual work. The principle of virtual work as applied to such cases may be expressed:

$$\sum Q\delta = \int F_Q e_0 \, ds + \int M_Q M_P \frac{ds}{EI} \tag{8.7}$$

where the axial strain e_0 includes the effect of both axial force and temperature.

Usually the entire structure can be split up into several portions over each of which F_Q, F_P, A, and t are constant. Likewise, to evaluate the second term on the right side of Eq. (8.7), the integration can be split up into several parts, each of which covers a portion over which the functions representing M_Q, M_P, and I remain of the same form. The portions used in evaluating the first term in Eq. (8.7) are not necessarily the same portions as those used to evaluate the second term. In any event, Eq. (8.7) may be represented more conveniently in the following form:

$$\sum Q\delta = \sum F_Q \, \Delta L + \sum \int M_Q M_P \frac{ds}{EI} \tag{8.8}$$

where
$$\Delta L = \int e_0 \, ds = \alpha_t L + \frac{F_P L}{AE}$$

The summation signs on the right side of Eq. (8.8) indicate that such terms must be summed up for all the separate portions of all members of the structure.

The application of Eq. (8.8) to a specific problem involves the techniques already illustrated in Examples 8.1 to 8.6. Example 8.7 will show how to arrange the computations. Most beam- or frame-deflection problems involve finding some component of the deflection of a point or the change in slope of a cross section. Sometimes, however, it is necessary to find the relative deflections of two adjacent cross sections such as a and a' in Fig. 8.14. The relative horizontal, vertical, or angular displacements at points a and a' can be found by selecting the virtual Q system shown in sketch A, B, or C, respectively.

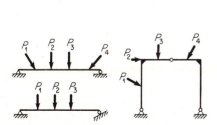

Figure 8.13 Structures subjected to both axial change in length and flexure

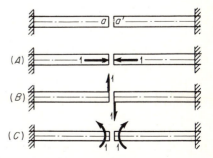

8.14 Q systems for obtaining relative deflections

Example 8.7 *Compute the change in slope of the cross section on the left side of the hinge at C due to the load shown.*

$$\sum Q\delta = \sum F_Q F_P \frac{L}{AE} + \sum \int M_Q M_P \frac{ds}{EI}$$

$$(1^{k\prime})(\alpha_{CL}^{\curvearrowright}) = \int_A^B M_Q M_P \frac{dy}{EI}$$

$$+ \int_B^C M_Q M_P \frac{dx}{EI} + \int_E^D + \int_D^C$$

$$+ \sum F_Q F_P \frac{L}{AE}$$

Note: $F_Q F_P L/AE$ can be evaluated for the same portions AB, BC, ED, and DC. Dashed line indicates lower fibers for applying beam convention to signs of M_Q and M_P.

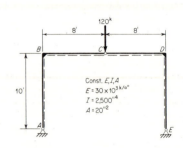

Const. E,I,A
$E = 30 \times 10^{3\,k/\sigma''}$
$I = 2,500^{''4}$
$A = 20^{''2}$

From A to B,

    $L = 10'$ $0 \rightarrow y \rightarrow 10$

    $F_Q = +\frac{1}{16}$ $M_Q = -\frac{y}{20}$

    $F_P = -60$ $M_P = -48y$

From B to C,

    $L = 8'$ $0 \rightarrow x \rightarrow 8$

    $F_Q = -\frac{1}{20}$ $M_Q = -\frac{1}{2} - \frac{x}{16}$

    $F_P = -48$ $M_P = -480 + 60x$

From E to D,

    $L = 10$ $0 \rightarrow y \rightarrow 10$

    $F_Q = -\frac{1}{16}$ $M_Q = -\frac{y}{20}$

    $F_P = -60$ $M_P = -48y$

From D to C,

    $L = 8$ $0 \rightarrow x \rightarrow 8$

    $F_Q = -\frac{1}{20}$ $M_Q = -\frac{1}{2} + \frac{x}{16}$

    $F_P = -48$ $M_P = -480 + 60x$

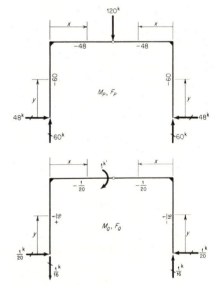

$$(1^{k\prime})(\alpha_{CL}^{\curvearrowright}) =$$

$$= \frac{1}{EI}\left[2\int_0^{10}\frac{-y}{20}(-48y)\,dy + \int_0^8\left(-\frac{1}{2}-\frac{x}{16}\right)(-480+60x)\,dx \right.$$

$$\left. + \int_0^8\left(-\frac{1}{2}+\frac{x}{16}\right)(-480+60x)\,dx \right] + \frac{2}{AE}(-\tfrac{1}{20})(-48)(8)$$

$$= \frac{1}{EI}\left[1.6y^3\Big|_0^{10} + 2(240x-15x^2)\Big|_0^8 \right] + \frac{38.4}{AE}$$

$$(1^{k\prime})(\alpha_{CL}^{\curvearrowright}) = \frac{3{,}520}{EI} + \frac{38.4}{AE} = \frac{3{,}520^{k2\prime3}}{[(30\times10^3)(144)]^{k\prime\prime2}\left[\dfrac{2{,}500}{144\times144}\right]^{\prime4}} + \frac{38.4^{k2\prime}}{[(30\times10^3)(144)]^{k\prime\prime2}\left(\dfrac{20}{144}\right)^{\prime2}}$$

$$\therefore\ \alpha_{CL} = +0.00676 + 0.000064 = \underline{\underline{+0.006824\ radian}} \qquad \therefore\ clockwise$$

Discussion:

In problems involving both bending and axial deformation, be particularly careful of units. Note that the contribution of the axial deformation term is only about 1 per cent of that due to bending deformation. This is more or less typical of the relative size of these two effects in frame-deflection problems. It is therefore usually permissible to neglect the effect of axial deformation in such cases.

A truss with riveted joints is essentially a rigid frame. In the discussion in Art. 4.3, it will be recalled that the members of such trusses are subjected to shear and bending moment as well as axial forces even when the loads are applied at the joints. However, as far as the stresses and strains in the members are concerned, a given riveted truss may be considered to be equivalent to a corresponding pin-connected truss, as shown in Fig. 8.15. This equivalent truss, however, is loaded not only with the given joint loads but also by couples on the ends of each bar, which are equal to the moments at the ends of the corresponding members of the riveted truss. More detailed analysis shows that, in most cases, these end couples by themselves produce very small bar forces in the members. In other words, the bar forces in the equivalent pin-connected truss are produced almost entirely by the joint loads. In the previous article, it is pointed out that the joint deflections of a pin-connected truss are a function only of the axial change in length of the members and do not depend on the bending deformation of such members. The deflection of the joints of the equivalent pin-connected truss under both the joint loads and the end couples is therefore essentially the same as that of an ideal pin-connected truss acted upon by only the joint loads. Hence, a riveted truss is assumed to be an ideal

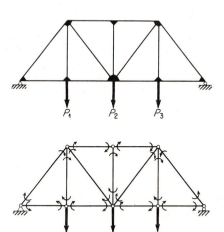

Figure 8.15   Deflections of truss with rigid joints

pin-connected truss when the deflection of the joints is computed. It should be realized, of course, that the bending of the members of riveted trusses does affect the deflection of points other than the joints.

At no place in the previous discussion has the case of a beam or a frame with a curved axis and varying cross sections been considered. The detailed consideration of such structures is beyond the scope of this book. If the curvature and variation of the cross sections are not great, the normal stresses in such structures may be assumed to be distributed linearly and the deflection of the structure can therefore be computed by applying Eq. (8.7). In such cases, the integrals on the right-hand side of this equation can seldom be evaluated exactly, and their values must be approximated by a summation process. For this purpose, the axis of the structure is divided into a number of short portions of equal length Δs. The values of F_Q, e_o, M_Q, M_P, and I are computed for the cross section at the center of each of these short portions. The products of $F_Q e_o \Delta s$ and $M_Q M_P \Delta s / EI$ can then be evaluated for each portion. The sum of these products for all the portions approximates the value of the right-hand side of Eq. (8.7). The accuracy of this summation process increases as the length of the individual portions is decreased.

8.8   Development of Moment-Area Theorems.   The moment-area theorems are often more convenient to use than the method of virtual work in the computation of slopes and deflections of beams and frames, particularly when the deformation is caused by concentrated rather than distributed loads. These theorems are based on a consideration of the geometry of the elastic curve of the beam and the relation between the rate of change of slope and the bending moment at a point on the elastic curve.

Referring to Fig. 8.16, consider a portion *ACB* of the elastic curve of a beam that was *initially straight* and *continuous* in the position $A_o B_o$ in its unstressed condition. Draw the tangents to the elastic curve at points *A* and *B*. The tangent at *A* intersects the vertical through *B* at *D*. The angle $\Delta \tau_{AB}$ is the change in slope between the tangents at points *A* and *B*. Recognize that the deflection and curvature have been exaggerated tremendously in this sketch. Actually, the inclination of any tangent to

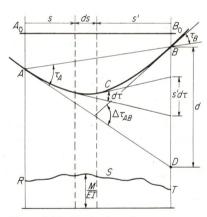

Figure 8.16   Derivation of moment-area theorems

the elastic curve is so small that an angle such as τ_A is approximately equal to its sine and its tangent and its cosine is approximately equal to unity.

Consider a differential element of this curve having a horizontal projection ds, and draw the tangents to the elastic curve at each end of this element. The change in slope between these tangents is the angle $d\tau$, and its value can be obtained by considering Fig. 8.17:

$$d\tau = \frac{(\sigma_c \, ds)/E}{c} = \frac{Mc}{EI} \frac{ds}{c} = \frac{M}{EI} ds$$

It is then evident that the total change in slope between the tangents at A and B is the sum of all the angles $d\tau$ for all the elements ds along the elastic curve ACB, or

$$\Delta\tau_{AB} = \int_A^B d\tau = \int_A^B \frac{M}{EI}\,ds \tag{8.9}$$

Let RST be the bending-moment diagram for the portion AB after it is modified by dividing every ordinate by the EI of the beam at that point. Such a diagram is

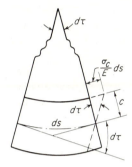

called the M/EI diagram. It is evident that the integral in Eq. (8.9) can be interpreted as the area under the M/EI diagram between A and B. Hence, from Eq. (8.9), we may state:

First moment-area theorem:

The change in slope of the tangents of the elastic curve between two points A and B is equal to the area under the M/EI diagram between these two points.†

In view of the fact that the deformation and slopes are actually small, it is evident from Fig. 8.16 that the intercept on the line BD (drawn normal to the unstrained position of the beam) between the tangents at the ends of the element ds can be written $s'\,d\tau$, and consequently

Figure 8.17 Differential change in slope

$$d = \int_A^B s'\,d\tau = \int_A^B \frac{M}{EI}\,s'\,ds \tag{8.10}$$

The last integral may be interpreted as the static moment, about an axis through point B, of the area under the M/EI diagram between points A and B. Therefore, from Eq. (8.10), we may state:

Second moment-area theorem:

The deflection of point B on the elastic curve from the tangent to this curve at point A is equal to the static moment about an axis through B of the area under the M/EI diagram between points A and B.†

Note that this deflection is measured in a direction normal to the original position of the beam.

The two theorems can be used directly to find the slopes and deflections of beams simply by drawing the moment diagram for the loads causing deformation and then computing the area and static moments of all or part of the corresponding M/EI diagram. This procedure is illustrated in Examples 8.8 and 8.9. It will

† Provided that within the portion AB of the beam no discontinuities (such as hinges) have been inserted.

become apparent from these examples that these computations can be facilitated by introducing some new ideas. The analogy based on these ideas, discussed in the next article, is called the **elastic-load method.**

These theorems can be extended without difficulty to members that were not initially straight. The consideration of such cases is beyond the scope of this book.

A sign convention could be formulated for the application of the moment-area theorems. This has not been done here for fear that it might be confused with the convention proposed in the next article for the elastic-load method. When applying the moment-area theorems, signs may be handled as shown in Examples 8.8 and 8.9. In these examples, the elastic curve is first sketched roughly showing the proper curvature as indicated by the signs of the bending-moment diagram. In the computation of a deflection (or angle) shown on this sketch, the contribution of a particular portion of the M/EI diagram (regardless of the sign of M in that portion) is considered as plus or minus, depending on whether it tends to increase or decrease, respectively, the magnitude of that deflection as shown on the assumed sketch. Thus, a positive final result confirms the sense shown for the deflection.

Example 8.8 *Using the moment-area theorems, compute the slope of the elastic curve at points a and m and the deflection at m.*

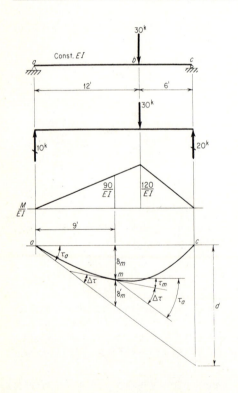

Since slopes are small, $\tan \tau_a = \tau_a$.

$$\therefore \tau_a = \frac{d}{18}$$

Applying second moment-area theorem gives

$$d = \frac{120}{EI}\left[(12\frac{1}{2})(6+4)+(6\frac{1}{2})(4)\right]$$

$$= \frac{8,640}{EI} \qquad \therefore \tau_a = \frac{480}{EI}$$

From sketch of elastic curve, $\tau_m = \tau_a - \Delta\tau$. *But by first moment-area theorem,*

$$\Delta\tau = \frac{90}{EI}\frac{9}{2} = \frac{405}{EI}$$

$$\therefore \tau_m = \frac{480-405}{EI} = \frac{75}{EI}$$

Also, from sketch, $\delta_m = 9\tau_a - \delta'_m$. *Applying second moment-area theorem gives*

$$\delta'_m = \frac{90}{EI}\frac{9}{2}(3) = \frac{1,215}{EI}$$

$$\therefore \delta_m = 9\frac{480}{EI} - \frac{1,215}{EI} = \frac{3,105}{EI}$$

Example 8.9 *Using the moment-area theorems, compute the deflections at points b and e and the slope at point c. Constant E and I.*

$$\tau_d = \frac{d_1}{20},$$

$$d_1 = \frac{1}{EI}\left\{(37.5)(20\tfrac{1}{2})(10)\right.$$

$$\left.-(25)(10\tfrac{1}{2})[10+(\tfrac{2}{3})(10)]\right\} = \frac{5,000}{3EI}$$

$$\therefore \tau_d = \frac{250}{3EI}$$

$$\delta_b = 10\tau_d + \delta_b',$$

$$\delta_b' = \frac{1}{EI}[(25)(10\tfrac{1}{2})(20\tfrac{2}{3})$$

$$-(37.5)(10\tfrac{1}{2})(10\tfrac{1}{3})] = \frac{625}{3EI}$$

$$\therefore \delta_b = \frac{2,500}{3EI} + \frac{625}{3EI} = \frac{3,125}{3EI} \text{ (down)}$$

$$\delta_e = 5\tau_d - \delta_e',$$

$$\delta_e' = \frac{1}{EI}[(25)(5\tfrac{1}{2})(10\tfrac{1}{3})] = \frac{625}{3EI}$$

$$\therefore \delta_e = \frac{1,250}{3EI} - \frac{625}{3EI} = \frac{625}{3EI} \text{ (up)}$$

$$\tau_c = \tau_d + \Delta\tau,$$

$$\Delta\tau = \frac{(25+12.5)(5)}{2EI} - \frac{(18.75)(5)}{2EI} = \frac{93.75}{2EI}$$

$$\therefore \tau_c = \frac{250}{3EI} + \frac{93.75}{2EI} = \frac{781.25}{6EI}$$

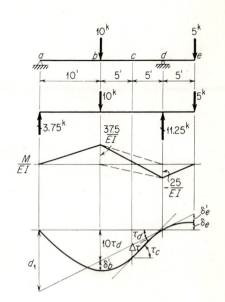

Discussion:

Note that it becomes more difficult to handle signs properly when a beam is bent by both plus and minus bending moment. To minimize these difficulties, it is advisable to sketch the elastic curve as accurately as possible. The curvature can be sketched correctly by simply following the bending-moment diagram: plus bending moment makes the elastic curve concave upward, and minus bending moment makes it concave downward. It will be found advantageous to handle these more difficult problems by the elastic-load method, for thereby the sign difficulty can be handled more or less automatically.

Note also that in a portion such as bd, where a straight-line portion of the moment diagram goes from plus to minus, it is often advantageous for computation purposes to replace the actual moment diagram by the dashed lines shown. In this case, the plus and minus triangles under the actual moment diagram are replaced by the plus triangle with an altitude of 37.5 and a base of 10 and the minus triangle with an altitude of 25 and also a base of 10. A little thought will verify that this procedure is legitimate for the computation of the net area or the net static moment of the area about any vertical axis.

8.9   Elastic-Load Method.   The ideas involved in the elastic-load method can be developed by considering the elastic curve ACB of a member AB which was originally straight and has been bent as shown in Fig. 8.18.   Let RST be the M/EI diagram.   Applying the second moment-area theorem gives:

$$d = \int_B^A s' \frac{M}{EI} ds$$

Then

$$\tau_A = \frac{d}{L} = \frac{1}{L} \int_A^B s' \frac{M}{EI} ds$$

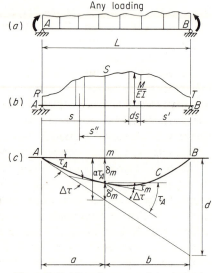

This equation simply states that τ_A is equal to the static moment about an axis through B of the area under the M/EI diagram between A and B divided by L. Note, however, that the form of this computation for τ_A seems very familiar to us. Suppose that we imagine that the M/EI diagram, RST, represents a distributed vertical load applied to a simple beam supported at points A and B, as shown by the dashed lines in Fig. 8.18b.   Then the computation for the vertical reaction at point A of this imaginary beam would yield a value exactly equal to the value of τ_A computed above.

This analogy can be carried still further if we consider the form of the computation for τ_m and δ_m.   Note that τ_m gives the slope of the tangent with reference to the direction of the chord AB of the elastic

Figure 8.18   Development of elastic-load method

curve and that δ_m gives the deflection of point m from the same chord AB. Consider first τ_m; then

$$\tau_m = \tau_A - \Delta\tau$$

But, according to the first moment-area theorem, $\Delta\tau = \int_A^m M/EI\ ds$, and hence

$$\tau_m = \tau_A - \int_A^m \frac{M}{EI} ds$$

Continuing to consider the M/EI diagram as the loading and τ_A as the reaction at point A on the imaginary beam, we see from this equation that τ_m is equal to the reaction τ_A minus the load applied between A and m.   In other words, τ_m may be interpreted as being the shear at point m of this imaginary beam.   Likewise,

$$\delta_m = \tau_A a - \delta'_m$$

But, from the second moment-area theorem, $\delta'_m = \int_A^m s''(M/EI)\,ds$, and hence

$$\delta_m = \tau_A a - \int_A^m s'' \frac{M}{EI}\,ds$$

To continue the analogy, from this equation it is apparent that δ_m may be interpreted as the bending moment at point m of the imaginary beam.

From these considerations, it can be concluded that the elastic curve of a beam AB is exactly the same as the bending-moment diagram for an imaginary simple end-supported beam of the same span AB which is loaded with a distributed transverse load equal to the M/EI diagram of actual beam AB. Further, the slope of the tangent to the elastic curve at any point is equal to the corresponding ordinate of the shear diagram for the imaginary beam AB loaded with the M/EI diagram. When used in this manner, the M/EI diagram is referred to as the **elastic load**, and therefore this procedure for computing the deflection and slope of a beam is called the **elastic-load method.** The shears and bending moments produced by the elastic load acting on the imaginary beam are often referred to as the **elastic shears** and the **elastic moments,** respectively.

By using this analogy, the problem of computing deflections and slopes of a beam is reduced to a procedure well known to a structural engineer. All he has to do is to compute the reactions, shear, and bending moment on an imaginary end-supported beam loaded with a distributed transverse load. The following examples illustrate the convenience of the elastic-load method.

The example used in developing these ideas has unyielding supports at points A and B. As a result, these points do not deflect, and therefore the chord of the elastic curve joining points A and B remains horizontal and coincident with the undeformed position of the axis of the beam. The slope τ_m and the deflection δ_m therefore give the true slope and deflection with reference to the original position of the beam. The elastic-load method may be applied, of course, to any portion AB of a beam whether the chord AB of the elastic curve remains fixed in position or not. Remember, however, that the shear and bending moment obtained by the elastic-load method give the slope and deflection *measured with reference to the chord AB*. If the chord has moved, the quantities so obtained do not give the true slope and deflection measured with reference to the original position of the beam until they are corrected for the effect of the chord movement.

In the general case where the chord may or may not move, the procedure of applying the elastic-load method may be summarized in the following statement:

The slopes and deflections of an elastic curve measured with reference to one of its chords AB are equal, respectively, to the (elastic) shears and (elastic) moments of an imaginary end-supported beam of span AB loaded with a distributed (elastic) load consisting of the M/EI diagram for that portion AB.†

---

† Provided that within the portion AB of the beam no discontinuities (such as hinges) have been inserted.

In order to take full advantage of the elastic-load method, *it is desirable to follow the same sign convention and principles as those used in drawing regular load, shear, and bending-moment diagrams.* Since upward loads are considered as positive in such computations, plus M/EI ordinates indicate upward loads. Plotting the shear and bending-moment diagrams for the imaginary beam according to the usual beam convention, plus bending moment would be plotted above the axis and minus values below. Plus bending moment on the imaginary beam therefore indicates deflections above the chord, and minus values indicate deflections below the chord. Likewise, plus shear on the imaginary beam indicates that the elastic curve slopes upward, proceeding from left to right, and minus shear that it slopes downward.

Example 8.10   *Compute the slopes and deflections of this beam.*

$$E = 30 \times 10^3 \text{ kips per sq. in.} \qquad I_1 = 600 \text{ in.}^4 \qquad I_2 = 300 \text{ in.}^4$$

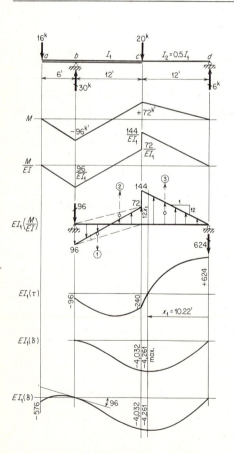

The chord *bd* of this beam remains horizontal. Slopes and deflections with reference to this chord are therefore the actual slopes and deflections. Use the elastic-load method to obtain these, using an imaginary beam supported at *b* and *d*. The reaction at *b* on this beam gives the true slope of the tangent to the elastic curve at point *b*. With this slope known, it is easy to apply the moment-area theorems directly to obtain slopes and deflections in the cantilever portion *ab*.

$$\sum M_b = 0, \,\curvearrowright +$$

$$
\begin{array}{llll}
(1) & (96)(6) = & (-576)(4) & = + \;\; 2{,}304 \\
(2) & (72)(6) = & (432)(8) & = - \;\; 3{,}456 \\
(3) & (144)(6) = & (864)(16) & = -13{,}824 \\
& = +720 & & \;\;\;\; -14{,}976 \\
\end{array}
$$

$$
\dfrac{}{24}
$$

$$= 624 \downarrow$$

$$\sum M_d = 0, \,\curvearrowright +$$

$$
\begin{array}{ll}
(576)(20) = & -11{,}520 \\
(432)(16) = & \;\;\;\; 6{,}912 \\
(864)(8) = & \;\;\;\; 6{,}912 \\
\end{array}
$$

$$\dfrac{2{,}304}{24} = 96 \downarrow$$

$$
\begin{aligned}
\tau_b = & \;\; -96 \\
& \;\; -576 \\
& \;\; +432 \\
\tau_e = & \;\; -240 \\
& \;\; +864 \\
\tau_d = & \;\; +624 \\
\end{aligned}
$$

$$0 = \delta_b$$

$$(-96)(12) = -1,152$$
$$(-576)(8) = -4,608$$
$$(+432)(4) = +1,728$$
$$\overline{ -4,032} = \delta_c$$
$$(-240)(12) = -2,880$$
$$(+864)(8) = +6,912$$
$$\overline{ 0} = \delta_d$$

Then at point a,

$$EI_1\delta_a = (96)(6) - (96)(3)(4)$$
$$\underline{\underline{= -576 \ (down)}}$$

Maximum occurs just to right of point c, where $\tau = 0$.

$$EI_1\tau = +624 - 12x_1\frac{x_1}{2} = 0 \qquad \therefore x_1^2 = 104 \qquad \underline{\underline{x_1 = 10.22'}}$$

$$EI_1\delta_{max} = -(624)(10.22) + \frac{(12)(10.22)^3}{(2)(3)} = -6,385 + 2,124 = \underline{\underline{-4,261 \ (down)}}$$

Substituting now for E and I_1,

$$\delta_{max} = \frac{-4,261^{k'^3}}{[(30)(10^3 \times 144)]^{k'^2}[600/(144)^2]'^4} = \underline{\underline{-0.0341 \ ft \ (down)}}$$

8.10 Application of Moment-Area Theorems and Elastic-Load Method to Deflection of Beams and Frames. To utilize the moment-area approach at its maximum efficiency, it is often desirable to combine the direct application of the two theorems with the use of the elastic-load procedure. This has already been illustrated in Example 8.10 and will be further illustrated in the examples that follow. To plan the method of attack in a given problem, it is advisable first to sketch the elastic curve of the structure. A certain amount of practice is necessary to develop proficiency in sketching such curves. However, even the novice can get the curvature correct by simply following the bending-moment diagram of the member and noting whether the bending is plus or minus.

Once the elastic curve has been sketched approximately, it then is easy to plan the solution. To obtain deflections with respect to a tangent, use the moment-area theorems directly. To obtain deflections with respect to a chord, use the elastic-load method. *In applying the elastic-load method as described here, the portion of the elastic curve of the actual beam corresponding to the span of the imaginary beam must never include an intermediate hinge.* At such hinges, there may be a sharp change in the slope of the elastic curve. This sudden change is not included in an elastic load that consists simply of the M/EI diagram. Such cases may be handled by an extension of the elastic-load method called the conjugate-beam method, discussed in Art. 8.11, or by the combined use of the moment-area theorems and the elastic-load method as illustrated in the following examples.

For the reasons just noted, it is likewise true that *the moment-area theorems cannot be applied between two points on the elastic curve if there is a hinge within that portion of the beam.*

Example 8.11   *Compute the maximum deflection of this beam.*

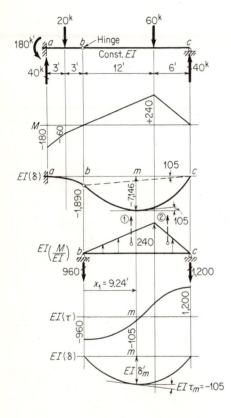

After studying the sketch of the elastic curve, it is apparent that δ_b can be computed by applying the second moment-area theorem to the portion ab. This deflection establishes the position of the chord bc. The deflections and slopes with reference to this chord can be obtained using the elastic-load method applied to an imaginary beam of span bc.

$$EI\delta_b = [(180)(\tfrac{3}{2})(5) + (60)(\tfrac{6}{2})(3)]$$
$$= 1{,}890 \; (\downarrow)$$

$$Rotation\ of\ chord\ bc = \frac{1{,}890}{18} = 105$$

$$(1)\quad (240)(6) = (1{,}440)(10) = 14{,}400$$
$$(2)\quad (240)(3) = \quad(720)\ (4) = \underline{\ 2{,}880}$$
$$\qquad\qquad\qquad\qquad\qquad\qquad 17{,}280$$
$$\qquad\qquad\qquad\qquad\qquad\qquad \overline{\ \ 18\ \ }$$
$$\qquad\qquad\qquad\qquad\qquad = 960 \downarrow$$

$$(1{,}440)(8) = 11{,}520$$
$$(720)(14) = \underline{10{,}080}$$
$$\qquad\qquad\quad 21{,}600$$
$$\qquad\qquad\quad \overline{\ \ 18\ \ } = 1{,}200 \updownarrow$$

The point of maximum δ occurs where the tangent to the elastic curve is horizontal. i.e., where the tangent slopes down to the right with respect to the chord bc, or

$$EI\tau_m = -105.$$

$$EI\tau_m = -960 + 20x_1\,\frac{x_1}{2} = -105$$

$$x_1^2 = 85.5$$
$$x_1 = 9.24'$$

$$EI\delta'_m = -(960)(9.24) + \frac{(20)(9.24)^3}{6}$$

$$= -6{,}225$$

$$EI\delta_m = -6{,}225 - (105)(18 - 9.24)$$
$$= -7{,}146$$

Discussion:

   Note that it is not permissible to apply the elastic-load method to an imaginary beam of span ac because of the presence of the hinge at b.

Example 8.12 *Compute the slopes and deflections of this beam.   Constant E and I.*

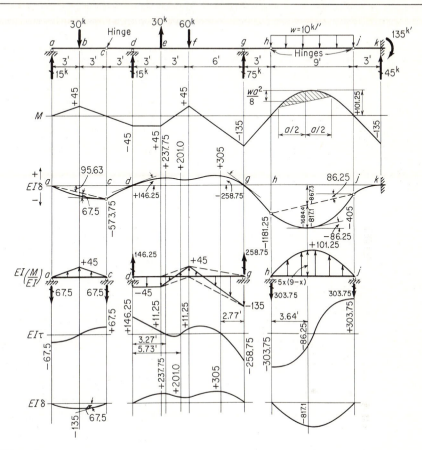

Discussion:

The chord *dg* of the elastic curve does not move, and therefore slopes and deflections determined with reference to it by applying the elastic-load method to the imaginary beam *dg* are the true slopes and deflections of the beam in this portion.   These computations are straightforward and produce the results shown for the location and magnitude of the maximum deflections in the portion *dg*.

The reactions of this imaginary beam establish the slope of the tangent of the elastic curve at *d* and *g*.   It is then easy to use the second moment-area theorem to compute the additional deflection of the hinges at *c* and *h* from these tangents.   The total deflections of these hinges from the original position of the beam are shown.

It is also easy to apply the second moment-area theorem to the cantilever portion *jk* and thus compute the deflection of the hinge at *j*.

In this manner, the position of the chords *ac* and *hj* is established, and it is now possible to compute the true slopes and deflections in these portions.   The position and magnitude of the maximum deflections in these portions are shown.   The slopes and deflections with respect to the chords *ac* and *hj* can be obtained by applying the elastic-load method to the imaginary beams of the same spans.

Note that there can be no point in the portion *ac* which deflects more than *c*, for the slope of the chord is greater than the slope of the tangent at *c* with respect to the chord.

The moment-area theorems and the elastic-load method can also be used advantageously in the computation of frame deflections. The frame deflections computed in this manner, however, do not include the effect of axial changes in length of the members. Fortunately, it is usually permissible to neglect the effect of axial deformation in most frame-deflection problems (see Example 8.7).

Example 8.13   *Compute the deflections of this frame.*

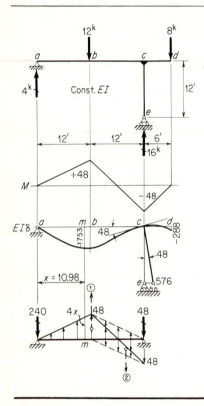

(1) $(48)(12) = (576)(^{12}\!/_{24}) = 288\downarrow$
(2) $(48)(6) \;\; = (288)(^{4}\!/_{24}) \;\; = \;\; 48\uparrow$
$\qquad\qquad\qquad\qquad\qquad\quad \overline{240\downarrow}$

(1) $(576)(^{12}\!/_{24})$ $\qquad\qquad = 288\downarrow$
(2) $(288)(^{20}\!/_{24})$ $\qquad\qquad = 240\uparrow$
$\qquad\qquad\qquad\qquad\qquad\quad \overline{48\downarrow}$

To find point of maximum deflection,

$$EI\tau_m = 0 = -240 + 4x\,\frac{x}{2}$$

$$x^2 = 120 \qquad x = 10.98$$

$$EI\delta_m = -(240)(10.98) + 4\,\frac{(10.98)^3}{6}$$
$$= -1,753(\downarrow)$$

$$EI\delta_d = (48)(6) - (48)(3)(4) = -288(\downarrow)$$

$$EI\delta_e = (48)(12) = 576\;(\rightarrow)$$

Since the joint at c is rigid, the tangents to the elastic curves of all members meeting at that joint rotate through the same angle. Since there is no bending moment in the column, the elastic curve is a straight line inclined at the same angle as the tangent of the elastic curve of the beam.

Example 8.14   *Compute the deflection at point e on this frame.*

$EI\tau_c = (360)(12)(^{16}\!/_{24}) = 2{,}880$
$EI\delta_e = (2{,}880)(12) + (360)\overline{(6)(8)}$
$\qquad\;\; = 51{,}840$
$\qquad\qquad\qquad\overline{(\leftarrow)}$

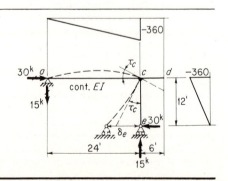

It should be apparent from these examples and the problems at the end of the chapter that the moment-area methods can be applied most advantageously when the moment diagram is composed of a series of straight lines, i.e., when the beam is acted upon by concentrated loads. In such cases, the area under the M/EI curve may be broken up into rectangles and triangles, and the necessary computations are quite simple. When the load is distributed, however, and the moment curve becomes a curved line, the computations become more difficult. When the load is uniformly distributed, the moment curve is parabolic and the M/EI diagram may be broken up into triangles, rectangles, and parabolic segments.[1] In the case of more complicated distributions, however, it is usually necessary to divide the M/EI diagram into a series of short portions each of which may be broken up into essentially triangles and rectangles. Sometimes it is desirable to use Simpson's rule to improve the results of approximate summation.

Newmark[2] developed a numerical procedure which is particularly useful for computing slopes and deflections due to bending of a beam with variable moment of inertia. He suggested replacing the distributed elastic load, defined by the M/EI diagram, by a series of equally spaced equivalent concentrated loads. Usually, adequate precision is obtained if the M/EI diagram is assumed to be straight between adjacent equivalent loads. For such an assumption, the equivalent loads at an end point a and at an interior point b can be computed from the following formulas, using the ordinates A, B, and C of the M/EI diagram shown in Fig. 8.19a:

$$P_a = \frac{h}{6}(2A + B) \qquad P_b = \frac{h}{6}(A + 4B + C)$$

Newmark's method is illustrated in Fig. 8.19 for a 36-ft simple end-supported beam with varying I, the M/EI diagram for which is shown in Fig. 8.19a. The elastic

[1] The following properties of parabolas are useful in handling parabolic elastic loads like those in the portion hj of the beam in Example 8.12, where $w = $ intensity of uniformly distributed load:

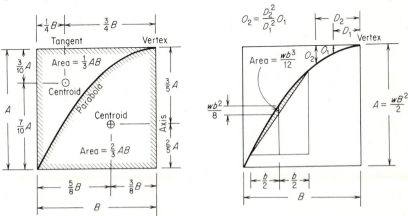

[2] N. M. Newmark, Numerical Procedure for Computing Deflections, Moments, and Buckling Loads, *Trans. ASCE*, vol. 108, p. 1161, 1943.

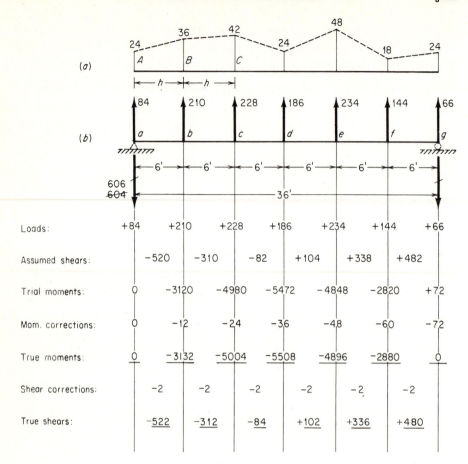

Figure 8.19 Newmark's method

load represented by this diagram is replaced by a series of concentrated loads spaced at 6-ft intervals and applied to the imaginary beam *ag*. The shears (slopes) and bending moments (deflections) for this imaginary beam can be computed in much the same manner as in Example 3.5. These numerical computations can be conveniently arranged as shown in Fig. 8.19. Note that the reactions of the imaginary beam need not be computed to start the calculations. Instead one of the reactions, say at *a*, may be assumed. Then, the corresponding shears and moments are calculated throughout the beam. If the assumed reaction is correct, the moment at *g* will calculate to be zero. If the moment at *g* is not zero, its erroneous value when divided by the span provides the basis for correcting the assumed reaction at *a* and the starting shear in panel *a-b*. The calculation of the remaining corrections of the moments and shears of the imaginary beam is self-explanatory.

8.11 Conjugate-Beam Method. From the examples of the previous article, it is apparent that deflection computations for any beam can be handled effectively

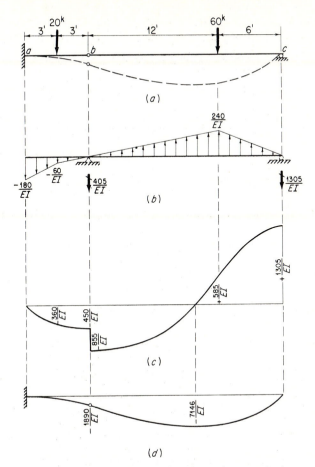

Figure 8.20   Development of conjugate-beam method

by a proper combination of the use of the moment-area theorems and of the elastic load method. The procedure involved in this combination, however, may be identified as nothing more than a slight extension and variation of the elastic-load method. This extension is called the **conjugate-beam method.**

In order to develop these new ideas, consider the beam used in Example 8.11 and shown in a somewhat different manner in Fig. 8.20. In Fig. 8.20a are shown the real beam and its loading. Also shown by the dashed line is the deflection curve, which has certain notable characteristics determined by the type of supports and by the hinge at *b*: (1) at support *a* both the deflection and slope of the elastic curve are zero; (2) at support *c* the deflection is zero but the elastic curve is free to assume any slope required; and (3) at hinge *b*, there can be a deflection, and, in addition, a sudden change in slope can occur between the left and right sides of the hinge.

The objective is to select a **conjugate beam** i.e., a corresponding fictitious beam, which has the same length as the real beam but is supported and detailed in such a manner that, *when the conjugate beam is loaded by the M/EI diagram of the real beam as an (elastic) load, the (elastic) shear in the conjugate beam at any location is equal to the slope of the real beam at the corresponding location and the (elastic) bending moment in the conjugate beam is equal to the corresponding deflection of the real beam.* Note that these slopes and deflections of the real beam are measured *with respect to its original position;* i.e., they are the true slopes and deflections. It is always possible to select the proper supports for the conjugate beam to achieve the desired objective by simply noting the known characteristics of the elastic curve of the real beam at its supports or at any special construction features, such as the hinge at *b.*

To illustrate the selection of the supports of the conjugate beam, consider the beam in Fig. 8.20. At *a*, there is neither slope nor deflection of the real beam; therefore, there must be neither shear nor moment at this point of the conjugate beam. That is, point *a* of the conjugate beam must be free and unsupported. At *c*, there is slope but no deflection of the real beam; therefore, there must be shear but no moment in the conjugate beam. That is, point *c* of the conjugate beam must be provided with the vertical reaction of a roller support. At *b*, there is deflection and possibly a discontinuous slope in the real beam; therefore, a vertical reaction must be provided to create the sudden change in shear in the conjugate beam, and it must be capable of resisting bending moment. That is, point *b* of the conjugate beam must also be provided with a roller support. Thus, the conjugate beam is loaded and supported as shown in Fig. 8.20*b.*

If the reactions, shears, and bending moments are now computed as for any statically determinate beam, the (elastic) shear and the (elastic) bending-moment diagrams can be drawn as shown in Fig. 8.20*c* and *d*, respectively. The ordinates of these diagrams give, respectively, the slope and the deflection at corresponding locations in the real beam, measured with respect to the original undeflected position. *The convention used for sign of the (elastic) load, (elastic) shear, and (elastic) bending moment is the usual beam convention as described at the end of Art. 8.9 regarding the elastic-load method.*

These and similar considerations lead to the rules shown in Fig. 8.21 for selecting the supports and other details of conjugate beams. Illustrations of selection of supports and details of typical conjugate beams are shown in Fig. 8.22. *Note that statically determinate real beams always have corresponding conjugate beams which are also statically determinate. Statically indeterminate real beams appear to have unstable conjugate beams. However, such conjugate beams turn out to be in equilibrium since they are stabilized by the elastic loading corresponding to the M/EI diagram for the corresponding real beam.* For example, the last case shown is the real beam clamped at both ends, and the corresponding conjugate beam is completely free and unsupported. For any loading (such as the centrally applied concentrated load shown), it will be found that the *M/EI* diagram corresponds to an elastic loading which is completely selfequalizing and requires no supporting reactions whatsoever. In fact, in such cases, the stabilizing conditions required of the elastic loading furnish one method of obtaining the necessary equations for finding the redundant moments for such indeterminate beams.

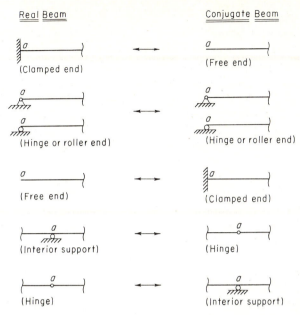

Figure 8.21 Selection of support and details of conjugate beams

When the conjugate-beam method is used to solve problems such as Example 8.10, 8.11, or 8.12, it will be found that exactly the same computations have been made as when the problems are solved using a combination of the moment-area theorems and the elastic-load method. However, the conjugate-beam approach furnishes a straightforward procedure for which a readily applied sign convention is available. As a result, one is less likely to make mistakes when following the conjugate-beam routine.

8.12 Williot-Mohr Method. When the method of virtual work is used to compute truss deflections, only one component of the deflection of a joint can be calculated at a time. To obtain the magnitude and direction of the true absolute movement of a joint, both horizontal and vertical components of its movement must be determined. Therefore, two separate applications of the method of virtual work are usually required in order to determine the resultant deflection of *each* truss joint. *One* graphical solution using the Williot-Mohr method, however, will determine the resultant deflection of *all* joints of the truss, and therefore this method obviously has a very important advantage in the solution of certain deflection problems.

The fundamentals of the Williot-Mohr method can be developed by considering the simple truss shown in Fig. 8.23. Assume that the changes of length of the members ΔL have been computed for the given condition of deformation, using Eq. (8.5a), (8.5b), or (8.5c). The deflected position of the truss can then be determined in the following manner as indicated by the dashed lines in Fig. 8.23a First, remove the pin at joint D and allow the change in length of member AB to take place. This will cause member DB to move parallel to itself as shown, and

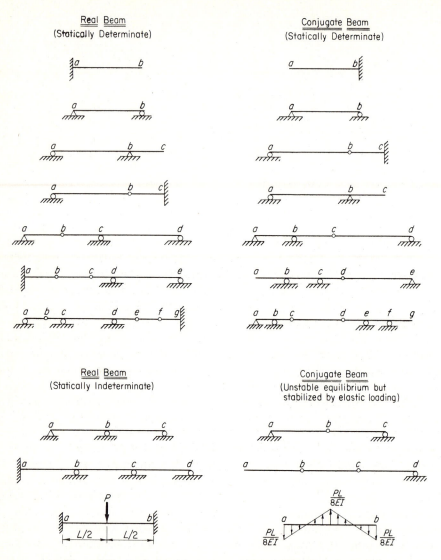

Figure 8.22 Typical real beams and corresponding conjugate beams

all points on this member will move horizontally to the right an amount equal to ΔL_{AB}. Now, upon allowing members AD and BD to change length, the D ends of each will move as shown if they remain connected to member AB at points A and B', respectively. Before pin D can be replaced and the truss connected together again, it is necessary to make the D ends of members AD and BD coincide again. This can be done by rotating AD about A and BD about B' until the arcs intersect. The deflection of any joint can then be determined from the original and deflected position of the joint.

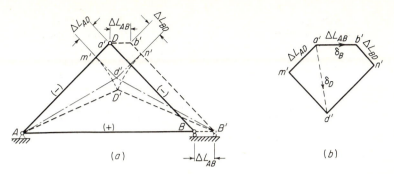

Figure 8.23 Kinematics of truss deflection

This procedure is straightforward but difficult to apply in practice because the deflections and changes in length are actually very much smaller than have been indicated in this sketch, so that a very large scale drawing is necessary in order to obtain any accuracy. Since the deformations are small, however, the angular rotation of any member is also small, in fact so small that it is permissible to assume that during the rotation of a member a point moves along the tangent drawn normal to the original direction of the member rather than along the true arc, as shown by the dot-dash lines in Fig. 8.23a. If ΔL had not been tremendously exaggerated in this sketch, the dashed and dot-dash lines would coincide for all practical purposes. Introducing this simplification makes it possible to obtain the joint deflections without drawing the entire lengths of the members because it is no longer necessary to draw the arcs about the centers of rotation.

The simplified diagram shown in Fig. 8.23b is similar to the portion of Fig. 8.23a marked with the same letters. It involves only the changes in length of the members and the tangents to the arcs of rotation and enables one to find the relative movements of the various joints. Such a diagram is called a **Williot diagram,** after the French engineer who suggested it. As before, imagine that the pin at D is temporarily removed and allow the changes in length to take place one at a time. Upon selecting a suitable scale for ΔL and designating the points in this diagram by the lowercase letters of the corresponding truss joints, the diagram is started by locating point a'. In this simple truss, it is known that joint A remains fixed in position and member AB remains horizontal. Joint B therefore moves horizontally to the right relative to A an amount equal to ΔL_{AB}, as indicated by the relative positions of points a' and b'. Owing to the shortening, ΔL_{AD}, the D end of member AD moves downward to the left parallel to AD with reference to joint A, as represented by the vector $\overline{a'm'}$; similarly, owing to the shortening, ΔL_{BD}, the D end of member BD moves downward to the right parallel to BD with reference to joint B, as represented by the vector $\overline{b'n'}$. In order to make the D ends coincide, member AD must be rotated about A and member BD about B. During these rotations, the D ends are assumed to move along the tangents as represented by the vectors $\overline{m'd'}$ and $\overline{n'd'}$, respectively. Note that these tangents are perpendicular to members AD and BD, respectively. The vectors $\overline{a'd'}$ and $\overline{a'b'}$ represent the

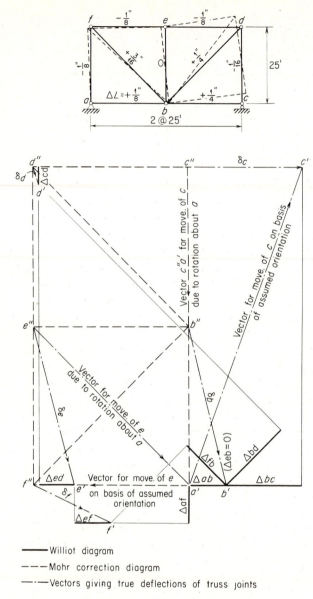

———— Williot diagram

– – – – Mohr correction diagram

–·–·– Vectors giving true deflections of truss joints

Figure 8.24   Williot-Mohr method

deflections of joints D and B, respectively, with respect to joint A. In this case, since joint A is actually fixed in position, these vectors represent the true absolute movements of these joints. Of course, the lengths of these vectors are measured to the same scale as that used in plotting ΔL.

The construction of the Williot diagram for a more elaborate truss is carried out in essentially the same manner. Temporarily, all members are assumed to be

disconnected.   Then, after the members are imagined to undergo their changes in
length, they are reassembled one at a time, and the Williot diagram is constructed
to show the resulting joint displacements.   In such cases, the true relative movement of the two ends of one particular bar are not known as was true of joints A and
B in the above simple case.   The Williot diagram can be drawn, however, on the
arbitrary assumption that some bar simply changes in length but does not rotate,
i.e., that the relative movement of the joints at the end of this bar is parallel to the
bar and equal to the change in its length.   When these two end joints have been
located, a third point corresponding to the third joint of a truss triangle formed by
these three joints can be located in the manner described above.   The remainder
of the construction can be carried out proceeding from joint to joint, always working with a triangle two joints of which have already been located, and locating
the new joint from these two points.   If the originally assumed orientation is correct, the vectors corresponding to certain known deflection conditions will be oriented correctly.   If these vectors are not consistent with the known conditions, the
Mohr correction diagram must be added to the Williot diagram.

The Williot diagram for the truss of Fig. 8.24 has been drawn on the assumption
that joint a is fixed in position and member ab fixed in direction.   If this orientation is correct, the vectors drawn from point a' to points b', c', d', etc., will be
the true deflections of joints b, c, d, etc., respectively.   Joint c on the truss, being a
roller support, can have no vertical deflection and is constrained to move horizontally.   On the basis of the assumed orientation, however, joint c moves upward to
the right as shown by the vector $\overline{a'c'}$, and the other joints have moved as indicated
to an exaggerated scale by the dashed lines on the line diagram of the truss.   This
means that the assumption that member ab remains fixed in direction is in error,
and it is now necessary to rotate the truss as a whole clockwise about a to bring
joint c back down on the support.   The amount of rotation required can be determined by knowing that the true deflection of c must be a horizontal vector.   This
true deflection is the resultant of vector $\overline{a'c'}$ and the vector representing the movement of c during the rotation of the truss about a.   During a small angular rotation
of the truss about a, joint c can be assumed to move along the tangent to the true
arc and hence normal to the line ac on the truss or, in this case, vertically.   If this
vector, giving the movement of c with reference to a during rotation, is added on
the diagram as the vertical vector $\overline{c''a'}$ drawn through a', then the resultant of the
vector $\overline{c''a'}$ ($\downarrow$) and the vector $\overline{a'c'}$($\nearrow$) must be the horizontal vector $\overline{c''c'}$ ($\rightarrow$),
which is the true deflection of joint c, and thus point c'' is located.

During the rotation of the truss about a,
not only joint c but all other joints may be
assumed to move normal to the radius from
the joint to the center of rotation a an
amount equal to that radius times the angle
of rotation.   In Fig. 8.25.a, the arrows on
the joints indicate the direction of the joint
movements as the truss is rotated clockwise about a through a small angle α.   If

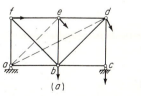

Figure 8.25 Mohr correction diagram

all these vectors are drawn to scale and acting toward point a'', as shown in Fig. 8.25b, the movement of joint b during rotation is represented by vector $\overline{b''a''}$, of joint c by $\overline{c''a''}$, of joint d by $\overline{d''a''}$, etc.   After these double-prime points have been connected as shown, consider any corresponding portions of these two figures such as triangles dac and $d''a''c''$.   Since $c''a''$ is perpendicular to ca and $d''a''$ is perpendicular to da, angle dac = angle $d''a''c''$.   Since $c''a''$ = αca and $d''a''$ = αda, then $c''a''/d''a''$ = ca/da.   Therefore, triangles dac and $d''a''c''$ are similar.   In this manner, it can be shown that Fig. 8.25a is similar to Fig. 8.25b.   Such considerations lead to the following conclusion:

If a rigid body is rotated about some center of rotation O through a small angle, and if the movements of two points i and k are plotted as vectors $\overline{i''O''}$ and $\overline{k''O''}$ toward point O", and if the shape of the body is then plotted to a scale with points i" and k" as a base, each line on this scale sketch will be perpendicular to the corresponding line on the actual body and the line drawn from any point on this scale sketch to the point O" will represent, as a vector, the movement of the corresponding point on the real body during the rotation about O.

This principle is the basis for the **Mohr correction diagram.**

In the example in Fig. 8.24, the Mohr correction diagram $a'b''c''d''e''f''$ is superimposed on the Williot diagram, using the vector $\overline{c''a'}$ as a basis, every line on it being at 90° to the corresponding line of the line diagram of the truss.   The vectors from the double-prime points to the pole a' give the movements of the corresponding joints during the rotation of the truss about joint a.   The true deflection of any joint can now be determined as the resultant of the vector from the corresponding double-prime point to a' and the vector from a' to the corresponding single-prime point, i.e., *the true deflection of any joint can be found in magnitude and direction from the vector drawn from the double-prime point on the Mohr correction diagram to the single-prime point on the Williot diagram.*

8.13   Application of Williot-Mohr Method.   The Williot diagram will be more compact and will require a smaller Mohr correction diagram if the orientation that is assumed to start the Williot diagram is reasonably correct.   In many cases, it is quite apparent that certain bars do not change direction very much.   For example, in the case of an end-support truss the center vertical remains essentially vertical and is therefore a good member to assume fixed in direction in starting the Williot diagram.   This is illustrated in Example 8.15.

It is also interesting to note that the Mohr correction diagram can be eliminated entirely if the relative movements of the two ends of a bar are computed first by the method of virtual work.   For example, consider the truss shown in Fig. 8.26. Joint a in this case does not move.   Consider one of the bars that meet at joint a, such as bar ab.   The relative movement of the two ends of this bar is the resultant of the relative movements parallel and perpendicular to the bar.   The relative movement of the two ends parallel to the bar is simply equal to its change in length, while the relative movement normal to the bar can be computed by the method of virtual work.   In this case, since point a does not move, all that is necessary is to compute the movement of point b normal to the bar, or vertically.

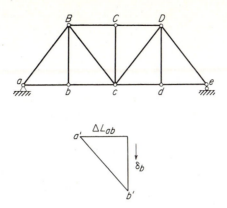

Figure 8.26   Correct orientation of Williot diagram

Once this has been done, these two vectors can be laid out as shown, and the vector $\overline{a'b'}$ then gives the true relative movements of joints a and b. The Williot diagram can now be drawn on the basis of the positions of points a' and b', locating first point B' from the triangle abB, etc. Since this diagram has been oriented correctly, the absolute movements of all the joints can be obtained from it directly without adding the Mohr correction diagram.

The Williot diagram can be drawn without difficulty as described in Art. 8.12 for any simple truss. It should be noted, however, that certain difficulties are encountered in applying these ideas to a compound truss such as that shown in Fig. 8.27. Suppose that the Williot diagram is started in this instance by assuming joint g fixed in position and bar gG fixed in direction. Points $F', f', e',$ and E' can be located without difficulty on the left side and similarly points $H', h', i',$ and I' on the right. It is impossible to proceed beyond these points, however, by using the previous-ly discussed procedures. One way of over-coming the difficulty is to recognize that the relative movement of joints C and e parallel

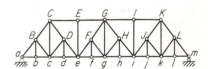

Figure 8.27   Compound truss

to the line CDe is simply equal to the sum of the changes in length of bars CD and De and, similarly, that the relative movement of joints c and e along line cde is equal to the sum of the changes in length of cd and de. On this basis, temporarily con-sider that there are no joints at D and d and that bars cD and Dd have been omitted and proceed as usual to locate points C' and c'. Having located these points, reinsert the omitted bars and backtrack to locate points D' and d'. If the work has been done correctly, it is now possible to locate e' from D' and d' and see whether or not it coincides with the existing location of point e'. Points $B', b',$ and a' can now be located without difficulty, and the remaining points to the right of iI can be located in the manner just described for the points to the left of Ee.

The application of the Williot-Mohr method to the case of a three-hinged arch requires a somewhat different technique, which is illustrated by Example 8.16.

Example 8.15   *Draw the Williot-Mohr diagram for this truss.*

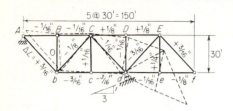

Assume point c fixed in position.
Assume bar cC fixed in direction.

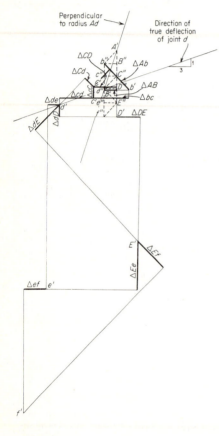

Point	Deflection	
	Horizontal	*Vertical*
A	0	0
B	0.063″ ←	0.130″ ↓
C	0.125″ ←	0.050″ ↓
D	0	0.145″ ↓
E	0.125″ →	0.700″ ↓
b	0.125″ →	0.130″ ↓
c	0.063″ ←	0.112″ ↓
d	0.250″ ←	0.083″ ↓
e	0.313″ ←	0.962″ ↓
f	0.438″ ←	1.513″ ↓

Discussion:

The vectors on the Williot diagram drawn from c′ to A′, B′, b′, C′, etc., give the movements of joints A, B, b, C, etc., it being assumed that joint c remains fixed in position and bar cC remains fixed in direction.   In such a case, the resulting joint deflections would be as shown to an exaggerated scale by the dashed lines on the line diagram.   Obviously, the assumed orientation is incorrect; for joint A has moved off the hinge support, and the movement of joint d is not parallel to the supporting surface.   Note, however, that the assumed orientation is not badly in error.

Joint A can be brought back to the hinge support by translating the truss as a rigid body parallel to the vector $\overline{A'c'}$ a distance equal to the scaled length of this vector.   During such a translation,

every joint would move the distance indicated by the vector $\overline{A'c'}$. After such translation, the re-
sultant movement of any joint, such as E, would be the vector sum of the vector $\overline{A'c'}$ and the vector
$\overline{c'E'}$, and equal to the vector $\overline{A'E'}$. Therefore, it is apparent that the resultant movements of the various
joints after joint A has been restored to its correct position are given by the vectors drawn from point
A' to B', b', C', c', D', d', etc. Obviously, the truss is still not oriented correctly, for the vector
$\overline{A'd'}$ does not indicate the proper direction for the movement of joint d. The truss must therefore
be rotated counterclockwise about A so that d moves the amount given by the vector $\overline{d''A'}$. Then the
total movement of joint d is the vector sum of vectors $\overline{d''A'}$ and $\overline{A'd'}$, that is, the vector $\overline{d''d'}$, which
is in the proper direction.

The Mohr correction diagram can then be completed as shown by the dotted lines, the line d''A'
being used as a base line. The final true resultant deflection of any joint is then given by the vector
drawn from the corresponding double-prime point to the corresponding single-prime point.

Example 8.16   *Find the deflections of this three-hinged arch, using the Williot-Mohr method.*

For left half:
Assume a fixed in position.
Assume aA fixed in direction.

For right half:
Assume e fixed in position.
Assume eE fixed in direction.

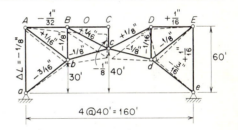

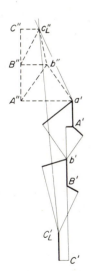

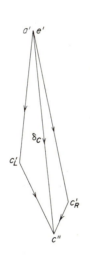

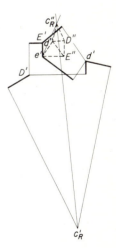

Discussion:
The presence of the hinge at point c introduces certain complications in applying the Williot-Mohr
method to this problem. After making the assumptions indicated, draw the Williot diagrams in
the usual manner for the two separate halves. On this basis, the vector $\overline{a'c_L'}$ on the left diagram
indicates a certain deflection for joint c. Similarly, the vector $\overline{e'c_R'}$ on the right diagram indicates a
different deflection for joint c. If the assumed orientations are correct, the deflection of joint c

will be the same whether determined from the left or the right diagram. Since these vectors are not equivalent, the left half must be rotated about a, causing joint c to move normal to the radius ac. Similarly, the right half is rotated about e, causing joint c to move normal to the radius ec. These rotations must be such as to produce the same resultant displacement of joint c for either half. They are determined as indicated in the center vector diagram, being given by the vector $\overline{c'_L c''}$ for the left half and $\overline{c'_R c''}$ for the right half. These vectors therefore establish the bases $\overline{c''_L a'}$ and $\overline{c''_R e'}$ for the Mohr correction diagrams for the left and right halves, respectively. The true deflections of the various joints can now be obtained from the vectors drawn from the double- to the single-prime points.

8.14 Bar-Chain Method. The bar-chain method is similar to the Williot-Mohr method in that by one application of the method we are able to compute the deflection of several joints of a truss simultaneously. This method was first suggested by H. Müller-Breslau and is essentially an adaptation of the elastic-load method, applied to trusses instead of beams.

To develop the fundamentals of this procedure, consider the case of the simple truss shown in Fig. 8.28a, where it is desired to compute the vertical component of the deflection of the lower-chord panel points. For the given cause of deformation, the changes in length of the members may be computed from Eq. (8.5a), (8.5b), or (8.5c). As a result of these changes in length, the angles of the truss triangles will change by certain small amounts, and the truss may change shape as indicated to an exaggerated scale by the dashed lines in Fig. 8.28b. The changes in the angles of the triangles can be computed rather easily by means of the expressions developed below. The bottom chord of this truss can be isolated and its deflected position drawn as shown in Fig. 8.28c. This suggests to us immediately that the problem of computing the deflected shape of this bottom chord resembles that of computing the elastic curve of a straight beam by the moment-area or elastic-load method. In the case of a beam, however, the elastic curve is a smooth, continuous curve the slope of which is continually changing, whereas the deflection curve of this bottom chord is composed of a series of straight lines which change slope only at the truss joints. This series of straight lines is called a *bar chain* and from this terminology the method gets its name.

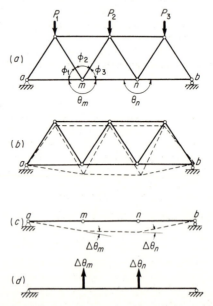

Figure 8.28 Fundamentals of bar-chain method

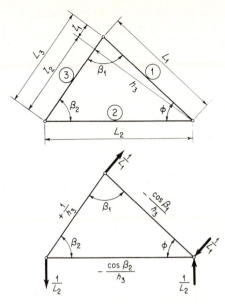

Figure 8.29   Angle changes of a truss triangle

In a beam, the change in slope between the tangents at the ends of a differential element dx of the elastic curve is equal to $M/EI\,dx$, which is the differential area under the M/EI diagram at that element.   At any joint m of the bar chain, however, the change in slope between the adjacent bars of the chain is equal to the change in the angle θ_m, that is, $\Delta\theta_m$, and between joints there is no change in slope.   When the moment-area method is applied to a bar chain, the M/EI diagram should be replaced, therefore, by a series of ordinates, one at each intermediate joint equal to $\Delta\theta$ at that joint.   Hence, the elastic load used for the bar chain consists of a series of concentrated loads such as those shown in Fig. 8.28d.   The technique of applying the moment-area or the elastic-load method to obtain the slope and deflection of the bar chain, however, is exactly the same as in the case of beams.   The angle θ should be considered as the angle included on the lower side between two adjacent bars of the chain. $\Delta\theta$ should be considered plus (and therefore an upward elastic load) when θ increases.

It is easy to compute the change in θ if the angle changes of the truss triangles are known. For example, at joint m in this case, $\Delta\theta_m$ is numerically equal but opposite in sense to the algebraic sum of the angle changes in angles ϕ_1, ϕ_2, and ϕ_3.   An expression to be used for computing such changes in the angles of the truss triangles will now be developed.

Consider the triangle shown in Fig. 8.29.   Suppose that we want to compute the change in angle ϕ caused by these bars increasing in length by ΔL_1, ΔL_2, and ΔL_3.   The increase $\Delta\phi$ in angle ϕ can be computed by the method of virtual work, using the Q system shown.   Let α_1 represent the clockwise rotation of bar 1 and α_2 the counter-clockwise rotation of bar 2.   Then, evaluating the external virtual work done by this Q system gives

$$\sum Q\delta = (1)(\alpha_1^{\curvearrowright}) + (1)(\alpha_2^{\curvearrowleft}) = (1)(\alpha_1^{\curvearrowright} + \alpha_2^{\curvearrowleft}) = 1\,\Delta\phi$$

since

$$\alpha_1^{\curvearrowright} + \alpha_2^{\curvearrowleft} = \text{increase in } \measuredangle\,\phi = \Delta\phi$$

Applying now the law of virtual work,

$$\sum Q\delta = \sum F_Q\,\Delta L$$

$$1\,\Delta\phi = \frac{1}{h_3}\Delta L_3 + \left(-\frac{\cos\beta_2}{h_3}\right)\Delta L_2 + \left(-\frac{\cos\beta_1}{h_3}\right)\Delta L_1 \qquad (a)$$

From the geometry of the triangle, the following relations exist:

$$h_3 = L_2 \sin \beta_2 \quad \text{and} \quad h_3 = L_1 \sin \beta_1 \tag{b}$$

Likewise, $$L_3 = l_1 + l_2 = h_3 \cot \beta_1 + h_3 \cot \beta_2 = h_3(\cot \beta_1 + \cot \beta_2)$$

$$\frac{1}{h_3} = \frac{\cot \beta_1 + \cot \beta_2}{L_3} \tag{c}$$

Substituting from Eqs. (b) and (c) in Eq. (a) gives

$$\Delta\phi = \frac{\Delta L_3}{L_3} (\cot \beta_1 + \cot \beta_2) - \frac{\Delta L_2}{L_2} \cot \beta_2 - \frac{\Delta L_1}{L_1} \cot \beta_1$$

Or this may be written in the following convenient form, noting that $\Delta L = eL$:

$$\Delta\phi = (e_3 - e_1) \cot \beta_1 + (e_3 - e_2) \cot \beta_2 \tag{8.11a}$$

Thus, the angle change $\Delta\phi$ is equal to the difference in the strain of the side opposite and one adjacent side multiplied by the cotangent of the angle between these sides plus the difference in the strain of the side opposite and the other adjacent side of the triangle multiplied by the cotangent of the angle between these two sides. When the strain is due to an axial-stress intensity σ, this equation can be expressed more conveniently as

$$E \Delta\phi = (\sigma_3 - \sigma_1) \cot \beta_1 + (\sigma_3 - \sigma_2) \cot \beta_2 \tag{8.11b}$$

A plus value of $\Delta\phi$ indicates an increase in angle ϕ and minus values a decrease.

After the angle changes of a simple truss have been computed by using either Eq. (8.11a) or (8.11b), $\Delta\theta$ can be computed at any joint of the bar chain. The moment-area method or the elastic-load procedure can then be applied to find the vertical components of the deflection of any bar chain that is *initially straight and horizontal*. This is illustrated in Example 8.17. It should be recalled that in applying the elastic-load method the shears and moments of the imaginary beam give the slopes and deflections of the deflection curve, measured with reference to the chord corresponding to the support points of this imaginary beam. Thus, in Example 8.17, where the chord *af* rotates, the true deflections can be obtained by drawing the known line of zero deflections through points *a* and *d* and correcting the deflections measured from the chord *af* in the manner indicated.

Example 8.17   *Using the bar-chain method, compute the vertical components of the deflection of the bottom-chord panel points.*

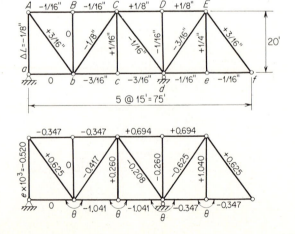

For convenience, multiply e by 10³ and divide the final results by 10³ to obtain true answers.

$\triangle$	$(e_3-e_1)\times 10^3$ (1)	cot β_1 (2)	(1)× (2)	$(e_3-e_2)\times 10^3$ (3)	cot β_2 (4)	(3)× (4)	$\Delta\phi\times$ 10^3	$\Delta\theta\times$ 10^3
abA	$-0.520-0 \quad=-0.520$	0	0	$-0.520-0.625=-1.145$	$1.\dot3$	-1.527	-1.527	
AbB	$-0.347-0.625=-0.972$	0.75	-0.729	$-0.347-0 \quad=-0.347$	0	0	-0.729	$+1.301$
BbC	$-0.347-0 \quad=-0.347$	0	0	$-0.347+0.417=+0.070$	0.75	$+0.052$	$+0.052$	
Cbc	$+0.260+0.417=+0.677$	$1.\dot3$	$+0.903$	$+0.260+1.041=+1.301$	0	0	$+0.903$	
bcC	$-0.417+1.041=+0.624$	0.75	$+0.468$	$-0.417-0.260=-0.677$	$1.\dot3$	-0.903	-0.435	$+0.435$
Ccd	$-0.208-0.260=-0.468$	$1.\dot3$	-0.624	$-0.208+1.041=+0.833$	0.75	$+0.625$	0	
cdC	$+0.260+1.041=+1.301$	0	0	$+0.260+0.208=+0.468$	$1.\dot3$	$+0.624$	$+0.624$	
CdD	$+0.694+0.208=+0.902$	0.75	$+0.676$	$+0.694+0.260=+0.954$	0	0	$+0.676$	-4.509
DdE	$+0.694+0.260=+0.954$	0	0	$+0.694+0.625=+1.319$	0.75	$+0.989$	$+0.989$	
Ede	$+1.040+0.625=+1.665$	$1.\dot3$	$+2.220$	$+1.040+0.347=+1.387$	0	0	$+2.220$	
deE	$-0.625+0.347=-0.278$	0.75	-0.208	$-0.625-1.040=-1.665$	$1.\dot3$	-2.220	-2.428	$+2.252$
Eef	$+0.625-1.040=-0.415$	$1.\dot3$	-0.553	$+0.625+0.347=+0.972$	0.75	$+0.729$	$+0.176$	

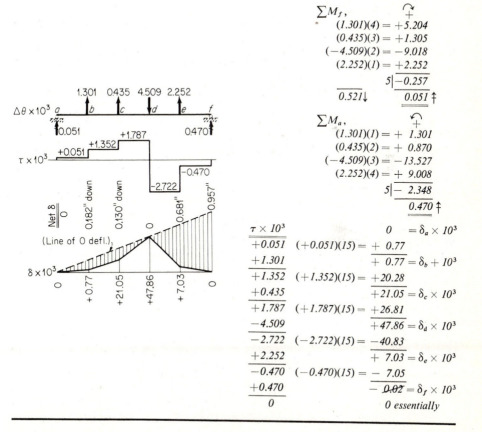

Using imaginary beam of span af,

$\sum M_f$, $\quad\curvearrowright +$

$(1.301)(4) = +5.204$
$(0.435)(3) = +1.305$
$(-4.509)(2) = -9.018$
$(2.252)(1) = +2.252$
$\qquad\qquad 5\overline{|-0.257}$
$\overline{0.521\downarrow} \qquad \overline{0.051}\,\updownarrow$

$\sum M_a$, $\quad\curvearrowleft +$

$(1.301)(1) = +\ 1.301$
$(0.435)(2) = +\ 0.870$
$(-4.509)(3) = -13.527$
$(2.252)(4) = +\ 9.008$
$\qquad\qquad 5\overline{|-\ 2.348}$
$\qquad\qquad\quad \overline{0.470}\,\updownarrow$

$\tau\times 10^3$		$0\quad=\delta_a\times 10^3$
$+0.051$	$(+0.051)(15) = +\ 0.77$	
$+1.301$		$+\ 0.77=\delta_b+10^3$
$+1.352$	$(+1.352)(15) = +20.28$	
$+0.435$		$+21.05=\delta_c\times 10^3$
$+1.787$	$(+1.787)(15) = +26.81$	
-4.509		$+47.86=\delta_d\times 10^3$
-2.722	$(-2.722)(15) = -40.83$	
$+2.252$		$+\ 7.03=\delta_e\times 10^3$
-0.470	$(-0.470)(15) = -\ 7.05$	
$+0.470$		$-\ \cancel{0.02}=\delta_f\times 10^3$
0		0 essentially

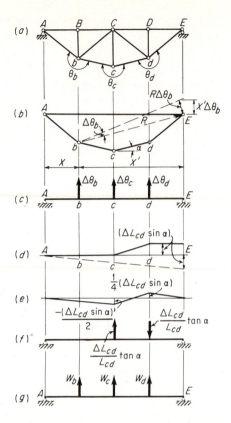

Figure 8.30 Bar chain (not initially straight)

Certain modifications must be introduced into this approach before it can be applied to a bar chain that is not initially straight and horizontal.   Suppose that we wish to compute the vertical components of the deflection of the lower-chord bar chain of the truss shown in Fig. 8.30a.   The effect of the changes in the θ angles on the deflection of this bar chain can be computed in the same manner as if the bar chain were initially horizontal.   This is evident when one considers the lower-chord bar chain shown in Fig. 8.30b.   Imagine that we allow only the angle θ_b to change and that at the same time we hold the portion of the bar chain to the left of joint b fixed in position. The vertical component of the movement of point E would then be equal to $x' \Delta\theta_b$, the same value as it would have been if the bar chain were initially straight.   Continuing in this manner, we can show that the vertical deflection of this lower-chord bar chain produced simply by the angle changes $\Delta\theta$ can be computed in the same manner as if the bar chain were initially horizontal, using the imaginary beam shown in Fig. 8.30c.

The total deflection of this lower-chord bar chain is produced, however, not only by the angle changes $\Delta\theta$ but also by the changes in length of the members of the bar chain themselves.   For example, consider the effect of an elongation of bar cd.   If all the joints of the bar chain were locked so that no angle changes could occur, and if the bars to the left of c were held fixed in position, the elongation of bar cd would cause all joints to the right of c to move vertically an amount equal to $\Delta L_{cd} \sin \alpha$, as indicated by the ordinates to the solid line in Fig. 8.30d.   In order to return point E to its proper position, the entire bar chain must be rotated about point A as a rigid body, causing the vertical displacements shown by the dashed line in Fig. 8.30d.   The net effect of the two displacements shown in Fig. 8.30d is shown in Fig. 8.30e. This displacement diagram would be the bending-moment diagram for the loads shown in Fig. 8.30f. In the same manner, it could be shown that the contribution of the change in length of any bar of the chain to the vertical deflection would be equal to the bending moment produced by two similar equal and opposite loads acting at the joints at the ends of that bar.

These considerations lead us to the following conclusion. To compute the total deflections of a bar chain that is not initially straight, the bending moments can be computed on an imaginary beam that is acted upon by a modified elastic loading as indicated in Fig. 8.30g. At any intermediate joint m of the bar chain, this loading is equal to the angle change $\Delta\theta_m$ modified by certain contributions proportional to the strains in the adjacent members of the bar chain. Let this modified elastic load at joint m be called W_m. Then

$$W_m = \Delta\theta_m - e_L \tan \alpha_L + e_R \tan \alpha_R \qquad (8.12)$$

where e_L, α_L = strain and initial slope with horizontal of adjacent member of bar chain on *left* of joint m

e_R, α_R = similar quantities for adjacent member on *right* of joint m

$\Delta\theta_m$ = change in angle θ_m, which, as defined above, is angle included on lower side between two adjacent members of chain

It is important for the proper sign convention to be adhered to in using this equation.

W_m is plus when indicating an upward elastic load
$\Delta\theta_m$ is plus when θ_m increases
e is plus when the member elongates
α is plus when the initial slope of the member is up to the right

In this application to a bar chain that is not initially straight, it is important to note that the shear produced on the imaginary beam by the elastic load has no significance. Of course, in cases where the bar chain is initially straight and horizontal, such shears give the slope of the members of the deflected bar chain. The elastic load for the general case of a polygonal bar chain as expressed in Eq. (8.12), however, was determined so that the bending moment on the imaginary beam gave the deflection of the bar chain, but no significance was attached to the shear produced by this load.

The bar-chain method can easily be applied in a similar manner to any simple truss. A suitable bar chain is selected joining those joints the vertical deflection of which is desired. For any such bar chain, the elastic loads can be computed from Eq. (8.12). In the special cases where the bar chain is initially straight and horizontal (as illustrated in Example 8.17), α will be zero for all bars and W_m equals simply $\Delta\theta_m$. Note that Eq. (8.12) cannot be used if the bar chain includes a vertical member since, then, $\alpha = 90°$ and $\tan \alpha = \infty$. It is never necessary to include such a member, however, since the difference in the vertical deflection of the ends of such a vertical is simply equal to the change in length of that member.

This method can also be applied to compound trusses such as that shown in Fig. 8.27. In such cases, however, it is necessary to insert imaginary bars between such joints as D and E, E and F, etc., so as to divide the truss into triangles for the computation of the angle changes. The changes in length of any one of these bars can be obtained by computing the relative deflection of the joints at the end of the bar by the method of virtual work. The method can likewise be applied to three-hinged arches such as that of Example 8.16. Here again, however, it is necessary to insert an imaginary bar between C and D before the angle changes can be computed. The change in length of this imaginary bar is equal to the relative deflection of joints C and D, which can likewise be computed by the method of virtual work.

8.15 Castigliano's Second Theorem. In 1879, Castigliano published the results of an elaborate research on statically indeterminate structures in which he used two theorems which bear his name.

The second theorem may be stated as follows:

Castigliano's second theorem:

In any structure the material of which is elastic and follows Hooke's law and in which the temperature is constant and the supports unyielding, the first partial derivative of the strain energy with respect to any particular force is equal to the displacement of the point of application of that force in the direction of its line of action.

In this statement, the words *force* and *displacement* should be interpreted also to mean *couple* and *angular rotation*, respectively. Further, it also is implied that, during the deformation of the structure, there is no appreciable change in its geometry. The application of this theorem is therefore limited to cases where it is legitimate to superimpose deflections.

To derive this theorem, consider any structure that satisfies the stated conditions, such as the beam shown in Fig. 8.31. Suppose that this beam is loaded gradually by the forces $P_1, P_2, \ldots, P_n$. Then, the external work done by these forces (let us call this $\mathscr{W}_E$) is some function of these forces. According to the principle of the conservation of

Figure 8.31 Derivation of Castigliano's second theorem

energy, we know that in any elastic structure at rest and in equilibrium under a system of loads, the internal work or strain energy stored in the structure is equal to the external work done by these loads during their gradual application. Designating the internal work or strain energy by $\mathscr{U}$, we therefore write

$$\mathscr{U} = \mathscr{W}_E = f(P_1, P_2, \ldots, P_n) \tag{a}$$

Suppose now that the force P_n is increased by a small amount dP_n; the internal work will be increased, and the new amount will be

$$\mathscr{U}' = \mathscr{U} + \frac{\partial \mathscr{U}}{\partial P_n} dP_n \tag{b}$$

The magnitude of the total internal work, however, does not depend upon the order in which the forces are applied; it depends only on the final value of these forces. Further, if the material follows Hooke's law, the deformation and deflection caused by the forces $P_1, P_2, \ldots, P_n$ and hence the work done by them are the same whether these forces are applied to a structure already acted upon by other forces or not, as long as the total stresses due to all causes remain within the elastic limit. If, therefore, the infinitesimal force dP_n is applied first and the forces $P_1, P_2, \ldots, P_n$ are applied later, the final amount of internal work will still be the same amount as that given by Eq. (*b*).

The force dP_n applied first produces an infinitesimal displacement $d\delta_n$, so that the corresponding external work done during the application of dP_n is a small quantity of the second order and can be neglected. If the forces $P_1, P_2, \ldots, P_n$ are now applied, the external work done just by them will not be modified owing to the presence of dP_n and hence will be equal to the value of $\mathscr{W}_E$ given by Eq. (a). However, during the application of these forces, the point of application of P_n is displaced an amount δ_n in the direction of the line of application of this force, and therefore dP_n does external work during this displacement equal to $dP_n \, \delta_n$. Let the total amount of external work done by the entire system during this loading sequence be $\mathscr{W}'_E$. Then,

$$\mathscr{W}'_E = \mathscr{W}_E + dP_n \, \delta_n \tag{c}$$

But, according to the principle of the conservation of energy, $\mathscr{W}'_E$ equals $\mathscr{U}'$, and therefore

$$\mathscr{W}_E + dP_n \, \delta_n = \mathscr{U} + \frac{\partial \mathscr{U}}{\partial P_n} \, dP_n \tag{d}$$

However, since $\mathscr{W}_E$ is equal to $\mathscr{U}$, Eq. (d) reduces to simply

$$\frac{\partial \mathscr{U}}{\partial P_n} = \delta_n \tag{8.13}$$

This latter equation is the mathematical statement of Castigliano's second theorem.

In order to use Castigliano's theorem, it is first necessary to develop suitable expressions for the strain energy stored, or the internal work done, by the stresses in a member. Consider first the case of the strain energy stored in a bar by an *axial force F* as this force gradually increases from zero to its final value. Consider a differential element of such a bar bounded by two adjacent cross sections, as shown in Fig. 8.32. Suppose that this element is acted upon by a force F_t, some intermediate value between zero and the final value of the force F. Suppose that this force is now increased by an amount dF_t, causing the element to change in length by an amount $\Delta(dL)_t$, where

$$\Delta(dL)_t = dF_t \, \frac{dL}{AE} \tag{e}$$

Neglecting quantities of the second order, the internal work done during the application of dF_t is equal to $F_t \, \Delta(dL)_t$, and therefore the total internal work $d\mathscr{U}$ done in this element during the increase of the force F from zero to its final value is

$$d\mathscr{U} = \int_0^F F_t \, \Delta(dL)_t = \int_0^F F_t \, \frac{dL}{AE} \, dF_t = \frac{F^2}{2AE} \, dL \tag{f}$$

For the entire member, the internal work will be the sum of the terms $d\mathscr{U}$ for all the elements dL, or

$$\mathscr{U} = \int_0^L \frac{F^2}{2AE} \, dL = \frac{F^2 L}{2AE} \tag{g}$$

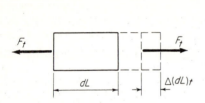

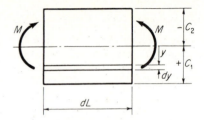

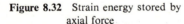

Figure 8.32 Strain energy stored by axial force

Figure 8.33 Strain energy stored by bending moment

For all the members of the structure, the internal work will be the sum of such terms for every bar of the structure, or

$$\text{Strain energy stored by axial forces} = \mathcal{U} = \sum \frac{F^2 L}{2AE} \qquad (8.14)$$

This equation can now be used to develop an expression for the strain energy stored in a beam by the stresses produced by a *bending moment* M. Consider a differential element of a beam having a length dL, as shown in Fig. 8.33. This element of the beam may be considered to be a bundle of little fibers each having a length dL, a depth dy, and a width normal to the plane of the paper of b. The axial force on such a fiber would be

$$F = \sigma b \, dy = \frac{My}{I} b \, dy \qquad (h)$$

The strain energy stored in all such fibers of the beam can be evaluated by applying Eq. (8.14), summing up the contributions of all the fibers across the element dL, and then summing up these quantities for all the elements in the length of the beam, or

$$\mathcal{U} = \int_0^L \int_{-c_2}^{c_1} \left(\frac{My}{I} b \, dy\right)^2 \frac{dL}{2(b \, dy)E} = \int_0^L \frac{M^2}{2EI} \int_{-c_2}^{c_1} \frac{y^2 b \, dy}{I} dL$$

$$= \int_0^L \frac{M^2}{2EI} dL \qquad (i)$$

since $\int_{-c_2}^{c_1} y^2 b \, dy = I$. Hence, for all the beam elements of the structure,

$$\text{Strain energy stored by bending moment} = \mathcal{U} = \sum \int \frac{M^2}{2EI} dL \qquad (8.15)$$

As discussed previously in Art. 8.5, it is usually permissible to neglect the strain energy stored by the shear stresses in a beam.

8.16 Castigliano's First Theorem.[1] To complete the discussion of Castigliano's theorems, his first theorem is also presented here although it is a method for expressing equilibrium conditions and for analyzing statically indeterminate structures rather than a method for computing deflections.

[1] See A. Castigliano, "Théorem de l'equilibre des systèmes élastiques et ses applications," translated by E. S. Andrews as "Elastic Stresses in Structures," Scott, Greenwood, 1919; or

Castigliano's first theorem:

In any structure the material of which is linearly or nonlinearly elastic and in which the temperature is constant and the supports are unyielding, the first partial derivative of the strain energy with respect to any particular deflection component is equal to the force applied at the point and in the direction corresponding to that deflection component.

This theorem can be derived in much the same fashion as the second theorem. Suppose a structure is in equilibrium under the action of the forces $P_1, P_2, \ldots, P_n$. These forces have done a certain amount of external work $\mathcal{W}_E$ and stored an equal amount of strain energy $\mathcal{U}$ in the structure. They likewise have produced certain deflections of the points of application of these forces, $\delta_1, \delta_2, \ldots, \delta_n$. If by varying the forces infinitesimal amounts the deflection δ_n is changed a small amount $d\delta_n$ while all other deflections $\delta_1, \delta_2, \ldots$ are held constant, the strain energy stored in the system will change to a value $\mathcal{U}'$, where

$$\mathcal{U}' = \mathcal{U} + \frac{\partial \mathcal{U}}{\partial \delta_n} d\delta_n \tag{a}$$

If the second-order contribution to the external work done by the differential force dP_n is neglected, the external work done on the structure will have increased to $\mathcal{W}'_E$ as a result of the introduction of the additional displacement $d\delta_n$, where

$$\mathcal{W}'_E = \mathcal{W}_E + P_n\, d\delta_n \tag{b}$$

Since $\mathcal{U}'$ must equal $\mathcal{W}'_E$, equating the right-hand sides of Eqs. (*a*) and (*b*) results in

$$\frac{\partial \mathcal{U}}{\partial \delta_n} = P_n \tag{c}$$

which is a mathematical statement of Castigliano's first theorem.

Application of the first theorem is not considered here. The reader is referred to Matheson's fine book referred to above or to papers by Argyris.[1] In order to use this theorem, it is apparent that alternative expressions for the strain energy will have to be developed. The alternative expressions must express the strain energy in terms of the deflections $\delta_1, \delta_2, \ldots, \delta_n$.

8.17 Computation of Deflections Using Castigliano's Second Theorem. Castigliano's second theorem is used principally in the analysis of statically indeterminate structures, although it is sometimes used to solve deflection problems. The technique of applying this method in the latter case is essentially the same as that of solving such problems by the method of virtual work. In fact, the following examples will demonstrate that the actual numerical computations involved in either of these methods are almost identical.

Example 8.18 illustrates how to apply this method when the deflection of the point of application of one of the loads is required. If this load has some numerical value, it should temporarily be denoted by a symbol, instead. Then, after the partial differentiation of the bending-moment expression has been performed, the symbol can be replaced by the given numerical value.

J. A. L. Matheson, "Hyperstatic Structures," 2d ed., vol. I, Butterworth's Scientific Publications, London, 1971.

Unfortunately there is considerable confusion in the literature regarding the nomenclature of Castigliano's theorems. Castigliano referred to his theorems as "theorem of the differential coefficients of the internal work, part I and part II." In this book, part I and part II are referred to, respectively, as Castigliano's first theorem and Castigliano's second theorem.

[1] J. H Argyris, Energy Theorems and Structural Analysis, *Aircr. Eng.*, October 1954, et seq.

Example 8.18   *Compute the vertical deflection of point b due to the load shown.*

$$\mathcal{U} = \sum \int \frac{M^2\, dx}{2EI}$$

but

$$\frac{\partial \mathcal{U}}{\partial P} = \delta_b^{\downarrow} = \sum \int M\, \frac{\partial M}{\partial P} \frac{dx}{EI}$$

From b to a,

$$0 < x < L \qquad M = -Px \qquad \frac{\partial M}{\partial P} = -x$$

Therefore,

$$\delta_b^{\downarrow} = \int_0^L (-Px)(-x)\, \frac{dx}{EI} = \left[\frac{Px^3}{3EI}\right]_0^L$$

whence

$$\delta_b = \frac{PL^3}{3EI}$$

Sometimes we wish to compute the deflection at a point where no load is acting.   In such cases, we may temporarily introduce an imaginary force (or couple) acting at the point and in the direction of the desired deflection component.   Then, after the partial differentiation of the strain energy has been performed, we let the introduced force equal zero and proceed with the numerical calculations.   In this way, we obtain the desired deflection produced by only the given loads. Example 8.19 illustrates this procedure.

By introducing suitable imaginary loads in this manner, it is possible to compute any desired deflection component.   The technique of selecting the appropriate imaginary loads in such cases is exactly the same as selecting the proper Q-force system in applying the method of virtual work. Example 8.20 illustrates this point and also demonstrates the close similarity between the solution by Castigliano's second theorem and that by the method of virtual work.

It should be noted that Castigliano's second theorem can be applied to any type of structure, whether beam, truss, or frame, the material of which behaves in a linearly elastic manner.   Its use is restricted, however, to cases where the deflection is caused by loads.   This theorem is not applicable to the computation of deflections produced by support settlements or temperature changes.

Example 8.19   *Compute the change in slope of the cross section at a due to the loads shown.*

Suppose that a couple M_1 was applied temporarily at a.   Considering this as part of the load system, we proceed as follows:

$$\mathcal{U} = \sum \int \frac{M^2\, dx}{2EI}$$

but

$$\frac{\partial \mathcal{U}}{\partial M_1} = \alpha_a^{\curvearrowright} = \sum \int M\, \frac{\partial M}{\partial M_1} \frac{dx}{EI}$$

From a to b,

$$0 < x < 10 \qquad M = M_1 + \left(7 - \frac{M_1}{20}\right)x \qquad \frac{\partial M}{\partial M_1} = \left(1 - \frac{x}{20}\right)$$

From d to c,

$$0 < x < 4 \qquad M = \left(13 + \frac{M_1}{20}\right)x \qquad \frac{\partial M}{\partial M_1} = \frac{x}{20}$$

From c to b,

$$4 < x < 10 \qquad M = \left(13 + \frac{M_1}{20}\right)x - 10(x - 4) \qquad \frac{\partial M}{\partial M_1} = \frac{x}{20}$$

Then

$$EI\frac{\partial \mathcal{U}}{\partial M_1} = EI\alpha_a = \int_0^{10} \left[M_1 + \left(7 - \frac{M_1}{20}\right)x\right]\left(1 - \frac{x}{20}\right)dx$$

$$+ \int_0^4 \left[\left(13 + \frac{M_1}{20}\right)x\right]\frac{x}{20}dx + \int_4^{10} \left[\left(13 + \frac{M_1}{20}\right)x - 10(x - 4)\right]\frac{x}{20}dx$$

In this equation, we may now let $M_1 = 0$ since it is an imaginary load.   Then

$$EI\alpha_a = \int_0^{10} 7x\left(1 - \frac{x}{20}\right)dx + \int_0^4 13x\frac{x}{20}dx + \int_4^{10}(3x + 40)\frac{x}{20}dx$$

$$= \left[\frac{7x^2}{2} - \frac{7x^3}{60}\right]_0^{10} + \left[\frac{13x^3}{60}\right]_0^4 + \left[\frac{3x^3}{60} + x^2\right]_4^{10}$$

$$= (350 - \frac{700}{6}) + \frac{208}{15} + \left[\frac{1,000 - 64}{20} + (100 - 16)\right]$$

or

$$\alpha_a = \frac{378}{EI}$$

If $E = (30 \times 10^3)^{k/''^2}$ and $I = 200$ in.^{4},

$$\alpha_a = \frac{378^{k/2}}{[(30 \times 10^3)(144)^{k/''2}[200/(144)^2]'^4} = 0.00907 \text{ radian}$$

Example 8.20   *Solve part a of Example 8.4 using Castigliano's second theorem.*

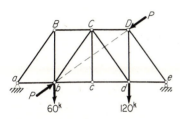

To solve for the relative deflection of joints b and D, add to the given load system the loads P acting as shown.   Then

$$\mathcal{U} = \sum \frac{F^2 L}{2AE}$$

and

$$\frac{\partial \mathcal{U}}{\partial P} = \delta_{b-D} = \sum \frac{FL}{AE}\frac{\partial F}{\partial P}$$

Using the bar forces computed in Example 8.4 for the 60- and 120-kip loads and noting that the forces due to the loads P are P times the forces caused by the unit loads in that problem, we may now proceed as follows:

Bar	L	A	$\dfrac{L}{A}$	F	$\dfrac{\partial F}{\partial P}$	$F\dfrac{L}{A}\dfrac{\partial F}{\partial P}$
Units	,	$''^2$	$'/''^2$	k	k/k	$k'/''^2$
bc	15	5	3	$+\,67.5\ -0.416P$	-0.416	$-\ 84.5$
cd	15	5	3	$+\,67.5\ -0.416P$	-0.416	$-\ 84.5$
CD	15	5	3	$-\,78.75-0.831P$	-0.831	$+197.0$
bC	25	2.5	10	$-\,18.75-0.695P$	-0.695	$+130$
Cd	25	2.5	10	$+\,18.75+0.695P$	$+0.695$	$+130$
dD	20	5	4	$+105\quad-0.555P$	-0.555	-233
Σ						$+\ 55.0$

Before evaluating the product in the last column of this tabulation, set P equal to zero since only the contribution of the constant portion of F need be included in the product. Thus,

$$\delta_{b-D} = \frac{+55}{E} = \frac{55^{k'/''^2}}{(30\times10^3)^{k'/''^2}} = +0.00183\ ft \qquad (together)$$

Note that those bars which have zero forces due to either the imaginary loads P or the given applied loads do not contribute to the products in the last column and therefore do not need to be included in the tabulation.

8.18 Deflection of Space Frameworks. The deflections of the joints of a space framework can be computed without difficulty by either the method of virtual work or Castigliano's second theorem. The expressions developed previously in order to apply these methods to planar trusses may likewise be applied to space frameworks (which are simply three-dimensional "trusses"), as illustrated in Example 8.21.

Example 8.21   *Compute the z component of the deflection of joint d due to the load shown. The cross-sectional area of all members is 2 sq in.   E $= 30 \times 10^3$ kips per sq in.*

Bar	Projection			L
	x	y	z	
ab	10	0	10	14.14
bc	10	0	10	14.14
ac	20	0	0	20
ad	10	20	5	22.91
bd	0	20	5	20.62
cd	10	20	5	22.91

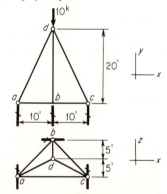

Using the method of virtual work,

$$\sum Q\delta = \sum F_Q\,\Delta L = \sum F_Q F_P \frac{L}{AE}$$

or

$$(1^k)(\delta_{dz}) = \frac{1}{AE}\sum F_Q F_P L$$

Bar	Components			F_Q
	X	Y	Z	
ab	+0.25	0	+0.25	+0.354
bc	+0.25	0	+0.25	+0.354
ac	−0.75	0	0	−0.75
ad	+0.5	+1	+0.25	+1.145
bd	0	−2	−0.5	−2.062
cd	+0.5	+1	+0.25	+1.145

Bar	Components			F_P
	X	Y	Z	
ab	+0.625	0	+0.625	+0.884
bc	+0.625	0	+0.625	+0.884
ac	+0.625	0	0	+0.625
ad	−1.25	−2.5	−0.625	−2.863
bd	0	−5	−1.25	−5.154
cd	−1.25	−2.5	−0.625	−2.863

Bar	L	F_Q	F_P	$F_Q F_P L$
Units	′	k	k	k²′
ab	14.14	+0.354	+0.884	+ 4.4
bc	14.14	+0.354	+0.884	+ 4.4
ac	20	−0.75	+0.625	− 9.4
ad	22.91	+1.145	−2.863	− 75.0
bd	20.62	−2.062	−5.154	+219.4
cd	22.91	+1.145	−2.863	− 75.0
Σ				+ 68.8

whence

$$(1^k)(\delta_{dz}) = \frac{+68.8^{k^2\prime}}{(2''^2)[(30\times 10^3)^{k/''^2}]}$$

$$\delta_{dz} = \underline{+0.00115\ ft}$$

8.19   Other Deflection Problems. All the examples in this chapter have illustrated the computation of the deflection of statically determinate structures. All the methods presented can be applied, however, to both statically determinate and indeterminate structures. Of course, the stress analysis of an indeterminate structure must be completed before the deformation of the elements of the structure can be defined. Once this has been done, however, the computations of the resulting deflections of the structure are essentially the same as if it were statically determinate. Several examples illustrating such computations are included in Art. 9.17.

In this chapter, the discussion of beam deflections has been limited to cases where the centroidal axis of the beam is straight and all the cross sections have axes of symmetry that lie in the same plane as the loads. Further, no discussion has been included concerning the deflections produced by shear. All such cases are beyond the scope of this book and are included in more advanced treatments of this subject. It should be noted, however, that either the method of virtual work or Castigliano's second theorem can be extended without difficulty to cover such cases. It is also possible to apply these methods to members that involve torsion.

8.20 Cambering of Structures. Cambering a structure consists in varying the unstressed shape of the members of the structure in such a manner that under some specified condition of loading the structure attains its theoretical shape. The purpose of such a procedure is twofold: (1) It improves the appearance of the loaded structure. (2) It ensures that the geometry of the loaded structure corresponds to the theoretical shape used in the stress analysis.

To illustrate this procedure, consider the problem of cambering a truss. In this case, the truss members are fabricated so that they are either longer or shorter than their theoretical lengths. Since the members receive their maximum stress under different positions of the live load, trusses cannot be cambered so as to assume their theoretical shape when each member receives its maximum stress. As a practical compromise, trusses are usually cambered so that they attain their theoretical shape under dead load, or under dead load plus some fraction of full live load over the entire structure.

To camber a truss exactly, we compute the change of length of each member under the stresses produced by the cambering load and then fabricate the compression members the corresponding amounts too long and the tension members the corresponding amounts too short. Then, when the truss is erected and subjected to its cambering load, it will deflect into its theoretical shape. The advantage of this exact method is that any truss so cambered can be assembled free of initial stress. The disadvantage is that most members are affected and the required changes in length are sometimes small and difficult to obtain.

The practical method of cambering trusses, therefore, is to alter only the lengths of the chord members. For example, if each top-chord member of an end-supported truss is lengthened $3/16$ in. for every 10 ft of its horizontal projection, this is equivalent to changing both top and bottom chords by one-half of this amount. This change in length is the same as that produced by a stress intensity of

$$(29{,}000{,}000)(3/32)(1/120) = 22{,}600 \text{ psi}$$

Since only the chords are being corrected, they must be overcorrected to allow for the contribution of the web members to the deflection. Assuming that the chords contribute 80 per cent of the deflection, the change in length specified above corresponds to that caused by a load which would produce a chord stress intensity of

$$(0.8)(22{,}600) = 18{,}000 \text{ psi}$$

In other words, the rule of thumb suggested above corresponds to a cambering load essentially equal to dead load plus full live and impact over the entire structure.

This approximate method of cambering can be applied without difficulty to statically determinate trusses. It must be applied with caution, however, to statically indeterminate trusses; otherwise, the assembled truss may have to be forced together, initial stresses thus being introduced into the truss.

8.21 Maxwell's Law of Reciprocal Deflections; Betti's Law. Maxwell's law is a special case of the more general Betti's law. Both laws are applicable to any type of structure, whether beam, truss, or frame. To simplify this discussion, however, these ideas will be developed by considering the simple truss shown in Fig. 8.34. Suppose that this truss is subjected to two separate and independent systems of forces, the system of forces P_m and the system of forces P_n. The P_m system develops the bar forces F_m in the various members of the truss, while the P_n system develops the bar forces F_n. Let us imagine two situations. First, suppose that the P_m system is at rest on the truss and that we then further deform the truss by applying the P_n system. As a second situation, suppose that just the reverse is true, i.e., that the P_n system is acting on the truss and that then we further deform the truss by applying the P_m system. In both situations, we may apply the law of virtual work and thereby come to a very useful conclusion known as Betti's law.

For purposes of this derivation, we shall assume that the supports of the structure are unyielding and that the temperature is constant. Also, let

δ_{mn} = deflection of point of application of one of the forces P_m (in direction and sense of this force) caused by application of P_n force system

δ_{nm} = deflection of point of application of one of the forces P_n caused by application of P_m force system

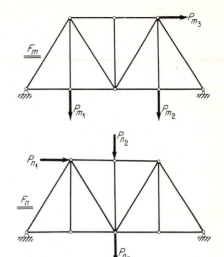

Figure 8.34   Derivation of Betti's law

Consider now the application of the law of virtual work to the first situation. In this case, the P_m force system is in the role of the virtual Q forces and will be given a ride as a result of the deformation caused by the P_n system. Thus, applying Eq. (8.5) gives

$$\sum P_m \delta_{mn} = \sum F_m \, \Delta L$$

where $\Delta L = F_n L / AE$, and thus

$$\sum P_m \delta_{mn} = \sum F_m F_n \frac{L}{AE} \tag{a}$$

In the second situation, however, the P_n force system will now be in the role of the virtual Q forces and will be given a ride as a result of the deformation caused by the P_m system. Then, applying Eq. (8.5) gives

$$\sum P_n \delta_{nm} = \sum F_n \, \Delta L$$

where $\Delta L = F_m L / AE$, and thus

$$\sum P_n \delta_{nm} = \sum F_n F_m \frac{L}{AE} \tag{b}$$

From Eqs. (a) and (b), it may be concluded that

$$\sum P_m \delta_{mn} = \sum P_n \delta_{nm} \tag{8.16}$$

which when stated in words is called Betti's law.

Betti's law:

In any structure the material of which is elastic and follows Hooke's law and in which the supports are unyielding and the temperature constant, the external virtual work done by a system of forces P_m during the deformation caused by a system of forces P_n is equal to the external virtual work done by the P_n system during the deformation caused by the P_m system.

Betti's law is a very useful principle and is sometimes called the **generalized Maxwell's law.** This suggests that Maxwell's law of reciprocal deflections can be derived directly from Betti's law.

Consider any structure such as the truss shown in Fig. 8.35. Suppose that the truss is acted upon first by a load P at point 1; then suppose that the truss is acted upon by a load of the same magnitude P but applied now at point 2. Let

δ_{12} = deflection of point 1 in direction *ab* due to a load P acting at point 2 in direction *cd*

δ_{21} = deflection of point 2 in direction *cd* due to a load P acting at point 1 in direction *ab*

Applying Betti's law to this situation gives

$$P\delta_{12} = P\delta_{21}$$

and therefore
$$\delta_{12} = \delta_{21} \tag{8.17}$$

which, can be stated in words, as follows:

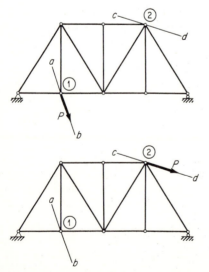

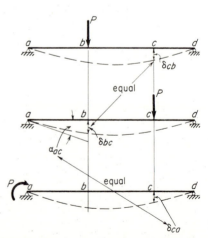

Figure 8.35 Development of Maxwell's law   **Figure 8.36** Application of Maxwell's law

Maxwell's law of reciprocal deflections:

In any structure the material of which is elastic and follows Hooke's law and in which the supports are unyielding and the temperature constant, the deflection of point 1 in the direction ab due to a load P at point 2 acting in a direction cd is numerically equal to the deflection of point 2 in the direction cd due to a load P at point 1 acting in a direction ab.

Maxwell's law is perfectly general and is applicable to any type of structure. This reciprocal relationship exists likewise between the rotations produced by two couples and also between the deflection produced by a couple of P and the rotation produced by a force P. This generality is illustrated by the beam in Fig. 8.36. That $\delta_{cb} = \delta_{bc}$ is a straightforward application of Maxwell's law. Note also that the rotation α_{ac} in radians produced by a force of P lb is numerically equal to the deflection δ_{ca} in feet produced by a couple of P ft-lb. In the latter case, be careful of the units.

It is important to become familiar with the subscript notations used in the above discussion to denote the deflections. *The first subscript denotes the point where the deflection is measured and the second subscript the point where the load causing the deflection is applied.*

8.22 Influence Lines for Deflections. Suppose that we wish to draw the influence line for the vertical deflection at point a on the beam in Fig. 8.37. The ordinates of such an influence line can be computed and plotted by placing a unit vertical load successively at various points along the beam and in each case computing the resulting vertical deflection of point a. In this manner, when the unit load is placed at any point m, it produces a deflection δ_{am} at point a; or when placed at some other point n, it produces a deflection δ_{an} at point a. Note, however, the advantage of applying Maxwell's law to this problem. If we simply place a unit vertical load at point a the deflections δ_{ma} and δ_{na} at points m and n will be equal, respectively, to δ_{am} and δ_{an} according to Maxwell's law. In other

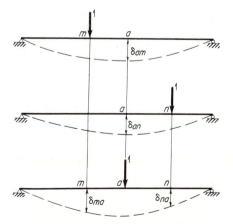

Figure 8.37 Influence lines for deflection

words, *the elastic curve of the beam when the unit load is placed at point a is the influence line for the vertical deflection at point a.* Thus by computing the ordinates of the elastic curve for one simple problem, we obtain the ordinates of the desired influence line.

This idea can be used to obtain the influence line for the deflection of any point on any structure. To obtain the influence line for the deflection of a certain point, simply place a unit load at that point, and compute the resulting elastic curve of the structure.

8.23   Problems for Solution

Problem 8.1   Using the method of virtual work, compute the vertical component of the deflection of joint *d* due to the load shown, for the structure of Fig. 8.38. Bar areas in square inches are shown in parentheses. $E = 30 \times 10^3$ kips per sq in.

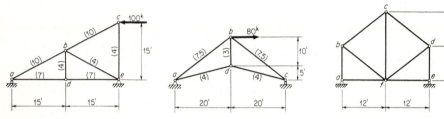

Figure 8.38   Problem 8.1   **Figure 8.39**   Problem 8.2   **Figure 8.40**   Problem 8.3

*Problem 8.2   Using the method of virtual work, compute the horizontal component of the deflection of joint *c* due to the load shown, for the structure of Fig. 8.39. Bar areas in square inches are shown in parentheses. $E = 30 \times 10^3$ kips per sq in.

*Problem 8.3   For the structure of Fig. 8.40, find the horizontal component of the deflection of joint *d* if bar *cf* is shortened 1 in.

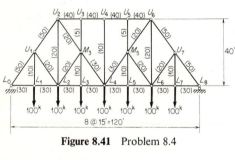

Figure 8.41   Problem 8.4

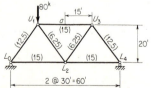

Figure 8.42   Problem 8.5

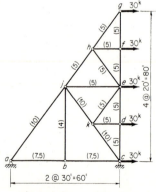

Figure 8.43   Problem 8.6

Problem 8.4 For the structure of Fig. 8.41, using the method of virtual work, determine the relative movement of joints M_3 and U_4 along the line $M_3 U_4$ (a) due to the loading shown; (b) due to a uniform increase in temperature of 50°F in the bottom chord. Bar areas in square inches are shown in parentheses. $E = 30 \times 10^3$ kips per sq in. $\alpha_t = 1/150,000$ per °F.

*Problem 8.5 For the structure of Fig. 8.42, using the method of virtual work, find the vertical deflection of point a due to the load shown. Cross-sectional areas of members in square inches are shown in parentheses. $E = 30 \times 10^3$ kips per sq in.

Problem 8.6 For the structure of Fig. 8.43, cross-sectional areas of members in square inches are shown in parentheses. $E = 30 \times 10^3$ kips per sq in. Compute the angular rotation of member bj due to the loads shown.

*Problem 8.7 For the structure of Fig. 8.44, E_1 of member $gf = 20 \times 10^3$ kips per sq in., and E_2 of all the other members $= 30 \times 10^3$ kips per sq in. Cross-sectional areas in square inches are shown in parentheses. Using the method of virtual work, (a) compute the vertical component of the deflection of point c due to the load shown. (b) If a turnbuckle in member gf were adjusted so as to shorten the member 0.5 in., what would be the vertical and horizontal components of the movement of point c due only to this adjustment?

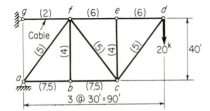

Figure 8.44 Problem 8.7

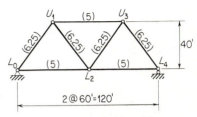

Figure 8.45 Problem 8.8

*Problem 8.8 Refer to Fig. 8.45. During repairs to the right abutment of this truss, it was necessary to support the truss temporarily at joint L_2 on a hydraulic jack. If the gross dead reaction at L_4 is 50 kips, compute the distance the jack must raise the truss at joint L_2 in order to free the support at L_4 and lift it 2 in. above its normal position. Cross-sectional areas of bars in square inches are shown in parentheses. $E = 30 \times 10^3$ kips per sq in.

*Problem 8.9 Refer to Fig. 8.46. Cross-sectional areas of members in square inches are shown in parentheses. E of member $ed = 20 \times 10^3$ kips per sq in. $= E_1$, and E of all other members $= 30 \times 10^3$ kips per sq in. $= E_2$. Using the method of virtual work, determine the direction and magnitude of the resultant deflection of joint d due to load shown.

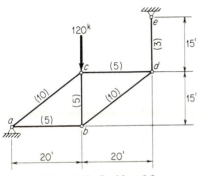

Figure 8.46 Problem 8.9

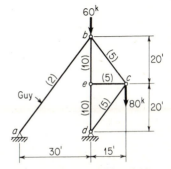

Figure 8.47 Problem 8.10

Problem 8.10 For the structure of Fig. 8.47, the cross-sectional areas of members in square inches are shown in parentheses. E of guy $= 20 \times 10^3$ kips per sq in. $= E_1$, and E of all other members $= 30 \times 10^3$ kips per sq in. $= E_2$.

(a) Compute the vertical component of the deflection of joint c due to the loads shown.

(b) How much would the length of the guy ab have to be changed, by adjusting a turnbuckle in the guy, to return joint c to its undeflected vertical elevation?

Problem 8.11 Using the method of virtual work, determine the vertical deflection and the change in slope of the cross section at point a of the beam of Fig. 8.48. E and I are both constant.

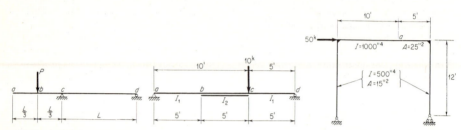

Figure 8.48 Problem 8.11

Figure 8.50 Problem 8.13

*__Problem 8.12__ For the girder of Fig. 8.49, using the method of virtual work, compute (a) the vertical deflection of point b due to the load shown and (b) the change in slope of the cross section at point d due to a uniformly distributed load of 2 kips per ft applied over the entire span.

$$E = 30 \times 10^3 \text{ kips per sq in.} \qquad I_1 = 300 \text{ in.}^4 \qquad I_2 = 500 \text{ in.}^4$$

Problem 8.13 Compute the vertical deflection of point a of the frame of Fig. 8.50, considering the effect of deformation due to both axial stress and bending. $E = 30 \times 10^3$ kips per sq in.

*__Problem 8.14__ For the structure of Fig. 8.51, I of member $ad = 3,456$ in.^{4}, and $E = 30 \times 10^3$ kips per sq in. Using the method of virtual work, compute the vertical component of the deflection of point b.

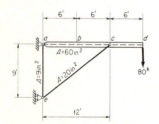

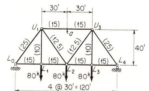

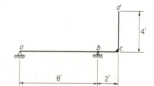

Figure 8.51 Problem 8.14   **Figure 8.52** Problem 8.15   **Figure 8.53** Problem 8.16

Problem 8.15 For the truss of Fig. 8.52, the cross-sectional areas of members in square inches are shown in parentheses. $E = 30 \times 10^3$ kips per sq in. Compute the relative deflection of point a and joint L_3 along the line joining them, due to the loads shown.

Problem 8.16 Refer to Fig. 8.53. Using the method of virtual work, compute the horizontal component of the deflection of point d due to the following support movements:

$$\begin{aligned}
\textit{Point } a\text{: Horizontal} &= 0.36 \text{ in. (to left)} \\
\text{Vertical} &= 0.48 \text{ in. (down)} \\
\textit{Point } b\text{: Vertical} &= 0.96 \text{ in. (down)}
\end{aligned}$$

*Problem 8.17   Refer to Fig. 8.54.

(*a*) Using the moment-area method, compute the vertical deflection of points *a*, *c*, and *d* and the slope of the elastic curve at points *b* and *c* in terms of *E* and *I*.

(*b*) If $E = 30 \times 10^3$ kips per sq in. and $I = 300$ in.⁴, compute from part *a* the vertical deflection of *a* in inches and the slope at *b* in radians.

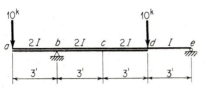

Figure 8.54   Problem 8.17

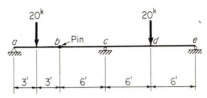

Figure 8.55   Problem 8.18

Problem 8.18   Refer to Fig. 8.55.   Using the elastic-load procedure and the moment-area theorems, find (*a*) the vertical deflection at 3-ft intervals along the beam in terms of *E* and *I* and (*b*) the position and magnitude of the maximum vertical deflection in span *ce*.   *E* and *I* are constant.

Problem 8.19   Refer to Fig. 8.56.   Using the moment-area method, find the position and magnitude of the maximum vertical deflection in member *bc* due to the load shown. $E = 30 \times 10^3$ kips per sq in.

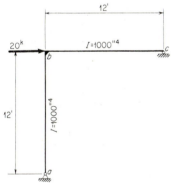

Figure 8.56   Problem 8.19

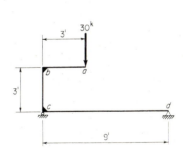

Figure 8.57   Problem 8.20

*Problem 8.20   Refer to Fig. 8.57.   *E* and *I* are constant throughout, $E = 30 \times 10^3$ kips per sq in., and $I = 432$ in.⁴   Compute the vertical component of the deflection of point *a* on the bracket attached to this beam due to the load shown.

Problem 8.21   Refer to Fig. 8.58.   *E* and *I* are constant, $E = 30 \times 10^3$ kips per sq in., and $I = 192$ in.⁴   Compute the location and magnitude of the maximum vertical deflection of the beam *ab*.

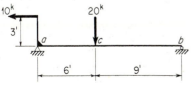

Figure 8.58   Problem 8.21

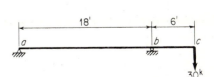

Figure 8.59   Problem 8.22

*Problem 8.22   Refer to Fig. 8.59. $E = 30 \times 10^3$ kips per sq in., and $I = 1,440$ in.[4]   If the support at b settled downward $\frac{1}{8}$ in., determine the location and magnitude of the maximum upward deflection in the portion of the beam ab resulting from both the load shown and the settlement.

Problem 8.23   Refer to Fig. 8.60.   Compute the position and magnitude of the maximum vertical deflection of this structure, using the moment-area theorems or their elastic-load adaptation. E and I are constant, $E = 30 \times 10^3$ kips per sq in., $I = 1,200$ in.[4]

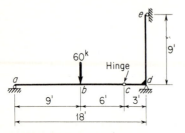

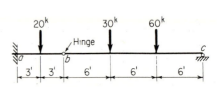

Figure 8.60   Problem 8.23   **Figure 8.61**   Problem 8.24

Problem 8.24   Refer to Fig. 8.61. $E = 30 \times 10^3$ kips per sq in., and $I = 576$ in.[4]   Compute the maximum vertical deflection of this beam, using the conjugate-beam method.

*Problem 8.25   Refer to Fig. 8.62.   Compute the position and magnitude of the maximum vertical deflection in this beam. E and I are constant, $E = 30 \times 10^3$ kips per sq in., $I = 432$ in.[4]

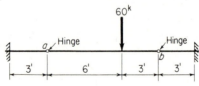

Figure 8.62   Problem 8.25

9

Stress Analysis of
Statically Indeterminate Structures

9.1 Introduction. During the twentieth century, the use of statically indeterminate structures has grown steadily. This is no doubt due to their economy and their increased rigidity under moving or movable loads. The details of reinforced-concrete and welded construction are such that structures of these types are usually wholly or partly continuous in their structural action and are therefore usually statically indeterminate. A knowledge of the analysis of indeterminate structures has thus become increasingly important as the use of these types of construction has become more extensive. Typical examples of indeterminate structures are continuous beams and trusses, two-hinged and hingeless arches, rigid-frame bridges, suspension bridges, and building frames.

Statically indeterminate structures differ from statically determinate ones in two important respects:

1. Their stress analysis involves not only their geometry but also their elastic properties such as modulus of elasticity, cross-sectional areas, and cross-sectional moments of inertia. Thus, to arrive at the final design of an indeterminate structure involves assuming preliminary sizes for the members, making a stress analysis of this design, redesigning the members for these stresses, making a new stress analysis of the revised design, redesigning again, reanalyzing, etc., until one converges on the final design.

2. In general, stresses are developed in indeterminate structures not only by loads but also by temperature changes, support settlements, fabrication errors, etc., and other factors that may affect the compatibility of the deformations of the elements of the structure.

In order to understand the stress analysis of indeterminate structures, it is imperative to understand first the fundamental difference between an unstable, a statically determinate and a statically indeterminate structure. For this reason, a short review of certain fundamentals is justified. Suppose that bar AB is supported only by a hinge support at A, as shown in Fig. 9.1a. If the 10-kip load shown is then applied to the bar, it will obviously rotate freely about the hinge at A. A bar supported in this manner will therefore be an **unstable structure.** By means of an additional roller support at B, however, as shown in Fig. 9.1b, we obviously prevent rotation about A and therefore have arranged the structure so that it is **stable;** by inspection we can see that it is also a **statically determinate structure.**

If we add still another roller support at C, as shown in Fig. 9.1c, we then have more reactions than the minimum required for static equilibrium and no longer have a statically determinate structure but, instead, a **statically indeterminate structure.** With this new arrangement, not only do we prevent the bar from translating or rotating as a rigid body, but also by the addition of the roller support at C we prevent this point from deflecting vertically.

This observation, however, immediately suggests a method by which we can compute the vertical reaction at C. If the support at C is temporarily removed, suppose that the 10-kip load acting on the beam supported simply by the remaining supports causes a downward deflection of 3 in. at C, as shown in Fig. 9.2a. If the support at C is now reestablished, it will have to provide an upward reaction in order to return point C to a position of zero deflection. Just how much this upward reaction will have to be can be determined by first finding how much a 1-kip load will deflect point C. If a 1-kip load acting at point C on the beam supported at A and B deflected this point $\frac{1}{2}$ in., as shown in Fig. 9.2b, then an upward force of 6 kips at point C, together with the 10-kip load, will produce a net deflection of zero at this point, as shown in Fig. 9.2c. The vertical reaction at point C of the beam shown in Fig. 9.1c must therefore be 6 kips acting upward. Once this reaction has been determined, the remaining reactions at A and B can easily be computed by statics.

9.2   Use of Deflection Superposition Equations[1] **in Analysis of Statically Indeterminate Structures.** Many simple indeterminate structures can be analyzed in the manner just discussed. Such an informal approach is confusing, however, in the case of more complicated structures. It is therefore desirable to develop a more formal and orderly procedure to facilitate the solution of the more complicated problems and also to handle the simple problems with maximum efficiency.

The ideas and philosophy behind a more orderly approach can be developed by considering a specific example such as the indeterminate beam shown in Fig. 9.3a, the supports of which are unyielding. This beam is statically indeterminate to the first degree, i.e., there is one more reaction component than the minimum necessary for static equilibrium. One of the reaction components may be considered as

[1] Authors have used a variety of names for this method, e.g., **Maxwell's method,** the **general method, the method of consistent displacements** as well as the **superposition-equation method.** Clerk Maxwell introduced this method in 1864. Independently in 1874, Otto Mohr developed the same method. In 1886, Heinrich Müller-Breslau published his variation and extensions of the work of Maxwell and Mohr.

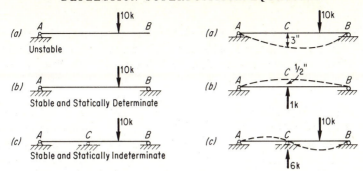

Figure 9.1   Classification of structures   **Figure 9.2**   Simple superposition of effects

being extra, or **redundant.** In this case, consider the vertical reaction at b as being the **redundant reaction.**

Suppose that we remove from the actual structure the vertical support at b and the vertical reaction X_b which it supplies. The statically determinate and stable cantilever beam that remains after the removal of this redundant support at b is called the **primary structure** (see Fig. 9.3b). We now subject this primary structure to the combined effect of the same loading as the actual structure plus the unknown redundant force X_b, as shown in Fig. 9.3c.

If the redundant force X_b acting on the primary structure has the same value as the vertical reaction at b on the actual structure, then the shear and bending moment at any point and the reactions at point a are the same for the two structures. If the conditions of stress in the actual structure and the primary structure are the same, then the conditions of deformation of the two structures must also be exactly the same. If the conditions of deformation of the two structures are the same and the deflections of the supports at point a on each are also identical, then the deflections at any other corresponding points must be the same. Therefore, since there is no vertical deflection of the support at b on the actual structure, the vertical deflection of point b on the primary structure, due to the combined action of the applied load and X_b, must also be equal to zero.

It is possible, however, to express this latter statement mathematically and thus obtain an equation from which the value of the unknown redundant X_b can be determined. Assuming the positive direction of the redundant X_b to be upward, let us introduce the following notation. Let

Δ_b = upward deflection of point b on primary structure due to all causes (see Fig. 9.3c)

Δ_{b0} = upward deflection of point b on primary structure due only to the applied load with the redundant removed, hereafter called condition $X = 0$ (see Fig. 9.3d)

Δ_{bb} = upward deflection of point b on primary structure due only to redundant X_b

It is impossible to compute Δ_{bb} until the magnitude of X_b is known. If, however, we also let

δ_{bb} = upward deflection of point b on primary structure due only to a unit upward load at b, hereafter called condition $X_b = +1$ kip (see Fig. 9.3e)

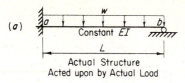

(a) Actual Structure
Acted upon by Actual Load

(b) Primary Structure

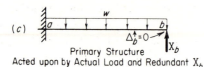

(c) Primary Structure
Acted upon by Actual Load and Redundant X_b

(d) Primary Structure
Acted upon by only Actual Load. Condition $X = 0$, M_0

(e) Primary Structure
Acted upon by unit value of X_b. Condition $X_b = +1^k$, M_b

Figure 9.3 Organization of analysis by deflection superposition equations

then we can say, *as long as the principle of superposition is valid,* that

$$\frac{\Delta_{bb}}{X_b} = \frac{\delta_{bb}}{1^k} \qquad (a)$$

Hence, $\qquad \Delta_{bb} = \frac{X_b}{1^k}\,\delta_{bb} \qquad \text{or} \qquad \Delta_{bb} = X_b\,\delta_{bb} \qquad (b)$

where, since we have replaced $X_b/1^k$ by simply X_b, *we henceforth must think of X_b as being dimensionless and representing simply the numerical value of X_b,* that is, the number of 1-kip forces represented by this redundant.[1] It is physical fact, however, that Δ_b, the total deflection due to all causes, is equal to the superposition of the

---

[1] Alternatively, we could have rearranged Eq. (a) as follows,

$$\Delta_{bb} = X_b \frac{\delta_{bb}}{1^k} \qquad \text{or} \qquad \Delta_{bb} = X_b\,d_{bb} \qquad (b')$$

where d_{bb} is called a **flexibility** (or **deflection**) **coefficient.** This coefficient d_{bb} denotes the upward vertical deflection of point b per unit upward force at point b and has units of distance per unit force. If we had elected to use this system, henceforth X_b would have been a dimensional number with force units.

contributions of the separate effects, namely, the applied load and the redundant X_b. Thus,

$$\Delta_b = \Delta_{b0} + \Delta_{bb} \qquad\qquad (c)$$

or, substituting from Eq. (b),

$$\Delta_b = \Delta_{b0} + X_b\delta_{bb} \qquad\qquad (9.1)$$

This equation is called a **superposition equation** for the deflection of point b on the primary structure.

Since Δ_b must be zero, we can solve Eq. (9.1) for X_b and obtain

$$X_b = -\frac{\Delta_{b0}}{\delta_{bb}} \qquad\qquad (9.2)$$

It is a simple matter, of course, to compute the numerical value of Δ_{b0} and δ_{bb} by any of the methods available for computing beam deflections. In substituting these values in Eq. (9.2), they should be considered plus when the deflections are upward as specified above. A plus value for X_b will then indicate that it acts upward; a minus value, that it acts downward. The organization of the numerical computations in problems such as this is illustrated by the examples in Art. 9.4.

9.3   General Discussion of Use of Deflection Superposition Equations in Analysis of Statically Indeterminate Structures. The method of attack suggested in Art. 9.2 is the most general available for the stress analysis of indeterminate structures. There are other methods, of course, that are definitely superior for certain specific types of structures, but there is no other one method that is as flexible and general as the method based on superposition equations. It is applicable to any structure, whether beam, frame, or truss or any combination of these simple types. It is applicable whether the structure is being analyzed for the effect of loads, temperature changes, support settlements, fabrication errors, or any other cause.

There is only one restriction on the use of this approach, namely, that the principle of superposition be valid. From the discussion in Art. 2.12, it will be recalled that the principle of superposition is valid unless the geometry of the structure changes an essential amount during the application of the loads or the material does not follow Hooke's law. All the other methods discussed here are likewise subject to this same limitation, however.

Before presenting numerical examples illustrating the application of this method to certain typical problems, it is desirable to emphasize and extend the ideas suggested in Art. 9.2. The student may have some difficulty understanding the somewhat generalized discussion that follows. After reading this article, however, absorbing as much as possible, and then studying the examples in Arts. 9.4 and 9.5, he should be able to understand the remainder of this article when he rereads it.

Suppose that we wish to analyze some particular indeterminate structure for any or all of the various causes which may produce stress in the structure. The structure may be of any type or statically indeterminate to any degree. In any case, the very first step in the analysis is to determine the degree of indeterminacy. Suppose the structure is indeterminate to the nth degree. We then select n redundant restraints, remove them from the structure, and replace them by the n redundant stress components they supply to the structure. All these n redundants X_a, X_b, ..., X_n will then be acting together with the applied loads on the *primary structure* that remains after the removal of the n restraints.

A wide choice of redundants is available in most problems, but it is usually possible to minimize the numerical calculations by choosing them judiciously. Some of the principles to follow in selecting the redundants are discussed in Art. 9.6. Suffice it to say for the present that the redundants must be selected so that the resulting primary structure is stable and statically determinate.[1]

Having selected the redundants, we can then reason that if the redundant forces acting on the primary structure have the same values as the forces supplied by the corresponding restraints on the actual structure, then the entire conditions of stress in the actual and primary structure are the same. As a result, the conditions of deformation of the two structures are the same. If, in addition, the deflections of the supports of the primary structure are the same as the deflections of the corresponding supports of the actual structure, then the deflections of any other corresponding points on the two structures must be identical. It may be concluded, therefore, that the deflections of the points of application of the n redundants on the primary structure must be equal to the specified deflections of the corresponding points on the actual structure.

If n superposition equations are written for these n deflections, they will involve among them all the n redundants as unknowns. Since all these n deflection equations must be satisfied simultaneously by the n unknowns, simultaneous solution of these equations leads to the values of the n unknown redundants.

In applying this method, it is convenient to have a definite notation for the various deflection terms involved in the solution. The total deflection of a point m on the primary structure due to all causes is denoted by Δ_m. The various supplementary deflections contributing to this total are identified, however, by two subscripts, the first of which denotes the point at which the deflection occurs and the second of which denotes the loading condition which causes that particular contribution. Thus, the deflection of a point m produced *on the primary structure* by certain typical causes would be denoted as follows:

Δ_m = total deflection of point m due to all causes

Δ_{m0} = deflection of point m due to condition $X = 0$, that is, applied load only with all redundants removed

Δ_{mT} = deflection of point m due to change in temperature

Δ_{mS} = deflection of point m due to *settlement of supports of primary structure*

Δ_{mE} = deflection of point m due to fabrication errors

δ_{ma} = deflection of point m due to condition $X_a = +1$ kip

δ_{mb} = deflection of point m due to condition $X_b = +1$ kip

δ_{mm} = deflection of point m due to condition $X_m = +1$ kip

etc.

---

[1] Occasionally it is advantageous to select a statically indeterminate and stable primary structure (see Art. 10.3).

Note that the deflections produced by unit values of the various redundants are denoted by a small delta (δ).

Any redundant may be arbitrarily assumed to act in a certain sense, thus establishing the positive sense of that redundant. Any deflection of the point of application of that redundant should then be measured along its line of action and should be considered positive when in the same sense as that assumed for the redundant.

Thus, using the above notation and sign convention, we have the following form for the n superposition equations involving the n redundants—one equation for the known total deflection of the point of application of each of the n redundants, on the primary structure:

$$\begin{aligned}
\Delta_a &= \Delta_{a0} + \Delta_{aT} + \Delta_{aS} + \Delta_{aE} + X_a\,\delta_{aa} + X_b\,\delta_{ab} + \cdots + X_n\,\delta_{an} \\
\Delta_b &= \Delta_{b0} + \Delta_{bT} + \Delta_{bS} + \Delta_{bE} + X_a\,\delta_{ba} + X_b\,\delta_{bb} + \cdots + X_n\,\delta_{bn} \\
&\cdots\cdots\cdots\cdots\cdots\cdots\cdots\cdots\cdots\cdots\cdots\cdots\cdots\cdots\cdots\cdots\cdots\cdots\cdots\cdots \\
\Delta_n &= \Delta_{n0} + \Delta_{nT} + \Delta_{nS} + \Delta_{nE} + X_a\,\delta_{na} + X_b\,\delta_{nb} + \cdots + X_n\,\delta_{nn}
\end{aligned} \tag{9.3}$$

If the known values of Δ_a, Δ_b, ..., Δ_n are substituted and the various deflection terms on the right-hand side of these equations are evaluated by *any* suitable method, then the n equations can be solved simultaneously[1] for the value of the redundants X_a, X_b, etc.[2]

Note that the values of X_a, X_b, etc., obtained from this solution are dimensionless. This is obviously so, since X_a, X_b, etc., must be pure numbers if all the terms in each of Eqs. (9.3) are to have the same deflection units. In order to assign units to these values for X_a, X_b, etc., it is necessary to note the units of the unit values of these redundants that were used in the computation of δ_{aa}, δ_{ab}, etc.

Note further that the redundants selected may be forces or couples. The deflections designated by Δ or δ may therefore represent a linear or angular deflection, depending on whether they represent the deflection of the point of application of a force or a couple. In the examples that follow, the positive direction of any particular deflection is indicated by the small arrow appended to the symbol, as $\Delta_a^{\uparrow}$, $\Delta_b^{\curvearrowright}$, $\Delta_c^{\rightarrow\leftarrow}$, etc.

9.4   Examples Illustrating Stress Analysis Using Deflection Superposition Equations. The following examples illustrate the application of the superposition-equation approach to the stress analysis of typical indeterminate structures acted upon by specified external loads.

---

[1] A convenient tabular method for solving simultaneous equations is discussed in connection with Example 9.19.

[2] If the alternative approach suggested in Eq. (*b'*) of the footnote of Art. 9.2 were used, the deflection superposition equations (9.3) could be rewritten in terms of the flexibility coefficients d_{ij}, where d_{ij} represents the deflection of the point of application of redundant X_i in the assumed direction of redundant X_i due to a unit value of redundant X_j.

$$\begin{aligned}
\Delta_a &= \Delta_{a0} + \Delta_{aT} + \Delta_{aS} + \Delta_{aE} + X_a d_{aa} + X_b d_{ab} + \cdots + X_n d_{an} \\
\Delta_a &= \Delta_{b0} + \Delta_{bT} + \Delta_{bS} + \Delta_{bE} + X_a d_{ba} + X_b d_{bb} + \cdots + X_n d_{bn} \\
&\cdots\cdots\cdots\cdots\cdots\cdots\cdots\cdots\cdots\cdots\cdots\cdots\cdots\cdots\cdots\cdots\cdots\cdots \\
\Delta_n &= \Delta_{n0} + \Delta_{nT} + \Delta_{nS} + \Delta_{nE} + X_a d_{na} + X_b d_{nb} + \cdots + X_n d_{nn}
\end{aligned} \tag{9.3'}$$

As noted previously, the redundants X in these equations would be dimensional numbers with appropriate force or couple units as the case may be.

After determining the degree of indeterminacy, one has considerable latitude in selecting the redundants.   However, *never make the mistake of selecting a statically determinate reaction component, bar force, shear, or moment, as a redundant.   If such a quantity is statically determinate, it is necessary for the stability of the structure; to remove it would leave an unstable primary structure.*   Such an error will never go undetected, however, since an attempted stress analysis of the primary structure would lead to impossible or inconsistent results.

The reader should consider alternative selections of the redundants in the following examples and try to decide whether in each case there is some other selection involving less numerical calculation than those used.   The question of selection of redundants is discussed further in Art. 9.6.

The positive sense of a redundant may be chosen arbitrarily.   The deflection of the point of application of this redundant should likewise be considered positive when in the same sense.   Consideration of any of these examples should make it apparent that the final interpretation of the direction of the redundants will not be affected by the initial choice of its positive sense.

In applying the superposition approach, any suitable method may be used to compute the various deflections involved.   Of course, as in any deflection problem, care must be exercised in handling units and signs.   *Note again that the values of X_a, X_b, etc., obtained from these solutions are simply numerical values, i.e., they are dimensionless.*   Such numerical values indicate, for example, that X_a is, say, 10.5 times larger than the unit load in condition $X_a = +1$ kip.   Thus, if this is a 1-kip unit load, X_a should be considered as 10.5 kips in stating the final results or in using it in subsequent computations.

The units will always be consistent if the same force and distance units are used throughout a given problem.   If, however, we mix up feet, inches, pounds, and kips in problems involving more than one type of deformation, we may run into difficulty.   Similar difficulty will be encountered if units are mixed in dealing with problems involving temperature changes, support settlements, etc., such as are discussed in Art. 9.5.

It should be noted that Δ_n, the total deflection of the point of application of a redundant X_n due to all causes, is almost always equal to zero.   In fact, the only case in which Δ_n would have a value other than zero would be when X_n represented a redundant reactive force or couple at a support on the actual structure, which underwent a movement in the direction of X_n.

Example 9.1   *Compute the reactions of this beam.*   *E and I are constant.*

Structure is indeterminate to first degree.
Select X_b as redundant.   Then

$$\Delta_b^1 = \Delta_{b0} + X_b\, \delta_{bb} = 0$$

Evaluate Δ_{b0} and δ_{bb} by method of virtual work.

$$\sum Q\,\delta = \sum \int M_Q M_P \frac{ds}{EI}$$

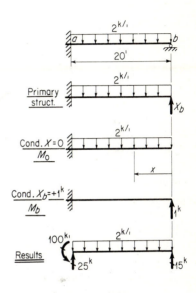

Δ_{b0}: $M_Q = M_b$ $M_P = M_0$

$$(1^k)(\Delta_{b0}^1) = \int M_b M_0 \frac{ds}{EI}$$

From b to a,

$$M_b = x \qquad M_0 = -x^2$$

$$(1^k)(\Delta_{b0}^1) = \int_0^{20} (x)(-x^2)\frac{dx}{EI}$$

$$\Delta_{b0} = \frac{-40{,}000^{k/3}}{EI}$$

δ_{bb}: $M_Q = M_b$ $M_P = M_b$

$$(1^k)(\delta_{bb}^1) = \int M_b^2 \frac{dx}{EI}$$

$$(1^k)(\delta_{bb}^1) = \int_0^{20} x^2 \frac{dx}{EI} \qquad \delta_{bb} = \frac{2{,}667^{k/3}}{EI}$$

Then

$$\frac{-40{,}000}{EI} + \frac{2{,}667}{EI} X_b = 0$$

$$\therefore X_b = +15 \qquad \therefore \text{ upward}$$

Then, by statics, the reactions at a are found to be
as shown.

Example 9.2 *Compute the reactions and draw the bending-moment diagram for this beam. E and I are constant.*

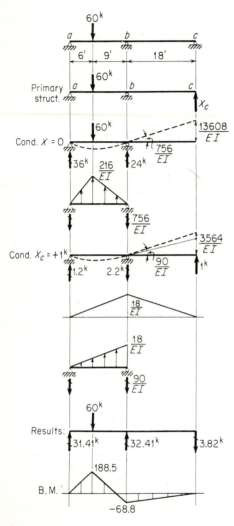

Structure is indeterminate to first degree. Select X_c as redundant. Then,

$$\Delta_c^1 = \Delta_{c0} + X_c \, \delta_{cc} = 0$$

Evaluate Δ_{c0} *and* δ_{cc} *by the elastic-load method.*

Δ_{c0}: *Compute first the reaction at b on imaginary beam.*

$$\begin{array}{ll}
(216)(3) \;\; = (648)(4) = & 2{,}592 \\
(216)(4.5) = (972)(9) = & 8{,}748 \\
\cline{2-2}
& 11{,}340 \\
& \overline{\phantom{11{,}340}} \;\; = 756 \\
& \quad 15
\end{array}$$

Then

$$EI\Delta_{c0} = (756)(18) = 13{,}608 \uparrow$$

$$\therefore \; \Delta_{c0} = \frac{+13{,}608^{k\prime3}}{EI}$$

δ_{cc}: *Compute first the reaction at b on imaginary beam.*

$$(\tfrac{2}{3})(18)(7.5) = 90$$

Then

$$EI\,\delta_{cc} = (90)(18) + (18)(9)(12) = 3{,}564 \uparrow$$

$$\therefore \; \delta_{cc} = \frac{+3{,}564^{k\prime3}}{EI}$$

Then

$$\frac{13{,}608}{EI} + \frac{3{,}564}{EI} \, X_c = 0$$

$$\therefore \; X_c = -\,3.82 \qquad \therefore \; down$$

The reaction at a and b can now be computed easily by statics. Likewise, the bending-moment diagram can be constructed without difficulty.

Example 9.3　*Compute the reactions and bar forces for this truss.　Cross-sectional areas in square inches are shown in parentheses.　$E = 30 \times 10^3$ kips per sq in.*

Indeterminate to first degree.　Select X_b as redundant.　Then

$$\Delta_b^\dagger = \Delta_{b0} + X_b \, \delta_{bb} = 0$$

Evaluate Δ_{b0} and δ_{bb} by the method of virtual work.

$$\sum Q \, \delta = \sum F_Q \, \Delta L = \sum F_Q F_P \frac{L}{AE}$$

$$\Delta_{b0}: \quad F_Q = F_b \quad F_P = F_0 \quad (1^k)(\Delta_{b0}^\dagger) = \frac{1}{E} \sum F_b F_0 \frac{L}{A}$$

$$\delta_{bb}: \quad F_Q = F_b \quad F_P = F_b \quad (1^k)(\delta_{bb}^\dagger) = \frac{1}{E} \sum F_b^2 \frac{L}{A}$$

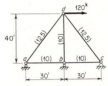

Primary struct. F

Cond. $X=0$ F_0

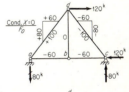

Cond. $X_b = +1^k$ F_b

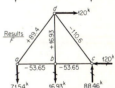

Results F

Bar	L	A	$\dfrac{L}{A}$	F_0	F_b	$F_b F_0 \dfrac{L}{A}$	$F_b^2 \dfrac{L}{A}$	$X_b F_b$	F
Units	$'$	$''^2$	$'/''^2$	k	k	$k^2 '/''^2$	$k^2 '/''^2$	k	k
ab	30	10	3	-60	-0.375	$+67.5$	$+0.422$	$+6.35$	-53.65
bc	30	10	3	-60	-0.375	$+67.5$	$+0.422$	$+6.35$	-53.65
ad	50	12.5	4	$+100$	$+0.625$	$+250$	$+1.563$	-10.6	$+89.4$
dc	50	12.5	4	-100	$+0.625$	-250	$+1.563$	-10.6	-110.6
bd	40	10	4	0	-1.0	0	$+4.0$	$+16.93$	$+16.93$
Σ						$+135$	$+7.970$		

$$(1^k)(\Delta_{b0}) = \frac{+135^{k^2 '/''^2}}{E} \qquad \Delta_{b0} = \frac{135^{k'/''^2}}{E}$$

$$(1^k)(\delta_{b0}) = \frac{+7.97^{k^2 '/''^2}}{E} \qquad \delta_{b0} = \frac{7.97^{k'/''^2}}{E}$$

$$\therefore \frac{135}{E} + \frac{7.97}{E} X_b = 0 \qquad X_b = \underline{-16.93} \qquad \therefore \text{ down}$$

The remaining reactions and the bar forces can now be computed by statics.　The bar forces can likewise be computed easily in a tabular manner by noting that by superposition

$$F = F_0 + X_b F_b$$

Example 9.4 *Compute the reactions of this truss. Cross-sectional areas in square inches are shown in parentheses.* $E = 30 \times 10^3$ *kips per sq in.*

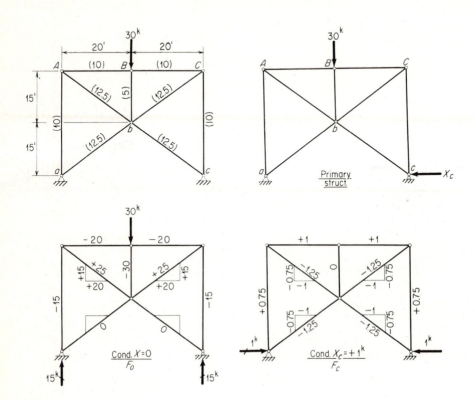

Indeterminate to first degree. Select X_c as redundant.

$$\Delta_c^{\leftarrow} = \Delta_{c0} + X_c \, \delta_{cc} = 0$$

Evaluate Δ_{c0} and δ_{cc} by the method of virtual work.

Δ_{c0}:　　　　　　$F_Q = F_c$　　$F_P = F_0$　　$(1^k)(\Delta_{c0}^{\leftarrow}) = \dfrac{1}{E} \sum F_c F_0 \dfrac{L}{A}$

δ_{cc}:　　　　　　$F_Q = F_c$　　$F_P = F_c$　　$(1^k)(\delta_{cc}^{\leftarrow}) = \dfrac{1}{E} \sum F_c^2 \dfrac{L}{A}$

Owing to symmetry of the load, structure, and redundant, only one-half of the bars need be included in the tabulation.

$$\frac{-272.5}{E} + \frac{19.875}{E} X_c = 0$$

$$\therefore X_c = +13.7 \qquad \therefore \leftrightarrow$$

Results:

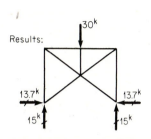

Bar	L	A	$\dfrac{L}{A}$	F_0	F_c	$F_0 F_c \dfrac{L}{A}$	$F_c^2 \dfrac{L}{A}$
Units	$'$	$''^2$	$'/''^2$	k	k	$k^2,/''^2$	$k^2,/''^2$
AB	20	10	2	-20	$+1.0$	-40	$+2.0$
ab	25	12.5	2	0	-1.25	0	$+3.125$
Aa	30	10	3	-15	$+0.75$	-33.75	$+1.688$
Ab	25	12.5	2	$+25$	-1.25	-62.5	$+3.125$
$\frac{1}{2}Bb$	7.5	5	1.5	-30	0	0	0
$\frac{1}{2}\Sigma$						-136.25	$+9.938$

$$\therefore \Delta_{co} = \frac{-272.5^{k,/''^2}}{E} \qquad \delta_{cc} = \frac{19.875^{k,/''^2}}{E}$$

Example 9.5   *Compute the bar forces in the members of this truss. E and A are constant for all members.*

This truss is statically determinate externally but indeterminate to the second degree with respect to its bar forces. Cut bars bC and Cd, and select their bar forces as the redundants. Then

$$\Delta_1^{\nwarrow} = \Delta_{10} + X_1 \delta_{11} + X_2 \delta_{12} = 0 \qquad (1)$$
$$\Delta_2^{\searrow} = \Delta_{20} + X_1 \delta_{21} + X_2 \delta_{22} = 0 \qquad (2)$$

By Maxwell's law, $\delta_{12} = \delta_{21}$, and by symmetry of the structure and in the selection of the redundants, $\delta_{11} = \delta_{22}$. Therefore, only four deflections need be computed: $\Delta_{10}, \Delta_{20}, \delta_{12},$ and δ_{11}. Using the method of virtual work, we find

$$(1^k)(EA\,\Delta_{10}) = \Sigma F_1 F_0 L = -2,040^{k2,}$$
$$(1^k)(EA\,\Delta_{20}) = \Sigma F_2 F_0 L = +760^{k2,}$$
$$(1^k)(EA\,\delta_{12}) = \Sigma F_1 F_2 L = +12.8^{k2,}$$
$$(1^k)(EA\,\delta_{11}) = \Sigma F_1^2 L = +86.4^{k2,}$$

Substituting these values in (1) and (2) and canceling out EA gives

$$-2,040 + 86.4X_1 + 12.8X_2 = 0$$
$$+760 + 12.8X_1 + 86.4X_2 = 0$$

Solving these simultaneously we have

$$X_1 = +25.5 \qquad X_2 = -12.6$$

In any bar, $\qquad F = F_0 + X_1 F_1 + X_2 F_2$

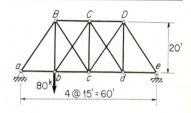

Primary struct.

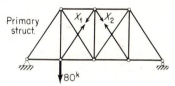

Cond. $X = 0$
F_0

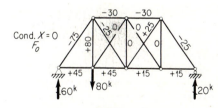

Cond. $X_1 = +1^k$
F_1

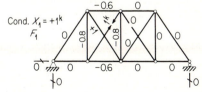

Bar	L	F_0	F_1	F_2	$F_0 F_1 L$	$F_0 F_2 L$	$F_1^2 L$	$F_1 F_2 L$	$F_1 X_1$	$F_2 X_2$	F
Units	'	k	k	k	$k^2{}'$	$k^2{}'$	$k^2{}'$	$k^2{}'$	k	k	k
bc	15	+45	−0.6	0	−405	0	+ 5.4	0	−15.3	0	+29.7
cd	15	+15	0	−0.6	0	−135	0	0	0	+ 7.6	+22.6
BC	15	−30	−0.6	0	+270	0	+5.4	0	−15.3	0	−45.3
CD	15	−30	0	−0.6	0	+270	0	0	0	+ 7.6	−22.4
bB	20	+80	−0.8	0	−1,280	0	+12.8	0	−20.4	0	+59.6
cC	20	0	−0.8	−0.8	0	0	+12.8	+12.8	−20.4	+10.1	−10.3
dD	20	0	0	−0.8	0	0	0	0	0	+10.1	+10.1
Bc	25	−25	+1	0	−625	0	+25.0	0	+25.5	0	+ 0.5
bC	25	0	+1	0	0	0	+25.0	0	+25.5	0	+25.5
cD	25	+25	0	+1	0	+625	0	0	0	−12.6	+12.4
Cd	25	0	0	+1	0	0	0	0	0	−12.6	−12.6
Σ					−2,040	+760	+86.4	+12.8			

Discussion:

In tabulating the computations in such a truss, it is unnecessary to include bars having neither an F_1 nor an F_2 bar force. Be sure to include the redundant bars, however. Why?

There may be some question about cutting truss bars as was done in this problem. Students sometimes worry about the two ends of a cut bar being unstable, since, as shown in the above line diagrams, there is nothing to prevent these two ends from rotating about the pin joints. To be strictly correct, we should show the cut in such cases as in Fig. 9.4. Since the shear and moment

Figure 9.4   Visualization of removing capacity to transmit axial force

in a straight bar hinged at both ends are statically determinate, the restraints of this type cannot be removed. Thus, when we say that we cut such a bar, we mean that we remove its capacity to carry axial force but retain its ability to carry shear and moment. This could be accomplished by cutting through the member and then inserting a telescopic device in the manner shown.

It is impractical, however, to show such a detail every time we remove the capacity to carry axial force from a bar that is hinged at its ends. In such cases, therefore, we shall imply such a detail but show the cut as has been done in the line diagrams in this example.

These remarks are likewise applicable to the tie rod in Example 9.8.

Example 9.6   *Compute the reactions and draw the bending-moment diagram for this beam. $E = 30 \times 10^3$ kips per sq in.*

This beam is statically indeterminate to the second degree. Select moments at supports as redundants. Then

$$\Delta_b^{\circlearrowright} = \Delta_{b0} + X_b\,\delta_{bb} + X_c\,\delta_{bc} = 0 \qquad (1)$$

$$\Delta_c^{\circlearrowright} = \Delta_{c0} + X_b\,\delta_{cb} + X_c\,\delta_{cc} = 0 \qquad (2)$$

Note $\delta_{bc} = \delta_{cb}$ by Maxwell's law. Use moment-area theorems to compute Δ_{b0}, Δ_{c0}, δ_{bb}, δ_{bc}, and δ_{cc}.

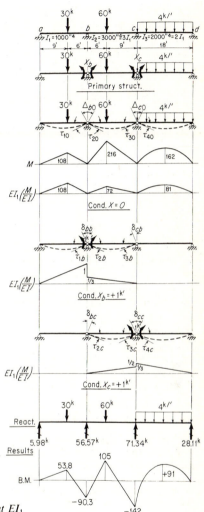

$$EI_1\tau_{10} = \frac{108}{15}\left[\frac{(9)(6)}{2} + \frac{(6)(11)}{2}\right] = 432$$

$$EI_1\tau_{20} = \frac{72}{15}\left[\frac{(9)(6)}{2} + \frac{(6)(11)}{2}\right] = 288$$

$$\therefore EI_1\,\Delta_{b0} = \underline{+720^{k\prime 2}}$$

$$EI_1\tau_{30} = \frac{72}{15}\left[\frac{(6)(4)}{2} + \frac{(9)(9)}{2}\right] = 252$$

$$EI_1\tau_{40} = \frac{2}{3}\frac{(81)(18)}{2} = 486$$

$$\therefore EI_1\,\Delta_{c0} = \underline{+738^{k\prime 2}}$$

$$EI_1\tau_{1b} = \frac{1}{15}\frac{(15)(10)}{2} = 5$$

$$EI_1\tau_{2b} = \frac{1}{3}\frac{(15)(10)}{(15)(2)} = \frac{5}{3}$$

$$\therefore EI_1\,\delta_{bb} = \underline{+6.67^{k\prime 2}}$$

$$EI_1\tau_{3b} = \frac{1}{3}\frac{(15)(5)}{(15)(2)} = \frac{5}{6}$$

$$\therefore EI_1\,\delta_{cb} = \underline{+0.83^{k\prime 2}}$$

$$EI_1\tau_{3c} = \frac{1}{3}\frac{(15)(10)}{(15)(2)} = \frac{5}{3}$$

$$EI_1\tau_{4c} = \frac{1}{2}\frac{(18)(12)}{(18)(2)} = 3$$

$$\therefore EI_1\,\delta_{cc} = \underline{+4.67^{k\prime 2}}$$

Substituting these values in (1) and (2) and canceling out EI_1

$$720 + 6.67X_b + 0.83X_c = 0$$
$$738 + 0.83X_b + 4.67X_c = 0$$

Solving these simultaneously, we have

$$X_b = -90.3 \qquad X_c = -142.0$$

With these moments known it is easy to isolate various portions of the beam and, by using statics, to compute shears, reactions, and bending moments, leading to the results shown.

Discussion:

When we select the support moments at b and c as the redundants, we remove the moment restraints from the actual structure at these points, i.e., we insert hinges at these points.

Note the manner in which the varying I is handled in this problem. One I is selected as a base, and all other I's are expressed in terms of the standard I. Once this is done, the actual values of the various I's need never be substituted in the solution of this problem.

Example 9.7 *Compute the reactions and draw the bending-moment diagram for this frame. E and I are constant. Consider only bending deformation.*

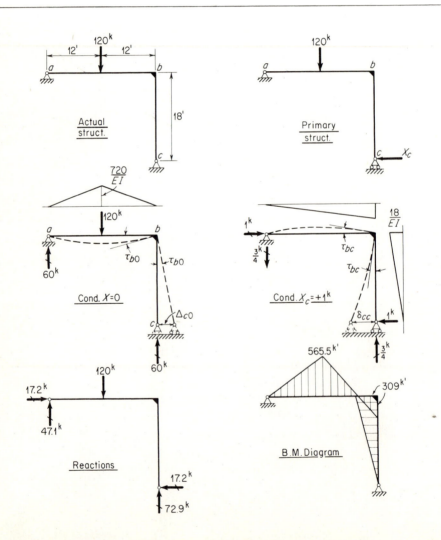

This frame is indeterminate to the first degree. Select X_c as the redundant. Then $\Delta_c = \Delta_{c0} + X_c \delta_{cc} = 0$. Compute Δ_{c0} and δ_{cc} by the moment-area theorems.

$$EI\tau_{b0} = (720)(12)\tfrac{12}{24} = 4,320 \qquad \therefore \ \Delta_{c0} = -\frac{4,320}{EI}(18) = \frac{-77,760^{k'3}}{EI}$$

$$EI\tau_{bc} = (18)(12)\tfrac{16}{24} = 144 \qquad \therefore \ \delta_{cc} = \frac{144}{EI}(18) + \frac{18}{EI}(9)(12) = +\frac{4,536^{k'3}}{EI}$$

Hence,

$$\frac{-77,760}{EI} + \frac{4,536}{EI}X_c = 0 \qquad X_c = +17.16 \qquad \therefore \ \leftrightarrow$$

The remaining computations can easily be done using only statics.

Example 9.8 *Compute the force in the tie rod of this structure.*

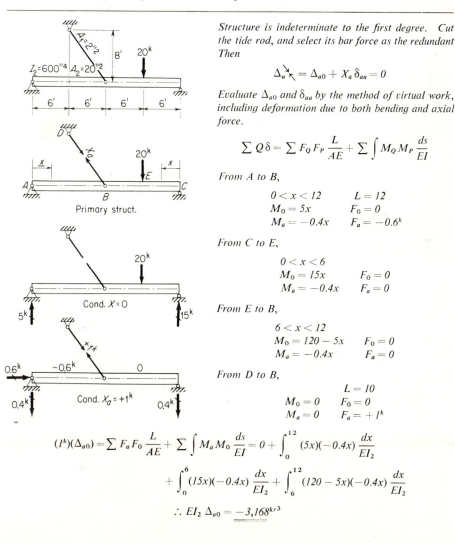

Primary struct.

Cond. $X=0$

Cond. $X_a = +1^k$

Structure is indeterminate to the first degree. Cut the tide rod, and select its bar force as the redundant Then

$$\Delta_a{\searrow} = \Delta_{a0} + X_a \delta_{aa} = 0$$

Evaluate Δ_{a0} and δ_{aa} by the method of virtual work, including deformation due to both bending and axial force.

$$\sum Q\delta = \sum F_Q F_P \frac{L}{AE} + \sum \int M_Q M_P \frac{ds}{EI}$$

From A to B,

$$0 < x < 12 \qquad L = 12$$
$$M_0 = 5x \qquad F_0 = 0$$
$$M_a = -0.4x \qquad F_a = -0.6^k$$

From C to E,

$$0 < x < 6$$
$$M_0 = 15x \qquad F_0 = 0$$
$$M_a = -0.4x \qquad F_a = 0$$

From E to B,

$$6 < x < 12$$
$$M_0 = 120 - 5x \qquad F_0 = 0$$
$$M_a = -0.4x \qquad F_a = 0$$

From D to B,

$$L = 10$$
$$M_0 = 0 \qquad F_0 = 0$$
$$M_a = 0 \qquad F_a = +1^k$$

$$(1^k)(\Delta_{a0}) = \sum F_a F_0 \frac{L}{AE} + \sum \int M_a M_0 \frac{ds}{EI} = 0 + \int_0^{12}(5x)(-0.4x)\frac{dx}{EI_2}$$

$$+ \int_0^6 (15x)(-0.4x)\frac{dx}{EI_2} + \int_6^{12}(120 - 5x)(-0.4x)\frac{dx}{EI_2}$$

$$\therefore \ EI_2 \Delta_{a0} = -3,168^{k'3}$$

$$(1^k)(\delta_{aa}) = \sum F_a^2 \frac{L}{AE} + \sum \int M_a^2 \frac{ds}{EI} = \frac{(-0.6)^2(12)}{EA_2} + \frac{(1)^2(10)}{EA_1}$$

$$+ 2 \int_0^{12} (-0.4x)^2 \frac{dx}{EI_2} \quad \therefore EI_2\,\delta_{aa} = +206.05^{k'3}$$

Since

$$\frac{I_2}{A_2} = \frac{600}{(144)^2} \frac{144}{20} = \left(\frac{30}{144}\right)^{'2} \quad and \quad \frac{I_2}{A_1} = \frac{600}{(144)^2} \frac{144}{2} = \left(\frac{300}{144}\right)^{'2}$$

then

$$\frac{-3,168}{EI_2} + \frac{206.05}{EI_2} X_a = 0 \quad \therefore X_a = +15.38 \quad \therefore \text{ tension}$$

If the deformation due to axial force has been neglected, we would have had

$$X_a = \frac{3,168}{184.32} = +17.2$$

9.5   Additional Examples Involving Temperature, Settlement, etc.   The following examples illustrate the application of the superposition-equation approach to the stress analysis of typical indeterminate structures subjected to temperature changes, support movements, fabrication errors, etc.

Fundamentally, these problems are no more difficult to handle than those involving simply the effect of loads. *Always remember, however, that the deflections* Δ_{aT}, Δ_{aS}, *etc., refer to deflections of points on the primary structure due to a change in temperature of the primary structure, due to the settlement of the supports of the primary structure, etc. These deflections when superimposed correctly with the contributions of each of the redundants must add up so that the total deflection of point a on the primary structure is equal to the known deflection of the corresponding point on the actual structure.*

Example 9.9   *Compute bar forces due to an increase of $60°F$ in the temperature at bars aB, BC, and Cd. No change in the temperature of any other bars. $\alpha_t = 1/150,000$ per $°F$. $E = 30 \times 10^3$ kips per sq in. Cross-sectional areas in square inches are shown in parentheses.*

This truss is indeterminate to the first degree. Cut bar bC, and select its bar force as the redundant. Then

$$\Delta_a^{\nwarrow} = \Delta_{aT} + X_a\,\delta_{aa} = 0$$

With the method of virtual work,

$$\Delta_{aT}: F_Q = F_a$$

$$(1^k)(\Delta_{aT}^{\nwarrow}) = \sum F_a \alpha_t tL = \alpha_t \sum F_a tL$$

$$\delta_{aa}: F_Q = F_P = F_a \qquad (1^k)(\delta_{aa}^{\nwarrow}) = \frac{1}{E} \sum F_a^2 \frac{L}{A}$$

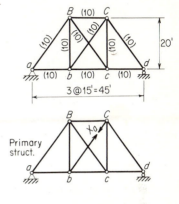

Bar	L	A	$\frac{L}{A}$	F_a	$F_a^2\frac{L}{A}$	t	$F_a tL$	$X_a F_a$
Units	$,$	$,^2$	$,/,^2$	k	$k^2,/,^2$	$°F$	$k°F'$	k
BC	15	10	1.5	−0.6	+0.54	+60	−540	− 7.5
bc	15	10	1.5	−0.6	+0.54	0	0	− 7.5
Bb	20	10	2	−0.8	+1.28	0	0	−10.0
Cc	20	10	2	−0.8	+1.28	0	0	−10.0
Bc	25	10	2.5	+1.0	+2.5	0	0	+12.5
bC	25	10	2.5	+1.0	+2.5	0	0	+12.5
Σ					+8.64		−540	

$$\Delta_{aT} = -(540)^{°F'}\left(\frac{1}{150,000}\right)^{/°F} = -0.0036 \, ft$$

$$\delta_{aa} = +\frac{8.64^{k'/,^2}}{(30 \times 10^3)^{k/,^2}} = +0.000288 \, ft$$

$$-0.0036 + 0.000288 X_a = 0 \qquad X_a = +12.5$$

$$\therefore tension$$

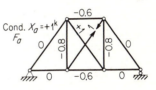

Example 9.10 *Compute the bar forces of the two-hinged trussed arch in Example 9.4 caused by forcing member AB into place even though it had been fabricated ⅛ in. too short.*

In this case,

$$\Delta_c^{\leftarrow} = \Delta_{cE} + X_c \delta_{cc} = 0$$

Using the method of virtual work, Δ_{cE} can be evaluated as

$$(1^k)(\Delta_{cE}^{\leftarrow}) = \sum F_c \Delta L_E = (+1^k)(-\tfrac{1}{96})' \qquad \Delta_{cE} = -0.0104'$$

From Example 9.4

$$\delta_{cc} = +\frac{19.875^{k'/,^2}}{E} = +\frac{19.875^{k'/,^2}}{30 \times 10^{3k/,^2}} = +0.0006625'$$

$$\therefore -0.0104' + 0.0006625' X_c = 0 \qquad \therefore X_c = +15.7 \qquad \therefore \leftarrow\!\rightarrow$$

The bar force in any member can then be computed, since $F = X_c F_c = 15.7 F_c$.

Example 9.11 *Compute the reactions and draw the bending-moment diagram for the beam in Example 9.6 due to the following support movements:*

Point a, 0.02 ft down   Point c, 0.05 ft down
Point b, 0.04 ft down   Point d, 0

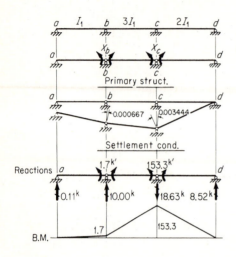

In this case,

$$\Delta_b^{\circlearrowright} = \Delta_{bS} + X_b\,\delta_{bb} + X_c\,\delta_{bc} = 0 \qquad (1)$$

$$\Delta_c^{\circlearrowright} = \Delta_{cS} + X_b\,\delta_{cb} + X_c\,\delta_{cc} = 0 \qquad (2)$$

From either the method of virtual work or the geometry of the adjacent sketch

$$\Delta_{bS} = -0.000667 \text{ radian}$$
$$\Delta_{cS} = -0.003444 \text{ radian}$$

From Example 9.6

$$\delta_{bb} = \frac{6.667^{k\prime 2}}{EI_1} \qquad \delta_{cc} = \frac{4.667^{k\prime 2}}{EI_1}$$

$$\delta_{bc} = \delta_{cb} = \frac{0.833^{k\prime 2}}{EI_1}$$

Upon substituting these values, Eqs. (1) and (2) become

$$6.667X_b + 0.833X_c = 0.000667EI_1$$
$$0.833X_b + 4.667X_c = 0.003444EI_1$$

from which, since

$$EI_1 = [(30 \times 10^3)(144)]^{k\prime 2}\left[\frac{1,000}{(144)^2}\right]^{\prime 4} = (0.2083 \times 10^6)^{k\prime 2}$$

$$X_b = +0.00000802EI_1 = \underline{+1.7}$$
$$X_c = +0.000736EI_1 \quad = \underline{+153.3}$$

The reactions and bending-moment diagram can now easily be computed.

Example 9.12 *Compute the bar forces in this truss due to the following support movements:*

Support at a, 0.24 in. down
Support at c, 0.48 in. down
Support at e, 0.36 in. down

Cross-sectional area in square inches shown in parentheses and E = 30 × 10³ kips per sq in.

Select the vertical reaction at c as the redundant:

$$\Delta_c^{\uparrow} = \Delta_{cs} + X_c\,\delta_{cc} = -0.04'$$

From the method of virtual work,

$\Delta_{cs}:$ $\sum Q\,\delta = 0$ *since $\Delta L = 0$ (see Example 8.3). Therefore,*

$$(1^k)\Delta_{cs}^! + (0.5^k)(0.02') + (0.5^k)(0.03') = 0$$
$$\therefore \Delta_{cs} = -0.025'$$

which checks the value obtained by geometry as shown on the sketch.

δ_{cc}:

$$(1^k)(\delta_{cc}^!) = \frac{1}{E}\sum F_c^2 \frac{L}{A}$$

Because of symmetry only one-half of the bars of the truss need be listed.

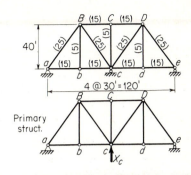

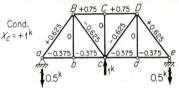

Primary struct.

Bar	L	A	$\dfrac{L}{A}$	F_c	$F_c^2\dfrac{L}{A}$	$X_c F_c$
Units	$'$	$''^2$	$'/''^2$	k	$k^2{}_,/''^2$	k
ab	30	15	2	-0.375	$+0.281$	$+26.0$
bc	30	15	2	-0.375	$+0.281$	$+26.0$
BC	30	15	2	$+0.75$	$+1.125$	-52.1
aB	50	25	2	$+0.625$	$+0.781$	-43.4
Bc	50	25	2	-0.625	$+0.781$	$+43.4$
$\frac{1}{2}\sum$						$+3.25$

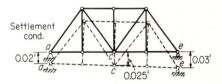

Cond. $X_c = +1^k$

$$\delta_{cc} = \frac{+6.50^{k'/''^2}}{(30\times10^3)^{k/''^2}} = +0.000216'$$

Settlement cond.

$$\therefore -0.025 + 0.000216X_c = -0.04'$$

$$X_c = \frac{-0.015}{0.000216} = -69.5 \qquad \therefore down$$

9.6   General Remarks Concerning Selection of Redundants.   From the previous discussion, we recognize that there is considerable latitude in selecting redundants, the only restriction being that they shall be selected so that a *stable* primary structure remains.   By proper selection of the redundants, however, we can minimize the numerical computations.   This objective can be achieved by adhering to the two following policies:

1. Take advantage of any symmetry of the structure.
2. Select the primary structure so that the effect of any of the various loading conditions is localized as much as possible.

Consideration of several alternative selections of the redundants for the continuous truss shown in Fig. 9.5 will illustrate the validity of these statements.   This structure is indeterminate to the second degree.   Any selection of the redundants will involve two equations of the following form:

$$\Delta_a = \Delta_{a0} + X_a\delta_{aa} + X_b\delta_{ab} = 0$$
$$\Delta_b = \Delta_{b0} + X_a\delta_{ba} + X_b\delta_{bb} = 0$$

(a)

Only the following five different deflection terms are involved, since δ_{ab} equals δ_{ba}:

$$(1)(\Delta_{a0}) = \sum F_a F_0 \frac{L}{AE} \qquad (1)(\delta_{aa}) = \sum F_a^2 \frac{L}{AE}$$

$$(1)(\Delta_{b0}) = \sum F_b F_0 \frac{A}{LE} \qquad (1)(\delta_{bb}) = \sum F_b^2 \frac{L}{AE} \qquad (b)$$

$$(1)(\delta_{ab}) = \sum F_a F_b \frac{L}{AE}$$

Before these terms can be evaluated, F_0, F_a, and F_b bar forces must be computed.

If the structure is symmetrical, and if symmetrical redundants are selected, the F_b forces can be obtained from the F_a forces by symmetry. Further, δ_{bb} will be equal to δ_{aa} in such a case, leaving only four deflection terms to be evaluated. The evaluation of these terms will involve less computation if the redundants are selected so as to restrict the effect of the various loading conditions to as few bars as possible. The latter will be true whether or not the structure is symmetrical.

All three alternative selections of the redundants shown in Fig. 9.5 take advantage of symmetry. The various loading conditions affect the portions of the structure indicated in each case. Comparison of these primary structures shows clearly that selection 3 is the best since it is most effective in localizing the effects of the various loading conditions.

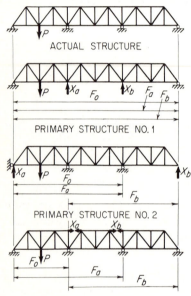

ACTUAL STRUCTURE

PRIMARY STRUCTURE NO. 1

PRIMARY STRUCTURE NO. 2

PRIMARY STRUCTURE NO. 3

Figure 9.5 Selection of redundants

There are several other items to be noted at this time. In the discussion of Example 9.5, it was pointed out that the shear and moment are statically determinate in a straight bar which is hinged at both ends. If such a bar is cut, only its axial force may be statically indeterminate and, therefore, considered as a redundant. On the other hand, the shear, moment, and axial force are often all statically indeterminate in a member that is rigidly connected at its ends to the rest of the structure. If such a member is cut, the shear, moment, and axial force may all be considered as redundants, provided that a stable structure remains when the restraints corresponding to all three of these elements are removed (as an illustration, see Example 9.13).

9.7 Analysis of Statically Indeterminate Structures Using Castigliano's Second Theorem; Theorem of Least Work.   The previous approach to the analysis of an indeterminate structure involved writing superposition equations for the deflections of the points of application of the redundants. Instead of doing this, however, expressions for these deflections can be set up using Castigliano's second theorem. The latter approach is actually very similar to the former. It is a somewhat more automatic procedure, however, and is therefore preferred by some students and engineers. Since Castigliano's theorem should really be limited to the computation of the deflections produced simply by loads on the structure, this method lacks the generality of the superposition-equation approach.

Consider, for example, the indeterminate beam shown in Fig. 9.3. After the degree of indeterminacy has been established and the redundant and the resulting primary structure have been selected, the deflection of the point of application of redundant X_b can be evaluated by using Castigliano's second theorem. In this particular case, only bending deformation is involved; therefore,

$$\mathcal{U} = \sum \int M^2 \frac{ds}{2EI}$$

but

$$\frac{\partial \mathcal{U}}{\partial X_b} = \Delta_b^{\uparrow} \tag{a}$$

Since point b on the actual structure does not deflect, Δ_b on the primary structure must equal zero. As a result,

$$\frac{\partial \mathcal{U}}{\partial X_b} = \sum \int M \frac{\partial M}{\partial X_b} \frac{ds}{EI} = 0 \tag{b}$$

However, M, being equal to the total bending moment on the primary structure due to all causes, may be expressed as being the superposition of the contribution of the applied load only and the contribution of the redundant X_b. Thus,

$$M = M_0 + X_b M_b \qquad \frac{\partial M}{\partial X_b} = M_b \tag{c}$$

Equation (b) therefore becomes

$$\sum \int M_0 M_b \frac{ds}{EI} + X_b \sum \int M_b^2 \frac{ds}{EI} = 0 \tag{d}$$

It is easy to evaluate these integrals for the primary structure and then solve for X_b.

If the superposition approach is used in this example, Δ_{b0} and δ_{bb} can be evaluated by the method of virtual work and found to be

$$(1)(\Delta_{b0}) = \sum \int M_0 M_b \frac{ds}{EI} \qquad (1)(\delta_{bb}) = \sum \int M_b^2 \frac{ds}{EI} \tag{e}$$

From Eqs. (e) it is immediately apparent that Eq. (d) is actually a statement that

$$\Delta_{b0} + X_b \delta_{bb} = 0 \tag{f}$$

Thus, if the method of virtual work is used as a basis for evaluating Δ_{b0} and δ_{bb} in the superposition-equation approach, the two methods are essentially identical.

The above illustration involves a structure that is statically indeterminate to only the first degree. In more highly indeterminate structures, the procedure is essentially the same. After selecting the n redundants and the resulting primary structure, express the displacement of the point of application of each redundant by n separate applications of Castigliano's second theorem. This will result in n simultaneous equations involving the n redundants, the value of which can then be obtained by simultaneous solution of the equations. The procedure for analyzing a multiply redundant frame in this manner is illustrated by Example 9.13.

In the example of Art. 9.8, equations comparable to Eq. (b) are evaluated in a slightly different manner. Thus it is possible to use Castigliano's theorem somewhat more automatically and effectively in certain problems. The procedure suggested in the above illustration may likewise be used to advantage in certain other problems. In cases of the latter type, however, there is no advantage in using the Castigliano approach instead of superposition equations.

If in the analysis of indeterminate structures the deflection of the point of application of a redundant is zero, then applying Castigliano's theorem as in Eq. (a) reduces to the statement that the first partial derivative of the strain energy with respect to that redundant is equal to zero. This is equivalent to stating that the value of the redundant must be such as to minimize the strain energy. This special case of Castigliano's second theorem is often called the **theorem of least work** and may be stated as follows.

Theorem of least work:

In a statically indeterminate structure, if there are no support movements and no change of temperature, the redundants must be such as to make the strain energy a minimum.

9.8   Example Illustrating Stress Analysis Using Castigliano's Second Theorem. The example that follows has been chosen primarily to illustrate the use of Castigliano's second theorem in the stress analysis of indeterminate structures. If, in such cases, we solved the problem by the superposition-equation approach, using the method of virtual work to evaluate the various deflection terms, we should find that the actual computations would be essentially the same as those involved in the Castigliano solution. One difference in the two approaches lies in the somewhat more automatic manner of setting up the solution when the Castigliano approach is used.

Another difference between the two approaches is also worthy of note. *In a Castigliano solution, the redundants carry their own units throughout a solution.* For example, in Eq. (a), X_b must be in kips if $\mathcal{U}$ is in kip-feet in order that the change in $\mathcal{U}$ divided by the change in X_b shall equal Δ_b in feet. As a result, if X_b is in kips, M_b must have units of kip-feet per kip if the units in Eqs. (c) are to be consistent. If it is recognized, therefore, that the redundants in a Castigliano solution carry their own units, dimensional checks should be consistent at all times.

Strictly speaking, Castigliano's theorem is applicable only when the deflection of the structure is caused by loads. It is possible, however, to handle the stress analysis of an indeterminate structure for the effect of temperature change, settlement of supports, etc., by proceeding as follows. Select the primary structure and temporarily remove all the redundants. Now allow the temperature change or settlement to take place on the primary structure. Compute the resulting displacements of the points of application of the redundants on this primary structure. Such computations may be performed by means of the method of virtual work or some other suitable method. Now apply the redundants. As they are applied, they must restore their points of application to their correct positions. Castigliano's theorem may be used to evaluate the deflections produced by the redundants. Substituting the previously computed values for the restoring deflections, we thereby obtain equations containing only the redundants as the unknowns, which can then be obtained by simultaneous solution of these equations. While this procedure is often not as straightforward as that of the superposition-equation approach, there are nevertheless instances where it may be used advantageously.

Example 9.13　*Solve this frame using Castigliano's theorem. Include the effect of deformation due to both axial force and bending.*

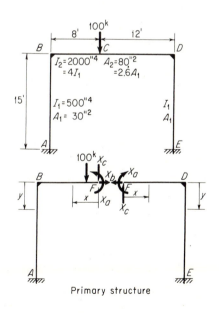

Primary structure

This frame is indeterminate to the third degree. Cut the girder at mid-span and select the moment, axial force, and shear as the three redundants X_a, X_b, and X_c.

$$\mathcal{U} = \sum \int M^2 \frac{ds}{2EI} + \sum \frac{F^2 L}{2AE}$$

But

$$\frac{\partial \mathcal{U}}{\partial X_a} = \Delta_a^{\curvearrowright} = 0$$

$$\frac{\partial \mathcal{U}}{X_b} = \Delta_b^{\leftarrow\rightarrow} = 0$$

$$\frac{\partial \mathcal{U}}{\partial X_c} = \Delta^{\updownarrow} = 0$$

Therefore, differentiating and canceling out E gives

$$\sum \int M \frac{\partial M}{\partial X_a} \frac{ds}{I} + \sum F \frac{L}{A} \frac{\partial F}{\partial X_a} = 0 \quad (1)$$

$$\sum \int M \frac{\partial M}{\partial X_b} \frac{ds}{I} + \sum F \frac{L}{A} \frac{\partial F}{\partial X_b} = 0 \quad (2)$$

$$\sum \int M \frac{\partial M}{\partial X_c} \frac{ds}{I} + \sum F \frac{L}{A} \frac{\partial F}{\partial X_c} = 0 \quad (3)$$

From F to C,

$$0 < x < 2 \quad M = X_a - xX_c \quad \frac{\partial M}{\partial X_a} = 1 \quad \frac{\partial M}{\partial X_b} = 0 \quad \frac{\partial M}{\partial X_c} = -x$$

$$L = 2' \quad F = X_b \quad \frac{\partial F}{\partial X_a} = 0 \quad \frac{\partial F}{\partial X_b} = 1 \quad \frac{\partial F}{\partial X_c} = 0$$

From C to B,

$$2 < x < 10 \quad M = X_a - xX_c - (100)(x-2) \quad \frac{\partial M}{\partial X_a} = 1 \quad \frac{\partial M}{\partial X_b} = 0 \quad \frac{\partial M}{\partial X_c} = -x$$

$$L = 8' \quad F = X_b \quad \frac{\partial F}{\partial X_a} = \frac{\partial F}{\partial X_c} = 0 \quad \frac{\partial F}{\partial X_b} = 1$$

From B to A,

$$0 < y < 15 \quad M = X_a - yX_b - 10X_c - 800 \quad \frac{\partial M}{\partial X_a} = 1 \quad \frac{\partial M}{\partial X_b} = -y \quad \frac{\partial M}{\partial X_c} = -10$$

$$L = 15' \quad F = -100 - X_c \quad \frac{\partial F}{\partial X_a} = \frac{\partial F}{\partial X_b} = 0 \quad \frac{\partial F}{\partial X_c} = -1$$

From F to D,

$$0 < x < 10 \quad M = X_a + xX_c \quad \frac{\partial M}{\partial X_a} = 1 \quad \frac{\partial M}{\partial X_b} = 0 \quad \frac{\partial M}{\partial X_c} = x$$

$$L = 10' \quad F = X_b \quad \frac{\partial F}{\partial X_a} = \frac{\partial F}{\partial X_c} = 0 \quad \frac{\partial F}{\partial X_b} = 1$$

From D to E,

$$0 < y < 15 \quad M = X_a - yX_b + 10X_c \quad \frac{\partial M}{\partial X_a} = 1 \quad \frac{\partial M}{\partial X_b} = -y \quad \frac{\partial M}{\partial X_c} = 10$$

$$L = 15' \quad F = X_c \quad \frac{\partial F}{\partial X_a} = \frac{\partial F}{\partial X_b} = 0 \quad \frac{\partial F}{\partial X_c} = 1$$

Setting up Eq. (1) gives

$$\int_0^2 (X_a - xX_c)(1) \frac{dx}{4I_1} + \int_2^{10} (X_a - xX_c - 100x + 200)(1) \frac{dx}{4I_1}$$

$$+ \int_0^{15} (X_a - yX_b - 10X_c - 800)(1) \frac{dy}{I_1} + \int_0^{10} (X_a + xX_c)(1) \frac{dx}{4I_1}$$

$$+ \int_0^{15} (X_a + 10X_c - yX_b)(1) \frac{dy}{I_1} = 0$$

Combining, canceling, and integrating, we obtain

$$35X_a - 225X_b = 12,800 \tag{1}$$

Setting up Eq. (2) gives

$$\int_0^{15} (X_a - yX_b - 10X_c - 800)(-y) \frac{dy}{I_1} + \int_0^{15} (X_a - yX_b + 10X_c)(-y) \frac{dy}{I}$$

$$+ X_b \frac{(1)(10)}{2.6A_1} + X_b \frac{(1)(10)}{2.6A_1} = 0$$

Again combining and canceling before integrating, we obtain

$$-225X_a + \left(2,250 + 7.5\,\frac{I_1}{A_1}\right)X_b = -90,000$$

However,

$$\frac{I_1}{A_1} = \frac{500}{(144)^2}\frac{144}{30} = \frac{50}{432}$$

and therefore

$$-225X_a + 2,250.87X_b = -90,000 \tag{2}$$

In a similar manner, Eq. (3) reduces to

$$\left(3,166.7 + 30\,\frac{I_1}{A_1}\right)X_c = -\left(125,866.7 + 1,500\,\frac{I_1}{A_1}\right)$$

$$3,170.14X_c = -126,040.2 \tag{3}$$

Solving Eqs. (1), (2), and (3),

$$X_a = +304.06^{k'}$$
$$X_b = -9.59^k$$
$$X_c = -39.76^k$$

If the effect of axial deformation were neglected, all the terms containing A would be omitted. Then, we should find

$$X_a = +304.0^{k'}$$
$$X_b = -9.60^k$$
$$X_c = -39.74^k$$

Discussion:

  These results indicate that the effect of axial deformation may be neglected in comparison with the effect of bending in the rigid-frame type of structure.

9.9   Development of the Three-Moment Equation. The **three-moment equation** was first presented in 1857 by the French engineer Clapeyron. This equation is a relationship that exists between the moments at three points in a continuous member. It is particularly helpful in solving for the moments at the supports of indeterminate beams.

  Designate three points on a continuous member as L, C, and R as shown in Fig. 9.6. Suppose that the moment of inertia is constant between point L and C and equal to I_L and likewise constant between C and R and equal to I_R.† The member is assumed to be straight initially, and the deflections from the original position are assumed to be δ_L, δ_C, and δ_R at points L, C, and R, respectively—all to be considered positive when upward as shown.

  Let the moments at these three points be M_L, M_C, and M_R. Bending moments are to be considered plus when causing tension on the lower fibers of a member. The bending-moment diagram for the portion LC or CR may be considered to be

---

  † Theoretically it is possible, though cumbersome, to include the effect of variable I in this development.

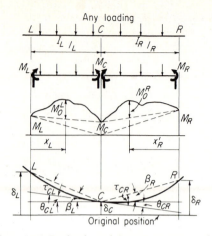

Figure 9.6 Derivation of three-moment equation

that resulting from superposition of three separate effects: the contribution of each of the end moments acting separately, which is given by the ordinates of the triangles indicated by the dashed lines, and the contribution of the applied load acting by itself with the end moments removed, which is given by the ordinates M_0^L in the portion LC and by M_0^R in the portion CR.

From the sketch of the elastic curve,

$$\theta_{CL} = \beta_L - \tau_{CL} \qquad \text{and} \qquad \theta_{CR} = \tau_{CR} - \beta_R$$

However, since the elastic curve is continuous through point C,

$$\theta_{CL} = \theta_{CR}$$

Hence,

$$\beta_L - \tau_{CL} = \tau_{CR} - \beta_R \tag{a}$$

Since these are all small angles, it is permissible to consider that

$$\beta_L = \frac{\delta_L - \delta_C}{l_L} \qquad \beta_R = \frac{\delta_R - \delta_C}{l_R} \tag{b}$$

If the bending-moment diagram were converted into an M/EI diagram, τ_{CL} and τ_{CR} could be evaluated easily by means of the second moment-area theorem,

$$\tau_{CL} = \frac{1}{EI_L\,l_L}\left(\frac{M_L\,l_L^2}{6} + \frac{M_C\,l_L^2}{3} + \int_0^{l_L} M_0^L x_L\,dx_L\right)$$

$$\tag{c}$$

$$\tau_{CR} = \frac{1}{EI_R\,l_R}\left(\frac{M_R\,l_R^2}{6} + \frac{M_C\,l_R^2}{3} + \int_0^{l_R} M_0^R x_R'\,dx_R'\right)$$

Let

$$(\mathscr{M}_0)_L = \int_0^{l_L} M_0^L x_L\,dx_L \qquad (\mathscr{M}_0)_R = \int_0^{l_R} M_0^R x_R'\,dx_R' \tag{9.4}$$

Substituting from Eq. (9.4) in Eqs. (c) and then from Eqs. (b) and (c) in Eq. (a), we obtain the so-called three-moment equation,

$$M_L \frac{l_L}{I_L} + 2M_C\left(\frac{l_L}{I_L} + \frac{l_R}{I_R}\right) + M_R \frac{l_R}{I_R} = -\frac{\mathscr{L}_0}{I_L} - \frac{\mathscr{R}_0}{I_R}$$

$$+ 6E\left[\frac{\delta_L}{l_L} - \delta_C\left(\frac{1}{l_L} + \frac{1}{l_R}\right) + \frac{\delta_R}{l_R}\right] \quad (9.5)$$

where the load terms are

$$\mathscr{L}_0 = +\frac{6(\mathscr{M}_0)_L}{l_L} \qquad \mathscr{R}_0 = +\frac{6(\mathscr{M}_0)_R}{l_R} \quad (9.6)$$

In using these equations, note particularly that:

1. M_L, M_C, and M_R are plus when causing tension on the lower fibers.
2. δ_L, δ_C, and δ_R are plus when upward from the original position.
3. $\mathscr{L}_0$ and $\mathscr{R}_0$ are load terms dependent on the applied load in the spans LC and CR, respectively.
4. The M_0 diagram for a member is the bending-moment diagram drawn for the member if it is assumed to be a simple end-supported beam. $(\mathscr{M}_0)_L$ represents the static moment of the area under this diagram, taken about an axis through the left end, while $(\mathscr{M}_0)_R$ represents the static moment about an axis through the right end. The sign of both these static moments depends simply on the sign of the ordinates of the M_0 diagram.

In the special case where $I_L = I_R = I$, Eq. (9.5) simplifies to

$$M_L l_L + 2M_C(l_L + l_R) + M_R l_R = -\mathscr{L}_0 - \mathscr{R}_0 + 6EI\left[\frac{\delta_L}{l_L} - \delta_C\left(\frac{1}{l_L} + \frac{1}{l_R}\right) + \frac{\delta_R}{l_R}\right] \quad (9.7)$$

The load terms in the cases of full uniform load and concentrated load are shown in Fig. 9.7.

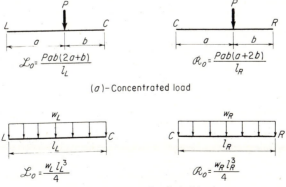

(a)-Concentrated load

(b)-Full uniformly distributed load

Figure 9.7   Load terms

9.10 Application of Three-Moment Equation. *The three-moment equation is applicable to any three points on a beam as long as there are no discontinuities, such as hinges in the beam within this portion.* In applying the equation to a continuous beam, if we select three successive support points as being L, C, and R, the deflection terms on the right-hand side of the equation will be equal either to zero or to the known movements of the support points. We thus obtain an equation involving the moments at the support points as the only unknowns.

In this manner, we can write an independent equation for any three successive support points along a continuous beam. We shall obtain n independent equations involving n unknown support moments, which can then be obtained from the simultaneous solution of these equations. There is a slight ambiguity in handling a fixed end on a continuous beam, but the technique of solving such problems is explained in the illustrative examples that follow.

The analysis of continuous beams by this method is straightforward. Be careful, however, to follow the sign convention noted in Art. 9.9. Likewise, be careful to use consistent units, particularly when there are movements of the supports.

Example 9.14   *Compute the moments at the support points of this beam. E and I are constant.*

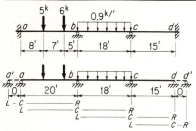

In this case, there are four unknown moments, M_a, M_b, M_c, and M_d. Four equations are therefore required. A fixed end may be handled by replacing it by an additional span of essentially zero length, as shown. The required equations can then be obtained by applying Eq. (9.7) four separate times, considering L, C, and R in turn, as indicated.

$$\delta_{a'} = \delta_a = \delta_b = \delta_c = \delta_d = \delta_{d'} = 0$$

Consider a', a, and b as L, C, and R.

$$\mathscr{L}_0 = 0 \qquad \mathscr{R}_0 = \frac{(5)(8)(12)(32)}{20} + \frac{(6)(15)(5)(25)}{20} = 1{,}330.5 \qquad M_{a'} = 0$$

$$\therefore 40M_a + 20M_b = -1{,}330.5 \tag{1}$$

Consider a, b, and c as L, C, and R.

$$\mathscr{L}_0 = \frac{(5)(8)(12)(28)}{20} + \frac{(6)(15)(5)(35)}{20} = 1{,}459.5 \qquad \mathscr{R}_0 = \frac{(0.9)(18)^3}{4} = 1{,}312.2$$

$$\therefore 20M_a + 76M_b + 18M_c = -2{,}771.7 \tag{2}$$

Consider b, c, and d as L, C, and R.

$$\mathscr{L}_0 = 1{,}312.2 \qquad \mathscr{R}_0 = 0 \qquad \therefore 18M_b + 66M_c + 15M_d = -1{,}312.2 \tag{3}$$

Consider c, d, and d' as L, C, and R.

$$\mathscr{L}_0 = \mathscr{R}_0 = 0 \qquad M_{d'} = 0 \qquad \therefore 15M_c + 30M_d = 0 \tag{4}$$

Solving Eqs. (1), (2), (3), and (4) simultaneously leads to the following values for the unknown moments:

$$M_a = -19.17^{k'} \qquad M_b = -28.20^{k'} \qquad M_c = -13.74^{k'} \qquad M_d = +6.87^{k'}$$

Example 9.15 *Compute the reactions and draw the bending-moment diagram for this beam due to the following support settlement:*

> *Support a rotates 0.005 radian clockwise*
> *Support b settles 0.0208 ft down*
> $E = 30 \times 10^3$ *kips per sq in.*

Consider a', a, and b as L, C, and R.

$$M_{a'} = 0 \qquad \delta_a = 0$$
$$\mathscr{L}_0 = \mathscr{R}_0 = 0 \qquad \delta_b = -0.0208'$$

As a result of rotation of support at a, δ_o'/l_L approaches +0.005. Applying Eq. (9.5) gives

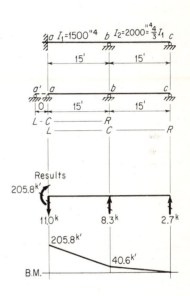

$$2M_a \frac{15}{I_1} + M_b \frac{15}{I_1} = (6E)(0.005) + 6E\left(-\frac{0.0208}{15}\right)$$

$$\therefore \; 30M_a + 15M_b = 0.02166EI_1 \qquad (1)$$

Consider a, b, and c as L, C, and R.

$$M_c = 0 \qquad \delta_b = -0.0208'$$
$$\delta_a = 0 \qquad \delta_c = 0$$

Applying Eq. (9.5) gives

$$M_a \frac{15}{I_1} + 2M_b \left(\frac{15}{I_1} + \frac{15}{\frac{4}{3}I_1}\right)$$

$$= (-6E)(-0.0208)(\tfrac{1}{15} + \tfrac{1}{15})$$

$$\therefore \; 15M_a + 52.5M_b = +0.01664EI_1 \qquad (2)$$

Solving Eqs. (1) and (2) simultaneously and substituting $EI_1 = 31.25 \times 10^4$ kip-ft², we have

$$M_a = +0.000658EI_1 = +205.8^{k'}$$
$$M_b = +0.0001291EI_1 = +40.6^{k'}$$

9.11 Development of the Slope-Deflection Equation. The **slope-deflection method** was presented by Prof. G. A. Maney in 1915 as a general method to be used in the analysis of rigid-joint structures. This method extended the use of equations, originally proposed by Manderla and Mohr, for computing secondary stresses in trusses. It is useful in its own right; and even more important, it provides an excellent means of introducing the method of moment distribution.

The following fundamental equations are derived by means of the moment-area theorems. Thus, these equations consider deformation caused by bending moment but neglect that due to shear and axial force. Since the effect of axial force and shear deformation on the stress analysis of most indeterminate beams and frames is very small, the error that results from using these equations as a basis for the slope-deflection method of analysis is also very small (for corroboration, see the results of Example 9.13). *The fundamental slope-deflection equation is an expression for the moment on the end of a member in terms of four quantities, namely, the rotation of the tangent at each end of the elastic curve of the member, the rotation of the chord joining the ends of the elastic curve, and the external loads applied to the member.* It is convenient in the application of this equation to use the following sign convention:

1. *Moments* acting on the *ends* of a *member* are *positive* when *clockwise*.

2. Let θ be the *rotation* of the *tangent* to the elastic curve at the end of a member referred to the *original direction* of the *member*. The angle θ is *positive* when the tangent to the elastic curve has rotated *clockwise* from its original direction.

3. Let ψ be the *rotation* of the *chord* joining the ends of the elastic curve referred to the *original direction* of the *member*. The angle ψ is positive when the chord of the elastic curve has rotated *clockwise* from its original direction.

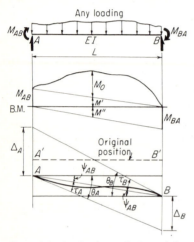

Figure 9.8 Derivation of slope-deflection equation

In designating the end moments, two subscripts will be used; these subscripts together designate the member under consideration, and the first one designates the end of the member to which the moment is applied. For example, M_{AB} designates the moment acting on the A end of member AB; M_{BA}, the moment on the B end of that member. The θ angles will be designated by one subscript indicating the end of the member. The ψ angles will be designated by two subscripts indicating the chord and, likewise, the member.

Using the above notation and convention, consider a member AB that has a constant E and I† throughout its length and that is initially straight. Suppose that this member is acted upon by the positive end moments M_{AB} and M_{BA} and any condition of an applied load, as shown in Fig. 9.8. Let AB be the elastic curve of this beam and $A'B'$ represent its original unstrained position. θ_A, θ_B, and ψ_{AB} are positive as shown.

The bending-moment diagram for this member may be considered to be the superposition of three separate effects: the contribution of each of the end moments acting separately, which is given by the ordinates of the triangular portions, M' and M'', and the contribution of the applied load acting by itself with the end moments removed, which is given by the ordinates M_0. In other words, the M_0 ordinates are the ordinates of the simple-beam bending-moment diagram for member AB, assuming temporarily that it acts as a simple end-supported beam. The total bending moment at any point will be the algebraic sum of M_0, M', and M'', but for this present derivation it is easier to consider these three contributions separately.

---

† It is theoretically possible, of course, to set up the slope-deflection equation considering the effect of variable I.

If the bending-moment diagram is converted into an M/EI diagram, Δ_A and Δ_B can be evaluated by the second moment-area theorem. Then

$$\Delta_A = -\frac{L^2}{6EI} M_{AB} + \frac{L^2}{3EI} M_{BA} - \frac{(\mathcal{M}_0)_A}{EI} \tag{a}$$

$$\Delta_B = \frac{L^2}{3EI} M_{AB} - \frac{L^2}{6EI} M_{BA} + \frac{(\mathcal{M}_0)_B}{EI} \tag{b}$$

in which $(\mathcal{M}_0)_A$ is the static moment about an axis through A of the area under the M_0 portion of the bending-moment diagram and $(\mathcal{M}_0)_B$ is a corresponding static moment about an axis through B.

Realizing that the angles and deformation shown in Fig. 9.8 are actually so small that an angle, its sine, and its tangent may all be considered equal, we see from the figure that

$$\frac{\Delta_A}{L} = \tau_B = \theta_B - \psi_{AB} \qquad \frac{\Delta_B}{L} = \tau_A = \theta_A - \psi_{AB} \tag{c}$$

Solving Eqs. (a) and (b) simultaneously for M_{AB} and M_{BA} and substituting in the resulting expressions for Δ_A/L and Δ_B/L from Eqs. (c), we obtain

$$M_{AB} = \frac{2EI}{L}(2\theta_A + \theta_B - 3\psi_{AB}) + \frac{2}{L^2}[(\mathcal{M}_0)_A - 2(\mathcal{M}_0)_B] \tag{d}$$

$$M_{BA} = \frac{2EI}{L}(2\theta_B + \theta_A - 3\psi_{AB}) + \frac{2}{L^2}[2(\mathcal{M}_0)_A - (\mathcal{M}_0)_B]$$

Up to this point, the condition of loading has not been defined, and Eqs. (d) are valid for any condition of transverse loading. The last term in brackets in each of these equations is a function of the type of loading, and it is important that its physical significance be recognized. Suppose that θ_A, θ_B, and ψ_{AB} are all equal to zero. Then the last terms of Eqs. (d) are, respectively, equal to the moment at the A end and the moment at the B end of the member. If, however, θ_A, θ_B, and ψ_{AB} are all equal to zero, it means physically that both ends of the member are completely fixed against rotation or translation and therefore that this member is what we call a **fixed-end beam**. These last terms of Eqs. (d) are therefore equal to the so-called **fixed-end moments**. Calling fixed-end moments FEM, we have

$$\text{FEM}_{AB} = \frac{2}{L^2}[(\mathcal{M}_0)_A - 2(\mathcal{M}_0)_B]$$

$$\tag{9.8}$$

$$\text{FEM}_{BA} = \frac{2}{L^2}[2(\mathcal{M}_0)_A - (\mathcal{M}_0)_B]$$

Substituting in Eqs. (d) from Eqs. (9.8), we obtain

$$M_{AB} = \frac{2EI}{L}(2\theta_A + \theta_B - 3\psi_{AB}) + \text{FEM}_{AB}$$

$$(9.9)$$

$$M_{BA} = \frac{2EI}{L}(2\theta_B + \theta_A - 3\psi_{AB}) + \text{FEM}_{BA}$$

Closer inspection of Eqs. (9.9) reveals that these two equations can be summarized by one general equation by calling the near end of a member N and the far end F. Also, if we let

$$K_{NF} = \textbf{stiffness factor} \text{ for member } NF = \frac{I_{NF}}{L_{NF}} \qquad (9.10)$$

then the fundamental slope-deflection equation can be written

$$M_{NF} = 2EK_{NF}(2\theta_N + \theta_F - 3\psi_{NF}) + \text{FEM}_{NF} \qquad (9.11)$$

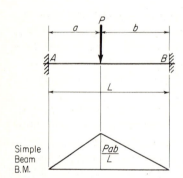

Figure 9.9　Concentrated load

Of course, the FEM can easily be determined for any given loading. If, in addition, the rotation of the tangent at each end and the rotation of the chord joining the ends of a member are known, the end moments in the member can easily be computed from Eq. (9.11). In Art. 9.12, the use of this equation in the solution of indeterminate beams and frames is discussed.

The FEM for any given loading can be evaluated by means of Eqs. (9.8), which involves computations such as the following regarding the area under the simple-beam bending-moment diagram:

Concentrated load (see Fig. 9.9):

$$(\mathcal{M}_0)_A = \frac{Pab}{L}\left[\frac{a}{2}\frac{2a}{3} + \frac{b}{2}\left(a + \frac{b}{3}\right)\right] = \frac{Pab}{6}(2a + b)$$

$$(\mathcal{M}_0)_B = \frac{Pab}{L}\left[\frac{b}{2}\frac{2b}{3} + \frac{a}{2}\left(b + \frac{a}{3}\right)\right] = \frac{Pab}{6}(2b + a)$$

$$\text{FEM}_{AB} = \frac{2}{L^2}\left[\frac{Pab}{6}(2a + b) - 2\frac{Pab}{6}(2b + a)\right] = -\frac{Pab^2}{L^2}$$

$$(9.12)$$

$$\text{FEM}_{BA} = \frac{2}{L^2}\left[2\frac{Pab}{6}(2a + b) - \frac{Pab}{6}(2b + a)\right] = +\frac{Pa^2b}{L^2}$$

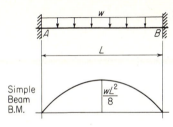

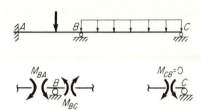

Figure 9.10 Full uniform load   **Figure 9.11** Isolation of joints

Full uniform load (see Fig. 9.10):

$$(\mathscr{M}_0)_A = (\mathscr{M}_0)_B = \frac{wL^4}{24}$$

$$\text{FEM}_{AB} = -\frac{wL^2}{12} \qquad \text{FEM}_{BA} = +\frac{wL^2}{12} \tag{9.13}$$

It will be noted that the proper signs of the FEM work out automatically from these calculations. In most cases, we know the direction of the end moments by inspection, and in this way we are usually able to verify the signs of the FEM.

Note again the sign convention to be used in applying the slope-deflection method. Note also that the above equations have been derived for a member that is initially straight and of constant E and I.

9.12 Application of Slope-Deflection Method to Beams and Frames. Consider first the application of the slope-deflection method to continuous-beam problems, such as the beam shown in Fig. 9.11, the supports of which are assumed to be unyielding. Think of this beam as being composed of two members AB and BC rigidly connected together at joint B. We could write expressions for the end moments at each end of each member, using Eq. (9.11). These four end moments M_{AB}, M_{BA}, M_{BC}, and M_{CB} would then be expressed in terms of the θ and ψ angles and the FEM, which could be computed from Eqs. (9.12) and (9.13).

Since the supports are unyielding, we know in this case that θ_A, ψ_{AB}, and ψ_{BC} are all equal to zero. Further, since members BA and BC are rigidly connected together at joint B, the tangent to the elastic curve at the B end of member AB must rotate with respect to its original direction algebraically the same amount θ_B as the tangent at the B end of member BC. Only the values of θ_B and θ_C are unknown, therefore, and involved in the expressions for the four end moments. If we could in some way find the values of θ_B and θ_C, we could then compute all the end moments; and, having them, we could compute by statics any other moment, shear, or reaction we desired. In other words, the stress analysis of this beam would be reduced to a problem in statics if we knew the values of θ_B and θ_C.

In this case, we are able to solve for these two unknowns by virtue of the fact that there are two convenient equations of statics which these end moments must satisfy. These equations are obtained by isolating joints B and C as shown in Fig. 9.11 and writing the equations $\sum M = 0$ for each of these joints. Thus,

From $\sum M_B = 0$: $M_{BA} + M_{BC} = 0$

From $\sum M_C = 0$: $M_{CB} = 0$

By substituting in these two equations the expressions for the end moments obtained by applying Eq. (9.11), we shall obtain two equations involving the two unknowns θ_B and θ_C. After solving these equations simultaneously for these unknowns, we shall then be able to compute the end moments and complete the stress analysis of the beam.

The actual numercial solution of such a problem is illustrated by Example 9.16. In Example 9.17 these ideas have been extended and applied to a case where the supports move.

Example 9.16 *Compute the reactions and draw the shear and bending-moment diagrams for this beam.   Supports are unyielding.*

Analyzing the θ and ψ angles, we have

$$\theta_a = 0 \qquad \psi_{ab} = \psi_{bc} = 0$$

$$\theta_b = ? \qquad \theta_c = ?$$

$$K_{ab} = \frac{I}{10} = K \qquad K_{bc} = \frac{3I}{15} = 2K$$

$$FEM_{ab} = -\frac{(20)(6)(4)^2}{(10)^2} = -19.2^{k'}$$

$$FEM_{ba} = +\frac{(20)(6)^2(4)}{(10)^2} = +28.8^{k'}$$

$$FEM_{bc} = -\frac{(2)(15)^2}{12} = -37.5^{k'}$$

$$FEM_{cb} = +37.5^{k'} \qquad M_{cd} = -25^{k'}$$

Using Eq. (9.11), write expressions for the end moments

$$M_{ab} = 2EK\theta_b - 19.2$$
$$M_{ba} = 4EK\theta_b + 28.8$$
$$M_{bc} = 8EK\theta_b + 4EK\theta_c - 37.5$$
$$M_{cb} = 8EK\theta_c + 4EK\theta_b + 37.5$$

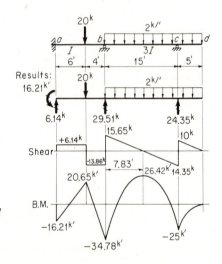

Isolate joints b and c; write the joint equations; and substitute for M_{ba}, M_{bc}, etc.

$\sum M_b = 0$	$M_{ba} + M_{bc} = 0$	$\therefore \quad 12EK\theta_b + 4EK\theta_c - 8.7 = 0 \qquad (1)$
$\sum M_c = 0$	$M_{cb} - 25 = 0$	$\therefore \quad 4EK\theta_b + 8EK\theta_c + 12.5 = 0 \qquad (2)$

Solving Eqs. (1) and (2), we find

$$EK\theta_b = +1.495 \qquad EK\theta_c = -2.31$$

Hence, substituting back in end-moment expressions gives

$$M_{ab} = +2.99 - 19.2 = -16.21^{k'} \qquad M_{bc} = +11.96 - 9.24 - 37.5 = -34.78^{k'}$$
$$M_{ba} = +5.98 + 28.8 = +34.78^{k'} \qquad M_{cb} = -18.48 + 5.98 + 37.5 = +25.0^{k'}$$

The remaining results can be computed by statics.

Discussion:

The cantilever portion cd adds no complications since the bending moment in this portion is statically determinate. The cantilever does, however, affect the joint equation at joint c. When isolating joints, assume unknown moments to be positive (i.e., to act clockwise on the end of the member and therefore counterclockwise on the end of the joint stub). Any known moment, however, should be shown acting with its known value in its known direction.

A convenient way of handling the K factor is to select one K factor as a standard and express all others in terms of this.

Units will always be consistent if all values are substituted in kip and foot units.

An alternative way of handling the cantilever effect is as follows. For moment equilibrium at any joint m with members connecting it to joints a, b, . . . , j,

$$M_{ma} + M_{mb} + \cdots + M_{mj} = 0$$

If any of these end moments are known, the proper sign and value can be substituted. For example, at joint c, $M_{cb} + M_{cd} = 0$, but $M_{cd} = -25$. Hence, $M_{cb} - 25 = 0$, which checks Eq. (2) above.

Note that the final results satisfy Eqs. (1) and (2). This check does not verify the work prior to setting up these equations, of course.

Example 9.17 *Compute the end moments for the beam in Example 9.16 caused only by the following support movements (no load acting):*

> *Support a, vertically 0.01 ft down, rotates 0.001 radian clockwise*
> *Support b, vertically 0.04 ft down*
> *Support c, vertically 0.0175 ft down*
> *Assume $E = 30 \times 10^3$ kips per sq in., $I = 1,000$ in.^{4}*

In this case,

$$\theta_a = +0.001 \qquad \psi_{ab} = \frac{0.04 - 0.01}{10} = +0.003 \qquad \psi_{bc} = -\frac{0.04 - 0.0175}{15} = -0.0015$$

but

$$\theta_b = ? \qquad and \qquad \theta_c = ?$$

There are no loads; therefore all FEMs are zero. Using Eq. (9.11) to write expressions for end moments,

$$M_{ab} = (2EK)(0.002 + \theta_b - 0.009) \quad = 2EK\theta_b - 0.014EK$$
$$M_{ba} = (2EK)(2\theta_b + 0.001 - 0.009) \quad = 4EK\theta_b - 0.016EK$$
$$M_{bc} = (2E)(2K)(2\theta_b + \theta_c + 0.0045) \quad = 8EK\theta_b + 4EK\theta_c + 0.018EK$$
$$M_{cb} = (2E)(2K)(2\theta_c + \theta_b + 0.0045) \quad = 4EK\theta_b + 8EK\theta_c + 0.018EK$$

From the joint equations,

$$\sum M_b = 0 \qquad M_{ba} + M_{bc} = 0 \qquad 12EK\theta_b + 4EK\theta_c + 0.002EK = 0 \tag{1}$$
$$\sum M_c = 0 \qquad \qquad M_{cb} = 0 \qquad 4EK\theta_b + 8EK\theta_c + 0.018EK = 0 \tag{2}$$

Solving Eqs. (1) and (2) simultaneously gives

$$\theta_b = +0.0007 \qquad \theta_c = -0.0026$$

Therefore, substituting back, we have

$$M_{ab} = -0.0126EK = -262.5^{k'} \qquad M_{bc} = 0.0132EK = +275.0^{k'}$$
$$M_{ba} = -0.0132EK = -275.0^{k'} \qquad M_{cb} = 0 = \qquad 0$$

since
$$EK = (30 \times 10^3)(144)\frac{1,000}{(144)^2(10)} = 20,833 \text{ kip-ft}$$

Discussion:
   In this type of problem, be particularly careful to insert the correct signs for the known θ and ψ angles.   Also be careful to keep units consistent.

Consider the rigid frames shown in Figs. 9.12a to d.   Suppose that we neglect the change in length of the members due to axial force, as is usually permissible in rigid frames, and consider only the effect of bending deformation.   With this assumption, it is easy to show in each of these four frames that the ψ angle is zero for every member (with the exception of the statically determinate cantilever portions, of course).   In Fig. 9.12b, for example, neglecting the axial change in length of members AB and BE, it is obvious that joint B cannot translate unless the supports move.   If joint B does not translate, we can reason in the same manner that joint C cannot translate.   The ψ angles must therefore be zero for all five members AB, BC, CD, BE, and CF.

At any particular joint, the θ angle will be the same for the ends of all members that are rigidly connected together at that joint.   In the case of this frame, therefore, there will be only four unknown θ angles, θ_A, θ_B, θ_C, and θ_D.   Hence, using

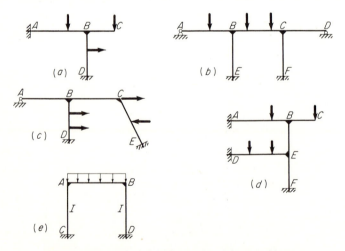

Figure 9.12   Typical rigid frames involving no sidesway

Eq. (9.11), we shall obtain expressions for the end moments involving these four θ angles as the only unknowns. Since we can write a joint equation $\Sigma M = 0$ at each joint where there is an unknown θ, we are able to obtain four equations containing the four θ angles as unknowns, in the same manner as in Example 9.16. After solving for the values of these θ angles, we can substitute back to obtain the values of the end moments, thus reducing the remainder of the stress analysis to a problem in statics.

In effect, we have shown that the slope-deflection solution of any frame in which there are no unknown ψ angles is essentially the same as that of a continuous beam. The solution of the frame in Fig. 9.12e is likewise in this category, provided that both the make-up and the loading are symmetrical. In this special case, the deflections will also be symmetrical and therefore there can be no horizontal deflection of the column tops. As a result, the ψ angles of all three members are zero.

In the most general type of rigid frame, both unknown θ and unknown ψ angles are involved even when we consider only the effect of bending deformation. In other words, there are both joint rotations and chord rotations, or sidesway. Several examples of such frames are shown in Fig. 9.15. Certain new ideas must be introduced to handle such problems. For this purpose, consider the frame shown in Fig. 9.13.

If the supports are unyielding, there are only two unknown θ angles, θ_B and θ_C. In this case, however, there is nothing that would prevent joints B and C from translating as well as rotating. Since we are neglecting deformation due to axial force, and since the chord rotations of the members are small, joint B translates essentially perpendicular to member AB, or, in this case, horizontally. Suppose that this movement is Δ. In the same manner, we should reason that joint C at the top of column CD must likewise translate horizontally. Since the axial change in length of BC is also neglected, the horizontal movement of C must also be Δ.

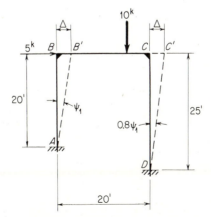

Figure 9.13 Frame involving sidesway

The deflected positions of the chords are shown by the dashed lines in Fig. 9.13. Note that these straight dashed lines indicate, not the elastic curve of the frame, but simply the *chords* of the elastic curve.   From the sketch it is apparent that

$$\psi_{BC} = 0 \qquad \psi_{AB} = \frac{\Delta}{20} = \psi_1$$

and therefore

$$\psi_{CD} = \frac{\Delta}{25} = \tfrac{4}{5}\,\psi_1$$

In this case, therefore, the ψ angles of the members of the frame can all be expressed in terms of one independent unknown, which we can call ψ_1.†

---

† In Example 9.19, this same technique is used on a more difficult problem to analyze the ψ angles assumed by the chords of the members and the relationships between these angles.   This same basic approach may be used to study the ψ angles of a frame of any complexity, such as those shown in Fig. 9.15, and may be outlined as follows:

1. Temporarily, consider all rigid joints in the indeterminate portion of the frame to be pinned joints and all fixed supports to be hinge supports.   Otherwise, consider the joint and support details of the structure to be as specified.   Now investigate whether or not this modified structure is statically and geometrically stable with respect to sidesway.   If one or more joints are free to translate without immediately encountering resistance, the structure is unstable and free to behave as a linkage or mechanism.   If such behavior of the modified structure is possible, corresponding chords of members of the actual frame may rotate through certain ψ angles.   In other words, in such a case sidesway of the actual structure is possible.

2. The degree of sidesway instability of the modified structure is equal to the number of roller (or link) supports which would have to be introduced in order to prevent movement of this structure as a linkage mechanism.   The position of such additional supports can be located by proceeding as follows.   Imagine that the members are disconnected from the supports and from one another at the joints.   Now reassemble the structure one member at a time starting from the supports. As this reassembly proceeds, it will become apparent that in order to prevent instability as a mechanism, auxiliary holding supports must be introduced at any joint which is not prevented from translating by at least two noncollinear bars connecting it to the stable structure already reassembled.

3. The potential ψ angles in the actual frame and the relationships between them can now be studied by translating successively each of the stabilizing supports which were introduced into the modified structure described in Step 2.   Each one of these supports may be translated an arbitrary amount and the resulting chord rotations thereby produced can then be evaluated.   One of these chord rotations produced by a particular support translation may be designated as the one independent ψ angle associated with that support translation and all other chord rotations involved expressed in terms of it.   The number of independent ψ angles involved in the actual frame is therefore equal to the degree of instability of the modified structure.   Note that the chord rotations of some members of a frame may be related to more than one of the independent ψ angles.

A rigid frame is said to have one degree of freedom with respect to sidesway associated with each independent ψ angle.   In Art. 9.14 it is shown that the organization of the moment-distribution solution of a frame is dependent on the proper recognition of the number of degrees of freedom with respect to sidesway.

When these procedures are applied to the frames shown in Fig. 9.15, the degree of instability of the corresponding modified structures (or the number of independent ψ angles of the actual structures) is found to be as follows: (*a*) 1, (*b*) 1, (*c*) 1, (*d*) 3, (*e*) 1, (*f*) 3, (*g*) 2, and (*h*) 2.

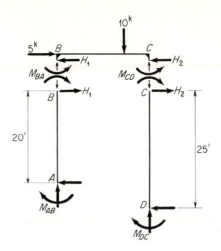

Figure 9.14 Isolation of columns and girder

Applying Eq. (9.11) in this case leads to expressions for the end moments that involve θ_B, θ_C, and ψ_1 as unknowns.   Two of the three equations required to solve for these unknowns can be obtained from the familiar joint equations at joints B and C.   The third independent equation of statics that the end moments must satisfy must be obtained in some other manner.   It can be derived as follows. Isolate each of the columns by cutting them out just below the girder and just above the foundation as shown in Fig. 9.14.   Likewise, isolate the girder by cutting it out just below the top of the columns.   Then, taking moments about the base of each isolated column

$$M_{AB} + M_{BA} + 20H_1 = 0 \tag{a}$$

$$M_{DC} + M_{CD} + 25H_2 = 0 \tag{b}$$

Likewise from $\sum F_x = 0$ on the girder,

$$H_1 + H_2 = 5 \tag{c}$$

Substituting for H_1 and H_2 in Eq. (c) from Eqs. (a) and (b), we obtain

$$M_{AB} + M_{BA} + 0.8M_{DC} + 0.8M_{CD} + 100 = 0 \tag{d}$$

which is the required third independent equation of statics.   From the two joint equations and this so-called **shear equation**, we are able to arrive at three equations from which we can solve for θ_B, θ_C, and ψ_1 and then to complete the solution as in previous problems.

Examples 9.18 and 9.19 illustrate the slope-deflection solution of certain typical frames where sidesway is involved.

Example 9.18 *Compute the end moments and draw the bending-moment diagram for the frame.*

Analyzing the θ and ψ angles, we have

$$\theta_A = \theta_D = \psi_{BC} = 0$$
$$\theta_B = ? \qquad \theta_C = ? \qquad \psi_{AB} = \psi_{CD} = \psi_1$$

$$K_{AB} = K_{CD} = \frac{I_1}{15} = K \qquad K_{BC} = \frac{4I_1}{20} = 3K$$

$$FEM_{BC} = -\frac{(100)(8)(12)^2}{(20)^2} = -288^{k'}$$

$$FEM_{CB} = +\frac{(100)(8)^2(12)}{(20)^2} = +192^{k'}$$

Using Eq. (9.11) gives

$$M_{AB} = 2EK\theta_B - 6EK\psi_1$$
$$M_{BA} = 4EK\theta_B - 6EK\psi_1$$
$$M_{BC} = 12EK\theta_B + 6EK\theta_C - 288$$
$$M_{CB} = 12EK\theta_C + 6EK\theta_B + 192$$
$$M_{CD} = 4EK\theta_C - 6EK\psi_1$$
$$M_{DC} = 2EK\theta_C - 6EK\psi_1$$

Joint B:

$$\sum M_B = 0 \qquad M_{BA} + M_{BC} = 0$$
$$\therefore 16EK\theta_B + 6EK\theta_C - 6EK\psi_1 - 288 = 0 \quad (1)$$

Joint C:

$$\sum M_C = 0 \qquad M_{CB} + M_{CD} - 500 = 0$$
$$\therefore 6EK\theta_B + 16EK\theta_C$$
$$-6EK\psi_1 - 308 = 0 \qquad (2)$$

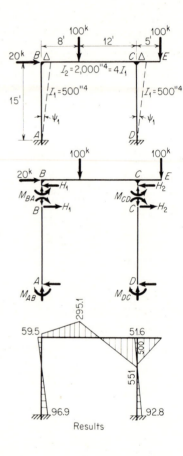

Results

Shear equation:

Col. AB: $M_{AB} + M_{BA} + 15H_1 = 0$
Col. DC: $M_{DC} + M_{CD} + 15H_2 = 0$
Girder: $H_1 + H_2 = 20$

$$\therefore M_{AB} + M_{BA} + M_{CD} + M_{DC} + 300 = 0$$

and $\therefore 6EK\theta_B + 6EK\theta_C - 24EK\psi_1 + 300 = 0$ (3)

Solving Eqs. (1), (2), and (3) gives

$$EK\theta_B = +18.63 \qquad EK\theta_C = +20.60 \qquad EK\psi_1 = +22.34$$

Then, $M_{AB} = -96.9^{k'}$ $M_{BC} = +59.2^{k'}$ $M_{CD} = -51.6^{k'}$
                     $M_{BA} = -59.5^{k'}$ $M_{CB} = +551.0^{k'}$ $M_{DC} = -92.8^{k'}$

Discussion:

Note that the dashed lines in the first sketch indicate the deflected position of the chords of the elastic curve of the columns, not the elastic curve itself.

When various portions of a frame are isolated as shown in the second sketch, all unknown end moments should be assumed to be positive, i.e., to act clockwise on the ends of the members. Shears and axial forces may be assumed to act in either direction, of course; but it having been assumed that a force such as H_1 acts in a certain direction on one isolated portion, it must be assumed that the force acts consistently on any subsequently isolated portions.

When plotting bending-moment diagrams for such frames, plot the ordinates on the side of the member that is in compression under the moment at that section of the frame.

Example 9.19   *Compute the end moments and draw the bending-moment diagram for this frame.*

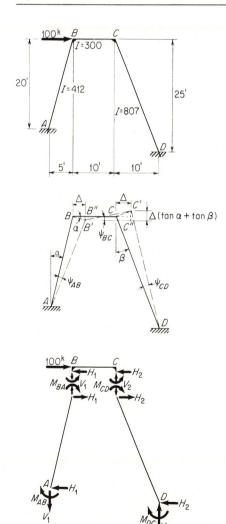

To analyze ψ-angle relationships, imagine that the members are temporarily disconnected at the joints and from the supports and then reassembled one by one. First connect member AB at the base. If the chord of this member did rotate an amount ψ_{AB}, the B end would move essentially perpendicular to AB along the path BB'. Translate member BC parallel to itself so that the B ends of AB and BC can be connected together at point B'. If the chord of BC were now rotated, the C end would move along a path $C''C'$ perpendicular to the original direction of BC. Likewise, if member CD were connected to the support at D and the chord were rotated, the C end would have to move along a path CC' perpendicular to CD. The C ends of members BC and CD could be joined at point C' where these paths intersect. Then

$$\psi_{AB} = \frac{BB'}{L_{AB}} = \frac{\Delta}{(\cos \alpha)L_{AB}} = \frac{\Delta}{20} = {}^{10}\!/_{13}\psi_1$$

$$\psi_{BC} = -\frac{C''C'}{10} = -\frac{\Delta}{10}(\tan \alpha + \tan \beta)$$

$$= -{}^{13}\!/_{200}\Delta = -\psi_1$$

$$\psi_{CD} = \frac{CC'}{L_{CD}} = \frac{\Delta}{(\cos \beta)L_{CD}} = \frac{\Delta}{25} = {}^{8}\!/_{13}\psi_1$$

Therefore all ψ angles can be expressed in terms of one unknown, ψ_1. Since $\theta_A = \theta_D = 0$, the independent unknowns are θ_B, θ_C, and ψ_1.

$$K_{AB} = {}^{412}\!/_{20.6} = 20 = K \qquad K_{BC} = {}^{300}\!/_{10} = 30 = 1.5K \qquad K_{CD} = {}^{807}\!/_{26.9} = 1.5K$$

All FEMs are zero since no loads are applied between joints. Applying Eq. (9.11) gives

$$M_{AB} = 2EK\theta_B - {}^{60}\!/_{13}EK\psi_1 \qquad\qquad M_{CB} = 3EK\theta_B + 6EK\theta_C + 9EK\psi_1$$
$$M_{BA} = 4EK\theta_B - {}^{60}\!/_{13}EK\psi_1 \qquad\qquad M_{CD} = 6EK\theta_C - {}^{72}\!/_{13}EK\psi_1$$
$$M_{BC} = 6EK\theta_B + 3EK\theta_C + 9EK\psi_1 \qquad M_{DC} = 3EK\theta_C - {}^{72}\!/_{13}EK\psi_1$$

Joint B:

$$\sum M_B = 0 \qquad M_{BA} + M_{BC} = 0 \qquad \therefore\ 10EK\theta_B + 3EK\theta_C + 4.385EK\psi_1 = 0 \tag{1}$$

Joint C:

$$\sum M_C = 0 \qquad M_{CB} + M_{CD} = 0 \qquad \therefore\ 3EK\theta_B + 12EK\theta_C + 3.462EK\psi_1 = 0 \tag{2}$$

Shear equation:

Col. AB: $\qquad\qquad \sum M_A = 0 \qquad M_{AB} + M_{BA} + 20H_1 - 5V_1 = 0 \qquad\qquad$ (a)

Col. CD: $\qquad\qquad \sum M_D = 0 \qquad M_{DC} + M_{CD} + 25H_2 - 10V_2 = 0 \qquad\qquad$ (b)

Girder: $\qquad\qquad \sum M_C = 0 \qquad M_{BA} + M_{CD} + 10V_1 = 0 \qquad\qquad\qquad$ (c)

$$\sum F_y = 0 \qquad V_1 = V_2 \qquad\qquad\qquad\qquad\qquad\quad \text{(d)}$$

$$\sum F_x = 0 \qquad H_1 + H_2 = 100 \qquad\qquad\qquad\qquad\quad \text{(e)}$$

Substitute in (a) and (b) for V_1, V_2 and H_2 from (c), (d), and (e); then eliminate H_1 from the modified Eqs. (a) and (b) and obtain

$$M_{AB} + 2.3M_{BA} + 2.1M_{CD} + 0.8M_{DC} + 2{,}000 = 0$$

or $\qquad\qquad\qquad 11.2EK\theta_B + 15EK\theta_C - 31.292EK\psi_1 = -2{,}000 \qquad\qquad$ (3)

Now solve Eqs. (1), (2), and (3) in the following tabular form:

Eq.	Operation	$EK\theta_B$ +	$EK\theta_C$ +	$EK\psi_1$	$= const \times 10^{-2}$	Check
1		+10	+ 3	+ 4.385	0	+17.385
2		+ 3	+ 12	+ 3.462	0	+18.462
3		+11.2	+15	−31.292	−20	−25.092
3′	(3)(0.893)	+10	+13.393	−27.939	−17.857	−22.404
3″	(3)(0.268)	+ 3	+ 4.018	− 8.382	− 5.357	− 6.721
4	1 − 3′		−10.393	+32.324	+17.857	+39.789
5	2 − 3″		+ 7.982	+11.844	+ 5.357	+25.183
4′	(4)(0.768)		− 7.982	+24.825	+13.715	+30.559
6	5 + 4′			+36.669	+19.072	+55.742
				+ 1.0	+ 0.5201	

$$7.982EK\theta_C = 5.357 - 6.159 = -0.802 \qquad EK\theta_C = -0.1005$$
$$3EK\theta_B = -5.357 + 4.359 + 0.404 = -0.594 \qquad EK\theta_B = -0.1980$$

The above results are for constants that are 0.01 of the actual constants. The actual results are 100 times the above.

$$EK\theta_B = -19.80 \qquad EK\theta_C = -10.05 \qquad EK\psi_1 = +52.01$$

The end moments are therefore

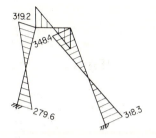

$$M_{AB} = -39.6 - 240.0 = -279.6^{k'}$$
$$M_{BA} = -79.2 - 240.0 = -319.2^{k'}$$
$$M_{BC} = -118.8 - 30.2 + 468.1 = +319.1^{k'}$$
$$M_{CB} = -59.4 - 60.4 + 468.1 = +348.3^{k'}$$
$$M_{CD} = -60.4 - 288.1 = -348.5^{k'}$$
$$M_{DC} = -30.2 - 288.1 = -318.3^{k'}$$

Discussion:

When analyzing the relations between the ψ angles, be sure to record the proper signs for them, thus indicating whether the chord rotates clockwise or counterclockwise.

In this problem, the solution of the simultaneous equations has been carried through in detail to illustrate a tabular procedure that is very convenient when there are three or more simultaneous equations to be solved. The important features of this procedure are as follows:

1. When eliminating an unknown, select the equation having the highest coefficient in that column. Operate on this equation so as to obtain several reduced versions of it, each version adjusted so as to match the coefficient of the unknown which is being eliminated in one of the original group of equations. Proceeding in this manner tends to minimize errors rather than to increase them.

2. Keep track of the operations performed on each equation, to facilitate checking.

3. In the check column, record the algebraic sum of all the coefficients and constant terms in each equation. Operate on this figure the same as on all other terms in an equation. After any operation, the new sum of coefficients and constant terms should be equal to the new figure in the check column. Note that this is a necessary check but is not sufficient to catch occasional compensating errors.

4. The order of magnitude of the constant should be adjusted so as be of the same order as the coefficients of the unknowns. This is done to make the check column most effective. After the solution of the equations is completed, the values of the unknowns can be adjusted to give the answers corresponding to the actual constants.

By this time, the student should have started comparing the advantages and disadvantages of the different methods of stress analysis. He should have begun to form some idea as to when to use one method and when to use another. For example, he should study the structures in Figs. 9.12 and 9.15 and decide in each case whether the approach using superposition equations or Castigliano's theorem is or is not superior to the slope-deflection method.

The amount of computation is more or less proportional to the square of the number of simultaneous equations involved in the solution. Generally speaking, the superior method is that involving the fewest unknowns. Thus, comparing the superposition-equation approach with the slope-deflection method consists largely in comparing the number of redundant stress components with the number of unknown θ and ψ angles.

9.13 Fundamentals of the Moment-Distribution Method. The moment-distribution method is an ingenious and convenient method of handling the stress

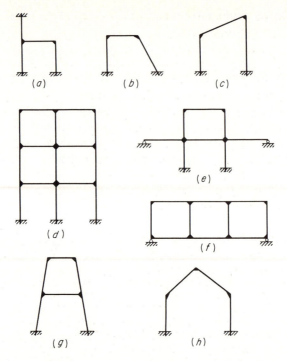

Figure 9.15 Typical rigid frames involving sidesway

analysis of rigid-joint structures.[1] All the methods discussed previously involve the solution of simultaneous equations, which constitutes a major part of the computational work when there are more than three or four unknowns. The method of moment distribution usually does not involve as many simultaneous equations and is often much shorter than any of the methods already discussed. It has the further advantage of consisting of a series of cycles, each converging on the precise final result; therefore the series can be terminated whenever one reaches the degree of precision required by the problem at hand.

If we refer to Eq. (9.11), the fundamental slope-deflection equation, we observe that the moment acting on the end of a member is the algebraic sum of four separate effects:

1. The moment due to the applied loads on the member if it were a fixed-end beam, i.e., the FEM

2. The moment caused by the rotation of the tangent to the elastic curve at the near end

3. The moment caused by the rotation of the tangent to the elastic curve at the far end

---

[1] The moment-distribution method was presented by Prof. Hardy Cross in an article in *Trans. ASCE*, vol. 96, pap. 1793, 1932, and in several prior publications, and is without a doubt one of the most important contributions to structural analysis in the twentieth century.

4. The moment caused by the rotation of the chord of the elastic curve joining the two ends of the member

Since the total end moment is the superposition of these four effects, it is suggested that it might be possible to allow them to take place separately and thus arrive at the total resulting from all four acting simultaneously.

To simplify the present discussion, we shall confine our attention to structures which are composed of prismatic members, i.e., members which have a constant I throughout their length, and also to structures in which there is no joint translation and in which, therefore, the ψ angles are all equal to zero.

Consider such a structure as that shown in Fig. 9.16. If the supports are unyielding, there is no joint rotation at a, d, c, or e but there will be some joint rotation at b when the load is applied. However, suppose that we first consider the unloaded structure and imagine that we temporarily apply an external clamp which locks joint b against rotation. Then if the load is applied, FEMs will be developed in member ab and can be computed from Eqs. (9.13).

The moment FEM_{ba} causes a counterclockwise moment on the locked joint b. If the clamp is released, this moment will cause the joint to rotate counterclockwise. When this joint rotates, certain moments are developed throughout the length of all members meeting at this joint. The joint will continue to rotate until sufficient

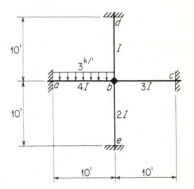

Figure 9.16 Development of moment-distribution ideas (joint rotation only)

end moments are developed in the b ends of these members to balance the effect of FEM_{ba}. Of course, simultaneously certain end moments have been developed in the far ends of these members. When equilibrium is established at joint b, the structure in this case will have attained its final deformed position and the total end moment at the ends of the various members will be the algebraic sum of the FEM and the moment caused by the rotation of joint b.

The procedure outlined above is essentially the moment-distribution method. It is convenient to adopt a certain terminology to simplify the description of the above procedure. We are already familiar with the term used to describe the end moments developed when the loads are applied to the structure with all joints locked against rotation—such moments are **fixed-end moments.** When a joint is unlocked, it will rotate if the algebraic sum of the FEMs acting on the joint does not add up to zero; this resultant moment acting on the joint is therefore called the **unbalanced moment.** When the unlocked joint rotates under this unbalanced moment, end moments are developed in the ends of the members meeting at the joint. These finally restore equilibrium at the joint and are called **distributed moments.** As the joint rotated, however, and bent these members, end moments were likewise developed at the far ends of each; these are called **carry-over moments.**

Before evaluating these various moments numerically, it is desirable to adopt a convenient sign convention. Three different conventions are in use in the literature, but the authors prefer to use the same convention as that previously suggested for the slope-deflection method, namely, *end moments are plus when they act clockwise on the ends of the members.* Note that such plus (clockwise) end moments are associated with corresponding counterclockwise moments applied to the stubs of the members attached to the joint (see Fig. 9.17).

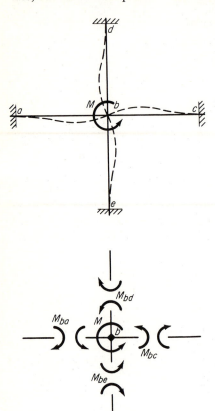

Expressions for the FEMs have already been developed in Art 9.11. The **unbalanced moment** acting on a locked joint n is simply the algebriac sum of the end moments at the n ends of all members that are rigidly connected to this locked joint.

Some further analysis is necessary to explain how to compute the distributed moments. Consider the structure in Fig. 9.16 under the condition where joint b is unlocked and allowed to rotate under the unbalanced moment M. The structure deforms as shown in Fig. 9.17 and develops distributed moments M_{ba}, M_{bd}, etc., which restore the equilibrium of joint b. These distributed moments, being unknown, are assumed to be positive and therefore to act clockwise on the ends of the members and counterclockwise on the joint. The unbalanced moment M is assumed to be the resultant of positive FEMs and therefore also acts counterclockwise on the joint. Since $\sum M_b = 0$,

Figure 9.17 Joint rotation

$$M_{ba} + M_{bc} + M_{bd} + M_{be} + M = 0 \qquad (a)$$

However, the distributed moments can be evaluated by means of Eq. (9.11), it being noted in this case that $\theta_a = \theta_d = \theta_c = \theta_e = 0$ and that all ψ angles are zero.

$$M_{ba} = 4EK_{ba}\theta_b \qquad M_{bc} = 4EK_{bc}\theta_b$$

$$M_{bd} = 4EK_{bd}\theta_b \qquad M_{be} = 4EK_{be}\theta_b \qquad (b)$$

By substituting in Eq. (*a*) from Eqs. (*b*), we can solve for θ_b, substitute this value back in Eqs. (*b*), and obtain expressions for M_{ba}, etc.   For example,

$$M_{ba} = \frac{-K_{ba}}{K_{ba} + K_{bd} + K_{bc} + K_{be}} M \quad \text{etc.} \tag{c}$$

or, in general, the distributed moment in any bar *bm* is given by

$$M_{bm} = -\frac{K_{bm}}{\sum\limits_b K} M \tag{d}$$

where the summation is meant to include all members meeting at joint *b*.   Let

$$DF_{bm} = \textbf{distribution factor} \text{ for } b \text{ end of member } bm = \frac{K_{bm}}{\sum\limits_b K} \tag{9.14}$$

Then
$$M_{bm} = -DF_{bm} M \tag{9.15}$$

Equation (9.15) may be interpreted as follows: *The **distributed moment** developed at the b end of member bm as joint b is unlocked and allowed to rotate under an unbalanced moment M is equal to the distribution factor DF_{bm} times the unbalanced moment M with the sign reversed.*

An expression for carry-over moment must also be developed. Consider any member, one end *b* of which has rotated θ_b, developing a distributed moment M_{bm} as shown in Fig. 9.18.   It should be noted that as joint *b* was unlocked and allowed

Figure 9.18 Carry-over moment

to rotate, the joint at *m* remained locked and therefore θ_m is equal to zero.   Since ψ_{bm} is also zero, applying Eq. (9.11) leads to the following:

$$M_{bm} = 4EK_{bm}\theta_b \quad \text{and} \quad M_{mb} = 2EK_{bm}\theta_b$$

Hence,
$$M_{mb} = \tfrac{1}{2}M_{bm} \tag{9.16}$$

*In words, the **carry-over moment** is equal to one-half of its corresponding distributed moment and has the same sign.*

We have now developed all the fundamental ideas and relations required to solve the simpler moment-distribution problems.   Note, however, that the above discussion is restricted to structures composed of prismatic members.   These ideas are extended to members having variable I in Art. 9.15.

9.14 Application of Moment Distribution to Beams and Frames.   For the first illustration of the application of moment distribution, consider the continuous beam shown in Fig. 9.19, the supports of which are unyielding.   In this case, the ψ angles of both members are zero, and support *a* is permanently locked against rotation.   Suppose that joints *b* and *c* are temporarily locked and the loads are applied, thereby developing the following FEMs:

$$\text{FEM}_{ab} = -9.6 \text{ kip-ft} \quad \text{FEM}_{bc} = -18.75 \text{ kip-ft}$$
$$\text{FEM}_{ba} = +14.4 \text{ kip-ft} \quad \text{FEM}_{cb} = +18.75 \text{ kip-ft}$$

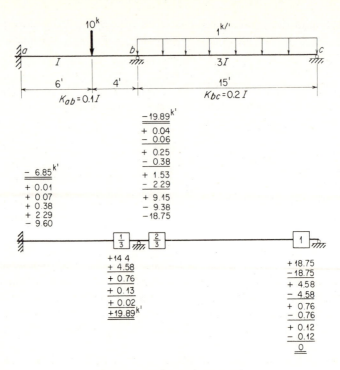

Figure 9.19   Illustrative example

In preparation for unlocking joints b and c and distributing the unbalanced moments, the stiffness factors K are computed, and from them the distribution factors at these joints are evaluated:

At b: $\sum K = 0.3I$ $DF_{ba} = \frac{1}{3}$ $DF_{bc} = \frac{2}{3}$

At c: $\sum K = 0.2I$ $DF_{cb} = 1$

These distribution factors are then recorded in the appropriate box on the working diagram in Fig. 9.19. All the computations for the end moments are recorded on this diagram, the numbers referring to a particular end moment being recorded in a column normal to the member and running away from the member on the side first encountered in proceeding clockwise about the joint. Such an arrangement of the computations is, of course, not imperative but will be found highly desirable for frames.

After recording the FEMs, joints b and c are successively unlocked and allowed to rotate gradually into their equilibrium positions. Upon unlocking joint c, it rotates under an unbalanced moment of $+18.75$ until a distributed moment of -18.75 is developed to restore equilibrium. Joint c is relocked in this new position, and a line is drawn under the -18.75 to indicate that the joint is now in equilibrium. As joint c rotated, a carry-over moment of one-half of the distributed moment was developed at the b end of member bc, or in this case -9.38.

If joint *b* is now unlocked, it will rotate under an unbalanced moment equal to the algebraic sum of the two fixed end moments at this joint and the above carry-over moment, or −13.73. Using the appropriate distribution factors, the distributed moments that restore equilibrium are computed and are found to be +4.58 and +9.15; they are recorded and underlined, indicating that the joint is now in equilibrium. Relocking this joint in this new position and recording the carry-over moments developed at *a* and *c* by the rotation of joint *b* lead to +2.29 and +4.58, respectively.

When we return to joint *c* and unlock it a second time, it will rotate under a unbalanced moment of +4.58, developing a balancing distributed moment of −4.58. Again half of this distributed moment is carried over to the *b* end of this member. Joint *c* is locked in its new position. When joint *b* is unlocked a second time, it will rotate under an unbalance of −2.29, developing distributed moments of +0.76 and +1.53, which in turn induce carry-over moments at *a* and *c* of +0.38 and +0.76.

We may proceed in this manner, unlocking first joint *c* and then joint *b*, until the effects are so small that we are willing to neglect them. This problem has been carried through more cycles than required for practical purposes simply to illustrate the procedure. After the moment-distribution procedure is completed, the final end moments are obtained by adding algebraically all the figures in the various columns.

The convergence in the above solution has been rather slow because joint *c* is actually a hinged end and continually throws back sizable carry-over moments to joint *b*. Whenever there is such a hinged end at the extremity of a structure, the convergence can be improved by modifying the above procedure as indicated in Fig. 9.20.

We start out as before, locking all joints against rotation, applying the loads, and developing the FEMs. Again as a first step, we unlock joint *c* and let it rotate and develop the distributed moment of −18.75, which then carries over −9.38 to

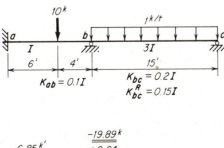

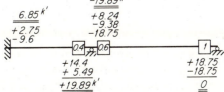

Figure 9.20 Illustrative example using reduced stiffness factor

joint b. At this point, we leave joint c unlocked so that it can rotate freely and hence can develop no end moment. Under these conditions, when we now proceed to joint b and unlock it, it will rotate under the unbalanced moment of -13.73 with the joint c unlocked instead of locked. Physically, this means that member bc is not so stiff as it was previously and hence does not take so much of the unbalanced moment. We shall now evaluate just how much its stiffness has been reduced.

Refer to Figs. 9.16 and 9.17 and the development of an expression for distributed moments. Suppose that the support at c were a roller or hinge support instead of a fixed support. Then, there would be a θ angle at c as well as b under the action of the unbalanced moment M, but M_{cb} would be equal to zero. Upon applying Eq. (9.11), the previous analysis would be modified as follows:

$$M_{cb} = 4EK_{bc}\theta_c + 2EK_{bc}\theta_b = 0$$

from which

$$\theta_c = -\frac{\theta_b}{2}$$

and then

$$M_{bc} = 4EK_{bc}\theta_b + 2EK_{bc}\theta_c = 3EK_{bc}\theta_b = (4E)(\tfrac{3}{4}K_{bc})(\theta_b) = 4EK_{bc}^{R}\theta_b$$

where

$$K_{bc}^{R} = \text{reduced stiffness factor} = \tfrac{3}{4}K_{bc} \tag{9.17}$$

Hence, our previous expressions for distribution factors may be used for this new case provided that a reduced stiffness factor K^R instead of the usual stiffness factor K is used for a member where the far end is hinged.[1]

We shall therefore revise the distribution factors at b, using the reduced stiffness factor for member bc:

At b: $\sum \text{eff } K = 0.25I$ $DF_{ba} = 0.4$ $DF_{bc} = 0.6$

As a result, distributed moments of $+5.49$ and $+8.24$ are developed to balance the unbalanced moment of -13.73. There is no carry-over to c, or course, since this joint was left unlocked. There is the usual carry-over, though, from b to a. In this particular case, our solution is now complete since joint c has not been unbalanced by the unlocking at b. All joints are therefore in equilibrium and will not be disturbed if the temporary locking device is permanently removed from joint b. Adding up the various columns will, in this case, now give us the exact end moments for this beam.

The revised procedure will not always give exact results in one cycle as it did in this problem, but in any case it will give more rapid convergence than the original

---

[1] Inasmuch as the effective stiffness factor at b for member bm may be either K_{bm} if the far end m is fixed or K_{bm}^{R} if the far end is hinged, it is desirable to rewrite Eq. (9.14) as

$$DF_{bm} = \text{distribution factor for } b \text{ end of member } bm = \frac{\text{eff } K_{bm}}{\sum_{b} \text{eff } K} \tag{9.14a}$$

where **eff K** designates the **effective stiffness factor** of a member, which may be equal to either K or K^R depending on whether the far end is fixed or hinged, respectively.

procedure.   Example 9.20 is designed to illustrate the application of the above ideas to a somewhat more elaborate structure that does not involve any unknown chord rotations.   In Example 9.21, these ideas are extended to cover a case where the ψ angles are not zero but their values are known.

Example 9.20   *Compute the end moments and draw the bending-moment diagram for this frame; unyielding supports.*

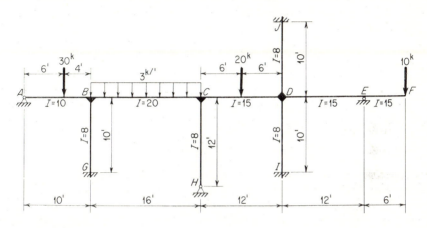

Distribution factors:

At B: $\sum$ **eff K** = 2.8	At C: $\sum$ **eff K** = 3.0	At D: $\sum$ **eff K** = 3.788
$DF_{BA} = 0.268$	$DF_{CB} = 0.417$	$DF_{DC} = 0.330$
$DF_{BC} = 0.446$	$DF_{CD} = 0.417$	$DF_{DJ} = 0.212$
$DF_{BG} = 0.286$	$DF_{CH} = 0.167$	$DF_{DE} = 0.248$
$\overline{1.000}$	$\overline{1.000}$	$DF_{DI} = 0.212$
		$\overline{1.002}$

Fixed-end moments:

$$FEM_{AB} = -\frac{(30)(6)(4)^2}{(10^2)} = -28.8^{k'} \qquad FEM_{BA} = +\frac{(30)(6)^2(4)}{(10)^2} = +43.2^{k'}$$

$$FEM_{BC} = -\frac{(3)(16)^2}{12} = -64^{k'} \qquad FEM_{CB} = +64^{k'}$$

$$FEM_{CB} = -\frac{(20)(12)}{8} = -30^{k'} \qquad FEM_{DC} = +30^{k'}$$

Cantilever moment:

$$M_{EF} = -(10)(6) = -60^{k'}$$

Moment distribution:

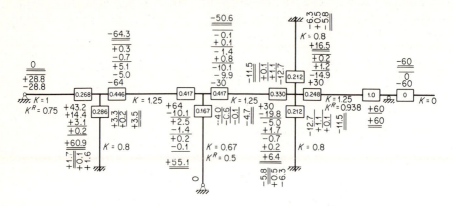

Final results:

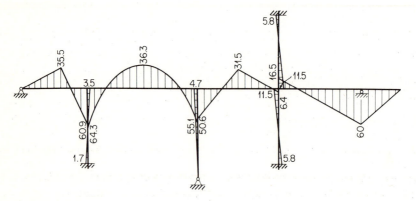

Discussion:

In this structure, when the loads are applied after all the joints have been locked against rotation, a moment will be developed at the E end of the cantilever, in addition to the usual FEMs at other points in the frame. The moment of 60 kip-ft contributes to the unbalanced moment at joint E, just as FEM would.

Note that the cantilever arm has no restraining effect on the rotation of joint E; that is, in effect, its stiffness factor is equal to zero. Any unbalanced moment is therefore carried entirely by the other members that meet at this joint.

Following the modified procedure, all hinged extremities must be unlocked first. This includes not only joints A and H but also E.

The remaining joints are unlocked in turn and gradually rotated into their equilibrium positions. If in doing this we start with the joint which has the largest unbalanced moment, the convergence of the solution will be somewhat more rapid, although the final results are independent of the order in which the joints are unlocked.

Example 9.21   *Solve Example 9.17, using moment distribution.*

Suppose that we temporarily lock all joints against rotation and that we then introduce the specified support movements, so that

$$\theta_a = +0.001 \ radian$$
$$\psi_{ab} = +0.003 \ radian$$
$$\psi_{bc} = -0.0015 \ radian$$

In effect, we have bent the members and developed certain initial end moments that can be evaluated by using Eq. (9.11), or

$$M_{ab} = (2E)(0.1I)(0.002 - 0.009)$$
$$= -0.0014EI = -292^{k'}$$
$$M_{ba} = (2E)(0.1I)(0.001 - 0.009)$$
$$= -0.0016EI = -334^{k'}$$
$$M_{bc} = (2E)(0.2I)(+0.0045)$$
$$= +0.0018EI = +376^{k'}$$
$$M_{cb} = +0.0018EI = +376^{k'}$$

since

$$EI = (30 \times 10^3)^{1.000}\!/_{144} = 208,333 \ kip\text{-}ft^2$$

The remainder of the solution is handled exactly the same as if these initial end moments were FEMs. Thus, the distribution, etc., is done exactly as it was in Fig. 9.20.

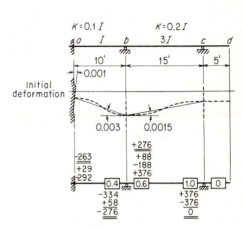

Once we understand the fundamental philosophy of moment distribution and the details of its application to the cases discussed above, it is quite easy to extend these ideas to more complicated cases involving sidesway, i.e., unknown ψ angles.[1] Consider, for example, the frame shown in Fig. 9.21. In this case, there is nothing to prevent a horizontal deflection of the column tops. In addition to rotation of joints B and C, we therefore have certain unknown chord rotations developed in the columns.

In order to handle this problem by moment distribution, we can break it down into two separate parts. First, suppose that we introduce a horizontal holding force R which prevents any horizontal movement of joint B. With the structure restrained in this manner, we can apply all the given loads and determine the resulting end moments by moment distribution just as we would for any structure that involved no sidesway. Having the end moments, we can then back-figure the holding force R by statics.[2] This part of the solution is called case A in Fig. 9.21.

---

[1] See footnote on p. 322 for description of a method of determining the number of independent unknown ψ angles. The number of degrees of freedom with respect to sidesway is equal to the number of independent unknown ψ angles.

[2] Suppose in a particular problem that the holding force R was found from statics to be zero. What does this mean regarding sidesway? Is it necessary to proceed with a case B solution?

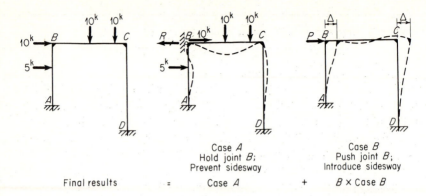

Figure 9.21   Moment distribution including effect of sidesway (joint translation)

In the second, or case B, part of the solution, we imagine that we lock the joints against rotation, push on the frame at joint B, and introduce some arbitrary horizontal displacement Δ. (*Note that the pushing force P in case B must be applied at the same point and along the same line of action as the holding force R in case A.*) We can then analyze the ψ angles thus introduced into the columns. Using Eq. (9.11), we can compute the corresponding initial end moments developed in the members. These end moments will be expressed in terms of $E\Delta$; but since Δ is arbitrary, we can let $E\Delta$ equal unity or any convenient amount and obtain numerical values for the initial end moments. If we then unlock, distribute, and carry over through several cycles at joints B and C, we shall arrive at a set of end moments for the frame. Again by statics we can back-figure the force P that has pushed the frame into this position and developed these end moments.

We are now in a position to superimpose the case A and case B parts of the solution so as to obtain the answers for the specified loads. To simplify the discussion of this superposition, let us assume that R was computed to be a force of 12 kips acting to the left and P a force of 6 kips acting to the right. We must take the case A moments as they are, since the external loads for this case are the same as the specified ones; but we can combine these results with any multiple of the case B results that we desire. Using the assumed values for R and P, if we take twice the case B loads and superimpose them on the case A loads, we shall obviously obtain the specified load system. The final answers for the end moments are therefore obtained by adding algebraically twice the case B moments to the case A moments.

Example 9.22 *Compute the end moments in this frame.*

At B: $K_{AB} = 15$ $DF_{BA} = 0.429$
         $K_{BD} = 20$ $DF_{BD} = 0.571$
         $\sum \text{eff } K = 35$

At D: $K_{DB} = 20$ $DF_{DB} = 0.426$
         $K_{DE}^{R} = 15$ $DF_{DE} = 0.319$
         $K_{DC}^{R} = 12$ $DF_{DC} = 0.255$
         $\sum \text{eff } K = 47$

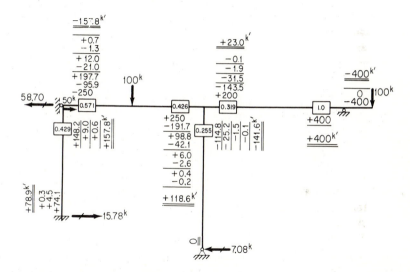

Case A: *Prevent sidesway.*   *Hold frame at joint B.*   *Apply all given loads to the structure.*

$$FEM_{BD} = -250 \qquad M_{EF} = -400$$
$$FEM_{DB} = +250$$

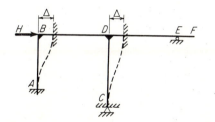

By isolating columns, horizontal reactions at A and C may be computed.   *Then, applying* $\sum H = 0$ *to the entire structure, we find the holding force at B to be 58.70 kips acting toward the left.*

Case B: *Introduce sidesway by pushing at point B, having first temporarily locked all joints (including C) against rotation. Compute initial end moments from Eq. (9.11).*

$$M_{AB} = M_{BA} = -(6E)(15)\frac{\Delta}{15} = -6E\Delta$$

$$M_{CD} = M_{DC} = -(6E)(16)\frac{\Delta}{20} = -4.8E\Delta$$

Let $E\Delta = 100$; *then*

$$M_{AB} = M_{BA} = -600$$
$$M_{CD} = M_{DC} = -480$$

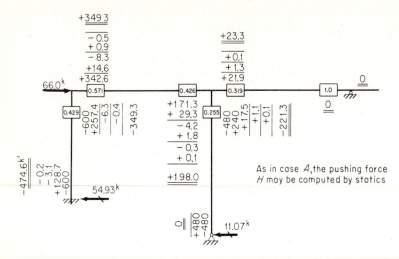

Combination of case A and case B: *Let B = factor by which we must multiply the case B solution. The superposition of B times the case B solution on the case A solution must be such that the modified pushing force from case B cancels the case A holding force at joint B, thereby leaving only the given loads acting on the structure. This cancellation may be expressed in equation form by considering forces acting to the right at joint B to be plus. Then*

$$-58.70 + 66B = 0 \quad or \quad B = 0.89$$
$$\therefore \ Final \ results = case \ A + (B)(case \ B)$$

$$M_{AB} = +78.9 - 421.5 \ = -342.6^{k'} \qquad M_{DC} = -141.6 - 196.7 = -338.3^{k'}$$
$$M_{BA} = +157.8 - 310.5 = -152.7^{k'} \qquad M_{DE} = +23.0 + 20.7 = +43.7^{k'}$$
$$M_{BD} = -157.8 + 310.5 = +152.7^{k'} \qquad M_{ED} = +400 + 0 = +400^{k'}$$
$$M_{DB} = +118.6 + 176.0 = +294.6^{k'}$$

Discussion:

When computing the initial end moments developed by a chord rotation, note that all joints, including hinged supports, are assumed to be locked against rotation. Note further that the stiffness factor K, not the reduced stiffness factor K^R, is used in these computations.

The reduced stiffness factor is used only to compute distribution factors and for no other purpose.

The end moments produced by any given horizontal load acting by itself at joint B can be computed by straight proportion from the case B part of the solution.

The ideas illustrated by Example 9.22 may be applied without modification to any frame involving only one independent ψ angle. It is also easy to extend the application of these ideas to frames involving more than one independent ψ angle. A frame having n independent ψ angles is said to have n **degrees of freedom with respect to sidesway.** For such a frame, the moment-distribution solution can be broken down into $n + 1$ separate cases: (1) a *case A* in which sidesway is completely prevented by introducing n holding forces; and (2) n separate sidesway cases, in each of which only *one* independent ψ angle is allowed to occur at a time by maintaining holding forces at $n - 1$ of the n joints held in case A while pushing at the one remaining joint held in case A.

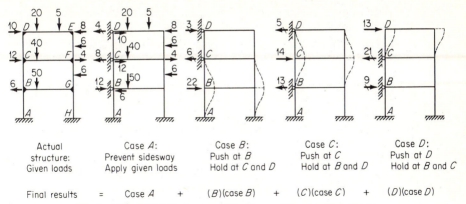

Figure 9.22   Multiple degrees of freedom with respect to sidesway

Note: Values noted above for pushing or holding forces have simply been assumed for illustrative purposes rather than being computed as would be done in solving an actual problem. Dashed-line deflected shapes correspond to the initial position of the frame where joint displacements have been introduced with the joints locked against rotation. The final elastic curve after the joints have been unlocked and have rotated into their final position corresponding to the completed moment distribution for each sidesway case has *not* been shown.

The result of these $n + 1$ cases can be superimposed in the same manner as used in Example 9.22. The factor by which each of the sidesway cases must be multiplied in this superposition can be found from the solution of n simultaneous equations. The n simultaneous equations are obtained by writing one equation for each of the n joints held in case A. For a joint j, such an equation simply states that the algebraic sum of the case A holding force at j plus a factor B times the holding or pushing force at j for the sidesway case B plus a factor C times the holding or pushing force at j for the sidesway case C, and so forth for each of the n sidesway cases, must be zero.

To illustrate this procedure for applying moment distribution to a frame having multiple degrees of freedom with respect to sidesway, consider the frame shown in Fig. 9.22. This frame has three independent ψ angles corresponding to the independent horizontal displacements of joints B, C, and D. The frame is therefore said to have three degrees of freedom with respect to sidesway. The moment-distribution solution is therefore broken down into $3 + 1$ separate cases: case A (sidesway prevented) and three sidesway cases (cases B, C, and D).

In case A after holding forces have been introduced at joints B, C, and D to prevent sidesway, all the given loads are applied and a regular no-sidesway moment-distribution solution is performed to obtain the end moments. Then by statics, the holding forces at B, C, and D are back-figured. For purposes of this discussion, these holding forces are assumed to have the values of 12, 8, and 4, respectively, and to be directed as shown in Fig. 9.22.

Then each of the three sidesway cases is considered separately. In case B, for example, joints C and D are held from displacing horizontally while joint B is displaced some arbitrary amount. If initially the joints are locked against rotation while this displacement is introduced, certain initial end moments are developed in members *AB*, *BC*, *GH*, and *GF*. These initial end moments can be computed for the arbitrary displacement as done above in case *B* of Example 9.22. A moment-distribution solution is performed for the end moments developed as a result of the introduction of these initial end moments. Then by statics the pushing force at *B* and the holding forces at *C* and *D* are back-figured. Also for purposes of the subsequent discussion, these forces are assumed to have the values and directions shown in Fig. 9.22 for case *B*. Similar solutions are conducted for the other two sidesway cases, case *C* and case *D*.

It now remains to find how much of each of the three sidesway cases should be superimposed on case *A* to obtain the final results for the actual given problem. In other words, we must obtain the factors *B*, *C*, and *D* by which cases *B*, *C*, and *D*, respectively, must be multiplied to obtain each of their contributions to the final results. These factors are obtained by solving three simultaneous equations, each of which is set up as follows. At joint *B*, for example, if forces to the right are considered as plus, we may write

$$12 + 22B - 13C + 9D = 0$$

At joint *C*:
$$-8 - 6B + 14C - 21D = 0$$

At joint *D*:
$$-4 + 3B - 5C + 13D = 0$$

The values of *B*, *C*, and *D* can be obtained from the simultaneous solution of these three equations. If we now multiply the external forces in case *B* by *B*, those in case *C* by *C*, and those in case *D* by *D*, and if we then superimpose these modified forces from these three cases on the forces in case *A*, we obtain simply the given loads (and their corresponding reactions). In other words, such a superposition has wiped out all the extra pushing and holding forces. Obviously, the final results (such as the end moments in the actual structure developed by the given loads) are then obtained by superimposing the results of the four cases, as indicated on Fig. 9.22.

It should be noted that one has considerable option as to how one introduces the arbitrary joint displacements in each of the sidesway solutions. The only requirement is that all the sidesway cases be independent of one another. For example, in case *B*, we could have displaced joints *B*, *C*, and *D* equal amounts horizontally. In case *C*, we could have held joint *B* and pushed *C* and *D* equal amounts. These new cases *B* and *C* could then have been combined with the previous cases *A* and *D*. A new set of simultaneous equations would be obtained from which new values of the factors *B*, *C*, and *D* would result. The superposition of the cases of this new solution would lead, however, to the same final results as were obtained from the first solution.

An alternative procedure that does not require simultaneous equations can be used for cases involving sidesway. In this method, we introduce joint displacements at the same time as we apply the loads that develop fixed end moments. After each cycle of distribution, we check the equilibrium equations and back-figure the joint loads. If the joint loads do not agree with the given ones, we introduce some additional joint displacements, carry through another cycle of distribution, check the forces again, etc.[1]

9.15 Moment-Distribution Method Applied to Nonprismatic Members. The fundamental philosophy of the moment-distribution procedure developed in Art. 9.13 is applicable to any beam or frame whether composed of members of constant or varying E and I. The expressions for fixed end moment, stiffness factor, and carry-over moment are derived there specifically, however, for members of constant E and I and are not applicable to nonprismatic members. We shall now develop new expressions for these quantities for *the case of originally straight members of varying E and I.*

Carry-over Factor C. The carry-over moment is the moment induced at a fixed end of a member when the opposite end is rotated by an end moment. It is convenient to express the carry-over moment at B as being equal to the applied moment at A multiplied by the *carry-over factor C_{AB}*, the order of the subscripts indicating the direction in which the effect is carried over, i.e., from A to B in this case, or

$$M_{BA} = C_{AB} M_{AB} \qquad (9.18)$$

From this equation, **the carry-over factor could be defined as the end moment induced at the fixed end of a member when the opposite end is rotated by an end moment of unity.** For a prismatic member, the carry-over factor is $\frac{1}{2}$. Note further that C_{AB} is equal to C_{BA} only when a member is symmetrical about its mid-point.

An outline of the procedure for computing the carry-over factor in any given case follows: Consider any member AB as shown in Fig. 9.23, and apply a moment of unity at A, thus inducing a moment C_{AB} at B. The resulting deflection of point A on the elastic curve from the tangent at B is equal to zero. Upon applying the second moment-area theorem, the static moment of the composite M/EI diagram taken about an axis through point A must therefore be equal to zero. C_{AB} may easily be evaluated from this equation.

True stiffness factor K'. Let us refer to Art. 9.13, where the evaluation of distributed moments is discussed, and consider that Figs. 9.16 and 9.17 now represent structures composed of nonprismatic members. In considering the effect of the rotation of joint B under the unbalanced moment M, we can no longer evaluate the distributed moment by Eq. (9.11). We can write, however,

$$M_{BA} = K'_{BA} \theta_B \qquad M_{BC} = K'_{BC} \theta_C \qquad \text{etc.} \qquad (a)$$

where K'_{BA}, K'_{BC}, etc., are called the **true stiffness factors** for the B end of member BA, for the B end of member BC, etc. From these equations, **the true stiffness factor K'_{BA} may be defined as the end moment required to rotate the tangent at the B end of member BA through a unit angle when the far end A is fixed.**

---

[1] Detailed discussion of this alternative procedure is beyond the scope of this book. For such information, the reader is referred to the more extensive treatment of the moment-distribution method found in the following textbooks. L. E. Grinter, "Theory of Modern Steel Structures," 3d ed., vol. 2, The Macmillan Company, New York, 1962; J. A. L. Matheson, "Hyperstatic Structures," 2d ed., vol. 1, Butterworth's Scientific Publications, London, 1971; J. I. Parcel and R. B. Moorman, "Analysis of Statically Indeterminate Structures," John Wiley & Sons, Inc., New York, 1955.

For more extensive treatments of the classical methods of analysis of statically indeterminate structures in general, the reader is referred to the following texts as well as those mentioned above: S. F. Borg and J. J. Gennaro, "Advanced Structural Analysis," D. Van Nostrand Co., Inc., Princeton, N.J., 1959; L. C. Maugh, "Statically Indeterminate Structures," 2d ed., John Wiley & Sons, Inc., New York, 1964; S. P. Timoshenko and D. H. Young, "Theory of Structures," 2d ed., McGraw-Hill Book Company, New York, 1965; C.-K. Wang, "Statically Indeterminate Structures," McGraw-Hill Book Company, Inc., New York, 1953; J. S. Kinney, "Indeterminate Structural Analysis," Addison-Wesley Publishing Company, Reading, Mass., 1957.

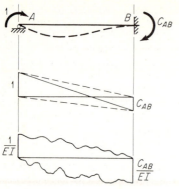 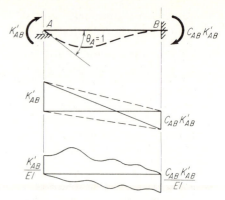

Figure 9.23   Carry-over factor   **Figure 9.24**   True stiffness factor

For a prismatic member, the true stiffness factor is $4EK$. Note that K, previously called simply the stiffness factor of a member, gives the relative value of the true stiffness factors for prismatic members. Perhaps K should have been called the relative stiffness factor, but this term is used so much that it is desirable to keep it as short as possible. Note further that, in general, a member has a different true stiffness factor at each end; only when the member is symmetrical will the true stiffness factor be the same at each end.

The true stiffness factor K'_{AB} for the A end of member AB shown in Fig. 9.24 can be computed as follows. By definition, an end moment of K'_{AB} applied at A will rotate the tangent at A through a unit angle when the far end B is fixed. Upon applying the first moment-area theorem, the net area under the composite M/EI diagram must therefore be equal to unity. If C_{AB} has previously been computed as described above, K'_{AB} can now be determined from this equation.

True reduced stiffness factor K'^R. The true reduced stiffness factor K'^R_{AB} may be defined as the end moment required to rotate the tangent at the A end through a unit angle when the far end is hinged. K'^R_{AB} can easily be computed by proceeding as follows, provided that K'_{AB}, C_{AB}, and C_{BA} have already been computed. Consider a bar AB, and suppose temporarily that we lock the B end against rotation. A moment of K'_{AB} applied at the A end would produce a rotation of A of unity and induce a carry-over moment at the far end of $C_{AB} K'_{AB}$. If we now lock the A end in this position and unlock the B end, the latter end will rotate under the unbalanced moment $C_{AB} K'_{AB}$ and develop a distributed moment of $-C_{AB} K'_{AB}$. As a result, a carry-over moment of $-C_{BA} C_{AB} K'_{AB}$ is induced at A. The member will now be in the deformed condition shown in Fig. 9.25, and the total end moment at A will be equal to the true reduced stiffness factor K'^R_{AB}. Hence,

$$K'^R_{AB} = K'_{AB}(1 - C_{BA} C_{AB}) \tag{9.19}$$

For a prismatic member where both C_{AB} and C_{BA} are equal to $\frac{1}{2}$, we see from Eq. (9.19) that K'^R_{AB} is equal to $\frac{3}{4} K'_{AB}$, as it should be.

Check relationship involving stiffness and carry-over factors. A useful relationship between stiffness and carry-over factors can be derived and used for checking purposes. Consider the member AB shown in Fig. 9.26, where it is acted upon by two separate loading systems I and II. Applying Betti's law, we find that

$$C_{AB} K'_{AB} = C_{BA} K'_{BA} \tag{9.20}$$

This expression proves very helpful when used as a check on the values of the carry-over and stiffness factors.

Distribution factors and distributed moments. If we complete the evaluation of distributed moments using the expressions in Eqs. (*a*) of this article and proceeding as in Art. 9.13, we obtain expressions for the distribution factor which are the same as Eqs. (9.14) or (9.14*a*) except that true stiffness factors K' are substituted for the stiffness factors K.

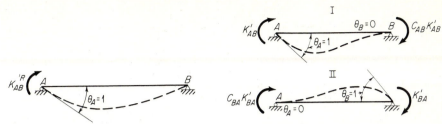

Figure 9.25 True reduced stiffness factor　　　　**Figure 9.26** Check relationship

Sidesway factor. It is likewise easy to develop expressions for the sidesway factor J that can be used to compute the initial end moments developed by a chord rotation. We can write the following expressions for such end moments:

$$M_{AB} = J_{AB}\psi_{AB} \qquad M_{BA} = J_{BA}\psi_{AB} \tag{9.21}$$

The sidesway factor J_{AB} **may be defined as the end moment developed at the A end of member AB by a chord rotation of unity, both ends of the member being locked against joint rotation.** For a prismatic member, the sidesway factor is $-6EK$. The sidesway factor will be the same at each end only when the member is symmetrical.

The sidesway factor for member AB in Fig. 9.27a may be expressed in terms of stiffness and carry-over factors by reasoning as follows. Suppose that the fixity is removed from the A and B ends temporarily and that the member is displaced as shown in Fig. 9.27b. Now lock the A end and rotate the B end until θ_B is restored to zero, thus developing end moments of $-K'_{BA}$ at B and $-C_{BA}K'_{BA}$ at A. Lock the B end in this position and rotate the A end until θ_A is likewise restored to zero, thus developing end moments of $-K'_{AB}$ at A and $-C_{AB}K'_{AB}$ at B. The member will now be in the same condition of deformation as shown in Fig. 9.27a, and the total end moments will be equal to the sidesway factor, or

$$\begin{aligned} J_{AB} &= -(K'_{AB} + C_{BA}K'_{BA}) \\ J_{BA} &= -(K'_{BA} + C_{AB}K'_{AB}) \end{aligned} \tag{9.22}$$

Fixed end moments. The fixed end moments in a nonprismatic member can be computed as follows. Consider member AB acted upon by any loading. Assume temporarily that the load is applied to this member acting as an end-supported beam. For this condition, compute the resulting rotations of the tangents θ_{AO} and θ_{BO}, using the moment-area theorems. Then if we imagine that we lock the B end and rotate the A end back to a zero slope, and then lock the A end in this position and rotate the B end back, the resulting end moments will be found to be

$$\begin{aligned} \text{FEM}_{AB} &= -K'_{AB}\theta_{AO} + C_{BA}K'_{BA}\theta_{BO} \\ \text{FEM}_{BA} &= +K'_{BA}\theta_{BO} - C_{AB}K'_{AB}\theta_{AO} \end{aligned} \tag{9.23}$$

General remarks. The above expressions are strictly applicable only to members the axes of which are originally straight. They may be used with satisfactory accuracy for members with slightly curved axes provided that the structure is arranged so that the axial thrust developed in such members is relatively small. Consideration of more accurate expressions for curved members is beyond the scope of this book.

Charts and tables have been prepared from which values of FEMs, stiffness, and carry-over factors can readily be obtained for typical haunched members encountered in practice.[1]

---

[1] For example, refer to "Handbook of Frame Constants," Portland Cement Association, Chicago, 1958.

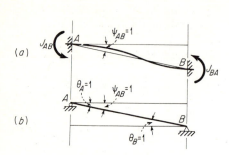

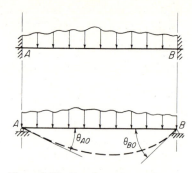

Figure 9.27 Sidesway factor

Figure 9.28 Supplementary computation relating to fixed end moment

9.16 Stress Analysis of Statically Indeterminate Space Frameworks. Statically indeterminate space frameworks may be analyzed by means of deflection superposition equations or Castigliano's theorem. Fundamentally, the stress analysis of such structures by these methods is accomplished in exactly the same manner as for an indeterminate truss. The detail of the analysis is more tedious, of course, for the stress analysis of the statically determinate primary structure is more complicated when it is a three-dimensional framework instead of a planar truss. All the ideas discussed above for indeterminate trusses are applicable here, but a detailed extension of these ideas to indeterminate space frameworks is not within the scope of this book.

9.17 Deflection of Statically Indeterminate Structures. Once the stress analysis of an indeterminate structure has been completed, the strains and the resulting deflections of the structure can be computed without difficulty by the same procedures as those discussed and applied to statically determinate structures in Chap. 8.

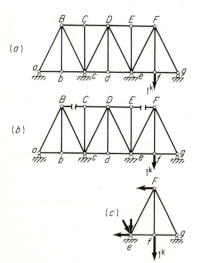

Figure 9.29 Deflections or statically indeterminate truss

For example, suppose that we have computed the bar forces and the corresponding changes in length ΔL of the members of the indeterminate truss shown in Fig. 9.29a, due to a given loading. Suppose that we then wish to compute the resulting vertical deflection of joint f by the method of virtual work. We could proceed in the usual manner and apply a unit vertical load at joint f of the actual structure shown in Fig. 9.29a. This load would cause F_Q forces in most of the bars of the indeterminate truss. After these forces had been computed, we could apply Eq. (8.5) and compute the required deflection of joint f. But this procedure, though straightforward, is laborious since most of the bars of the truss are involved in the computations.

Deflection computations for indeterminate structures can be simplified considerably if we simply recognize that the redundants of such structures have been computed so that the stresses, strains, and deflections of the primary and actual structures are identical. Deflections can therefore be computed by using the statically determinate primary structure rather than the actual indeterminate structure. It is much easier to use the primary structure, of course. For example, in the above structure, suppose that we use the primary structure shown in Fig. 9.29b and proceed to compute the vertical deflection of joint f by the method of virtual work. In this case, the unit load at joint f would develop F_Q forces in only five bars, namely, ef, fg, eF, fF, and Fg. As a result, when

Eq. (8.5) is applied, only these five bars contribute to the sum of $F_Q \, \Delta L$, the computations being thus tremendously reduced in this particular case. Naturally the ΔL's used for these five bars in these computations are the ΔL's defined for these bars as members of the actual indeterminate truss shown in Fig. 9.29a.

To some, a descriptive explanation such as is given above is not adequate, and they prefer a mathematical demonstration. Although it will not be done here, it can be proved mathematically that the results are identical whether the unit Q load is considered to be applied to the actual indeterminate structure or to the statically determinate primary structure.

Physically, it seems obvious that the deflections computed for the right span of the actual structure must be identical with the deflections of this portion isolated as a simple span and considered to be acted upon by any given external loads and also by external forces corresponding to the bar forces in the members cut by the isolating section as shown in Fig. 9.29. These ideas are even more obvious from a physical viewpoint in the case of the deflections of a continuous beam. In such a structure, it seems perfectly natural to isolate each span and compute its deflections by considering it as a simple end-supported beam acted upon by any loads which act in that span and also by end moments which are equivalent to the bending moments at the corresponding support points of the actual continuous beam.

9.18 Secondary Stresses in Trusses. From Chap. 4 it will be recalled that the elementary stress analysis of a truss is based upon the following assumptions:

1. The members are connected together at their ends by frictionless pin joints.

2. All external loads and reactions (including the weight of the members) are applied to the truss at the joints.

3. The centroidal axes of all members are straight, coincide with the lines connecting the joint centers, and lie in a plane that also contains the lines of action of all the loads and reactions.

A stress analysis based on these assumptions leads to the determination of so-called **primary stresses.**

Stresses caused by conditions not considered in the primary stress analysis are called **secondary stresses.** Of these the most important are caused by the fact that the joints are rigid, and hence the members are not free to change their relative directions when the truss is deformed. Several classical methods[1] are available for making approximate analyses for secondary stresses. This problem can likewise be solved very efficiently by means of moment distribution. The latter method is the only one that will be discussed here.

A truss with rigid joints is actually a rigid frame. Theoretically, we could analyze such a frame by the use of deflection superposition equations or Castigliano's theorem, considering deformation due to both bending and axial force. Such structures would be so highly statically indeterminate, however, that it would not be practical to carry through an exact stress analysis. We have already noted that the deflection of the joints is primarily a function of the axial forces in the members since the bending of the members has only a second-order effect on their axial change in length. Further, it can be shown that the axial forces in the members of a truss with rigid joints are essentially the same as those in an ideal pin-jointed truss, i.e., that the presence of shears and moments in the members has a small effect on the axial forces.[2] This suggests that the axial forces and the resulting joint deflections can be computed, assuming the truss to be pin-jointed. If the joint deflections are known, all the ψ angles of the members can be computed and the remainder of the problem solved either by slope deflection or moment distribution, in the same manner as for any frame in which only the joint rotations or θ angles are unknown.

---

[1] A very comprehensive paper on these methods was presented by Cecil Vivian von Abo, *Trans. ASCE*, vol. 89, 1926.

[2] See J. I. Parcel and R. B. B. Moorman, "Analysis of Statically Indeterminate Structures," chap. 9, John Wiley & Sons, Inc., New York, 1955.

The solution of a secondary-stress problem using moment distribution can be outlined as follows:

1. Compute the bar forces assuming the truss to be pin-jointed.
2. Compute the joint deflections and the corresponding chord rotations of the members. This can be done conveniently by means of the Williot method or the angle-change procedure used in Example 9.23.
3. Compute the initial end moments developed if all the joints are locked against rotation and then the above joint deflections and chord rotations introduced. Likewise, compute any unbalanced moments that may act on the joints as a result of eccentricities of the bar.
4. Distribute and carry over through as many cycles as is necessary. The final end moments thus obtained will be the first approximation of the secondary end moments, from which the secondary stresses can be computed. From these end moments, the shears in the members can be computed as well and also, by statics, new values of the bar forces in the members.
5. If these new values of the bar forces are markedly different from those computed in step 1 repeat steps 2, 3, and 4, and thus arrive at a second approximation for the secondary stresses.

This procedure is illustrated by Example 9.23.

Example 9.23   *Compute the secondary stresses in the members of this truss.*

Bar	L	A	I	c	K	Make-up
	$''$	$''^2$	$''^4$	$''$	$''^3$	
ab						
bc	300	18.0	174.9	6.375	0.583	4 ⌊s 6 × 3½ × ½
Bb	336	15.88	153.4	6.375	0.456	4 ⌊s 6 × 3½ × 7⁄16
Bc	450.4	13.68	131.4	6.375	0.292	4 ⌊s 6 × 3½ × 3⁄8
Cc	336	11.44	78.9	5.375	0.235	4 ⌊s 5 × 3 × 3⁄8
aB	450.4	27.68	960.9	5.739 / 9.699	2.134	2[s 15″ − 33.9# / 1Pl 18 × 7⁄16
BC	300	26.55	922.7	5.921 / 9.453	3.076	2[s 15″ − 33.9# / 1Pl 18 × 3⁄8

Notes: 1. Members ab, bc, Bb, Bc, and Cc assembled so:
2. Working lines of members aB and BC lie on centerline of channels.

Step 1: *Primary bar forces and stress intensities shown on line diagram.*
Step 2: *Compute ψ angles using angle changes computed by Eq. (8.11b).*

Angle	$f_3 - f_1$	$\cot \beta_1$	$f_3 - f_2$	$\cot \beta_2$	First term	Second term	$E\,\Delta\gamma$
B-a-b	+11.71 + 13.50 = +25.21	1.12		0	+28.25	0	+28.25
b-B-a	+13.82 + 13.50 = +27.32	0.893		0	+24.40	0	+24.40
a-b-B	−13.50 − 13.82 = −27.32	0.893	−13.50 − 11.71 = −25.21	1.12	−24.40	−28.25	−52.65
c-B-b	+13.82 − 9.11 = +4.71	0.893		0	+ 4.21	0	+ 4.21
B-b-c	+ 9.11 − 11.71 = −2.60	1.12	+ 9.11 − 13.82 = −4.71	0.893	− 2.91	− 4.21	− 7.12
b-c-B		0	+11.71 − 9.11 = +2.60	1.12	0	+ 2.91	+ 2.91
C-B-c		0	0 −9.11 = − 9.11	1.12	0	−10.20	−10.20
c-C-B	+ 9.11 − 0 = +9.11	1.12	+ 9.11 + 12.50 = +21.61	0.893	+10.20	+19.30	+29.50
B-c-C		0	−12.50 − 9.11 = −21.61	0.893	0	−19.30	−19.30

Then,

$$
\begin{aligned}
E\psi_{Cc} &= \quad 0 & E\psi_{aB} &= +47.91 \\
E\Delta_{BcC} &= -19.30 & E\Delta_{aBb} &= +24.40 \\
E\psi_{Bc} &= +19.30 & E\psi_{Bb} &= +23.51 \\
E\Delta_{Bcb} &= + 2.91 & E\Delta_{bBc} &= + 4.21 \\
E\psi_{bc} &= +16.39 & E\psi_{Bc} &= +19.30 \\
E\Delta_{Bbc} &= - 7.12 & E\Delta_{cBC} &= -10.20 \\
E\psi_{Bb} &= +23.51 & E\psi_{Bc} &= +29.50 \\
E\Delta_{abB} &= -52.65 & E\Delta_{BCc} &= +29.50 \\
E\psi_{ab} &= +76.16 & E\psi_{Cc} &= \quad 0 \\
E\Delta_{Bab} &= +28.25 \\
E\psi_{aB} &= +47.91
\end{aligned}
$$

In this case, ψ_{Cc} is known to equal zero by symmetry. All these ψ angles are therefore oriented correctly. However, any ψ angle may be assumed to be zero and all other ψ angles computed to correspond. This is obviously permissible since it means in effect that the truss has been rotated as a rigid body which does not alter the condition of stress.

Step 3: *Compute initial end moments and eccentric moments.*

Joint *a***:**

$$M_e = +(374)(2.20) = +823^{k''}$$

Joint *B***:**

$$
\begin{aligned}
M_e &= -(374)(2.20) + (332)(1.953) \\
&= -174^{k''}
\end{aligned}
$$

Bar	K	$E\psi$	$-6EK\psi$
ab	0.583	+76.16	−266.0$^{k''}$
bc	0.583	+16.39	− 57.3$^{k''}$
aB	2.134	+47.91	−613.5$^{k''}$
BC	3.076	+29.50	−544.0$^{k''}$
Bb	0.456	+23.51	− 64.3$^{k''}$
Cc	0.235	0	0
Bc	0.292	+19.30	− 33.8$^{k''}$

Step 4: *Distribute, carry over, and obtain secondary end moments. These end moments are underlined. With the section moduli known, it is simple to compute the secondary stress intensities from these.*

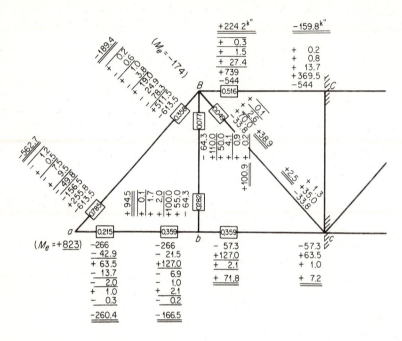

Note in this case that because of symmetry only half the truss needs to be considered; further, it is known, therefore, that joints C and c do not rotate. These joints never have to be unlocked.

9.19   Supplementary Remarks—Symmetrical and Antisymmetrical Loads, Elastic Center, and Column Analogy. In presenting an elementary discussion of any subject, the writer must draw a rather arbitrary line between the fundamental and advanced aspects of the field, since there are always marginal topics that may or may not be included. It is desirable, however, to mention some of these topics briefly so as to suggest transitional reading material for those who intend to study further.

Here and there in previous discussions we have mentioned and used symmetry to a limited extent. Whenever we are dealing with a symmetrical structure, we should always be on the alert to utilize symmetry as effectively as possible. For example, consider the closed-ring type of structure shown in Fig. 9.30. Strictly speaking, this structure is indeterminate to the third degree. If both the frame and the loading are symmetrical about both the *x* and the *y* axis, however we can reason from symmetry that the shear is zero and the axial force is 6 kips compression at the mid-point cross section of the girder. Only the moment remains unknown. Because of symmetry, therefore, only one statically indeterminate quantity remains, instead of three.

The gain from symmetry in the above situation suggests that we can gain considerable simplification even when we have an *unsymmetrical load* acting on a *symmetrical structure*. Consider such a situation as that shown in Fig. 9.31. Suppose that this frame (called a **Vierendeel truss**) is symmetrical about the vertical axis only. Consider the solution of the frame under the given unsymmetrical load as shown in Fig. 9.31a. If the slope-deflection method is used, there is a total of 11 unknowns, 8θ and 3ψ angles. Suppose that the load is broken down into two separate

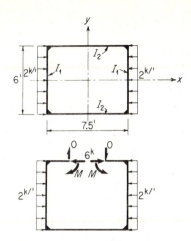

Figure 9.30 Symmetrical structure, symmetrically loaded

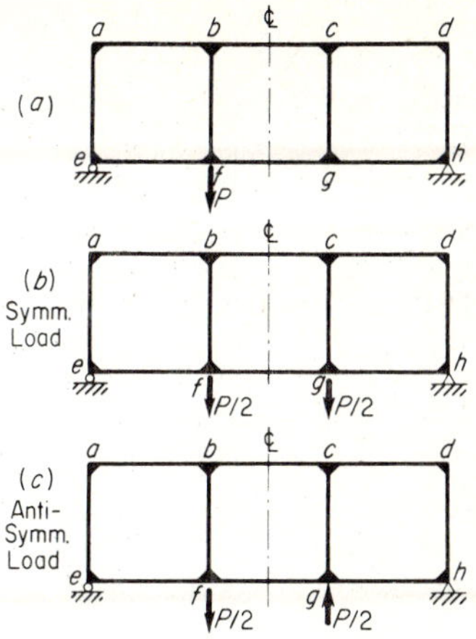

Figure 9.31 Use of symmetrical and antisymmetrical loads

systems, one a symmetrical system shown in Fig. 9.31b and one an antisymmetrical system shown in Fig. 9.31c. Obviously, the sum of these two systems is equal to the given load, and therefore the sum of the results for the two separate systems is equal to the results for the given load.

Let us now compare the computational work of analyzing the structure for the given unsymmetrical load with the sum of the computational works for the symmetrical and antisymmetrical loads. For the case of the symmetrical loads, there are only five independent unknowns (four θ angles and one ψ angle) since, by symmetry,

$$\theta_a = -\theta_d \qquad \theta_b = -\theta_c \qquad \theta_e = -\theta_h \qquad \theta_f = -\theta_g$$
$$\psi \text{ cols} = \psi_{fg} = 0 \qquad \psi_{ef} = -\psi_{gh}$$

For the case of the antisymmetrical loads, there are six independent unknowns (four θ and two ψ angles) since in this case, by antisymmetry,

$$\theta_a = \theta_d \qquad \theta_b = \theta_c \qquad \theta_e = \theta_h \qquad \theta_f = \theta_g \qquad \psi_{ef} = \psi_{gh}$$

and, in addition, ψ_{fg} is opposite but related to ψ_{ef}. Since the computational work is roughly proportional to the square of the number of unknowns, using symmetrical and antisymmetrical loads almost cuts the computational work in half. This idea is very useful when we are dealing with symmetrical structures and deserves further study.[1]

The so-called **elastic center** is another useful idea that we have not discussed.[2] The elastic-center technique is applicable to the closed-ring type of structures; a frame such as that shown in Fig. 9.32 would be classified as this type if we consider the ground as being the closing side of the ring. We can analyze this structure by means of deflection superposition equations, selecting the redundants as shown in Fig. 9.32b, which is the same selection as that used in Example 9.13. Proceeding with this solution involves three simultaneous equations of the form

$$\Delta_{ao} + X_a\,\delta_{aa} + X_b\,\delta_{ab} + X_c\,\delta_{ac} = 0$$
$$\Delta_{bo} + X_a\,\delta_{ab} + X_b\,\delta_{bb} + X_c\,\delta_{bc} = 0$$
$$\Delta_{co} + X_a\,\delta_{ac} + X_b\,\delta_{bc} + X_c\,\delta_{cc} = 0$$

[1] J. S. Newell, Symmetric and Anti-symmetric Loadings, *Civ. Eng.*, April 1939, pp. 249–251; W. L. Andrée, "Das B = U Verfahren," R. Oldenbourg-Verlag, Munich, 1919.

[2] W. M. Fife and J. B. Wilbur, "Theory of Statically Indeterminate Structures," pp. 114–120, McGraw-Hill Book Company, New York, 1937.

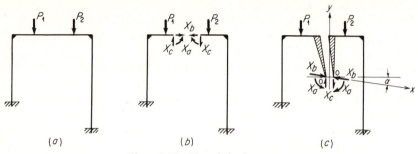

Figure 9.32 Use of elastic center

If we can select the three redundants in such a manner that δ_{ab}, δ_{ac}, and δ_{bc} are all equal to zero, then each of these equations will contain only one redundant and we will not have to solve them simultaneously. We can do this if we apply the redundants at some point o as shown in Fig. 9.32c, where the two coordinates of point o and the inclination α of the x axis are selected so as to make δ_{ab}, δ_{ac}, and δ_{bc} all equal to zero simultaneously. The redundants acting at point o are assumed to be connected by two rigid (nondeformable) arms to the two sides of the cut in the girder. Note that the two sets of redundants in Fig. 9.32b and c are statically equivalent but do not have the same values, of course.

The detailed calculations involved in using this elastic-center procedure are found to be similar to computing static moments and products and moments of inertia of areas. Professor Hardy Cross recognized this and suggested organizing the computations as in computing the stresses in a column subjected to combined bending and direct stress. He called this suggested procedure **column analogy.**[1]

9.20 Experimental Structural Analysis. In 1948, when this book was first published, the literature available on experimental stress analysis was limited. Three chapters on the use of experimental models in the analysis of structures were therefore included in the first edition and continued in the second edition in 1960. This material is not being included here, however, because in these last three decades a new division of the technical literature has developed to provide coverage of this growing field of engineering activity.

The Society for Experimental Stress Analysis (SESA) was founded in 1943. Its fine journal, *Engineering Mechanics*, furnishes an excellent source that should be consulted when one is searching for information on experimental techniques applicable either to structures or structural models or to field or laboratory testing. SESA has also published a number of useful books and monographs, e.g., "Handbook of Experimental Stress Analysis," 1950; "Experimental Techniques in Fracture Mechanics," 1974; and "Manual on Experimental Stress Analysis," 2d ed., 1975.

Ideally engineers would like to judge the validity and precision of their mathmatical structural theories by comparing theoretical predictions with actual behaviors during the testing of full-scale structures under service conditions. Unfortunately, however, experimental investigations are severely limited by the technical and economic difficulties encountered in testing full-scale structures and

[1] H. Cross and N. D. Morgan, "Continuous Frames of Reinforced Concrete," John Wiley & Sons, Inc., New York, 1932.

structural elements.   Only on rare occasions has it been feasible to perform a
sufficient number of full-scale tests to provide good statistical averages of the test
data and also to furnish an adequate statistical description of the loading conditions
and the material and geometrical dimensions of the structure.   Since prototype
structures are not readily studied, the alternate approach of investigating small-
scale replicas of these structures and elements is being developed with the aim of
establishing the experimental approach in its proper place in the fields of structural
engineering research, education, and design.

Laboratory testing and the use of models have increased steadily in the last
fifty years.   Today, these experimental methods are not only extremely important
tools in structural research and development but also provide an important supplement to the mathematical methods used in the actual design of structures.   Euro-
pean engineers particularly, such as Nervi, Oberti, Torroja, and Rocha, have often
used models in arriving at their designs.   Perhaps the most widely publicized use
of models in this latter respect has been in connection with the design of most of
the important suspension bridges erected in recent years.   But the use of models
in the design of large concrete dams is also well known.   Further evidence of the
importance of experimental stress analysis is furnished by the well-equipped
laboratories that have been established by several governmental agencies.   There
are also many academic institutions that possess fine structural laboratories established primarily for educational and research purposes.

Experimental analysis of structural problems encountered in either research or
actual design may be used for one or more of three reasons: (1) mathematical
analysis of the problem concerned is virtually impossible; (2) the analysis, though
possible, is so complex and tedious that the model analysis offers an advantageous
shortcut; (3) the importance of the problem is such that verification of the mathematical solution by model test is warranted.   The stress distribution in an irregu-
larly shaped member may be investigated by the use of a model for the first reason;
a model test may serve as the basis for the analysis of a complex building frame or
of a shell roof for the second reason; a model study of the proposed design of a
suspension bridge may come under the third classification.

The objective of structural tests usually belongs in one of the following four
categories: (1) determination of the resultant of the internal stresses on cross
sections of the test member; (2) determination of the distribution of stresses over
such cross sections; (3) determination of critical or buckling loads; or (4) the
analysis of the dynamic characteristics and response of the test member.

9.21   Problems for Solution

*Problem 9.1   Find all the reactions of the structure shown in Fig. 9.33 using deflection superposition equations as the basis for the solution. $L/A = 1$ for all members.

Problem 9.2   Compute the bar forces in the members of the truss shown in Fig. 9.34, using the
deflection superposition-equation method.   Cross-sectional areas of members, in square inches,
are shown in parentheses.

*Problem 9.3   Using deflection superposition equations, compute the bar force in the tie rod
ad of the structure of Fig. 9.35.   Cross-sectional areas of the members, in square inches, are
shown in parentheses.

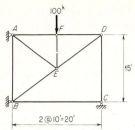

Figure 9.33   Problem 9.1

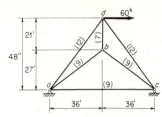

Figure 9.34   Problem 9.2

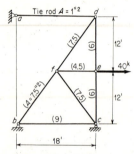

Figure 9.35   Problem 9.3

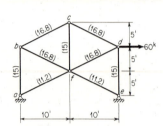

Figure 9.36   Problem 9.4

*Problem 9.4   Using deflection superposition equations, compute the bar force in all members of the arch of Fig. 9.36.   Cross-sectional areas in square inches are shown in parentheses. $E = 30 \times 10^3$ kips per sq in., $\alpha_t = 1/150,000$ per °F.

*Problem 9.5   Find the horizontal component of the right reaction of the arch in Prob. 9.4,. due to each of the following conditions: (a) Increase of temperature of 50°F in bars ab, bc, cd, de. No change in remaining bars.   (b) Bars bc and cd are ¼ in. too short and bar cf is ⅛ in. too long owing to errors in fabrication, and it was necessary to force them into place.   (c) The supports settle as follows:

Left support: Vertical = 0.48 in. down   horizontal = 0.24 in. to left
Right support: Vertical = 0.24 in. down   horizontal = 0.36 in. to right

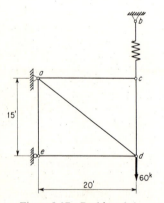

Figure 9.37   Problem 9.6

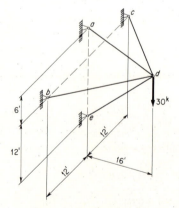

Figure 9.38   Problem 9.7

Problem 9.6 Compute the horizontal component of the deflection of joint d due to the load acting on the structure shown in Fig. 9.37. The stiffness of spring bc is such that it elongates 0.036 in. under a tensile force of 1 kip. For all other members, $A = 2$ sq in. and $E = 30,000$ kips per sq in.

Problem 9.7 All joints of the structure shown in Fig. 9.38 are frictionless ball-and-socket joints. Joints b, c, and d lie in the same horizontal plane. For all members, $A = 3$ sq in. and $E = 30,000$ kips per sq in. Compute the bar forces in all members due to the 30-kip vertical load shown.

Problem 9.8 Consider the beam shown in Fig. 9.39.

(*a*) Estimate in some reasonable manner (which you should describe) the reactions developed in this beam by the loads shown. Be sure that your estimated reactions at least satisfy static equilibrium. Draw the shear and bending-moment diagrams for your solution, and sketch the elastic curve for the beam.

(*b*) Using the deflection superposition equations as a basis for the solution, analyze the beam for the load shown and draw the shear and bending-moment diagrams corresponding to your results.

(*c*) Analyze the beam only for the effect of a vertical settlement of support b of 0.24 in. and a clockwise rotation of support a of 0.005 radian. $E = 30,000$ kips per sq in.

(*d*) The values which you estimated in part *a* for the reactions will not, in general, give consistent deflections for the structure. For example, instead of the supports being immovable as intended in this part of the solution, your values may imply certain support movements in addition to the given loads. What movements are implied by your estimate?

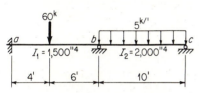

Figure 9.39   Problem 9.8

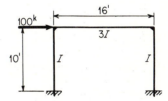

Figure 9.40   Problem 9.9

*__*Problem 9.9__* Draw the bending-moment diagram for the frame shown in Fig. 9.40. Use Castigliano's theorem as a basis for the solution. Consider only the effect of bending deformation.

*__*Problem 9.10__* Compute the reactions of the structure shown in Fig. 9.41. Neglect deformation due to axial force. E and I are constant.

*__*Problem 9.11__* Compute the reactions on the structure shown in Fig. 9.42. Neglect deformation due to axial force. E and I are constant.

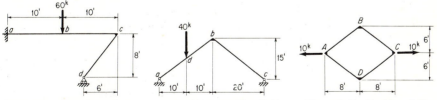

Figure 9.41   Problem 9.10   **Figure 9.42**   Problem 9.11   **Figure 9.43**   Problem 9.12

*__*Problem 9.12__* Draw the shear and bending-moment diagrams for member AB of the frame shown in Fig. 9.43. Neglect deformation due to axial force.

*__*Problem 9.13__* Referring to Fig. 9.44; draw the bending-moment diagram for the beam AB, using Castigliano's theorem.

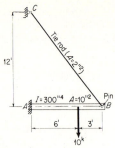

Figure 9.44 Problem 9.13

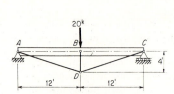

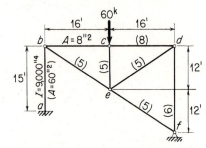

Figure 9.45 Problem 9.14

Problem 9.14   Referring to Fig. 9.45, compute the axial forces in the members of the king-post truss, due to the load shown.   Also, draw the bending-moment diagram for member AC.

Member AC: $A = 12$ sq in. $I = 432$ in.[4]
Member AD: $A = 3$ sq in.
Member DC: $A = 3$ sq in.
Member BD: $A = 2$ sq in.

*Problem 9.15   In the structure shown in Fig. 9.46, the cross-sectional areas of members, in square inches, are shown in parentheses. $E = 30 \times 10^3$ kips per sq in., I of beam $= 4,000$ in.[4] Compute the force in the tie rod, which is connected to the beam and the truss by hinged ends.

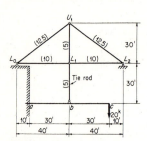

Figure 9.46 Problem 9.15

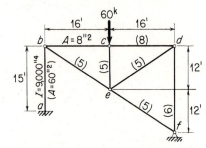

Figure 9.47 Problem 9.16

Problem 9.16   Compute the reactions of the structure shown in Fig. 9.47.

Problem 9.17   Discuss each of the structures shown in Fig. 9.48 with respect to stability and statical indeterminany.   If indeterminate, indicate what you would select for redundants.   For the right structure, what can you reason about the stress analysis by applying principles of symmetry and antisymmetry?

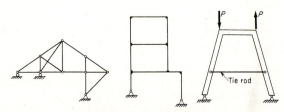

Figure 9.48 Problem 9.17

Problem 9.18   Consider the two-span continuous beam used in Prob. 9.8.   Suppose that the moments of inertia in each span were doubled by the addition of flange plates.   What effect would this modification of the structure have on the following results previously computed in the solution of that problem?

(a) The bending moments computed for the given loading.

(b) The bending moments computed for the given support movements.

(c) The bending deflections computed for the given support movements.

(d) The bending stress intensities computed for the given support movements.

*Problem 9.19   Refer to Fig. 9.49.   Compute the end moments in all the members of this frame, using the slope-deflection method.   Draw the moment diagrams for members *AB* and *BD*.

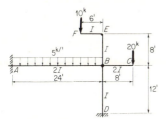

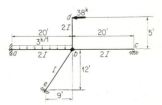

Figure 9.49   Problem 9.19   **Figure 9.50**   Problem 9.20

*Problem 9.20   Refer to Fig. 9.50.   Compute the end moments in all members of this frame, using the slope-deflection method.   Draw the shear and moment diagrams for member *ab*.

Problem 9.21   Solve parts *b* and *c* of Prob. 9.8 using the slope-deflection method.

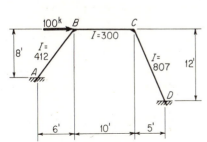

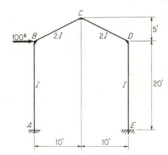

Figure 9.51   Problem 9.22   **Figure 9.52**   Problem 9.23

*Problem 9.22   Using the slope-deflection method, find all the end moments and support reactions of the frame shown in Fig. 9.51.

Problem 9.23   Using the slope-deflection method, find all the end moments and support reactions of the frame shown in Fig. 9.52.

Problem 9.24   Assume that the structure shown in Fig. 9.53 is symmetrical about its vertical centerline.

(a) Indicate the independent θ and ψ angles, and write the appropriate equations of statics, assuming that the loading is symmetrical about the vertical centerline.

(b) Do likewise for an antisymmetrical loading.

*Problem 9.25   Refer to Fig. 9.54.   Using moment distribution, compute the end moments in the members of this frame, and draw the shear and moment diagrams for member *AB*.

Problem 9.26   Solve parts *b* and *c* of Prob. 9.8 using moment distribution.

*Problem 9.27   Refer to Fig. 9.55.   Find the end moments of this frame using the moment-distribution method.

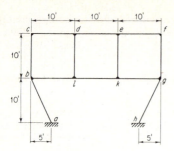

Figure 9.53   Problem 9.24

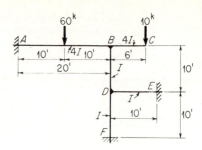

Figure 9.54   Problem 9.25

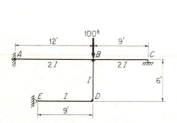

Figure 9.55   Problem 9.27

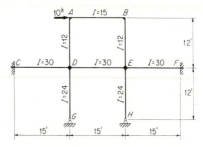

Figure 9.56   Problem 9.28

Problem 9.28   Compute the end moments in the members of the frame shown in Fig. 9.56 using the moment-distribution method.

Problem 9.29   Refer to Fig. 9.57.   Find the end moments in the members of this frame using the moment-distribution method.

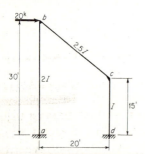

Figure 9.57   Problem 9.29

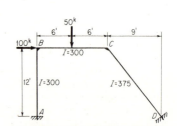

Figure 9.58   Problem 9.30

Problem 9.30   Refer to Fig. 9.58.   Find the end moments in this frame using the moment-distribution method.

Problem 9.31   Consider the structure shown in Fig. 9.59.

(a) Using the moment-distribution method, compute the end moments produced in all members by the load shown.

(b) Using the results of part a, describe how you would compute the horizontal deflection of C.

(c) How large a horizontal force would now have to be applied at joint C in order to push it back to its original position?

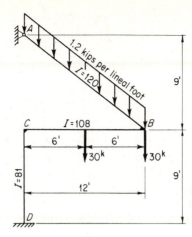

Figure 9.59   Problem 9.31

Problem 9.32   Consider the structure shown in Fig. 9.60.

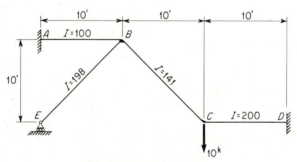

Figure 9.60   Problem 9.32

(a) Using the moment-distribution method, compute the end moments produced in all members by the load shown.

(b) Suppose that this structure were altered by pin-connecting joint B to a support 12 ft above it by means of a tie rod. Compute the axial force developed in this tie rod by the 10-kip vertical load at joint C. Neglect the effect of the axial change in length of the tie rod.

(c) Describe how you could incorporate the effect of the axial change of length of the tie rod on the moment-distribution solution.

Problem 9.33   Assuming that the structures shown in Fig. 9.61 are unsymmetrical and using suitable sketches to describe your ideas, indicate the various cases into which you would divide the moment-distribution solution of each of the structures.

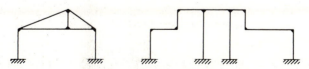

Figure 9.61   Problem 9.33

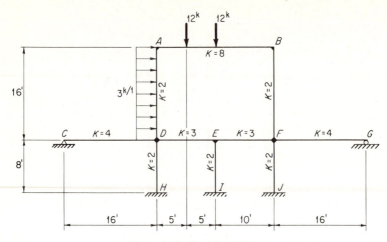

Figure 9.62   Problem 9.34

Problem 9.34   Utilizing principles of symmetry and antisymmetry, compute the end moments in all members of the frame shown in Fig. 9.62 by the moment-distribution method.

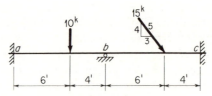

Figure 9.63   Problem 9.35

Problem 9.35   Consider the beam shown in Fig. 9.63, in which $I = 500$ in.^{4}, $A = 30$ sq in., and $E = 30 \times 10^3$ kips per sq in.   Compare the solution of this problem by the theorem of three moments, slope deflection, and superposition equations.   (*Hints:* To what degree is the beam indeterminate?   Can you find the horizontal reactions by three moments or slope deflection? Which method do you consider to be superior?   How many simultaneous equations are involved in each solution?)

Problem 9.36   Determine the number of independent θ and ψ angles in each of the structures shown in Figs. 9.12 and 9.15.

10

Influence Lines for Statically Indeterminate Structures

10.1 Introduction. In Chap. 9, various methods of analyzing indeterminate structures were discussed; in all cases, however, the structures were subjected to some particular condition of loading. Often it is necessary to analyze an indeterminate structure for the effect of a movable or a moving load. In such cases, it is convenient to prepare influence lines or influence tables for the various stress components, since from these we can determine both how to load the structure to produce a maximum effect and likewise the magnitude of this maximum effect.

In Chap. 5, the preparation and use of influence lines for statically determinate structures were discussed. In such cases, we found that after a little practice we could draw an entire influence line by figuring a few key ordinates and connecting them with straight lines. Influence lines for indeterminate structures cannot be drawn so easily, however, since in general they are curved lines or, at best, a series of chords of a curved line.

Fortunately, influence lines of the latter type occur quite frequently. When an indeterminate structure is loaded at the panel points by floor beams supporting stringers which act as simple beams between the floor beams, it is easy to show that the various influence lines for the indeterminate structure are straight lines between panel points. If, however, the stringers are continuous over several floor beams, the influence lines are curved lines between panel points, though in most practical cases the departure of these curves from a straight line between panel points is rather slight. If the moving load does not act through a stringer-floor-beam system and can be applied to the structure at any point along its length, the influence lines are likewise curved lines

The first step in preparing influence lines for the various stresses in an indeterminate structure is to determine the influence lines for the redundants. Once this has been done, the influence lines for any other reaction—bar force, shear, or moment—can be computed by statics.

10.2 Influence Line by Successive Positions of Unit Load. In Chap. 5, it was pointed out that the ordinates of an influence line for some particular quantity can always be obtained by placing a unit load successively at each possible load point on the structure and computing the value of that quantity for each of these positions. This same procedure can be followed to obtain the influence lines for the redundants of an indeterminate structure. Doing so means solving a number of problems in just the same way as in Chap. 9.

Offhand this might seem a long and tedious process. It will be found, however, that the computations can be organized very efficiently, as illustrated in Example 10.1. Moreover, once the influence lines for the redundants have been obtained, the influence lines for all other stress components can be obtained very easily, simply by superimposing quantities already computed. In Example 10.1, once the influence-line ordinates have been computed for the redundants X_1 and X_2, the ordinates for all other bar forces can be computed in tabular form by means of the relation

$$F = F_0 + X_1 F_1 + X_2 F_2$$

Note that this is very easy, for F_0 bar forces have already been computed for all bars for every position of the unit load. While it takes longer to compute simply the influence lines for the redundants of an indeterminate truss by this method, influence lines for all the bar forces and reactions can be obtained essentially as fast by this method as by any other.

Example 10.1 *Assuming A to be constant for all bars, prepare influence lines for the forces in the redundant bars of this truss.*

Select forces in bars CD and FG as redundants. Compute these redundants for a unit load at any joint n, using deflection superposition equations.

$$\Delta_1^{\rightarrow\leftarrow} = \Delta_{1n} + X_1\delta_{11} + X_2\delta_{12} = 0 \quad (1)$$
$$\Delta_2^{\rightarrow\leftarrow} = \Delta_{2n} + X_1\delta_{21} + X_2\delta_{22} = 0 \quad (2)$$

By symmetry, $\delta_{11} = \delta_{22}$ *and by Maxwell's law,* $\delta_{12} = \delta_{21}$, *Equations (1) and (2) therefore become*

$$\Delta_{1n} + X_1\delta_{11} + X_2\delta_{12} = 0 \quad (1a)$$
$$\Delta_{2n} + X_1\delta_{12} + X_2\delta_{11} = 0 \quad (2a)$$

When the method of virtual work is used to evaluate these deflection terms, the constants, of the structure are

$$(I^k)(\delta_{11}) = \frac{1}{EA}\sum F_1^2 L$$

$$(I^k)(\delta_{12}) = \frac{1}{EA}\sum F_1 F_2 L$$

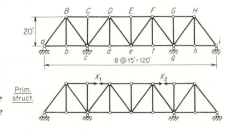

while the load terms are

$$(1^k)(\Delta_{1n}) = \frac{1}{EA}\sum F_1 F_n L$$

$$(1^k)(\Delta_{2n}) = \frac{1}{EA}\sum F_2 F_n L$$

Since EA is constant for all bars and the right-hand sides of Eqs. (1a) and (2a) are zero, for convenience we can let EA = 1 for the rest of the solution. By symmetry, X_1 due to a unit load at b is equal to X_2 due to a unit load at h, etc. As a result, if we compute both X_1 and X_2 for a unit load placed successively at points b, d, and e, we shall have enough information to plot the complete influence lines for both X_1 and X_2.

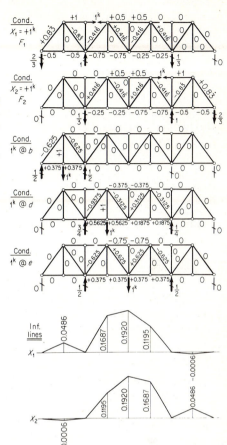

Constants of structure:

Bar	L	F_1	F_2	$F_1 F_2 L$	$F_1^2 L$
ac	30	−0.5	0	0	+ 7.5
ce	30	−0.75	−0.25	+5.625	+ 16.875
eg	30	−0.25	−0.75	+5.625	+ 1.875
BD	30	+1.0	0	0	+ 30.0
DF	30	+0.5	+0.5	+7.5	+ 7.5
aB	25	+0.83̇	0	0	+ 17.35
Bc	25	−0.83̇	0	0	+ 17.35
cD	25	−0.416̇	+0.416̇	−4.34	+ 4.34
De	25	+0.416̇	−0.416̇	−4.34	+ 4.34
eF	25	−0.416̇	+0.416̇	−4.34	+ 4.34
Fg	25	+0.416̇	−0.416̇	−4.34	+ 4.34
Σ				+1.39	+115.81

$$\therefore \delta_{11} = +115.81 \qquad \delta_{12} = +1.39$$

Load terms:

Unit load in left span

			At b	
Bar	L	F_1	F_n	F_1F_nL
ac	30	−0.5	+0.375	− 5.625
aB	25	+0.83	−0.625	−13.02
Bc	25	−0.83	−0.625	+13.02
Σ				− 5.625

Note:
When load is at b, $\Delta_{2b} = 0$.
When load is at e, $\Delta_{1e} = \Delta_{2e}$.

Unit load in center span

				At d			At e	
Bar	L	F_1	F_2	F_n	F_1F_nL	F_2F_nL	F_n	F_1F_nL
ce	30	−0.75	−0.25	+0.563	−12.66	− 4.22	+0.375	− 8.44
eg	30	−0.25	−0.75	+0.188	− 1.41	− 4.22	+0.375	− 2.81
DF	30	+0.5	+0.5	−0.375	− 5.63	− 5.63	−0.75	−11.25
cD	25	−0.416	+0.416	−0.938	+ 9.78	− 9.78	−0.625	+ 6.50
De	25	+0.416	−0.416	−0.313	− 3.26	+ 3.26	+0.625	+ 6.50
eF	25	−0.416	+0.416	+0.313	− 3.26	+ 3.26	+0.625	− 6.50
Fg	25	+0.416	−0.416	−0.313	− 3.26	+ 3.26	−0.625	− 6.50
Σ					−19.70	−14.07		−22.50

Solution of equations:

Eq.	*Operation*	X_1 +	X_2	= constant, unit load at: b	d	e	*Check*
(1)		+115.81	+ 1.39	+5.625	+19.70	+22.50	+165.025
(2)		+ 1.39	+115.81	+0	+14.07	+22.50	+153.77
(1′)	(1)(0.01200)	+ 1.39	+ 0.017	+0.068	+ 0.236	+ 0.270	+ 1.980
(3)	(2) − (1′)		+115.793	−0.068	+13.834	+22.230	+151.790
			+ 1.0	−0.00059	+ 0.1195	+ 0.1920	+ 1.3108
			− 1.39	+0.00082	− 0.1661	− 0.2669	
		+115.81		+5.6258	+19.5339	+22.2331	
		+ 1.0		+0.0486	+ 0.1687	+ 0.1920	

10.3 Müller-Breslau's Principle for Obtaining Influence Lines. Müller-Breslau's principle provides a very convenient method of computing[1] influence lines and likewise is the basis for certain indirect methods of model analysis.

[1] Likewise, by applying Müller-Breslau's principle, it is often possible to sketch the shape of an influence line roughly and thus determine, accurately enough for design purposes, how to load the structure so as to produce a maximum effect.

Müller-Breslau's principle:

The ordinates of the influence line for any stress element (such as axial force, shear, moment, or reaction) of any structure are proportional to those of the deflection curve which is obtained by removing the restraint corresponding to that element from the structure and introducing in its place a corresponding deformation into the primary structure which remains.

This principle is applicable to any type of structure, whether beam, truss, or frame, statically determinate or indeterminate. *In the case of indeterminate structures, this principle is limited to structures the material of which is elastic and follows Hooke's law.* This limitation is not particularly important, however, since the vast majority of practical cases fall into this category.

The validity of this principle can be demonstrated in the following manner. For this purpose, consider the two-span continuous beam shown in Fig. 10.1a. Suppose that the influence line for the vertical reaction at a is required. The influence line can be plotted after the reaction has been evaluated for a unit vertical load applied successively at various points n along the structure, each evaluation being carried out as follows. Temporarily remove the roller support at a from the actual structure, leaving the primary structure shown in Fig. 10.1b. Suppose that this primary structure is acted upon by the unit load at a point n and a vertical upward force R_a at point a. If this force has the same value as the vertical reaction at point a on the actual structure, then the stresses—and hence the deformation—of the primary structure will be exactly the same as those of the actual structure. The elastic curve of the primary structure under such conditions will therefore be as indicated in Fig. 10.1b, the vertical deflection at point a being zero.

Suppose that we now consider the primary structure to be acted upon by simply a vertical force F at point a. In this case, the primary structure will deflect as shown in Fig. 10.1c. Thus we have considered the primary structure under the action of two separate and distinct force systems: (1) the forces in sketch b and (2) those in sketch c. Applying Betti's law to this situation, we write

$$(R_a)(\Delta_{aa}) + (1)(\Delta_{na}) = (F)(0)$$

and therefore

$$R_a = -\frac{\Delta_{na}}{\Delta_{aa}}(1) \qquad (10.1)$$

where the same notation is used to designate these deflections as that in the deflection superposition equation method.

From this equation, it is apparent that the reaction R_a when the unit vertical load is at point n is proportional to the deflection Δ_{na} at that point. The shape of the influence line for R_a is therefore the same as the shape of the elastic curve of the structure when it is acted upon by a force F at point a. The magnitude of the influence-line ordinate at any point n can be obtained by dividing the deflection at that point on this elastic curve by the deflection at point a. In this manner, we have demonstrated that influence lines can be obtained in the manner outlined by Müller-Breslau's principle.

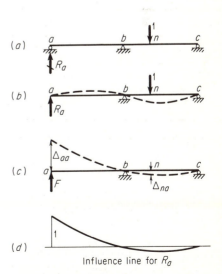

Figure 10.1   Müller Breslau principle

In a similar manner, the validity of this principle can be demonstrated for any stress element of any structure. In the general case, Eq. (10.1) can be written as follows for any stress element X_a:

$$X_a = -\frac{\Delta_{na}}{\Delta_{aa}} 1 \tag{10.2}$$

It is important to note the sign convention of this equation, namely, X_a is plus when in the same sense as the introduced deflection Δ_{aa}, and Δ_{na} is plus when in the same sense as the applied unit load, the influence of which is given by the ordinates of the influence line. Note further that X_a may represent either a force or a couple. If X_a is a force, the corresponding Δ_{aa} is a linear deflection; but if X_a is a couple, the corresponding Δ_{aa} is an angular rotation.

It is also important to note that the magnitude of any influence-line ordinate is independent of the magnitude of the force F which must be applied to introduce the deflection Δ_{aa} into the primary structure. *For computation of the influence-line ordinates, F is usually taken as unity and Δ_{aa} and Δ_{na} are computed to correspond.*

Examples 10.2 and 10.3 illustrate the application of this procedure to continuous beams. *When applying this method, note that, if the actual structure is indeterminate to more than the first degree, the primary structure which remains after the removal of a redundant is still a statically indeterminate structure.* This does not cause any real difficulty however. It simply means that before the deflections Δ_{aa} and Δ_{na} can be computed, the statically indeterminate primary structure must be analyzed by one of the methods of Chap. 9.

Example 10.2 *Prepare an influence line for the vertical reaction at b of this beam.*

Applying Müller-Breslau's principle, we may select the primary structure shown and write

$$X_b^\uparrow = -\frac{\Delta_{nb}^\downarrow}{\Delta_{bb}^\uparrow} (1^k)$$

where the plus direction of the various terms is shown.

If, for convenience, we decide to call Δ_{nb} plus when up, we must reverse the sign of the right-hand side of this equation and write

$$X_b^\uparrow = \frac{\Delta_{nb}^\uparrow}{\Delta_{bb}^\uparrow} (1^k)$$

By the second moment-area theorem,

$$EI\Delta_{nb} = 20\,\frac{x}{2}\frac{2x}{3} + (20 - x)\frac{x}{2}\frac{x}{3}$$

$$= \frac{x^2}{6} (60 - x)$$

$$EI\Delta_{bb} = \frac{(20)^2}{6} (60 - 20) = \frac{8,000}{3}$$

$$\therefore X_b^\uparrow = \frac{x^2(60 - x)}{16,000}$$

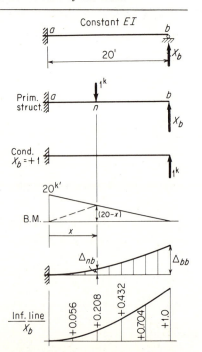

Example 10.3 *Prepare an influence line for the moment at support b of this beam.*

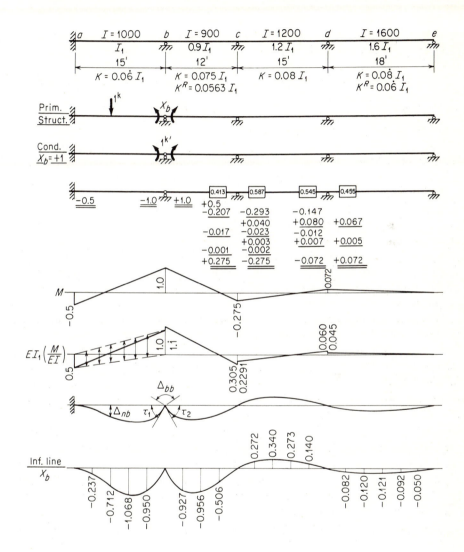

Using Müller-Breslau's principle,

$$X_b \supset \subset = -\frac{\Delta_{nb}^{\downarrow}}{\Delta_{bb} \supset \subset}(1^k) \quad or \quad X_b \supset \subset = +\frac{\Delta_{nb}^{\uparrow}}{\Delta_{bb} \supset \subset}(1^k)$$

if we reverse the sign convention for Δ_{nb}.

Computing Δ_{nb} and Δ_{bb} by the moment-area or elastic-load procedure gives

$$EI_1\tau_1 = (0.5)(7.5) = 3.75 \qquad\qquad \therefore EI_1\Delta_{bb} = 3.75 + 3.83 = 7.58$$

$$EI_1\tau_2 = (1.1)(6)(\tfrac{2}{3}) - (0.305)(6)(\tfrac{1}{3}) = 3.834$$

Span *ab*:

$$EI_1\Delta_{nb}^1 = \frac{x}{15}\frac{x^2}{6} - \frac{1}{2}\frac{x}{2}\frac{2x}{3} - \left(\frac{1}{2} - \frac{x}{30}\right)\frac{x^2}{6} = \frac{x^3}{60} - \frac{x^2}{4}$$

$$\therefore X_b = \frac{1}{7.58}\left(\frac{x^3}{60} - \frac{x^2}{4}\right)$$

Span *bc*:

$$EI_1\Delta_{nb}^1 = -3.834x + 1.\dot{1}\frac{x^2}{3} + (1.\dot{1} - 0.0928x)\frac{x^2}{6} - 0.02542\frac{x^3}{6}$$

$$= -3.834x + 0.556x^2 - 0.0197x^3$$

$$\therefore X_b = \frac{1}{7.58}(-3.834x + 0.556x^2 - 0.0197x^3)$$

etc., for spans cd and de. From these equations, the ordinates every 3 ft are found to be as noted.

10.4 Influence Lines Obtained by Superposition of the Effects of Fixed-End Moments.

This method is very useful for determining influence lines for the end moments of indeterminate beams or frames. It involves using moment distribution to determine the separate effects of each of the FEMs of the various loaded members. These separate effects can then be superimposed to give the total end moments. When properly organized as illustrated in Example 10.4, it is a very effective method of obtaining influence lines for the end moments.

Essentially this method consists of the following steps:

1. Apply a FEM of unity at one end of a member. Using the moment-distribution procedure, compute the resulting end moments in all members. Repeat this process for each end of every member that can have a fixed end moment developed by the applied loads.

2. Compute the FEMs developed by a unit load placed in turn at each of the various load points.

3. Combine the data from steps 1 and 2 to find the end moments throughout caused by a unit load at each of the various load points of the structure.

Of course, when the influence lines have been computed for the end moments, other influence lines can be computed by statics.

When computing step 1 of the above procedure, note that actually only one moment distribution solution is required for each joint of the frame. This may be done for an FEM of unity in any member at a joint. Once this is done, the effect of an FEM of unity in any of the other members meeting at that joint can be determined by inspection. This is evident in the solution of Example 10.4.

Example 10.4 *By superimposing the effects of fixed moments, prepare an influence table for the end moments of the frame used in Example 9.22. Give ordinates every 5 ft along girder BDEF.*

In this case, there can be a sidesway of the columns. To consider this we need to combine a case A and a case B type of solution.

Use the case B end moments from Example 9.22, dividing the data there by 60.

Step 1: *Find the end moments caused by an FEM = +1 in ends of girder members.*

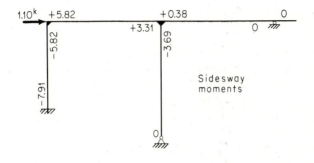

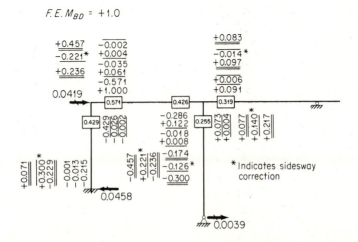

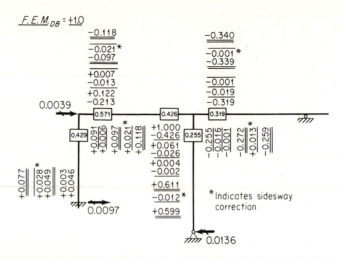

These are the only actual moment-distribution solutions that need to be carried through. The end moments caused by $FEM_{DE} = +1.000$ can be written by inspection simply by shifting the $+1.000$ figure from the DB to the DE column of figures. Likewise, the end moments due to either $FEM_{ED} = +1.0$ or $FEM_{EF} = +1.0$ will be equal to one-half those due to $FEM_{DE} = -1$, with the exception of the ends ED and EF.

Summary of end moments caused by FEM = +1 at the various points:

$FEM = +1$ at point	M_{AB}	M_{BD}	M_{DB}	M_{DE}	M_{DC}	M_{ED}
BD	+0.071	+0.236	−0.300	+0.083	+0.217	0
DB	+0.077	−0.118	+0.599	−0.340	−0.259	0
DE	+0.077	−0.118	−0.401	+0.660	−0.259	0
ED	−0.038	+0.059	+0.201	−0.330	+0.129	0
EF	−0.038	+0.059	+0.201	−0.330	+0.129	−1.0

Step 2: FEM due to unit load at various load points:

Load at	FEM developed at end:				
	BD	DB	DE	ED	EF
1	−2.8125	+0.9375			
2	−2.50	+2.50			
3	−0.9375	+2.8125			
4			−2.8125	+0.9375	
5			−2.50	+2.50	
6			−0.9375	+2.8125	
F					−4.00

Step 3:   **Influence table for end moments:**

Load at:	Factor × [end moment corresponding to FEM = +1 at point (−)]	M_{AB}	M_{BD}	M_{DB}	M_{DE}	M_{DC}	M_{ED}
1	−2.8125 × BD	−0.199	−0.664	+0.844	−0.233	−0.610	
	+0.9375 × DB	+0.072	−0.111	+0.561	−0.318	−0.243	
		−0.127	−0.775	+1.405	−0.551	−0.853	0
2	−2.50 × BD	−0.177	−0.590	+0.750	−0.207	−0.543	
	+2.50 × DB	+0.192	−0.295	+1.498	−0.850	−0.648	
		+0.015	−0.885	+2.248	−1.057	−1.191	0
3	−0.9375 × BD	−0.067	−0.221	+0.281	−0.078	−0.203	
	+2.8125 × DB	+0.216	−0.332	+1.682	−0.956	−0.729	
		+0.149	−0.553	+1.963	−1.034	−0.932	0
4	−2.8125 × DE	−0.216	+0.332	+1.128	−1.852	+0.729	
	+0.9375 × ED	−0.036	+0.054	+0.188	−0.308	+0.120	
		−0.252	+0.386	+1.316	−2.160	+0.849	0
5	−2.50 × DE	−0.192	+0.295	+1.002	−1.648	+0.648	
	+2.50 × ED	−0.095	+0.145	+0.500	−0.820	+0.320	
		−0.287	+0.440	+1.502	−2.468	+0.968	0
6	−0.9375 × DE	−0.072	+0.111	+0.376	−0.619	+0.243	
	+2.8125 × ED	−0.107	+0.163	+0.565	−0.928	+0.360	
		−0.179	+0.274	+0.941	−1.547	+0.603	0
F	−4.0 × EF	+0.152	−0.236	−0.804	+1.320	−0.516	+4.0

10.5   Problems for Solution

Problem 10.1   (*a*) Assuming L/A to be constant for all bars, prepare an influence line for the force in bar *CD* of the truss shown in Fig. 10.2.

(*b*) Using the data from part *a*, likewise prepare influence lines for the force in bars *Bc* and *bc*.

Problem 10.2   Prepare influence lines for the forces in bars *bC* and *Cd* of the truss shown in Example 9.5.

Problem 10.3   Using Müller-Breslau's principle, prepare influence lines for (*a*) the moment at point *c* of the beam shown in Example 10.3; (*b*) the vertical rection at point *b* of this same beam.

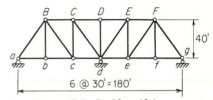

Figure 10.2   Problem 10.1

Problem 10.4   Using the method described in Art. 10.4 prepare influence lines for the end moments of the beam shown in Example 10.3.

11

Work and Energy Methods

11.1 Introduction. Several work and energy methods have already been presented and used for computing deflections and analyzing statically indeterminate structures. The method of virtual work was developed in Arts. 8.3 to 8.7, and Castigliano's first and second theorems were derived in Arts. 8.16 and 8.15, respectively.

Several more energy relationships will be needed for the development of the methods of systematic analysis to be discussed in Part II. Specifically, the **principle of stationary total potential energy** and the **principle of stationary total complementary potential energy** will be utilized in these subsequent discussions. Before developing these two new principles, it will be helpful to review some basic definitions. Then some of the previous developments regarding work and energy will also be reviewed, after which these concepts will be expanded to obtain the additional principles needed in the following chapters.

11.2 Some Basic Definitions and Concepts. Perhaps it should only be necessary at this point to suggest that the reader review his elementary dynamics, concentrating particularly on the basic concepts and definitions introduced during the elementary applications of Newton's laws of motion. However, those ideas that are particularly pertinent to the present developments in this chapter will be restated here for ready reference.

A **force** may be defined as any action that tends to change the state of motion (or rest) of the body to which it is applied. When the point of application of an active force moves, then the force is said to do **work** $\mathscr{W}$ equal to the product of the force and the lineal displacement of its point of application in the direction of the force.[1]

---

[1] In this case, the words *force* and *displacement* are used in a generalized sense which may also be interpreted to mean *couple* and *rotational displacement*, respectively.

If a force is *very* gradually applied to a body, there will be essentially only static deformation and displacement of the body, with no essential accelerations or changes of velocity involved. In such cases, due to the deformation of the body, the work done by the force is stored up in the body by a particular form of potential energy referred to as **strain energy**. If, on the other hand, a force is *not* gradually applied, accelerations and changes of velocity, as well as deformation of the body, will be produced and the work done by the force may be converted into both strain energy and a change in the kinetic energy of the body.

Most often we associate **potential energy** $\mathscr{V}$ with the capability (or potential) of a weight to do work. We measure such potential by selecting arbitrarily some convenient datum plane such as shown in Fig. 11.1a and then computing $\mathscr{V}_o$, the initial potential energy (or the initial potential) of the weight, to be the product of D_o, the initial distance of the weight from the datum, and W_1, the force of gravity acting on the weight.

Actually, of course, any active force P, whether it acts vertically or horizontally or in any direction, has the potential for doing work, and therefore it has potential energy. As in the case of the weight, any force P has an initial potential energy $\mathscr{V}_o$ equal to PD_o, where D_o is the distance to a convenient datum measured in the sense of, and along the line of action of, the force.

If the weight W_1 in Fig. 11.1b were pushed off of its supporting ledge, it would fall freely. As it fell, the weight would steadily acquire more and more kinetic energy, the amount of such energy at any instant being exactly equal to the potential energy $W_1 D$ it had lost between that instant and the start of its fall. On the other hand, consider the situation shown in Fig. 11.1c, where the weight is very gradually transferred from the ledge to the supporting spring. It can be imagined that this gradual tranfer is accomplished with the assistance of a friendly genie who at the start of the transfer process bears the entire load of the weight. But as the transfer proceeds, the spring gradually deflects and absorbs part of the load of the weight. Finally, the spring has deflected enough for the resisting force of the spring to carry the entire load and the genie can be discharged for the time being. Depending on whether the force-deflection characteristics of the spring are linear or nonlinear, the load-transfer process can be depicted by Fig. 11.1d or e, respectively.

During the gradual assumption of the weight W_1 by the spring, the weight has *lost* potential energy by the amount of $-W_1\Delta_1$. Note that this amount is numerically equal to the area of the rectangle *Oabc* in either Fig. 11.1d or e. On the other hand, the area *Obc* under force-deflection curve *Ob* in either of these figures represents the strain energy stored in the spring. During the gradual loading of the spring, the resisting force R of the spring at any displacement is equal and opposite to the net load; i.e., the difference between the weight and the portion of the weight being supported by the genie. The curve *Ob* therefore could also represent the net load-deflection curve; and for such an interpretation the area *Obc* would be equal to the work done by the net load during the gradual loading of the spring.

The area *Oab* above the curve *Ob* represents that part of the potential energy lost by the weight that is *not* transformed into strain energy stored in the spring.

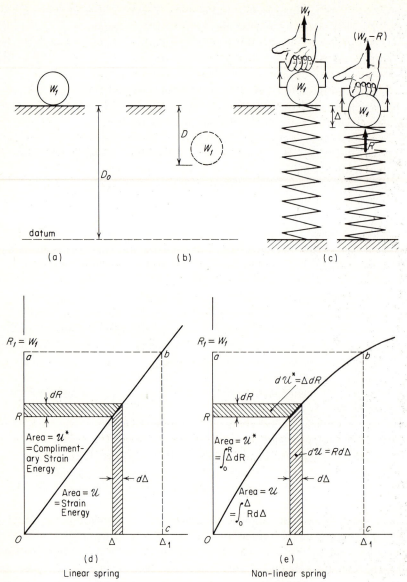

Figure 11.1 Work and energy relationships

In this case, the area Oab represents the work done by the genie in partially holding back the weight so that the net load on the spring gradually increases from zero to the final value W_1.

In the parlance of structural mechanics, the work done by the genie and therefore lost to the system is called the **complementary work** done by the net load. Likewise, interpreting Fig. 11.1d and e as they are actually drawn to represent the force-deflection behavior of the spring, the areas Obc above the curves are said to represent the **complementary strain energy** of the spring.

11.3 Principle of Stationary Total Potential Energy. There is a general principle of rigid-body mechanics that can be represented by Fig. 11.2, in which identical balls are shown resting on three different types of surfaces representing the cases of stable, neutral, and unstable equilibrium, respectively. In each case, a small horizontal displacement of the ball would not be accompanied by any essential vertical displacement. Therefore, for such small horizontal displacements of the balls from their equilibrium positions shown, there is no change in the potential energies associated with the weights W_1. Reasoning from situations like that represented in Fig. 11.2a, it can be stated that the potential energy of a system has a stationary value when the system is in equilibrium and this value is a minimum when the equilibrium is stable.

These considerations suggest the validity of the principle of stationary total potential energy for deformable structural systems. For such a system, the total potential energy of the system consists in part of the potential of the active applied loads $\mathscr{V}$, the remainder consisting of the strain energy stored in the structure $\mathscr{U}$; both these parts can be expressed mathematically in terms of the relevant independent displacements of the system. If the system is in stable equilibrium under the loads, the total potential energy of the system must be at a minimum. Therefore, for small displacements of the system from the equilibrium position, there must be no significant changes in the total potential energy $\mathscr{U} + \mathscr{V}$, or

$$\delta_D(\mathscr{U} + \mathscr{V}) = 0 \qquad (11.1)$$

where the subscript D has been added to the variation symbol δ to emphasize that only deformations and displacements are to be varied. When the following minimizing conditions are expressed mathematically, n equations are obtained, from which the n displacements Δ can be computed, thereby defining the deflected position of the system in its equilibrium position under the active applied loads:

$$\frac{\partial}{\partial \Delta_1}(\mathscr{U} + \mathscr{V}) = 0, \qquad \frac{\partial}{\partial \Delta_2}(\mathscr{U} + \mathscr{V}) = 0, \qquad \dots, \frac{\partial}{\partial \Delta_n}(\mathscr{U} + \mathscr{V}) = 0 \quad (11.2)[1]$$

In the setup of these equations, the strain energy of the system can be evaluated from expressions such as those developed in Art. 8.15. The potential of any load P_j can be expressed as $\mathscr{V}_{oj} - P_j\Delta_j$, and such quantities may be summed for all loads of the system to obtain the potential of the entire load system:

$$\mathscr{V} = \sum^{n} (\mathscr{V}_{oj} - P_j\Delta_j) \qquad (11.3)$$

in which $\mathscr{V}_{oj}$ is a constant that represents the potential (with respect to some convenient datum) of the load P_j in its original position on the unloaded, and hence undeformed, system. With respect to the process of applying Eq. (11.2), note that

---

[1] Basically these equations represent force-equilibrium conditions in the directions of the displacements Δ.

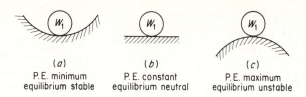

Figure 11.2   Stationary Potential Energy

the value of $\mathscr{V}_{oj}$ need not in fact be computed for each load since these values of $\mathscr{V}_{oj}$, being constants, contribute nothing when partially differentiated in turn with respect to each of the deflections, Δ.

Although we shall not do so formally in this presentation, the principle of total potential energy we have been considering can be rigorously derived by an application of the theorem of virtual deformations, discussed in Art. 11.4. The total potential energy principle has been expressed mathematically by Eqs. (11.1) and (11.2), and may be stated in words as follows:

Principle of stationary total potential energy:

Of all geometrically compatible deformation states of a structural system that also satisfy the deflection boundary conditions, those that also satisfy the force equilibrium requirements give stationary and minimum values to the total potential energy.

It should be noted that this principle is valid theoretically for all structural systems, linear or otherwise, and that even large deformations causing significant changes in geometry can, in principle, be handled. Admittedly, however, cases involving large deflections and/or nonlinear behavior result in very cumbersome solutions which may be impractical to complete.

It is also particularly important to recognize that the principle of stationary total potential energy is not based in any way on the concept of the conservation of energy. If it were, the principle would have to account not only for the various forms of potential energy possessed by the system but also for the kinetic energy involved and any losses or diversions of energy that might occur. The principle of stationary total potential energy is simply a very useful relationship regarding all the potential energy involved in the system. The various terms of the relationship can often be evaluated without much difficulty, and the solution of the resulting equations leads to useful results.

11.4 Virtual-Work Theorems. Several virtual-work relationships have already been discussed. Bernoulli's principle of virtual work for rigid bodies was developed in Art. 8.3. The principle of virtual work for deformable bodies was developed in Art. 8.4 and converted into a convenient form for use in deflection computations in Art. 8.5. This latter principle of virtual work is restated here for convenient reference.

Principle of virtual work for deformable bodies:

If a deformable body is in equilibrium under a virtual Q-force system and remains in equilibrium while the body is subjected to a small and compatible deformation, the external virtual work done by the external Q-forces acting on the body is equal to the internal virtual work of deformation done by the internal Q stresses.

In order to visualize this statement mathematically, it is convenient to think of this principle in the form applicable to frameworks,

virtual-force system

$$\sum Q\delta = \sum F_Q \Delta L \tag{11.4}$$

deformations caused by actual forces

Two important corollaries can be reasoned from this principle of virtual work. The first may be stated as follows:

Theorem of virtual forces for deformable bodies:

If, during any statically possible virtual variation from the equilibrium state of the virtual Q-force system, the external virtual work done by the virtual variation of the external Q forces is equal to the internal virtual work done by the virtual variation of the internal Q stresses, then the unchanging displacements and strains of the deformation system to which the deformable body is being subjected are internally compatible and externally consistent with the support constraints imposed on the body.

Mathematically this theorem may be expressed as follows:

variation of virtual-force system

$$\sum \delta_F(Q)\, \delta = \sum \delta_F(F_Q)\, \Delta L \tag{11.5}$$

∴ compatible deformation

where the subscript F has been added to the variation symbol δ to emphasize that only the external Q forces and the internal Q stresses are to be varied. The second corollary may be stated as follows:

Theorem of virtual deformations for deformable bodies:

If, during any consistent virtual variation of a compatible deformation system to which a deformable body is subjected, the external virtual work done by the external Q forces during the virtual variations of the deformations is equal to the internal virtual work done by the internal Q-stresses during the variation, then the external Q forces, which have remained constant during the variation of the deformations, are in a state of equilibrium with the internal Q stresses.

Mathematically this second theorem may be expressed as follows:

$$\therefore \text{ virtual-force system in equilibrium}$$

$$\sum Q \, \delta_D(\delta) = \sum F_Q \, \delta_D(\Delta L) \qquad (11.6)$$

variation of deformation system

where the subscript D has been added to the variation symbol δ to emphasize that only deformations and displacements are to be varied.

It should be noted that the principle of virtual work and its two corollaries are valid for any type of deformation no matter what its cause and for linear or nonlinear material. The only limitation is that the deformations should not be so large that the equilibrium conditions of the structure need to be altered to include the effects of these large displacements.

11.5 Principle of Stationary Total Complementary Potential Energy. Like potential energy, the complementary potential energy of a system can often be readily evaluated. Also, as in the case of potential energy, there is a useful relationship involving the complementary potential energy of a structural system. This complementary energy relationship is also *not* based on the concept of the conservation of energy. The relationship may be stated as follows:

Principle of stationary total complementary potential energy:

Of all force states satisfying the equilibrium equations and the force boundary conditions, those that also satisfy the geometric compatibility requirements give stationary and minimum values to the total complementary potential energy.

Mathematically this relationship involving π^*, the total complementary potential energy, which is equal to the sum of the complementary strain energy $\mathcal{U}^*$ and the complementary potential energy of the active applied loads $\mathcal{V}^*$, may be written

$$\delta_F(\pi^*) = \delta_F(\mathcal{U}^* + \mathcal{V}^*) = 0 \qquad (11.7)$$

where the subscript F has been added to the variation symbol δ to emphasize that only the external forces and internal stresses are to be varied. Basically this relationship leads to statements of the geometric compatibility conditions.

The complementary strain energy $\mathcal{U}^*$ can be evaluated in a manner similar to the evaluation of the strain energy $\mathcal{U}$. Whereas $\bar{\mathcal{U}}$, the density, i.e., the strain energy per unit volume, of the strain energy $\mathcal{U}$, is computed from the expression

$$\bar{\mathcal{U}} = \int \sigma \, d\varepsilon \qquad (11.8)$$

the density $\bar{\mathcal{U}}^*$ of the complementary strain energy $\mathcal{U}^*$ is computed from the reciprocal relationship

$$\bar{\mathcal{U}}^* = \int \varepsilon \, d\sigma \qquad (11.9)$$

Comparisons of these expressions with similar quantities for the spring shown on Figs. 11.1d and e reveals that the computation of $\bar{\mathcal{U}}$ is associated so to speak with the area *under* the curve and the computation of $\bar{\mathcal{U}}^*$ is associated with the area *above* the curve. Applying Eq. (11.9) to a particular structural member would lead to the value of its complementary strain energy.

In somewhat similar fashion, the value of the complementary potential energy $\mathcal{V}^*$ of an active load P can be determined. The work $\mathcal{W}$ done by such a load moving through a displacement Δ in the sense of its line of action is equal to $\int P \, d\Delta$. On the other hand, the complementary work $\mathcal{W}^*$ done by such a load, as the load varies is equal to $\int \Delta dP$. However, from the definition of the potential energy (or potential) of a load, it is apparent that the load *loses* potential energy in an amount equal to the work it does as it is displaced. Therefore, the variation of the potential energy $\delta_D \mathcal{V}$ as the displacement varies is equal to the *negative* of the variation of the work $\delta_D \mathcal{W}$, or

$$\delta_D \mathcal{V} = -\delta_D \mathcal{W} = -P \, d\Delta \qquad (11.10)$$

But in an analogous manner, if the loads instead of the displacements are varied, the variation of the complementary potential energy $\delta_F \mathcal{V}^*$ is equal to the negative of the variation of the complementary work $\delta_F \mathcal{W}^*$, or

$$\delta_F \mathcal{V}^* = -\delta_F \mathcal{W}^* = -\Delta dP \qquad (11.11)$$

Equation (11.7), the mathematical expression of the principle of stationary total complementary potential energy, follows directly from an application of the theorem of virtual forces. Referring to Eq. (11.5) and the statement of this theorem, if the Q-force system mentioned there is considered to be the actual P-force system causing deformation of the structure, then the external work done is equal to the variation of the complementary work $\delta_F \mathcal{W}^*$ and the internal virtual work is equal to the variation of the complementary strain energy $\delta_F \mathcal{U}^*$, or

$$\delta_F \mathcal{W}^* = \delta_F \mathcal{U}^*$$

From which,

$$\delta_F (\mathcal{U}^* - \mathcal{W}^*) = 0$$

Substituting in this from Eq. (11.11) leads to

$$\delta_F (\mathcal{U}^* + \mathcal{V}^*) = 0$$

which confirms the relationship expressed in Eq. (11.7).

11.6 Complementary-Strain-Energy Theorem. In 1889, ten years after Castigliano published his well-known first and second theorems, Engesser published a paper on statically indeterminate structures in which he stated what we shall call the complementary-strain-energy-theorem.

Complementary-strain-energy theorem:

In any structure the material of which is linearly or nonlinearly elastic, the first partial derivative of the complementary strain energy $\mathscr{U}^*$ with respect to any particular force P_n is equal to the displacement δ_n of the point of application of that force in the direction of its line of action.

Mathematically, this theorem may be expressed as follows:

$$\frac{\partial \mathscr{U}^*}{\partial P_n} = \delta_n \tag{11.12}$$

This theorem is closely related to Castigliano's second theorem, which, of course, involves ordinary strain energy and is restricted to linearly elastic materials.

11.7   References

For additional information on Engesser's theorem refer to Matheson's book. For additional discussion of work and energy methods in general, Matheson and the other four following references are recommended:

1. Matheson, J. A. L.: "Hyperstatic Structures," 2d ed., vol. I, Butterworth's Scientific Publications, London, 1971.

2. Argyris, J. H., and S. Kelsey: "Energy Theorems and Structural Analysis," Butterworth's Scientific Publications, London, 1960.

3. Przemieniecki, J. S.: "Theory of Matrix Structural Analysis," McGraw-Hill Book Company, New York, 1968.

4. Crandall, S. H.: "Engineering Analysis," McGraw-Hill Book Company, 1956.

5. Oden, J. T.: " Mechanics of Elastic Structures," McGraw-Hill Book Company, New York, 1967.

Part II

Introduction to Systematic Structural Analysis

One objective of Part II is to familiarize the reader with the inputs and outputs of automated structural-analysis procedures. The other is to explain how the descriptive data of a linear structural-analysis problem can be systematically transformed into the quantities describing the structural response.

Chapter 12 is designed to accomplish the first objective. Article 12.1 is a review of the basics of matrix algebra. Articles 12.2 to 12.6 deal with the quantitative descriptions of the geometry, the material, the deflection boundary conditions, and the force boundary conditions of structures. Then follows a discussion of the quantities describing the structural response and the methods of transforming the basic data into the description of structural response.

Chapters 13 to 16 deal with the second objective. In Chap. 13, the transformation of the description of a vector (such as position, deflection, and force vectors) from one cartesian coordinate system to another is first established. Then the relationships between kinematically equivalent deflections and between statically equivalent forces are developed.

Chapter 14 develops flexibility and stiffness relationships between the vertex forces and the vertex deflections of a structural element with two or more vertices. The relationships between the descriptions of element flexibility and element stiffness relations in more than one cartesian coordinate system are also established.

In Chap. 15, the systematic analysis by the displacement method is detailed. Economical generation of the matrices involved is discussed, and methods of solving the equations for the unknown deflection components are given.

Chapter 16 details the systematic analysis by the force method. Systematic generation of the force and moment-equilibrium equations of the nodes in terms of the vertex forces of the elements, and the automatic identification of the redundants from these equations are presented. A comparison of systematic-analysis costs by the displacement and force methods is also given.

Additional Symbols for Part II

$\mathbf{A}^m_{(0)}$	$n^m e$ by $Ne - b$ matrix relating $\mathbf{u}_{(1)}$ to $\mathbf{w}^m$
$\mathbf{A}^m_{(1)}$	$n^m e$ by b matrix relating $\mathbf{u}_{(20)}$ to $\mathbf{w}^m$
$\mathbf{B}^m$	$n^m e$ by $\jmath^m$ matrix relating $\mathbf{s}^m$ to $\mathbf{q}^m$
$\mathbf{\bar{B}}$	Ne by $\jmath + b$ matrix relating $\mathbf{z} = [\mathbf{s}^T \mathbf{p}_r^T]^T$ to $\mathbf{\bar{p}}$
$\mathbf{B}^0$	$Ne - b$ by $\jmath$ matrix relating $\mathbf{s}$ to $\mathbf{p}^0$
$\mathscr{B}$	$\jmath$ by $\jmath + b - Ne$ matrix relating $\mathbf{x}$ to $\mathbf{s}$
$\mathscr{B}_0$	$\jmath$ by $Ne - b$ matrix relating $\mathbf{p}^0$ to $\mathbf{s}$
b	number of deflection constraints, i.e., number of reactions
$\mathscr{C}$	compliance matrix of material
$\mathbf{c}_i$	description of position vector at point i
$\mathscr{D}$	material matrix of material
$\mathbf{d}_i$	description of displacement vector of point i
e	degrees of freedom per node (vertex)
e'	number of rigid body degrees of freedom
$\mathbf{F}^m$	$\jmath^m$ by $\jmath^m$ flexibility matrix of the mth element
$\mathbf{\bar{F}}$	$\jmath + b - Ne$ by $\jmath + b - Ne$ flexibility matrix associated with $\mathbf{x}$ directions
f^m	number of interelement force boundary conditions of mth element
$\mathbf{H}$	b by Ne coefficient matrix of $\mathbf{u}$ in the deflection constraint equations
$\mathbf{H}_{(1)} = \mathscr{H}^T$	Ne by $Ne - b$ matrix relating $\mathbf{u}_{(1)}$ to $\mathbf{u}$
$\mathbf{H}_{(0)} = \mathscr{H}_0^T$	Ne by b matrix relating $\mathbf{u}_{(20)}$ to $\mathbf{u}$
$\mathbf{h}_0$	b by 1-constant matrix in the deflection constraint equations
$\mathbf{I}_{j_g^m}$	Ne by e matrix denoting the j_g^mth column partition of N by N partitioned identity matrix of order Ne
$\mathbf{I}_g^m$	$n^m e$ by e matrix denoting the gth column partition of n^m by n^m partitioned identity matrix of order $n^m e$
$\mathbf{I}_*^m$	$\jmath$ by $\jmath^m$ matrix denoting the mth column partition of m by m partitioned identity matrix of order $\jmath$
$\mathbf{j}^m$	n^mth-order column matrix listing the node labels of the vertices of mth element
j_g^m	gth entry of $\mathbf{j}^m$, namely, the node label of the gth vertex of mth element
$\mathbf{K}^m$	$\jmath^m$ by $\jmath^m$ *stiffness matrix* of mth element *in the statically determinate state*
$\mathscr{K}^m$	$n^m e$ by $n^m e$ *free-free stiffness matrix* of mth element
$\mathbf{\bar{K}}$	$Ne - b$ by $Ne - b$ *overall stiffness matrix*
$\mathbf{L}$	matrix of direction cosines of local axes; (α, β) entry of $\mathbf{L}$ is the cosine of the angle between βth local axis and αth overall axis
M	total number of elements in the structure

$\mathbf{m}_i$	description of moment vector at point i (internal or external)
N	total number of nodes in the structure
$\mathbf{N}^m$	$n^m e$ by Ne matrix relating $\mathbf{u}$ to $\mathbf{w}^m$
$\mathscr{N}$	*connectivity matrix*
n^m	number of vertices in the mth element
$\mathbf{n}_i$	description of actual force vector at point i (internal or external)
$\mathbf{P}$	permutation matrix
$\mathbf{p}$	Neth-order column matrix listing external forces in directions of $\mathbf{u}$ (ith partition, $\mathbf{p}_i$, is of order e)
$\mathbf{p}_0$	Neth-order column matrix listing prescribed external concentrated loads in directions of $\mathbf{u}$ (ith partition, $\mathbf{p}_{0i}$, is of order e)
$\mathbf{p}^*$	Neth-order column matrix listing unprescribed external forces (i.e., reactions) in directions of $\mathbf{u}$ (ith partition, $\mathbf{p}_i^*$, is of order e)
$\mathbf{p}_r$	bth-order column matrix listing reactions
$\mathbf{p}^0$	$Ne - b$th-order column matrix listing prescribed external forces in directions of $\mathbf{u}_{(1)}$
$\bar{\mathbf{p}}_0$	$Ne - b$th-order *overall load* column matrix (right-hand side of equilibrium equations in terms of $\mathbf{u}_{(1)}$)
$\bar{\mathbf{p}}$	Neth-order column matrix listing known forces resolved in directions of $\mathbf{u}$ (ith partition, $\bar{\mathbf{p}}_i$, is of order e)
$\not\!p$	prescribed distributed surface force
$\mathbf{q}^m$	$n^m e$th-order column matrix listing *actual vertex forces* of mth element (gth partition, $\mathbf{q}_g^m$, is of order e)
$\mathbf{q}_0^m$	$n^m e$th-order column matrix listing *fixed vertex forces* of mth element (gth partition, $\mathbf{q}_{0g}^m$, is of order e); negative of q_0^m is the *free-free load matrix of mth element*
$\mathbf{q}^{0m}$	$n^m e$th-order column matrix listing vertex forces which are statically equivalent (in any convenient way) to element loads in the mth element (gth partition, $\mathbf{q}_g^{0m}$, is of order e)
$\mathbf{s}$	$\jmath$th-order column matrix listing *all of the element forces*
$\mathbf{s}^m$	$\jmath^m$th-order column matrix listing the *element forces of the mth element*, i.e., mth partition of $\mathbf{s}$
$\mathbf{s}^{0m}$	$\jmath^m$th-order column matrix listing the *fixed vertex forces of mth element*
$\jmath$	order of $\mathbf{s}$, i.e., $\displaystyle \jmath = \sum_{m=1}^{M} \jmath^m$
$\jmath^m$	order of $\mathbf{s}^m$, i.e., $\jmath^m = n^m e - e' - \not\!f^m$
$\mathbf{T}_{ji}$	e by e *deflection transfer matrix* from point i to point j (deflections described in the same coordinate system); inverse transpose is the *force transfer matrix*
$\mathscr{T}_{ji}$	e by e *general deflection transfer matrix* from point i to point j (deflections are described in different coordinate systems); inverse transpose is the *general force transfer matrix*
$\mathbf{u}$	Neth-order column matrix listing actual deflections of all nodes (ith partition, $\mathbf{u}_i$, is of order e)

$\mathbf{u}_{(1)}$	$Ne - b$th-order column matrix listing the independent components of $\mathbf{u}$
$\mathbf{u}_{(2)}$	bth-order column matrix listing the dependent components of $\mathbf{u}$
$\mathbf{u}_{(20)}$	bth-order column matrix listing prescribed deflections in directions of $\mathbf{u}_{(2)}$
$\mathbf{u}_0$	Neth-order column matrix listing prescribed deflections in directions of $\mathbf{u}$ (ith partition, $\mathbf{u}_{0i}$, is of order e)
$\mathbf{v}^m$	∂^mth-order column matrix listing *element deformations* of mth element (actual vertex deflections in directions of $\mathbf{s}^m$)
$\mathbf{v}_0^m$	∂^mth-order column matrix listing *element deformations* caused by element loads in the mth element
$\bar{\mathbf{v}}_0$	$\partial + b - Ne$th-order column matrix denoting deformations in directions of $\mathbf{x}$
$\mathbf{w}^m$	$n^m e$th-order column matrix listing actual vertex deflections in directions of $\mathbf{q}^m$ (gth partition of $\mathbf{w}^m$, $\mathbf{w}_g^m$, lists the actual vertex deflections in directions of $\mathbf{q}_g^m$)
$\mathbf{x}$	$\partial + b - Ne$th-order column matrix listing the *redundants*
δ	variation symbol
ε	sixth-order column matrix listing independent components of strains at a point
σ	sixth-order column matrix listing independent components of stress at a point
σ_x	description of stress vector acting on a unit area with normal in x direction
λ	bth-order column matrix listing the Lagrange multipliers
$\boldsymbol{\Gamma}^m$	∂^m by $n^m e$ matrix relating $\mathbf{w}^m$ to $\mathbf{v}^m$
$\boldsymbol{\theta}_i$	description of rotation vector of point i

12

Describing Problems to a Digital Computer

12.1 Basic Matrix Algebra and Notation. In Part II, the **systematic analysis of structures** will be discussed. The term systematic analysis is introduced here to mean the type of analysis where systematized computational representations and routines are used to minimize the effect of the complexities of the structure on the application of the pertinent structural-analysis procedures. As a result of such systematic organization of the computations, algebra rather than geometry becomes the dominant mathematical tool used in the analysis. The difference between *geometry* and *analytical geometry* is analogous to the difference between *ordinary structural analysis* and *systematic structural analysis*.

The need for systematic analysis arises primarily from the use of digital computers in structural-analysis problems. In describing a problem to a digital computer, one is forced to use arrays of numbers rather than sketches. Once the description of the problem is in the form of arrays, there is no reference to the sketch of the structure during the manipulation of these arrays until all the results are obtained. This is why matrix algebra plays a very important role in the systematic structural analysis. In this article, the important definitions and rules and the notation for matrix algebra are summarized.

12.1.1 Definitions. A **matrix** is an array of elements arranged in rows and columns which, as a whole, obey certain algebraic rules detailed in this article. Unless otherwise specified, a matrix element is a single scalar quantity, although, in general, it may be a complex number, a symbol, or another matrix. The rows of a matrix are labeled with positive consecutive integers from top to bottom, starting with 1 at the top. Likewise, the columns of a matrix are labeled with consecutive integers from left to right, starting with 1 at the left. A matrix element carries as the first subscript its row number and as the second subscript its column number. As a general rule matrices are denoted by boldface roman letters.

Lightface italic letters are used for scalars. In particular, a lightface lowercase italic letter with two subscripts refers to an element of a matrix denoted by the same letter in uppercase boldface roman type. For example, the elements of a rectangular matrix are displayed within brackets:

$$\mathbf{A} = \begin{bmatrix} a_{11} & a_{12} & \cdots & a_{1n} \\ a_{21} & a_{22} & \cdots & a_{2n} \\ \cdots\cdots\cdots\cdots\cdots \\ a_{m1} & a_{m2} & & a_{mn} \end{bmatrix} \tag{12.1}$$

The number of rows in a matrix is called the **row order** of the matrix. The number of columns is called the **column order.** Matrix $\mathbf{A}$ in Eq. (12.1) may be referred to as either a matrix of row order m and column order n or an m by n matrix. When there is only one row in a matrix, it is called a **row matrix** or sometimes simply a **row.** When there is only one column in a matrix, it is called a **column matrix** or simply a **column.** Column matrices are denoted by bold face lower case roman letters. Matrix $\mathbf{A}$ in Eq. (12.1) may be considered to consist of n column matrices. Since these column matrices belong to matrix $\mathbf{A}$, their name should contain the same letter but lowercase bold face roman type. Therefore, $\mathbf{A}$ may be represented as

$$\mathbf{A} = [\mathbf{a}_1 \quad \mathbf{a}_2 \quad \cdots \quad \mathbf{a}_n] \tag{12.2}$$

where the subscripts of the column matrix symbols $\mathbf{a}$ correspond to the column numbers in $\mathbf{A}$. The number of elements in a column matrix is called the **order of the column.** Braces rather than brackets are used to display column matrices such as $\mathbf{x}$, a column matrix of order m, shown as

$$\mathbf{x} = \begin{Bmatrix} x_1 \\ x_2 \\ \vdots \\ x_m \end{Bmatrix} \tag{12.3}$$

Note that an element of a column matrix is named with the same letter as the column but lowercase lightface italic type and carries a subscript showing the place in the column.

When we interchange the corresponding columns and rows of a matrix, the resulting array is called the **transpose** of the original matrix. For example, the transpose of matrix $\mathbf{A}$ in Eq. (12.1) may be denoted $\mathbf{A}^T$ and displayed as

$$\mathbf{A}^T = \begin{bmatrix} a_{11} & a_{21} & \cdots & a_{m1} \\ a_{12} & a_{22} & \cdots & a_{m2} \\ \cdots\cdots\cdots\cdots\cdots \\ a_{1n} & a_{2n} & \cdots & a_{mn} \end{bmatrix} \tag{12.4}$$

or, from Eq. (12.2), this matrix may be displayed as

$$\mathbf{A}^T = \begin{Bmatrix} \mathbf{a}_1^T \\ \mathbf{a}_2^T \\ \vdots \\ \mathbf{a}_n^T \end{Bmatrix} \tag{12.5}$$

where superscript T shows the transposition operation. The transpose of the column matrix $\mathbf{x}$ in Eq. (12.3) may be shown as the row matrix $\mathbf{x}^T$

$$\mathbf{x}^T = [x_1 \quad x_2 \quad \cdots \quad x_m] \tag{12.6}$$

In order to save space, column matrices are often printed in the form of their corresponding row transposes.

If all elements of a matrix are zeros, it is called a **zero matrix** or **null matrix.** When the row order and the column order of a matrix are the same, the array is called a **square matrix.** The diagonal line established by the matrix elements with repeated subscripts, such as a_{11}, a_{22}, etc., is called the **main diagonal.** Matrix elements on the main diagonal are called **diagonal elements,** and the remaining elements are called **off-diagonal elements.** A square matrix in which all the off-diagonal elements are zero is called a **diagonal matrix.** Let k_i denote the ith diagonal entry of a diagonal matrix $\mathbf{K}$; then we use the notation

$$\mathbf{K} = \operatorname{diag} k_i \tag{12.7}$$

to indicate that $\mathbf{K}$ is diagonal.

A diagonal matrix in which all the diagonal elements are equal to unity is called an **identity matrix** or a **unit matrix.** Identity matrices are shown by letter $\mathbf{I}$, and the columns of identity matrices are shown by letter $\mathbf{i}$ with appropriate subscript. Let $\mathbf{I}$ denote an identity matrix of order n. This matrix may be displayed as

$$\mathbf{I} = \begin{bmatrix} 1 & 0 & \cdots & 0 \\ 0 & 1 & \cdots & 0 \\ \cdots\cdots\cdots\cdots \\ 0 & 0 & \cdots & 1 \end{bmatrix} \tag{12.8}$$

or alternately as

$$\mathbf{I} = [\mathbf{i}_1 \quad \mathbf{i}_2 \quad \cdots \quad \mathbf{i}_n] \tag{12.9}$$

The (i, j) element of $\mathbf{I}$ may be denoted by the **Kronecker delta** δ_{ij} rather than i_{ij}. For the elements of $\mathbf{I}$, note that

$$\delta_{ij} = 0 \quad (i \neq j) \quad \text{and} \quad \delta_{ii} = 1 \tag{12.10}$$

If the columns of an identity matrix are shuffled, the resulting matrix is called a **permutation matrix.** Permutation matrices are denoted by $\mathbf{P}$ and a notation used in the subscript to describe the shuffling:

$$\mathbf{P}_{\alpha_1, \ldots, \alpha_n} = [\mathbf{i}_{\alpha_1} \quad \mathbf{i}_{\alpha_2} \quad \cdots \quad \mathbf{i}_{\alpha_n}] \tag{12.11}$$

where $\alpha_1, \alpha_2, \ldots, \alpha_n$ is some permutation of $1, 2, \ldots, n$.

A diagonal matrix in which all the diagonal elements are equal to the same scalar is called a **scalar matrix.**

12.1.2 Determinants. A determinant is an important scalar, and it is related only to square arrays. Let $\mathbf{A}$ denote a square matrix of order n. The determinant of $\mathbf{A}$, shown by det $\mathbf{A}$, is defined as

$$\det \mathbf{A} = \sum \operatorname{sgn}(\alpha_1 \alpha_2 \cdots \alpha_n) a_{1\alpha_1} a_{2\alpha_2} \cdots a_{n\alpha_n} \tag{12.12}$$

where each product $a_{1\alpha_1} a_{2\alpha_2} \cdots a_{n\alpha_n}$ consists of one and only one number from each row and column of $\mathbf{A}$ and where the sum extends over all possible **permutations** $\alpha_1\alpha_2 \cdots \alpha_n$ of the numbers 1, 2, ..., n. If the permutation $\alpha_1\alpha_2 \cdots \alpha_n$ is *even*, sgn $(\alpha_1\alpha_2 \cdots \alpha_n)$ should be taken as $+1$, and if the permutation is *odd*, it should be taken as -1. A permutation is called **even** if an even number of interchanges is required to obtain the original order; otherwise it is called **odd**. Note that there are $n!$ different permutations of numbers 1, 2, ..., n.

As an example, suppose matrix $\mathbf{A}$ is of order 3. The possible number of different permutations of 1, 2, 3 is $3! = 6$. These permutations are 1, 2, 3; 2, 3, 1; 3, 1, 2; 1, 3, 2; 2, 1, 3; and 3, 2, 1. Note that the first three permutations are even and the last three are odd. Then by the definition given in Eq. (12.12), we may write

$$\det \mathbf{A} = a_{11}a_{22}a_{33} + a_{12}a_{23}a_{31} + a_{13}a_{21}a_{32} - a_{11}a_{23}a_{32} - a_{12}a_{21}a_{33} - a_{13}a_{22}a_{31}$$

$$(12.13)$$

where the second subscripts in each triple product are one of the six permutations of 1, 2, 3.

For a general square matrix, we rarely use the definition given in Eq. (12.12) for the computation of its determinant, but there are cases where this definition is very useful. For example, consider the diagonal matrix defined by Eq. (12.7). Here, out of possible $n!$ terms of the summation in Eq. (12.12), since only one may not be zero, we can write

$$\det (\operatorname{diag} k_i) = k_1 k_2 \cdots k_n \qquad (12.14)$$

The definition given in Eq. (12.12) is also useful in computing the determinants of upper and lower triangular matrices. A square matrix with zero for all the off-diagonal elements below the main diagonal is called an **upper triangular matrix** and may be displayed as

$$\mathbf{U} = \begin{bmatrix} u_{11} & u_{12} & \cdots & u_{1n} \\ 0 & u_{22} & \cdots & u_{2n} \\ \cdots & \cdots & \cdots & \cdots \\ 0 & \cdots & \cdots & u_{nn} \end{bmatrix} \qquad (12.15)$$

Since only one of the $n!$ terms in the formal definition of det $\mathbf{U}$ may be nonzero, we can write

$$\det \mathbf{U} = u_{11}u_{22} \cdots u_{nn} \qquad (12.16)$$

Similarly, for the lower triangular matrix $\mathbf{L}$, we can write

$$\det \mathbf{L} = l_{11}l_{22} \cdots l_{nn} \qquad (12.17)$$

From the basic definition of the determinant given in Eq. (12.12), the following properties can easily be proved:

1. $\det \mathbf{A} = \det \mathbf{A}^T$.

2. Interchanging any two rows (columns) of $\mathbf{A}$ changes the sign of its determinant.

3. If two rows (columns) of **A** are identical, then det **A** $= 0$.

4. If the elements of a row (column) of **A** are all zero, then det **A** $= 0$.

5. The determinant of the matrix which is obtained by multiplying all the elements of a row (column) of **A** with a constant c is c det **A**.

6. The determinant of the matrix which is obtained by adding any c multiple of a row (column) of **A** to any other row (column) is still det **A**.

In practice, by applying property 6 we first reduce a given matrix into an equivalent triangular form and then apply Eqs. (12.16) and (12.17) to compute its determinant. This is called **pivotal condensation.**

Another practical way of computing the determinant of a square matrix is by using **Laplace expansion** of the determinant. For the Laplace expansion, we need two new concepts. Let **A** denote a square matrix of order n. Associated with any one element a_{ij} of matrix **A**, we define a scalar m_{ij} such that it is the determinant of the array which is obtained from **A** by deleting its ith row and jth column. The determinant m_{ij} thus obtained is called the **minor** of element a_{ij}. The **cofactor** associated with a_{ij} is shown by c_{ij} and defined by

$$c_{ij} = (-1)^{i+j} m_{ij} \tag{12.18}$$

Using the definition given in Eq. (12.12), we can show that

$$\det (A)\delta_{ij} = \sum_{k=1}^{n} a_{ik} c_{jk} \tag{12.19}$$

or

$$\det (A)\delta_{ij} = \sum_{k=1}^{n} a_{kj} c_{ki} \tag{12.20}$$

where δ_{ij} is the Kronecker delta as defined in Eqs. (12.10). Equations (12.19) and (12.20) are called the **Laplace expansion of the determinant.** In words, these equations state that the determinant of the matrix is equal to the sum of the products of the elements of any row (column) by their cofactors; however, the sum of the products of the elements of a row (column) by the cofactors of any other row (column) is always zero.

The matrix consisting of the cofactors c_{ij} is called the **matrix of cofactors C.** If the determinant of a square matrix is zero, such a matrix is called **singular.** A square matrix is called **nonsingular** if its determinant is not zero.

12.1.3 Rank of a matrix. Let **A** denote an n by m matrix. Consider all possible square matrices formed from **A** by deleting rows and columns. The **rank** of **A** is defined as the order of the highest-order nonsingular matrix which can be established in this way. For example, the ranks of the matrices

$$\begin{bmatrix} 1 & -2 & -1 \\ 2 & -4 & -2 \end{bmatrix} \qquad \begin{bmatrix} -2 & 1 & 0 \\ 1 & 2 & 3 \end{bmatrix} \qquad \begin{bmatrix} 1 & 3 \\ 2 & 1 \end{bmatrix} \tag{a}$$

are 1, 2, and 2, respectively. If **A** is a square matrix of order n, it is nonsingular only if its rank is also n. The rank of a null matrix is zero. The rank of an identity matrix of order n is also n.

12.1.4   Matrix equalities.   Two matrices are equal to each other if the elements in corresponding positions are equal.   For example, if

$$A = B \tag{12.21}$$

this implies that

$$a_{ij} = b_{ij} \qquad \begin{aligned} i &= 1, \ldots, m \\ j &= 1, \ldots, n \end{aligned} \tag{12.22}$$

Note that in order for two matrices to be equal to each other, their row and column orders must be equal.   If both sides of a matrix equation are subjected to the same **matrix operation,** the equality sign survives this operation.   Allowable matrix operations are defined in Arts. 12.1.5 to 12.1.13.   In a matrix equality, one or both sides may consist of matrix expressions instead of single matrices.   A **matrix expression** consists of one or more matrices operated on by one or more matrix operations.

12.1.5   Matrix multiplied by a scalar.   Let A denote an m by n matrix, and let s denote a scalar.   The product of A by s, which may be indicated by sA, is also an m by n matrix.   If B denotes the product matrix, then

$$B = sA \tag{12.23}$$

and the definition of a matrix multiplied by a scalar can be stated as

$$b_{ij} = sa_{ij} \qquad \begin{aligned} i &= 1, \ldots, m \\ j &= 1, \ldots, n \end{aligned} \tag{12.24}$$

which means that to multiply a matrix by a scalar amounts to multiplying every element of the matrix by the same scalar.   Note that if an identity matrix is multiplied by a scalar, the resulting matrix is a scalar matrix.   Note also that from the basic definition of the determinant given by Eq. (12.12) we may reason that if A is a square matrix of order n, then

$$\det sA = s^n \det A \tag{12.25}$$

If we replace s by $1/s$, we obtain the definition of a matrix divided by a scalar from Eqs. (12.23) and (12.24).

12.1.6   Matrix addition and subtraction.   Let A and B denote two matrices both of row order m and column order n.   The sum of these matrices, which may be denoted by $A + B$ or $B + A$, is also an m by n matrix.   Let C denote the sum matrix; then

$$C = A + B \tag{12.26}$$

and the definition of matrix addition can be stated as

$$c_{ij} = a_{ij} + b_{ij} \qquad \begin{aligned} i &= 1, \ldots, m \\ j &= 1, \ldots, n \end{aligned} \tag{12.27}$$

Since this definition is the same for either $B + A$ or $A + B$ matrix addition is said to be **commutative.**   If more than two matrices are to be added, the sum can be obtained by first adding any two and then adding the remaining matrices one by one in any sequence.   From this we can see that the matrix addition is **associative,** since

$$(\mathbf{A} + \mathbf{B}) + \mathbf{D} = \mathbf{A} + (\mathbf{B} + \mathbf{D}) \tag{12.28}$$

Matrix subtraction is a special case of matrix addition, since

$$\mathbf{E} = \mathbf{A} - \mathbf{B} = \mathbf{A} + (-1)\mathbf{B} \tag{12.29}$$

From Eqs. (12.23) and (12.24) we can see that

$$e_{ij} = a_{ij} - b_{ij} \tag{12.30}$$

Note again that matrix addition or subtraction is possible only if the matrices involved are of the same row and column orders.

12.1.7   Matrix multiplied by a column.   Let A denote an m by n matrix, and let **x** denote a column matrix of order n.   The **postmultiplication** of **A** by **x**, denoted by **Ax**, yields a product that is a column matrix of order m.   Let **b** denote this product. Then we can write

$$\mathbf{Ax} = \mathbf{b} \tag{12.31}$$

and the definition of a rectangular matrix postmultiplied by a column matrix is

$$b_i = \sum_{j=1}^{n} a_{ij} x_j \qquad i = 1, \ldots, m \tag{12.32}$$

Note that in order for this multiplication to be possible, the column order of **A** must be equal to the row order of **x**.

If we display **x** as

$$\mathbf{x}^T = [x_1 \quad x_2 \quad \cdots \quad x_n] \tag{12.33}$$

and use the definition of **A** as in Eq. (12.2), we can also write

$$\mathbf{Ax} = x_1 \mathbf{a}_1 + \cdots + x_n \mathbf{a}_n = \sum_{j=1}^{n} x_j \mathbf{a}_j \tag{12.34}$$

Let $\mathbf{i}_j$ denote the jth column of the nth-order identity matrix.   From Eq. (12.34) we can see that

$$\mathbf{Ai}_j = \mathbf{a}_j \qquad j = 1, \ldots, n \tag{12.35}$$

which enables us to single out any one column of **A**.

Note that the **premultiplication** of **A** by **x**, namely, **xA**, cannot be defined, since the column order of **x** is not equal to the row order of **A**.   Therefore we state that matrix multiplication is *not commutative.*

From Eq. (12.34), we can see that the order of column matrix **Ax** is equal to the row order of **A**.   As an example, let **A** and **x** be given as

$$\mathbf{A} = \begin{bmatrix} 1 & -2 & 3 \\ 2 & 1 & 5 \end{bmatrix} \qquad \mathbf{x}^T = [4 \quad 6 \quad 7] \tag{b}$$

Then for the $\mathbf{Ax}$ product we write

$$\mathbf{Ax} = \begin{bmatrix} 1 & -2 & 3 \\ 2 & 1 & 5 \end{bmatrix} \begin{Bmatrix} 4 \\ 6 \\ 7 \end{Bmatrix} = \begin{Bmatrix} (1)(4) + (-2)(6) + (3)(7) \\ (2)(4) + (1)(6) + (5)(7) \end{Bmatrix} = \begin{Bmatrix} 13 \\ 49 \end{Bmatrix} \tag{c}$$

12.1.8   Matrix multiplied by matrix. This multiplication is the generalization of the matrix multiplied by column. Let $\mathbf{A}$ denote an m by n matrix, and let $\mathbf{B}$ denote an n by p matrix. The postmultiplication of $\mathbf{A}$ by $\mathbf{B}$ denoted $\mathbf{AB}$, will yield a definable product provided that the column order of the first matrix, $\mathbf{A}$, is equal to the row of the second matrix, $\mathbf{B}$. Since these orders are both equal to n in this case, the product $\mathbf{AB}$ is said to be conformable. Let $\mathbf{C}$ denote the product matrix. Then we can write

$$\mathbf{C} = \mathbf{AB} \tag{12.36}$$

and the definition of matrix multiplied by matrix can be given as

$$c_{ij} = \sum_{k=1}^{n} a_{ik} b_{kj} \qquad \begin{aligned} i &= 1, \ldots, m \\ j &= 1, \ldots, p \end{aligned} \tag{12.37}$$

To calculate the (i, j) element of $\mathbf{C}$, the ith row of $\mathbf{A}$ is multiplied by the jth column of $\mathbf{B}$. This operation may be sketched as follows:

$$\begin{bmatrix} c_{11} & \cdots & c_{1j} & \cdots & c_{1p} \\ & & & & \\ c_{i1} & \cdots & \boxed{c_{ij}} & \cdots & c_{ip} \\ & & & & \\ c_{m1} & \cdots & c_{mj} & \cdots & c_{mp} \end{bmatrix} = \begin{bmatrix} a_{11} & \cdots & a_{1n} \\ & & \\ a_{i1} & \cdots & a_{in} \\ & & \\ a_{m1} & \cdots & a_{mn} \end{bmatrix} \begin{bmatrix} b_{11} & \cdots & b_{1j} & \cdots & b_{1p} \\ & & & & \\ b_{n1} & \cdots & b_{nj} & \cdots & b_{np} \end{bmatrix} \tag{d}$$

$$(m \text{ by } p) \qquad\qquad (m \text{ by } n) \qquad\qquad (n \text{ by } p)$$

We observe from these definitions that the row order of the product matrix is equal to that of the multiplier (the first matrix) and the column order of the product matrix is equal to that of the multiplicand (the second matrix). The definition also tells us why the column order of the first matrix must equal the row order of the second matrix. If p is not equal to m, the premultiplication of $\mathbf{A}$ by $\mathbf{B}$, namely, the product $\mathbf{BA}$, is not possible. Even if p is equal to m, although now $\mathbf{BA}$ is conformable, one can easily see from the definition in Eq. (12.37) that, in general,

$$\mathbf{BA} \neq \mathbf{AB} \tag{12.38}$$

In other words, matrix multiplication is noncommutative. As an example we perform the following multiplication:

$$\begin{bmatrix} 5 & 0 & 3 \\ 2 & 1 & 4 \end{bmatrix} \begin{bmatrix} 2 & 4 \\ 1 & 3 \\ 7 & 0 \end{bmatrix} = \begin{bmatrix} (5)(2) + (0)(1) + (3)(7) & (5)(4) + (0)(3) + (3)(0) \\ (2)(2) + (1)(1) + (4)(7) & (2)(4) + (1)(3) + (4)(0) \end{bmatrix}$$

$$= \begin{bmatrix} 31 & 20 \\ 33 & 11 \end{bmatrix} \tag{e}$$

Note that if the second matrix had an additional column and we reversed the order of multiplication of the matrices, the multiplication would not even be possible.

Suppose that **A** is an m by n matrix, **B** is an n by r matrix, and **C** is an r by s matrix, in which case, their product **ABC** is possible. We can then easily verify that the matrix multiplication is *associative*, namely,

$$\mathbf{A}(\mathbf{BC}) = (\mathbf{AB})\mathbf{C} \tag{12.39}$$

Suppose **D** is an m by n matrix and **E** and **F** are n by r matrices such that **DE** and **DF** products are conformable. In such a case, again we can easily verify that

$$\mathbf{D}(\mathbf{E} + \mathbf{F}) = \mathbf{DE} + \mathbf{DF} \tag{12.40}$$

i.e., matrix multiplication is *distributive*.

Let **A** denote an m by n matrix. The reader may verify the following important rules:

1. Postmultiplying **A** by the identity matrix of order n does not change its value. Likewise premultiplying **A** by the identity matrix of order m does not change its value.

2. Postmultiplying **A** by a diagonal matrix of order n amounts to multiplying the columns of **A** by the corresponding diagonal elements of the diagonal matrix, namely,

$$\mathbf{A} \operatorname{diag} d_i = [d_1\mathbf{a}_1 \quad \cdots \quad d_n\mathbf{a}_n] \tag{12.41}$$

Similarly, premultiplying **A** by a diagonal matrix of order m amounts to multiplying the rows of **A** by the corresponding diagonal elements of the diagonal matrix.

3. Postmultiplying **A** by a permutation matrix of order n reorders the columns of **A**. Similarly, premultiplying **A** by the transpose of a permutation matrix of order m reorders the rows of **A**. In both cases the reordering is in accordance with the subscript of the permutation matrix [see Eq. (12.11)].

4. Postmultiplying **A** by a null matrix of row order n or premultiplying it by a null matrix of column order m produces a null matrix.

5. Postmultiplying **A** by the jth column of the nth-order identity matrix singles out the jth column of **A** [see Eq. (12.35)]. Similarly, premultiplying **A** by the transpose of the ith column of the mth-order identity matrix singles out the ith row of **A**.

6. Postmultiplying **A** by the jth column of the nth-order identity matrix and premultiplying by the transpose of the ith column of the mth-order identity matrix singles out the a_{ij} entry of **A**, namely,

$$\mathbf{i}_i'^T \mathbf{A} \mathbf{i}_j = a_{ij} \tag{12.42}$$

where the prime indicates that $\mathbf{i}_i'$ belongs to the mth order identity matrix.

12.1.9 Matrix inversion. This operation is applicable only to nonsingular square matrices. It resembles division in ordinary algebra. Let **A** denote an nth-order nonsingular square matrix. The inverse of **A** is denoted by $\mathbf{A}^{-1}$, which

is also a unique nth-order nonsingular square matrix.   It has the following important property:

$$AA^{-1} = A^{-1}A = I \tag{12.43}$$

where I is the nth-order identity matrix.   The definition of A^{-1} may be given as
follows.   Consider the cofactor matrix C associated with A.   Matrix C by definition is a unique matrix.   We can express the **inverse matrix** as

$$A^{-1} = \frac{1}{\det A} C^T \tag{12.44}$$

Since Eqs. (12.19) and (12.20) may be written in matrix notation as

$$C^T A = A C^T = \text{diag} (\det A) \tag{12.45}$$

the inverse matrix as defined by Eq. (12.44) does indeed satisfy Eqs. (12.43).
Since C is unique, A^{-1} is also unique.   In practice we seldom use Eq. (12.44) in
obtaining the inverse matrix.   There are many numerical algorithms by which the
inverse matrix can be computed more economically (one such is explained in
Art. 16.3).

We can demonstrate the use of the inverse matrix by the following example.
Suppose we wish to obtain an explicit expression for the nth-order column matrix x
from

$$Ax = b \tag{12.46}$$

where A is a nonsingular matrix of order n.   By premultiplying both sides of the
equality by A^{-1} and then using Eq. (12.43) and rule 1 listed in Art. 12.1.8 we can
write the desired expression as

$$x = A^{-1}b \tag{12.47}$$

Note that if Eq. (12.46) shows a set of linear simultaneous equations in x, Eq. (12.47)
is the explicit expression for its solution.

12.1.10   Differentiation of a matrix.   Let A denote an m by n matrix such that
every element of the matrix is a scalar function of independent variables x and y.
The partial derivative of matrix A with respect to x is also an m by n matrix.   Let
B denote the derivative matrix; then we may write

$$\frac{\partial}{\partial x} A = B \tag{12.48}$$

and the definition of the differentiation is given by

$$b_{ij} = \frac{\partial}{\partial x} a_{ij} \qquad i = 1, \ldots, m \tag{12.49}$$
$$j = 1, \ldots, n$$

If the matrix elements of A are the functions of only x, we use Eqs. (12.48) and
(12.49) with ordinary differentiation.

12.1.11   Differentiation of a matrix expression.   The rules for the differentiation of ordinary algebraic expressions also cover the matrix expressions.   Here we should be careful not to change the order of matrices in a product, since matrix multiplication is not commutative.   For example, let $\mathbf{A}$ and $\mathbf{B}$ denote two matrices such that product $\mathbf{AB}$ is defined, and let us assume that the elements of both matrices are functions of independent variables x and y.   The x partial derivative of product $\mathbf{AB}$ can be written

$$\frac{\partial}{\partial x}(\mathbf{AB}) = \left(\frac{\partial}{\partial x}\mathbf{A}\right)\mathbf{B} + \mathbf{A}\left(\frac{\partial}{\partial x}\mathbf{B}\right) \qquad (12.50)$$

If the elements of $\mathbf{A}$ and $\mathbf{B}$ are functions of only x, we would use ordinary differentiation in Eq. (12.50).

12.1.12   Matrix integration.   Let $\mathbf{A}$ denote an m by n matrix such that every element of the matrix is a continuous function of x.   The integral of $\mathbf{A}$ with respect to x is also an m by n matrix.   Let this matrix be denoted by $\mathbf{B}$.   Then we can write

$$\mathbf{B} = \int \mathbf{A}\, dx \qquad (12.51)$$

The definition of the integration is given by

$$b_{ij} = \int a_{ij}\, dx \qquad i = 1, \ldots, m \qquad (12.52)$$
$$j = 1, \ldots, n$$

If the elements of $\mathbf{A}$ are functions of independent variables x, y, and z, we represent the product of $dx\, dy\, dz$ by the volume element dv and write Eqs. (12.51) and (12.52) with dv instead of dx.

12.1.13   Transposition of a matrix product.   Let $\mathbf{A}$, $\mathbf{B}$, and $\mathbf{C}$ be conformable matrices such that product $\mathbf{ABC}$ is defined.   If we are interested in the transpose of the product, we can verify that

$$(\mathbf{ABC})^T = \mathbf{C}^T\mathbf{B}^T\mathbf{A}^T \qquad (12.53)$$

namely, that the transpose of a matrix product is the product of transposes in reverse order.

12.1.14   Inverse of a matrix product.   Let $\mathbf{A}$, $\mathbf{B}$, and $\mathbf{C}$ denote nonsingular square matrices of order n.   It can be shown that

$$(\mathbf{ABC})^{-1} = \mathbf{C}^{-1}\mathbf{B}^{-1}\mathbf{A}^{-1} \qquad (12.54)$$

namely, that the inverse of a nonsingular matrix product is the product of inverses in the reverse order.

12.1.15   Symmetric matrices.   These are a subset of square matrices.   Let $\mathbf{A}$ denote a square matrix of order n.   If

$$\mathbf{A} = \mathbf{A}^T \qquad (12.55)$$

holds, we call matrix **A** a **symmetric matrix.** Note that in a symmetric matrix, elements in (i, j) and (j, i) positions are algebraically equal to each other, as in

$$\mathbf{A} = \begin{bmatrix} 1 & 2 & 3 \\ 2 & 4 & 5 \\ 3 & 5 & 6 \end{bmatrix} \qquad\qquad (f)$$

From the definitions, we can see that the sum of symmetric matrices is also symmetric, whereas the product of symmetric matrices is not symmetric unless the matrices in the product are commutative. The latter situation arises, for example, when we deal with diagonal matrices. The matrix product of two or more diagonal matrices is commutative, and the resulting matrix is also diagonal. Note that a diagonal matrix is a special symmetric matrix. We can also see that the derivative and the integral of a symmetric matrix are also symmetric. From Eq. (12.44) we can observe that the inverse of a symmetric nonsingular matrix is also symmetric. In storing symmetric matrices, we may consider only one-half the matrix including the main diagonal elements.

12.1.16 Skew-symmetric matrices. These matrices also constitute a subset of square matrices. Let **B** denote a square matrix of order n. If

$$\mathbf{B} = -\mathbf{B}^T \qquad\qquad (12.56)$$

we call matrix **B** a **skew-symmetric matrix.** Since $\mathbf{B} + \mathbf{B}^T = \mathbf{0}$, the main-diagonal elements of a skew-symmetric matrix are always zero, namely,

$$b_{ii} = 0 \qquad i = 1, \ldots, n \qquad\qquad (12.57)$$

but for other elements of such matrices

$$b_{ij} = -b_{ji} \qquad \begin{aligned} i &= 1, \ldots, n \\ j &= 1, \ldots, n \end{aligned} \qquad\qquad (12.58)$$

It can be shown that a skew-symmetric matrix is always singular. An example of skew-symmetric matrix is

$$\mathbf{B} = \begin{bmatrix} 0 & 1 & 2 \\ -1 & 0 & 3 \\ -2 & -3 & 0 \end{bmatrix} \qquad\qquad (g)$$

In storing skew-symmetric matrices we may consider only one-half of the matrix, excluding the main-diagonal elements. The sum of skew-symmetric matrices is also skew-symmetric, whereas the product of skew-symmetric matrices is not a skew-symmetric matrix.

12.1.17 Symmetric and skew-symmetric components of a square matrix. Any square matrix can be decomposed into one symmetric and one skew-symmetric matrix of the same order. Let **C** denote a square matrix of order n. Let matrices **A** and **B** be such that

$$\mathbf{A} = \tfrac{1}{2}(\mathbf{C} + \mathbf{C}^T) \qquad\qquad (12.59)$$

and

$$\mathbf{B} = \tfrac{1}{2}(\mathbf{C} - \mathbf{C}^T) \qquad\qquad (12.60)$$

Since $(C + C^T)^T = C + C^T$, matrix A is symmetric, and since $(C - C^T)^T = -(C - C^T)$, matrix B is skew-symmetric. As an example, consider the matrix C,

$$C = \begin{bmatrix} 1 & 3 & 5 \\ 1 & 4 & 8 \\ 1 & 2 & 6 \end{bmatrix} \tag{h}$$

Application of Eqs. (12.59) and (12.60) will yield symmetric and skew-symmetric components of Eq. (h) exactly the same as in A and B displayed in Eqs. (f) and (g), respectively.

12.1.18 Partitioned matrices and columns. If the elements of a matrix are also matrices, we call the matrix **partitioned** provided that the elements on the same row (or column) have the same row (or column) order. The matrix elements of a partitioned matrix are called **submatrices.** Let A denote a partitioned matrix. The submatrix of A in the (i, j) position is shown as A_{ij}. We display the partitioned matrix A as follows:

$$A = \begin{bmatrix} A_{11} & A_{12} & \cdots & A_{1n} \\ A_{21} & A_{22} & \cdots & A_{2n} \\ \vdots & \vdots & \vdots & \vdots \\ A_{m1} & A_{m2} & \cdots & A_{mn} \end{bmatrix} \tag{12.61}$$

The dashed lines are called the **partitioning lines.** All submatrices between two successive vertical partitioning lines are of the same column order, and all submatrices between two successive horizontal partitioning lines are of the same row order. A scalar element of submatrix A_{ij} in position (α, β) in the submatrix will be denoted by $a_{i\alpha j\beta}$. For example, the element in the $(1, 2)$ position of submatrix $A_{3,4}$ will be shown by $a_{3,1,4,2}$. When symbolic labels are used, Latin letters will be used for partition labels and Greek letters for positions in the submatrices. We shall use the symbolism of A_j to denote the jth partition column of the partitioned matrix A. With this symbolism, we can redisplay A of Eq. (12.61) as

$$A = [A_1 \quad A_2 \quad \cdots \quad A_n] \tag{12.62}$$

When we want to refer to the βth column of the jth column partition of A, we shall use $a_{j\beta}$. Suppose the range of β is e; then we display the jth partition column of A as

$$A_j = [a_{j1} \quad a_{j2} \quad \cdots \quad a_{je}] \tag{12.63}$$

Let I denote a partitioned identity matrix with n column and row partitions. We display this identity matrix as

$$I = [I_1 \quad I_2 \quad \cdots \quad I_n] \tag{12.64}$$

Suppose the jth column partition of I is of column order e. Then I_j consists of $n - 1$ zero submatrices and one identity matrix of order e, which is the jth submatrix of I_j. Suppose the column partitioning of A in Eq. (12.61) is the same as the row partitioning of I in Eq. (12.64). We can see from the definition of matrix multiplication that

$$AI_j = A_j \tag{12.65}$$

which states that postmultiplying the partitioned matrix $\mathbf{A}$ with $\mathbf{I}_j$ amounts to singling out the jth column partition of $\mathbf{A}$.   Let $\mathbf{I}'$ denote the mth-order partitioned identity matrix such that its column partitioning matches the row partitioning of $\mathbf{A}$ as given by Eq. (12.61).   Let $\mathbf{I}'_i$ denote the ith column partition of $\mathbf{I}'$.   From the matrix multiplication rules, we see that

$$\mathbf{I}'_i{}^T\mathbf{A} = (\mathbf{A}^T)^T_i \tag{12.66}$$

namely, premultiplying the partitioned matrix $\mathbf{A}$ with the transpose of the ith column partition of $\mathbf{I}'$ singles out the ith row partition of $\mathbf{A}$.   From Eqs. (12.65) and (12.66) we can also write

$$\mathbf{I}'_i{}^T\mathbf{A}\mathbf{I}_j = \mathbf{A}_{ij} \tag{12.67}$$

That is, premultiplying $\mathbf{A}$ with the transpose of the ith column partition of the mth-order partitioned identity matrix $\mathbf{I}'$ and postmultiplying it with the jth column partition of the nth-order partitioned identity matrix $\mathbf{I}$ singles out the $\mathbf{A}_{ij}$ element. The reader can also verify that

$$\mathbf{A} = \sum_{i=1}^{m} \sum_{j=1}^{n} \mathbf{I}'_i \mathbf{A}_{ij} \mathbf{I}^T_j \tag{12.68}$$

a relationship that is exploited extensively in the systematic analysis of structures. The operation in Eq. (12.68) is called the **direct sum.**   With this definition, we can state that a matrix is the direct sum of its elements.

If the elements of a column matrix are also column matrices, we call the column matrix partitioned.   The elements of a partitioned column matrix are called sub-column matrices.   Let $\mathbf{x}$ denote a partitioned column matrix.   The ith subcolumn of $\mathbf{x}$ is shown as $\mathbf{x}_i$.   We display the partitioned column matrix $\mathbf{x}$ as

$$\mathbf{x} = \begin{Bmatrix} \mathbf{x}_1 \\ \vdots \\ \mathbf{x}_n \end{Bmatrix} \tag{12.69}$$

The αth scalar element of subcolumn $\mathbf{x}_i$ is denoted by $x_{i\alpha}$.   For example, the second element of the fifth subcolumn of $\mathbf{x}$ is shown as $x_{5,2}$.   When symbolic labels are used, we shall use Latin letters for the labels of subcolumns and Greek letters for the labels of the elements of a subcolumn.   Note that some confusion may arise in interpreting $x_{5,2}$, since it may also mean the $(5, 2)$ element of matrix $\mathbf{X}$. The proper interpretation should be clear from the context in which the symbol appears.

If we want to obtain the transpose of a partitioned matrix, we can first treat the matrix as if it were unpartitioned and interchange the rows and columns as in the usual transposition process.   Then we reestablish the partitioning lines in this transposed matrix and recognize that each of its submatrices is the transpose of the corresponding submatrix in the original matrix.   Therefore, the $\mathbf{A}_{ij}$ submatrix at the (i, j) position of a partitioned matrix **A** appears with its transpose in the (j, i) position of $\mathbf{A}^T$. With this, the transpose of **A** given in Eq. (12.61) can be written

$$\mathbf{A}^T = \begin{bmatrix} \mathbf{A}^T_{11} & \mathbf{A}^T_{21} & \cdots & \mathbf{A}^T_{m1} \\ \mathbf{A}^T_{12} & \mathbf{A}^T_{22} & \cdots & \mathbf{A}^T_{m2} \\ \vdots & & & \vdots \\ \mathbf{A}^T_{1n} & \mathbf{A}^T_{2n} & \cdots & \mathbf{A}^T_{mn} \end{bmatrix} \tag{12.70}$$

The transpose of a partitioned column matrix is obtained similarly. The jth subcolumn $\mathbf{x}_j$ of the partitioned column matrix **x** appears with its transpose at the jth position of $\mathbf{x}^T$. For example, the transpose of **x** given by Eq. (12.69) can be displayed as

$$\mathbf{x}^T = [\mathbf{x}^T_1 \quad \cdots \quad \mathbf{x}^T_n] \tag{12.71}$$

We shall use the formalism of Eq. (12.71) rather than that of Eq. (12.69) in displaying a partitioned column matrix because it is easier to print.

In the product of two partitioned matrices the number and spacings of the vertical partitioning lines in the first matrix should be the same as the number and the spacings of the horizontal partitioning lines in the second matrix. The number and the spacings of the horizontal partitioning lines in the product matrix are the same as those in the first matrix, and the number and the spacings of the vertical partitioning lines in the product matrix are the same as those in the second matrix.

12.1.19   Linear forms. Let **a** and **x** denote two column matrices of order n. The product $\mathscr{W}$ of these two matrices in the form of

$$\mathscr{W} = \mathbf{x}^T \mathbf{a} \tag{12.72}$$

is called the **linear form** of **x**. Note that $\mathscr{W}$ is a scalar. We usually assume that column matrix **a** is a constant column, and column matrix **x** represents the variables. When $n = 2$; we note that $\mathscr{W}$ represents a plane in the space of x_1, x_2, and $\mathscr{W}$. When $n = 1$, $\mathscr{W}$ represents a straight line in the planar space of x_1 and $\mathscr{W}$. In general we say that $\mathscr{W}$ in Eq. (12.72) represents a plane in $(n + 1)$-dimensional space. Physically, if **a** represents the magnitudes of a list of forces and **x** lists the displacements in the directions of these forces, we can consider $\mathscr{W}$ as the work done by the forces.

12.1.20   Quadratic forms. Let **A** denote a symmetric square matrix of order n. Let **x** denote a column of order n. The scalar defined by

$$\mathscr{U} = \tfrac{1}{2}\mathbf{x}^T \mathbf{A} \mathbf{x} \tag{12.73}$$

is called the **quadratic form** of **x**. When $n = 1$, $\mathscr{U}$ shows a parabola passing through the origin in $x_1, \mathscr{U}$ space. If $a_{11} > 0$, the parabola is such that $\mathscr{U}$ is always larger than zero regardless of the value of x_1. However, if $a_{11} < 0$, the parabola is such that it is always less than zero regardless of the value of x_1. Of course, if $a_{11} = 0$, $\mathscr{U}$ is always zero. When $n = 2$, $\mathscr{U}$ shows a quadratic surface in $x_1, x_2, \mathscr{U}$ space. Depending upon the values of elements a_{11}, $a_{12} = a_{21}$, and a_{22}, the quadratic surface may be such that $\mathscr{U}$ may be always greater than or less than zero, or it may assume both positive and negative values as we assign values to x_1 and x_2.

In general, we say that $\mathcal{U}$ as defined in Eq. (12.73) shows a quadratic surface in the $(n + 1)$-dimensional space of $x_1, x_2, \ldots, x_n$ and $\mathcal{U}$. We classify a quadratic form with its sign. If $\mathcal{U}$ is always larger than zero for all possible values of $x_1, \ldots, x_n$ and zero only if $x_1 = x_2 = \cdots = x_n = 0$, the quadratic form and the matrix $\mathbf{A}$ are called **positive definite**. If $\mathcal{U}$ is always larger than or equal to zero for all possible values of $x_1, \ldots, x_n$, we call the quadratic form and the matrix $\mathbf{A}$ **positive**. The distinction between positive and positive definite quadratic forms is that a positive quadratic form may be zero even if one or more x_i components are not zero. Similarly, if $\mathcal{U}$ is always less than or equal to zero and zero only when all x_i components are zero, we call the quadratic form and the matrix $\mathbf{A}$ **negative definite**. If $\mathcal{U}$ is always less than or equal to zero for all values of x_i components, we call the quadratic form and the matrix $\mathbf{A}$ **negative**. If $\mathcal{U}$ can assume both negative and positive values as we assign arbitrary values to $\mathbf{x}$, the quadratic form and the matrix $\mathbf{A}$ are called **indefinite**. Positive and positive definite quadratic forms and matrices play a very important role in linear structural analysis.

From the foregoing definitions it is understood that if a matrix is positive or positive definite, the quadratic form related to the matrix is also positive or positive definite. There are various ways of testing the positive definiteness of a symmetric matrix. One of them states that if all the **leading principal minors** of a symmetric matrix are larger than zero, the matrix is positive definite. Namely, if $\mathbf{A}$ is positive definite, then

$$a_{11} > 0, \det \begin{bmatrix} a_{11} & a_{12} \\ a_{21} & a_{22} \end{bmatrix} > 0, \ldots, \det \mathbf{A} > 0 \qquad (12.74)$$

We see from Eq. (12.73) that if $\mathbf{A}$ is positive definite, then $a_{ii} > 0$, $i = 1, \ldots, n$. However, when $a_{ii} > 0$, $i = 1, \ldots, n$, the matrix is not necessarily positive definite. As discussed in Art. 15.5, we can decompose a real symmetric matrix into the product of a lower triangular matrix with its transpose. In such cases, according to Eq. (12.74), we can observe that in order for matrix $\mathbf{A}$ to be positive definite it is sufficient for all the diagonal elements of its triangular decomposition to be larger than zero.

12.1.21   Remarks about the notation in Part II.   In systematic structural analysis, we are usually confronted with matrices which are either too large or somehow modular either in **structural elements** or in certain special points called **nodes.** Such characteristics encourage us to partition the matrices whenever possible. If partitioning is governed by the *structural elements*, we shall indicate this by placing the label as a *superscript*. On the other hand, if the partitioning is done by the *nodes*, we shall show this by using the lowercase *Latin* letters as the node labels in the *subscripts*. The positions of the scalar elements of a submatrix or a subcolumn will be shown by Greek letters in the subscripts. Since there are many indices and their ranges are not constant, the summation convention of repeated indices will not be used in Part II for the sake of clarity. All summations will be explicitly indicated by the summation symbols where the index ranges will be clearly displayed.

Right-handed cartesian coordinate systems will be used throughout.   When there is more than one coordinate system, one will be selected as the **overall coordinate system** and the others will be referred to as the **local coordinate systems**.   The quantities described in local coordinate systems will be marked by a prime symbol whenever feasible.   Greek letters will be used in the subscripts whenever references are made to the coordinate directions.

12.1.22   Problems for Solution

Problem 12.1   Show that the set of linear equations

$$\begin{aligned}
a_{11}x_1 + a_{12}x_2 + \cdots + a_{1n}x_n &= c_1 \\
a_{21}x_1 + a_{22}x_2 + \cdots + a_{2n}x_n &= c_2 \\
&\cdots\cdots\cdots\cdots\cdots\cdots\cdots\cdots \\
a_{n1}x_1 + a_{n2}x_2 + \cdots + a_{nn}x_n &= c_n
\end{aligned}$$
(12.75)

can conveniently be written in matrix notation as

$$\mathbf{Ax} = \mathbf{c}$$
(12.76)

and the solution expressed as

$$\mathbf{x} = \mathbf{A}^{-1}\mathbf{c}$$
(12.77)

Then deduce the rule which states that in order for the linear system in Eq. (12.75) to have a unique solution, $\mathbf{A}$ should be nonsingular.

Problem 12.2   Using the definition given in Eq. (12.12), prove that if two rows (or columns) of a square array are the same, the value of its determinant is zero.   Then prove that if any multiple of a row (or column) is added to any other row (or column), the value of the determinant does not change.   (Such operations are called the **elementary operations.**)

*Problem 12.3   Given the fourth-order square matrix

$$\mathbf{A} = \begin{bmatrix} 4 & 1 & 3 & 5 \\ 4 & 2 & 6 & 10 \\ -8 & 0 & 4 & 5 \\ -4 & 3 & 1 & 3 \end{bmatrix}$$
(i)

compute det $\mathbf{A}$ by first transforming the array into an upper triangular form by the second rule of Prob. 12.2 (which is rule 6 in Art. 12.1.2) and then using Eq. (12.16).

Problem 12.4   Let $\mathbf{A}$ denote a nonsingular matrix such that

$$\mathbf{A}^T = \mathbf{A}^{-1}$$
(12.78)

Such matrices are called **orthogonal.**   Prove that the identity matrix $\mathbf{I}$ and the permutation matrix defined in Eq. (12.11) are orthogonal matrices.

*Problem 12.5   Prove that

$$(\operatorname{diag} \mathbf{K}_i)^{-1} = \operatorname{diag} \mathbf{K}_i^{-1}$$
(12.79)

Specialize this for diagonal matrices which are not partitioned.

Problem 12.6   Consider the quadratic form given by Eq. (12.73).   Suppose $\mathbf{A}$ and $\mathbf{x}$ are partitioned conformably into n partitions.   Show that

$$\mathscr{U} = \frac{1}{2} \sum_{i=1}^{n} \sum_{j=1}^{n} \mathbf{x}_i^T \mathbf{A}_{ij} \mathbf{x}_j$$
(12.80)

Assuming that the partitioning line spacings are constant and equal to e, show also that

$$\mathscr{U} = \frac{1}{2} \sum_{i=1}^{n} \sum_{j=1}^{n} \sum_{\alpha=1}^{e} \sum_{\beta=1}^{e} x_{i\alpha} a_{i\alpha j\beta} x_{j\beta}$$
(12.81)

Write a computer procedure to evaluate $\mathscr{U}$ from Eq. (12.81).

12.2   Geometry of Structure.   The geometry of the structure is defined by the volume of those material particles which carry stresses to withstand the applied loads.   This volume is usually too complicated for analytical or geometrical description.   The structural analysis may be very costly unless the geometry can be idealized as simple forms.   Fortunately, manufacturing, transportation, and construction costs and aesthetic and performance constraints have forced the geometry of structures into configurations which are modular in simple volumes, where usually one or two dimensions are much smaller than the remaining, as in beams, bars, plates, and their curved counterparts.   Long experience with these modules, called **structural elements,** has taught us that the general behavior of the material particles in the short dimension can usually be described with acceptable accuracy by means of simple laws.   For example, we have the Euler-Navier rule for beams, which states that *plane sections remain plane after deformation.*   This enables us to represent a beam by its axis and its cross-sectional properties, and the analysis becomes simplified since the behavior of particles in a cross section can easily be expressed in terms of the displacements and the rotations of the particle on the beam axis (see Fig. 12.1).   The Kirchhoff-Love assumption for plates and shells states that *normals to the middle surface remain normal after deformation.* With this, we can represent a thin structural element by its middle surface and its thickness and simplify the analysis by paying attention only to the behavior of the particles on the middle surface, since the behavior of the particles in a thickness direction can be expressed in terms of the behavior of the particle in the middle of the thickness (see Fig. 12.2).   Of course there are structural elements, such as foundation blocks, which are neither thin nor long; however, they are usually simple in geometry.

12.2.1   Overall coordinate system, nodes, elements, element vertices and labels. Consider the structures composed of bars, beams, and columns.   They constitute the majority of engineering structures and are the main subject of ordinary structural analysis.   In such structures, the structural elements are in contact with each

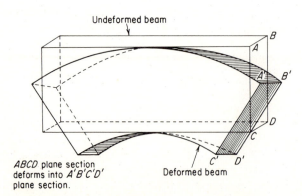

Figure 12.1   Plane-sections-remain-plane assumption

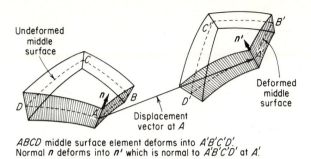

ABCD middle surface element deforms into A'B'C'D'.
Normal n deforms into n' which is normal to A'B'C'D' at A'.

Figure 12.2   Normals-remain-normal assumption

other at special points, called **nodes**. The points of a structural element which are in contact with the nodes are called **vertices.** The geometry of the structure is defined by the coordinates of nodal points in an overall coordinate system and by the elemental information listing for each element the node labels of its vertices and its cross-sectional data. If a group of basic elements is considered as a structural element, its constituents can be defined in a similar fashion.

If the structure is a planar one, the overall coordinate system is selected such that its xy plane is coincident with the plane of the structure. If the structure has one or more planes of geometrical symmetry, we try to overlap them with as many coordinate planes as possible. In order to minimize numerical round-off errors, we choose the origin of the overall coordinate system in the vicinity of the geometric center of the structure. For each structural element we may choose a *local coordinate system* such that the first axis may extend from the first vertex toward the second vertex. The first vertex is the one whose node label is listed first in the list of node labels of the element vertices. In planar structures, the third local axis is always coincident with the third overall axis, and the second local axis is in the plane of the structure and directed such that the local coordinate system is a right-handed one. In nonplanar structures, the second and the third local axes are taken as the principal axes of the cross section at the first vertex. To fully define such a coordinate system, we usually give the overall coordinates of some **auxiliary point** on the second local axis as part of the elemental information and assume that the second local axis is directed from the first vertex toward this point (see Fig. 12.3). The cross-sectional data usually consist of the area, principal moments of inertia, and torsional constant for uniform structural elements where the center of the cross section is coincident with the shear center.

In Fig. 12.4 a space-frame structure with eight uniform beam elements, six rigid joints, and three fixed supports is shown. In nonvertical elements, the principal axis of the cross section producing the larger moment of inertia is placed horizontally. In the vertical elements, the same kind of axis is parallel to the horizontal element. The figure shows the overall coordinate system and the labels of nodes and elements. It is clear from the preceding paragraph that it is imperative to label all nodes with positive integers starting from 1, called **node labels,** and all elements

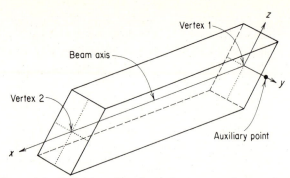

Figure 12.3 Definition of local coordinate system in space beams

with a different set of positive integers, again starting from 1, called **element labels.** To use positive integers as node and element labels is peculiar to systematic analysis. It is a natural result of our wish to use the matrix algebra in the analysis as much as possible. The structure in Fig. 12.4 contains six nodes, and there are $6! = 720$ possible ways of labeling these nodes. For the present structure any one of these possibilities may be chosen. However, in general, one should try to label the nodes in such a way that the difference between the highest and the lowest label in a **node set** is as small as possible. The term **node set** refers to a node and all other nodes connected with it by an element. The reasons for this will be seen during the discussion of the displacement method. Similarly, since there are eight structural elements, there are $8! = 40,320$ possible ways of labeling these elements. We may choose any one of these, since the labeling of elements is not important, as we shall see later.

In the systematic analysis of structures, during the preparation of the description of the problem in the form of numerical arrays, it is always necessary to have a clear sketch of the structure showing the overall coordinate system, the node labels, and the element labels. With such a **key sketch,** it is a straightforward exercise to generate the arrays describing the geometry. For example, let **c**

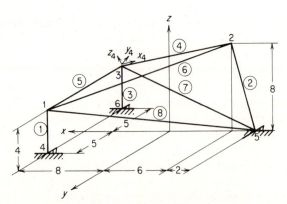

Figure 12.4 Key sketch of a space frame

denote a partitioned column matrix containing overall coordinates of the nodes such that $\mathbf{c}_i$ lists the description of the **position vector** of node i. Since the components of the position vector are the coordinates of its tip point, $\mathbf{c}_i$ actually lists the x, y, z coordinates of the ith node. For example, $c_{i\alpha}$ would show the y coordinate of node 3 when $i = 3$ and $\alpha = 2$. We may define

$$\mathbf{c}^T = [\mathbf{c}_1^T \quad \mathbf{c}_2^T \quad \mathbf{c}_3^T \quad \mathbf{c}_4^T \quad \mathbf{c}_5^T \quad \mathbf{c}_6^T]$$

and determine each of the subcolumns from the key sketch. For example, $\mathbf{c}_2^T = [-6 \quad 0 \quad 8]$. To complete the description of the geometry, we must provide the elemental information. Let us assume that for the structure in Fig. 12.4 all elements are alike as far as cross-sectional data are concerned. Then we have to provide the numerical values of the cross-sectional area, the two moments of inertia, and the torsional constant. The remainder of the elemental information consists of a partitioned column matrix $\mathbf{j}$, where the mth partition lists the node labels of the vertices of the mth element, that is,

$$\mathbf{j}^T = [\mathbf{j}^{1T} \quad \mathbf{j}^{2T} \quad \mathbf{j}^{3T} \quad \mathbf{j}^{4T} \quad \mathbf{j}^{5T} \quad \mathbf{j}^{6T} \quad \mathbf{j}^{7T} \quad \mathbf{j}^{8T}],$$

and from the key sketch, for example, $\mathbf{j}^{4T} = [3 \quad 2]$. Note that matrices $\mathbf{c}$ and $\mathbf{j}$, including the cross-sectional data, complete the definition of the geometry.

For example, if we are interested in the length L of an element, say the mth one, we simply write

$$L^m = \|\mathbf{c}_{j_2^m} - \mathbf{c}_{j_1^m}\| = \sqrt{\sum_{\alpha=1}^{3} (c_{j_2^m \alpha} - c_{j_1^m \alpha})^2} \tag{12.82}$$

Suppose we would like to compute the length of the fourth element. Since $j_2^4 = 2$ and $j_1^4 = 3$ and $\mathbf{c}_2^T = [-6 \quad 0 \quad 8]$ and $\mathbf{c}_3^T = [8 \quad -5 \quad 4]$.

$$L^4 = \|\mathbf{c}_2 - \mathbf{c}_1\| = \sqrt{(-6-8)^2 + (0+5)^2 + (8-4)^2} = 15.3948$$

In this way we can compute all pertinent information related to the element geometry.

12.2.2　Local coordinate systems. It is very important that the arrays defining the geometry of the structure be sufficient for the computation of the direction cosines of local coordinate axes in the overall coordinate system. The local coordinate systems play an important role in describing the forces acting on the elements and the thermal gradients. A local coordinate system is uniquely determined in the overall coordinate system by the coordinates of its origin and the direction cosines of its axes. The direction cosines can be arranged to define the **matrix of direction cosines,** which is a third-order square matrix in which the first, second, and third columns contain the direction cosines of the first, second, and third local axes, respectively. We have already indicated that in beam-type elements the first local axis is taken in the direction of the element axis and the other two local axes are taken as the principal axes of the cross section at the local origin.

Before we get into the details of computation of the direction cosines of local axes, let us recall the handling of three-dimensional vector operations in terms of the vector components in cartesian coordinates.   Let the column matrix $\mathbf{c}_i$ denote the cartesian description of a three-dimensional vector related to a certain point i:

$$\mathbf{c}_i^T = [c_{i1} \quad c_{i2} \quad c_{i3}] \tag{12.83}$$

Let L_c denote the length of this vector.   As in Eq. (12.82), we may express the length of the vector in terms of its cartesian components as

$$L_c = (c_{i1}^2 + c_{i2}^2 + c_{i3}^2)^{1/2} \tag{12.84}$$

Let $\hat{\mathbf{c}}_i$ denote the column matrix listing the components of the unit vector in the direction of the given vector.   For $\hat{\mathbf{c}}_i$, we write

$$\hat{\mathbf{c}}_i^T = \frac{\mathbf{c}_i^T}{L_c} = \left[\frac{c_{i1}}{L_c} \quad \frac{c_{i2}}{L_c} \quad \frac{c_{i3}}{L_c} \right] \tag{12.85}$$

where the components listed in $\hat{\mathbf{c}}_i$ are the **direction cosines** of the vector described by $\mathbf{c}_i$.

Let $\mathbf{d}_i$ denote the cartesian description of another vector related to point i:

$$\mathbf{d}_i^T = [d_{i1} \quad d_{i2} \quad d_{i3}] \tag{12.86}$$

Denoting the length of this vector by L_d and the cartesian description of its unit vector by $\hat{\mathbf{d}}_i$, similar to Eqs. (12.84) and (12.85), we write

$$L_d = (d_{i1}^2 + d_{i2}^2 + d_{i3}^2)^{1/2} \tag{12.87}$$

and

$$\hat{\mathbf{d}}_i = \frac{\mathbf{d}_i}{L_d} \tag{12.88}$$

Let d_{ic} denote the component of the vector described by $\mathbf{d}_i$ in the direction of the vector described by $\mathbf{c}_i$ and θ denote the angle between these two vectors.   We can easily verify that

$$d_{ic} = \hat{\mathbf{c}}_i^T \mathbf{d}_i \tag{12.89}$$

and

$$\cos \theta = \hat{\mathbf{c}}_i^T \hat{\mathbf{d}}_i \tag{12.90}$$

Let $\mathbf{e}_i$ denote the description of the vector which is the cross product (or vector product) of the vectors described by $\mathbf{c}_i$ and $\mathbf{d}_i$.   It can be shown that

$$\mathbf{e}_i = \tilde{\mathbf{c}}_i \mathbf{d}_i \tag{12.91}$$

where

$$\tilde{\mathbf{c}}_i = \begin{bmatrix} 0 & -c_{i3} & c_{i2} \\ c_{i3} & 0 & -c_{i1} \\ -c_{i2} & c_{i1} & 0 \end{bmatrix} \tag{12.92}$$

Note that $\tilde{\mathbf{c}}_i$ is a third-order skew-symmetric matrix.   Since its determinant is zero, it has no inverse and it has the following property

$$\tilde{\mathbf{c}}_i^T = -\tilde{\mathbf{c}}_i \tag{12.93}$$

which can easily be verified by Eq. (12.92).   Note that in order to preserve the strong relationship between the components of $\mathbf{c}_i$ and $\tilde{\mathbf{c}}_i$, we are using a lowercase

Latin letter with the tilde for the designation of the third-order square matrix $\tilde{c}_i$. In a way, the tilde tells us that the components of the multiplier vector in a vector product should be arranged in accordance with Eq. (12.92) if we want to obtain the cartesian description of the vector product as the result of a matrix multiplied by a column, as shown in Eq. (12.91). Since in a vector product the sign changes when we interchange the multiplier with the multiplicand, the reader may verify that

$$\tilde{d}_i c_i = -\tilde{c}_i d_i \tag{12.94}$$

Now let us obtain the direction cosines of the axes of the local coordinate system of element 4 of the structure in Fig. 12.4. The origin of this coordinate system is located at the first vertex of element 4, that is, at node 3. Let the third-order matrix L denote the matrix of direction cosines. To indicate that this matrix belongs to the local coordinate system at the first vertex of element 4, we could label this matrix as L_1^4. However, at this time, let us suppress the subscript and superscript. Since the first local axis is coincident with the element axis and heads from vertex 1 to vertex 2, observing that $j_1^4 = 3$ and $j_2^4 = 2$, we write the first column of L as

$$l_1 = (c_2 - c_3)/\|c_2 - c_3\| \tag{a}$$

noting that direction cosines are components of the unit vector. In order to obtain the direction cosines of the second local axis, we need, as additional element data, the overall coordinates of a point on the positive side of the second local axis. Let c_p denote the description of the position vector of such a point. Then

$$l_2 = (c_p - c_3)/\|c_p - c_3\| \tag{b}$$

which is the second column of L. Since the coordinate axes are mutually orthogonal and the cosine of the angle between these axes is zero, we obtain

$$l_1^T l_2 = 0 \tag{c}$$

as a result of Eq. (12.90). The direction cosines of the third local axis can now be obtained as the description of the vector product of the unit vectors on the first and the second axes:

$$l_3 = \tilde{l}_1 l_2 \tag{d}$$

The matrix of the direction cosines can now be displayed as

$$L = [l_1 \quad l_2 \quad l_3] \tag{12.95}$$

Since the unit vectors on the coordinate axes are mutually orthogonal, we have, utilizing the properties of the Kronecker delta δ,

$$l_\alpha^T l_\beta = \delta_{\alpha\beta} \tag{12.96}$$

implying that

$$L^T L = I$$

or

$$L^{-1} = L^T \tag{12.97}$$

In words, we state that the matrix of direction cosines $\mathbf{L}$ is an **orthogonal matrix.** The numerical value of $\mathbf{L}$ for element 4 can be obtained by using $\mathbf{c}_2^T = [-6 \quad 0 \quad 8]$, $\mathbf{c}_3^T = [8 \quad -5 \quad 4]$, $\mathbf{c}_p^T = [13 \quad 9 \quad 4]$, and Eqs. (a), (b), and (d) as

$$\mathbf{L} = \begin{bmatrix} -0.9094 & 0.3363 & -0.2447 \\ 0.3248 & 0.9417 & 0.0874 \\ 0.2598 & 0.0000 & -0.9657 \end{bmatrix} \qquad (e)$$

The reader may verify this. Note that the overall coordinates of an auxiliary point on the second local axis, in the present case [13   9   4], should be part of the elemental information. It is possible to reduce the auxiliary information to a single scalar. However, we shall not dwell on such special schemes here.

The type of structures we have considered so far can be assembled from structural elements which are line segments in their idealized form. Recalling the Euler-Navier assumptions about the behavior of such structural elements, we observe that in order to describe the behavior of the assembled structure all we have to know is the behavior of the particles located at the vertices of each element. That is, a finite amount of information suffices to describe the behavior of the whole structure. Such structures are called **discrete structures** and are the main concern of ordinary structural analysis. All trusses and frames fall into the category of discrete structures. Other structures, however, require infinite information to describe their behavior fully. Such structures include plates, shells, and solids. They are called the **structures of two- and three-dimensional continua.**

12.2.3 The finite elements in a continuum. Thanks to the use of digital computers in structural analysis, and the efforts of structural engineers, we now have reliable approximate methods to handle continuum structures as if they were discrete structures. These methods are usually referred to as the **finite-element methods.** They are, in a way, based on simplifying assumptions analogous to the plane-sections-remain-plane or normals-remain-normal assumptions, which may be stated as follows: *the behavior of material particles in a sufficiently small neighborhood may be approximated by the behavior of a few.* Of course, the smaller the neighborhood the better the approximation.

Consider, for example, a floor slab in a building. Since the dimension in the thickness direction is much smaller than in the other two, such a structure can be treated as a transversely loaded plate where the normals-remain-normal after deformation assumption is valid. Now suppose that we are interested in experimentally obtaining the stress on the upper face of the slab due to a given load. We can place strain-gage rosettes with a certain pattern on the upper face, load the slab, and make the measurements. With proper calibration, these measurements yield the stresses at the points where the strain gages were attached. After the experiment if we are asked to give the stresses at a point where there was no strain gage, we can use the results of three strain-gage rosettes which are closest to the point and contain it in the triangle established by the strain gages. The required stresses can be obtained approximately by linear interpolation. If the triangle containing the point is small, we are confident that the error of interpolation is negligibly small. This is the theory of finite-element methods. We place a mesh to cover the material

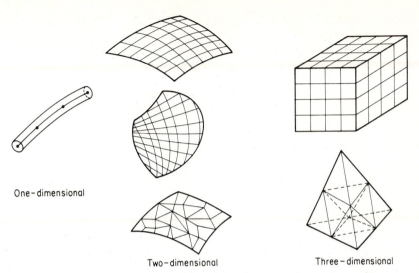

Figure 12.5 Various finite-element meshes

volume of the structure and pay attention to the behavior of the particles at the mesh points only. Whenever we are interested in the behavior of a particle which is not on a mesh point, we express its behavior by interpolation in terms of those of the nearby mesh points.

In thin structures, such as plates and shells, we usually use triangular or quadrilateral mesh elements to cover the middle surface. In solid structures we may use tetrahedral or hexahedral mesh elements to make up the material volume. Each mesh element is called a **finite element,** and the mesh points are called nodes. The points of a finite element which are coincident with the nodes are called vertices. The reason for using simple geometrical figures as finite elements is to simplify their definition. Straight lines and planes are easier to define than curves and surfaces. To prevent ambiguity during interpolation, finite elements are always taken as convex figures without overlaps or gaps. In Fig. 12.5 several finite elements and finite-element meshes are shown.

By introducing a finite-element mesh to represent the material volume of the structures of two- and three-dimensional continua, we can proceed to describe their geometry in exactly the same way discussed earlier for the structures composed of bars and beams. The only difference here is that the elements have more than two vertices. We must have an additional rule to order the vertices. In line elements we choose either one of the two vertices as the first vertex, and the remaining automatically becomes the second vertex. For more complicated elements, we may also choose any one of the vertices as the first vertex and prescribe a rule for determining the second, third, etc., vertices, if any, of the element. For example, in thin structures, we first select the general direction of the middle-surface normals by imagining an arrowhead on one of the line segments representing thickness and assuming that all the other thickness line segments are directed similarly. Then given a triangular, or quadrilateral element, we select any one of the vertices as the

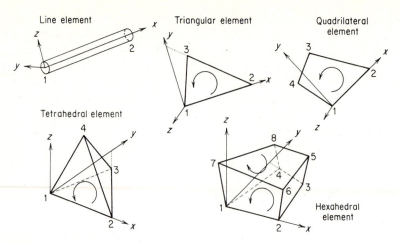

Figure 12.6   Vertex ordering and local-coordinate-system selection in various finite elements

first vertex, and label the other vertices by going around the element in counter-clockwise direction about the normal with the right-hand rule.   For tetrahedral and hexahedral elements, we may pretend to stand inside the elements.   We may select any one of the vertices of the face on which we are standing as the first vertex and order the other vertices of this face with the right-hand rule about the vector joining our feet to our head.   If the element is a tetrahedron, the last vertex becomes the fourth one.   If the element is an hexahedral element, we may take the vertex space diagonally across the first vertex as the fifth vertex and order again the remaining vertices with the right-hand rule.   Once we have a definite rule for ordering the vertices, it is a comparatively easy matter to select a local coordinate system for the element.   For example we may choose the origin of the local coordinate system as the first vertex.   The local x axis may be taken as the line joining the first vertex to the second vertex.   Let $\mathbf{l}_1$ denote the description of the unit vector on local x axis.   We describe the unit vector on the local z axis as $\mathbf{l}_3 = \tilde{\mathbf{l}}_1 \mathbf{c}^\circ / \|\tilde{\mathbf{l}}_1 \mathbf{c}^\circ\|$, where $\mathbf{c}^\circ$ is the description of the vector joining the first vertex to the third vertex.   The description of the unit vector on the local y axis can be obtained from $\mathbf{l}_2 = \tilde{\mathbf{l}}_3 \mathbf{l}_1$.   To complete the definition of geometry in thin structures, for each triangular and quadrilateral element we should also give the thickness of the element.   In Fig. 12.6 the vertex ordering and the selection of element local coordinate system as described here are shown for triangular, quadrilateral, tetrahedral, and hexahedral elements.

It is obvious from the foregoing discussions that we can span the material volume of any structure by establishing a finite-element mesh using line segments, triangular, quadrilaterial, tetrahedral, and hexahedral elements and define the geometry of the material volume by reference to this mesh   As detailed earlier for the space frame of Fig. 12.4, after establishing the finite-element mesh, we can select a suitable overall coordinate system and label the nodes and the finite elements.   With these labels we can generate a partitioned matrix $\mathbf{c}$ containing the overall coordinates of the nodes, where c_i contains the overall coordinates of node i, and a partitioned matrix $\mathbf{j}$ containing the node labels of the vertices of the elements, where $\mathbf{j}^m$ lists the node labels of the vertices of the mth element. In addition to matrices $\mathbf{c}$ and $\mathbf{j}$, we have to provide the area, the moments of inertia about the principal axes, the torsional constant and the coordinates of an auxiliary point on the local y axis for the cross sections of line elements, and the thicknesses for the triangular and quadrilateral elements in order to complete the description of the geometry in the form of numerical arrays.

When curvatures exist in the geometry of a structural element, we usually decrease the size of the element so that the effect of curvature becomes negligible. This will be our policy in the present treatment. However, at the expense of handling additional data, it is of course possible to carry the actual curvatures and their rates of change throughout the solution.

Given any structure, the selection of node points is very important. We usually locate the nodes with the objective of creating zones of uniformity in the material volume. Furthermore, when there is a geometric singularity, we identify it with a node. That is why, in trusses and frames, all joints are natural nodes. Likewise, load singularities, e.g., point loads, and point supports are also desirable node locations. Other than these, the **density of nodes,** i.e., the number of nodes per unit volume of material, is a function of the rate of change of geometry, material, and loading with the space. A rapid change should imply a high node density. The same applies to internal quantities such as stresses, strains, and displacements. If we have a priori knowledge about their variations in the material volume of the structure, we can adjust the nodal densities to their variations. For example, in general, we can expect rapid variations in internal quantities in the vicinity of load and geometrical singularities.

The measuring units of geometrical quantities are in length units. For example, distances are measured in length units, areas in length^2, volumes in length^3,

Table 12.1   Basic Geometric Information of Line Elements

	Planar truss	Planar frame	Planar gridwork	Space truss	Space frame
Node label, first vertex	×	×	×	×	×
Second vertex	×	×	×	×	×
Cross-sectional area	×	×	×	×	×
Cross-sectional moment of inertia, about local y axis			×		×
About local z axis		×			×
Torsional constant of cross section			×		×
Additional data to determine local y axis					×

moments of inertia in length^4, and small angles in units length per length. We may choose for length units from any suitable system; however, we must be consistent throughout the analysis problem. If we choose meters to express the span of a structure, we must continue to use them for other geometrical quantities, e.g.,use m^2 for areas, m^3 for volumes, m^4 for moments of inertia, etc.

In Table 12.1 we summarize the elemental information for uniform bar- and beam-type structural elements, and in Table 12.2 the elemental information for the finite elements of two- and three-dimensional continua are given. Only the elemental information pertaining to the geometrical definitions is included in these tables.

Table 12.2   Basic Geometric Information of Elements of Two- and Three-dimensional Continua

	Triangular element	Quadrilateral element	Tetrahedral element	Hexahedral element
Node label, first vertex	×	×	×	×
Second vertex	×	×	×	×
Third vertex	×	×	×	×
Fourth vertex		×	×	×
Fifth vertex				×
Sixth vertex				×
Seventh vertex				×
Eighth vertex				×
Thickness	×	×		

12.2.4   Problems for Solution

*Problem 12.7   Give explicit expressions for the column matrices **j** and **c** of the structure in Fig. 12.4.

*Problem 12.8   What is the length of element 8 in Fig. 12.4?

Problem 12.9   Referring to Fig. 12.4, compute the matrix of direction cosines of the local coordinate system of element 8. Assume that the overall coordinates of the auxiliary point for local y axis are $[-3 \quad -16 \quad 0]$, and the first vertex is at node 5.

Problem 12.10   A triangular face of a geodesic dome is defined by nodes 15, 45, and 3, which are coincident with the first, second, and third vertices, respectively. The descriptions of the position vectors of these nodes, in the overall coordinate system, are $\mathbf{c}_{15}^T = [10 \quad 10 \quad 25]$, $\mathbf{c}_{45}^T = [18.16 \quad 14.08 \quad 29.08]$, and $\mathbf{c}_3^T = [14.08 \quad 5.92 \quad 33.16]$. Compute the matrix of direction cosines of the local axes of the triangular element.

Problem 12.11   Compute the angles of the triangle defined in Prob. 12.10.

* Problem 12.12   What is the area of the triangular face defined in Prob. 12.10?

Problem 12.13 In order to quantify the topology of the element connections in an N-node structure, we can define an Nth-order square matrix $\mathscr{N}$ as

$$\mathscr{N} = \sum_{m=1}^{M} \sum_{k=1}^{n^m} \sum_{l=1}^{n^m} \mathbf{i}_{j_k^m} \mathbf{i}_{j_l^m}^T \tag{12.98}$$

where M = number of elements

$\quad n^m$ = number of vertices in mth element

$\quad \mathbf{i}_i$ = ith column of the Nth-order identity matrix $\mathbf{I}$.

Show that:

(a) $\mathscr{N}$ is symmetrical.

(b) The value of n_{ij}, $i \neq j$, is the number of elements containing nodal line ij.

(c) The value of n_{ii} is the number of elements meeting at node i.

When matrix $\mathscr{N}$ is generated as a binary matrix to distinguish its zero entries from the nonzero ones, it is called the **connectivity matrix of the whole structure.**

12.3 Material. In describing a structure, in addition to the description of the geometry of material volume, the quantitative description of the material itself is also required. The quantitative description of the material includes the numerical definition of its unit mass, its thermal-expansion coefficients, and its stress-strain relations, i.e., its constitutive law. In general, the properties of a material may change from material particle to material particle, and at a material point they may change from direction to direction. If the properties are not changing from point to point, the material is called **homogeneous;** otherwise it is called **nonhomogeneous.** If the material properties at a point are not dependent on direction, the material is called **isotropic;** otherwise it is called **anisotropic.** If an anisotropic material exhibits one or two planes of symmetry in its behavior, it is called **orthotropic;** otherwise it is called a **general anisotropic** material. The material properties may be functions of time and temperature, and depending upon the case, we talk about **time-dependent, time-independent, temperature-dependent,** or **temperature-independent** materials. If the temperature in the material is a function of its stresses, the material is called **thermomechanically coupled;** otherwise it is called **thermomechanically uncoupled.** If the material particles assume their original configuration when the loads are removed, the material is called **elastic;** otherwise it is called **inelastic.** If the stresses in the material are linear functions of the strains, the material is called **linear;** otherwise it is called **nonlinear.** The methods of systematic structural analysis enable us to solve the structural-analysis problems involving materials with any one or all of these qualifiers. We shall confine ourselves, however, to materials which are thermomechanically uncoupled, time- and temperature-independent, linear elastic, and homogeneous in a given structural element. This subclass of materials to which we confine ourselves will be referred simply as the material, without qualifiers. The qualifiers will be invoked when we need to make a point about them. Since a structure may consist of many elements, each of which can be of different material, as far as the totality of the structure is concerned, the material may be piecewise homogeneous, i.e., homogeneous in an element but nonhomogeneous as we go from one element to another.

12.3.1 Quantification of mechanical properties. In our earlier studies of strength of materials, we found that we need four constants to describe a linear isotropic material: the unit mass ρ, the thermal expansion coefficient α, Young's modulus E, and the shear modulus G. We need ρ for the computation of unit weight, α for the computation of thermal strains, and E and G for the computation of normal and shearing stresses when lineal and angular strains are known. Let

$$\lambda = \text{earth's gravitational acceleration}$$
$$\Delta t = \text{temperature increase}$$
$$g = \text{unit weight}$$
$$\epsilon = \text{lineal strain}$$
$$\gamma = \text{angular strain}$$
$$\sigma = \text{normal stress}$$
$$\tau = \text{shearing stress}$$
$$\epsilon_0 = \text{thermal strain}$$

Then we can write

$$g = \lambda\rho \tag{12.99}$$

$$\epsilon_0 = \alpha\,\Delta t \tag{12.100}$$

$$\sigma = E(\epsilon - \epsilon_0) \tag{12.101}$$

$$\tau = G\gamma \tag{12.101'}$$

These are the familiar relationships used in beam- and bar-type structural elements where there is only one lineal direction which is of importance, i.e., the axial direction, and where we are usually interested in only the change of right angles with the element axis. Actually, at a material particle there are three lineal strains, ϵ_x, ϵ_y, ϵ_z, in the directions of three mutually orthogonal axes, x, y, z, and three angular strains, γ_{xy}, γ_{xz}, γ_{yz}, related to the three-right angles (see Fig. 12.8). Likewise, at this material particle we can define three normal stresses, σ_x, σ_y, σ_z, and three shear stresses, τ_{xy}, τ_{xz}, τ_{yz} (Fig. 12.8). Equations (12.99) to (12.101) therefore represent a very special case and are insufficient for the general case.

The general case is of course the one represented by **Hooke's law** or its inverse. In general, in order to describe the material properties without ambiguity, we select a reference coordinate system. The local coordinate system of a structural element is usually taken as the **material coordinate system** in which the material is described. In Fig. 12.7, the preferred orientations of coordinate systems relative to the structural elements are shown.

In systematic structural analysis, we usually list the independent components of the stresses at a point as a column matrix like

$$\boldsymbol{\sigma}^T = [\sigma_x \quad \sigma_y \quad \sigma_z \quad \tau_{xy} \quad \tau_{xz} \quad \tau_{yz}] \tag{12.102}$$

We call $\boldsymbol{\sigma}$ the **stress column matrix** for the state of stress at the material particle. Likewise, we list the independent components of the strains at the material particle as a column, matrix

$$\boldsymbol{\epsilon}^T = [\epsilon_x \quad \epsilon_y \quad \epsilon_z \quad \gamma_{xy} \quad \gamma_{xz} \quad \gamma_{yz}] \tag{12.103}$$

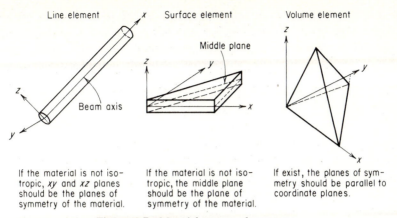

Line element   Surface element   Volume element

Middle plane

Beam axis

If the material is not iso-   If the material is not iso-   If exist, the planes of sym-
tropic, xy and xz planes   tropic, the middle plane   metry should be parallel to
should be the planes of   should be the plane of   coordinate planes.
symmetry of the material.   symmetry of the material.

Figure 12.7   Material axes vs. element axes

and call it the **strain column matrix** for the state of strain at the material particle. Note that the ordering of the components of $\boldsymbol{\sigma}$ and $\boldsymbol{\epsilon}$ is such that the corresponding components carry the same subscripts. Denoting the components of the displacement vector of a material particle by u, v, and w, using Fig. 12.8 and the basic definitions for lineal and angular strains, we can write

$$\boldsymbol{\epsilon}^T = [u_{,x} \quad v_{,y} \quad w_{,z} \quad u_{,y} + v_{,x} \quad u_{,z} + w_{,x} \quad v_{,z} + w_{,y}] \qquad (12.104)$$

where a comma in the subscript indicates partial differentiation with respect to the quantity following it.

If the unit cube in Fig. 12.8 is subjected to a temperature increase of Δt, the **induced-thermal-strains column matrix** $\boldsymbol{\epsilon}_0$ can be expressed as

$$\boldsymbol{\epsilon}_0^T = \Delta t [\alpha_x \quad \alpha_y \quad \alpha_z \quad \alpha_{xy} \quad \alpha_{xz} \quad \alpha_{yz}] \qquad (12.105)$$

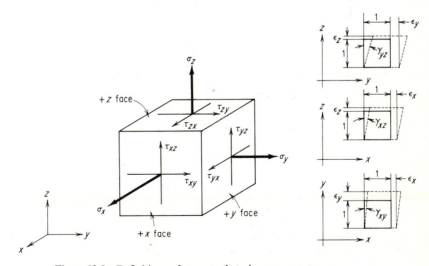

Figure 12.8   Definitions of stress and strain components

where the first three entries are the thermal-expansion coefficients in the x, y, z directions and the last three entries are the thermal coefficients for angular strains. For isotropic materials $\alpha_x = \alpha_y = \alpha_z = \alpha$ and $\alpha_{xy} = \alpha_{xz} = \alpha_{yz} = 0$.

The relationship between the matrices describing the state of stress and the state of strain at a material particle is called the **constitutive law** of the material. Since we are confined to linear elastic materials, the most general linear relationship may be stated as

$$\epsilon = \mathscr{C}\sigma + \epsilon_0 \tag{12.106}$$

where $\mathscr{C}$ is a symmetric positive definite matrix of order 6, called the **compliance matrix** of the material. Equation (12.106) is nothing but the **generalized Hooke's law**. The elements of the compliance matrix can be measured experimentally in the laboratory. When the material is isotropic, it can be displayed as

$$\mathscr{C} = \frac{1}{E} \begin{bmatrix} 1 & -v & -v & 0 & 0 & 0 \\ & 1 & -v & 0 & 0 & 0 \\ & & 1 & 0 & 0 & 0 \\ & & & 2(1+v) & 0 & 0 \\ & \text{Symmetric} & & & 2(1+v) & 0 \\ & & & & & 2(1+v) \end{bmatrix} \tag{12.107}$$

where E is **Young's modulus** and v is **Poisson's ratio**. We may recall that E and v are related to the **shear modulus** G

$$G = \frac{E}{2(1+v)} \tag{12.108}$$

The inverse of the compliance matrix is called the **material matrix** $\mathscr{D}$:

$$\mathscr{C}^{-1} = \mathscr{D} \tag{12.109}$$

For isotropic materials, $\mathscr{D}$ can be expressed, by inverting $\mathscr{C}$ from Eq. (12.107), as

$$\mathscr{D} = \frac{E}{(1+v)(1-2v)} \begin{bmatrix} 1-v & v & v & 0 & 0 & 0 \\ & 1-v & v & 0 & 0 & 0 \\ & & 1-v & 0 & 0 & 0 \\ & & & \dfrac{1-2v}{2} & 0 & 0 \\ & \text{Symmetric} & & & \dfrac{1-2v}{2} & 0 \\ & & & & & \dfrac{1-2v}{2} \end{bmatrix} \tag{12.110}$$

Note that when the material is isotropic, the 21 independent components of the material matrix (or the compliance matrix) are expressible in terms of two constants, for example, E and v. From Eq. (12.106) we can solve for $\boldsymbol{\sigma}$ to obtain

$$\boldsymbol{\sigma} = \mathscr{D}\boldsymbol{\epsilon} - \mathscr{D}\boldsymbol{\epsilon}_0 \qquad (12.111)$$

where we have made use of Eq. (12.109).

In Table 12.3 the number of constants to describe the mechanical and thermo-mechanical properties of various elastic materials are given. In describing a structure by arrays of numbers, for each structural element we should list its material constants. Since most engineering structures contain a small number of different materials, we usually list the material constants with type numbers, and in the description of the structural element we use only the material type number of its material, instead of a long list of constants.

During the discussion of Eq. (12.106), we stated that the compliance matrix is symmetric and positive definite. Naturally, the inverse of the compliance matrix, i.e., the material matrix, is also symmetric and positive definite, since these properties of square matrices are not lost during inversion. Symmetry and positive definiteness are very favorable properties, and their existence in structural-analysis problems has its roots in the fact that the material matrix carries these properties. For further information about elastic materials, see Ref. 2.

The stresses are measured in force per length^2 units. Strains are dimensionless quantities which may be considered to be in units of length per length. The entries of material matrix $\mathscr{D}$ are measured in units of force per length^2, and those of $\mathscr{C}$ are measured in units of length^2 per force. The units of the thermal expansion

Table 12.3 Material Constants for Various Elastic Materials

Type of material	Unit mass	Thermal-expansion coefficients	Independent constants defining material matrix	Total number of material constants
Isotropic	ρ	α	E, G	4
Orthotropic with symmetries about xy and xz planes	ρ	$\alpha_x, \alpha_y, \alpha_z$	$d_{11}, d_{12}, d_{13}, d_{22}, d_{23}, d_{33}, d_{44},$ d_{55}, d_{66}	13
Orthotropic with symmetry about xy planes	ρ	$\alpha_x, \alpha_y, \alpha_z, \alpha_{xy}$	$d_{11}, d_{12}, d_{13}, d_{14}, d_{22}, d_{23}, d_{24},$ $d_{33}, d_{34}, d_{44}, d_{55}, d_{56}, d_{66}$	18
Anisotropic	ρ	$\alpha_x, \alpha_y, \alpha_z, \alpha_{xy}, \alpha_{xz}, \alpha_{yz}$	$d_{11}, d_{12}, d_{13}, d_{14}, d_{15}, d_{16}, d_{22},$ $d_{23}, d_{24}, d_{25}, d_{26}, d_{33}, d_{34}, d_{35},$ $d_{36}, d_{44}, d_{45}, d_{46}, d_{55}, d_{56}, d_{66}$	28

coefficients are (length per length) per degrees of temperature. The units of the unit mass are

$$\text{Force} \Big/ \left(\frac{\text{length}}{\text{time}^2} \times \text{length}^3\right)$$

and those of the accelerations are length per time^2. We must be consistent in length, temperature, force, and time units throughout the systematic analysis.

12.3.2 Problems for Solution

Problem 12.14   Stresses at a point of a material are given as

$$\sigma^T = [-1{,}200 \quad 400 \quad -900 \quad 125 \quad -300 \quad -1{,}200]$$

in a local coordinate system (x, y, z). Sketch the directions on a unit cube where the faces are parallel to the coordinate planes.

Problem 12.15   Strains at a material particle are given as

$$\epsilon^T = 10^{-3}[-1 \quad 2 \quad -3 \quad -2 \quad 1 \quad -4]$$

in a local coordinate system (x, y, z). Sketch the deformed configuration of a unit cube where the undeformed faces are parallel to the coordinate planes.

Problem 12.16   Display the complete compliance matrix of an isotropic material where Young's modulus is $E = 8 \times 10^6$ and the shear modulus is $G = 3 \times 10^6$.

Problem 12.17   Show that the compliance matrix of an isotropic elastic material is positive definite only if $\nu < 0.5$.

Problem 12.18   Obtain the material matrix of the material of Prob. 12.16 by first inverting the compliance matrix, and then using Eq. (12.110).

Problem 12.19   Consider the material matrix of an orthotropic material with one plane of symmetry, as given in Table 12.3.   If one is given that $d_{55} = 6 \times 10^6, d_{66} = 4 \times 10^6, d_{56} = 5 \times 10^6$, show that the material matrix is not positive definite and therefore such a material cannot exist.

*Problem 12.20   The **spherical stress** and the **spherical strain** at a point are defined as $\sigma_m = (\sigma_x + \sigma_y + \sigma_z)/3$ and $\epsilon_m = (\epsilon_x + \epsilon_y + \epsilon_z)/3$, respectively.   If one writes $\sigma_m = 3K\epsilon_m$, K is called **bulk modulus.**   What is the value of the bulk modulus of the material in Prob. 12.16?

12.4 Deflection Boundary Conditions. Given a structure, we usually have a priori information about the displacements or rotations of some of the material particles, such as those on the support points.   In fact, we reserve the term **structure** for bodies where the rigid-body movement is prevented, which means that we have at least statically determinate supports.   Therefore the definition of a structure is not complete unless its support conditions are clearly stated.   In this section we discuss how we can express the support conditions of structures in the form of numerical arrays, since in the systematic analysis of structures we can no longer use the geometrical symbolism of the ordinary structural analysis for the supports.

12.4.1 Deflection degrees of freedom at a point. Let us consider a material particle located at a node point, say i.   When the structure is loaded, the particle will move to another point.   The vector which joins node i to the new location of the material particle is called the **displacement vector.** Suppose we attached a certain line segment to the particle before the loads are applied.   The line segment will assume a different orientation when the particle moves to its new location. The angle between the original orientation of the line segment and its final position

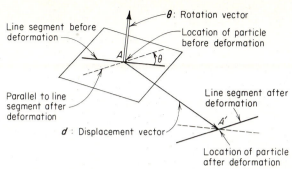

Figure 12.9   Displacement and rotation vectors of a particle

is called **rotation.** Since rotations are very small angles, we can represent a rotation with a vector which is perpendicular to the plane established by the original and final orientations of the line segment. This vector is called the **rotation vector** of the particle. The magnitude of the rotation vector is the rotation, and its direction is such that it defines the actual rotation direction by the right-hand rule (see Fig. 12.9).

Let $\mathbf{d}_i$ and $\boldsymbol{\theta}_i$ denote the descriptions of the displacement and the rotation vectors of a material particle located at node i, respectively in a cartesian coordinate system. When we want to refer to both the displacement and the rotation of a particle, we shall use the term **deflection.** Let the column matrix $\mathbf{u}_i$ list the deflections of the particle at point i. We display $\mathbf{u}_i$ as

$$\mathbf{u}_i^T = [\mathbf{d}_i^T \quad \boldsymbol{\theta}_i^T] \tag{12.112}$$

which is of order 6, that is, three displacement components and three rotation components. Each one of the six components of $\mathbf{u}_i$ is called a **degree of freedom** of the particle at i.

Depending upon the type of the structure, some of the components of $\mathbf{u}_i$ may always be zero throughout the range of i. For example, in trusses the universal joints and the condition that the bars are loaded only at their vertices ensure that at every particle $\boldsymbol{\theta}$ is zero if we ignore the rigid-body rotations of the bars. In planar structures undergoing in-plane deformations, e.g., a planar frame loaded in its plane, the displacement vector is always in the plane of the structure, and the rotation vector, if it exists, is always perpendicular to the plane of the structure. In planar structures undergoing out-of-plane deformations, e.g., the gridwork in Fig. 12.10 loaded normal to its plane, the displacement vector is normal to the plane of the structure, and the rotation vector lies in the plane of the structure.

When certain components of $\mathbf{u}_i$ are identically zero for all i, we usually exclude them completely when we count the number of degrees of freedom at a particle point. Referring to the directions defined by the axes of a suitably selected overall coordinate system, we have the degrees of freedom at a point in various structures given in Table 12.4. From now on, we shall assume that $\mathbf{u}_i$ lists only the nonzero components. For example, if i is the label of a node of a planar frame placed in the overall xy plane, we display $\mathbf{u}_i$ as $\mathbf{u}_i^T = [d_{i1} \quad d_{i2} \quad \theta_{i3}]$. Similarly, the deflections of the ith node of a gridwork frame placed in the overall xy plane will be displayed

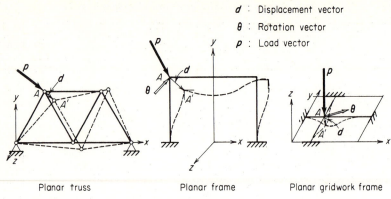

Figure 12.10 Deformations of planar trusses, frames, and gridworks

as $\mathbf{u}_i^T = [d_{i3} \quad \theta_{i1} \quad \theta_{i2}]$, and those of a space truss are $\mathbf{u}_i^T = [d_{i1} \quad d_{i2} \quad d_{i3}]$, etc. Using d and θ symbolism in these displays has the advantage of clearly indicating the exact type of deflection components involved. However, for purposes of systematizing the analysis procedures, it is often desirable to utilize just one symbol, u, to denote all deflection components, whether they are displacements or rotations. When one symbol is used for convenience in an analysis, the nature of the deflection components, i.e., whether they are displacements or rotations or both, should be identified at the beginning. In Part II, this identification is by means of Table 12.4.

12.4.2 Constraints on deflections. In describing the support conditions of a structure quantitatively, we use the same coordinate system which we referred to in describing the geometry. Let $\mathbf{u}$ denote a partitioned column matrix where the ith partition is $\mathbf{u}_i$, that is, the subcolumn matrix listing the deflections of the ith node. If the total number of nodes in the structure is N and the number of degrees of freedom per node is e, then the number of components of $\mathbf{u}$ is Ne. We can obtain the value of e from Table 12.4, according to the type of the structure. For example, consider the planar truss shown in Fig. 12.11. From Table 12.4 we observe that $e = 2$, and from the figure we have $N = 8$. Therefore, $\mathbf{u}$ is of order $(8)(2) = 16$. We may display matrix $\mathbf{u}$ of this structure as: $\mathbf{u}^T = [u_{1,1} \quad u_{1,2} \quad \cdots \quad u_{8,2}]$. This

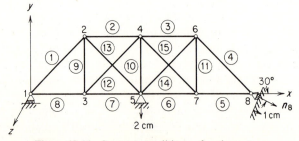

Figure 12.11 Support conditions of a planar truss

Table 12.4   Degrees of Freedom at a Point in Various Structures

	Displacements[†] along overall axes			Rotations[†] about overall axes			Total number of degrees of freedom at a point[‡]
	x	y	z	x	y	z	
Planar trusses	1	2					2
Space frameworks	1	2	3				3
Planar frames	1	2				3	3
Space frames	1	2	3	4	5	6	6
Gridwork frames			1	2	3		3
Plane-stress problems	1	2					2
Plane-strain problems	1	2					2
Plate-bending problems			1	2	3		3
Shells of revolution, membrane	1	2					2
Membrane and bending	1	2				3	3
General shells, membrane	1	2	3				3
Membrane and bending	1	2	3	4	5	6	6
Solids of revolution	1	2					2
General solids	1	2	3				3

† Numbers are the entry numbers of the component in column matrix $\mathbf{u}_i$.

‡ Order of the column matrix $\mathbf{u}_i$ listing the nonzero deflection components.

list shows us that there are 16 deflection components to fully describe the deformed configuration of the structure uniquely. However, because of the supports, we have a knowledge on some of these components. From the figure we observe that node 1 can move only parallel to x axis, node 5 cannot move in x direction and is subjected to a support settlement of 2 units in the negative y direction, and node 8 can move only parallel to the inclined support plane, which itself settles 1 unit in the direction of its normal.

The mathematical description of these support conditions is called **constraints on deflections.** These constraints of the truss in Fig. 12.11 can be written

$$
\begin{aligned}
u_{1,2} &= 0 \\
u_{5,1} &= 0 \\
u_{5,2} &= -2 \\
\mathbf{d}_8^T u_8 &= 1
\end{aligned} \qquad (a)
$$

where $\mathbf{d}_8$ is the column matrix listing the components of the displacement vector of node 8, n_8 is the column matrix listing the components of the unit normal vector of the inclined support plane at node 8, and the equation itself expresses the fact that the displacement component in the normal direction is equal to the settlement of the inclined plane. Using $\mathbf{d}_8^T = [u_{8,1} \quad u_{8,2}]$ and $n_8^T = [0.866 \quad -0.5]$, we rewrite Eqs. ($a$) as

$$
\begin{aligned}
u_{1,2} &= 0 \\
u_{5,1} &= 0 \\
u_{5,2} &= -2 \\
0.866 u_{8,1} &- 0.500 u_{8,2} = 1
\end{aligned}
\qquad (b)
$$

which are basically linear equations indicating the constraints imposed on the components of $\mathbf{u}$ by the supports. These equations may also be displayed as

$$
\begin{bmatrix}
\cdot & 1 & \cdot & \cdot & \cdot & \cdot & \cdot & \cdot & \cdot & \cdot & \cdot & \cdot & \cdot & \cdot & \cdot & \cdot \\
\cdot & \cdot & \cdot & \cdot & \cdot & \cdot & \cdot & \cdot & 1 & \cdot & \cdot & \cdot & \cdot & \cdot & \cdot & \cdot \\
\cdot & \cdot & \cdot & \cdot & \cdot & \cdot & \cdot & \cdot & \cdot & 1 & \cdot & \cdot & \cdot & \cdot & \cdot & \cdot \\
\cdot & \cdot & \cdot & \cdot & \cdot & \cdot & \cdot & \cdot & \cdot & \cdot & \cdot & \cdot & \cdot & \cdot & 0.866 & -0.5
\end{bmatrix}
\begin{Bmatrix}
u_{1,1} \\ u_{1,2} \\ u_{2,1} \\ u_{2,2} \\ \vdots \\ u_{8,1} \\ u_{8,2}
\end{Bmatrix}
=
\begin{Bmatrix}
\cdot \\ \cdot \\ -2 \\ 1
\end{Bmatrix}
\qquad (b')
$$

where the zero entries are shown by boldface dots. These constraint equations on deflections are also called the **deflection boundary conditions.**

For the general case the deflection boundary conditions can be expressed in the same form as Eqs. (b'), or

$$
\mathbf{H}\mathbf{u} = \mathbf{h}_0 \qquad (12.113)
$$

where $\mathbf{H}$ and $\mathbf{h}_0$ are known matrices. If the number of deflection constraints is denoted by b, $\mathbf{H}$ is a b by Ne matrix and $\mathbf{h}_0$ is a column matrix of order b. For the structure of Fig. 12.11, from the display in Eqs. (b'), we see that $\mathbf{H}$ is a 4 by 16 matrix and $\mathbf{h}_0$ is a column matrix of order 4. We also observe that these matrices contain very few nonzero entries.

12.4.3 Dependent and independent deflection components. Consider the truss example of Fig. 12.11 and its deflection constraints shown in Eqs. (b). We can explicitly express 4 of the 16 deflection components in terms of the remaining 12 components. For example, we see that we can express either $u_{8,1}$ or $u_{8,2}$ in terms of the other from the last of Eqs. (b). In numerical operations, in order to minimize the possible **round-off errors,** when we have options, we prefer to divide by numbers of larger magnitude. With this criterion, we choose to solve $u_{8,1}$ and write from Eqs. (b)

$$
\begin{aligned}
u_{1,2} &= 0 \\
u_{5,1} &= 0 \\
u_{5,2} &= -2 \\
u_{8,1} &= 0.577 u_{8,2} + 1.155
\end{aligned}
\qquad (c)
$$

or equivalently

$$
\begin{Bmatrix} u_{1,2} \\ u_{5,1} \\ u_{5,2} \\ u_{8,1} \end{Bmatrix}
=
\begin{bmatrix}
\cdot \;\; \cdot \;\; \cdot \;\; \cdot \;\; \cdot \;\; \cdot \;\; \cdot \;\; \cdot \;\; \cdot \;\; \cdot \;\; \cdot & \cdot \\
\cdot \;\; \cdot \;\; \cdot \;\; \cdot \;\; \cdot \;\; \cdot \;\; \cdot \;\; \cdot \;\; \cdot \;\; \cdot \;\; \cdot & \cdot \\
\cdot \;\; \cdot \;\; \cdot \;\; \cdot \;\; \cdot \;\; \cdot \;\; \cdot \;\; \cdot \;\; \cdot \;\; \cdot \;\; \cdot & \cdot \\
\cdot \;\; \cdot \;\; \cdot \;\; \cdot \;\; \cdot \;\; \cdot \;\; \cdot \;\; \cdot \;\; \cdot \;\; \cdot \;\; \cdot & 0.577
\end{bmatrix}
\begin{Bmatrix} u_{1,1} \\ u_{2,1} \\ \vdots \\ u_{4,2} \\ u_{6,1} \\ u_{6,2} \\ u_{7,1} \\ u_{7,2} \\ u_{8,2} \end{Bmatrix}
+
\begin{Bmatrix} \cdot \\ \cdot \\ -2 \\ 1.155 \end{Bmatrix}
\qquad (c')
$$

The deflection components appearing on the left side of these equations are called the **dependent deflection components,** and those appearing at the right are called the **independent deflection components.**

We can apply the above procedure in a more formal way to change the format of Eq. (12.113). If b denotes the number of constraint equations, obviously, $b < Ne$. We can solve b deflection components in terms of the others, from Eq. (12.113). Let $\mathbf{u}_{(2)}$ denote the column matrix listing the b dependent deflection components, and let $\mathbf{u}_{(1)}$ denote the column matrix listing the $Ne\text{-}b$ independent deflection components. Let the columns of $\mathbf{H}$ associated with the components of $\mathbf{u}_{(1)}$ and $\mathbf{u}_{(2)}$ be denoted by $\mathbf{H}'_{(1)}$ and $\mathbf{H}_{(2)}$, respectively. With these, we can rewrite Eq. (12.113) as

$$\mathbf{H}'_{(1)}\mathbf{u}_{(1)} + \mathbf{H}_{(2)}\mathbf{u}_{(2)} = \mathbf{h}_0 \qquad (12.114)$$

Note that $\mathbf{H}_{(2)}$ is a square matrix of order b. It may have an inverse if its columns are linearly independent. Assuming that $\mathbf{H}_{(2)}^{-1}$ exists, we can solve $\mathbf{u}_{(2)}$ from Eq. (12.114):

$$\mathbf{u}_{(2)} = -\mathbf{H}_{(2)}^{-1}\mathbf{H}'_{(1)}\mathbf{u}_{(1)} + \mathbf{H}_{(2)}^{-1}\mathbf{h}_0 \qquad (12.115)$$

Defining

$$\mathbf{H}_{21} = -\mathbf{H}_{(2)}^{-1}\mathbf{H}'_{(1)} \qquad (12.116)$$

and

$$\mathbf{u}_{(20)} = \mathbf{H}_{(2)}^{-1}\mathbf{h}_0 \qquad (12.117)$$

we can rewrite Eq. (12.115) as

$$\mathbf{u}_{(2)} = \mathbf{H}_{21}\mathbf{u}_{(1)} + \mathbf{u}_{(20)} \qquad (12.118)$$

which is the alternate form of Eq. (12.113), and corresponds to the display in Eqs. (c'). In most of the analysis problems, $\mathbf{H}_{(2)}$ is almost a diagonal matrix, and therefore it is just as easy to write the deflection boundary conditions in the form of Eq. (12.118) [instead of Eq. (12.113)]. We already demonstrated this when we converted Eqs. (b) into Eqs. (c). In the more complicated cases, if the need arises, one can use the systematic procedure explained in Art. 16.3.1 to reduce Eqs. (12.113) into the format of Eqs. (12.118).

Since the dependent and independent components of the deflections listed in $\mathbf{u}$ are clearly identified, we always prefer to define the deflection boundary conditions in the format of Eq. (12.118).   How to transfer this information to the digital computer is discussed in the Art. 12.4.4.

12.4.4   The DBC input units.   As exemplified by Eqs. (c) for the truss problem, in actual structures, both matrices H_{21} and $u_{(20)}$ in Eq. (12.118) contain very few nonzero entries.   Such matrices, called **sparse matrices,** are defined economically by listing only the nonzero entries by their row and column designations.   For example, consider the matrix displayed in Eqs. (c').   We can define this matrix completely and uniquely if we say that it is zero throughout with only one nonzero element 0.577 at (4,12), where 4 is the row designation and 12 is the column designation of the nonzero element.   Referring to Eq. (12.118) we observe that the row designation of a nonzero element of H_{21} is the same as the row designation of the entry of $u_{(2)}$ which is on the same row.   This entry of $u_{(2)}$ can also be identified by its node and degree-of-freedom numbers, which are called the **index pair** of the component.   On the other hand, the column designation of the nonzero entry of H_{21} is the same as the row number of the component of $u_{(1)}$ which multiplies it.   This component of $u_{(1)}$ can also be identified by its node and degree-of-freedom numbers, i.e., with its index pair.   Therefore it may be more convenient to define the nonzero entry of H_{21} by two index pairs and a constant.   For example, the nonzero entry of the sparse matrix displayed in Eqs. (c') can be defined as

$$(8,1) \quad (8,2) \quad 0.577$$

instead of (4,12) and 0.577, as earlier.   Note that the **first index pair** (8,1) corresponds to row 4 and the **second index pair** (8,2) corresponds to column 12.   Also note that the first index pair is the index pair corresponding to the dependent deflection component and the second index pair is the index pair corresponding to the independent deflection component.   A set of numbers consisting of two index pairs and a constant constitute one **deflection-boundary-condition input unit** (DBC input unit).

We can also use DBC input units to define the entries of the $u_{(20)}$ column matrix. Here, although an entry can be defined by its row number and its value alone, for the sake of uniformity, we may still choose to define the entries of $u_{(20)}$ by two index pairs and a constant by repeating the first index pair twice.   For example, for the third entry of the displayed column on the right side of Eqs. (c'), we write

$$(5,2) \quad (5,2) \quad -2$$

Note that the first index pair (5,2) indicates the third row, and its repetition as the second index pair shows that an entry of $u_{(20)}$ is under question.   Since the index pairs of the nonzero entries of H_{21} are always different, identical first and second index pairs in a DBC input unit clearly show that the DBC input unit belongs to $u_{(20)}$ rather than H_{21}.   Another important point regarding the DBC input units of $u_{(20)}$ is that we have also to prepare DBC input units for those zero entries of $u_{(20)}$ associated with dependent deflection components which have zero rows in H_{21}, such as (1,2) and (5,1) entries of $u_{(20)}$ in the display in Eqs. (c').   This is

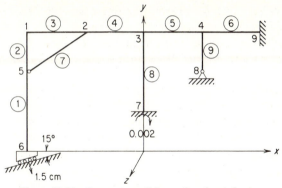

Figure 12.12   Support conditions of a planar frame

a convenient way of transmitting the deflection constraints of the type of first two equations of Eqs. (c) to the computer.

We may list the DBC input units, as defined here, in any order, since each DBC input unit is an independent and unambiguous piece of information. For example, the DBC input units corresponding to H_{21} and $u_{(20)}$ matrices of the truss example of Fig. 12.11 can be displayed as

$$
\begin{array}{ccc}
(1,2) & (1,2) & 0 \\
(5,1) & (5,1) & 0 \\
(5,2) & (5,2) & -2 \\
(8,1) & (8,1) & 1.155 \\
(8,1) & (8,2) & 0.577
\end{array}
\qquad (d)
$$

which can be verified from Eqs. (c) and (c′). These DBC input units conveniently and economically give a quantitative representation of the support conditions of the truss.

In Arts. 12.4.5 to 12.4.7 the use of DBC input units in various deflection constraint situations is illustrated by examples.

12.4.5 Planar frame example. Let us describe the support conditions of the planar frame shown in Fig. 12.12. The degrees of freedom at a node, say ith one, consist of $[d_{i1} \quad d_{i2} \quad \theta_{i3}]$, where d_{i1} and d_{i2} are the displacement components of the ith node in the directions of the overall x axis and overall y axis, respectively, and θ_{i3} is the rotation at node i about the overall z axis. The list of deflections of the whole structure is $u^T = [u_1^T \quad \cdots \quad u_9^T]$, where each partition is of order 3, namely, $u_i^T = [u_{i1} \quad u_{i2} \quad u_{i3}]$, $i = 1, 2, \ldots, 9$. It is understood that $u_{i1} = d_{i1}$, $u_{i2} = d_{i2}$, and $u_{i3} = \theta_{i3}$. Since the number of degrees of freedom per node is 3 and there are 9 nodes, the order of u is 27. Looking at the figure, we can write the support conditions as

$$
\begin{array}{ll}
u_{6,3} = \quad 0 & u_{8,1} = 0 \\
d_6^T n_6 = \quad 1.5 & u_{8,2} = 0 \\
u_{7,1} = \quad 0 & u_{9,1} = 0 \\
u_{7,2} = \quad 0 & u_{9,2} = 0 \\
u_{7,3} = -0.002 & u_{9,3} = 0
\end{array}
\qquad (e)
$$

Noting that $\mathbf{d}_6^T = [u_{6,1} \quad u_{6,2}]$ and from the figure $n_6^T = [0.259 \quad -0.966]$, we can write the second of Eqs. (e) as

$$0.259u_{6,1} - 0.966u_{6,2} = 1.5 \tag{f}$$

or, solving for $u_{6,2}$,

$$u_{6,2} = 0.268u_{6,1} - 1.553 \tag{g}$$

Replacing the second of Eqs. (e) by its equivalent in Eq. (g), we obtain the base equations from which the following DBC input units can readily be obtained:

$$
\begin{array}{lll}
(6,3) & (6,3) & 0 \\
(6,2) & (6,1) & 0.268 \\
(6,2) & (6,2) & -1.553 \\
(7,1) & (7,1) & 0 \\
(7,2) & (7,2) & 0 \\
(7,3) & (7,3) & -0.002 \\
(8,1) & (8,1) & 0 \\
(8,2) & (8,2) & 0 \\
(9,1) & (9,1) & 0 \\
(9,2) & (9,2) & 0 \\
(9,3) & (9,3) & 0
\end{array}
\tag{h}
$$

These are the support conditions in numerical form. The fact that the moment is zero at node 8 is not considered a deflection boundary condition. Likewise we do not treat the zero-moment condition at vertex on node 5 of element 5-2 as a deflection boundary condition. These do not tell us anything about the deflections at the nodes. We shall see the handling of such cases during our discussion of force boundary conditions in Art. 12.5.

12.4.6 Deflection behavior of symmetrical structures. We call a structure symmetrical with respect to a plane when its geometry, material, and support conditions are symmetrical with respect to the same plane. The solution of many linear problems of structural analysis of symmetrical structures can be greatly simplified by utilizing the following **symmetry-antisymmetry theorems:**

Theorem 1

Any loading of a structure symmetrical with respect to a plane can be decomposed into components symmetrical and antisymmetrical with respect to the same plane, as depicted in Fig. 12.13.

Theorem 2

A structure symmetrical with respect to a plane subjected to loading symmetrical with respect to the same plane undergoes deformations symmetrical with respect to the plane of symmetry, and the material particles in the plane of symmetry remain in the plane with rotations possible only about the normal of the plane (Fig. 12.14a).

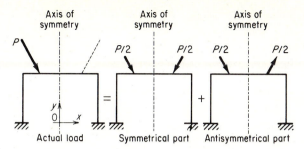

Figure 12.13 Decomposition of the loading of a symmetrical structure into symmetrical and antisymmetrical components

Theorem 3

A structure symmetrical with respect to a plane subjected to loading antisymmetrical with respect to the same plane undergoes deformations antisymmetrical with respect to the plane of symmetry of the structure, and the material particles in the plane of symmetry leave the plane along its normal direction with rotations possible only about the lines parallel to the plane (see Fig. 12.14b).

These theorems enable us to consider only a symmetrical half of the structure during the analysis. We first apply the symmetrical component of the loading on the symmetrical half of the structure with deflection boundary conditions on the plane of symmetry determined by Theorem 2 and obtain the symmetrical part of the response. Next we apply the antisymmetrical component of the loading on the symmetrical half of the structure with deflection boundary conditions on the plane of symmetry determined by Theorem 3 and obtain the antisymmetrical component of the response. Then we apply superposition to obtain the complete response.

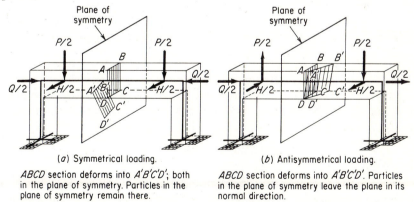

(*a*) Symmetrical loading.

(*b*) Antisymmetrical loading.

ABCD section deforms into *A'B'C'D'*; both in the plane of symmetry. Particles in the plane of symmetry remain there.

ABCD section deforms into *A'B'C'D'*. Particles in the plane of symmetry leave the plane in its normal direction.

(Note that symmetry, if any, of the structure about its own midplane is not involved in these considerations.)

Figure 12.14 Symmetrical and antisymmetrical deformations at particles in the plane of symmetry of a symmetrical structure

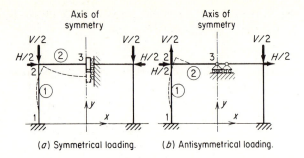

(a) Symmetrical loading.   (b) Antisymmetrical loading.

Figure 12.15 Key sketches of a symmetrical frame for symmetrical and antisymmetrical loadings

Planar structures such as those shown in Figs. 12.11 and 12.12 are actually three-dimensional structures whose geometry, material, support conditions and loading are symmetrically disposed with respect to z-positions of their particles. Since the *xoy* midplanes of such structures are planes of symmetry of structure and loading, we may treat these as **planar structures** for certain computational purposes. According to the terminology of symmetry and antisymmetry, a planar structure being deflected in its plane is, in fact, an example of a symmetrical structure being deformed symmetrically (with respect to that plane). But such a structure deflecting out of its midplane illustrates a symmetrical structure being antisymmetrically deformed.

Often planar structures are also symmetrical with respect to one or more planes normal to their midplanes, as illustrated by the planar frame in Fig. 12.13. Note that, when such a structure has more than one plane of symmetry, the intersection of two if its planes of symmetry is called an **axis of symmetry**.

Using Theorem 2, we can say that when a structure with an axis of symmetry is loaded symmetrically about the axis, it will undergo deformations symmetrical about the axis and the particles on the axis of symmetry will move only along this axis without rotation. An example of this is a symmetrical planar frame undergoing symmetrical deformations. When a structure with an axis of symmetry is loaded antisymmetrically about this axis, it will undergo antisymmetrical deformations about this axis such that the particles on the axis of symmetry will not move; however, they can rotate about the axis. This can be deduced by using Theorem 3. For example, the antisymmetrical deformations of a symmetric gridwork frame fall in the domain of this deduction. In fact, one can deduce many other interesting conclusions by using Theorems 2 and 3.

12.4.7 Examples of the use of DBC input units. As an example of the use of DBC input units let us consider the symmetrical planar frame shown in Fig. 12.15a. Since this frame is subjected to symmetrical loads, it will undergo symmetrical deformations. In the analysis we consider only one symmetrical half of the structure, say the left half, where the nodes are labeled. Note that we also label the point on the axis of symmetry. The degrees of freedom of a node in the overall coordinate system are $[d_{i1} \quad d_{i2} \quad \theta_{i3}]$, which can also be denoted as $[u_{i1} \quad u_{i2} \quad u_{i3}]$; that is, $u_{i1} = d_{i1}$ is the displacement along overall x axis; $u_{i2} = d_{i2}$, is the displacement along overall y axis; and $u_{i3} = \theta_{i3}$ is the rotation about the overall z axis.

Since the particles on the axis of symmetry can move only along the axis of symmetry, we can list the DBC input units as follows:

$$
\begin{array}{cccccc}
(1,1) & (1,1) & 0 & (3,1) & (3,1) & 0 \\
(1,2) & (1,2) & 0 & (3,3) & (3,3) & 0 \\
(1,3) & (1,3) & 0 & & &
\end{array}
\qquad (i)
$$

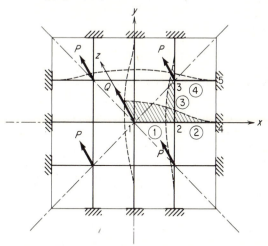

Figure 12.16   Key sketch of a symmetrical gridwork frame for symmetrical loading

The independent degree-of-freedom directions are the ones associated with (2,1), (2,2), (2,3), and (3,2).

In the case of this symmetrical planar frame undergoing antisymmetrical deformations, shown in Fig. 12.15b, the particle originally on the axis of symmetry rotates and moves normal to this axis; therefore we have the following DBC input units:

$$
\begin{array}{cccccc}
(1,1) & (1,1) & 0 & (3,2) & (3,2) & 0 \\
(1,2) & (1,2) & 0 & & & \\
(1,3) & (1,3) & 0 & & &
\end{array}
\qquad (j)
$$

Note that here the loading is symmetrical with respect to the xy plane but antisymmetrical with respect to the yz plane. According to Theorems 2 and 3, this means that the particles on the axis of symmetry will remain and rotate in xy plane; however, they should move normal to the yz plane.

Let us now consider the gridwork frame shown in Fig. 12.16. The frame itself is symmetrical about the xy, yz, and xz planes. However, the loading is symmetrical about the xz, and yz planes, and, being the loading of a gridwork frame, it can be considered antisymmetrical with respect to the xy plane. Because of the square plan of the gridwork, both the loading and the structure are symmetrical about the planes containing the z axis and the diagonal directions. Therefore the response of the structure can be obtained by confining the analysis to one-eighth of the structure. In the figure, only the nodes of such a portion are labeled. For the analysis we take only one-half of the loads which are on the planes of symmetry. Since the deformations will be antisymmetrical about the xy plane, the degrees of

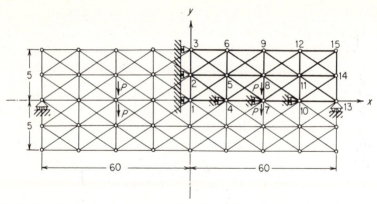

Figure 12.17 Key sketch of a planar truss undergoing deformations which can be approximated using the **plane sections remain plane** assumption

freedom relative to the selected overall coordinate system are $[d_{i3} \quad \theta_{i1} \quad \theta_{i2}]$ or $[u_{i1} \quad u_{i2} \quad u_{i3}]$, where $u_{i1} = d_{i3}$ is the displacement along the overall z-axis, $u_{i2} = \theta_{i1}$ is the rotation about the overall x axis, and $u_{i3} = \theta_{i2}$ is the rotation about the overall y axis. Since the loading is symmetrical about the planes passing through the z axis and including the x axis, y axis, and the diagonals, the deformations will also be symmetrical about these planes. Node 1, being on the z axis, can move only along z but cannot rotate. Node 2, being on the xz plane, will remain and rotate in the xz plane, and node 3, being in the plane of z and the diagonal, will remain and rotate in this plane. Therefore, we can list the DBC input units as follows:

$$
\begin{array}{ccccccc}
(1,2) & (1,2) & 0 & \quad & (4,1) & (4,1) & 0 \\
(1,3) & (1,3) & 0 & & (4,2) & (4,2) & 0 \\
(2,2) & (2,2) & 0 & & (4,3) & (4,3) & 0 \\
(3,2) & (3,3) & -1 & & (5,1) & (5,1) & 0 \\
 & & & & (5,2) & (5,2) & 0 \\
 & & & & (5,3) & (5,3) & 0
\end{array}
\qquad (k)
$$

where the DBC input units related to nodes 4 and 5 are the result of the fact that the structure is clamped around the boundary.

Before we leave the subject of handling the deflection boundary conditions by means of arrays, let us see how we can express any known information about the variation of deflections from node to node. In Fig. 12.17 a planar truss panel is shown supported along the major axis of symmetry, which is taken as the overall x axis. The structure is also symmetrical about the y axis, and, as a planar structure subjected to in-plane deformations, is also symmetrical about the xy plane. Suppose the loading in the xy plane is symmetrical about the y axis and confined to the points of the x axis. Because of the symmetries, we can confine the analysis to the quarter of the structure comprising the labeled nodes in the figure. The deflection degrees of a node are $[d_{i1} \quad d_{i2}]$ or $[u_{i1} \quad u_{i2}]$, where $u_{i1} = d_{i1}$ is the displacement component along the overall x axis and $u_{i2} = d_{i2}$ is the displacement component along the overall y axis. Since there are 15 nodes, the order of column matrix **u** is 30. Suppose the loading is antisymmetrical about the xz plane, as shown in the figure. Then, by the theorems on symmetry, we observe that the nodes on the y axis can move only along the y axis and the nodes on x axis will have to move in the y direction, excluding node 13, which is a support point. In other words, deflection components $u_{1,1}$, $u_{2,1}$, $u_{3,1}$, $u_{4,1}$, $u_{7,1}$, $u_{10,1}$, $u_{13,1}$, and $u_{13,2}$ are dependent, and all are prescribed as zero. The remaining $30 - 8 = 22$ deflection components appear to be independent. However, we see that the height-span ratio of the truss panel is less than $\frac{1}{10}$, which implies that we can use the plane-sections-remain-plane assumption of the elementary beam theory. This is, in fact, an a priori knowledge of the variation of the deflections of the nodes. The conditions implied by this assumption are

$$u_{i,1} = 0.500(u_{i-1,1} + u_{i+1,1}) \qquad i = 2, 5, 8, 11, 14 \tag{l}$$

which state that deflection components $u_{5,1}$, $u_{8,1}$, $u_{11,1}$, and $u_{14,1}$ are also dependent. We can exploit beam theory further and assume that the height of the truss will not change. This leads to

$$u_{i+1,2} = u_{i+2,2} = u_{i,2} \qquad i = 1, 4, 7, 10, 13 \tag{m}$$

which implies that deflection components $u_{2,2}$, $u_{3,2}$, $u_{5,2}$, $u_{6,2}$, $u_{8,2}$, $u_{9,2}$, $u_{11,2}$, $u_{12,2}$, $u_{14,2}$, and $u_{15,2}$ are also dependent. Thus the conditions in Eqs. (l) and (m) reduce the number of independent components from 22 to $22 - 4 - 10 = 8$. We can list the DBC input units for this case as follows:

(1,1)	(1,1)	0	(5.1)	(6.1)	0.5	(2,2)	(1,2)	1	
(2,1)	(2,1)	0	(8.1)	(9.1)	0.5	(3,2)	(1,2)	1	
(3,1)	(3,1)	0	(11.1)	(12.1)	0.5	(5,2)	(4,2)	1	
(4,1)	(4,1)	0	(14.1)	(15.1)	0.5	(6,2)	(4,2)	1	
(7,1)	(7,1)	0				(8,2)	(7,2)	1	
(10,1)	(10,1)	0				(9,2)	(7,2)	1	(n)
(13,1)	(13,1)	0				(11,2)	(10,2)	1	
(13,2)	(13,2)	0				(12,2)	(10,2)	1	
						(14,2)	(14,2)	0	
						(15,2)	(15,2)	0	

where the second and the third columns correspond to the DBC input units of conditions in Eqs. (l) and (m), respectively.

With this last example, it is clear that the DBC input units can represent not only the support conditions but also all other known conditions related to deflections. In fact, the term deflection boundary condition is misleading if we note that nodes such as 5, 8, and 11 in Fig. 12.17 are not on the actual boundary of the structure. A better term would be simply deflection condition. However, we shall continue to use deflection boundary condition with the understanding that it actually means deflection condition and implies that the concept of boundary has been generalized.[1]

---

[1] For further information about the deflection boundary conditions and their processing, see Ref. 3.

12.4.8 Problems for Solution

* Problem 12.21 In an indeterminate space framework, node 15 is located on a roller support which can roll freely over the plane whose unit normal in the overall coordinate system is given as $n^T = [0.816 \quad 0.408 \quad 0.408]$. The support plane is subjected to a settlement which is expressed in the overall coordinate system by $d_0 = [2 \quad 1 \quad -1]$. What are the DBC input units associated with this support?

Problem 12.22 The regular hexagonal gridwork frame shown in Fig. 12.18 is simply supported on the outer nodes and subjected to transverse loads such that it undergoes out-of-plane deformations which are symmetrical with respect to the planes established by the z axis and the 1-4, 1-6, and 1-6' lines. What are the DBC input units related to nodes 1, 2, 3, 4, 5, and 6?

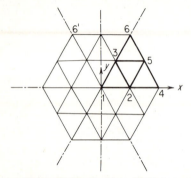

Figure 12.18 Problem 12.22

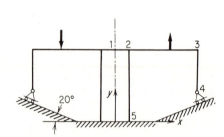

Figure 12.19 Problem 12.23

Problem 12.23 The planar-frame structure shown in Fig. 12.19 is symmetrical with respect to the yz plane and subjected to antisymmetrical loads with respect to the same plane such that the frame is to deform in the xy plane only. What are the DBC input units related to nodes 1, 4, and 5?

* Problem 12.24 The rigid space frame shown in Fig. 12.20 is symmetrical with respect to xz and yz planes and subjected to a loading which is symmetrical with respect to the xz plane but antisymmetrical with respect to the yz plane. What are the DBC input units at nodes 1 and 2?

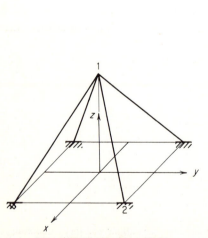

Figure 12.20 Problem 12.24

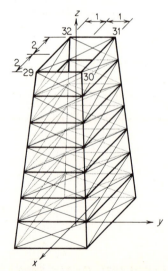

Figure 12.21 Problem 12.25

Problem 12.25   The three-dimensional space framework shown in Fig. 12.21 is deforming such that the plane sections-remain plane assumption is valid for planes parallel to the xy plane. Write the DBC input units associated with nodes 29, 30, 31, and 32.

12.5   Force Boundary Conditions.   This term is used to represent all conditions related to the forces acting on the structure. As discussed briefly in the last paragraph of Art. 12.4.7, the term *boundary* should be understood to include all points where the force conditions are asserted. According to the character of the force boundary conditions, they can be classified into two main categories: those which affect the overall rigidity of the structure and those which do not. As an example of the first category consider the zero-moment condition at the vertex at node 5 of element 5-2 in the frame structure of Fig. 12.12. On the other hand, in the same structure, the fact that node 1 is subjected to a concentrated load may be an example of the second category of force boundary conditions. The first-category conditions amount to prescribing the interparticle forces, whereas the second-category conditions merely prescribe the forces acting on the particles as an external agent. Special measures must be taken during the construction of the structure to create conditions of the first category, such as placing hinges and inserting sliding mechanisms inside the structure. The second-category conditions develop only after the structure is constructed. We shall call the first category **interelement-force boundary conditions** and the second category **loading-force boundary conditions.** In Art. 12.5 we shall study the method of describing interelement- and loading-force boundary conditions in the form of numerical arrays.

12.5.1   Interelement-force boundary conditions.   Consider node 5 of the planar-frame structure shown in Fig. 12.12. Three frame elements meet at this node. The connection of element 5-2 to the node is different from those of the other two, as indicated geometrically by a hinge symbol in the figure. The hinge symbol means that no moment will be transmitted from the element 5-2 to the node or vice versa. Actually, in structural analysis three basic symbols are used to indicate zero conditions for moment, axial force, and transverse shear force. Shown also in Fig. 12.22 are several combinations of the basic three symbols. These symbols have been devised for planar beam- and bar-type structural elements. To use these symbols in three-dimensional structures consisting of bars and beams usually creates ambiguity, since we may have to indicate zero condition in any of the two possible bending moments or transverse shear forces or try to show that the torsional moment is zero. We may have additional difficulties with these symbols in

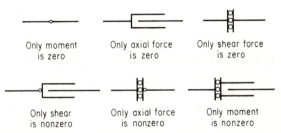

Figure 12.22   Geometric representations for various zero interelement-force conditions

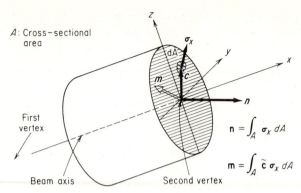

Figure 12.23   Forces at a vertex of a space frame

the structures of two- and three-dimensional continua. Actually special inter-element-force devices are rarely used in a continuum, since the implementation of such conditions during construction is even harder than the graphical representations of these conditions. Continuum structures are usually constructed monolithically, so that no force discontinuities are encountered between interelement boundaries. We shall therefore confine our attention to interelement-force boundary conditions in bar- and beam-type structures where the interelement boundaries can be idealized as discrete points.

Conditions that can be prescribed between a support point and a structural element are not considered to be part of the interelement-force boundary conditions. For example, in Fig. 12.12, at node 8, a zero-moment condition exists between the support and the element; however, this is not an interelement-force boundary condition since it involves only one element. Prescribed moments such as at node 8 are treated as loading-force boundary conditions.

In Fig. 12.23, the cross section at the positive face of a line element is shown. A positive face is the one at which the outer normal of the beam is in the same direction as the local x axis. The local coordinate system is such that the origin is at the area center (which is also considered to be the shear center), and local y and z axes are the principal axes of the cross section. Let $\mathbf{c}^T = [0 \quad y \quad z]$ denote the description of the position vector of a point in the cross section and

$$\boldsymbol{\sigma}_x^T = [\sigma_x \quad \tau_{xy} \quad \tau_{xz}]$$

denote the stresses at this point. On the cross section, the sum of the stresses is called the **stress resultant,** and the moment of the stresses about the origin is called the **stress couple.** Let $\mathbf{n}$ denote the stress resultant and $\mathbf{m}$ denote the stress couple of the cross section. From their definitions, we can write

$$\mathbf{n} = \int_A \boldsymbol{\sigma}_x \, dA \tag{12.119}$$

and

$$\mathbf{m} = \int_A \tilde{\mathbf{c}} \boldsymbol{\sigma}_x \, dA \tag{12.120}$$

where A is the cross-sectional area and $\tilde{\mathbf{c}}$ is as defined by Eq. (12.92) from the components of $\mathbf{c}$. Let the components of $\mathbf{n}$ and $\mathbf{m}$ with respect to the coordinate axes be denoted by n_x, n_y, n_z and m_x, m_y, m_z, respectively, namely $\mathbf{n}^T = [n_x, n_y, n_z]$, and $\mathbf{m}^T = [m_x, m_y, m_z]$. Of course, strictly speaking, the internal stress resultant n_x is the equilibrant of the normal force, and n_y and n_z, similarly, of the shear forces, while the internal stress couple m_x is the equilibrant of the torsional moment, and m_y and m_z of the corresponding bending moments imposed on the cross section. Let $\mathbf{q}$ denote a column matrix such that

$$\mathbf{q}^T = [\mathbf{n}^T \quad \mathbf{m}^T] \tag{12.121}$$

Column matrix $\mathbf{q}$ is called the **resisting forces** of the cross section, although some of the components of $\mathbf{q}$ are actually moments. The word *force* here is used in a generalized sense. Note that there is one-to-one correspondence between the components of $\mathbf{q}$ and the components of the deflection column matrix $\mathbf{u}$ [see Eq. (12.112)] of the centroid of the cross section, since $\mathbf{u}^T = [\mathbf{d}^T \quad \boldsymbol{\theta}^T]$, where $\mathbf{d}$ is the description of the displacement vector and $\boldsymbol{\theta}$ is the description of the rotation vector. We observe that in line elements, if a particle has six deflection degrees of freedom, there are six force components in the corresponding directions.

In Art. 12.2.1 we defined the terms vertex and node and the method of identifying such points via labels. In Art. 12.1.21 we indicated that superscripts are used for element labels and subscripts for node and vertex labels. With these points in mind, consider the section created by a cut between a node and a vertex of an element attached to the node. The forces acting on the element at this section are called **vertex forces.** To identify the description of the vertex forces in a suitably selected cartesian coordinate system, we append to the letter designation of the column matrix listing the vertex-force components a subscript and a superscript defining the location of the section by the vertex number of the related element. For example, the forces at the ith vertex of the mth element can be shown by $\mathbf{q}_i^m$, and the αth component of this column matrix is $q_{i\alpha}^m$. When the element coordinate system is used in expressing the forces at the positive face of a beam element, we can call the components of the column matrix by their related engineering names, such that q_{i1}^m is the normal force at the ith vertex of the mth element, q_{i4}^m is the torsional moment at the same point, etc.

During the construction of the structure, using the special devices sketched in Fig. 12.22, we can reduce some of the components of the column listing the forces at a vertex to zero. For example, in trusses we use pin connections which assure us that no moments develop within the limits of idealization. The order of the column matrix listing the forces at a vertex is usually 6, that is, three force components and three moment components. By special loading and special devices, we can consistently reduce certain components of the force column matrix at all vertices of all elements to zero. For example, in planar frames, choosing the plane of the structure as a plane of symmetry for both the structure and its loading, we can ensure that in the coordinate system of the beam, which is taken such that local z axis is perpendicular to the plane of the structure, the third, fourth, and fifth components of the column matrix of forces will be zero throughout. Then we can eliminate these zero components from the column matrix of the forces. For planar frames we can write $\mathbf{q}^T = [n_x \quad n_y \quad m_z]$. Note that this corresponds to the deflection column matrix of the vertex point, that is, $\mathbf{u}^T = [d_x \quad d_y \quad \theta_z]$. In planar gridwork frames, we can choose the plane of the gridwork as the plane of symmetry of the structure and the plane of antisymmetry of its loads, thus ensuring that in the coordinate system of the element, which is selected such that local z axis is perpendicular to the plane of the structure, the first, second, and sixth components of the column matrix of forces will be zero throughout. Deleting these zero components, we can write $\mathbf{q}^T = [n_z \quad m_x \quad m_y]$ for the column matrix of forces at a point. Note that this corresponds to the column matrix of deflections at a vertex point, that is, $\mathbf{u}^T = [d_z \quad \theta_x \quad \theta_y]$. From the similarities between force and deflection components, we observe that the order of the column matrix of forces without the zero components is equal to the number of degrees of freedom at a point of that structure.

Table 12.4 can be used to obtain the number of degrees of freedom at a point of various structures. For example, for planar trusses in the xy plane, the description of the deflections at a vertex point in the overall coordinate system is $\mathbf{u}^T = [d_x \quad d_y]$, from which it can be deduced that the vertex forces in the overall coordinate system should be $\mathbf{q}^T = [n_x \quad n_y]$. Since in truss elements, a zero-moment condition prevails throughout the bar, we are aware that the vector corresponding to $\mathbf{q}$ is in the direction of the bar axis. Similarly, in three-dimensional trusses, the column matrix of deflections at a vertex point as described in the overall coordinate system is $\mathbf{u}^T = [d_x \quad d_y \quad d_z]$; therefore the column matrix of vertex forces should be $\mathbf{q}^T = [n_x \quad n_y \quad n_z]$. Here, too, the vector corresponding to $\mathbf{q}$ should be in the direction of the bar axis, since there cannot be any moment at any point along the bar axis. When, as in the case of zero end moments for the elements of these trusses, the vertex-force components are consistently zero, it is unnecessary to invoke special interelement-force boundary conditions. We can recognize such conditions by merely prescribing the appropriate degrees of freedom for points of the structure. We confine the use of specific interelement-force boundary conditions to those atypical connections which differ from the joining arrangements utilized throughout most of a particular structure, such as the connection of element 5-2 at node 5 of the structure shown in Fig. 12.12.

We may list the zero vertex forces at the vertices of an element as part of the elemental information. Actually such a list constitutes the interelement-force boundary conditions of that element. We provide such a list only if there are interelement-force boundary conditions related to the element. We select at a vertex where there is special vertex-to-node connection mechanism a coordinate system such that a binary column matrix can list zero and nonzero force components. For example, if a component is prescribed as zero, we indicate this with a 0; otherwise we indicate it with a 1. We order these 0s and 1s with the order of the degrees of freedom of the vertex relative to the coordinate system at the vertex. For example, suppose we have a space-frame structure where the number of degrees of freedom at a point is six, the first three denoting the displacement degrees of freedom and the last three denoting the rotational degrees of freedom. Suppose a beam element of this structure is connected so that at the second vertex the torsional

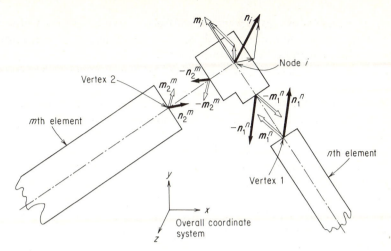

Figure 12.24 Free-body diagram of a node

moment is zero. The binary column matrix of this vertex is $[1 \quad 1 \quad 1 \quad 0 \quad 1 \quad 1]^T$, which can be represented by the octal number 73. We can use the scalar octal number 73 as the quantity representing the interelement-force boundary condition of this element at its second vertex. Now consider the 5-2 element of the planar frame of Fig. 12.12, where the first vertex is such that there is no moment to develop there. Since the number of degrees of freedom at a point of this structure is three, the column matrix of the vertex forces is of the form of $[n_x \quad n_y \quad m_z]$. Then the binary column matrix representing the connection at the first vertex would be $[1 \quad 1 \quad 0]^T$, which corresponds to octal number 6. In order to have the binary representation of the interelement-force boundary conditions meaningful, we must have a definite rule which enables us to orient the coordinate system at the vertex relative to the coordinate system of the element. For line elements, the rule consists of taking the coordinate systems at the vertices parallel to the coordinate system of the element. For elements with more than two vertices, additional rules may be required.

12.5.2 Loading-force boundary conditions related to nodes. These are the conditions which define the forces acting on the nodes as an external agent. They are the prescribed forces at the nodes. In general a node i may be subjected to a concentrated force $\mathbf{n}_i$ and a concentrated moment $\mathbf{m}_i$, as shown in Fig. 12.24. To describe these in array form, we use the overall coordinate system and the node labels employed in the key sketch of the structure. For node i we define $\mathbf{p}_i^T = [\mathbf{n}_i^T \quad \mathbf{m}_i^T]$, which is the column matrix listing the **concentrated loads of node** i. In general the order of $\mathbf{p}_i$ is 6, and the components are such that

$$\mathbf{p}_i^T = [n_{ix} \quad n_{iy} \quad n_{iz} \quad m_{ix} \quad m_{iy} \quad m_{iz}].$$

However, because of constraints on loading, some of the components of $\mathbf{p}_i$ may be zero for all i. For example, consider the planar frames in the overall xy plane undergoing in-plane deformations. This situation requires that $n_{iz} = m_{ix} = m_{iy} = 0$

for all i.   Then we choose not to write the consistently zero components and declare the column matrix of concentrated loads at node i as $\mathbf{p}_i^T = [n_{ix} \quad n_{iy} \quad m_{iz}]$, which is of order 3.   The number of degrees of freedom for points of this type of structures is also three (see Table 12.4).   Consider now the planar gridwork frames in the overall xy plane undergoing out-of-plane deformations.   Since the loading should be treated as antisymmetrical with respect to the overall xy plane in this type of structures, at a node, say the ith, we should have $n_{ix} = n_{iy} = m_{iz} = 0$.   Then we declare its column matrix of concentrated loads as $\mathbf{p}_i^T = [n_{iz} \quad m_{ix} \quad m_{iy}]$, the order of which is also equal to the number of degrees of freedom for points of this type of structures.   Likewise, in planar trusses located in the overall xy plane, $\mathbf{p}_i^T = [n_{ix} \quad n_{iy}]$, and in space trusses, $\mathbf{p}_i^T = [n_{ix} \quad n_{iy} \quad n_{iz}]$.   We observe that the order of column matrix $\mathbf{p}_i$ is equal to the number of degrees of freedom at node i.

Let $\mathbf{p}$ denote a partitioned column matrix such that its ith partition is $\mathbf{p}_i$. Column matrix $\mathbf{p}$ is called the **column matrix of the nodal loads.**   The definition of the loading-force boundary conditions related to the nodes is equivalent to defining the column matrix of the nodal loads.   Column matrix $\mathbf{p}$ is usually a sparse column matrix unless all nodes are loaded in all degree-of-freedom directions.   As usual, we can define the column matrix of the nodal loads by its nonzero entries. Suppose only the entry corresponding to the αth component at node i is nonzero, with a value, say, of p; then we can define the whole column matrix $\mathbf{p}$ by

$$(i, \alpha) \quad p \qquad\qquad (a)$$

where the parenthetical quantity is the index pair corresponding to the component number of the nonzero entry of $\mathbf{p}$ and p is the numerical value of this nonzero component.   We shall call the quantity comprising an index pair and a value the **concentrated-load-input unit.**   If there is more than one nonzero entry in $\mathbf{p}$, then for each such entry we define one concentrated-load-input unit.   We need as many concentrated-load-input units as there are nonzero components of $\mathbf{p}$.   We may list the concentrated-load-input units in any order.   The concentrated-load-input unit as an entity uniquely defines the loading-force boundary conditions related to the nodes.

12.5.3   Loading-force boundary conditions related to elements.   All prescribed loading conditions related to the structural elements fall into this category; e.g., body forces, prescribed surface tractions, concentrated loads acting at points other than nodes, and thermal loads are all loading-force boundary conditions related to the elements.   Whenever possible, we prefer to describe these forces in the coordinate system of the element (see Fig. 12.25).   In systematic structural analysis, we are usually primarily interested in the behavior of the whole structural system rather than the detailed behavior of the structural element.   Therefore, whenever we can, we should use the nodal equivalents of the loads applied directly to the elements.   We discuss this point further in later chapters.   Here we are focusing on the descriptions of uniform loading conditions in the element, such as uniformly distributed body forces, uniformly distributed surface tractions, uniform temperature increase, uniform temperature-increase gradients, etc.   The discussion of each one of these uniform loading conditions of an element follows.

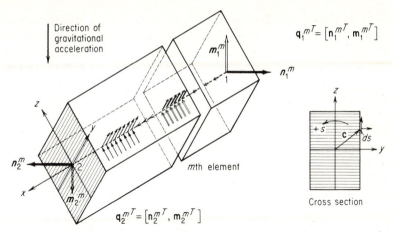

Figure 12.25   Free-body diagram of a space-frame element

Body Forces. A typical example of body forces is gravity loads. They are measured in force per length^3 units. For example, let ρ denote the unit mass of the material, and let **a** denote the cartesian description of the earth's gravitational acceleration, that is, $\mathbf{a}^T = [a_x \;\; a_y \;\; a_z]$. Then the description of the body force **g** of the element may be given as

$$\mathbf{g} = \rho \mathbf{a} \tag{12.122}$$

We usually describe the acceleration vector in the overall coordinate system. Since the orientation of the element coordinate system relative to the overall coordinate system is known, the description of acceleration loads in local or overall coordinate systems can be obtained with equal ease. Sometimes we handle the response of a structure to earthquake loads statically by subjecting the structure to an imaginary constant earthquake acceleration. Let **a*** denote the cartesian description of this acceleration; then the body force due to this hypothetical earthquake can be expressed as **a***ρ, which is the direct generalization of Eq. (12.122). The definition of this type of body forces is therefore complete when the description of the acceleration vector in the overall coordinate system and the unit mass of the material are known. For other types of body forces, which are uniform in the element, we need the three components of the body force along the axes of the element coordinate system.

Surface Tractions. These are the prescribed forces acting on the free surface of an element. They are usually measured in force per length^2 units. The exception is bar- and beam-type elements, where we talk about their sum and the sum of their moments, considering the circumference of the cross section. The measuring units of such sums are force per length and force-length per length, respectively. On the free edges of thin structural elements, the same sums are performed across the thickness. The summed tractions are called **line loads.** A line load can have as many independent components as the number of degrees of freedom at a point of the structural element. For example, in planar-frame elements, we may talk about axial, transverse, and moment components of the line loads. Since the surface tractions are usually considered to be uniform for an element, they are defined by as many constants as the number of degrees of freedom, representing the components along the axes of the element coordinate system. The component of the pre-scribed surface traction along the normal of the free surface is also called **pressure.** The pressure loading of structural elements is much more common than the loadings associated with the tangential components of the surface tractions. Frictionless soil pressure, hydrostatic pressure, and atmospheric pressure are examples of pressure loading. Friction on the skin of a pile and friction caused by the steady flow of a viscous fluid in a conduit are examples of the loading caused by the tangential components of surface tractions.

Thermal Loads. These loads are identified by a **temperature change** in the material. We know that the change of temperature induces strains in the material [see Eq. (12.105)]. If the material particles are not free to move, thermal stress will develop in the structure. A temperature change is assumed to be positive if it corresponds to a temperature increase. Considering only uniform loading condi-tions in the structural elements, one scalar for each element, representing the temperature change, is usually enough to represent the thermal-loading conditions of that element. In thin structural elements, where the degrees of freedom at a point also include the rotational ones, change in the temperature gradient in the direction of thickness is also required. Let Δt denote the temperature increase, and let the local z axis be in the direction of the thickness. The gradient of temperature increase in the thickness direction may be shown as $\Delta t_{,z}$, where the comma in the subscript indicates partial differentiation with respect to z. If $\Delta t_{,z}$ is not constant, we use an average value. The units of temperature increase are degrees of temperature, and the units of temperature gradient are degrees of temperature per length. In planar-gridwork frames, assuming that the local z axis is normal to the plane of the structure, the thermal loads are defined simply as $\Delta t_{,z}$, and there is no need to prescribe Δt. In planar frames where the local z axes are normal to the plane of structure, the thermal loads are defined by pres-cribing Δt and $\Delta t_{,y}$. For truss structures, only Δt is required. In space-frame structures Δt, $\Delta t_{,y}$, and $\Delta t_{,z}$ are necessary for the definition of the thermal loads. Note that thermal gradients are always relative to the local axes. Of course, when a structural-analysis problem involves thermal loads, the descriptions of the material should include the thermal-expansion coefficients. Temperature increase and its gradients are part of the elemental information. By scaling **shrinkage** with the thermal-expansion coefficient, we can treat it as a temperature decrease. **Fabrication-error** problems in bar- and beam-type structures can be handled as a temperature-increase problem by scaling the extra length of the element by its nominal length and the thermal-expansion coefficient and using the resulting quantity as a temperature increase.

12.5.4   Problems for Solution

Problem 12.26   Referred to the local coordinate system, the descriptions of the stress resultant and the couple of the positive face at $x = 15$ of a space beam are $\mathbf{n}^T = [70 \quad -4 \quad 14]$ and $\mathbf{m}^T = [25 \quad -40 \quad 60]$. The cross section of the beam is a circle with a radius of 2 units. Compute the column matrix of stresses $\boldsymbol{\sigma}$ at points defined by $\mathbf{c}_1^T = [15 \quad 2 \quad 0]$, $\mathbf{c}_2^T = [15 \quad 0 \quad 0]$ and $\mathbf{c}_3^T = [15 \quad \sqrt{2} \quad \sqrt{2}]$, all in the local coordinate system. Sketch the components of $\mathbf{n}$ and $\mathbf{m}$.

Problem 12.27 A space-beam element is to be connected at its second vertex to a node such that the octal-number representation of the connection is 33. Sketch the connection.

Problem 12.28 The connections of the first and the second vertices of a planar-frame element to the nodes are represented by octal numbers 4 and 5, respectively. Sketch the connections.

*Problem 12.29 In a submerged structure, a horizontal circular cylindrical beam element of 0.20 m diameter is located 10 m below the freshwater surface. The local y axis of the element is also horizontal. What is the description of the pressure-loading vector of this space-frame element in the element coordinate system, and what are the units of its components?

*Problem 12.30 The temperature of a beam element during construction was 25°C. The beam has a rectangular cross section of 0.20 by 0.50 m. The long side of the cross section is parallel to the local y axis. It is expected that the temperature at the corner of the cross section in the positive coordinate quadrant will be 55°C and the corner diagonally across will be heated to 35°C. The other two corners will have new temperatures which are equal. Assuming that the temperature variation in the cross section will be linear, what are the values of Δt, $\Delta t_{,y}$, and $\Delta t_{,z}$, and what are their units?

12.6 Description of Structural Response. In preceding articles we studied how to describe the inputs of a structural-analysis problem in the form of arrays. In this section we shall extend the same principles to a description of the output of the analysis. Before we do so, let us identify the basic components of the structural analysis itself. In Fig. 12.26 the analysis is defined by three basic blocks: structure, excitation, and response. A structure is uniquely defined by its geometry, material, deflection boundary conditions (excluding support settlements), and interelement-force boundary conditions. Note that all these evolve during the construction of the structure and are responsible for its rigidity. Once the structure is defined, we are interested in its behavior under loads. This is the objective of structural

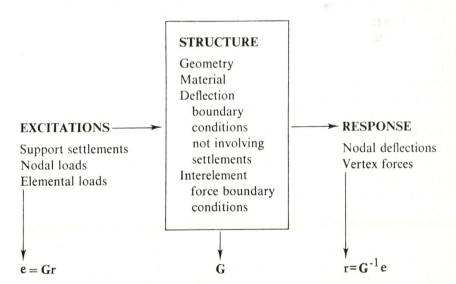

Figure 12.26 Basic components of structural analysis

analysis, i.e., given the structure, to determine its behavior under any loading. Here we shall use **excitation** to represent any or all of the possible loadings and **response** for the quantities describing the deformations and the internal stresses of the structure as a result of the loading.

We may consider the structure as an entity transforming the excitations into response. The excitations consist of loading-force boundary conditions, i.e., nodal and elemental loads, and of those deflection boundary conditions representing support settlements. For example, for a building subjected to the weight of its nonstructural components the excitation is the weight of such components, whereas for a building subjected to foundation movement the excitation is the settlement of the foundations. In the preceding articles, we discussed the method of quantifying the structure and its excitations without making any distinction between the two. However, we must clearly differentiate the information defining the structure from the information defining its excitations. We may repeat the analysis for a given structure many times, each time using a different excitation. If the data defining the structure are clearly separated from the data defining the excitations, for each analysis we need provide only the data defining the excitation after we have defined the structure once at the beginning.

The information defining the structure should be just enough to determine its intrinsic properties. It turns out that the geometry, the material properties, the deflection boundary conditions with zero support settlements, and the interelement-force boundary conditions are the quantities which enable us to determine these properties. We must have complete knowledge of these quantities before the start of the analysis. Sometimes we have this knowledge only in a statistical sense, such as knowing Young's modulus with its **mean** and **standard deviation.** Since the quantities defining the structure are the result of some measurements, our knowledge about them is, in general, always statistical. Through **quality control** during manufacturing and construction, we can minimize the statistical uncertainties to a level where the data defining the structure may be considered deterministic rather than statistical. If this happens, we call the structural system **deterministic;** otherwise it is **probabilistic.** We should, of course, never confuse the term deterministic with the term determinate. We use the latter term for the structures where the internal stresses can be computed from the equilibrium requirements without referring to the material properties. For example, we may talk about the deterministic determinate structural systems when the data defining the determinate structure are not statistical.

We shall confine ourselves to deterministic structural systems. During our discussion of the material of the structure, we stated that we would confine ourselves to time-independent, linearly elastic materials. The structures of such materials exhibit a linear relationship between excitations and the responses provided that the deformed geometry of the structure is almost the same as its original geometry, i.e., the deflections are small. The smallness of deflections can be achieved either by high rigidity in the structure or by the small magnitudes in the excitations. We shall limit ourselves to those structural-analysis problems where the excitations are linearly related to the response. Let **e** denote the algebraic representation of the excitations and **r** denote the algebraic representation of the response. The linear relationship between **e** and **r** may be shown as

$$\mathbf{G}\mathbf{r} = e \qquad\qquad (12.123)$$

where **G** may be considered as the algebraic representation of the structure. We shall call the structure **linear** when the related analysis problem can be cast in the form of Eq. (12.123); otherwise it will be called **nonlinear**. In a linear structure, **G** is said to represent the intrinsic properties of the structure. Most of the effort in systematic structural analysis goes into the systematic generation of **G** from the arrays defining the structure.

The excitations of the structure may consist of the loading-force boundary conditions and those deflection boundary conditions defining the support settlements. The loading-force boundary conditions are nodal loads and elemental loads. The excitations of a structure may or may not depend on time. We shall assume throughout that they are time-independent. Such excitations are called **static loading.** Static loading implies that the loads are applied very slowly, so as to avoid developing inertial forces or friction forces between the material particles. Except for a very few special cases, it is extremely difficult to assess quantitatively the excitations to which a structure may be subjected during its lifetime. Excitations are therefore often defined by statistical estimates involving the maximum, minimum, mean, a distribution law, and the standard deviation. We classify the excitations as deterministic or probabilistic according as their definitions are absolute or statistical. For example, gravity loads are deterministic, whereas wind loads, earthquake loads, etc., are probabilistic. In linear structures, as far as analysis is concerned, probabilistic loads pose no special problem once the statistical inference in assessing them is complete. This is due to the fact that the statistical characteristics of the excitations are preserved during linear transformations of the type of Eq. (12.123); for example, maximum load according to some criterion will produce maximum response with the same criterion. With this in mind, we shall confine ourselves to time-independent deterministic excitations throughout. In earlier articles, we showed the method of describing such excitations in the form of arrays. Part of the effort in systematic structural analysis goes into the systematic generation of an algebraic column matrix, in the form of **e** in Eq. (12.123), from the arrays defining the excitations.

12.6.1 Quantities describing the response. Structural response should now be viewed within the context of the structure and the excitations as described in the previous paragraphs. The response of a linearly behaving structure to a deterministic static loading consists of the deflections and internal stresses corresponding to its deformed configuration. The deflections are actually those of the nodes. In other words, the partitioned column matrix **u**, where the ith partition $\mathbf{u}_i$ lists the deflection components of the ith node, is the deflection response of the structure. From our discussion of the deflection boundary conditions we know that in order to determine the deflection response of the structure we must compute the independent deflection components first and then evaluate the dependent ones using the DBC input units. The term deflection response actually includes the deflections of all material particles.   However, it is a comparatively easy task to compute the deflections in an element once we know the deflections of its vertices.

In structural elements of bar and beam types, we support the element at its vertices and apply the computed vertex deflections as support settlements together with the prescribed element loads.   The analysis of the element under such loading will yield the response at the particles which are not on the nodes.   In the elements of structures of two- and three-dimensional continua, we can obtain the deflection response at the material particles not located at the vertices by the same interpolation rule which is used during the quantitative description of the intrinsic properties of the structure.

The stress response of the structure consists of the quantities defining the stress state at every material particle.   In the systematic analysis of bar- and beam-type structures, we consider the stress response to be completely defined when we determine the vertex forces of every element, since the stress resultants and couples at other cross sections can be obtained by means of simple calculations.   In the structures of two- and three-dimensional continua, the description of the stress response is a little more involved.   In such structures, we consider the stress response to be defined when we determine stress states at one or more strategically located points in each element.   We compute the stress state at a point by first computing the strains from the deflections of nearby nodes by interpolation and then using them in the stress-strain relations in the form of Eq. (12.111).

In the following chapters, the treatment of systematic analysis is limited to structures where the interelement boundaries consist of discrete points, as in the case of structures built by bars and beams.   However, no limit is fixed for the number of vertices in an element.   By doing so, excluding a detailed computation of stress response and the problems of convergence, we in effect include the systema-

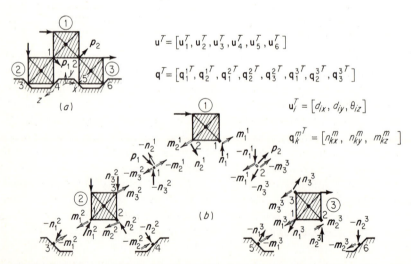

Figure 12.27 Response quantities of a six-node idealization of a planar frame: (*a*) key sketch; (*b*) unassembled form

tic analysis of structures of two- and three-dimensional continua. Denoting the list of deflections of node i by $\mathbf{u}_i$, and denoting the list of forces at vertex k of the mth element by $\mathbf{q}_k^m$, we shall consider the response computations complete when the numerical values of $\mathbf{u}_i$ and $\mathbf{q}_k^m$ are known for all i, m, and k. For example, see Fig. 12.27.

12.6.2 Problems for Solution

Problem 12.31 Consider the structure given in Fig. 12.4. What are the orders of the partitioned column matrix of deflections $\mathbf{u}$ and partitioned column matrix of vertex forces $\mathbf{q}$ of the whole structure? List these column matrices explicitly, using $u_{i\alpha}$ for the αth component of the deflection at node i and $q_{k\alpha}^m$ for the αth component of the forces at the kth vertex of element m.

Problem 12.32 What is the order of the deflection column matrix $\mathbf{u}$ of a 30-node, and 45-bar space-truss structure? What is the order of the vertex-force column matrix $\mathbf{q}$ of the same structure? Note that the order of $\mathbf{q}$ can be decreased if one takes into account the fact that in a bar element vertex forces are equal and opposite and they are in the direction of the bar axis, due to element equilibrium. We use $\mathbf{s}$ to denote the list of vertex forces which is contracted by the condition of element equilibrium. What is the order of column matrix $\mathbf{s}$ of this structure?

Problem 12.33 Using the definition given for column matrix $\mathbf{s}$ in Prob. 12.32, compute the order of $\mathbf{s}$ of the structure shown in Fig. 12.4. Display matrix $\mathbf{s}$.

Problem 12.34 What are the orders of the column matrix of deflections $\mathbf{u}$ and the column matrix of vertex forces $\mathbf{q}$ and $\mathbf{s}$ (see Prob. 12.32 for definition) of a planar gridwork frame consisting of 16 nodes and 24 beams? Display these matrices.

Problem 12.35 Consider Eq. (12.123), which represents the excitation-response relationship of a linearly behaving structure loaded statically. The matrix $\mathbf{G}$ is always a nonsingular square matrix; therefore its inverse $\mathbf{G}^{-1}$ exists. We can rewrite Eq. (12.123) as

$$\mathbf{r} = \mathbf{G}^{-1}\mathbf{e} \qquad (12.124)$$

where $\mathbf{G}^{-1}$ may be called the **transfer function** of the structure. Note that once the transfer function is available, the computation of the response for any excitation amounts to a matrix multiplication. Let n denote the order of $\mathbf{G}$. If $\mathbf{G}$ is a full matrix, it is known that the algebraic operations, specifically the number of multiplications, in computing $\mathbf{G}^{-1}$ are normally proportional to n^3 and the number of multiplications in computing $\mathbf{r}$ as $\mathbf{G}^{-1}\mathbf{e}$ is proportional to n^2. If the order of $\mathbf{G}$ is doubled, what would be the number of multiplications for obtaining $\mathbf{G}^{-1}$ and $\mathbf{G}^{-1}\mathbf{e}$ as multiples of the corresponding number of multiplications before doubling the order?

12.7 Methods for Computing the Response.

Let us discuss briefly the methods by which we can compute the response from the arrays defining the structure and its excitations. We assume that the structure is discrete, so that its response can also be described in the form of arrays. Furthermore both the structural system and its excitations are deterministic, and the structure is behaving linearly. Under statical loading, the relationship between the excitations and the response is a linear algebraic one, as depicted by Eq. (12.123), where $\mathbf{G}$ represents the intrinsic properties of the structure, $\mathbf{e}$ is the excitations, and $\mathbf{r}$ represents the response. Physically, Eq. (12.123) represents two conditions, **force equilibrium** and **deflection compatibility.** The force-equilibrium conditions are the result of the requirement that every element be in static equilibrium under external loads and its vertex forces and that every node be in static equilibrium under nodal loads and the vertex forces of those elements meeting at the node. The deflection-compatibility conditions are the results of the requirement that the deformations of the structural elements be geometrically compatible with the nodal deflections.

12.7.1   Introduction to various methods.   Let $\mathbf{q}$ denote the column matrix listing vertex forces of each vertex of each element such that $\mathbf{q}_k^m$ would be the partition of $\mathbf{q}$ listing the vertex forces at the kth vertex of the mth element.   Let $\mathbf{u}$ denote the column matrix listing the nodal deflections such that the ith partition of $\mathbf{u}$ would be the nodal deflections of node i, namely, $\mathbf{u}_i$.   We can generate Eq. (12.123) with the understanding that $\mathbf{r}^T = [\mathbf{q}^T \quad \mathbf{u}^T]$.   Such a procedure is called the **hybrid method.** In such methods the two components of the response, namely, the nodal deflections and the vertex forces, are obtained simultaneously.   Although appearing attractive, these methods are very uneconomical since they involve very large sets of linear simultaneous equations.   Note that, in this case, the order of $\mathbf{r}$ is the sum of orders of $\mathbf{q}$ and $\mathbf{u}$.   In Part II, we exclude the hybrid methods from our discussions.

During the analysis process, we may focus our attention first on the computation of nodal deflections $\mathbf{u}$ by expressing the force-equilibrium requirements in terms of deflections.   Such an approach is called the **displacement method.**   In displacement methods, Eq. (12.123) represents the equilibrium conditions and $\mathbf{r}$ represents the independent components of deflection column $\mathbf{u}$.   After the computation of the components of $\mathbf{r}$, we proceed to compute the dependent components of $\mathbf{u}$ and then the column of vertex forces $\mathbf{q}$.   The last two steps of the method involve only simple computations such as matrix multiplications, whereas in the first step we have to invert the coefficient matrix of $\mathbf{r}$.   Chapter 15 covers systematic implementation of this method.

The alternative method in systematic analysis is the **force method.**   Here we focus our attention first on the computation of vertex forces.   To do this we first express the equilibrium equations in terms of the vertex forces.   If the structure is statically indeterminate, there will be fewer equations than the number of unknown vertex forces.   We generate the missing equations from the geometric compatibility conditions.   In the force method, therefore, Eq. (12.123) represents partly force equilibrium requirements and partly the geometric-compatibility requirements and $\mathbf{r}$ lists the unknown vertex forces.   After the computation of vertex forces, we proceed to compute the nodal deflections with simple calculations amounting to matrix multiplications.   In the first step of the method, however, we must invert the coefficient matrix of unknown vertex forces.   Chapter 16 covers systematic implementation of the force method.

Since systematic handling of both the force method and the displacement method involves the transformation of descriptions of forces and deflections and the vertex-force–vertex-deflection relations of structural elements, we shall discuss these topics in Chaps. 13 and 14, respectively, before we describe the systematic applications of the force and the displacement methods.

12.7.2   Problems for Solution

*Problem 12.36   Consider a structure consisting of N nodes and M elements, each with n vertices.   Suppose the number of degrees of freedom per point is e.   Assume that the numbers of known components of the column matrix of deflections $\mathbf{u}$ and the column matrix of vertex forces $\mathbf{q}$ (or $\mathbf{s}$, as defined in Prob. 12.32) are negligibly small relative to the orders of their respective matrices.   What is the order of the column matrix of deflections $\mathbf{u}$?   Consider the jth partition of $\mathbf{q}$, namely, $\mathbf{q}^j$, which lists the forces at the n vertices.   We can define a column matrix $\mathbf{s}^j$ to list the forces at the first $n - 1$ vertices as the independent components of the vertex forces of the jth element, since when we know s^j, we can compute q_n^j by the force-equilibrium equations of the element (see Prob. 12.32). What is the order of column matrix s (which is a partitioned column matrix, where the jth partition is s^j)? What is the order of the combined-response matrix $[u^T \quad s^T]$?

*Problem 12.37 Consider Probs. 12.35 and 12.36. Suppose we use the hybrid method of analysis, where the response r is considered to be $[u^T \quad s^T]$. Assume that the number of multiplications for computing s from u is negligible compared with that of computing u in the displacement method. In computing the response to a given excitation, what is the ratio of total number of multiplications of the hybrid method to that of the displacement method as a function of M, N, n, and e (obtain separate expressions for computing both the transfer function G^{-1} and the response $G^{-1}e$)?

12.8 References

1. Schwarz, H. R., H. Rutishauser, and E. Stiefel: "Numerical Analysis of Symmetric Matrices," Prentice-Hall, Inc., Englewood Cliffs, N.J., 1973.

2. Sokolnikoff, I. S.: "Mathematical Theory of Elasticity," 2d ed., McGraw-Hill Book Company, New York, 1956.

3. Utku, S.: "ELAS75 Computer Program, User's Manual," *Duke Univ. Sch. Eng. Struct. Mech. Ser.* 10, Durham, N.C., 1971.

4. Hildebrand, F. B.: "Methods of Applied Mathematics," 2d ed., chap. 1, Prentice-Hall, Inc., Englewood Cliffs, N.J., 1965.

13

Transformation Matrices

13.1 **Transformation of Descriptions of Position Vectors.** The vector c_i joining an origin to a point is called the **position vector** of that point (Fig. 13.1). Let O denote the origin and i the point; then the vector c_i joining O to i, that is, $\overrightarrow{Oi}$, is the position vector of point i. In numerical operations, we refer to vectors by their components relative to a coordinate system located at point O. If in such a coordinate system the coordinates of point i are x, y, and z, the **description** of the position vector may be given as $c_i^T = [x \quad y \quad z]$, which is a matrix of order 3. To be precise we should refer to $[x \quad y \quad z]$ as the description of the position vector in the selected coordinate system, not as the position vector itself. We may ignore the distinction between a vector and its description so long as we can maintain a one-to-one correspondence between the two. However, sometimes, given a vector, we may use more than one coordinate system to describe it. When this happens, we have more than one description of the same vector. In the Arts. 13.1 to 13.3 we deal with the various descriptions of a given vector and establish the rules existing between the descriptions, assuming that the coordinate systems are all right-handed and cartesian.

13.1.1 **Relationship between cartesian descriptions.** In Fig. 13.1, an origin O, a point i, and two right-handed cartesian coordinate systems, both with origins at O, are shown. For the sake of argument, let us call the coordinate system (x, y, z) the overall coordinate system and the coordinate system (x', y', z') the local coordinate system. Let us denote the descriptions of the position vector of point i in the overall and local coordinate systems by c_i and c_i', respectively. If the coordinates of point i in the overall coordinate system are x_i, y_i, z_i and its coordinates in the local coordinate system are x_i', y_i', z_i', we may state that $c_i^T = [x_i \quad y_i \quad z_i]$ and $c_i'^T = [x_i' \quad y_i' \quad z_i']$. We would like to know the relationship between c_i and c_i'. We assume here that the orientation of the local coordinate system relative to the overall coordinate system is known. In Art. 12.2.2 we observed that the knowledge of the orientation can be quantified by the third-order matrix of direction

448

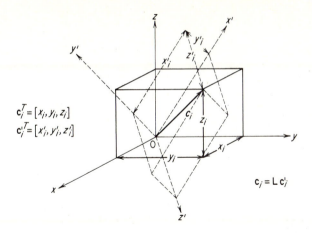

Figure 13.1 Position vector of a point and transformation of its descriptions

cosines **L**, where the columns are descriptions of the unit vectors of local axes in the overall coordinate system. In other words, l_1, l_2, and l_3 are the descriptions of the unit vectors along the first, second, and third local axes, respectively, in the overall coordinate system. We repeat Eq. (12.95) here for ready reference.

$$\mathbf{L} = [l_1 \quad l_2 \quad l_3] \tag{13.1}$$

We assume that **L** is known. Consider the matrix $x_i' l_1$. This is the description in overall coordinates of the vector on the first local axis with a length of x_i'. Similarly $y_i' l_2$ and $z_i' l_3$ are the descriptions in the overall coordinate system of the vectors on the second and third local axes with lengths y_i' and z_i', respectively. Since the sum of $x_i' l_1$, $y_i' l_2$, and $z_i' l_3$ is the description of the position vector itself, expressed in the overall coordinate system, we can write

$$\mathbf{c}_i = x_i' l_1 + y_i' l_2 + z_i' l_3 = \begin{bmatrix} l_1 & l_2 & l_3 \end{bmatrix} \begin{Bmatrix} x_i' \\ y_i' \\ z_i' \end{Bmatrix} \tag{13.2}$$

or noting that $\mathbf{c}_i'^T = [x_i' \quad y_i' \quad z_i']$ and using Eq. (13.1), we have

$$\mathbf{c}_i = \mathbf{L}\mathbf{c}_i' \tag{13.3}$$

This relationship enables us to compute the description of the position vector in the overall coordinate system when its description in the local coordinate system is known. As discussed earlier, **L** is an orthogonal matrix, and therefore its inverse is equal to its transpose [see Eqs. (12.96) and (12.97) and the related discussions]. Therefore, we can also write

$$\mathbf{c}_i' = \mathbf{L}^T\mathbf{c}_i \tag{13.4}$$

which enables us to obtain the description of the position vector in the local coordinate system when its description in the overall coordinate sytem is known.

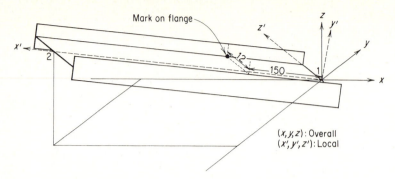

Figure 13.2 Finding the overall coordinates of a point on the beam

As an example, consider the wide-flange steel-beam element shown in Fig. 13.2. This beam is positioned such that its first vertex is located at the origin, its second vertex is located at a point with overall coordinates $[-748.331\ -1{,}122.497\ 374.166]$, and its principal axis with the larger moment of inertia at the cross section at vertex 1 passes through the point with overall coordinates $[0.\ \ 31.623\ \ 94.868]$. It is assumed that this point on the principal axis is on the positive side of the second local axis of the beam. The height of the beam is 24 units, and the flange width is 12 units. On the beam, 150 units away from the first vertex, there is a mark at the middle of the flange with a positive z' coordinate. Suppose we are interested in finding the overall coordinates of this mark. If we denote the description of the position vector of this mark in the local coordinate system of the beam by $\mathbf{c}'$, according to the given information, we can write

$$\mathbf{c}'^T = [150 \quad 0 \quad 12] \tag{a}$$

As described in Art. 12.2.2, we can obtain the matrix of direction cosines from the given information as

$$\mathbf{L} = \begin{bmatrix} -0.5345 & 0 & -0.8452 \\ -0.8018 & 0.3162 & +0.5071 \\ 0.2673 & 0.9487 & -0.1690 \end{bmatrix} \tag{b}$$

Then, using Eq. (13.3), we can obtain the overall coordinates of the mark as

$$\mathbf{c}^T = (\mathbf{L}\mathbf{c}')^T = [-90.3202 \quad -114.1825 \quad 38.0608] \tag{c}$$

where $\mathbf{c}$ is the description of the position vector of the mark in the overall coordinate system. The reader should verify this.

In dealing with planar structures, we usually place the structure in the overall xy plane and orient local coordinate systems so that the local and overall z axes are coincident. In such cases, we can display the matrix of direction cosines of the local coordinate systems as

$$\mathbf{L} = \left[\begin{array}{cc|c} \times & \times & 0 \\ \times & \times & 0 \\ \hline 0 & 0 & 1 \end{array} \right] \tag{d}$$

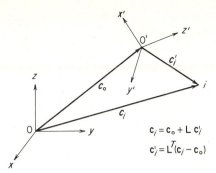

Figure 13.3 Relation between the position vectors when the coordinate origins are not coincident

where $\times$ denotes entries which are not necessarily zero. Since $\mathbf{L}$ is a diagonal partitioned matrix, its inverse is also in the same form, because $(\text{diag } \mathbf{A}_i)^{-1} = \text{diag } \mathbf{A}_i^{-1}$. The form of $\mathbf{L}$ in Eq. (d) indicates that there is no coupling between the first two and the last components of a position vector during transformation. When we confine ourselves to the points of xy plane, the position vectors will have only two nonzero components and their transformations will be governed by the upper left diagonal submatrix in Eq. (d). So long as we deal with the position vectors in xy plane, this second-order submatrix may be considered as the matrix of direction cosines of the local axes in xy plane. Equations (13.3) and (13.4) are also valid, the orders of $\mathbf{c}_i$, $\mathbf{c}_i'$, and $\mathbf{L}$ being 2.

Sometimes we may refer to a point by two position vectors originating from two different origins O and O'. Let the descriptions of these vectors in their respective coordinate systems be denoted by $\mathbf{c}_i$ and $\mathbf{c}_i'$, and let $\mathbf{c}_0$ denote the description of vector c_0 in the coordinate system located at O (see Fig. 13.3). Let $\mathbf{L}$ denote the matrix of direction cosines of the coordinate system at O' relative to the coordinate system at O. Note that the matrix of direction cosines is not a function of the locations of coordinate origins. With the help of Fig. 13.3 and Eq. (13.3) we can write

$$\mathbf{c}_i = \mathbf{c}_0 + \mathbf{L}\mathbf{c}_i' \tag{13.5}$$

and, solving for $\mathbf{c}_i'$, we obtain

$$\mathbf{c}_i' = \mathbf{L}^T(\mathbf{c}_i - \mathbf{c}_0) \tag{13.6}$$

Note that these equation reduce to Eqs. (13.3) and (13.4) when $\mathbf{c}_0$ is a null matrix, i.e., when O and O' are coincident.

13.1.2 Problems for Solution

*Problem 13.1 The dimensions of the rectangular cross section of a uniform beam are 20 and 40 units of length. The first and the second vertices of the beam are placed at nodes 45 and 17. The descriptions of the position vectors of these nodes in the overall coordinate system are $\mathbf{c}_{45}^T = [10 \quad 20 \quad 30]$ and $\mathbf{c}_{17}^T = [100 \quad 500 \quad 200]$. The principal axis of the cross section with larger moment of inertia is taken as the local y axis. The local origin is at the first vertex. The overall coordinates of a point on the positive side of the local y axis are $[58 \quad 11 \quad 30]$. What are the local and overall coordinates of the corner with positive local y and z coordinates, in the mid-section of the beam?

*Problem 13.2   The first, second, and third vertices of a triangular element are placed on nodes with overall coordinates [5 15 25], [40 -20 13], and [-5 6 -4], respectively. Find the local and the overall coordinates of the center of the circle passing through the vertices of the triangle.

Problem 13.3   Find the matrix of direction cosines of the local axes of a tetrahedral element where the first, second, third, and fourth vertices are placed on nodes 17, 41, 37, and 26.   The descriptions of the position vectors of these nodes in the overall coordinate system are

$$\mathbf{c}_{17}^T = [10 \quad -5 \quad 0], \mathbf{c}_{41}^T = [-10 \quad -2 \quad 10], \mathbf{c}_{37}^T = [0 \quad 7 \quad 5], \text{ and } \mathbf{c}_{26}^T = [4 \quad 3 \quad 6].$$

Problem 13.4   The first four vertices of a hexahedral element are located at nodes with overall coordinates [0 5 3], [5 0 7], [0 15 6], and [4 7 8], respectively.   Find the matrix of direction cosines of the local coordinate system of the element and then compute the local co-ordinates of the first four vertices.

Problem 13.5   The description of the boundary curve of an elliptical plate in the overall coordinate system is given as $(x/5)^2 + (y/4)^2 = 1$.   We would like to choose a local coordinate system at the boundary point in the first quadrant with $x = 3$, so that the first local axis is in the direction of the outer normal at this point and the third local axis is in the direction of the overall z axis.   What is the matrix of direction cosines of the local coordinate system at this point? (*Hint:* If the equation of a curve is $f(x, y) = 0$, the normal at point x_0, y_0 on the curve is parallel to the vector described by $[f_{,x} \quad f_{,y}]_{x_0, y_0}$, where a comma in the subscript indicates partial differentiation with respect to the quantity following it and the subscripts on the bracket indicate that the partial derivatives are to be evaluated at point x_0, y_0.)

Problem 13.6   The description of the boundary surface of a three-dimensional ellipsoidal solid in the overall coordinate system is given by $(x/5)^2 + (y/4)^2 + (z/6)^2 = 1$.   Find the matrix of directions cosines of the local coordinate system located at the first-quadrant boundary point which is on the space bisector of this quadrant.   The local coordinate system is such that the first local axis is in the direction of the outer normal of the solid at the boundary point and the second local axis is in the plane established by the outer normal and the overall z axis and heads in the same general direction as the overall z axis.   (*Hint:*   With the notation of the hint of Prob. 13.5, the description of a vector in the normal direction of surface $f(x, y, z) = 0$ at point x_0, y_0, z_0 on the surface may be expressed as $[f_{,x} \quad f_{,y} \quad f_{,z}]_{x_0, y_0, z_0}$.)

13.2   Transformation of Descriptions of Displacement and Rotation Vectors.

As discussed in Art. 12.4.1, the displacement and the rotation vectors of a material particle are the vectorial quantities which define the deformed configuration of the particle relative to its original configuration.   Referring to Fig. 13.4, let A denote a material particle and i a unit vector passing through A before deformations and let A' and i' denote the corresponding entities after deformations.   Vector $\overrightarrow{AA'}$ is called the displacement vector d, and vector $i \times i'$ is called the rotation vector θ.

13.2.1   Transformation of cartesian descriptions of deflections at a point.   Con-sider two coordinate systems located at point A, both cartesian and right-handed. Let us call one of the coordinate systems the overall coordinate system and the other the local coordinate system.   Let d denote the description of the displacement vector in the overall coordinate system and d' its description in the local coordinate system.   Similarly, let θ denote the description of the rotation vector in the overall coordinate system and θ' denote its description in the local coordinate system. Let the matrix of direction cosines of the local coordinate system relative to the

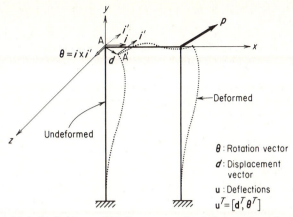

Figure 13.4 Definitions of displacement and rotation vectors of a point

overall coordinate system be denoted by **L**. With the reasoning which led us to Eq. (13.3), we can write

$$\mathbf{d} = \mathbf{L}\mathbf{d}' \tag{13.7}$$

and

$$\boldsymbol{\theta} = \mathbf{L}\boldsymbol{\theta}' \tag{13.8}$$

Since **L** is an orthogonal matrix, $\mathbf{L}^T = \mathbf{L}^{-1}$ and therefore from the above equations we can obtain

$$\mathbf{d}' = \mathbf{L}^T\mathbf{d} \tag{13.9}$$

and

$$\boldsymbol{\theta}' = \mathbf{L}^T\boldsymbol{\theta} \tag{13.10}$$

We can use these equations in converting the descriptions of displacement and rotation vectors from one coordinate system to another. Note that the transformation of the descriptions of displacement and rotation vector is independent of the locations of the origins of overall and local coordinate systems since **L** is independent of the locations of the origins.

In Art. 12.4.1, we used deflection as the combined name of displacements and rotations of a particle. Let **u** list the deflection components of point A in the overall coordinate system and **u**' list those in the local coordinates system, such that

$$\mathbf{u}^T = [\mathbf{d}^T \quad \boldsymbol{\theta}^T] \tag{13.11}$$

and

$$\mathbf{u}'^T = [\mathbf{d}'^T \quad \boldsymbol{\theta}'^T] \tag{13.12}$$

Using Eq. (13.1), we can obtain the relationship between **u** and **u**':

$$\mathbf{u} = \mathbf{R}\mathbf{u}' \tag{13.13}$$

where the **deflection transformation matrix R** can be displayed as

$$\mathbf{R} = \begin{bmatrix} \mathbf{L} & \mathbf{0} \\ \mathbf{0} & \mathbf{L} \end{bmatrix} \tag{13.14}$$

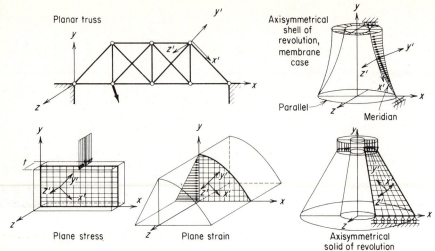

Figure 13.5   Structures with two lineal degrees of freedom per point

Note that $\mathbf{0}$ is the null square matrix of order 3. Since $\mathbf{R}$ is a diagonal matrix and its diagonal submatrices are orthogonal, $\mathbf{R}$ is also orthogonal, namely,

$$\mathbf{R}^{-1} = \mathbf{R}^T \tag{13.15}$$

It then follows from Eq. (13.13) that

$$\mathbf{u}' = \mathbf{R}^T \mathbf{u} \tag{13.16}$$

Equations (13.13) and (13.16) can be used in transforming the descriptions of deflections of a point from one coordinate system to another.

13.2.2 Specialization of relations for various groups of structures. As summarized in Table 12.4, the order of the column matrix of deflections $\mathbf{u}$ at a point varies from structure to structure since some of the components of the deflections are identically zero because of the nature of the structure. By properly selecting the local coordinate systems, we can ensure that a zero component of $\mathbf{u}$ will also be a zero component in $\mathbf{u}'$. Taking Table 12.4 as a guide, we can display the column matrix of deflections $\mathbf{u}$ at a point explicitly with its components in the overall coordinate system and the corresponding transformation matrix in terms of $l_{\alpha\beta}$ coefficients, representing the cosine of the angle between the βth local axis and the αth overall axis. This is done in the following paragraphs.

In planar trusses, plane-stress and plane-strain problems, membrane shells of revolution undergoing axisymmetrical deformations, and solids of revolution undergoing axisymmetrical deformations we may consider that the structure is idealized in the overall xy plane and that the local coordinate systems are such that the third local axis is always taken parallel to, and in the direction of, the overall z axis. In axisymmetrical structures we may take the axis of symmetry as the overall y axis. These types of structures are sketched in Fig. 13.5 Referred to the overall coordinate system, the column matrix of deflections at a point of such structures may be displayed as

$$\mathbf{u}^T = [d_x \quad d_y] \tag{13.17}$$

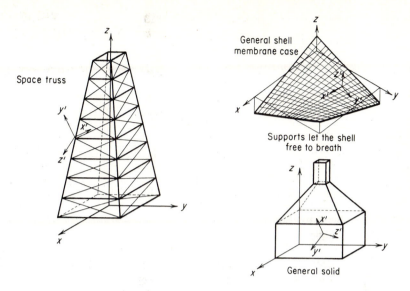

Figure 13.6   Structures with three lineal degrees of freedom per point

where d_x and d_y are the components of the displacement vector along the overall axes. In these structures, the number of degrees of freedom at a point is two. The order of the column matrix of deflections is likewise two. Since the respective deflection components are zero, deleting the third through the last rows and columns of **R** of Eq. (13.14), gives the deflection transformation matrix **R** for these structures as

$$\mathbf{R} = \begin{bmatrix} l_{11} & l_{12} \\ l_{21} & l_{22} \end{bmatrix} \tag{13.18}$$

where $l_{\alpha\beta}$ is the cosine of the angle between βth local axis and αth overall axis. It follows from Eq. (13.16) that $\mathbf{u}' = \mathbf{R}^T\mathbf{u}$.

According to Table 12.4, space trusses, membrane cases of general shells, and general solids have similar deflection components at a point. These structures are sketched in Fig. 13.6. Referred to the overall coordinate system, the column matrix of deflections at a point in these structures can be displayed as

$$\mathbf{u}^T = [d_x \quad d_y \quad d_z] \tag{13.19}$$

where d_x, d_y, and d_z are the components of the displacement vector in the overall coordinate system. The order of the column matrix of deflections and the number of degrees of freedom at a point are both 3. Since the respective deflection components are zero, deleting the last three rows and columns in **R** of Eq. (13.14) gives the deflection transformation matrix of these structures as

$$\mathbf{R} = \begin{bmatrix} l_{11} & l_{12} & l_{13} \\ l_{21} & l_{22} & l_{23} \\ l_{31} & l_{32} & l_{33} \end{bmatrix} \tag{13.20}$$

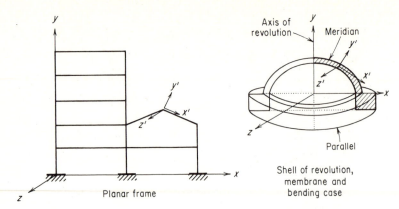

Figure 13.7 Structures with two lineal and one angular degree of freedom per point

where $l_{\alpha\beta}$ is the cosine of the angle between the βth local axis and the αth overall axis. It follows from Eq. (13.16) that $\mathbf{u'} = \mathbf{R}^T\mathbf{u}$.

In the case of planar frames, membranes, and bending shells of revolution undergoing axisymmetrical deformations, we may consider that the overall xy plane contains the idealized structure. If such is the case, the overall y axis can be taken as the axis of revolution. We can select the local coordinate systems in these structures so that the third local axis is always parallel to, and in the direction of, the overall z axis. These structures are sketched in Fig. 13.7. From Table 12.4, the column matrix of deflections in the overall coordinate system can be displayed as

$$\mathbf{u}^T = [d_x \quad d_y \quad \theta_z] \tag{13.21}$$

where d_x and d_y are displacement components along the overall x and y axes and θ_z is the rotation about the overall z axis. The order of the column matrix of deflections at a point is three; likewise the number of degrees of freedom at a point is also three. Since the respective deflection components are zero, deleting the third, fourth, and fifth rows and columns of $\mathbf{R}$ of Eq. (13.14) gives the deflection transfer matrix in these structures as

$$\mathbf{R} = \begin{bmatrix} l_{11} & l_{12} & 0 \\ l_{21} & l_{22} & 0 \\ 0 & 0 & 1 \end{bmatrix} \tag{13.22}$$

where $l_{\alpha\beta}$ is the cosine of the angle between the βth local axis and the αth overall axis and again it follows from Eq. (13.16) that $\mathbf{u'} = \mathbf{R}^T\mathbf{u}$.

In planar gridwork frames and plate-bending problems we select the overall xy plane as the plane of the structure and choose our local coordinate systems so that the third local axis is always parallel to, and in the direction of, the overall z axis. According to Table 12.4, these structures have similar deflection components at a point. The sketches of these structures are given in Fig. 13.8. Referred to the overall coordinate system, the column matrix of deflections at a point can be displayed as

$$\mathbf{u}^T = [d_z \quad \theta_x \quad \theta_y] \tag{13.23}$$

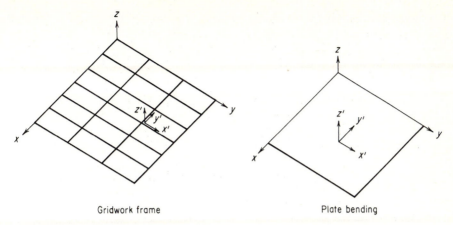

Gridwork frame   Plate bending

Figure 13.8   Structures with one lineal and two angular degrees of freedom per point

where d_z is the displacement component along the overall z axis and θ_x and θ_y are the rotation components about overall x and y axes.   Here, too, the order of the column matrix of deflections and the number of degrees of freedom at a point are both 3.   Since the respective deflection components are zero, deleting the first, second, and sixth rows and columns of $\mathbf{R}$ in Eq. (13.14) gives the deflection transformation matrix for these structures as

$$\mathbf{R} = \begin{bmatrix} 1 & 0 & 0 \\ 0 & l_{11} & l_{12} \\ 0 & l_{21} & l_{22} \end{bmatrix} \tag{13.24}$$

where $l_{\alpha\beta}$ is the cosine of the angle between βth local axis and αth overall axis and it follows from Eq. (13.16) that $\mathbf{u}' = \mathbf{R}^T\mathbf{u}$.

Finally we consider space-frame structures and the membrane and bending cases of general shells.   These structures are sketched in Fig. 13.9.   The number of degrees

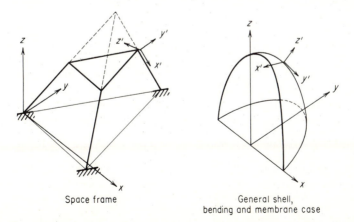

Space frame   General shell,
                                 bending and membrane case

Figure 13.9   Structures with three lineal and three angular degrees of freedom per point

of freedom at a point in these structures is six.   Referred to the overall coordinate system the column matrix of deflections can be displayed as

$$\mathbf{u}^T = [d_x \quad d_y \quad d_z \quad \theta_x \quad \theta_y \quad \theta_z] \tag{13.25}$$

where d_x, d_y, d_z are the components of the displacement vector and θ_x, θ_y, θ_z are the components of the rotation vector.   The deflection transformation matrix is as in Eq. (13.14), which can be displayed explicitly as

$$\mathbf{R} = \begin{bmatrix} l_{11} & l_{12} & l_{13} & & & \\ l_{21} & l_{22} & l_{23} & & & \\ l_{31} & l_{32} & l_{33} & & & \\ & & & l_{11} & l_{12} & l_{13} \\ & & & l_{21} & l_{22} & l_{23} \\ & & & l_{31} & l_{32} & l_{33} \end{bmatrix} \tag{13.26}$$

where $l_{\alpha\beta}$ is the cosine of the angle between the βth local axis and αth overall axis and by Eq. (13.16) $\mathbf{u}' = \mathbf{R}^T\mathbf{u}$.

The repetitiousness of the last five paragraphs will be obvious to the reader. Each refers to a figure and two displayed equations for $\mathbf{u}$ and $\mathbf{R}$.   The figures contain one discrete structure and one or more structures of two- or three-dimensional continua which behave similarly to the discrete structure.   The figures also show relative orientations of the structure and its local and overall coordinate systems. In systematic structural analysis, the features detailed in these figures and displayed equations are not referred to during the description of general procedures, which we shall discuss in Chaps. 15 and 16.   However, the above paragraphs become extremely important when we start the actual computations.

13.2.3   Problems for Solution

Problem 13.7   In a space beam, the columns of the matrix of direction cosines of the local axes are given as

$l_1^T = [0.4 \quad 0.7 \quad 0.5916]$, $l_2^T = [-0.8682 \quad 0.4962 \quad 0]$, and $l_3^T = [-0.2936 \quad -0.5136 \quad 0.8062]$.

The descriptions of the first and the second vertex deflections in the overall coordinate system are $10^{-3}[40 \quad -20 \quad 18 \quad 2 \quad -1 \quad -6]$ and $10^{-3}[3 \quad -25 \quad -12 \quad 3 \quad 5 \quad -1]$, respectively.   Compute the vertex deflections in the local coordinate system of the element.   What is the change in the length of the beam?   How can you compute the length change of the beam directly?

Problem 13.8   The displacement vectors of the vertices in the triangular element of Prob. 13.2 are described in the overall coordinate system as $\mathbf{d}_1^T = 10^{-2}[3 \quad -2 \quad -1]$, $\mathbf{d}_2^T = 10^{-2}[1 \quad 2 \quad 3]$, and $\mathbf{d}_3^T = 10^{-2}[-1 \quad 0 \quad 4]$.   What are the descriptions of the vertex displacement vectors in the local coordinate system of the element?   What are the length changes in the sides of the triangular element?   What is the rotation of the triangle as a rigid body about the local z axis?   (*Hint:* Assume $\mathbf{d}_1$ as translation and find θ_z.]

*Problem 13.9   In linear thin-shell theory, the rotation vector at a middle-surface point is always tangent to the undeformed middle surface.   Suppose that the middle surface is defined in the overall coordinate system by the equation of $f(x, y, z) = 0$ and the description of the rotation vector at a middle surface point is $\mathbf{\theta}^T = [\theta_x \quad \theta_y \quad \theta_z]$, also in the overall coordinate system.   What is the condition which states that $\mathbf{\theta}$ is tangent to the middle surface?   Suppose the boundary of the solid mentioned in Prob. 13.6 is a shell.   Obtain the DBC input units corresponding to the above condition at node 15, which is located at the origin of the local coordinate system worked out in Prob. 13.6.

*Problem 13.10   Consider a spherical truss where the nodes are placed at the intersections of parallels and meridians (like latitudes and longitudes) of the sphere. Suppose that the overall y axis is the axis of symmetry of the structure and the meridional plane containing 0 and 180° longitudes is the overall xy plane. The origin of the overall coordinate system is at the center of the sphere. The truss is undergoing axisymmetrical deformations. Node 10 is placed at 0° longitude and 60° latitude, and the description of its displacement vector in the overall coordinates is $\mathbf{d}_{10}^T = 10^{-2}[-3 \quad 7 \quad 0]$. What is the description of the displacement vector of node 25 which is on 60° latitude and 30° longitude, in the overall coordinates? Because of the axial symmetry, components of displacements of node 25 and node 10 are interdependent. Write down the DBC input units expressing the displacements of node 25 in terms of those of node 10.

Problem 13.11   The first and the second vertex of a gridwork-frame element are placed at nodes 5 and 10, respectively. The descriptions of the position and deflection vectors of these nodes in the overall coordinate system are $\mathbf{c}_5^T = [5 \quad 10 \quad 0]$, $\mathbf{u}_5^T = 10^{-3}[12 \quad -3 \quad 1]$, $\mathbf{c}_{10}^T = [12 \quad 35 \quad 0]$, and $\mathbf{u}_{10}^T = 10^{-3}[2 \quad 4 \quad -3]$. What are the vertex deflections in the local coordinate system of the element? What is the twist angle of the vertex 2 cross section relative to that of vertex 1?

Problem 13.12   The first and the second vertex of a planar-frame element are at nodes 11 and 47, respectively. In the overall coordinate system, the position vectors of the nodes are described as $\mathbf{c}_{11}^T = [5 \quad 2 \quad 0]$ and $\mathbf{c}_{47}^T = [25 \quad 12 \quad 0]$. The description of the deflections of node 11 in the overall coordinate system is $\mathbf{u}_{11}^T = 10^{-3}[10 \quad 4 \quad 7]$. If the matrix of **element deformations** is given by $10^{-3}[-17 \quad 14 \quad 2]$, what is the description of the deflections of node 47 in the overall coordinate system? Element deformations are defined as the relative movements of the cross section at vertex 2 relative to the cross section at vertex 1 provided that these movements are expressed in the local coordinate system of the element (see Art. 13.4.3 for a more precise definition).

13.3   Transformation of Descriptions of Force and Moment Vectors.

During our discussion of force boundary conditions in Art. 12.5, we talked about concentrated forces and the concentrated moments. These entities occur in structural analysis either as external agents loading the structure or as internal quantities representing the normal or moment sums of the stresses in the cross sections of elements, usually at the vertices. Concentrated forces and concentrated moments are vectorial quantities; i.e., they involve a magnitude and a direction. Here we shall study the rules governing the transformation of their descriptions from one cartesian coordinate system to another.

13.3.1   Transformation of cartesian descriptions of forces at a point.

In Fig. 13.10 a concentrated force and a concentrated moment are shown acting at a point A, where two right-handed cartesian coordinate systems are located. Let us call one of these coordinate systems the overall coordinate system and the other the local coordinate system. Let $\mathbf{n}$ and $\mathbf{m}$ denote the descriptions of the force and the moment vectors in the overall coordinate system, respectively. Let the corresponding descriptions in the local coordinate system be $\mathbf{n}'$ and $\mathbf{m}'$, and let $\mathbf{L}$ denote the matrix of direction cosines of the local coordinate system in the overall coordinate system. Using the reasoning which led us to Eq. (13.3), we may write

$$\mathbf{n} = \mathbf{Ln}' \tag{13.27}$$

and

$$\mathbf{m} = \mathbf{Lm}' \tag{13.28}$$

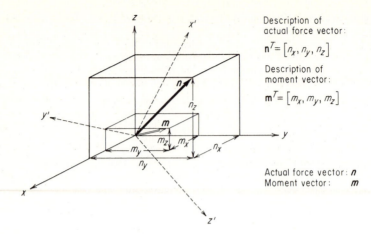

Description of actual force vector:

$$\mathbf{n}^T = [n_x, n_y, n_z]$$

Description of moment vector:

$$\mathbf{m}^T = [m_x, m_y, m_z]$$

Actual force vector: $\mathbf{n}$
Moment vector: $\mathbf{m}$

Figure 13.10   Actual force and moment vectors at a point and their cartesian descriptions

and using the orthogonality of $\mathbf{L}$, namely, $\mathbf{L}^{-1} = \mathbf{L}^T$, we have

$$\mathbf{n}' = \mathbf{L}^T\mathbf{n} \tag{13.29}$$

and

$$\mathbf{m}' = \mathbf{L}^T\mathbf{m} \tag{13.30}$$

We can use these equations in converting the descriptions of concentrated force and moment vectors from one coordinate system to another. Since $\mathbf{L}$ is independent of the locations of coordinate origins, for the transformation formulas listed above, it is not important whether the origins of the coordinate systems and the point of application of the force and moment vectors are coincident with each other or not.

In Art. 12.5 we introduced the term forces for the combined name of the components of the concentrated force and moment vectors acting at a point. Let $\mathbf{q}$ denote the description of the forces at a point in the overall coordinate system such that

$$\mathbf{q}^T = [\mathbf{n}^T \quad \mathbf{m}^T] \tag{13.31}$$

and similarly let $\mathbf{q}'$ denote, the description of the same forces in the local coordinates

$$\mathbf{q}'^T = [\mathbf{n}'^T \quad \mathbf{m}'^T] \tag{13.32}$$

From Eqs. (13.27) and (13.28) the relationship between $\mathbf{q}$ and $\mathbf{q}'$ can be written compactly as

$$\mathbf{q} = \mathbf{R}\mathbf{q}' \tag{13.33}$$

where $\mathbf{R}$ is as defined in Eq. (13.14). According to Eq. (13.15), $\mathbf{R}$ is orthogonal; therefore

$$\mathbf{q}' = \mathbf{R}^T\mathbf{q} \tag{13.34}$$

We observe that the transformation rules for the descriptions of forces at a point are identical with those of the deflections. Indeed this is true even if the number

of degrees of freedom at a point is less than six, as in many structures studied in
the second half of Art. 13.2.　If in a structure certain components of the deflections at a point are identically zero, the corresponding components in the forces
are also identically zero.　This follows the rules of symmetry in Art. 12.4.6.

13.3.2 Specialization of relationship for various groups of structures.　Consider,
for example, the structures shown in Fig. 13.5, where the structures are idealized
in the overall xy plane and local coordinate systems are such that the third local
axis is always parallel to, and in the direction of, the overall z axis.　In these
structures, all components of the forces at a point but the first two are identically
zero, and the description of the forces in the overall coordinate system is

$$\mathbf{q}^T = [n_x \quad n_y] \tag{13.35}$$

where n_x and n_y are the actual force components along the overall x and y axes.
It follows from the rule for selecting the local coordinate systems that $\mathbf{q}'$ also has
similar components.　The relation between $\mathbf{q}$ and $\mathbf{q}'$ is as in Eqs. (13.33) and (13.34),
and $\mathbf{R}$ is as displayed in Eq. (13.18).　Note that in these structures the number of
degrees of freedom at a point is two, and both the force matrix $\mathbf{q}$ and the deflection matrix $\mathbf{u}$ are of order 2.　Moreover, the components of $\mathbf{q}$ are the actual force
components in the coordinate directions, and the components of $\mathbf{u}$ are the displacements in the same directions.　The measuring units of the force components in
planar trusses are in force units; in plane-stress and plane-strain problems they are
also in force units; in axially symmetric solids and membranes they are in units of
force per unit length of parallel.　Axially symmetric structures undergoing axially
symmetric deformations cannot have concentrated loads because of the constraints
imposed by the symmetry conditions.　In these structures, instead of concentrated
loads, we have line loads which are uniformly distributed loads along parallels.

For the structures shown in Fig. 13.6, the components of the forces at a point in
the overall coordinates system can be displayed as

$$\mathbf{q}^T = [n_x \quad n_y \quad n_z] \tag{13.36}$$

where n_x, n_y, n_z are actual force components along the overall coordinate axes.
The description of the forces in the local coordinates, that is, $\mathbf{q}'$, has a similar form.
The relation between $\mathbf{q}$ and $\mathbf{q}'$ is as shown in Eqs. (13.33) and (13.34), where the
transformation matrix is as displayed in Eq. (13.20).　In these structures, the number of degrees of freedom at a point is three, and the matrices for forces and deflections are of order 3.　Components of $\mathbf{q}$ are measured in force units.　Comparing Eq. (13.36) with Eq. (13.19), we observe that corresponding components of
$\mathbf{q}$ and $\mathbf{u}$ are actual force and displacements along the same coordinate direction.

In the structures of Fig. 13.7, the components of the forces at a point in the overall coordinate system are

$$\mathbf{q}^T = [n_x \quad n_y \quad m_z] \tag{13.37}$$

where n_x, n_y are actual force components along the overall x and y axes and m_z
is the moment about the overall z axis.　Since in these structures we select the
local coordinate systems such that the local third axis is always parallel to, and in

the direction of, the overall z axis, the description in local coordinate systems, $\mathbf{q}'$, is in similar form. The transformation between $\mathbf{q}$ and $\mathbf{q}'$ can be performed by Eqs. (13.33) and (13.34) provided that the transformation matrix $\mathbf{R}$ is taken as displayed by Eq. (13.22). In these structures the number of degrees of freedom at a point is three, and so are the orders of the column matrices for its forces and deflections. Comparing Eqs. (13.37) and (13.21), we observe that when a component of the force matrix is an actual force, the corresponding deflection component is a displacement in the same direction, whereas when the component of the force matrix is a moment, the corresponding component of the deflection matrix is a rotation in the same direction. In planar frames actual force components are measured in force units, and moments are measured in force-length units. In shells of revolution undergoing axially symmetric deformations, due to symmetry requirements, forces are line forces which are uniformly distributed along the parallels; therefore, in these structures the actual force components are measured in units of force per unit length of parallel, and the moments are measured in units of force-length per unit length of parallel.

In structures of Fig. 13.8, the components of the forces at a point in the overall coordinate system can be displayed as

$$\mathbf{q}^T = [n_z \quad m_x \quad m_y] \tag{13.38}$$

where n_z is the actual force component along the overall z axis and m_x and m_y are moment components about the overall x and y axes. Since in these structures we select the local coordinate systems so that the local third axis is always parallel to, and in the direction of, the overall z axis, the description of the forces at a point $\mathbf{q}'$ in local coordinate systems is in similar form. Transformation between $\mathbf{q}$ and $\mathbf{q}'$ is accomplished by Eqs. (13.33) and (13.34) provided that the transformation matrix $\mathbf{R}$ is taken as displayed by Eq. (13.24). In these structures the number of degrees of freedom at a point is three, and the deflection and force vectors are of order 3. Comparing Eqs. (13.38) and (13.23), we observe that the corresponding components in these column matrices are either an actual force and a displacement or a moment and a rotation, measured in the same direction. Actual forces are measured in force units, and moments are measured in force-length units.

Finally in the structures of Fig. (13.9), the matrix for forces at a point in the overall coordinate system can be displayed as

$$\mathbf{q}^T = [n_x \quad n_y \quad n_z \quad m_x \quad m_y \quad m_z] \tag{13.39}$$

where n_x, n_y, n_z are the components of actual force along the overall coordinate axes and m_x, m_y, m_z are the moment components about the overall coordinate axes. The description in local coordinate systems, $\mathbf{q}'$, has the same form. Transformation between $\mathbf{q}$ and $\mathbf{q}'$ is accomplished by Eqs. (13.33) and (13.34), and the transformation matrix $\mathbf{R}$ is as displayed in Eq. (13.26). Comparing Eq. (13.39) with Eq. (13.25), we observe that the corresponding components of $\mathbf{q}$ and $\mathbf{u}$ are either actual force and displacement or moment and rotation, effective in the same direction. Actual forces are measured in force units, and moments are measured in force-length units.

To recapitulate, we observe that the order of the column matrix for forces at a point is equal to the number of degrees of freedom at that point and the transformation of the description of the forces at a point uses the same **R** matrix which would be used in the transformation of the description of the deflections at the same point.

13.3.3   Problems for Solution

*Problem 13.13   Consider the space beam mentioned in Prob. 13.7. The description of the forces acting on vertex 2 of the element is expressed in the overall coordinate system as

$$q_2^T = [120 \quad 6 \quad -75 \quad 63 \quad -35 \quad 4].$$

What are the normal force, the shear forces, the torsional moment, and the bending moments at vertex 2?

Problem 13.14   The column matrix listing the forces at the second vertex of the gridwork frame mentioned in Prob. 13.11 is expressed in the overall coordinate system as $q_2^T = [100 \quad 35 \quad -40]$. Compute the numerical values of the transverse shear force, the torsional moment, and the bending moment acting on vertex 2 and show them on a sketch of the cross section at the second vertex.

*Problem 13.15   At the first vertex of the planar beam of Prob. 13.12, the column matrix listing the vertex forces in the overall coordinate system is $q_1^T = [62 \quad 25 \quad 30]$. Compute the numerical values and sketch graphically the shear force, the normal force, and the bending moment acting on the cross section of the first vertex.

Problem 13.16   At the second vertex of a space beam the axial force is 30 kN creating compression, and shearing forces are 20 and 40 kN, both creating negative shearing stresses in the cross section. The bending moments are 130 and 105 kN·m, and both create compression at the first-quadrant portion of the cross section. The torsional moment is 80 kN·m and causes negative shears at the first-quadrant points in planes normal to y. The first and the second vertices of the beam are located at nodes 3 and 15, respectively. The descriptions of the position vectors of these nodes in the overall coordinate system are $c_3^T = [5 \quad 10 \quad 15]$, and $c_{15}^T = [15 \quad 20 \quad -30]$. The description of the position vector of an auxiliary point on the positive side of the local y axis is $c_0^T = [5 \quad 55 \quad 25]$ in the overall coordinate system. What is the description of the forces acting at vertex 2 in the overall coordinate system?

Problem 13.17   Consider the spherical truss of Prob. 13.10. Since the truss is undergoing axially symmetrical deformations, its loading is also axially symmetric. The description of the forces at node 10 is given as $[10 \quad -15 \quad 0]$ in the overall coordinate system. What is the description of the forces at node 25 in the overall coordinate system?

Problem 13.18   A curved space beam is clamped at vertex 1 and subjected to a force which is described by the matrix $q_2^T = [50 \quad 10 \quad 15 \quad 2 \quad -15 \quad 75]$ in the overall coordinate system. The description of the position vector of the points on the beam axis is $c^T = [x \quad y \quad z]$, where x, y, z are the overall coordinates, defined by $x = 3t + 5t^3$, $y = t + t^2$, and $z = 10 \sin t$, such that $t = 0$ corresponds to vertex 1 and $t = 1$ corresponds to vertex 2. The dimensionless parameter t is called the **curve parameter**. The local x axis is tangent to the beam axis and heads away from the beam at vertex 2. The local y axis at vertex 2 heads toward the curvature center of the curve at this point. What are the normal force, the shear forces, the bending moments, and the torsional moment acting on the beam at vertex 2? (*Hint:* If **c** is the cartesian description of the position vector of a point on the curve, $\dot{c}/\|\dot{c}\|$ is the description of the unit tangent vector of the curve at this point heading in the increased t direction and $\dfrac{d}{dt}\left(\dfrac{\dot{c}}{\|\dot{c}\|}\right)$ is the description of a vector heading toward the center of the curvature at the same point. In these definitions a dot shows differentiation once with respect to t. A derivative of a column matrix is obtained by differentiating its components. The derivative of the description of unit tangent vector with respect to the arc length of the curve is the description of its **curvature vector**. The magnitude of the curvature vector is the **curvature** of the curve at the point.)

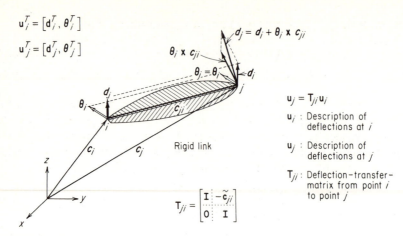

$$\mathbf{u}_i^T = [\mathbf{d}_i^T, \boldsymbol{\theta}_i^T]$$

$$\mathbf{u}_j^T = [\mathbf{d}_j^T, \boldsymbol{\theta}_j^T]$$

$$\boldsymbol{\theta}_i \times c_{ji}$$

$$\boldsymbol{\theta}_j = \boldsymbol{\theta}_i$$

$$d_j = d_i + \boldsymbol{\theta}_i \times c_{ji}$$

Rigid link

$$\mathbf{u}_j = \mathbf{T}_{ji}\mathbf{u}_i$$

$\mathbf{u}_i$: Description of deflections at i

$\mathbf{u}_j$: Description of deflections at j

$\mathbf{T}_{ji}$: Deflection-transfer-matrix from point i to point j

$$\mathbf{T}_{ji} = \begin{bmatrix} \mathbf{I} & -\tilde{\mathbf{c}}_{ji} \\ \mathbf{0} & \mathbf{I} \end{bmatrix}$$

Figure 13.11   Equivalent deflections as the deflections of the vertices of a rigid link

13.4   Kinematically Equivalent Deflections.   Consider two distinct points i and j. If the deflections of these points are related like those of similar points on a rigid link joining them, we call such deflections **kinematically equivalent**.   In a set of kinematically equivalent deflections, knowing deflections at one point enables us to compute all related deflections.   For example, in the rigid link joining point i to point j, if we know the deflections of point i, we can compute the deflections not only of point j but of all points of the rigid link.

13.4.1   Deflection transfer matrix.   In Fig. 13.11, two distinct points i and j, the rigid link joining these points, and an overall coordinate system are shown. Let $\mathbf{u}_i$ and $\mathbf{u}_j$ list the deflections of points i and j in the coordinate system of the figure, and let $\mathbf{c}_i$ and $\mathbf{c}_j$ denote the descriptions of the position vectors of these points in the same coordinate system.   Denoting the descriptions of the displacement vectors by $\mathbf{d}$ and those of rotation vectors by $\boldsymbol{\theta}$ and using the labels of the points as subscripts, we observe that

$$\mathbf{u}_i^T = [\mathbf{d}_i^T \quad \boldsymbol{\theta}_i^T] \tag{13.40}$$

$$\mathbf{u}_j^T = [\mathbf{d}_j^T \quad \boldsymbol{\theta}_j^T] \tag{13.41}$$

Denoting the description of the position vector of point j relative to point i by $\mathbf{c}_{ji}$, we write

$$\mathbf{c}_{ji} = \mathbf{c}_j - \mathbf{c}_i \tag{13.42}$$

We would like to express the relationship between $\mathbf{u}_i$ and $\mathbf{u}_j$ in matrix notation. From rigid-body dynamics we know that

$$d_j = d_i - c_{ji} \times \boldsymbol{\theta}_i \tag{13.43}$$

$$\boldsymbol{\theta}_j = \boldsymbol{\theta}_i$$

which, with the help of Eq. (12.91), can be written in matrix notation as

$$\mathbf{d}_j = \mathbf{d}_i - \tilde{\mathbf{c}}_{ji}\boldsymbol{\theta}_i \tag{13.44}$$

$$\boldsymbol{\theta}_j = \boldsymbol{\theta}_i \tag{13.45}$$

We know that these equations are correct for small rotations and express the fact that if one end of a rigid link is subjected to a displacement and a rotation, the other end of the link undergoes the same rotation; however, its displacement is some amount more than that of the first end. The additional displacement is proportional to the length of the rigid link and its rotation. Defining

$$\mathbf{T}_{ji} = \begin{bmatrix} \mathbf{I} & -\tilde{\mathbf{c}}_{ji} \\ \mathbf{0} & \mathbf{I} \end{bmatrix} \tag{13.46}$$

and using Eqs. (13.40) and (13.41), we can cast Eqs. (13.44) and (13.45) into a single matrix equation

$$\mathbf{u}_j = \mathbf{T}_{ji}\mathbf{u}_i \tag{13.47}$$

which enables us to compute the kinematically equivalent deflections at point j from the deflections of point i. The matrix $\mathbf{T}_{ji}$ is called the **deflection transfer matrix** from point i to point j. Note that when $\mathbf{u}_i$ is of order 6, $\mathbf{T}_{ji}$ is also of order 6, its diagonal partitions are identity matrices of order 3, and the lower left partition is a zero matrix of order 3. The upper right partition is a skew-symmetric matrix of order 3 which is derivable from the description of the position vector of point j relative to point i using Eq. (12.92). In other words, the deflection transfer matrix is known when we know the relative locations of points i and j. By using the deflection transfer matrix from point j to point i we can write the inverse relationship as

$$\mathbf{u}_i = \mathbf{T}_{ij}\mathbf{u}_j \tag{13.48}$$

which implies that

$$\mathbf{T}_{ij}^{-1} = \mathbf{T}_{ji} \tag{13.49}$$

We can easily verify the last equation by observing that $\mathbf{T}_{ij}\mathbf{T}_{ji} = \mathbf{I}$ when we use the definition in Eq. (13.46) and note that $\tilde{\mathbf{c}}_{ij} = -\tilde{\mathbf{c}}_{ji}$.

13.4.2　General deflection transfer matrix. So far we have assumed that the descriptions of the deflections and position vectors are all in the same coordinate system. Now let us suppose that $\mathbf{u}_j$ and $\mathbf{c}_{ji}$ are described in the overall coordinate system and the deflections of point i are given in a local coordinate system as $\mathbf{u}_i'$. We are interested in obtaining the kinematically equivalent deflections to $\mathbf{u}_i'$ at point j expressed in the overall coordinate system. For this purpose we can use Eq. (13.47), but we must first express $\mathbf{u}_i'$ deflections in the overall coordinate system. The knowledge of the matrix of direction cosines of the local coordinate system $\mathbf{L}$ enables us to generate the transformation matrix $\mathbf{R}_i$ as discussed in Art. 13.2.1. and then, using Eq. (13.13), we write $\mathbf{u}_i = \mathbf{R}_i\mathbf{u}_i'$, where $\mathbf{u}_i$ is the description of $\mathbf{u}'_i$ in the overall coordinate system. Then, using Eq. (13.47), we find the required deflections

$$\mathbf{u}_j = \mathbf{T}_{ji}\,\mathbf{R}_i\,\mathbf{u}'_i \tag{13.50}$$

Now let us suppose that the description of the position vector of point j relative to point i and also the deflections at i are both expressed in the local coordinate system and we are interested in obtaining the kinematically equivalent deflections to $\mathbf{u}'_i$ at point j in the overall coordinate system. Let $\mathbf{c}'_{ji}$ denote the description of the position vector in local coordinate system and $\mathbf{T}'_{ji}$ the related deflection transfer matrix in the local coordinate system. Matrix $\mathbf{T}'_{ji}$ can be obtained from the right-hand side of Eq. (13.46) by using $\mathbf{c}'_{ji}$ components. For the equivalent deflection at point j in the local coordinate system, we can write $\mathbf{u}'_j = \mathbf{T}'_{ji}\,\mathbf{u}'_i$ by using Eq. (13.47). Since $\mathbf{u}_j = \mathbf{R}_j\,\mathbf{u}'_j$, the required deflections can be expressed as

$$\mathbf{u}_j = \mathbf{R}_j\,\mathbf{T}'_{ji}\,\mathbf{u}'_i \tag{13.51}$$

In obtaining the kinematically equivalent deflection at point j in the overall coordinate system, we use Eq. (13.50) providing that the deflections at point i are expressed in the local coordinate system, and we use Eq. (13.51) if both the deflections and the description of the position vector of point i are expressed in the local coordinate system. Since $\mathbf{R}_j = \mathbf{R}_i = \mathbf{R}$, from the comparison of Eqs. (13.50) and (13.51) we obtain

$$\mathbf{T}_{ji}\,\mathbf{R} = \mathbf{R}\mathbf{T}'_{ji} = \mathscr{T}_{ji} \tag{13.52}$$

or, using Eqs. (13.46) and (13.14), we get

$$\tilde{\mathbf{c}}_{ji}\,\mathbf{L} = \mathbf{L}\,\tilde{\mathbf{c}}'_{ji} \tag{13.53}$$

which leads to

$$\tilde{\mathbf{c}}'_{ji} = \mathbf{L}^T\tilde{\mathbf{c}}_{ji}\,\mathbf{L} \tag{13.54}$$

Equation (13.54) is also known from the cartesian descriptions of vector products of vectors in three-dimensional space. Using the definitions in Eqs. (13.46) and (13.14) in Eqs. (13.52), we have

$$\mathscr{T}_{ji} = \begin{bmatrix} \mathbf{L} & -\tilde{\mathbf{c}}_{ji}\mathbf{L} \\ \mathbf{0} & \mathbf{L} \end{bmatrix} = \begin{bmatrix} \mathbf{L} & -\mathbf{L}\tilde{\mathbf{c}}'_{ji} \\ \mathbf{0} & \mathbf{L} \end{bmatrix} \tag{13.55}$$

which can be used in Eqs. (13.50) and (13.51) to give

$$\mathbf{u}_j = \mathscr{T}_{ji}\mathbf{u}'_i \tag{13.56}$$

Note that when points i and j are coincident, $\mathscr{T}_{ji}$ reduces to $\mathbf{R}$ and it is identical to $\mathbf{T}_{ji}$ when the coordinate systems are parallel. The inverse relationship may be written

$$\mathbf{u}'_i = \mathscr{T}'_{ij}\mathbf{u}_j \tag{13.57}$$

Comparing this with Eq. (13.56), we observe that

$$\mathscr{T}'_{ij} = \mathscr{T}_{ji}^{-1} \tag{13.58}$$

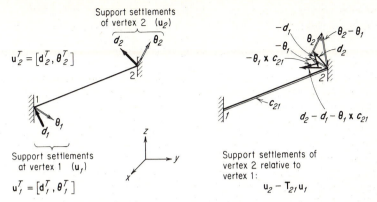

Figure 13.12 Computation of element deformations from the deflections of its vertices

We can easily verify this, noting that the matrix of direction cosines of the overall coordinate system in the local coordinate system is $\mathbf{L}^T$ and $\tilde{\mathbf{c}}_{ij} = -\tilde{\mathbf{c}}_{ji}$. Indeed $\mathscr{T}'_{ij}\mathscr{T}_{ji} = \mathbf{I}$. We shall call $\mathscr{T}_{ji}$ the **general deflection transfer matrix** from the local descriptions at node i to the overall descriptions at node j.

13.4.3. Computation of element deformations. Deflection transfer matrices are very useful in computing the deformations of the structures subjected to rigid-body movements in addition to the usual loadings. For example, consider the space-beam element which is clamped at both vertices, as shown in Fig. 13.12. Suppose the supports of this structure settle so that in the coordinate system of the figure the descriptions of the settlements are $\mathbf{u}_1$ and $\mathbf{u}_2$ at vertices 1 and 2, respectively. If we consider $\mathbf{u}_1$ representing the rigid-body movement, we can compute its equivalent at vertex 2 as $\mathbf{T}_{21}\mathbf{u}_1$, and the movement of vertex 2 relative to vertex 1 can be expressed as $\mathbf{u}_2 - \mathbf{T}_{21}\mathbf{u}_1$. In fact, the stresses developed in the structure when vertex 1 is subjected to settlement $\mathbf{u}_1$ and vertex 2 is subjected to $\mathbf{u}_2$ are identical with those when vertex 1 is not settling and vertex 2 is settling with $\mathbf{u}_2 - \mathbf{T}_{21}\mathbf{u}_1$. The relative settlements, when expressed in the coordinate system of the element, are called the **element deformations.**

Deflection transfer matrices are also useful in determining the **force transfer matrices,** which are the subject of Art. 13.5.

13.4.4   Problems for Solution

Problem 13.19 Compute the column matrix representing the element deformations of the space-beam element of Prob. 13.7. Assume the beam length as 10 units.

Problem 13.20 Resolve Prob. 13.12 by using deflection transfer matrices.

Problem 13.21 In planar frames, the local and overall coordinate systems are selected as shown in Fig. 13.7, and deflections at a point are expressed as shown in Eq. (13.21). Display explicitly the third-order deflection transfer matrices $\mathbf{T}_{ji}$, $\mathscr{T}_{ji}$ and their inverses for planar frames.

Problem 13.22 Figure 13.8 shows the local and overall coordinate systems of gridwork frames, and Eq. (13.23) displays the deflections at a point. Display explicitly the third-order deflection transfer matrices $\mathbf{T}_{ji}$, $\mathscr{T}_{ji}$ and their inverses for gridwork frames.

Problem 13.23   Let $\mathbf{d}_i^T = [d_{ix} \quad d_{iy} \quad d_{iz}]$ and $\mathbf{c}_i^T = [c_{ix} \quad c_{iy} \quad c_{iz}]$ denote the cartesian descriptions of vectors $\boldsymbol{d}_i$ and $\boldsymbol{c}_i$. Let $\boldsymbol{i}_x$, $\boldsymbol{i}_y$, and $\boldsymbol{i}_z$ denote the unit vectors of the cartesian coordinate axes. We know that, for right-hand coordinate systems

$$\boldsymbol{c}_i \times \boldsymbol{d}_i = \det \begin{bmatrix} \boldsymbol{i}_x & \boldsymbol{i}_y & \boldsymbol{i}_z \\ c_{ix} & c_{iy} & c_{iz} \\ d_{ix} & d_{iy} & d_{iz} \end{bmatrix} \tag{13.59}$$

holds. Show that the cartesian description of $\boldsymbol{c}_i \times \boldsymbol{d}_i$ is $\tilde{\mathbf{c}}_i \mathbf{d}_i$, where $\tilde{\mathbf{c}}_i$ is as defined by Eq. (12.92). Let $\mathbf{c}_i'$ and $\mathbf{d}_i'$ denote the descriptions of $\boldsymbol{c}_i$ and $\boldsymbol{d}_i$ in another cartesian coordinate system the matrix of direction cosines of which is given as $\mathbf{L}$. We know that $\mathbf{c}_i = \mathbf{L}\mathbf{c}_i'$ and $\mathbf{d}_i = \mathbf{L}\mathbf{d}_i'$. Prove that $\tilde{\mathbf{c}}_i' = \mathbf{L}^T \tilde{\mathbf{c}}_i \mathbf{L}$.

*Problem 13.24   In order to cover the structural elements with more than two vertices, we generalize the definition of the matrix of element deformations as the list of vertex deflections in local coordinates when the rigid-body movements are eliminated. Consider the triangular element of Prob. 13.8, which has six degrees of freedom in rigid-body movements, i.e., the three translations along the overall axes and the three rotations about them. Suppose all components of $\mathbf{d}_1'$, the last two components of $\mathbf{d}_2'$, and the last component of $\mathbf{d}_3'$ are due to rigid-body movement. Compute the nonzero components of the matrix of element deformations as a column matrix of order 3.

13.5   **Statically Equivalent Forces.**   A statically equivalent force system is one that has the same resultant as the given force system. Statically equivalent force systems are often used for studying the equilibrium relations between external and internal or internal and internal forces of a structure. Given a force system, one can devise infinitely many other force systems which are statically equivalent to the given system. In finding a statically equivalent force system for the actual internal stresses at a section of a structure, the selection of an appropriate one is a very important problem. In this article, we are interested only in establishing the relationships between the statically equivalent forces. The problem of feasibility of a selected statically equivalent force system in the context of an actual structural analysis problem is beyond our present discussion.

A good example of statically equivalent forces is the set of internal forces developed in the cross sections of a cantilevered beam subjected to a transverse concentrated force at the free end, as shown in Fig. 13.13. Note that the internal

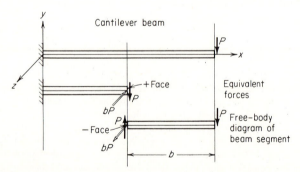

Figure 13.13   Stress resultant and couple of a cross section as equivalent forces to a given load in a cantilever beam

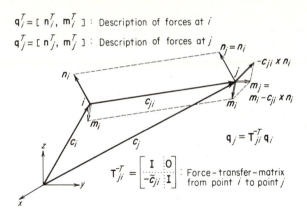

Figure 13.14   Equivalent forces at points i and j

forces, i.e., the bending moment and the transverse shear force on the positive face of a section are statically equivalent to the external tip force, whereas the internal forces on the negative face of a section are the static equilibrants of the external tip force. Note also that the internal forces on the positive faces of any two sections of this beam are statical equivalents of each other. As this example shows, we use the statically equivalent force concept extensively to develop internal-force diagrams of beams, such as normal force, shear force, and bending-moment diagrams. We use the same concept also in defining the stress resultants and stress couples which are statical equivalents of the actual stresses of a cross section, as in Eqs. (12.119) and (12.120). In Art. 13.5 we shall confine ourselves to establishing the relationships between statically equivalent force systems in matrix notation.

13.5.1   Force transfer matrix. Let i and j denote two distinct points, and let $\mathbf{q}_i$ and $\mathbf{q}_j$ denote the descriptions of the forces acting at these points in an overall coordinate system. In Fig. 13.14, the points, their position vectors $\mathbf{c}_i$ and $\mathbf{c}_j$, the force vectors, $\mathbf{n}_i$ and $\mathbf{n}_j$, and the moment vectors $\mathbf{m}_i$ and $\mathbf{m}_j$ acting at these points are shown. Let $\mathbf{c}$, $\mathbf{n}$, and $\mathbf{m}$, with proper subscripts, denote the descriptions of the position, the force, and the moment vectors in the overall coordinate system. By definition [see Eq. (12.121)]

$$\mathbf{q}_i^T = [\mathbf{n}_i^T \quad \mathbf{m}_i^T] \tag{13.60}$$

and

$$\mathbf{q}_j^T = [\mathbf{n}_j^T \quad \mathbf{m}_j^T] \tag{13.61}$$

Let $\mathbf{c}_{ji}$ denote the position vector of point j relative to point i, and let $\mathbf{c}_{ji}$ denote the description of this vector. As in Eq. (13.42), we can write

$$\mathbf{c}_{ji} = \mathbf{c}_j - \mathbf{c}_i \tag{13.62}$$

If $\mathbf{q}_i$ and $\mathbf{q}_j$ are the descriptions of statically equivalent forces, from statics we know that

$$\mathbf{n}_j = \mathbf{n}_i \tag{13.63}$$

$$\mathbf{m}_j = \mathbf{m}_i - \mathbf{c}_{ji} \times \mathbf{n}_i \tag{13.64}$$

which, with the help of Eq. (12.91), can be written in matrix notation

$$\mathbf{n}_j = \mathbf{n}_i \tag{13 65}$$

$$\mathbf{m}_j = \mathbf{m}_i - \tilde{\mathbf{c}}_{ji}\mathbf{n}_i \tag{13.66}$$

Observing from Eq. (13.46) that

$$\mathbf{T}_{ji}^{-T} = \begin{bmatrix} \mathbf{I} & \mathbf{0} \\ -\tilde{\mathbf{c}}_{ji} & \mathbf{I} \end{bmatrix} \tag{13.67}$$

we can express Eqs. (13.65) and (13.66) compactly as

$$\mathbf{q}_j = \mathbf{T}_{ji}^{-T}\,\mathbf{q}_i \tag{13.68}$$

where superscript $-T$ refers to the inverse transpose and the definitions of Eqs. (13.60) and (13.61) are also used. The matrix $\mathbf{T}_{ji}^{-T}$ is called the **force transfer matrix** from point i to point j. It can be used in finding the statically equivalent force components at point j when the description of the forces at point i is available. Note that the force descriptions in both the column matrices $\mathbf{q}_i$ and $\mathbf{q}_j$ are in the same coordinate system. We also observe that the force transfer matrix from point i to point j is the inverse transpose of the deflection transfer matrix from point i to point j. In Eq. (13.68) considering point i as point j and point j as point i, we can write the inverse relationship as

$$\mathbf{q}_i = \mathbf{T}_{ij}^{-T}\,\mathbf{q}_j \tag{13.69}$$

which imples that $\mathbf{T}_{ji} = \mathbf{T}_{ij}^{-1}$, as stated in (13.49).

13.5.2   General force transfer matrix.   In deriving Eqs. (13.67) to (13.69) we assumed that the descriptions listed in column matrices $\mathbf{c}_i$, $\mathbf{c}_j$, $\mathbf{q}_i$, and $\mathbf{q}_j$ are all expressed in the same coordinate system. Now suppose that the force system at point i is given in a local coordinate system by column matrix $\mathbf{q}_i'$ whereas the position vector of point i relative to point j is described in the overall coordinate system. We would like to obtain the description of the equivalent force system at point j in the overall coordinate system. For this purpose, we may first use Eq. (13.33) as $\mathbf{q}_i = \mathbf{R}_i\mathbf{q}_i'$ to obtain the description of the force system at point i in the overall coordinate system and then use this in Eq. (13.68) to write

$$\mathbf{q}_j = \mathbf{T}_{ji}^{-T}\mathbf{R}_i\mathbf{q}_i' \tag{13.70}$$

If the description of the position vector of point j relative to point i and also the force system at point i are both given in the local coordinate system and we are asked to obtain the description of the equivalent force system at point j in the overall coordinate system, we can first express the equivalent force system at point j from the force transfer matrix $\mathbf{T}_{ji}'^{-T}$ in the local coordinate system as $\mathbf{T}_{ji}'^{-T}\mathbf{q}_i'$ and then transform this into the overall coordinate system:

$$\mathbf{q}_j = \mathbf{R}_j\mathbf{T}_{ji}'^{-T}\mathbf{q}_i' \tag{13.71}$$

Since $\mathbf{R}_j = \mathbf{R}_i = \mathbf{R}$ and $\mathbf{R}^{-T} = \mathbf{R}$, the comparison of Eqs. (13.70) and (13.71) leads to

$$\mathbf{T}_{ji}^{-T}\mathbf{R}^{-T} = \mathbf{R}^{-T}\mathbf{T}_{ji}'^{-T} \tag{13.72}$$

Noting that the inverse transpose of a matrix product is the product of inverse transposes of its factors in the same order (provided that inverses exist), we can rewrite the last equation and compare it with Eq. (13.52) to get

$$(\mathbf{T}_{ji}\mathbf{R})^{-T} = (\mathbf{R}\mathbf{T}'_{ji})^{-T} = \mathscr{T}_{ji}^{-T} \tag{13.73}$$

where $\mathscr{T}_{ji}$ is as defined by Eqs. (13.55). Using these equations, we can rewrite Eqs. (13.70) and (13.71) compactly as

$$\mathbf{q}_j = \mathscr{T}_{ji}^{-T}\mathbf{q}'_i \tag{13.74}$$

We can use this equation to obtain the description of the force system at point j in the overall coordinate system equivalent to a given force system at point i which is described in a local coordinate system. Matrix $\mathscr{T}_{ji}^{-T}$ is called the **general force transfer matrix** from the local descriptions at point i to overall descriptions at point j. We observe that the general force transfer matrix from point i to point j is the inverse transpose of the general deflection transfer matrix from point i to point j. The inverse relationship can be written

$$\mathbf{q}'_i = \mathscr{T}_{ij}'^{-T}\mathbf{q}_j \tag{13.75}$$

The comparison of this equation with Eq. (13.74) leads to Eq. (13.58), which can easily be verified, as discussed in Art. 13.4.2.

13.5.3   Statically equilibrant forces at a point. Force transfer matrices are very useful in obtaining the internal-force diagrams of beam elements. They are also very useful in generating force equilibrium equations. For example, let $\mathbf{q}_i$, $i = 1, \ldots, n - 1$, denote the descriptions in an overall coordinate system of $n - 1$ force systems acting at $n - 1$ distinct points. The statical equivalent of these $n - 1$ force systems at point n can be expressed as $\sum_{i=1}^{n-1} \mathbf{T}_{n,i}^{-T}\mathbf{q}_i$. Let $\mathbf{q}_n$ denote the description of the force system at point n, which is in statical equilibrium with the $n - 1$ force systems. Then one can write

$$\mathbf{q}_n = -\sum_{i=1}^{n-1} \mathbf{T}_{n,i}^{-T}\mathbf{q}_i \tag{13.76}$$

which is called the **statically equilibrant force system** at point n of the $n - 1$ force systems.

13.5.4   Problems for Solution

Problem 13.25 Let $\mathbf{q}_i$, $\mathbf{q}_j$, and $\mathbf{q}_k$ denote the descriptions in the overall coordinate system of the forces acting at points i, j, and k of a rigid body. It is known that these forces are in static equilibrium. What is the matrix equation of equilibrium in terms of the transfer matrices to point i? Suppose we apply a deflection system $\mathbf{u}_i$ at point i. Show that the work done by the forces is zero.

Problem 13.26 Consider the space beam of Prob. 13.18. Let points 1 and 2 of the beam be located at the vertices. For points 1 and 2, the values of the curve parameter t are 0 and 1, respectively. Let t_0, $0 < t_0 < 1$, denote the value of t at point 3 of the beam axis. Assume that the local coordinate system at point 3 is such that the local x axis is tangent to the beam axis at point 3 and heading in the increasing t direction and the local y axis is heading toward the center of curvature of the beam axis at point 3. Display explicitly $\mathscr{T}_{32}'^T$, the general force transfer matrix in the local coordinates. Using $t_0 = 1$ and $t_0 = 0$, obtain $\mathscr{T}_{22}'^T$ and $\mathscr{T}_{12}'^T$ matrices and compute the cross-sectional forces at + faces of points 2 and 1.

Problem 13.27 Using $\mathscr{T}_{32}'^T$, the general force transfer matrix in local coordinates of Prob. 13.26, obtain explicit expressions in terms of t_0 for the normal force, the shear forces, the torsional moment, and the bending moments acting on the positive faces and plot them as a function of t_0 to obtain the internal-force diagrams.

Problem 13.28 A gridwork-frame structure is subjected to forces at nodes 3, 15, and 27. The descriptions of these forces in the overall coordinate system are

$$\mathbf{q}_3^T = [-12 \quad 4 \quad 2], \quad \mathbf{q}_{15}^T = [-7 \quad 3 \quad 5], \quad \text{and} \quad \mathbf{q}_{27}^T = [19 \quad 26 \quad 36]$$

[see Eq. (13.38) for the meanings of the components]. The descriptions of the position vectors of these nodes in the overall coordinate system are

$$\mathbf{c}_3^T = [1 \quad 2 \quad 0], \quad \mathbf{c}_{15}^T = [3 \quad 4 \quad 0], \quad \text{and} \quad \mathbf{c}_{27}^T = [4 \quad 1 \quad 0].$$

Using the force transfer matrices, show that these forces are self-equilibrating.

*****Problem 13.29** A planar-cantilever-frame structure is clamped at node 1 and subjected to forces at nodes 3 and 5. The position vectors in the overall coordinate system are $\mathbf{c}_1^T = [0 \quad 0 \quad 0]$, $\mathbf{c}_3^T = [3 \quad 5 \quad 0]$, and $\mathbf{c}_5^T = [6 \quad 15 \quad 0]$. The force vectors in the overall coordinate system are $\mathbf{q}_3^T = [10 \quad 15 \quad -40]$ and $\mathbf{q}_5^T = [2 \quad -20 \quad 15]$ [see Eq. (13.37) for the meanings of these components]. Using the force transfer matrices to node 1, compute the reactions at the clamped support.

Problem 13.30 Prove that the statement in Eq. (13.58) is true.

Problem 13.31 Prove that the statement in Eq. (13.49) is true.

13.6 References

1. Meriam, J. L.: "Statics," 2d ed., John Wiley & Sons, Inc., New York, 1971.

2. Hall, A. S., and R. W. Woodhead: "Frame Analysis," chap. 6, John Wiley & Sons, Inc., New York, 1961.

14

Stiffness and Flexibility Relations of Structural Elements

14.1 Flexibility Relations for Elements of Discrete Structures. Previously we defined a discrete structure as one where interelement boundaries consist of a finite number of discrete points. Several examples are given in Fig. 14.1. Familiar discrete structures are two- and three-dimensional frames and frameworks and gridwork frames. In these familiar discrete structures, the elements are in contact with each other at points called **nodes**. The points of an element at which the element is attached to the nodes are called **vertices**. Familiar discrete structures can be idealized as the assembly of **structural elements with two vertices.** As shown in Fig. 14.1, we can conceive many other discrete structures where an element may have more than two vertices.

Even a familiar discrete structure may be thought of consisting of structural elements with more than two vertices. For example, in Fig. 14.2a a compound truss is shown. Each structurally stable unit of this truss may be considered as an individual **structural element**. In Fig. 14.2b, the elements and their vertices and the nodes of the compound truss are displayed. Note that out of 32 joints of the truss, only 12 are considered nodes. Of course, if we wish, we can consider all of the joints as nodes, in which case the elements of the structure are the familiar bar elements. The choice between the two idealizations depends merely on what we consider as the building blocks of the structure. In large structures, the choice also depends on the cost of analysis. From the short discussion in Art. 12.7, we know that the cost of analysis grows rapidly with the number of nodes and total number of vertices. The idealization in Fig. 14.2b involves 12 nodes and 20 vertices altogether, whereas to idealize the structure by assuming that each joint is a node would involve 32 nodes and 122 vertices. Conceiving of a structure as consisting of structurally stable smaller structures is called **substructuring.**

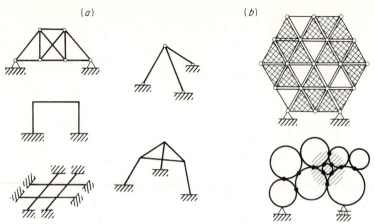

Figure 14.1   Examples of discrete structures: (*a*) familiar types; (*b*) plausible types

The behavior of a structure depends on the behavior of its structural elements and on how these elements are brought together. We quantify the behavior of the structural elements by means of their **flexibility** or **stiffness** relations. These relations merely express quantitatively the dependence of **vertex forces** q_k^m and **vertex deflections** w_k^m (see Fig. 14.2*d*) on each other and the **element loads** (see Art. 12.5.3). In Chap. 14, we shall discuss the flexibility and stiffness relations of the structural elements.

In this article, we give the fundamentals related to the flexibility relations of elements with two or more vertices and confine ourselves to discrete structures. We illustrate these relations with those of the two-vertex elements of the familiar discrete structures, i.e., those of beams and bars.

In discussing the fundamentals, we assume that the vertices are labeled with positive integers as described in Art. 12.2.1. Unless otherwise indicated, the local coordinate system will be used for reference purposes. In the typical element we shall be studying, there are n vertices and the number of degrees of freedom per vertex is e, such that

$$n > 1 \quad \text{and} \quad e > 1 \tag{14.1}$$

The order of the column matrix listing the deflections of the vertices of the element is denoted by r, which is

$$r = ne \tag{14.2}$$

We shall assume throughout that the deformations are small and that the material is linearly elastic with at least two planes of symmetry, the axis of the element being coincident with the intersection of these planes if an axis exists. If the element contains initial stresses, their effect on the flexibility properties will be ignored. With this background information in mind, we shall develop the basic information for the flexibility properties and element loads.

14.1.1   Definitions.   As we embark on a general discussion including multi-vertex elements, it is important to correlate what will follow with related material

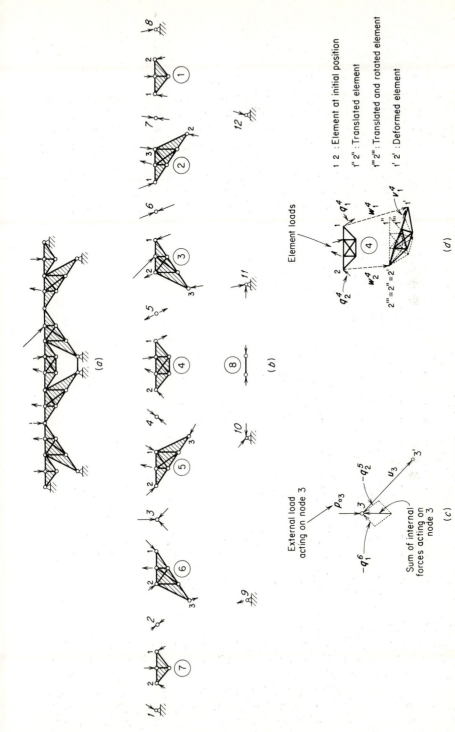

Figure 14.2 Two- and three-vertex element idealization of a planar truss: (*a*) the structure and its loading; (*b*) node, element, and vertex labels and the free-body diagrams of the nodes and elements; (*c*) internal and external forces of a node and its displacement vector; (*d*) free-body diagram of an element and its rigid-body movements and deformation

in Chaps. 12 and 13, where two-vertex elements were often used to illustrate basic definitions. In Art. 12.4 we used u to denote deflections and in Art. 12.5 q to denote the resultant of internal stresses on internal sections of a structure. In Art. 12.5.1 we introduced the terms **vertex forces** and **vertex deflections**, and we became quite specific in using the symbols $\mathbf{q}$ and $\mathbf{u}$, respectively, to denote the column matrices describing these two quantities. In Fig. 12.24 we featured the vertex forces acting on the end cross sections of two-vertex elements. In the following discussions, however, we shall use q (or s) and w (or v) instead of q and u, respectively, to designate the internal forces and deflections at the vertices of the multivertex elements. Moreover, we shall visualize the vertex forces acting on such elements at the actual locations of attachment to the corresponding nodes, without regard to where such points may be located on the configuration of the element. Sometimes vertices are at the ends of element members; but often in multivertex elements they are not and instead are at locations such as shown in Figs. 14.2b and 14.3c.

Consider the free-body diagram of an isolated element of an actual structure after the loads are applied. Such an element is shown in Figs. 14.2d and 14.3c. Whenever we consider the element as a free body, we shall refer to it as an **element in free state**. The list of forces acting at the vertices of the mth element is denoted by the partitioned column matrix $\mathbf{q}'^m$, where the ith partition $\mathbf{q}_i'^m$ is a matrix listing the forces in the degree-of-freedom directions of the ith vertex. Since in this chapter we shall be discussing the properties of a single element quantified in its own local coordinate system, we shall drop the prime and the superscript m and use $\mathbf{q}$ and $\mathbf{q}_i$ for $\mathbf{q}'^m$ and $\mathbf{q}_i'^m$, respectively. The column matrix $\mathbf{q}$ is called the **complete list of vertex forces** of the loaded element in free state. The list of actual deflections of the vertices in the directions of the components of $\mathbf{q}$ will be shown by $\mathbf{w}$. The reader may refer to Fig. 14.2d for a visualization of vertex forces and vertex deflections.

The forces listed in $\mathbf{q}$ are caused by two independent agents, the element loads, discussed in Art. 12.5.3, and the vertex deflections listed in column matrix $\mathbf{w}$. If we want to find the portion of the components of $\mathbf{q}$ directly caused by the element loads only, we take the element, lock all its vertices against deflection to ensure that $\mathbf{w} = \mathbf{0}$, and then load it with its element loads. The reactions thereby developed in the locked vertices are denoted by $\mathbf{q}_0$, and they are equal to the portions of the components of $\mathbf{q}$ which are caused simply by the element loads. Whenever we consider the element with all its vertices locked against deflection, we shall refer to it as the **element in kinematically determinate state**. Note that if at a point of the structure the number of degrees of freedom is e, we lock a joint by restraining the deflections only in the directions associated with the e degrees of freedom. The column matrix $\mathbf{q}_0$ will be referred to as the **complete list of fixed vertex forces of the element.**

If we subtract from the total vertex forces $\mathbf{q}$ the portions $\mathbf{q}_0$ caused by simply the element loads (with $\mathbf{w} = \mathbf{0}$), the remainders $(\mathbf{q} - \mathbf{q}_0)$ will be the portions associated with only the vertex deflections listed in $\mathbf{w}$. We may reason that the vertex forces $\mathbf{q} - \mathbf{q}_0$ of the element result only from the relative deflection (rather than the

absolute deflection) of the vertices of the element.   This conclusion can be justified as follows.   Support the isolated element shown in Fig. 14.3d in a so-called
statically determinate state by fixing it at the last vertex n.[1]   Note that this
statical determinancy refers only to the external support state since the multi-
vertex element may still be internally indeterminate.   Now imagine that we move
this isolated element as a rigid body by translating and rotating its fixed support
through the deflections $\mathbf{w}_n$ specified for vertex n.   Such rigid-body motion encounters no resistance, produces no deformation within the element, and induces
no vertex forces at the other $n - 1$ vertices of the element.   However, only the
vertex n is now in its proper deflected position, while the other $n - 1$ vertices have
undergone the rigid-body movements $\mathscr{T}_{in}\mathbf{w}_n$, $i = 1, \ldots, n'$, where $\mathscr{T}_{in}$ is the general
deflection transfer matrix defined in Art. 13.4.2.[2]   In order to move the other
$n - 1$ vertices into the final vertex positions defined by $\mathbf{w}$, the vertex forces $\mathbf{q} - \mathbf{q}_0$
may be applied and allowed to move these $n - 1$ vertices by the remaining amounts
defined by $\mathbf{v}_i$, where

$$\mathbf{v}_i = \mathbf{w}_i - \mathscr{T}_{in}\mathbf{w}_n \qquad i = 1, \ldots, n' \tag{14.4}$$

Note that physically $\mathbf{v}_i$ represents the relative movements of vertex i with respect
to the support at vertex n.

The vertex deflections $\mathbf{v}_i$, which have been imposed on the element by the vertex
forces $\mathbf{q} - \mathbf{q}_0$, produce deformations of the element identical with those called
element deformations in Art. 13.4.3.   In this and subsequent discussions, we shall
use the term **element deformations** and the symbol $\mathbf{v}$ to denote the vertex deflections
of an element relative to the supports of the element in its *statically determinate
state*.   The element deformations will be described by the partitioned column
matrix $\mathbf{v}$.   Note that $\mathbf{v}$ has n' partitions, where $n' = n - 1$.   All n' partitions are
of the order e except that if $e' > e$, the n'th partition will be of the order $2e - e'$.

Actually in the subsequent considerations of the isolated element in its *statically
determinate state*, we shall deal with the following two separate contributions to
the *total* element deformations represented by $\mathbf{v}$: (1) the element deformations $\mathbf{v}_0$
caused by applying the **element loads** (see Art 12.5.3) to such an element and (2) the
element deformations $\mathbf{v}'$ caused by the vertex forces exerted on the element at its

---

[1] Choosing such a fixed support is not an appropriate selection in some cases for defining the
statically determinate state of the element.   Consider, for example, the element shown in Fig. 14.2d,
where fixing the element at vertex 2 prevents rigid-body displacements but will not prevent rotation
of the element.   To prevent rotation, we would have to introduce one more restraint at vertex 1.
Other statically determinate supports will be discussed in Art. 14.2.6.

To generalize, let e' denote the rigid-body degrees of freedom.   When $e' > e$, provision of only
e restraints at vertex n is insufficient to prevent rigid-body motion of the element.   In such cases,
$e' - e$ more restraints must be provided at vertex $n - 1$ to complete the selection of an adequate
statically determinate support system for the element.   Note that the restrained element in the
statically determinate state is left with r' degrees of freedom, where

$$r' = r - e' \tag{14.3}$$

[2] Note that for situations like those described in the preceding footnote, the computation of
rigid-body movements must be altered to correspond.   For example, how should these computations be modified in order to cover fully the element shown in Fig. 14.2d?

first n' vertices by the neighboring nodes. We shall refer to these latter vertex forces as the **element forces** and list in the column matrix s their components in the direction of the corresponding components of $\mathbf{v}'$. At these first n' vertices, we shall define the element forces as

$$\mathbf{s}_i = \mathbf{q}_i \qquad i = 1, \ldots, n' \tag{14.5}$$

where $\mathbf{s}_i$ and $\mathbf{q}_i$ are the ith partitions of s and q, respectively.

Previously in this article, we considered the element *in a free state*. For such an element, we should note that the vertex forces $\mathbf{q} - \mathbf{q}_0$ associated with the deflections w are in static equilibrium. Thus,

$$\sum_{i=1}^{n} \mathscr{T}_{ni}^{-T}(\mathbf{q}_i - \mathbf{q}_{0i}) = 0 \tag{14.6}$$

where $\mathscr{T}_{ni}^{-T}$ is the general force transfer matrix from vertex i to vertex n (see Art. 13.5.2). Defining

$$\mathbf{q}_n^0 = \sum_{i=1}^{n} \mathscr{T}_{ni}^{-T}\mathbf{q}_{0i} \tag{14.7}$$

and using this definition and the ones in Eq. (14.5) in Eq. (14.6), we can solve for $\mathbf{q}_n$ as

$$\mathbf{q}_n = \mathbf{q}_n^0 - \sum_{i=1}^{n'} \mathscr{T}_{ni}^{-T}\mathbf{s}_i \tag{14.8}$$

Equations (14.5) and (14.8) established the relationship between the element forces s and the vertex forces q.

For the element in the statically determinate state, the total element deformations v due to the combined effect of element loads and element forces can be expressed as

$$\mathbf{v} = \mathbf{v}_0 + \mathbf{v}' \tag{14.9}$$

Note that $\mathbf{v}_i$, the ith partition of the partitioned column matrix v representing the total element deformations, is defined in terms of the absolute vertex deflections w by Eq. (14.4). Note also that in the absence of element loads $\mathbf{v} = \mathbf{v}'$.

14.1.2 Element flexibility matrix. Consider the element in the statically determinate state and in the absence of the element loads. In this configuration, the forces acting on the first n' vertices are listed in the column matrix s and the associated deflections listed in the column matrix $\mathbf{v}'$. We display these partitioned columns as

$$\mathbf{s}^T = [\mathbf{s}_1^T \quad \cdots \quad \mathbf{s}_{n'}^T] \tag{14.10}$$

and

$$\mathbf{v}'^T = [\mathbf{v}_1'^T \quad \cdots \quad \mathbf{v}_{n'}'^T] \tag{14.11}$$

where

$$\mathbf{s}_i^T = [s_{i,1} \quad \cdots \quad s_{i,e}] \tag{14.12}$$

and

$$\mathbf{v}_i'^T = [v_{i,1}' \quad \cdots \quad v_{i,e}'] \tag{14.13}$$

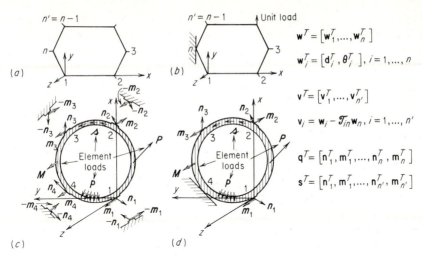

Figure 14.3 Typical elements and their local coordinate systems: (*a*) element in free state; (*b*) element in statically determinate state; (*c*) element in free state (see Fig. 14.1*b*); (*d*) element in statically determinate state (see Fig. 14.1*b*)

We observe that when the element is in statically determinate state with no element loads, the existence of $\mathbf{v}'$ is due to $\mathbf{s}$ or vice versa. Since we assume linear behavior, there must be a simple linear relationship between the components of $\mathbf{s}$ and the components of $\mathbf{v}'$. Let $f_{i,\alpha,j,\beta}$ denote the αth deflection component at vertex i due to a unit force at vertex j in direction β of the element in a statically determinate state with no element loads or other vertex forces present (see Fig. 14.3*b*). The scalar $f_{i,\alpha,j,\beta}$ is the deflection influence coefficient associated with directions identified by the index pairs (i, α) and (j, β). If the element consists of one or more line elements, we can use Eqs. (14.48) and (14.49) to obtain the influence coefficients. Otherwise, the theorem of virtual forces [see Eq. (11.5)] or the complementary strain energy theorem [see Eq. (11.12)] can be used to obtain the influence coefficients. We can use the principle of superposition to obtain the deflection components listed in $\mathbf{v}'$ when all the force components listed in $\mathbf{s}$ are acting simultaneously. When the force component $s_{j,\beta}$ is acting alone in the direction defined by the index pair (j, β), the deflection in the direction defined by the index pair (i, α) is $s_{j,\beta} f_{i,\alpha,j,\beta}$; the deflection when all the forces are acting simultaneously is

$$v'_{i,\alpha} = \sum_{j=1}^{n'} \sum_{\beta=1}^{e} f_{i,\alpha,j,\beta}\, s_{j,\beta} \tag{14.14}$$

Considering all i and α and using Eqs. (14.10) to (14.13), we have

$$\mathbf{v}' = \mathbf{F}\mathbf{s} \tag{14.15}$$

where $\mathbf{F}$ is the matrix of influence coefficients. Note that the entry of $\mathbf{F}$ on the row corresponding to index pair (i, α) and the column corresponding to index pair (j, β) is $f_{i,\alpha,j,\beta}$. Since thanks to the Maxwell's law of reciprocal deflections,

$f_{(i, \alpha), (j, \beta)} = f_{(j, \beta), (i, \alpha)}$, $\mathbf{F}$ is symmetrical; that is, $\mathbf{F} = \mathbf{F}^T$. Matrix $\mathbf{F}$ is called the **flexibility matrix** of the element described in the coordinate system of the element. In Eq. (14.15), $\mathbf{v}'$ and $\mathbf{s}$ are partitioned column matrices. With the rules discussed at the end of Art. 12.1.18, we can partition $\mathbf{F}$ so that the $\mathbf{F}_{ij}$ partition of $\mathbf{F}$ consists of the influence coefficients related with $\mathbf{s}_j$ forces and $\mathbf{v}'_i$ deflections.

The total complementary strain energy of the element is numerically equal to the complementary work done on the first n' vertices of the element as element forces listed in $\mathbf{s}$ build up from zero to their final values. With this we can write

$$\mathcal{U}^* = \tfrac{1}{2}\mathbf{s}^T\mathbf{F}\mathbf{s} \tag{14.16}$$

which, in terms of scalar force components, corresponds to

$$\mathcal{U}^* = \frac{1}{2} \sum_{i=1}^{n'} \sum_{\alpha=1}^{e} \sum_{j=1}^{n'} \sum_{\beta=1}^{e} s_{i, \alpha}\, f_{i, \alpha, j, \beta}\, s_{j, \beta} \tag{14.16'}$$

Since $\mathbf{s}$ lists the forces at the first n' vertices of the element in the statically determinate state, $\mathcal{U}^* \geq 0$, being zero only if all the components of $\mathbf{s}$ are zero. This implies that $\mathcal{U}^*$ is a positive definite quadratic form of $\mathbf{s}$ and $\mathbf{F}$ is a positive definite matrix and therefore $\mathbf{F}^{-1}$ exists (see Art. 12.1.20). If we apply the complementary strain-energy theorem to $\mathcal{U}^*$ as given by Eq. (14.16), the partial derivatives of $\mathcal{U}^*$ with respect to the components of $\mathbf{s}$, that is, the **Jacobian matrix** of $\mathcal{U}^*$, will give us the deflections in the directions of the components of $\mathbf{s}$, namely,

$$\mathbf{v}'^T = \left[\frac{\partial \mathcal{U}^*}{\partial s_{1,\,1}} \quad \cdots \quad \frac{\partial \mathcal{U}^*}{\partial s_{n',\,e}} \right] = \mathcal{U}^*_{,\mathbf{s}} \tag{14.17}$$

Since $\mathbf{v}' = \mathbf{F}\mathbf{s}$, from Eq. (14.17), we can write

$$\mathcal{U}^*_{,\mathbf{s}} = \mathbf{F}\mathbf{s} \tag{14.18}$$

By partial differentiation of both sides with respect to the components of $\mathbf{s}$ and showing this with another comma in the subscript we obtain

$$\mathcal{U}^*_{,,\mathbf{s}} = \mathbf{F} \tag{14.19}$$

which states that $\mathbf{F}$ is the **Hessian matrix** of $\mathcal{U}^*$. In other words, the flexibility matrix is the matrix of second derivatives of the scalar function $\mathcal{U}^*$ representing the total complementary strain energy expressed in terms of the forces. This is a very important conclusion, since in many problems we may know $\mathcal{U}^*$, in which case it is relatively easy to obtain the flexibility matrix $\mathbf{F}$ from Eq. (14.19).

To illustrate the development of the flexibility matrix $\mathbf{F}$ for an element, consider the uniform space-beam element given in Fig. 14.4a. Let A denote the cross-sectional area, I_{yy} and I_{zz} the moments of inertia, and J the torsional constant. We assume that the axis of shear centers is coincident with the x axis. Ignoring the shear effects, we would like to express the flexibility matrix of the element. For this case $n = 2$, $n' = 1$, and $e = 6$; therefore the flexibility matrix is of order 6. The first column of $\mathbf{F}$ is denoted by $\mathbf{f}_1$ and lists the deflections at vertex 1 when a unit force is applied at vertex 1 in the first-degree-of-freedom direction. We observe that $\mathbf{f}_1^T = [L/AE \quad 0 \quad 0 \quad 0 \quad 0 \quad 0]$. The second column of $\mathbf{F}$ is $\mathbf{f}_2$, which can be computed as $\mathbf{f}_2^T = [0 \quad L^3/3EI_{zz} \quad 0 \quad 0 \quad 0 \quad -L^2/2EI_{zz}]$. The third column of $\mathbf{F}$ is $\mathbf{f}_3$, which lists the vertex deflections at vertex 1 due to a unit load in the z direction. We can verify that $\mathbf{f}_3^T = [0 \quad 0 \quad L^3/3EI_{yy} \quad 0 \quad L^2/2EI_{yy} \quad 0]$. If we continue in this fashion, the flexibility matrix of the space-beam element is expressed as

$$
\mathbf{F} =
\begin{bmatrix}
\dfrac{L}{EA} & \cdot & \cdot & \cdot & \cdot & \cdot \\[2ex]
\cdot & \dfrac{L^3}{3EI_{zz}} & \cdot & \cdot & \cdot & -\dfrac{L^2}{2EI_{zz}} \\[2ex]
\cdot & \cdot & \dfrac{L^3}{3EI_{yy}} & \cdot & \dfrac{L^2}{2EI_{yy}} & \cdot \\[2ex]
\cdot & \cdot & \cdot & \dfrac{L}{GJ} & \cdot & \cdot \\[2ex]
\cdot & \cdot & \dfrac{L^2}{2EI_{yy}} & \cdot & \dfrac{L}{EI_{yy}} & \cdot \\[2ex]
\cdot & -\dfrac{L^2}{2EI_{zz}} & \cdot & \cdot & \cdot & \dfrac{L}{EI_{zz}}
\end{bmatrix}
\tag{14.20}
$$

corresponding to element forces $\mathbf{s}^T = [n_{1x} \quad n_{1y} \quad n_{1z} \quad m_{1x} \quad m_{1y} \quad m_{1z}]$ as shown in Fig. 14.4a. In the expression of $\mathbf{F}$, E is Young's modulus and G is the shear modulus.

The flexibility matrices of other familiar line elements can be obtained by specializing the one in Eq. (14.20) (see Prob. 14.1). If we express the total complementary strain energy $\mathcal{U}^*$ of the space-beam element in terms of the forces at vertex 1, we obtain the same matrix as in Eq. (14.20) by the rule shown in Eq. (14.19) (see Prob. 14.2).

14.1.3 Element deformations caused by element loads. In Art. 14.1.1 we defined the element deformations caused by element loads as the deflections of the first n' vertices when the element in the statically determinate state is subjected to its element loads. We defined the element loads in Art. 12.5.3. In Art. 14.1.1, we listed the element deformations caused by element loads in the partitioned column matrix $\mathbf{v}_0$. We display $\mathbf{v}_0$ as

$$
\mathbf{v}_0^T = [\mathbf{v}_{0_1}^T \quad \cdots \quad \mathbf{v}_{0_{n'}}^T]
\tag{14.21}
$$

where

$$
\mathbf{v}_{0_i}^T = [v_{0_{i,\,1}} \quad \cdots \quad v_{0_{i,\,e}}]
\tag{14.22}
$$

After loading the element in the statically determinate state with its element loads, we can compute the components of $\mathbf{v}_0$ either by the complementary-strain-energy theorem or by the theorem of virtual forces. If the element consists of one or more line elements, we can use Eqs. (14.48) and (14.49) to compute the deflections.

As an example, consider the space beam shown in Fig. 14.4b in the statically determinate state with element loads consisting of uniform thermal loads Δt, $\Delta t_{,y}$, $\Delta t_{,z}$, pressure loads $\not\!p_x$, $\not\!p_y$, $\not\!p_z$, a concentrated load at location x_0 with components $P_x, P_y, P_z, M_x, M_y, M_z$, a fabrication error of ΔL (if ΔL is positive, it means that the beam is manufactured with a length of $L + \Delta L$, instead of L), and a shrinkage of α_s (which means that after the shrinkage the beam's length will be $L - \alpha_s L$). Using Eqs. (14.48) and (14.49), the reader can verify that the element deformations listed in $\mathbf{v}_0$ corresponding to this loading are

$$
\mathbf{v}_0 = \begin{Bmatrix} v_{0_{1,1}} \\ v_{0_{1,2}} \\ v_{0_{1,3}} \\ v_{0_{1,4}} \\ v_{0_{1,5}} \\ v_{0_{1,6}} \end{Bmatrix} = \begin{Bmatrix} \dfrac{1}{2}\dfrac{\not\!p_x L^2}{EA} + \dfrac{P_x}{EA}(L - x_0) - \Delta t\,\alpha L + \alpha_s L - \Delta L \\[2mm] \dfrac{\not\!p_y L^4}{8EI_{zz}} + \dfrac{P_y}{6EI_{zz}}(L - x_0)^2(2L + x_0) - \tfrac{1}{2}\Delta t_{,y}\,\alpha L^2 - \dfrac{M_z(L^2 - x_0^2)}{2EI_{zz}} \\[2mm] \dfrac{\not\!p_z L^4}{8EI_{yy}} + \dfrac{P_z}{6EI_{yy}}(L - x_0)^2(2L + x_0) - \tfrac{1}{2}\Delta t_{,z}\,\alpha L^2 + \dfrac{M_y(L^2 - x_0^2)}{2EI_{yy}} \\[2mm] \dfrac{M_x}{GJ}(L - x_0) \\[2mm] \dfrac{\not\!p_z L^3}{6EI_{yy}} + \dfrac{P_z(L - x_0)^2}{2EI_{yy}} - \Delta t_{,z}\,\alpha L + \dfrac{M_y(L - x_0)}{EI_{yy}} \\[2mm] -\dfrac{\not\!p_y L^3}{6EI_{zz}} - \dfrac{P_y(L - x_0)^2}{2EI_{zz}} + \Delta t_{,y}\,\alpha L + \dfrac{M_z(L - x_0)}{EI_{zz}} \end{Bmatrix}
$$

$$(14.23)$$

where E = Young's modulus

$\qquad G$ = shear modulus

$\qquad \alpha$ = thermal-expansion coefficient

$\qquad A$ = cross-sectional area

$\qquad J$ = torsional constant

I_{yy}, I_{zz} = moments of inertia

and where in the computations the shear effects are ignored. In the calculations, the pressure load is considered as a uniformly distributed load. Equation (14.23) can easily be developed for triangular, parabolic, or any other kind of distributed load.

14.1.4 Flexibility relation of element. In Art. 14.1.1, we showed by Eq. (14.9) that the element deformations listed in $\mathbf{v}$ are the sum of two contributions, i.e., the contributions of the element loads (those listed in column matrix $\mathbf{v}_0$) and the contributions of the element forces $\mathbf{s}$ (those listed in column matrix $\mathbf{v}'$). Substituting $\mathbf{v}'$ from Eq. (14.15) into (14.9), we can write

$$\mathbf{v} = \mathbf{Fs} + \mathbf{v}_0 \qquad (14.24)$$

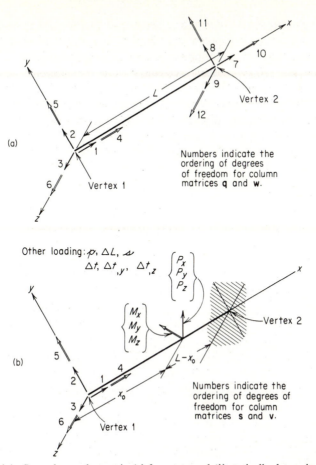

Figure 14.4 Space-beam element in (*a*) free state and (*b*) statically determinate state

which is referred to as the **flexibility relation of the element.** Note that when we know the material and the geometry of the element, by selecting a suitable coordinate system, we can obtain the matrix $\mathbf{F}$ numerically; and in addition if the element loads are present, we can compute the components of $\mathbf{v}_0$. In the flexibility relation of the element, $\mathbf{F}$ represents the intrinsic properties of the element, and $\mathbf{v}_0$ represents the effect of prescribed element loads. Column matrix $\mathbf{s}$ may be considered to list the independent variables, and $\mathbf{v}$ to be the response of the element. Note that in the absence of element loads, $\mathbf{v}_0 = 0$, and the response of the element depends solely upon the element forces listed in $\mathbf{s}$, namely, the forces exerted on the element by its neighbors at its first n' vertices. The element-flexibility relation is one of the concise ways of representing element material, geometry, and loads.

14.1.5 Alternate for element-flexibility relation (total complementary potential energy of element). The element-flexibility relation shown in Eq. (14.24) is a matrix relationship. The information contained in this relationship can also be expressed by means of a scalar function, i.e., the total complementary potential energy of the element.   In order to be able to apply the stationary principle related to the total complementary potential energy whenever the need arises, we must express the total complementary potential energy of the element π^* in terms of the force components satisfying the force-equilibrium requirements (see Art. 11.5). For such force components we may use the ones listed in the element forces column **s**.   Indeed for any arbitrary values of the components of **s**, the force equilibrium of the element is always guaranteed, since the restraining forces $\mathbf{q}_n$ [see Eq. (14.8)] are capable of keeping the **s** force system in equilibrium.   In other words, if we choose the element forces listed in **s** as independent force components, the equilibrium requirement of the element as a free body imposes a condition only on $\mathbf{q}_n$ and not on the components of **s**.

After establishing the fact that the components of **s** are independent as far as the force equilibrium of the element is concerned, we can now express the total complementary potential energy π^* of the element in terms of the components of **s**. We can do this in two ways, as shown below.   Both ways lead to the same expression.

In the first way, we use the definition of the total complementary energy given in Art. 11.5, namely, $\pi^* = \mathscr{U}^* + \mathscr{V}^*$.   The expression for $\mathscr{U}^*$ is available from Eq. (14.16). According to Eq. (14.15), if we start from the element in a kinematically determinate state without the element loads, to develop the **s** forces at the first n' vertices we must move these vertices the amounts listed in **v′**.   According to Eq. (14.9), in order to develop the final values of **s** forces after moving the first n' vertices the amounts listed in **v**, we must start the element in a kinematically determinate state with support settlements in the first n' vertices as listed in $-\mathbf{v}_0$.   If we start from this state, after moving the vertices the amounts listed in **v**, we would have effectively $-\mathbf{v}_0 + \mathbf{v} = \mathbf{v}'$ deformations, which are just enough to create the final values of the forces listed in **s**.   This argument indicates that we can treat the $-\mathbf{v}_0$ deflections as prescribed deflections.   Then by Eq. (11.11), the potential $\mathscr{V}^*$ of the prescribed deflections becomes

$$\mathscr{V}^* = -\mathbf{s}^T(-\mathbf{v}_0) = +\mathbf{s}^T\mathbf{v}_0 \tag{14.25}$$

Now, using $\mathscr{U}^*$ from Eq. (14.16) and $\mathscr{V}^*$ from Eq. (14.25) in $\pi^* = \mathscr{U}^* + \mathscr{V}^*$, we can write

$$\pi^* = \tfrac{1}{2}\mathbf{s}^T\mathbf{F}\mathbf{s} + \mathbf{s}^T\mathbf{v}_0 \tag{14.26}$$

which is the total complementary potential energy of the element.

In the second way, we need not consider the $-\mathbf{v}_0$ deflections as the prescribed deflections of the element.   Instead we argue that there are no prescribed deflections in the element; however, the force-deflection relationships at the first n' vertices of the element are as in Eq. (14.24).   If there are no prescribed deflections, we must have

$$\mathscr{V}^* = 0 \tag{14.27}$$

Since the total complementary strain energy is numerically equal to the complementary work done on the boundaries where the forces are not prescribed, namely, at the first n' vertex points, we can write

$$\mathscr{U}^* = \int_0^s d\mathbf{s}^T \mathbf{v} = \int_0^s d\mathbf{s}^T (\mathbf{Fs} + \mathbf{v}_0) = \tfrac{1}{2}\mathbf{s}^T\mathbf{Fs} + \mathbf{s}^T\mathbf{v}_0 \qquad (14.28)$$

Now, using $\mathscr{V}^*$ and $\mathscr{U}^*$ from Eqs. (14.27) and (14.28) in $\pi^* = \mathscr{U}^* + \mathscr{V}^*$, we obtain

$$\pi^* = \tfrac{1}{2}\mathbf{s}^T\mathbf{Fs} + \mathbf{s}^T\mathbf{v}_0 \qquad (14.29)$$

which is identical to Eq. (14.26).

Note that the complementary-strain-energy expression of Eq. (14.16) excludes the complementary strain energy associated with $\mathbf{v}_0$ deflections, whereas the complementary-strain-energy expression in Eq. (14.28) includes the complementary strain energy associated with $\mathbf{v}_0$ deflections. If we apply the complementary-strain-energy theorem to Eq. (14.16) as the deflections in the directions of the components of $\mathbf{s}$, we obtain $\mathbf{v}'$ [see Eq. (14.17)]; whereas if we apply this theorem to the expression in Eq. (14.28), we obtain $\mathbf{v}$ as in Eq. (14.24). The complementary-strain-energy expression in Eq. (14.28) is preferable, since it relieves us from the burden of remembering whether the element is subjected to initial straining. Whatever the interpretation we associate with $\mathscr{U}^*$ and $\mathscr{V}^*$, we observe that we have a unique expression for the total complementary potential energy of the element. The expression for π^* concisely represents the material, geometry, and loads of the element in the form of a scalar function, which is an alternate form for the element-flexibility relation.

14.1.6   Imposition of interelement-force boundary conditions. In Art. 12.5.1 we saw that in connecting the elements of a structure with each other we can use special devices to ensure that certain components of the vertex forces of an element are zero. If a certain force component is zero at every vertex of the element, we can handle this situation by considering the number of degrees of freedom per point, i.e., the quantity e, 1 less than its nominal value. For example, if the space beam is connected to the structure such that it can deform only in the xy plane, then the value of e is 3 instead of 6 and the matrix $\mathbf{F}$ and the column $\mathbf{v}_0$ are of order 3. Since the three degrees of freedom of the planar-beam points are d_x, d_y, and θ_z, by deleting the third, the fourth, and the fifth rows and columns of $\mathbf{F}$ in Eq. (14.20), we can write the flexibility matrix corresponding to the *planar beam* as

$$\mathbf{F} = \begin{bmatrix} \dfrac{L}{EA} & \cdot & \cdot \\[2ex] \cdot & \dfrac{L^3}{3EI_{zz}} & -\dfrac{L^2}{2EI_{zz}} \\[2ex] \cdot & -\dfrac{L^2}{2EI_{zz}} & \dfrac{L}{EI_{zz}} \end{bmatrix} \qquad (14.30)$$

By deleting the third, fourth, and fifth rows of $\mathbf{v}_0$ in Eq. (14.23) we can write for the corresponding element deformations

$$
\mathbf{v}_0 = \left\{
\begin{array}{l}
\dfrac{p_x L^2}{2EA} + \dfrac{P_x(L - x_0)}{EA} - \Delta t\, \alpha L + \alpha_s L - \Delta L \\[2mm]
\dfrac{p_y L^4}{8EI_{zz}} + \dfrac{P_y(L - x_0)^2(2L + x_0)}{6EI_{zz}} - \tfrac{1}{2}\Delta t_{,y}\, \alpha L^2 - \dfrac{M_z(L^2 - x_0^2)}{2EI_{zz}} \\[2mm]
-\dfrac{L^3 p_y}{6EI_{zz}} - \dfrac{P_y(L - x_0)^2}{2EI_{zz}} + \Delta t_{,y}\, \alpha L + \dfrac{M_z(L - x_0)}{EI_{zz}}
\end{array}
\right\} \quad (14.31)
$$

If we define

$$\mathbf{s}^T = [\,n_{1x} \quad n_{1y} \quad m_{1z}\,] \tag{14.32}$$

and

$$\mathbf{v}^T = [\,d_{1x} \quad d_{1y} \quad \theta_{1z}\,] \tag{14.33}$$

and use $\mathbf{F}$ and $\mathbf{v}_0$ from Eqs. (14.30) and (14.31), the flexibility relations of the planar-beam element are as in Eqs. (14.24) or (14.26).

In *gridwork frames*, the value of e is also 3, and by deleting the first, second, and sixth rows and columns of $\mathbf{F}$ in Eq. (14.20), we obtain its flexibility matrix as

$$
\mathbf{F} =
\begin{bmatrix}
\dfrac{L^3}{3EI_{yy}} & \cdot & \dfrac{L^2}{2EI_{yy}} \\[2mm]
\cdot & \dfrac{L}{GJ} & \cdot \\[2mm]
\dfrac{L^2}{2EI_{yy}} & \cdot & \dfrac{L}{EI_{yy}}
\end{bmatrix}
\tag{14.34}
$$

and upon deleting the first, second, and sixth rows of $\mathbf{v}_0$ in Eq. (14.23), the corresponding column matrix of element deformations becomes

$$
\mathbf{v}_0 = \left\{
\begin{array}{l}
\dfrac{p_z L^4}{8EI_{yy}} + \dfrac{P_z(L - x_0)^2(2L + x_0)}{6EI_{yy}} - \tfrac{1}{2}\Delta t_{,z}\, \alpha L^2 + \dfrac{M_y(L^2 - x_0^2)}{2EI_{yy}} \\[2mm]
\dfrac{M_x}{GJ}(L - x_0) \\[2mm]
\dfrac{p_z L^3}{6EI_{yy}} + \dfrac{P_z(L - x_0)^2}{2EI_{yy}} - \Delta t_{,z}\, \alpha L + \dfrac{M_y(L - x_0)}{EI_{yy}}
\end{array}
\right\} \quad (14.35)
$$

If we define

$$\mathbf{s}^T = [\,n_{1z} \quad m_{1x} \quad m_{1y}\,] \tag{14.36}$$

and

$$\mathbf{v}^T = [\,d_{1z} \quad \theta_{1x} \quad \theta_{1y}\,] \tag{14.37}$$

and use $\mathbf{F}$ and $\mathbf{v}_0$ as defined above, the flexibility relations of the gridwork-frame element remain the same as in Eqs. (14.24) and (14.26).

In *planar and space frameworks*, the values of e are 2 and 3, respectively. However, in these structures we consider only the axial forces and displacements. Therefore, in a bar element of either the planar truss or the space framework we should consider only the axial degree of freedom. From Eq. (14.20), by deleting all rows and columns but the first we obtain

$$\mathbf{F} = \frac{L}{AE} \tag{14.38}$$

and from Eq. (14.23), by deleting all rows but the first we have

$$\mathbf{v}_0 = \frac{\mu_x L^2}{2EA} + \frac{P_x(L - x_0)}{AE} - \Delta t \, \alpha L + \alpha_s L - \Delta L \tag{14.39}$$

Note that both $\mathbf{F}$ and $\mathbf{v}_0$ are of order 1. Defining

$$\mathbf{s} = n_{1x} \quad \text{and} \quad \mathbf{v} = d_{1x} \tag{14.40}$$

we can use Eqs. (14.24) and (14.26) as the flexibility relations of the bar element.

A more interesting case of interelement boundary conditions develops if a vertex-force component is prescribed as zero but not at every vertex of the element (see, for example, element 5-2 in Fig. 12.12). In such elements, during the ordering of the vertices, we make sure that the nth vertex contains no prescribed forces. Then we generate $\mathbf{F}$ and $\mathbf{v}_0$ ignoring the zero vertex force. Finally, we delete the row and the column in $\mathbf{F}$ corresponding to the zero vertex force and delete the row in $\mathbf{v}_0$ corresponding to the zero vertex force. As an example consider the 5-2 beam of the planar frame in Fig. 12.12. Note that $n = 2$, $e = 3$. The matrices $\mathbf{F}$ and $\mathbf{v}_0$ are as in Eqs. (14.30) and (14.31) in the coordinate system of the beam when vertex 2 is clamped. Since the special mechanism at the vertex on node 5 enforces

$$m_{1z} = 0 \tag{14.41}$$

we have

$$\mathbf{F} = \begin{bmatrix} \dfrac{L}{EA} & \cdot \\ \cdot & \dfrac{L^3}{3EI_{zz}} \end{bmatrix} \tag{14.42}$$

$$\mathbf{v}_0 = \left\{ \begin{array}{l} \dfrac{\mu_x L^2}{2EA} + \dfrac{P_x(L - x_0)}{EA} - \Delta t \, \alpha L + \alpha_s L - \Delta L \\[4mm] \dfrac{\mu_y L^4}{8EI_{zz}} + \dfrac{P_y(L - x_0)^2(2L + x_0)}{6EI_{zz}} - \tfrac{1}{2}\Delta t_{,y} \, \alpha L^2 - \dfrac{M_z(L^2 - x_0^2)}{2EI_{zz}} \end{array} \right\} \tag{14.43}$$

$$\mathbf{s}^T = [n_{1x} \quad n_{1y}] \tag{14.44}$$

and

$$\mathbf{v}^T = [d_{1x} \quad d_{1y}] \tag{14.45}$$

The flexibility relations of this element are as in Eqs. (14.24) and (14.26) provided we use $\mathbf{F}$, $\mathbf{v}_0$, $\mathbf{s}$, and $\mathbf{v}$ from Eqs. (14.42) to (14.45) above. If it is not possible to have a vertex without any prescribed vertex force, we select the vertex with only one prescribed vertex-force component as the nth one. Ignoring the fact that one component of force at this vertex is zero, we proceed to generate $\mathbf{F}$ and $\mathbf{v}_0$ in the usual way. When we obtain the flexibility relations as in Eqs. (14.24) and (14.26), we attach to these relations a constraint on $\mathbf{s}$ representing the fact that a component is prescribed as zero. Suppose component β of the force at vertex n is prescribed as zero. We can write the constraint equation from Eq. (14.8) as

$$q_{n\beta} = 0 = q_{n\beta}^0 - \sum_{i=1}^{n'} (\mathcal{T}_{ni}^{-T} \mathbf{s}_i)_\beta \tag{14.46}$$

If more than one component is prescribed as zero, we shall have more than one constraint of this type. For an application, see Prob. 14.9.

To recapitulate our discussions in this article, we observe that the flexibility relation of a discrete element in matrix form is as in Eq. (14.24), where the element flexibility matrix $\mathbf{F}$ and the element-deformation matrix corresponding to element loads $\mathbf{v}_0$ can be obtained by deflection computations for the element in a statically determinate state. The total complementary potential energy of the element is given in Eq. (14.26), which expresses the flexibility relation of the element in the form of a scalar function. If the support of the statically determinate state involves a prescribed vertex force, $\mathbf{F}$ and $\mathbf{v}_0$ can be generated by ignoring it; however, a constraint equation of the type of Eq. (14.46) should be included with Eqs. (14.24) and (14.26).

14.1.7   Problems for Solution

Problem 14.1   In Art. 14.1 after deriving $\mathbf{F}$ and $\mathbf{v}_0$ of uniform space-beam element by ignoring the shear effects in Eqs. (14.20) and (14.23), we specialized them in Eqs. (14.30) to (14.40) for planar beams, gridwork beams, and bars. Obtain $\mathbf{F}$ and $\mathbf{v}_0$ for these structures from the basic definitions of the flexibility matrix and the matrix of element deformations without neglecting shear effects.

Problem 14.2   For planar-beam, gridwork-frame, and bar elements, express the total complementary strain energy in terms of the forces at vertex 1, and then obtain the flexibility matrix by Eq. (14.19). Do not neglect the shear effects.

Problem 14.3   The moment of inertia I_{zz} of a planar-beam element is I_1 at vertex 1 and I_2 at vertex 2. Between the vertices, I_{zz} varies linearly, that is, $I_{zz} = I_1 + (I_2 - I_1)x/L$. What are the flexibility matrix $\mathbf{F}$ and the element deformations $\mathbf{v}_0$ corresponding to a concentrated load P_y at mid-length of the element? The beam is otherwise uniform, and the cross-sectional area is A. Ignore the shear effects.

Problem 14.4   Consider a curved beam referred to its natural coordinate system (see, for example, Prob. 13.18). Let s denote the arc length measured along the beam axis from vertex 1. Let r_y and r_z denote the radii of curvature of the beam axis at a point in the xy and xz planes, respectively. These quantities are positive if on the positive face the local y and z axes are heading toward the center of curvatures. Let

N = normal force
Q_y, Q_z = shear forces in y- and z-axis directions
M_x = torsional moment
M_y, M_z = bending moment about y and z axes
A = cross-sectional area for normal force

A_{sy}, A_{sz} = cross-sectional areas for shear forces in y- and z-axis directions
J = torsional constant
I_{yy}, I_{zz} = principal moment of inertia
E = Young's modulus
G = shear modulus
L = length of beam measured along axis

It can be shown that the total complementary strain energy of the beam is

$$\mathscr{U}^* = \frac{1}{2} \int_0^L \left(\frac{N^2}{AE} + \frac{Q_y^2}{GA_{sy}} + \frac{Q_z^2}{GA_{sz}} + \frac{M_x^2}{GJ} + \frac{M_y^2}{EI_{yy}} + \frac{M_z^2}{EI_{zz}} - 2\frac{M_y N}{AEr_z} + 2\frac{M_z N}{AEr_y} \right) ds \quad (14.47)$$

and the first variation of $\mathscr{U}^*$ can be obtained from this equation as

$$\delta\mathscr{U}^* = \int_0^L \left(\frac{N}{AE} \delta N + \frac{Q_y}{GA_{sy}} \delta Q_y + \frac{Q_z}{GA_{sz}} \delta Q_z + \frac{M_x}{GJ} \delta M_x \right.$$

$$\left. + \frac{M_y}{EI_{yy}} \delta M_y + \frac{M_z}{EI_{zz}} \delta M_z - \frac{\delta M_y N + M_y \delta N}{AEr_z} + \frac{\delta M_z N + M_z \delta N}{AEr_y} \right) ds \quad (14.48)$$

Note that if $\delta N, \delta Q_y, \delta Q_z, \delta M_x, \delta M_y$, and δM_z are the internal forces associated with a unit force to establish a virtual-force system, the deflection u in the direction of the unit load and caused by the loading of $N, Q_y, Q_z, M_x, M_y, M_z$ is

$$u = \delta\mathscr{U}^* \quad (14.49)$$

The terms related to the radii of curvature are negligible if the ratio of the radius of curvature to the beam height in that direction is more than 10. Specialize Eq. (14.48) for planar circular uniform curved beams in the xy plane. Obtain the flexibility matrix $\mathbf{F}$ and element deformations $\mathbf{v}_0$ corresponding to uniform radial pressure p_r of the beam shown in Fig. 14.5. Assume $r/d > 10$, ignore normal-force and shear effects, and use the coordinate system shown for the vertex forces and deflections at vertex 1.

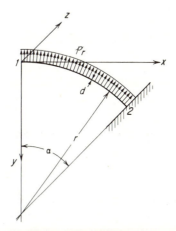

Figure 14.5 Problem 14.4

Problem 14.5   Obtain the inverse of $\mathbf{F}$ in Eq. (14.20) in the following way. First reorder $\mathbf{F}$ by row and column interchanges so that it corresponds to $\mathbf{s}^T = [n_{1x} \quad m_{1x} \quad n_{1y} \quad m_{1z} \quad n_{1z} \quad m_{1y}]$ and $\mathbf{v'}^T = [d_{1x} \quad \theta_{1x} \quad d_{1y} \quad \theta_{1z} \quad d_{1z} \quad \theta_{1y}]$. Then noting that the new $\mathbf{F}$ is diagonal with diagonal submatrices of order 1, 1, 2, and 2, obtain the inverse with $(\text{diag } \mathbf{F}_i)^{-1} = \text{diag } \mathbf{F}_i^{-1}$. Finally reorder diag $\mathbf{F}_i^{-1}$ by column and row interchanges to correspond to

$$\mathbf{s}^T = [n_{1x} \quad n_{1y} \quad n_{1z} \quad m_{1x} \quad m_{1y} \quad m_{1z}] \quad \text{and} \quad \mathbf{v}'^T = [d_{1x} \quad d_{1y} \quad d_{1z} \quad \theta_{1x} \quad \theta_{1y} \quad \theta_{1z}].$$

*Problem 14.6   Consider the planar beam described for Eqs. (14.30) to (14.33). What is v_0 corresponding to the distributed load $p_y = p_1 + (p_2 - p_1)x/L$ acting alone?

*Problem 14.7   Solve Prob. 14.6 for $p_y = x(L - x)4p_0/L^2$ acting alone.

Problem 14.8   Use Eqs. (14.41) to (14.45) to obtain the fixed vertex forces of the uniform beam element shown in Fig. 14.6.

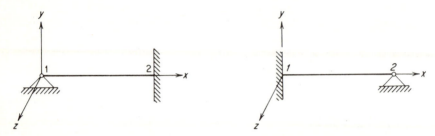

Figure 14.6   Problem 14.8   **Figure 14.7**   Problem 14.9

Problem 14.9   Suppose we have the planar beam of Prob. 14.8 as shown in Fig. 14.7. Use Eqs. (14.30) to (14.33) and the constraint in Eq. (14.46) to obtain the fixed vertex forces. Compare the results with those of Prob. 14.8.

14.2   Stiffness Relations for Elements of Discrete Structures.

In Art. 14.1 the flexibility relations for the elements of discrete structures were developed. In this article we discuss stiffness relations. Earlier we defined a discrete structure as one where interelement boundaries consist of a finite number of discrete points. Several examples were given in Fig. 14.1. Familiar discrete structures are two- and three-dimensional frames and frameworks and gridwork frames. In these familiar discrete structures, each element contains exactly two vertices, and the elements are in contact with each other only at these vertices. As shown in Fig. 14.1, we can imagine many other discrete structures where an element may have more than two vertices. We shall first discuss the fundamentals related with the stiffness relations of elements with two or more vertices. Later we shall special-ize our discussion for the elements of familiar discrete structures. We shall assume throughout that the deformations are small and that the material is linearly elastic with at least two planes of symmetry, the axis of the element always being coincident with the intersection of these planes if an axis exists. If the element contains initial stresses, their effect on the stiffness properties will be ignored.

14.2.1   Definitions.

In Art. 14.1.1, we gave the definitions of the free state, the kinematically determinate state, and the statically determinate state of an element. When we consider the loaded element in the free state, there may be two sets of forces acting on the element: the element loads and the vertex forces. These forces are in statical equilibrium. In Fig. 14.3a, a typical element is shown in the free state with its local coordinate system. There are n vertices, and the degrees of freedom per vertex is e, so that Eq. (14.1) holds. If the element label is j, the list of vertex forces in the free state can be shown by the partitioned column matrix $\mathbf{q}'^j$, where the ith partition $\mathbf{q}_i'^j$ lists the forces at the ith vertex. Note that the prime indicates that the vertex forces are described in the local coordinate system of the element. Since we shall be discussing a single element, as we did in Art. 14.1, from now on we shall drop the prime and the superscript from the descriptions of forces and deflections. Thus, instead of $\mathbf{q}'^j$ and $\mathbf{q}_i'^j$, we shall be using $\mathbf{q}$ and $\mathbf{q}_i$. The list of actual deflections of the vertices in the directions of the components of $\mathbf{q}$ will be shown by $\mathbf{w}$, which is also a partitioned column matrix like $\mathbf{q}$.

Another alternative is to support the isolated element in a statically determinate state by fixing it at the last vertex n, as shown in Fig. 14.3b. If we then move this support through the proper deflections $\mathbf{w}_n$ specified for vertex n, we produce rigid-body movements throughout the element, which are readily computed at the other n' vertices. The remainder of the vertex deflections of these n' vertices are movements relative to the support at n, which are called **element deformations** and represented by the partitioned column matrix $\mathbf{v}$ defined by Eqs. (14.4). We have already discussed these matters in Arts. 13.4.2, 13.4.3, and 14.1.1. In Art. 14.1.1 we defined the vertex forces in the direction of the components of $\mathbf{v}$ as the element forces and listed them in the partitioned column matrix $\mathbf{s}$. The relationship between the partitions of $\mathbf{q}$ and $\mathbf{s}$ is defined in Eqs. (14.5) and (14.8). Note that the orders of $\mathbf{w}$ and $\mathbf{q}$ are r and the orders of $\mathbf{v}$ and $\mathbf{s}$ are r', with r and r' being defined in Eqs. (14.2) and (14.3), respectively.

Considering the element in free state, we observe that the vertex forces $\mathbf{q}$ are caused by two agents: the element loads, which were discussed in Art. 12.5.3, and the vertex deflections listed in $\mathbf{w}$. The portion of $\mathbf{q}$ caused by the element loads alone, i.e., when $\mathbf{w} = \mathbf{0}$, is called the **complete list of the fixed vertex forces** and is listed in the partitioned column matrix $\mathbf{q}_0$. These forces are the reactions developed at the vertices when the element is subjected to its element loads in the kinematically determinate state, i.e., the state of the element when $\mathbf{w} = \mathbf{0}$. Note that $\mathbf{q}_0$ forces are in statical equilibrium with the element loads whereas $-\mathbf{q}_0$ forces are statically equivalent to the element loads. The matrix $-\mathbf{q}_0$ is called the **free-free element load matrix**. The column matrix established by the components of $\mathbf{q}_0$, which are in the directions of $\mathbf{v}$, is called the **fixed vertex forces**, and denoted by $\mathbf{s}_0$. If the statically determinate state is chosen as shown in Fig. 14.3b, the relationship between partitions of $\mathbf{q}_0$ and $\mathbf{s}_0$ can be stated as

$$\mathbf{s}_{0_i} = \mathbf{q}_{0_i} \qquad i = 1, \ldots, n' \tag{14.50}$$

and from the overall equilibrium of the element in the free state and subjected to the element loads and $\mathbf{s}_0$ forces at its first n' vertices, with the help of Eq. (14.8),

$$\mathbf{q}_{0_n} = \mathbf{q}_n^0 - \sum_{i=1}^{n'} \mathcal{T}_{ni}^{-T} \mathbf{s}_{0_i} \tag{14.51}$$

where $\mathbf{q}_n^0$ is the description of the forces located at vertex n, which equilibrate all the element loads.

With the definitions given above, in the remainder of this article we develop the basic information related to the stiffness properties of the element. We shall do this in two stages. In the first stage we discuss the stiffness properties of the element in the statically determinate state, relying heavily on the discussion of Art. 14.1. In the next stage we discuss the stiffness properties of the element in the free state.

14.2.2 Stiffness relations of the element in the statically determinate state. Consider the element in the statically determinate state as shown in Fig. 14.3b. In Art. 14.1.4 we discussed the flexibility relation of the element and established Eq. (14.24), repeated here:

$$\mathbf{v} = \mathbf{Fs} + \mathbf{v}_0 \qquad (14.52)$$

where $\mathbf{F}$ is the flexibility matrix, $\mathbf{v}_0$ is the column matrix of element deformations caused by the element loads, and $\mathbf{s}$ and $\mathbf{v}$ are as defined in Art. 14.2.1, namely, the list of forces and associated deflections at the first n' vertices. Since $\mathbf{F}$ is positive definite, $\mathbf{F}^{-1}$ exists and therefore we can solve for $\mathbf{s}$ from Eq. (14.52)

$$\mathbf{s} = \mathbf{F}^{-1}\mathbf{v} + (-\mathbf{F}^{-1}\mathbf{v}_0) \qquad (14.53)$$

From Eq. (14.52) we observe that the second term on the right of Eq. (14.53) is the value of $\mathbf{s}$ when $\mathbf{v} = \mathbf{0}$. In Eq. (14.50), such values of $\mathbf{s}$ are defined as $\mathbf{s}_0$. Therefore we can write

$$\mathbf{s}_0 = -\mathbf{F}^{-1}\mathbf{v}_0 \qquad (14.54)$$

Defining

$$\mathbf{K} = \mathbf{F}^{-1} \qquad (14.55)$$

we can rewrite Eqs. (14.53) and (14.54) as

$$\mathbf{s} = \mathbf{Kv} + \mathbf{s}_0 \qquad (14.56)$$

and

$$\mathbf{s}_0 = -\mathbf{Kv}_0 \qquad (14.57)$$

The matrix $\mathbf{K}$ is called the **stiffness matrix of the element in the statically determinate state.** The relationship in Eq. (14.56) is the **stiffness relation of the element in the statically determinate state.** We have already defined $\mathbf{s}_0$ in Art. 14.2.1 as the fixed vertex forces of the element.

The definition equations (14.55) and (14.57) are also the equations from which $\mathbf{K}$ and $\mathbf{s}_0$ are computed. Since, by the discussions of Art. 14.1, we now know how to compute $\mathbf{F}$ and $\mathbf{v}_0$ with or without shear effects and curvature effects, using Eqs. (14.55) and (14.57), we can also compute $\mathbf{K}$ and $\mathbf{s}_0$ with and without shear effects and curvature effects.

Sometimes the inversion involved in Eq. (14.55) is easily performed, as in the case of **uniform space beam** without the shear effects. The flexibility matrix $\mathbf{F}$ for this case is given explicitly in Eq. (14.20). Following the instructions given in

Prob. 14.5, we can obtain the stiffness matrix of the element in the statically determinate state as

$$
\mathbf{K} = \begin{bmatrix}
\dfrac{EA}{L} & \cdot & \cdot & \cdot & \cdot & \cdot \\[4mm]
\cdot & \dfrac{12EI_{zz}}{L^3} & \cdot & \cdot & \cdot & \dfrac{6EI_{zz}}{L^2} \\[4mm]
\cdot & \cdot & \dfrac{12EI_{yy}}{L^3} & \cdot & -\dfrac{6EI_{yy}}{L^2} & \cdot \\[4mm]
\cdot & \cdot & \cdot & \dfrac{GJ}{L} & \cdot & \cdot \\[4mm]
\cdot & \cdot & -\dfrac{6EI_{yy}}{L^2} & \cdot & \dfrac{4EI_{yy}}{L} & \cdot \\[4mm]
\cdot & \dfrac{6EI_{zz}}{L^2} & \cdot & \cdot & \cdot & \dfrac{4EI_{zz}}{L}
\end{bmatrix} \tag{14.58}
$$

By substituting $\mathbf{v}_0$ from Eq. (14.23) and $\mathbf{K}$ from Eq. (14.58) into Eq. (14.57), the fixed vertex forces of the element corresponding to the element loads considered in Art. 14.1.3 can be obtained as follows:

$$
\mathbf{s}_0 = \begin{Bmatrix} s_{0_{1,1}} \\[3mm] s_{0_{1,2}} \\[3mm] s_{0_{1,3}} \\[3mm] s_{0_{1,4}} \\[3mm] s_{0_{1,5}} \\[3mm] s_{0_{1,6}} \end{Bmatrix} = \begin{Bmatrix}
-\tfrac{1}{2}p_x L - P_x \dfrac{L - x_0}{L} + \Delta t\, \alpha EA - \alpha_s EA + \dfrac{\Delta L}{L} EA \\[4mm]
-\tfrac{1}{2}p_y L - 2P_y \left(\dfrac{L - x_0}{L}\right)^2 \dfrac{L + 2x_0}{2L} + \dfrac{6M_z}{L}\left(1 - \dfrac{x_0}{L}\right)\dfrac{x_0}{L} \\[4mm]
-\tfrac{1}{2}p_z L - 2P_z \left(\dfrac{L - x_0}{L}\right)^2 \dfrac{L + 2x_0}{2L} - \dfrac{6M_y}{L}\left(1 - \dfrac{x_0}{L}\right)\dfrac{x_0}{L} \\[4mm]
-M_x \dfrac{L - x_0}{L} \\[4mm]
\tfrac{1}{12}p_z L^2 + P_z L\left(\dfrac{L - x_0}{L}\right)^2 \dfrac{x_0}{L} + \Delta t_{,z}\, \alpha EI_{yy} - M_y\left(1 - \dfrac{x_0}{L}\right)\left(1 - 3\dfrac{x_0}{L}\right) \\[4mm]
-\tfrac{1}{12}p_y L^2 - P_y L\left(\dfrac{L - x_0}{L}\right)^2 \dfrac{x_0}{L} - \Delta t_{,y}\, \alpha EI_{zz} - M_z\left(1 - \dfrac{x_0}{L}\right)\left(1 - 3\dfrac{x_0}{L}\right)
\end{Bmatrix} \tag{14.59}
$$

The corresponding $\mathbf{v}$ and $\mathbf{s}$ columns are

$$
\mathbf{v}^T = \begin{bmatrix} d_{1x} & d_{1y} & d_{1z} & \theta_{1x} & \theta_{1y} & \theta_{1z} \end{bmatrix} \tag{14.60}
$$

and

$$
\mathbf{s}^T = \begin{bmatrix} n_{1x} & n_{1y} & n_{1z} & m_{1x} & m_{1y} & m_{1z} \end{bmatrix} \tag{14.61}
$$

For *uniform planar beams*, defining s and v as in Eqs. (14.32) and (14.33) and inverting **F** in Eq. (14.30), we have

$$
\mathbf{K} = \begin{bmatrix} \dfrac{EA}{L} & \cdot & \cdot \\[2ex] \cdot & \dfrac{12EI_{zz}}{L^3} & \dfrac{6EI_{zz}}{L^2} \\[2ex] \cdot & \dfrac{6EI_{zz}}{L^2} & \dfrac{4EI_{zz}}{L} \end{bmatrix}
\tag{14.62}
$$

and using this and $\mathbf{v}_0$ from Eq. (14.31) in Eq. (14.57) gives

$$
\mathbf{s}_0 = \left\{ \begin{array}{l} -\tfrac{1}{2}f_x L - P_x \dfrac{L - x_0}{L} + \Delta t\, \alpha EA - \alpha_s EA + \dfrac{\Delta L}{L} EA \\[3ex] -\tfrac{1}{2}f_y L - 2P_y \left(\dfrac{L - x_0}{L}\right)^2 \dfrac{L + 2x_0}{2L} + \dfrac{6M_z}{L}\left(1 - \dfrac{x_0}{L}\right)\dfrac{x_0}{L} \\[3ex] -\tfrac{1}{12}f_y L^2 - P_y L \left(\dfrac{L - x_0}{L}\right)^2 \dfrac{x_0}{L} - \Delta t_{,y}\, \alpha EI_{zz} - M_z\left(1 - \dfrac{x_0}{L}\right)\left(1 - 3\dfrac{x_0}{L}\right) \end{array} \right\}
\tag{14.63}
$$

Note that **K** and $\mathbf{s}_0$ of the planar uniform beams can also be obtained by deleting the third, fourth, and fifth rows and columns of **K** and $\mathbf{s}_0$ given in Eqs. (14.58) and (14.59). Similarly for a *uniform-gridwork-frame element*, defining s and v as in Eqs. (14.36) and (14.37) and inverting **F** in Eq. (14.34), we obtain

$$
\mathbf{K} = \begin{bmatrix} \dfrac{12EI_{yy}}{L^3} & \cdot & -\dfrac{6EI_{yy}}{L^2} \\[2ex] \cdot & \dfrac{GJ}{L} & \cdot \\[2ex] -\dfrac{6EI_{yy}}{L^2} & \cdot & \dfrac{4EI_{yy}}{L} \end{bmatrix}
\tag{14.64}
$$

and using this together with $\mathbf{v}_0$ from Eq. (14.35) in Eq. (14.57) gives

$$
\mathbf{s}_0 = \left\{ \begin{array}{l} -\tfrac{1}{2}f_z L - 2P_z \left(\dfrac{L - x_0}{L}\right)^2 \dfrac{L + 2x_0}{2L} - \dfrac{6M_y}{L}\left(1 - \dfrac{x_0}{L}\right)\dfrac{x_0}{L} \\[3ex] -M_x \dfrac{L - x_0}{L} \\[3ex] \tfrac{1}{12}f_z L^2 + P_z L \left(\dfrac{L - x_0}{L}\right)^2 \dfrac{x_0}{L} + \Delta t_{,z}\, \alpha EI_{yy} - M_y\left(1 - \dfrac{x_0}{L}\right)\left(1 - 3\dfrac{x_0}{L}\right) \end{array} \right\}
\tag{14.65}
$$

Again we observe that $\mathbf{K}$ and $\mathbf{s}_0$ of the uniform gridwork frame can also be obtained by deleting the first, second, and sixth rows and columns of $\mathbf{K}$ and $\mathbf{s}_0$ given by Eqs. (14.58) and (14.59). Finally, by defining $\mathbf{s}$ and $\mathbf{v}$ as in Eqs. (14.40) we can write for the *uniform bar element*

$$\mathbf{K} = \frac{EA}{L} \tag{14.66}$$

and

$$\mathbf{s}_0 = -\tfrac{1}{2}\rho_x L - P_x \frac{L - x_0}{L} + \Delta t\, \alpha EA - \alpha_s EA + \frac{\Delta L}{L} EA \tag{14.67}$$

where both the matrices are of order 1.

In cases where we cannot invert matrix $\mathbf{F}$ explicitly, it is always preferable first to define $\mathbf{F}$ and $\mathbf{v}_0$ numerically and then obtain $\mathbf{K}$ and $\mathbf{s}_0$ by numerical inversion.

We close this article by observing that the stiffness relation of the element in the statically determinate state is a matrix relation and is one of the concise ways of representing the material, geometry, and loads of the element.

14.2.3 Alternate for element-stiffness relation in the statically determinate state (total potential energy of the element).

The information contained in the matrix relationship discussed in Art. 14.2.2 can also be expressed by means of a scalar function, i.e., the total potential energy of the element. In order to be able to apply the stationary principle related to the total potential energy whenever the need arises, we must express the total potential energy π in terms of the deflection components satisfying the geometric-compatibility requirements (see the definition given in Art. 11.3). Obviously, the components of $\mathbf{v}$ do satisfy the geometric-compatibility requirements as far as the element is concerned. We can express the total potential energy of the element in terms of the components of $\mathbf{v}$ in two ways. Both lead us to the same expression, as shown below.

In the first way, we use the definition of the total potential energy given in Art. 11.3, that is, $\pi = \mathcal{U} + \mathcal{V}$, where $\mathcal{U}$ is the total strain energy of the element and $\mathcal{V}$ is the potential of the prescribed forces. Since the strain energy is numerically equal to the work done on the boundaries where the displacements are not prescribed, namely, at the first n' vertices, we can write

$$\mathcal{U} = \int_0^{\mathbf{v}} d\upsilon^T(\mathbf{s} - \mathbf{s}_0) = \int_0^{\mathbf{v}} d\upsilon^T \mathbf{K}\upsilon = \tfrac{1}{2}\mathbf{v}^T\mathbf{K}\mathbf{v} \tag{14.68}$$

where use is made of Eq. (14.56) and the symmetry property of $\mathbf{K}$. From the discussions in Arts. 14.1.1 and 14.1.2 we know that $-\mathbf{q}_0$ forces are the statical equivalents of the prescribed element loads. Therefore, by Eq. (11.10), we can write

$$\mathcal{V} = -\mathbf{0}^T(-\mathbf{q}_0)_n - \sum_{i=1}^{n'} \mathbf{v}_i^T(-\mathbf{q}_{0_i}) = \sum_{i=1}^{n'} \mathbf{v}_i^T \mathbf{s}_{0_i} = +\mathbf{v}^T\mathbf{s}_0 \tag{14.69}$$

where we used Eqs. (14.50). Now, using $\mathcal{U}$ from Eq. (14.68) and $\mathcal{V}$ from Eq. (14.69) in $\pi = \mathcal{U} + \mathcal{V}$, we obtain

$$\pi = \tfrac{1}{2}\mathbf{v}^T\mathbf{K}\mathbf{v} + \mathbf{v}^T\mathbf{s}_0 \tag{14.70}$$

which is the total potential energy of the element in terms of the element deformations $\mathbf{v}$.

In the second way, we consider that the element is not subjected to any prescribed loads but that its force-deformation relationship at the vertices is as given by Eq. (14.56). Then, since there are no prescribed forces, according to Eq. (11.10), we have

$$\mathscr{V} = 0 \qquad (14.71)$$

The total strain energy is numerically equal to the work done on the boundaries where the deflections are not prescribed, namely, at the first n' vertices:

$$\mathscr{U} = \int_0^{\mathbf{v}} dv^T\mathbf{s} = \int_0^{\mathbf{v}} dv^T(\mathbf{K}v + \mathbf{s}_0) = \tfrac{1}{2}\mathbf{v}^T\mathbf{K}\mathbf{v} + \mathbf{v}^T\mathbf{s}_0 \qquad (14.72)$$

With these, the total-potential-energy expression becomes

$$\pi = \mathscr{U} + \mathscr{V} = \tfrac{1}{2}\mathbf{v}^T\mathbf{K}\mathbf{v} + \mathbf{v}^T\mathbf{s}_0 \qquad (14.73)$$

which is identical to the one in Eq. (14.70).

Note that the strain-energy expression in Eq. (14.68) excludes the strain energy associated with the element loads, whereas the strain-energy expression in Eq. (14.72) includes the strain energy associated with the element loads. If we apply Castigliano's first theorem to the one in Eq. (14.68), as forces in the directions of the components of $\mathbf{v}$, we obtain $\mathbf{s} - \mathbf{s}_0$, whereas if we apply this theorem to the expression in Eq. (14.72), we obtain $\mathbf{s}$. Whatever interpretation we associate with $\mathscr{U}$ and $\mathscr{V}$, we observe that we have a unique expression for the total potential energy of the element. The expression for π concisely represents the material, geometry, and loads of the element in the form of a scalar function of element deformations. It is an alternate form of the element-stiffness relation.

14.2.4 Stiffness relations of the element in the free state. The stiffness relations developed for the element in Arts. 14.2.2 and 14.2.3 are valid only for the associated statically determinate state. In this article we shall develop the stiffness relations of the element in the free state. These latter stiffness relations are useful, since no identification with a related statically determinate support state is necessary. In Arts. 14.1.1 and 14.2.1 we defined the vertex deflections $\mathbf{w}$ and element deformations $\mathbf{v}$ and established the relationship between the two in Eqs. (14.4). We can rewrite these equations compactly as

$$\mathbf{v} = \boldsymbol{\Gamma}\mathbf{w} \qquad (14.74)$$

where
$$\boldsymbol{\Gamma} = \begin{bmatrix} \mathbf{I} & \mathbf{0} & \cdots & -\mathscr{T}_{1,n} \\ \mathbf{0} & \mathbf{I} & \cdots & -\mathscr{T}_{2,n} \\ \cdots\cdots\cdots\cdots\cdots\cdots \\ \mathbf{0} & \mathbf{0} & \mathbf{I} & -\mathscr{T}_{n',n} \end{bmatrix} \qquad (14.75)$$

In the last equation $\mathbf{I}$ and $\mathbf{0}$ are identity and null matrices of order e, respectively.

If we substitute $\mathbf{v}$ from Eq. (14.74) into Eq. (14.68), for the strain energy of the element in free state, we obtain

$$\mathscr{U} = \tfrac{1}{2}\mathbf{w}^T\mathbf{\Gamma}^T\mathbf{K}\mathbf{\Gamma}\mathbf{w} \tag{14.76}$$

or defining

$$\mathscr{K} = \mathbf{\Gamma}^T\mathbf{K}\mathbf{\Gamma} \tag{14.77}$$

which is called the **free-free stiffness matrix of the element,** we have

$$\mathscr{U} = \tfrac{1}{2}\mathbf{w}^T\mathscr{K}\mathbf{w} \tag{14.78}$$

In Arts. 14.1.1 and 14.2.1 we saw that $-\mathbf{q}_0$ forces are statical equivalents of the element loads. Since the potential of statically equivalent forces are the same, for $\mathscr{V}$ of the element loads we can write

$$\mathscr{V} = +\mathbf{w}^T\mathbf{q}_0 \tag{14.79}$$

Note that from Eqs. (14.50) we can determine the first n' partitions of $\mathbf{q}_0$ from knowledge of $\mathbf{s}_0$. The nth partition of $\mathbf{q}_0$ can be computed from Eq. (14.51), which was derived by considering the force equilibrium of the element subjected to the element loads and $\mathbf{s}_0$ forces. Whereas $\mathbf{q}_0$ forces are called the complete list of fixed vertex forces of the element, $-\mathbf{q}_0$ forces are called the **free-free element-load matrix,** even though these $-\mathbf{q}_0$ forces are *not* the element loads themselves but only statically equivalent to them.

Given an element and its loads, we can calculate its free-free stiffness matrix and the free-free load matrix as follows. First, as explained in Art. 14.1, we compute its $\mathbf{F}$ and $\mathbf{v}_0$ matrices. Then, using Eqs. (14.55) and (14.57), we compute $\mathbf{K}$ and $\mathbf{s}_0$ from $\mathbf{F}$ and $\mathbf{v}_0$. Following this, using Eqs. (14.50) and (14.51), we obtain $\mathbf{q}_0$ from $\mathbf{s}_0$ and element loads, and using Eq. (14.77) we obtain $\mathscr{K}$ from $\mathbf{K}$.

Now, substituting $\mathscr{U}$ from Eq. (14.78) and $\mathscr{V}$ from Eq. (14.79) into the total-potential-energy expression $\pi = \mathscr{U} + \mathscr{V}$, we obtain the total potential energy of the element in terms of $\mathbf{w}$ deflections as

$$\pi = \tfrac{1}{2}\mathbf{w}^T\mathscr{K}\mathbf{w} + \mathbf{w}^T\mathbf{q}_0 \tag{14.80}$$

Let $\mathbf{K}_{ij}$ denote the partition of $\mathbf{K}$ in the (i, j) position. Using $\mathbf{K}$ in partitioned form and $\mathbf{\Gamma}$ from Eq. (14.75) in Eq. (14.77), after multiplication we obtain for $\mathscr{K}$ an explicit expression in terms of partitions $\mathbf{K}_{ij}$, $i = 1, \ldots, n', j = 1, \ldots, n'$, and the general deflection transfer matrices $\mathscr{T}_{in}$, $i = 1, \ldots, n'$. We can display this explicit expression of $\mathscr{K}$ as

$$\mathscr{K} = \begin{bmatrix} \mathbf{K}_{11} & \cdots & \mathbf{K}_{1n'} & -\sum\limits_{j=1}^{n'}\mathbf{K}_{1j}\mathscr{T}_{jn} \\ & \cdots & \cdots & \cdots \cdots \cdots \\ & & \mathbf{K}_{n'n'} & -\sum\limits_{j=1}^{n'}\mathbf{K}_{n'j}\mathscr{T}_{jn} \\ \text{Symmetric} & & & \sum\limits_{i=1}^{n'}\sum\limits_{j=1}^{n'}\mathscr{T}_{in}^T\mathbf{K}_{ij}\mathscr{T}_{jn} \end{bmatrix} \tag{14.81}$$

We observe that the first n' by n' partitions of $\mathscr{K}$ are identical to those of $\mathbf{K}$.

Let $\mathbf{k}_t$ denote the tth column of $\mathbf{K}$, and let $\mathscr{k}_t$ denote the tth column of $\mathscr{K}$. We can interpret $\mathbf{k}_t$ by Eq. (14.56) as the forces developed in the first n' vertices of the element when a support settlement of unity happens in the tth direction of the element in the kinematically determinate state. The force developed at vertex n can be computed as $-\sum_{i=1}^{n'} \mathscr{T}_{ni}^{-T} \mathbf{k}_{it}$, where $\mathbf{k}_{it}$ is the ith partition of $\mathbf{k}_t$. Noting that $\mathscr{T}_{ni}^{-T} = \mathscr{T}_{in}^{T}$, we observe from Eq. (14.81) that

$$\mathscr{k}_{it} = \mathbf{k}_{it} \qquad i = 1, \ldots, n' \tag{14.82}$$

and

$$\mathscr{k}_{nt} = - \sum_{i=1}^{n'} \mathscr{T}_{ni}^{-T} \mathbf{k}_{it} \tag{14.83}$$

It follows that $\mathscr{k}_t$ lists forces developed at the n vertices of the element when a support settlement of unity takes place in the tth direction of the element in kinematically determinate state, and these forces are in static equilibrium. Then we state that the *entries in a column of the free-free stiffness matrix $\mathscr{K}$ are forces in static equilibrium and they respresent the reactions at the vertices of the element in the kinematically determinate state when a unit support settlement takes place in the associated direction.* The latter half of this statement offers us an alternative method for computing $\mathscr{K}$.

Since Eq. (14.83) is true for all t, that is, $1 \leq t \leq r = ne$, it tells us that the last e rows (or columns) of $\mathscr{K}$ are linearly dependent on the first $r' = r - e$ rows (or columns); therefore its rank is r' (we assume that $e = e'$; see Art. 14.1.1), although matrix $\mathscr{K}$ is of order r. We conclude that $\mathscr{K}$ does not have an inverse (the order of a square matrix must be equal to its rank for its inverse to exist). We can see this alternately as follows. The total strain energy of the element can be expressed either in terms of $\mathbf{v}$, as in Eq. (14.68), or in terms of $\mathbf{w}$, as in Eq. (14.78), namely,

$$\mathscr{U} = \tfrac{1}{2}\mathbf{v}^T \mathbf{K} \mathbf{v} = \tfrac{1}{2}\mathbf{w}^T \mathscr{K} \mathbf{w} \tag{14.84}$$

The first part of these equations shows that $\mathscr{U}$ is always larger than or equal to zero and is zero only if $\mathbf{v} = \mathbf{0}$. Then it follows that the quadratic form at the extreme right is always larger than or equal to zero and is to zero for nonzero $\mathbf{w}$ [note from Eq. (14.4) that for $\mathbf{v} = \mathbf{0}$ we can have $\mathbf{w} \neq \mathbf{0}$], indicating that $\mathscr{K}$ is only positive but not positive definite. Positive matrices are always singular. This fact should always be kept in mind.

We next discuss the stiffness relations of the element in the free state. For this we shall take advantage of the alternate possibilities for interpretation of the linear portion of the total potential energy π, as discussed in Art. 14.2.3 for the element in the statically determinate state. We have already used one interpretation in deriving Eq. (14.80). In the other interpretation we assume that

$$\mathscr{V} = 0 \tag{14.85}$$

Then

$$\mathscr{U} = \pi = \tfrac{1}{2}\mathbf{w}^T \mathscr{K} \mathbf{w} + \mathbf{w}^T \mathbf{q}_0 \tag{14.86}$$

According to Castigliano's first theorem, the partial derivative of the strain energy with respect to a deflection component gives the force component in the direction of the deflection. Let $\mathbf{q}$ denote the forces in the directions of $\mathbf{w}$. By differentiating in Eq. (14.86) with respect to the components of $\mathbf{w}$ we obtain

$$\mathbf{q} = \mathscr{K}\mathbf{w} + \mathbf{q}_0 \tag{14.87}$$

the **stiffness relations of the element in free state.** Note that in Eq. (14.87) $\mathbf{q}$ lists the vertex forces at the n vertices of the element due to element loads plus the vertex forces associated with the vertex movements $\mathbf{w}$. When $\mathbf{w} = \mathbf{0}$, the element is in the kinematically determinate state and the vertex forces are nothing but the locked reactions $\mathbf{q}_0$ associated with the element loads. In the absence of element loads, the vertex forces are merely a linear function of vertex deflections $\mathbf{w}$, as seen from Eq. (14.87). We observe that the stiffness relations of the element in the free state are characterized by the free-free stiffness matrix $\mathscr{K}$ and the free-free load matrix $-\mathbf{q}_0$.

As an example let us obtain explicit expressions for $\mathscr{K}$ and $-\mathbf{q}_0$ for the **uniform-space-beam element** we have been studying. We have expressions in Eqs. (14.58) and (14.59) for $\mathbf{K} = \mathbf{K}_{1,1}$ and $\mathbf{s}_0 = \mathbf{s}_{0_1}$ of this beam. Since $n = 2$, we have to determine only one general deflection transfer matrix, namely, $\mathscr{T}_{1,2}$. Since we choose the coordinate system at vertex 2 parallel to the one at vertex 1, the matrix of direction cosines of the coordinate axes at vertex 2 relative to the one at vertex 1 is the identity matrix, namely, $\mathbf{L} = \mathbf{I}$, and therefore, according to Eq. (13.14), $\mathbf{R}$ is an identity matrix itself, and with Eq. (13.52) $\mathscr{T}_{1,2} = \mathbf{T}_{1,2}$. The definition of the $\mathbf{T}_{1,2}$ matrix can be obtained from Eq. (13.46). For the beam of length L, $\mathbf{c}_{1,2} = \mathbf{c}_1 - \mathbf{c}_2$ [see Eq. (13.42)] and $\mathbf{c}_1 = 0$, and $\mathbf{c}_2^T = [L \quad 0 \quad 0]$; then $\mathbf{c}_{1,2}^T = -[L \quad 0 \quad 0]$. For the $-\tilde{\mathbf{c}}_{1,2}$ matrix appearing in the definition [Eq. (13.46)], we can write from Eq. (12.92)

$$-\tilde{\mathbf{c}}_{1,2} = \begin{bmatrix} \cdot & \cdot & \cdot \\ \cdot & \cdot & -L \\ \cdot & L & \cdot \end{bmatrix} \tag{14.88}$$

Using this in Eq. (13.46), we obtain

$$\mathbf{T}_{1,2} = \begin{bmatrix} 1 & \cdot & \cdot & \cdot & \cdot & \cdot \\ \cdot & 1 & \cdot & \cdot & \cdot & -L \\ \cdot & \cdot & 1 & \cdot & L & \cdot \\ \cdot & \cdot & \cdot & 1 & \cdot & \cdot \\ \cdot & \cdot & \cdot & \cdot & 1 & \cdot \\ \cdot & \cdot & \cdot & \cdot & \cdot & 1 \end{bmatrix} \tag{14.89}$$

Note that to premultiply a matrix with $\mathbf{T}_{1,2}^T$ amounts to keeping the matrix the same except that the $-L$ multiple of the second row is added to the sixth row and the L multiple of the third row is added to the fifth row. Similarly, postmultiplying a matrix with $\mathbf{T}_{1,2}$ amounts to keeping the matrix the same except that the $-L$ multiple of the second column is added to the sixth column and L multiple of the third column is added to the fifth column. From Eq. (14.81), noting that $n = 2$, we write

$$\mathscr{K} = \begin{bmatrix} \mathbf{K}_{1,1} & \mathstrut & -\mathbf{K}_{1,1}\mathbf{T}_{1,2} \\ \hline -\mathbf{T}_{1,2}^T \mathbf{K}_{1,1} & \mathstrut & \mathbf{T}_{1,2}^T \mathbf{K}_{1,1}\mathbf{T}_{1,2} \end{bmatrix} \tag{14.90}$$

Substituting $\mathbf{K}_{1,1}$ from Eq. (14.58) (note that $\mathbf{K} = \mathbf{K}_{1,1}$) and using $\mathbf{T}_{1,2}$ from Eq. (14.89), we obtain for the uniform-space-beam element

$$
\mathscr{K} = \begin{bmatrix}
\dfrac{EA}{L} & \cdot & \cdot & \cdot & \cdot & \cdot & -\dfrac{EA}{L} & \cdot & \cdot & \cdot & \cdot & \cdot \\
& \dfrac{12EI_{zz}}{L^3} & \cdot & \cdot & \cdot & \dfrac{6EI_{zz}}{L^2} & \cdot & -\dfrac{12EI_{zz}}{L^3} & \cdot & \cdot & \cdot & \dfrac{6EI_{zz}}{L^2} \\
& & \dfrac{12EI_{yy}}{L^3} & \cdot & -\dfrac{6EI_{yy}}{L^2} & \cdot & \cdot & \cdot & -\dfrac{12EI_{yy}}{L^3} & \cdot & -\dfrac{6EI_{yy}}{L^2} & \cdot \\
& & & \dfrac{GJ}{L} & \cdot & \cdot & \cdot & \cdot & \cdot & -\dfrac{GJ}{L} & \cdot & \cdot \\
& & & & \dfrac{4EI_{yy}}{L} & \cdot & \cdot & \cdot & \dfrac{6EI_{yy}}{L^2} & \cdot & \dfrac{2EI_{yy}}{L} & \cdot \\
& & & & & \dfrac{4EI_{zz}}{L} & \cdot & -\dfrac{6EI_{zz}}{L^2} & \cdot & \cdot & \cdot & \dfrac{2EI_{zz}}{L} \\
& & & & & & \dfrac{EA}{L} & \cdot & \cdot & \cdot & \cdot & \cdot \\
& & & & & & & \dfrac{12EI_{zz}}{L^3} & \cdot & \cdot & \cdot & -\dfrac{6EI_{zz}}{L^2} \\
& & \text{Symmetric} & & & & & & \dfrac{12EI_{yy}}{L^3} & \cdot & \dfrac{6EI_{yy}}{L^2} & \cdot \\
& & & & & & & & & \dfrac{GJ}{L} & \cdot & \cdot \\
& & & & & & & & & & \dfrac{4EI_{yy}}{L} & \cdot \\
& & & & & & & & & & & \dfrac{4EI_{zz}}{L}
\end{bmatrix}
$$

$$(14.91)$$

In order to obtain $\mathbf{q}_0$, we must first find the $\mathbf{q}_2^0$ matrix which is the statical equilibrant of element loads at vertex 2. Since only the concentrated loads at x_0 and the uniformly distributed loads will have a statical equilibrant, we can easily verify that in the coordinate system of the element we have

$$
\mathbf{q}_2^0 = - \left\{ \begin{array}{l}
P_x + \not{p}_x L \\[4pt]
P_y + \not{p}_y L \\[4pt]
P_z + \not{p}_z L \\[4pt]
M_x \\[4pt]
M_y + P_z(L - x_0) + \dfrac{\not{p}_z L^2}{2} \\[4pt]
M_z - P_y(L - x_0) - \dfrac{\not{p}_y L^2}{2}
\end{array} \right\}
\qquad (14.92)
$$

Using $\mathbf{q}_2^0$ from this equation, $\mathbf{s}_0 = \mathbf{s}_{0_1}$ from Eq. (14.59), and $\mathbf{T}_{2,1}^{-T}$ from Eq. (14.89) (note that $\mathbf{T}_{2,1}^{-T} = \mathbf{T}_{1,2}^{T}$) from Eq. (14.51) for $n = 2$ we obtain $\mathbf{q}_{0_2}$. Since $\mathbf{q}_{0_1} = \mathbf{s}_{0_1}$ and $\mathbf{q}_0^T = [\mathbf{q}_{0_1}^T \quad \mathbf{q}_{0_2}^T]$, the matrix $-\mathbf{q}_0$ corresponding to the loads described above Eq. (14.23) can be displayed as

$$
-\mathbf{q}_0 = \left\{
\begin{array}{l}
+\tfrac{1}{2}\!\rho_x L + P_x \dfrac{L - x_0}{L} - \Delta t\, \alpha EA + \alpha_s EA - \dfrac{\Delta L}{L} EA \\[2.5ex]
+\tfrac{1}{2}\!\rho_y L + 2P_y \left(\dfrac{L - x_0}{L}\right)^2 \dfrac{L + 2x_0}{2L} - \dfrac{6M_z(L - x_0)x_0}{L^3} \\[2.5ex]
+\tfrac{1}{2}\!\rho_z L + 2P_z \left(\dfrac{L - x_0}{L}\right)^2 \dfrac{L + 2x_0}{2L} + \dfrac{6M_y(L - x_0)x_0}{L^3} \\[2.5ex]
+ M_x \dfrac{L - x_0}{L} \\[2.5ex]
-\dfrac{\rho_z}{12} L^2 - P_z L \left(\dfrac{L - x_0}{L}\right)^2 \dfrac{x_0}{L} - \Delta t_{,z}\,\alpha EI_{yy} + M_y \dfrac{L - x_0}{L}\dfrac{L - 3x_0}{L} \\[2.5ex]
+\dfrac{\rho_y}{12} L^2 + P_y L \left(\dfrac{L - x_0}{L}\right)^2 \dfrac{x_0}{L} + \Delta t_{,y}\,\alpha EI_{zz} + M_z \dfrac{L - x_0}{L}\dfrac{L - 3x_0}{L} \\[2.5ex]
+\tfrac{1}{2}\!\rho_x L + P_x \dfrac{x_0}{L} + \Delta t\, \alpha EA - \alpha_s EA + \dfrac{\Delta L}{L} EA \\[2.5ex]
+\tfrac{1}{2}\!\rho_y L + P_y \left(1 - 2\dfrac{L - x_0}{L}\dfrac{L - x_0}{L}\dfrac{L + 2x_0}{2L}\right) + \dfrac{6M_z(L - x_0)x_0}{L^3} \\[2.5ex]
+\tfrac{1}{2}\!\rho_z L + P_z \left(1 - 2\dfrac{L - x_0}{L}\dfrac{L - x_0}{L}\dfrac{L + 2x_0}{2L}\right) - \dfrac{6M_y(L - x_0)x_0}{L^3} \\[2.5ex]
+ M_x \dfrac{x_0}{L} \\[2.5ex]
+\dfrac{\rho_z}{12} L^2 + P_z L \dfrac{L - x_0}{L}\left(1 - \dfrac{L + 2x_0}{L}\dfrac{L - x_0}{L} + \dfrac{L - x_0}{L}\dfrac{x_0}{L}\right) \\[2.5ex]
\quad + M_y \left(1 - 6\dfrac{L - x_0}{L}\dfrac{x_0}{L} - \dfrac{L - x_0}{L}\dfrac{L - 3x_0}{L}\right) + \Delta t_{,z}\,\alpha EI_{yy} \\[2.5ex]
-\dfrac{\rho_y}{12} L^2 - P_y L \dfrac{L - x_0}{L}\left(1 - \dfrac{L + 2x_0}{L}\dfrac{L - x_0}{L} + \dfrac{L - x_0}{L}\dfrac{x_0}{L}\right) \\[2.5ex]
\quad + M_z \left(1 - 6\dfrac{L - x_0}{L}\dfrac{x_0}{L} - \dfrac{L - x_0}{L}\dfrac{L - 3x_0}{L}\right) - \Delta t_{,y}\,\alpha EI_{zz}
\end{array}
\right.
\qquad (14.93)
$$

We can display the vertex deflections w associated with $\mathcal{K}$ and $\mathbf{q}_0$ of Eqs. (14.91) and (14.93) as

$$\mathbf{w}^T = [d_{1x} \quad d_{1y} \quad d_{1z} \quad \theta_{1x} \quad \theta_{1y} \quad \theta_{1z} \quad d_{2x} \quad d_{2y} \quad d_{2z} \quad \theta_{2x} \quad \theta_{2y} \quad \theta_{2z}] \quad (14.94)$$

where d stands for displacements, θ for rotations, the first subscript for vertex numbers, and the second subscript for component directions.

We can obtain the free-free stiffness matrix and the free-free load matrix of a *uniform planar beam* from those of a uniform space beam by deleting the third, fourth, fifth, ninth, tenth, and eleventh rows and columns in Eqs. (14.91) and (14.93). Alternately, we observe that the deflection transfer matrix in the xy plane can be obtained from Eq. (14.89) by deleting the third, fourth, and fifth rows and columns as

$$\mathbf{T}_{1,2} = \begin{bmatrix} 1 & \cdot & \cdot \\ \cdot & 1 & -L \\ \cdot & \cdot & 1 \end{bmatrix} \quad (14.95)$$

Using $\mathbf{T}_{1,2}$ from this equation and $\mathbf{K} = \mathbf{K}_{1,1}$ from Eq. (14.62) in Eq. (14.90), we obtain the free-free stiffness matrix of the uniform-planar-beam element as

$$\mathcal{K} = \begin{bmatrix} \dfrac{EA}{L} & \cdot & \cdot & -\dfrac{EA}{L} & \cdot & \cdot \\[2ex] & \dfrac{12EI_{zz}}{L^3} & \dfrac{6EI_{zz}}{L^2} & \cdot & -\dfrac{12EI_{zz}}{L^3} & \dfrac{6EI_{zz}}{L^2} \\[2ex] & & \dfrac{4EI_{zz}}{L} & \cdot & -\dfrac{6EI_{zz}}{L^2} & \dfrac{2EI_{zz}}{L} \\[2ex] & & & \dfrac{EA}{L} & \cdot & \cdot \\[2ex] \text{Symmetric} & & & & \dfrac{12EI_{zz}}{L^3} & -\dfrac{6EI_{zz}}{L^2} \\[2ex] & & & & & \dfrac{4EI_{zz}}{L} \end{bmatrix} \quad (14.96)$$

Using $\mathbf{T}_{1,2}$ from Eq. (14.95), $\mathbf{s}_0 = \mathbf{s}_{0_1}$ from Eq. (14.63), and $\mathbf{q}_2^{0T} = -[P_x + \rlap{/}{p}_x L \quad P_y + \rlap{/}{p}_y L \quad M_z - P_y(L-x) - \rlap{/}{p}_y L^2/2]$ in Eq. (14.51), we can obtain the free-free load matrix of the element corresponding to the element loads given above Eq. (14.23) as

$$
-\mathbf{q}_0 = \left\{
\begin{array}{l}
+\tfrac{1}{2}\rlap{/}{p}_x L + P_x \dfrac{L - x_0}{L} - \Delta t\,\alpha EA + \alpha_s E A - \dfrac{\Delta L}{L}\,EA \\[2ex]
+\tfrac{1}{2}\rlap{/}{p}_y L + 2P_y\left(\dfrac{L - x_0}{L}\right)^2 \dfrac{L + 2x_0}{2L} - \dfrac{6M_z(L - x_0)x_0}{L^3} \\[2ex]
+\dfrac{\rlap{/}{p}_y}{12} L^2 + P_y L\left(\dfrac{L - x_0}{L}\right)^2 \dfrac{x_0}{L} + \Delta t_{,y}\,\alpha EI_{zz} + M_z \dfrac{L - x_0}{L}\dfrac{L - 3x_0}{L} \\[2ex]
+\tfrac{1}{2}\rlap{/}{p}_x L + \dfrac{P_x x_0}{L} + \Delta t\,\alpha EA - \alpha_s EA + \dfrac{\Delta L}{L}\,EA \\[2ex]
+\tfrac{1}{2}\rlap{/}{p}_y L + P_y\left(1 - 2\dfrac{L - x_0}{L}\dfrac{L - x_0}{L}\dfrac{L + 2x_0}{2L}\right) + \dfrac{6M_z(L - x_0)x_0}{L^3} \\[2ex]
-\dfrac{\rlap{/}{p}_y}{12} L^2 - P_y L \dfrac{L - x_0}{L}\left(1 - \dfrac{L + 2x_0}{L}\dfrac{L - x_0}{L} + \dfrac{L - x_0}{L}\dfrac{x_0}{L}\right) \\[2ex]
\quad + M_z\left(1 - 6\dfrac{L - x_0}{L}\dfrac{x_0}{L} - \dfrac{L - x_0}{L}\dfrac{L - 3x_0}{L}\right) - \Delta t_{,y}\,\alpha EI_{zz}
\end{array}
\right\}
\tag{14.97}
$$

The vertex deflections $\mathbf{w}$ associated with $\mathcal{K}$ and $\mathbf{q}_0$ of Eqs. (14.96) and (14.97) can be displayed as

$$
\mathbf{w}^T = [d_{1x} \quad d_{1y} \quad \theta_{1z} \quad d_{2x} \quad d_{2y} \quad \theta_{2z}]
\tag{14.98}
$$

where d is used for displacements and θ for rotations.

The free-free stiffness matrix and the free-free load matrix of a *uniform-gridwork-frame-element* can be obtained by deleting the first, second, sixth, seventh, eighth, and twelfth rows and columns of $\mathcal{K}$ and $-\mathbf{q}_0$ of Eqs. (14.91) and (14.93). Alternately, we can use Eq. (14.90) with $\mathbf{K} = \mathbf{K}_{1,1}$ of Eq. (14.64) and $\mathbf{T}_{1,2}$, as shown below:

$$
\mathbf{T}_{1,2} = \begin{bmatrix} 1 & . & L \\ . & 1 & . \\ . & . & 1 \end{bmatrix}
\tag{14.99}
$$

which is easily obtained from $\mathbf{T}_{1,2}$ of Eq. (14.89) by deleting the first, second, and sixth rows and columns. For the free-free load matrix, we can use Eq. (14.51) with $\mathbf{s}_0 = \mathbf{s}_{0_1}$ of Eq. (14.65) and $\mathbf{q}_2^{0^T} = -[P_z + \mathscr{l}_z L \quad M_x \quad M_y + P_z(L - x) + \mathscr{l}_z L^2/2]$. Either method leads to

$$
\mathscr{K} =
\begin{bmatrix}
\dfrac{12EI_{yy}}{L^3} & \cdot & -\dfrac{6EI_{yy}}{L^2} & -\dfrac{12EI_{yy}}{L^3} & \cdot & -\dfrac{6EI_{yy}}{L^2} \\[2ex]
 & \dfrac{GJ}{L} & \cdot & \cdot & -\dfrac{GJ}{L} & \cdot \\[2ex]
 & & \dfrac{4EI_{yy}}{L} & \dfrac{6EI_{yy}}{L^2} & \cdot & \dfrac{2EI_{yy}}{L^2} \\[2ex]
 & & & \dfrac{12EI_{yy}}{L^3} & \cdot & \dfrac{6EI_{yy}}{L^2} \\[2ex]
 & \text{Symmetric} & & & \dfrac{GJ}{L} & \cdot \\[2ex]
 & & & & & \dfrac{4EI_{yy}}{L}
\end{bmatrix}
\tag{14.100}
$$

and

$$
-\mathbf{q}_0 =
\left\{
\begin{array}{l}
+\tfrac{1}{2}\mathscr{l}_z L + 2P_z\left(\dfrac{L - x_0}{L}\right)^2 \dfrac{L + 2x_0}{2L} + \dfrac{6M_y(L - x_0)x_0}{L^3} \\[2.5ex]
+ M_x \dfrac{L - x_0}{L} \\[2.5ex]
-\dfrac{\mathscr{l}_z}{12}L^2 - P_z L\left(\dfrac{L - x_0}{L}\right)^2 \dfrac{x_0}{L} - \Delta t_{,z}\alpha EI_{yy} + M_y \dfrac{L - x_0}{L}\dfrac{L - 3x_0}{L} \\[2.5ex]
+\tfrac{1}{2}\mathscr{l}_z L + P_z\left(1 - 2\dfrac{L - x_0}{L}\dfrac{L - x_0}{L}\dfrac{L + 2x_0}{2L}\right) - \dfrac{6M_y(L - x_0)x_0}{L^3} \\[2.5ex]
+ M_x \dfrac{x_0}{L} \\[2.5ex]
+\dfrac{\mathscr{l}_z}{12}L^2 + P_z L \dfrac{L - x_0}{L}\left(1 - \dfrac{L + 2x_0}{L}\dfrac{L - x_0}{L} + \dfrac{L - x_0}{L}\dfrac{x_0}{L}\right) \\[2.5ex]
\quad + M_y\left(1 - 6\dfrac{L - x_0}{L}\dfrac{x_0}{L} - \dfrac{L - x_0}{L}\dfrac{L - 3x_0}{L}\right) + \Delta t_{,z}\alpha EI_{yy}
\end{array}
\right\}
\tag{14.101}
$$

The vertex deflections $\mathbf{w}$ associated with $\mathcal{K}$ and $\mathbf{q}_0$ of Eqs. (14.100) and (14.101) are

$$\mathbf{w}^T = [d_{1z} \quad \theta_{1x} \quad \theta_{1y} \quad d_{2z} \quad \theta_{2x} \quad \theta_{2y}] \tag{14.102}$$

where d denotes displacements and θ denotes rotations.

The free-free stiffness matrix and the free-free load matrix of a *bar element* can be obtained by deleting the second, third, fourth, fifth, sixth, eighth, ninth, tenth, eleventh, and twelfth rows and columns of $\mathcal{K}$ and $-\mathbf{q}_0$ of Eqs. (14.91) and (14.93). Or alternately, we can use Eq. (14.90) with $\mathbf{K} = \mathbf{K}_{1,1}$ of Eq. (14.66) and $\mathbf{T}_{1,2}$:

$$\mathbf{T}_{1,2} = \begin{bmatrix} 1 & \cdot \\ \cdot & 1 \end{bmatrix} \tag{14.103}$$

which can be obtained from $\mathbf{T}_{1,2}$ of Eq. (14.89) by deleting the third, fourth, fifth, and sixth rows and columns. For the free-free load matrix, we can use Eqs. (14.51) with $\mathbf{s}_0 = \mathbf{s}_{0_1}$ of Eq. (14.67) and $-\mathbf{q}_2^0 = P_x + \not{p}_x L$. Either method leads to

$$\mathcal{K} = \begin{bmatrix} \dfrac{EA}{L} & -\dfrac{EA}{L} \\[2mm] -\dfrac{EA}{L} & \dfrac{EA}{L} \end{bmatrix} \tag{14.104}$$

and

$$-\mathbf{q}_0 = \left\{ \begin{array}{l} +\tfrac{1}{2}\not{p}_x L + P_x \dfrac{L - x_0}{L} - \Delta t\, \alpha EA + \alpha_s EA - \dfrac{\Delta L}{L} EA \\[3mm] +\tfrac{1}{2}\not{p}_x L + P_x \dfrac{x_0}{L} + \Delta t\, \alpha EA - \alpha_s EA + \dfrac{\Delta L}{L} EA \end{array} \right\} \tag{14.105}$$

The vertex deflections $\mathbf{w}$ associated with $\mathcal{K}$ and $\mathbf{q}_0$ of Eqs. (14.104) and (14.105) can be displayed as

$$\mathbf{w}^T = [d_{1x} \quad d_{2x}] \tag{14.106}$$

where d denotes displacement.

14.2.5 Imposition of interelement-force boundary conditions. We discussed in Art 12.5.1 the interelement-force boundary conditions which may arise during connection of structural elements with each other. Using special devices, the construction engineer can ensure that certain force components will be zero under all loadings. In Art. 14.2.4 if a certain force component is identically zero for all vertices of all elements, we simply delete the associated row and column in the stiffness matrix and the corresponding entry in the column representing the element loads. We have done this for the element both in the statically determinate state and in the free state. In the earlier discussions of Art. 14.2 we included the stiffness relations of the element in the statically determinate state so that we could develop the ideas concerning the stiffness relations in the free state in an orderly manner, taking full advantage of the reader's basic structural knowledge. Historically, the stiffness relations in the statically determinate state have been exploited to establish a duality between the force method and the displacement method. In the present treatment they will be used only to obtain the stiffness relations in the free state from the flexibility relations or vice versa. For this reason, the treatment of incidental interelement-force boundary conditions will be confined to the stiffness relations in the free state.

Consider element 5-2 of the planar-frame structure shown in Fig. 12.12. If it were not for the hinge at the vertex placed on node 5, the free-free stiffness matrix $\mathscr{K}$ and the free-free load matrix $-\mathbf{q}_0$ of this element in its own coordinate system would be given by Eqs. (14.96) and (14.97). The presence of the hinge, however, induces the constraint that no moment can exist at the first vertex of the element (note that the element is introduced as the 5-2 element, indicating that the first vertex is the one at node 5). In other words, the hinge tells us that the third component of $\mathbf{q}$ in Eq. (14.87) is prescribed as zero. We can still use $\mathscr{K}$ and $-\mathbf{q}_0$ of Eqs. (14.96) and (14.97) for this element provided that we also carry the constraint of $q_{1,3} = 0$ in our treatments. It is preferable, however, to handle the constraint at this level; hence we do not carry it as a separate entity in the treatments of Chap. 15. In the following paragraphs we shall show how this is done in the general case of a n-vertex element with e degrees of freedom at each vertex.

Given a structural element, we first obtain its stiffness relations in the free state, ignoring the interelement-force boundary conditions. Suppose that $\mathscr{K}$ and $-\mathbf{q}_0$ denote the free-free stiffness matrix and the free-free load matrix of the element, respectively, generated without considering any interelement-force boundary conditions. These matrices are of order $r = ne$, and the corresponding stiffness relation of the element is given by Eq. (14.87). For the present discussion we shall consider the matrix and the column matrices of this equation without any partitions, namely,

$$\mathbf{q}^T = [q_1 \quad q_2 \quad \cdots \quad q_r] \tag{14.107}$$

$$\mathbf{w}^T = [w_1 \quad w_2 \quad \cdots \quad w_r] \tag{14.108}$$

$$\mathbf{q}_0^T = [q_{0_1} \quad q_{0_2} \quad \cdots \quad q_{0_r}] \tag{14.109}$$

and the jth column of $\mathscr{K}$

$$\pmb{k}_j^T = [k_{1j} \quad k_{2j} \quad \cdots \quad k_{rj}] \qquad \text{where } 1 \le j \le r \tag{14.110}$$

Let us suppose that the tth component of the list of vertex forces is prescribed as zero by means of a special device, that is,

$$q_t = 0 \tag{14.111}$$

This condition implies that the tth scalar equation of Eq. (14.87) is

$$0 = [k_{t1} \quad \cdots \quad k_{tt} \quad \cdots \quad k_{tr}]\mathbf{w} + q_{0_t} \tag{14.112}$$

which can be written

$$0 = \pmb{k}_t^T \mathbf{w} + q_{0_t} \tag{14.113}$$

We can solve w_t out of this equation and eliminate it from the rest of Eq. (14.87). Since the coefficients of w_t in Eq. (14.87) are listed in k_t, we can premultiply both sides of Eq. (14.113) by $(1/k_{tt})k_t$ and subtract the resulting matrix equation from Eq. (14.87) to obtain

$$\mathbf{q} = \left(\mathscr{K} - \frac{1}{k_{tt}} k_t k_t^T \right) \mathbf{w} + \left(\mathbf{q}_0 - \frac{q_{0_t}}{k_{tt}} k_t \right) \tag{14.114}$$

We can observe from this equation that tth scalar equation is indeed

$$q_t = \left(k_t^T - \frac{k_{tt}}{k_{tt}} k_t^T \right) \mathbf{w} + \left(q_{0_t} - \frac{q_{0_t}}{k_{tt}} k_{tt} \right) = 0 \tag{14.115}$$

and the coefficients of w_t in Eq. (14.114) are zero, since the tth column of the co-efficient matrix of $\mathbf{w}$ in Eq. (14.114) is

$$\left(k_t - \frac{1}{k_{tt}} k_t k_{tt} \right) = \mathbf{0} \tag{14.116}$$

These statements ensure that Eq. (14.114) will produce $q_t = 0$ for all values of $\mathbf{w}$ and the new stiffness relations will not be affected by the values of w_t. Let $\mathscr{K}^*$ denote the free-free stiffness matrix and $-\mathbf{q}_0^*$ denote the free-free load matrix of the element after the interelement-force boundary condition is imposed. Then

$$\mathscr{K}^* = \mathscr{K} - \frac{1}{k_{tt}} k_t k_t^T \tag{14.117}$$

and

$$\mathbf{q}_0^* = \mathbf{q}_0 - \frac{q_{0_t}}{k_{tt}} k_t \tag{14.118}$$

With this notation, the stiffness relation in Eq. (14.114) can be rewritten as

$$\mathbf{q} = \mathscr{K}^* \mathbf{w} + \mathbf{q}_0^* \tag{14.119}$$

which is the stiffness relation of the element in the free state after the interelement-force boundary condition is imposed.

If there is more than one interelement force-boundary condition, we use the above procedure for each of the interelement-force boundary conditions, one at a time, in any order, each time using the latest-obtained free-free stiffness matrix and the free-free load matrix. The advantage of this procedure is that for incidental interelement-force boundary conditions the orders of $\mathscr{K}^*$ and $-\mathbf{q}_0^*$ remain the same yet contain the constraint built into them. Comparing Eq. (14.119) with Eq. (14.87), we observe that the structure of the stiffness relations remains the same and therefore the energy expressions will not be altered in form. Note that when all other vertex-deflection components are known, we can obtain the value of w_t from Eq. (14.113)

As an example, let us obtain $\mathscr{K}^*$ and $-\mathbf{q}_0^*$ matrices of element 5-2 in the planar-frame structure shown in Fig. 12.12. Suppose that $\mathscr{K}$ and $-\mathbf{q}_0$ matrices of the element before the $q_3 = 0$ constraint is imposed are as given by Eqs. (14.96) and (14.97) in the coordinate system of the element. We note that in this case

$$t = 3 \qquad k_{33} = \frac{4EI_{zz}}{L} \qquad \text{and} \qquad k_3^T = \begin{bmatrix} 0 & \dfrac{6EI_{zz}}{L^2} & \dfrac{4EI_{zz}}{L} & 0 & \dfrac{-6EI_{zz}}{L^2} & \dfrac{2EI_{zz}}{L} \end{bmatrix}$$

Using these together with $\mathscr{K}$ from Eq. (14.96) in Eq. (14.117), we obtain

$$\mathscr{K}^* = \begin{bmatrix} \dfrac{EA}{L} & \cdot & \cdot & -\dfrac{EA}{L} & \cdot & \cdot \\[2mm] & \dfrac{3EI_{zz}}{L^3} & \cdot & \cdot & -\dfrac{3EI_{zz}}{L^3} & \dfrac{3EI_{zz}}{L^2} \\[2mm] & & \cdot & \cdot & \cdot & \cdot \\[2mm] & & & \dfrac{EA}{L} & \cdot & \cdot \\[2mm] & \text{Symmetric} & & & \dfrac{3EI_{zz}}{L^3} & -\dfrac{3EI_{zz}}{L^2} \\[2mm] & & & & & \dfrac{3EI_{zz}}{L} \end{bmatrix} \qquad (14.120)$$

and using them together with $-\mathbf{q}_0$ from Eq. (14.97) in Eq. (14.118), we find

$$-\mathbf{q}_0^* = \left\{ \begin{array}{l} + \tfrac{1}{2} p_x L + P_x \dfrac{L - x_0}{L} - \Delta t\, \alpha EA + \alpha_s EA - \dfrac{\Delta L}{L} EA \\[3mm] + \tfrac{3}{8} p_y L + P_y \left(\dfrac{L - x_0}{L}\right)^2 \dfrac{2L + x_0}{2L} - \tfrac{3}{2} M_z \dfrac{L^2 - x_0^2}{L^3} - \dfrac{3}{2}\dfrac{\Delta t_{,y}\, \alpha EI_{zz}}{L} \\[3mm] \qquad\qquad\qquad\qquad 0 \\[3mm] + \tfrac{1}{2} p_x L + P_x \dfrac{x_0}{L} + \Delta t\, \alpha AE - \alpha_s AE + \dfrac{\Delta L}{L} AE \\[3mm] + \tfrac{5}{8} p_y L + P_y \left[1 - \left(\dfrac{L - x_0}{L}\right)^2 \dfrac{2L + x_0}{2L} \right] + \dfrac{3}{2} M_z \dfrac{L^2 - x_0^2}{L^3} + \dfrac{3}{2}\dfrac{\Delta t_{,y}\, \alpha EI_{zz}}{L} \\[3mm] - \tfrac{1}{8} p_y L^2 - P_y L \dfrac{L - x_0}{L} \left(1 - \dfrac{2L + x_0}{2L} \dfrac{L - x_0}{L} \right) \\[3mm] \qquad\qquad + M_z \left(1 - \dfrac{3}{2}\dfrac{L^2 - x_0^2}{L^2} \right) - \tfrac{3}{2} \Delta t_{,y}\, \alpha EI_{zz} \end{array} \right\}$$

$$(14.121)$$

where the loading is defined above Eq. (14.23).

14.2.6 Obtaining flexibility relations from the stiffness relations in the free state. In the foregoing discussions we have confined ourselves to the method where the element flexibility relations are obtained first and the stiffness relations are obtained from the flexibility relations.   As we shall see in Art. 14.3, sometimes we can obtain the stiffness relations in the free state directly, without ever referring to the flexibility relations.   Given the stiffness relations of the element in free state, how can we obtain the flexibility relations of the element from the stiffness relations?   In this article we deal with this question.

Assume that the free-free stiffness matrix $\mathcal{K}$ and the free-free load matrix $-\mathbf{q}_0$ of the element are available, so that its stiffness relations as stated by Eq. (14.87) are completely defined.   Note that Eq. (14.87) enables us to compute the vertex forces of the element when vertex deflections $\mathbf{w}$ are known.   We can see from the equation that when $\mathbf{w} = \mathbf{0}$, vertex forces are simply $\mathbf{q}_0$.   We have seen [see Eq. (14.84) and the discussion following it] that when $\mathbf{w}$ represents a rigid-body movement, the vertex forces are still $\mathbf{q}_0$.   In other words, $\mathcal{K}\mathbf{w} = \mathbf{0}$ if $\mathbf{w}$ represents a rigid-body movement.   We can therefore assign zero values to enough components of $\mathbf{w}$ to eliminate rigid-body movements and thus place the element on statically determinate supports.   In space structures we assign zero values to six suitably selected components of $\mathbf{w}$ to prevent three rigid-body translations and three rigid-body rotations.   In planar structures, we assign zero values to three suitably selected components of $\mathbf{w}$ to prevent two rigid-body translations and one rigid-body rotation.

Once we decide which components of $\mathbf{w}$ are to be made zero in Eq. (14.87), we place zero at those entries of $\mathbf{w}$.   We may choose to delete the zeros and the associated columns in $\mathcal{K}$.   We may delete the scalar equations of Eq. (14.87) corresponding to the zero deflection directions.   The remaining equations are of the type of Eq. (14.56); that is, they represent the stiffness relations of the element in the statically determinate state.   Suppose Eq. (14.56) represents the remainder of the scalar equations of Eq. (14.87) after the rows and the columns corresponding to the zero components of $\mathbf{w}$ are removed.   If we selected proper components of $\mathbf{w}$ to be zero, thus preventing the rigid-body movements, $\mathbf{K}$ of Eq. (14.56) would be positive definite, implying that $\mathbf{K}^{-1}$ exists.   Thus solving $\mathbf{v}$ from Eq. (14.56) is possible, and so we can write

$$\mathbf{v} = \mathbf{K}^{-1}\mathbf{s} - \mathbf{K}^{-1}\mathbf{s}_0 \qquad (14.122)$$

Comparing this with the flexibility relations of the element given in Eq. (14.24), we observe that

$$\mathbf{F} = \mathbf{K}^{-1} \qquad (14.123)$$

and

$$\mathbf{v}_0 = -\mathbf{K}^{-1}\mathbf{s}_0 \qquad (14.124)$$

which are already stated, indirectly, by Eqs. (14.55) and (14.54), respectively.   In other words, once we establish the stiffness relations of the element in the statically determinate state, the flexibility relations are obtained by inversion, as shown by Eqs. (14.123) and (14.124).

Let e denote the number of components of $\mathbf{w}$ to be made zero in order to prevent any rigid-body movement of the element.   Let r denote the order of $\mathbf{w}$ (which is also the order of $\mathcal{K}$). The number of possibilities of zero-component selection is equal to the number of e combinations of r objects, namely

$$\binom{r}{e}^{\dagger} = \frac{r!}{(r-e)!\,e!}$$

Although certain combinations may not be capable of preventing the rigid-body movements of the element, the number of acceptable combinations is very large, even for the line elements. For example, consider the space-beam element for which $r = 12$ and $e = 6$. The number of statically determinate states for this beam can be as large as $12!/6!\,6! = 924$. Although some of these cannot prevent the rigid-body movements, and therefore are not acceptable, there are still very many acceptable ones. In Art. 14.1.1 we adopted the policy of choosing the degree-of-freedom directions of the last vertex as the directions in which the deflections are made zero for the statically determinate state. Actually such a choice is one of many other acceptable possibilities. The superiority of this choice over many others lies in the fact that it can easily be identified. Whenever possible we shall stick to this policy. As an example, consider the uniform space beam and its $\mathcal{K}$ and $-\mathbf{q}_0$ matrices, as given by Eqs. (14.91) and (14.93). From these we may obtain matrices $\mathbf{F}$ and $\mathbf{v}_0$ to identify its flexibility relations as follows. We choose the last six components of $\mathbf{w}$ as the components to be made zero for the statically determinate state. Deleting the last six rows and columns of $\mathcal{K}$ and deleting the last six entries of $-\mathbf{q}_0$, we obtain $\mathbf{K}$ and $-\mathbf{s}_0$, which are already listed in Eqs. (14.58) and (14.59). Then using these in Eqs. (14.123) and (14.124), we obtain $\mathbf{F}$ and $\mathbf{v}_0$, which are already given in Eqs. (14.20) and (14.23). For the inversion of $\mathbf{K}$ we may take advantage of the hint given in Prob. 14.5 for $\mathbf{F}$, since the same hint is also applicable to $\mathbf{K}$.

Finally we can use the deleted equations of the stiffness relations in the free state in computing the reactions of the element in the statically determinate state.

One can fully exploit the procedure explained here for obtaining the flexibility relations from the stiffness relations of the element in free state, especially for the elements of a continuum.

14.2.7   Problems for Solution

Problem 14.10   From $\mathbf{F}$ and $\mathbf{v}_0$ of the beam in Prob. 14.3, obtain the stiffness matrix $\mathbf{K}$ and fixed-vertex-forces column matrix $\mathbf{s}_0$ of the beam in the statically determinate state.

Problem 14.11   From $\mathbf{K}$ and $\mathbf{s}_0$ of the beam in Prob. 14.10, obtain the free-free stiffness matrix $\mathcal{K}$ and the free-free load matrix $-\mathbf{q}_0$.

Problem 14.12   From $\mathbf{F}$ and $\mathbf{v}_0$ of the curved planar beam of Prob. 14.4, obtain its $\mathbf{K}$ and $\mathbf{s}_0$.

Problem 14.13   From $\mathbf{K}$ and $\mathbf{s}_0$ of the curved beam in Prob. 14.12, compute its $\mathcal{K}$ and $-\mathbf{q}_0$.

*Problem 14.14   Obtain $\mathbf{s}_0$ corresponding to $\mathbf{v}_0$ of Prob. 14.6, and then obtain $-\mathbf{q}_0$, the free-free load matrix, from $\mathbf{s}_0$.

*Problem 14.15   Obtain $\mathbf{s}_0$ corresponding to $\mathbf{v}_0$ of Prob. 14.7, and then obtain $-\mathbf{q}_0$, the free-free load matrix, from $\mathbf{s}_0$.

Problem 14.16   Suppose the beam in Prob. 14.11 has the interelement-force boundary condition such that no moment can develop at vertex 1. Obtain $\mathcal{K}^*$ and $-\mathbf{q}_0^*$ from $\mathbf{K}$ and $\mathbf{s}_0$.

*Problem 14.17   What is $\mathbf{q}_0^*$ associated with the beam and its loading in Prob. 14.14 if the beam has the interelement-force boundary condition such that no shear exists at vertex 1?

---

† Binomial coefficient symbol.

*Problem 14.18   What is $\mathbf{q}_0^*$ for the beam and its loading in Prob. 14.15 if the beam has the interelement-force boundary condition such that no moment exists at vertex 1?

Problem 14.19   Prove that when $\mathbf{w}$ represents a rigid-body movement, $\mathscr{K}\mathbf{w} = \mathbf{0}$.

Problem 14.20   Let $\mathscr{K}$ and $-\mathbf{q}_0$ denote the free-free stiffness and load matrices of a structural element.   By preventing rigid-body movements from $\mathscr{K}$ and $-\mathbf{q}_0$ we can obtain $\mathbf{K}$ and $\mathbf{s}_0$ of the element in the statically determinate state, which in turn yield $\mathbf{F}$ and $\mathbf{v}_0$.   Since there are various ways of supporting the element in a statically determinate fashion, obviously there are various expressions for $\mathbf{K}$ and $\mathbf{s}_0$ or for $\mathbf{F}$ and $\mathbf{v}_0$.   Show that given an element and its load, all $\mathbf{F}$s associated with different statically determinate support conditions have the same total complementary strain energy $\mathscr{U}^* = \frac{1}{2}\mathbf{s}^T\mathbf{F}\mathbf{s}$ for a given $\mathbf{q}$ vertex force.   Also show that $\mathbf{K}$s associated with different statically determinate support states have the same total strain energy $\mathscr{U} = \frac{1}{2}\mathbf{v}^T\mathbf{K}\mathbf{v}$ for a given $\mathbf{w}$.

14.3   Stiffness Relations for Elements of Continuum Structures.

An element of a continuum structure is different from an element of a discrete structure in the following respect.   Although an element of a discrete structure is connected with other elements only at its vertex points, an element of a continuum structure is connected with the other elements along the surfaces between its vertices.   The difference can be best illustrated by the following example.   Consider a rectangular sheet of steel plate.   We can construct the plate as an assembly of many triangular plate elements.   If we ask a welder to weld the triangular elements only at their vertices and not in between, we shall obtain a discrete structure with gaps along the nodal lines.   Along these gaps, stresses and deflections will be, in general, discontinuous when the welded structure is subjected to loads.   On the other hand, we can ask the welder to weld the triangular elements along their sides to create a monocoque structure without any gaps.   In this structure there will be no discontinuities in stresses and deflections along the nodal lines when the structure is loaded.   This difference makes it conceptually very difficult to extend the ideas we have developed for the stiffness relations of discrete elements to those of the continuum elements.   Note that even the term stiffness relations may appear meaningless when we use it for the elements of a continuum structure.   This is because in the foregoing discussions we obtained the stiffness relations from the flexibility relations, which were obtained by applying unit concentrated loads at the vertices of the element on statically determinate supports located at the vertices.   However, if we consider the free-free stiffness matrix $\mathscr{K}$ and the free-free load matrix $-\mathbf{q}_0$ as quantities helpful in defining the total potential energy of the element, we can easily extend the ideas developed for the stiffness properties of discrete elements to the elements of continuum structures.   Once we achieve this, the procedures of Chap. 15, which were historically developed for discrete structures, become applicable for analysis problems of continuum structures.   With a small additional effort, the reader can apply the methods of basic discrete structural analysis to solve the problems of continuum structures, which were until very recently in the domain of applied mathematics.

Ignoring the details of the computation, we see that the total potential energy of a continuum element can be expressed as

$$\pi \approx \tfrac{1}{2}\mathbf{w}^T\mathscr{K}\mathbf{w} + \mathbf{w}^T\mathbf{q}_0 \tag{14.125}$$

where $\mathbf{w}$ = list of vertex deflections

$\quad\mathscr{K}$ = free-free stiffness matrix of the element

$\quad-\mathbf{q}_0$ = free-free load matrix

Note that the total-potential-energy expression in Eq. (14.125) is very similar to that in Eq. (14.80). The difference is the fact that the expression in Eq. (14.125) is approximate and the error is a function of the size of the element relative to the overall structure. By applying the first theorem of Castigliano to the expression in Eq. (14.125) the vertex forces $\mathbf{q}$ in the directions of $\mathbf{w}$ can be obtained from Eq. (14.125) by differentiation with respect to $\mathbf{w}$ as

$$\mathbf{q} \approx \mathscr{K}\mathbf{w} + \mathbf{q}_0 \qquad (14.126)$$

which is the stiffness relation of the continuum element in the free-free state.

14.4 Flexibility Relations for Elements of Continuum Structures. The difficulties encountered in the stiffness relations of the continuum elements were briefly stated in Art. 14.3. These difficulties are even more extensive for the flexibility relations of continuum structures. However, once such relations are established, the procedures of Chap. 16, which were historically developed for the discrete structures, become immediately applicable for the analysis problems of continuum structures. There are various approaches for obtaining the flexibility relations of the continuum elements. Probably the most straightforward is to obtain them from the stiffness relations of the continuum element.

Suppose that $\mathscr{K}$ and $\mathbf{q}_0$ of Eqs. (14.125) and (14.126) are available. Then by following the procedure explained in Art. 14.2.6 one can obtain the approximate expressions for the element-flexibility matrix $\mathbf{F}$ and the element deformations $\mathbf{v}_0$ corresponding to the element loads. With these, as in Eq. (14.26), one can write the total complementary potential energy π^* of the element in terms of independent force components $\mathbf{s}$ which satisfy the force-equilibrium requirements as

$$\pi^* \approx \tfrac{1}{2}\mathbf{s}^T\mathbf{F}\mathbf{s} + \mathbf{s}^T\mathbf{v}_0 \qquad (14.127)$$

Assuming that $\mathscr{V}^* = 0$ and $\mathscr{U}^* = \pi^*$, we can apply the second theorem of Castigliano to the expression in Eq. (14.127) to express deflections $\mathbf{v}$ in the directions of the independent force components $\mathbf{s}$ of the element. By differentiation with respect to $\mathbf{s}$, from Eq. (14.127), we obtain

$$\mathbf{v} \approx \mathbf{F}\mathbf{s} + \mathbf{v}_0 \qquad (14.128)$$

which is the flexibility relation of the element in the statically determinate state.

14.5 Transformations of Stiffness and Flexibility Relations. In Arts. 14.1 to 14.4 we discussed the stiffness and flexibility relations of elements of discrete and continuum structures. In expressing the stiffness and flexibility relations we systematically used the local coordinate system of the element. Before these relationships are used, it is generally necessary to express them in the overall coordinate system of the structure. In this article we discuss the transformation of these relations when we change the coordinate system.

Because the transformation operations are easy to describe, we shall assume that the vertex forces and the vertex deflections are ordered such that they are partitionable with the vertices. In discussing the flexibility and the stiffness relations of elements of discrete structures we used column matrices listing the vertex forces and deflections which were partitionable with the vertices. However, if we have matrices which are partitionable with the degrees of freedom, they can easily be reordered by means of permutation matrices such that the partitioning is with vertices again. In our developments in Arts. 14.1 to 14.4, since the only coordinate system used was that of the element, we did not use primes over the symbols denoting the matrices, contrary to our convention in Art. 12.1.21. In this article, since there are at least two coordinate systems, we shall use a prime in a symbol denoting a matrix with entries in the local coordinate system and no prime if the matrix elements are in the overall coordinate system. Since in the stiffness and flexibility relations of structural elements the approximation sign is the only difference between the expressions for discrete structures and for those of the continuum structures, we shall make no distinction between the elements of discrete structures and the elements of continuum structures. In fact this distinction is not needed in most of the remaining treatments of Part II.

14.5.1   The general case of an element in the statically determinate state. Let $\mathbf{s}'$, $\mathbf{v}'$, $\mathbf{s}_0'$, and $\mathbf{v}_0'$ denote the element forces, the element deformations, the fixed vertex forces, and the element deformations due to element loads, respectively, referred to the local coordinate system of the element. Let $\mathbf{K}'$ and $\mathbf{F}'$ denote the stiffness and the flexibility matrix of the element, both described in the local coordinate system of the element. We assume that all these matrices are partitionable with vertices. With these notations, we can rewrite the total potential energy π and the total complementary potential energy π^* from Eqs. (14.70) and (14.29), respectively, by restoring the primes which were ignored initially, as

$$\pi = \tfrac{1}{2}\mathbf{v}'^T\mathbf{K}'\mathbf{v}' + \mathbf{v}'^T\mathbf{s}_0' \tag{14.129}$$

and

$$\pi^* = \tfrac{1}{2}\mathbf{s}'^T\mathbf{F}'\mathbf{s}' + \mathbf{s}'^T\mathbf{v}_0' \tag{14.130}$$

Let $\mathbf{s}_i'$, $\mathbf{v}_i'$, $\mathbf{s}_{0_i}'$, and $\mathbf{v}_{0_i}'$ denote the element forces, the element deformations, the fixed vertex forces, and the element deformations due to element loads, all at the ith vertex. Let $\mathbf{F}_{ij}'$ and $\mathbf{K}_{ij}'$ denote the partitions of $\mathbf{F}'$ and $\mathbf{K}'$, respectively, at the (i, j) position. Noting that the ranges of i and j are both n' (since we assume that the element has n vertices and it is supported at the nth vertex, leaving us $n' = n - 1$ vertices to include in the treatment), we can rewrite Eqs. (14.129) and (14.130) as

$$\pi = \frac{1}{2}\sum_{i=1}^{n'}\sum_{j=1}^{n'}\mathbf{v}_i'^T\mathbf{K}_{ij}'\mathbf{v}_j' + \sum_{i=1}^{n'}\mathbf{v}_i'^T\mathbf{s}_{0_i}' \tag{14.131}$$

and

$$\pi^* = \frac{1}{2}\sum_{i=1}^{n'}\sum_{j=1}^{n'}\mathbf{s}_i'^T\mathbf{F}_{ij}'\mathbf{s}_j' + \sum_{i=1}^{n'}\mathbf{s}_i'^T\mathbf{v}_{0_i}' \tag{14.132}$$

Let $\mathbf{v}_i$ denote kinematically equivalent deflections to $\mathbf{v}_i'$ acting at point i. Point i may or may not be the same as vertex i. With Eq. (13.57) we can write

$$\mathbf{v}_i' = \mathscr{T}_{ii}'\mathbf{v}_i \tag{14.133}$$

where $\mathscr{T}_{ii}'$ is the general deflection transfer matrix from point i to point i expressed in the local coordinate system. Substituting $\mathbf{v}_i'$ from Eq. (14.133) into Eq. (14.131) we can write

$$\pi = \frac{1}{2}\sum_{i=1}^{n'}\sum_{j=1}^{n'}\mathbf{v}_i^T\mathscr{T}_{ii}'^T\mathbf{K}_{ij}'\mathscr{T}_{jj}'\mathbf{v}_j + \sum_{i=1}^{n'}\mathbf{v}_i^T\mathscr{T}_{ii}'^T\mathbf{s}_{0_i}' \tag{14.134}$$

Similarly, let $\mathbf{s}_i$ denote statically equivalent forces of $\mathbf{s}_i'$ acting at point i. Point i may or may not be the same as vertex i. With the help of Eq. (13.75) we can write

$$\mathbf{s}_i' = \mathscr{T}_{ii}'^{-T} \mathbf{s}_i \tag{14.135}$$

where $\mathscr{T}_{ii}'$ is the general deflection transfer matrix as defined above. Substituting $\mathbf{s}_i'$ from Eq. (14.135) into Eq. (14.132), we obtain

$$\pi^* = \frac{1}{2} \sum_{i=1}^{n'} \sum_{j=1}^{n'} \mathbf{s}_i^T \mathscr{T}_{ii}'^{-1} \mathbf{F}_{ij}' \mathscr{T}_{jj}'^{-T} \mathbf{s}_j + \sum_{i=1}^{n'} \mathbf{s}_i^T \mathscr{T}_{ii}'^{-1} \mathbf{v}_{0i}' \tag{14.136}$$

By defining

$$\mathbf{K}_{ij} = \mathscr{T}_{ii}'^{T} \mathbf{K}_{ij}' \mathscr{T}_{jj}' \tag{14.137}$$

and

$$\mathbf{s}_{0i} = \mathscr{T}_{ii}'^{T} \mathbf{s}_{0i}' \tag{14.138}$$

we can rewrite Eq. (14.134) as

$$\pi = \frac{1}{2} \sum_{i=1}^{n'} \sum_{j=1}^{n'} \mathbf{v}_i^T \mathbf{K}_{ij} \mathbf{v}_j + \sum_{i=1}^{n'} \mathbf{v}_i^T \mathbf{s}_{0i} \tag{14.139}$$

or denoting by $\mathbf{v}$, $\mathbf{s}_0$, and $\mathbf{K}$ the matrices with partitions $\mathbf{v}_i$, $\mathbf{s}_{0i}$, and $\mathbf{K}_{ij}$, we can rewrite Eq. (14.139) as

$$\pi = \tfrac{1}{2} \mathbf{v}^T \mathbf{K} \mathbf{v} + \mathbf{v}^T \mathbf{s}_0 \tag{14.140}$$

Note that

$$\mathbf{K} = \mathrm{diag}\,(\mathscr{T}_{ii}'^{T}) \mathbf{K}' \, \mathrm{diag}\,(\mathscr{T}_{jj}') \tag{14.141}$$

and

$$\mathbf{s}_0 = \mathrm{diag}\,(\mathscr{T}_{ii}'^{T}) \mathbf{s}_0' \tag{14.142}$$

which follow from Eqs. (14.137) and (14.138). We define

$$\mathbf{F}_{ij} = \mathscr{T}_{ii}'^{-1} \mathbf{F}_{ij}' \mathscr{T}_{jj}'^{-T} \tag{14.143}$$

and

$$\mathbf{v}_{0i} = \mathscr{T}_{ii}'^{-1} \mathbf{v}_{0i}' \tag{14.144}$$

and rewrite Eq. (14.136) as

$$\pi^* = \frac{1}{2} \sum_{i=1}^{n'} \sum_{j=1}^{n'} \mathbf{s}_i^T \mathbf{F}_{ij} \mathbf{s}_j + \sum_{i=1}^{n'} \mathbf{s}_i^T \mathbf{v}_{0i} \tag{14.145}$$

or denoting by $\mathbf{s}$, $\mathbf{v}_0$, and $\mathbf{F}$ the matrices with partitions $\mathbf{s}_i$, $\mathbf{v}_{0i}$, and $\mathbf{F}_{ij}$, we can rewrite Eq. (14.145) compactly as

$$\pi^* = \tfrac{1}{2} \mathbf{s}^T \mathbf{F} \mathbf{s} + \mathbf{s}^T \mathbf{v}_0 \tag{14.146}$$

Note also that

$$\mathbf{F} = \mathrm{diag}\,(\mathscr{T}_{ii}'^{-1}) \mathbf{F}' \, \mathrm{diag}\,(\mathscr{T}_{jj}'^{-T}) \tag{14.147}$$

and

$$\mathbf{v}_0 = \mathrm{diag}_i\,(\mathscr{T}_{ii}'^{-1}) \mathbf{v}_{0i}' \tag{14.148}$$

which follows from Eqs. (14.143) and (14.144). Thus we have established expressions for the total potential energy and the total complementary potential energy of the element in Eqs. (14.140) and (14.146), respectively, in terms of element deformations and element forces applied and described at the ends of rigid bars which are welded to the element vertices. Note that the ends of the rigid bars may be considered new vertices of the element. Note also that at each vertex we may have a different coordinate system. The lengths and the directions of the rigid bars and the orientations of the coordinate systems are all represented in the general deflection transfer matrices.

14.5.2 Special cases of an element in the statically determinate state. As discussed in Art. 13.4.2, the definitions of the general deflection transfer matrices are very simple, and these matrices have the useful property that $\mathscr{T}_{ij} = \mathscr{T}_{ji}^{-1}$. We observe that the stiffness and flexibility matrices of the element with rigid bars in the statically determinate state are as in Eqs. (14.141) and (14.147), respectively. And the corresponding fixed vertex forces and element deformations are defined in Eqs. (14.142) and (14.148), respectively. If we want to express these quantities with zero lengths of rigid bars, it is sufficient to make both subscripts of the general deflection transfer-matrix the same. In that case, with Eq. (13.55), we observe that $\mathscr{T}'_{i_i}$ matrices become $\mathbf{R}_i^T$ matrices, which are established by the matrix of direction cosines of the local coordinate system of the element relative to the coordinate system at vertex i. In Art. 13.2.2 we gave $\mathbf{R}$ matrices for various types of structures. In this case for the stiffness matrix and fixed vertex forces of the element in the statically determinate state, we can write, from Eqs. (14.141) and (14.142),

$$\mathbf{K} = \text{diag } (\mathbf{R}_i)\mathbf{K}' \text{ diag } (\mathbf{R}_j^T) \qquad (14.149)$$

and

$$\mathbf{s}_0 = \text{diag } (\mathbf{R}_i)\mathbf{s}'_0 \qquad (14.150)$$

For the flexibility matrix and the element deformations of the element in the statically determinate state we write, from Eqs. (14.147) and (14.148),

$$\mathbf{F} = \text{diag } (\mathbf{R}_i)\mathbf{F}' \text{ diag } (\mathbf{R}_j^T) \qquad (14.151)$$

and

$$\mathbf{v}_0 = \text{diag } (\mathbf{R}_i) \mathbf{v}'_0 \qquad (14.152)$$

In practice, we are usually confronted with the task of expressing $\mathbf{K}'$, $\mathbf{s}'_0$, $\mathbf{F}'$, and $\mathbf{v}'_0$ in the overall coordinate system. In such a case

$$\mathbf{R}_i = \mathbf{R}_j = \mathbf{R} \qquad \text{for all } i \text{ and } j \qquad (14.153)$$

Then it follows from Eqs. (14.149) to (14.152) that

$$\mathbf{K} = \text{diag } (\mathbf{R})\mathbf{K}' \text{ diag } (\mathbf{R}^T) \qquad (14.154)$$

$$\mathbf{s}_0 = \text{diag } (\mathbf{R})\mathbf{s}'_0 \qquad (14.155)$$

$$\mathbf{F} = \text{diag } (\mathbf{R})\mathbf{F}' \text{ diag } (\mathbf{R}^T) \qquad (14.156)$$

and

$$\mathbf{v}_0 = \text{diag } (\mathbf{R})\mathbf{v}'_0 \qquad (14.157)$$

which are the transformation relations to express $\mathbf{K}'$, $\mathbf{s}_0'$, $\mathbf{F}'$, and $\mathbf{v}_0'$ (quantities defined in the local coordinate system of the element) in the overall coordinate system.

14.5.3   General case of an element in the free state.   The transformation relations for the free-free stiffness matrix and the free-free load matrix can be obtained in the same way.   Let $\mathcal{K}'$ and $-\mathbf{q}_{0_i}'$ denote the free-free stiffness matrix and the free-free load matrix of the element described in the local coordinate system.   Let $\mathbf{w}'$ denote the list of all vertex deflections, again in the local coordinate system.   With this notation we can write the total potential energy of the element, from Eqs. (14.80) or (14.125), as

$$\pi = \tfrac{1}{2}\mathbf{w}'^T \mathcal{K}'\mathbf{w}' + \mathbf{w}'^T \mathbf{q}_0' \tag{14.158}$$

Let $\mathbf{w}_i'$ and $\mathbf{q}_{0_i}'$ denote the partitions of $\mathbf{w}'$ and $\mathbf{q}_0'$ corresponding to the ith vertex, and let $\mathcal{K}_{ij}'$ denote the (i, j) partition of $\mathcal{K}'$.   Noting that the ranges of i and j are both n, we can rewrite Eq. (14.158) as

$$\pi = \frac{1}{2}\sum_{i=1}^{n}\sum_{j=1}^{n} \mathbf{w}_i'^T \mathcal{K}_{ij}'\mathbf{w}_j' + \sum_{i=1}^{n} \mathbf{w}_i'^T \mathbf{q}_{0_i} \tag{14.159}$$

Let $\mathbf{w}_i$ denote a kinematically equivalent deflection of $\mathbf{w}_i'$ acting at point i, which may or may not be the same as vertex i.   With Eq. (13.57) we write

$$\mathbf{w}_i' = \mathcal{T}_{ii}'\mathbf{w}_i \tag{14.160}$$

where $\mathcal{T}_{ii}'$ is the general deflection transfer matrix from point i to point i, expressed in the local coordinate system.   Substituting $\mathbf{w}_i'$ from Eq. (14.160) into Eq. (14.159), we obtain

$$\pi = \frac{1}{2}\sum_{i=1}^{n}\sum_{j=1}^{n} \mathbf{w}_i^T \mathcal{T}_{ii}'^T \mathcal{K}_{ij}'\mathcal{T}_{jj}'\mathbf{w}_j + \sum_{i=1}^{n} \mathbf{w}_i^T \mathcal{T}_{ii}'^T \mathbf{q}_{0_i}' \tag{14.161}$$

We define

$$\mathcal{K}_{ij} = \mathcal{T}_{ii}'^T \mathcal{K}_{ij}'\mathcal{T}_{jj}' \tag{14.162}$$

and

$$\mathbf{q}_{0_i} = \mathcal{T}_{ii}'^T \mathbf{q}_{0_i}' \tag{14.163}$$

and rewrite Eq. (14.161) as

$$\pi = \frac{1}{2}\sum_{i=1}^{n}\sum_{j=1}^{n} \mathbf{w}_i^T \mathcal{K}_{ij}\mathbf{w}_j + \sum_{i=1}^{n} \mathbf{w}_i^T \mathbf{q}_{0_i} \tag{14.164}$$

Or denoting by $\mathbf{w}$, $\mathbf{q}_0$, and $\mathcal{K}$ the matrices with partitions $\mathbf{w}_i$, $\mathbf{q}_{0_i}$, and $\mathcal{K}_{ij}$, we can express Eq. (14.164) as

$$\pi = \tfrac{1}{2}\mathbf{w}^T \mathcal{K}\mathbf{w} + \mathbf{w}^T \mathbf{q}_0 \tag{14.165}$$

Note that

$$\mathcal{K} = \operatorname{diag}(\mathcal{T}_{ii}'^T)\mathcal{K}' \operatorname{diag}(\mathcal{T}_{jj}') \tag{14.166}$$

and

$$\mathbf{q}_0 = \operatorname{diag}(\mathcal{T}_{ii}'^T)\mathbf{q}_0' \tag{14.167}$$

which follow from Eqs. (14.162) and (14.163). Matrices $\mathscr{K}$ and $-\mathbf{q}_0$ are the free-free stiffness matrix and the free-free load matrix of the element at the ends of the rigid bars which are welded to the element at its vertices. The lengths and the directions of the rigid bars, as well as the coordinate systems for deflections at their free ends, are all included in the general deflection transfer matrices.

14.5.4 Special cases of element in free state. If the lengths of the rigid bars are zero $\mathscr{T}'_{ii}$ matrices become $\mathscr{T}'_{ii} = \mathbf{R}_i^T$ and we have, from Eqs. (14.166) and (14.167),

$$\mathscr{K} = \text{diag } (\mathbf{R}_i)\mathscr{K}' \text{ diag } (\mathbf{R}_j^T) \tag{14.168}$$

and

$$\mathbf{q}_0 = \text{diag } (\mathbf{R}_i)\mathbf{q}_0' \tag{14.169}$$

In practice we are usually confronted with the task of expressing $\mathscr{K}'$ and $\mathbf{q}_0'$ in the overall coordinate system. For such cases, the statement in Eq. (14.153) is true, and we can rewrite Eqs. (14.168) and (14.169) as

$$\mathscr{K} = \text{diag } (\mathbf{R})\mathscr{K}' \text{ diag } (\mathbf{R}^T) \tag{14.170}$$

and

$$\mathbf{q}_0 = \text{diag } (\mathbf{R})\mathbf{q}_0' \tag{14.171}$$

As a concluding remark, we should point out that the stiffness relations in Eqs. (14.56), (14.87), and (14.126) and the flexibility relations in Eqs. (14.24) and (14.128) remain the same in form, provided we use the transformed quantities in them and also interpret the left-hand sides as the transformed quantities.

14.5.5 Problems for Solution

Problem 14.21 If the total potential energy of a bar element in the overall coordinate system is given as in Eq. (14.165), where $\mathbf{w}^T = [d_{1x} \quad d_{1y} \quad d_{1z} \quad d_{2x} \quad d_{2y} \quad d_{2z}]$, d denoting displacements, show that

$$\mathscr{K} = \text{diag } (\mathbf{l}_i)\mathscr{K}' \text{ diag } (\mathbf{l}_j^T) \tag{14.172}$$

and

$$\mathbf{q}_0 = \text{diag } (\mathbf{l}_i)\mathbf{q}_0' \tag{14.173}$$

where $\mathscr{K}'$ and $\mathbf{q}_0'$ are as given in Eqs. (14.104) and (14.105) and $\mathbf{l}_1 = \mathbf{l}_2 = \mathbf{l}$ lists the direction cosines of the local x axis in the overall coordinate system. Show that for planar trusses the order of $\mathbf{l}$ is 2 and for space trusses the order of $\mathbf{l}$ is 3.

Problem 14.22 Assume that $\mathscr{K}'$ and $\mathbf{q}_0'$ of a gridwork frame element are as given in Eqs. (14.100) and (14.101). If the local x axis of the element is parallel to the overall y axis, what are $\mathscr{K}$ and $\mathbf{q}_0$?

Problem 14.23 Assume that $\mathscr{K}'$ and $\mathbf{q}_0'$ of a planar-beam element are given as in Eqs. (14.96) and (14.97). If the local x axis of the element is parallel to the negative of the overall y axis, what are $\mathscr{K}$ and $\mathbf{q}_0$?

Problem 14.24 Suppose the gridwork element of Prob. 14.22 is part of the reinforcing ribs of a slab with its middle plane coincident with the overall xy plane. The vertices of the gridwork frame are attached to the nodes in the middle plane of the slab by means of rigid bars whenever the vertices of the gridwork-frame element are not coincident with the nodes. Referring to the overall coordinate system, suppose that the second vertex of the gridwork-frame element is attached to a node with position vector $\mathbf{c}^T = [x + a \quad y + b \quad 0]$, where x and y are the coordinates of the second vertex. There is no offset between the first vertex and its attachment node. What are $\mathscr{K}$ and $\mathbf{q}_0$ matrices of the gridwork-frame element relative to the attachment nodes if its vertices are extended to these nodes by rigid bars?

*Problem 14.25 The first and the second vertices of a planar-frame element are at points i and j, which have position vectors $\mathbf{c}_i^T = [1 \quad 5 \quad 0]$ and $\mathbf{c}_j^T = [11 \quad 8 \quad 0]$. The frame element is to be attached to nodes i and j, which have position vectors $\mathbf{c}_i^T = [2 \quad 6 \quad 0]$ and $\mathbf{c}_j^T = [12 \quad 9 \quad 0]$. The position vectors are described in the overall coordinate system. Suppose $\mathscr{K}'$ and $\mathbf{q}_0'$ of the element are known. What transformation matrices $\mathscr{T}_{li}$ and $\mathscr{T}_{jj}$ will produce $\mathscr{K}$ and $\mathbf{q}_0$ in the overall coordinate system relative to points i and j from $\mathscr{K} = \mathrm{diag}\,(\mathscr{T}_{li}^T)\mathscr{K}'\,\mathrm{diag}\,(\mathscr{T}_{jj})$ and $\mathbf{q}_0 = \mathrm{diag}\,(\mathscr{T}_{li}^T)\mathbf{q}_0'$?

14.6   References

1. Utku, Ş.: "ELAS: A General Purpose Computer Program for the Equilibrium Problems of Linear Structures," vol. II, "Documentation of the Program," *Calif. Inst. Technol. Jet Propul. Lab. Tech. Rep.* 32–1240, Pasadena, Calif., 1969.

2. Przemieniecki, J. S.: "Theory of Matrix Structural Analysis," chaps. 5 and 7, McGraw-Hill Book Company, New York, 1968.

15

Systematic Analysis by the Displacement Method

15.1   General Remarks.   In this chapter we shall study systematic analysis by the displacement method.   No distinction will be made between discrete structures and continuum structures.   We shall assume that a key sketch of the structure to be analyzed is already prepared, as detailed in Art. 12.2.1.   The total number of elements in the structure will be denoted by M, and the total number of nodes will be denoted by N, throughout.   The number of degrees of freedom at a node e will be assumed constant for a given structure.   The degree-of-freedom directions are those implied by the overall coordinate system.   For example, for a space-frame structure $e = 6$.   Table 12.4 should be consulted for the values of e and the associated degree-of-freedom directions in various structures.

15.1.1   Definitions.   The number of vertices in the mth element will be shown by n^m.   As stated in Art. 12.1.21, superscripts will be used exclusively for the element labels.   Given an element, say the mth one, we shall assume that $\mathscr{K}^m$, $\mathbf{q}_0^m$, and $\mathbf{j}^m$ are available.   Here $\mathscr{K}^m$ is the free-free stiffness matrix and $-\mathbf{q}_0^m$ is the free-free load matrix of the mth element, both of which are described in the overall coordinate system in such a way that they are partitionable by vertices. We also assume that the interelement-force boundary-conditions, if any, are already imposed on $\mathscr{K}^m$ and $\mathbf{q}_0^m$ as described in Art. 14.2.5.   Column matrix $\mathbf{j}^m$ is the list of node labels of the vertices of the mth element, which are involved in $\mathscr{K}^m$ and $\mathbf{q}_0^m$.   In the strict finite-element sense, $\mathbf{j}^m$ lists the actual vertices of the mth element.   However, for the elements of continuum structures, if the interpolation rule exceeds the boundaries of the element, $\mathbf{j}^m$ contains the labels of those nodes which are involved in the interpolation rule.   Considering such possibilities, we interpret n^m as the order of $\mathbf{j}^m$.

Note that, with these definitions, the orders of $\mathscr{K}^m$ and $\mathbf{q}_0^m$ are $n^m e$.   In practice, one usually refers to standard computer procedures to generate $\mathscr{K}^m$ and

$\mathbf{q}_0^m$ in the form required here. These procedures can be prepared with the information provided in Chaps. 12 and 14. We observe that the generation of $\mathcal{K}^m$ requires the overall coordinates of the nodes listed in $\mathbf{j}^m$, the material matrix $\mathcal{D}^m$, and the interelement-force boundary conditions of the element. The generation of $\mathbf{q}_0^m$ again requires the overall coordinates of the nodes listed in $\mathbf{j}^m$, the material constants and the thermal coefficients of the element, the element-loading information, and the interelement-force boundary conditions, if any. For this reason, a single procedure can be used to generate both $\mathcal{K}^m$ and $\mathbf{q}_0^m$. Excluding the computation of stresses from the calculated deflections, the base information for the generation of $\mathcal{K}^m$ and $\mathbf{q}_0^m$ is not needed for the procedures of this chapter.

We shall assume that the deflection boundary conditions of the structure are available as in either Eq. (12.118) or (12.113). We also assume that the column matrix listing the nonzero components (in the overall coordinate system) of the concentrated forces (which may include moments) acting at the nodes is available. From such a matrix we can create a column matrix $\mathbf{p}_0$ such that its ith partition $\mathbf{p}_{0i}$ would list the prescribed concentrated forces at node i. Note that the order $\mathbf{p}_{0i}$ is e.

We shall denote by $\mathbf{u}$ the list of deflections of all nodes such that the ith partition, $\mathbf{u}_i$, lists the deflection components of the ith node. Note that the order of $\mathbf{u}_i$ is e and the order of $\mathbf{u}$ is eN. The deflection and the force at the gth vertex of the mth element will be shown by $\mathbf{w}_g^m$ and $\mathbf{q}_g^m$, respectively. Both these matrices are of order e. We shall use $\mathbf{w}^m$ to denote the vertex deflections and $\mathbf{q}^m$ to denote the list of vertex forces of the mth element. Note that the orders of $\mathbf{w}^m$ and $\mathbf{q}^m$ are both $n^m e$ and that $\mathbf{w}_g^m$ and $\mathbf{q}_g^m$ are the gth partitions of these columns. Note also that $\mathbf{u}_i$, $\mathbf{w}_g^m$, and $\mathbf{q}_g^m$ are all described in the overall coordinate system. With this notation we can write

$$\mathbf{q}^m = \mathcal{K}^m \mathbf{w}^m + \mathbf{q}_0^m \tag{15.1}$$

as discussed in Arts. 14.2.4 and 14.3 [see, for example, Eqs. (14.87) and (14.126)]. We note that the stiffness relation in Eq. (15.1) is only approximate if the mth element is part of a two- or three-dimensional continuum. In this relation the known information is $\mathcal{K}^m$ and $\mathbf{q}_0^m$.

In place of Eq. (15.1), we can use the total potential energy to convey the same information. Let π^m denote the total potential energy of the mth element; then we can write

$$\pi^m = \tfrac{1}{2}\mathbf{w}^{m^T} \mathcal{K}^m \mathbf{w}^m + \mathbf{w}^{m^T} \mathbf{q}_0^m \tag{15.2}$$

as described in Arts. 14.2.4, 14.3, 14.5.3, and 14.5.4 [see, for example, Eqs. (14.80), (14.125), and (14.165)]. Here, too, the expression for the total potential energy of the mth element is only approximate if the element is part of a two- or three-dimensional continuum. Although we know $\mathcal{K}^m$ and $\mathbf{q}_0^m$ numerically, the numerical value of π^m depends on $\mathbf{w}^m$, that is, the list of vertex deflections of the mth element. At this stage $\mathbf{w}^m$ is not known.

In the following articles of this chapter we develop procedures to compute the unknown deflections from the known information first and then to compute the stresses from the deflections.

15.1.2   Problems for Solution

*Problem 15.1   Consider the building bent shown in Fig. 7.8b.   Prepare the key sketch of the structure.   What is e?   What are the values of N, M, and n^m, $m = 1, \ldots, M$?   What is column matrix $\mathbf{j}$ (mth partition of $\mathbf{j}$ is $\mathbf{j}^m$)?   What are the deflection boundary conditions in the form of Eq. (12.118)?   Write down the DBC input units.

Problem 15.2   Repeat Prob. 15.1 for the structure of Fig. 7.18.

Problem 15.3   Repeat Prob. 15.1 for the structure in Fig. 9.53, considering the symmetry and antisymmetry mentioned in Prob. 9.24.

Problem 15.4   Repeat Prob. 15.1 for the structure of Fig. 9.62, considering the symmetry and antisymmetry mentioned in Prob. 9.34.

Problem 15.5   Consider the key sketch given in Fig. 15.2 for the plane-stress problem.   Repeat Prob. 15.1 for this structure.   Do the same for the structure in Fig. 15.1.

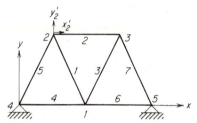

Figure 15.1   Key sketch of a planar truss

Problem 15.6   Suppose the sketch given in Fig. 15.2 is the plan view of a thin shell undergoing membrane and bending action.   The z coordinates of the nodes are proportional with the distance in xy plane from node 39.   Repeat Prob. 15.1 for this shell, assuming only membrane action is present.

15.2   Relations Between Deflections; Compatibility.

In this article we shall study the relations which exist between the several deflection components.   For example, the column matrix of nodal deflections $\mathbf{u}$ and the column matrices of vertex deflections $\mathbf{w}^m$, $m = 1, \ldots, M$ convey the same information; therefore a certain compatibility relationship must exist between $\mathbf{u}$ and $\mathbf{w}$.   Also, because of deflection boundary conditions, $\mathbf{u}$ must satisfy certain constraints in the form of Eq. (12.113).

15.2.1   Relation between vertex deflections w and nodal deflections u.

Consider node 39 in the key sketch given in Fig. 15.2.   We observe that elements 54, 55, 66, and 67 meet at this node.   Suppose that the vertices of these elements are listed such that $\mathbf{j}^{54^T} = [32 \quad 38 \quad 39]$, $\mathbf{j}^{55^T} = [32 \quad 39 \quad 40]$, $\mathbf{j}^{66^T} = [38 \quad 46 \quad 39]$, and $\mathbf{j}^{67^T} = [39 \quad 46 \quad 40]$.   From this information we see that the third vertex of element 54, the second vertex of element 55, the third vertex of element 66, and the first vertex of element 67 are all connected to node 39, and consequently, the deflections of these vertices and the deflections of node 39 are the same.   Since we express all deflections in the overall coordinate system, we can see that $\mathbf{u}_{39} = \mathbf{w}_3^{54} = \mathbf{w}_2^{55} = \mathbf{w}_3^{66} = \mathbf{w}_1^{67}$.   Consider now the key sketch shown in Fig. 12.12.   At node 5 of the structure elements 1, 2, and 7 meet.   Suppose that the vertices of these elements are listed such that $\mathbf{j}^{1^T} = [5 \quad 6]$, $\mathbf{j}^{2^T} = [1 \quad 5]$, and $\mathbf{j}^{7^T} = [5 \quad 2]$.   From

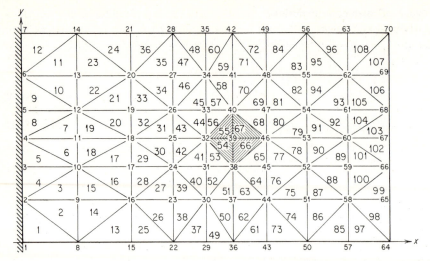

Figure 15.2   Key sketch of a plate problem

this information we observe that $\mathbf{u}_5 = \mathbf{w}_1^1 = \mathbf{w}_2^2 = \mathbf{w}_1^7$. Note that due to the inter-element-force boundary condition, $u_{5,3} \neq w_{1,3}^7$. However, since the interelement-force boundary condition is already imposed on $\mathcal{K}^7$ and $\mathbf{q}_0^7$, we can take $u_{5,3} = w_{1,3}^7$ without altering the stiffness relations of element 7 (see Art. 14.2.5). We conclude that the *vertices placed on the same node have the same deflections as those of the node*. With this in mind, we would like to express $\mathbf{w}^m$ in terms of $\mathbf{u}$, using the information in $\mathbf{j}^m$.

Consider the identity matrix $\mathbf{I}$, which is of the same order as $\mathbf{u}$, that is, of order eN. Since $\mathbf{u}$ is partitionable with the nodes, we consider that $\mathbf{I}$ is also partitionable with the nodes. We use $\mathbf{I}_{ij}$ to denote the partition of $\mathbf{I}$ which is in the (i, j) position. We use $\mathbf{I}_j$ to denote the jth partition column of $\mathbf{I}$. Note that $\mathbf{I}_j$ is of row order eN and column order e and that it is full of zeros, excluding the jth partition, which is an identity matrix of order e. For example, if $e = 3$, $N = 6$, for $j = 4$, we display $\mathbf{I}_4$ as

$$\mathbf{I}_4^T = \begin{bmatrix} \cdot & \cdot & \cdot & | & \cdot & \cdot & \cdot & | & \cdot & \cdot & \cdot & | & 1 & \cdot & \cdot & | & \cdot & \cdot & \cdot & | & \cdot & \cdot & \cdot \\ \cdot & \cdot & \cdot & | & \cdot & \cdot & \cdot & | & \cdot & \cdot & \cdot & | & \cdot & 1 & \cdot & | & \cdot & \cdot & \cdot & | & \cdot & \cdot & \cdot \\ \cdot & \cdot & \cdot & | & \cdot & \cdot & \cdot & | & \cdot & \cdot & \cdot & | & \cdot & \cdot & 1 & | & \cdot & \cdot & \cdot & | & \cdot & \cdot & \cdot \end{bmatrix} \tag{15.3}$$

Since $\mathbf{I}_j$ is of the same order as $\mathbf{u}$, we observe that

$$\mathbf{u}_j = \mathbf{I}_j^T \mathbf{u} \tag{15.4}$$

holds (see Art. 12.1.18). In other words, premultiplying $\mathbf{u}$ by $\mathbf{I}_j^T$ singles out its jth partition.

Now consider the identity matrix $\mathbf{I}^m$ which is of the same order as $\mathbf{w}^m$, that is, of order en^m. Since $\mathbf{w}^m$ is partitionable with vertices, we assume that $\mathbf{I}^m$ is also partitionable with vertices. We denote the partition of $\mathbf{I}^m$ which is at the (g, h) position by $\mathbf{I}_{g,h}^m$. The gth partition column of $\mathbf{I}^m$ will be denoted by $\mathbf{I}_g^m$, which is of row order en^m and column order e, and it is full of zeros, excluding the gth partition, which is an identity matrix of order e. If we displayed $\mathbf{I}_g^{m^T}$, it would look similar to the matrix displayed in Eq. (15.3) when $e = 3$, $n^m = 6$, and $g = 4$. We observe that

$$\mathbf{w}^m = \sum_{g=1}^{n^m} \mathbf{I}_g^m \mathbf{w}_g^m \tag{15.5}$$

is a true statement [see Eq. (12.68) for the general case]. Noting that the node label of the gth vertex of the mth element is j_g^m, we can write

$$\mathbf{w}_g^m = \mathbf{u}_{j_g^m} \tag{15.6}$$

Since

$$\mathbf{u}_{j_g^m} = \mathbf{I}_{j_g^m}^T \mathbf{u} \tag{15.7}$$

substituting $\mathbf{u}_{j_g^m}$ from Eq. (15.7) into Eq. (15.6) and then substituting $\mathbf{w}_g^m$ from this equation into Eq. (15.5), we obtain

$$\mathbf{w}^m = \sum_{g=1}^{n^m} \mathbf{I}_g^m \mathbf{I}_{j_g^m}^T \mathbf{u} \tag{15.8}$$

Defining

$$\mathbf{N}^m = \sum_{g=1}^{n^m} \mathbf{I}_g^m \mathbf{I}_{j_g^m}^T \tag{15.9}$$

we can rewrite Eq. (15.8) as

$$\mathbf{w}^m = \mathbf{N}^m \mathbf{u} \tag{15.10}$$

which expresses the list of vertex deflections of the mth element in terms of the list of deflections of all nodes.

Note that in order to generate $\mathbf{N}^m$, knowledge of the quantities e, N, and $\mathbf{j}^m$ is sufficient. For example, $\mathbf{N}^7$ of the seventh element of the structure in Fig. 12.12 can be displayed as shown below, noting that $e = 3$, $n = 9$, and $\mathbf{j}^{7^T} = [5 \quad 2]$:

$$\mathbf{N}^7 = \begin{bmatrix} \cdots & \cdots & \cdots & \cdots & 1.. & \cdots & \cdots & \cdots & \cdots \\ \cdots & \cdots & \cdots & \cdots & .1. & \cdots & \cdots & \cdots & \cdots \\ \cdots & \cdots & \cdots & \cdots & ..1 & \cdots & \cdots & \cdots & \cdots \\ \cdots & 1.. & \cdots & \cdots & \cdots & \cdots & \cdots & \cdots & \cdots \\ \cdots & .1. & \cdots & \cdots & \cdots & \cdots & \cdots & \cdots & \cdots \\ \cdots & ..1 & \cdots & \cdots & \cdots & \cdots & \cdots & \cdots & \cdots \end{bmatrix} \tag{15.11}$$

We can denote

$$\begin{bmatrix} . & . & . \\ . & . & . \\ . & . & . \end{bmatrix} = \mathbf{0} \quad \text{and} \quad \begin{bmatrix} 1 & . & . \\ . & 1 & . \\ . & . & 1 \end{bmatrix} = \mathbf{1} \tag{15.12}$$

and rewrite Eq. (15.11) as a binary matrix

$$\mathbf{N}^7 = \begin{bmatrix} 0 & 0 & 0 & 0 & 1 & 0 & 0 & 0 & 0 \\ 0 & 1 & 0 & 0 & 0 & 0 & 0 & 0 & 0 \end{bmatrix} \tag{15.13}$$

We observe that $\mathbf{N}^7$ is basically a zero matrix except that it has one **1** at every row in columns indicated by $\mathbf{j}^7$ (in the present case, since $\mathbf{j}^{7^T} = [5 \quad 2]$, **1**s are located at the fifth column in the first row and at the second column in the second row). Note that orders of **0** and **1** are both e and that the row order of $\mathbf{N}^7$ is equal to the order of $\mathbf{j}^7$ when $\mathbf{N}^7$ is expressed as a binary matrix. In practice, an $\mathbf{N}^m$ matrix is never generated and stored in the computer's memory. Whenever needed, it can be created from knowledge of e, N, and $\mathbf{j}^m$. If we consider $\mathbf{w}^m$ as the mth partition of $\mathbf{w}$, that is,

$$\mathbf{w}^T = [\mathbf{w}^{1^T} \quad \cdots \quad \mathbf{w}^{M^T}] \tag{15.14}$$

we can write Eq. (15.10) to include all elements as

$$\mathbf{w} = \mathbf{Nu} \tag{15.15}$$

which relates the nodal deflections $\mathbf{u}$ to the vertex deflections $\mathbf{w}$ of the whole structure. In Eq. (15.15) N is partitioned so that its mth-row partition is $\mathbf{N}^m$.

15.2.2 Relations between dependent and independent components of u. As discussed in Art. 12.4, the presence of yielding and/or unyielding supports, symmetry, and antisymmetry conditions or some other simplifying assumptions on the variation pattern of deflections means that the column matrix of nodal deflections $\mathbf{u}$ must satisfy certain linear constraints of the type of Eq. (12.113), reproduced below:

$$\mathbf{Hu} = \mathbf{h}_0 \tag{15.16}$$

Matrices $\mathbf{H}$ and $\mathbf{h}_0$ are known. Let the row order of $\mathbf{H}$, that is, the number of constraints, be denoted by b. Given a structural-analysis problem, usually $b < Ne$. If $b = Ne$, we can solve all the deflections from Eq. (15.16) directly, without the structural behavior. Of course, the case of $b > Ne$ is impossible, since the number of constraints cannot be larger than what they are constraining. The interesting case is when $b < Ne$. Here we can solve b components of $\mathbf{u}$ in terms of the *remaining ones*. We call the solved components the **dependent deflection components** and the remaining ones the **independent deflection components.**

Let $\mathbf{u}_{(1)}$ denote the independent deflection components, and let $\mathbf{u}_{(2)}$ denote the dependent deflection components. Note that the order of $\mathbf{u}_{(2)}$ is b and the order of $\mathbf{u}_{(1)}$ is $Ne - b$. Given a structural-analysis problem, by a simple inspection we can easily determine which components of $\mathbf{u}$ can be taken as dependent and which as independent. As shown in Art. 12.4.3, we can recast the constraint conditions of Eq. (15.16) as in Eq. (12.118), reproduced below:

$$\mathbf{u}_{(2)} = \mathbf{H}_{(21)}\mathbf{u}_{(1)} + \mathbf{I}'\mathbf{u}_{(20)} \tag{15.17}$$

Here matrices $\mathbf{H}_{(21)}$ and $\mathbf{u}_{(20)}$ are easily defined by the DBC input units of the problem (see Art. 12.4.4), and $\mathbf{I}'$ is an identity matrix of order b. Considering the identity

$$\mathbf{u}_{(1)} = \mathbf{I}\mathbf{u}_{(1)} + \mathbf{0}\mathbf{u}_{(20)} \tag{15.18}$$

(where $\mathbf{I}$ is the identity matrix of order $Ne - b$ and $\mathbf{0}$ is a $Ne - b$ by b null matrix) in conjunction with Eq. (15.17), we can write

$$\left\{\begin{matrix} \mathbf{u}_{(1)} \\ \hline \mathbf{u}_{(2)} \end{matrix}\right\} = \left[\begin{matrix} \mathbf{I} \\ \hline \mathbf{H}_{(21)} \end{matrix}\right]\mathbf{u}_{(1)} + \left[\begin{matrix} \mathbf{0} \\ \hline \mathbf{I'} \end{matrix}\right]\mathbf{u}_{(20)} \tag{15.19}$$

The matrix on the left is, in general, a shuffled form of $\mathbf{u}$. By using the appropriate permutation matrix $\mathbf{P}$, we can write

$$\mathbf{u} = \mathbf{P}\left\{\begin{matrix} \mathbf{u}_{(1)} \\ \mathbf{u}_{(2)} \end{matrix}\right\} \tag{15.20}$$

Therefore, if we premultiply both sides of Eq. (15.19) by $\mathbf{P}$ and use Eq. (15.20), we obtain

$$\mathbf{u} = \mathbf{H}_{(1)}\mathbf{u}_{(1)} + \mathbf{H}_{(0)}\mathbf{u}_{(20)} \tag{15.21}$$

where

$$\mathbf{H}_{(1)} = \mathbf{P}\left[\begin{matrix} \mathbf{I} \\ \mathbf{H}_{(21)} \end{matrix}\right] \tag{15.22}$$

and

$$\mathbf{H}_{(0)} = \mathbf{P}\left[\begin{matrix} \mathbf{0} \\ \mathbf{I'} \end{matrix}\right] \tag{15.23}$$

Note that Eq. (15.21) expresses the complete list of nodal deflections $\mathbf{u}$ as a linear function of the independent deflections $\mathbf{u}_{(1)}$. Note also that matrices $\mathbf{H}_{(1)}$, $\mathbf{H}_{(0)}$, and $\mathbf{u}_{(20)}$ can be systematically defined by means of the DBC input units of the problem (see Art. 12.4.4).

15.2.3　Vertex deflections w in terms of independent deflections $\mathbf{u}_{(1)}$. We can now proceed to express the vertex deflections of an element in terms of the independent deflection components $\mathbf{u}_{(1)}$. We substitute $\mathbf{u}$ from Eq. (15.21) into Eq. (15.10) to obtain

$$\mathbf{w}^m = \mathbf{N}^m\mathbf{H}_{(1)}\mathbf{u}_{(1)} + \mathbf{N}^m\mathbf{H}_{(0)}\mathbf{u}_{(20)} \tag{15.24}$$

which expresses the list of vertex deflections of the mth element in terms of the independent deflection components $\mathbf{u}_{(1)}$ and prescribed support settlements $\mathbf{u}_{(20)}$. We substitute $\mathbf{u}$ from Eq. (15.21) into Eq. (15.15) to obtain

$$\mathbf{w} = \mathbf{A}_{(1)}\mathbf{u}_{(1)} + \mathbf{A}_{(0)}\mathbf{u}_{(20)} \tag{15.25}$$

where

$$\mathbf{A}_{(1)} = \mathbf{N}\mathbf{H}_{(1)} \tag{15.26}$$

and

$$\mathbf{A}_{(0)} = \mathbf{N}\mathbf{H}_{(0)} \tag{15.27}$$

In Eq. (15.25) the complete list of vertex deflections is expressed in terms of the independent deflection components $\mathbf{u}_{(1)}$ and the prescribed deflection components $\mathbf{u}_{(20)}$. Defining

$$\mathbf{A}_{(1)}^m = \mathbf{N}^m\mathbf{H}_{(1)} \tag{15.28}$$

and

$$\mathbf{A}_{(0)}^m = \mathbf{N}^m\mathbf{H}_{(0)} \tag{15.29}$$

we can rewrite Eq. (15.24) as

$$\mathbf{w}^m = \mathbf{A}_{(1)}^m\mathbf{u}_{(1)} + \mathbf{A}_{(0)}^m\mathbf{u}_{(20)} \tag{15.30}$$

which expresses $\mathbf{w}^m$ compactly in terms of $\mathbf{u}_{(1)}$ and $\mathbf{u}_{(20)}$.

15.2.4 Problems for Solution

Problem 15.7 Consider the truss shown in Fig. 12.11. The DBC input units of the truss are as in Eq. (*d*) of Art. 12.4. What are e, N, and n^m, $m = 1, \dots, M$ values for this truss? Identify the components of $\mathbf{u}_{(1)}$ and $\mathbf{u}_{(2)}$ column matrices. Display $\mathbf{H}_{(21)}$ and $\mathbf{u}_{(20)}$. What is the permutation matrix $\mathbf{P}$ such that Eq. (15.20) holds? Display the $\mathbf{H}_{(1)}$ and $\mathbf{H}_{(0)}$ matrices. Let $\mathbf{u}_0 = \mathbf{H}_{(0)}\mathbf{u}_{(20)}$, and consider the augmented matrix $[\mathbf{H}_{(1)} \quad \mathbf{u}_0]$. In many problems this augmented matrix is such that it contains at the most only one nonzero element per row. When this happens, we can express the whole augmented matrix by two column matrices, each of the order of Ne, one containing the column numbers of the nonzero entries, the other the values of the nonzero entries. In other words, if we call the first column matrix $\mathbf{n}_0$ and the second one $\mathbf{a}_0$, n_{0_i} is the column number and a_{0_i} is the value of the nonzero entry in the ith row of the augmented matrix. If in the present problem, the augmented matrix can be expressed by two such column matrices, $\mathbf{n}_0$ and $\mathbf{a}_0$, display them.

Problem 15.8 Repeat Prob. 15.7 for the structure shown in Fig. 12.12. The DBC input units of this structure are given in Eqs. (*h*) of Art. 12.4.

Problem 15.9 Repeat Prob. 15.7 for the structure shown in Fig. 12.15*a*. The DBC input units of this structure are given in Eqs. (*i*) of Art. 12.4. Due to symmetry, note that we are interested in only one-half of the structure.

*Problem 15.10 Repeat Prob. 15.7 for the structure shown in Fig. 12.15*b*. The DBC input units of this structure are given in Eqs. (*j*) of Art. 12.4. Due to antisymmetry, note that we are interested in only one-half of the structure.

Problem 15.11 Repeat Prob. 15.7 for the structure shown in Fig. 12.16. The DBC input units of this structure are given in Eqs. (*k*) of Art. 12.4. Note that we are interested in only one-eighth of the structure.

Problem 15.12 Display matrix $\mathbf{N}^m$ in binary form [see Eqs. (15.12) and (15.13)] for $m = 12$ and $m = 4$ in the structure of Fig. 12.11. What are the orders of $\mathbf{0}$ and $\mathbf{1}$?

Problem 15.13 Display matrix $\mathbf{N}^m$ in binary form [see Eqs. (15.12) and (15.13)] for $m = 4$ and $m = 9$ in the structure of Fig. 12.12. What are the orders of $\mathbf{0}$ and $\mathbf{1}$?

Problem 15.14 Display $\mathbf{N}^m$ in binary form [see Eqs. (15.12) and (15.13)] for $m = 3$ and $m = 4$ in the structure of Fig. 12.16. What are the orders of $\mathbf{0}$ and $\mathbf{1}$?

Problem 15.15 Display $\mathbf{N}^m$ in binary form [see Eqs. (15.12) and (15.13)] for $m = 54$, $m = 66$, and $m = 3$ in the structure of Fig. 15.2. What are the orders of $\mathbf{0}$ and $\mathbf{1}$?

15.3 Governing Equations for Deflections.

Once the elemental matrices $\mathcal{K}^m$ and $\mathbf{q}_0^m$, $m = 1, \dots, M$, and the prescribed nodal forces $\mathbf{p}_{0_i}$, $i = 1, \dots, N$, are available in the overall coordinate system and we have the mesh-topology information $\mathbf{j}$ which enables us to relate vertex deflections $\mathbf{w}$ to the nodal deflections $\mathbf{u}$, as in Eq. (15.15), we can obtain the governing equations for the deflections in various ways, depending upon the form of the deflection boundary conditions, i.e., whether in the form of Eq. (15.16) or in the form of (15.21).

If the deflection boundary conditions are available in the form of Eq. (15.21), we can use a method based on either the **principle of stationary total potential energy (with substitution)** or the **theorem of virtual deformations** to obtain the governing equations for the independent deflections $\mathbf{u}_{(1)}$. After the computation of $\mathbf{u}_{(1)}$ from these equations, we can use Eq. (15.21) to compute the complete list of nodal deflections $\mathbf{u}$. If the deflection boundary conditions are given in the form of Eq. (15.16), we can use a method based on the **principle of stationary total potential energy (with Lagrange multipliers)** to obtain the governing equations for the complete list of nodal deflections $\mathbf{u}$. Since the effort necessary to convert the deflection boundary conditions from the format of Eq. (15.16) into the format of Eq. (15.21) is minimal (see the discussion in Art. 12.4.3), in order to minimize the order of equations to be solved for the deflections, we always prefer the methods related to the format of the boundary conditions, as in Eq. (15.21). Because of its simplicity the method utilizing the principle of stationary total potential energy (with substitution) may be preferred over the virtual-work method, although these two methods yield the same set of equations. In the following articles we detail these methods, in the order of preference.

15.3.1 Method utilizing the principle of stationary total potential energy (with substitution). This method requires that the deflection boundary conditions be available in the form of Eq. (15.21). The complete list of nodal concentrated loads $\mathbf{p}_0$ and the mesh-topology information $\mathbf{j}$ are also assumed available from the input data. As discussed in Art. 15.2.1, matrices $\mathbf{N}^m$, $m = 1, \ldots, M$, that is, $\mathbf{N}$ of Eq. (15.10), are derivable from $\mathbf{j}$.

In this method, we first write the total potential energy of the structure in terms of the vertex deflections $\mathbf{w}^m$ and nodal deflections $\mathbf{u}$, then express the energy in terms of the independent deflections components $\mathbf{u}_{(1)}$, and finally apply the principle of stationary total potential energy to obtain the governing equations for $\mathbf{u}_{(1)}$.

Let π denote the total potential energy of the structure. Noting that the potential of prescribed nodal forces $\mathbf{p}_0$ is $-\mathbf{u}^T\mathbf{p}_0$ [see Eq. (11.10)], we can write

$$\pi = \sum_{m=1}^{M} \pi^m - \mathbf{u}^T\mathbf{p}_0 \tag{15.31}$$

or, substituting π^m from Eq. (15.2), we have

$$\pi = \frac{1}{2}\left(\sum_{m=1}^{M} \mathbf{w}^{m^T}\mathcal{K}^m\mathbf{w}^m\right) + \left(\sum_{m=1}^{M} \mathbf{w}^{m^T}\mathbf{q}_0^m\right) - \mathbf{u}^T\mathbf{p}_0 \tag{15.32}$$

Using Eq. (15.10) in Eq. (15.32) we can express π in terms of $\mathbf{u}$ alone

$$\pi = \frac{1}{2}\mathbf{u}^T\left(\sum_{m=1}^{M} \mathbf{N}^{m^T}\mathcal{K}^m\mathbf{N}^m\right)\mathbf{u} + \mathbf{u}^T\left(\sum_{m=1}^{M} \mathbf{N}^{m^T}\mathbf{q}_0^m\right) - \mathbf{u}^T\mathbf{p}_0 \tag{15.33}$$

Since the components of $\mathbf{u}$ are not independent but subject to the constraints shown in Eq. (15.21), we can eliminate $\mathbf{u}$ from Eq. (15.33) by using $\mathbf{u}$ from Eq. (15.21)

$$\pi = \frac{1}{2}(\mathbf{H}_{(1)}\mathbf{u}_{(1)} + \mathbf{u}_0)^T\left(\sum_{m=1}^{M} \mathbf{N}^{m^T}\mathcal{K}^m\mathbf{N}^m\right)(\mathbf{H}_{(1)}\mathbf{u}_{(1)} + \mathbf{u}_0)$$

$$- (\mathbf{H}_{(1)}\mathbf{u}_{(1)} + \mathbf{u}_0)^T\left(\mathbf{p}_0 - \sum_{m=1}^{M} \mathbf{N}^{m^T}\mathbf{q}_0^m\right) \tag{15.34}$$

where the form of Eq. (15.21)

$$\mathbf{u} = \mathbf{H}_{(1)}\mathbf{u}_{(1)} + \mathbf{u}_0 \tag{15.35}$$

with

$$\mathbf{u}_0 = \mathbf{H}_{(0)}\mathbf{u}_{(20)} \tag{15.36}$$

is used. We call $\mathbf{u}_0$ the **column matrix of prescribed deflections.** Noting that

$$\mathbf{u}_0^T\left(\sum_{m=1}^{M} \mathbf{N}^{m^T}\mathcal{K}^m\mathbf{N}^m\mathbf{H}_{(1)}\right)\mathbf{u}_{(1)} = \mathbf{u}_{(1)}^T\left(\sum_{m=1}^{M} \mathbf{H}_{(1)}^T\mathbf{N}^{m^T}\mathcal{K}^m\mathbf{N}^m\right)\mathbf{u}_0 \tag{15.37}$$

since $\mathscr{K}^m = \mathscr{K}^{m^T}$, we can rewrite Eq. (15.34) compactly as

$$\pi = \tfrac{1}{2}\mathbf{u}_{(1)}^T \overline{\mathbf{K}} \mathbf{u}_{(1)} - \mathbf{u}_{(1)}^T \overline{\mathbf{p}}_0 + \text{const} \tag{15.38}$$

where

$$\overline{\mathbf{K}} = \sum_{m=1}^{M} \mathbf{H}_{(1)}^T \mathbf{N}^{m^T} \mathscr{K}^m \mathbf{N}^m \mathbf{H}_{(1)} \tag{15.39}$$

and

$$\overline{\mathbf{p}}_0 = \mathbf{H}_{(1)}^T \mathbf{p}_0 - \sum_{m=1}^{M} \mathbf{H}_{(1)}^T \mathbf{N}^{m^T} (\mathbf{q}_0^m + \mathscr{K}^m \mathbf{N}^m \mathbf{u}_0) \tag{15.40}$$

and the constant is the sum of all remaining terms which do not contain $\mathbf{u}_{(1)}$. Matrix $\overline{\mathbf{K}}$ is called the **overall stiffness matrix** of the structure, and $\overline{\mathbf{p}}_0$ is called the **overall load matrix** of the structure in the directions of $\mathbf{u}_{(1)}$.   Using the definitions given in Eqs. (15.28), (15.29), and (15.36), we can also write

$$\overline{\mathbf{K}} = \sum_{m=1}^{M} \mathbf{A}_{(1)}^{m^T} \mathscr{K}^m \mathbf{A}_{(1)}^m \tag{15.41}$$

and

$$\overline{\mathbf{p}}_0 = \mathbf{H}_{(1)}^T \mathbf{p}_0 - \sum_{m=1}^{M} \mathbf{A}_{(1)}^T (\mathbf{q}_0^m + \mathscr{K}^m \mathbf{A}_{(0)}^m \mathbf{u}_{(20)}) \tag{15.42}$$

From Eq. (15.41) we observe that $\overline{\mathbf{K}}$ is symmetric, i.e.,

$$\overline{\mathbf{K}} = \overline{\mathbf{K}}^T \tag{15.43}$$

According to the principle of stationary total potential energy (see Art. 11.3), $\delta\pi = 0$ for the true $\mathbf{u}_{(1)}$.[†]   Hence, keeping Eq. (15.43) in mind, from Eq. (15.38) by differentiation with respect to the components of $\mathbf{u}_{(1)}$ we obtain

$$\overline{\mathbf{K}}\mathbf{u}_{(1)} = \overline{\mathbf{p}}_0 \tag{15.44}$$

which is the governing equation for $\mathbf{u}_{(1)}$.

Note that, according to Eq. (15.40), $\overline{\mathbf{p}}_0 = 0$ if the structural elements are not loaded, i.e., if $\mathbf{q}_0^m = 0$, $m = 1, \ldots, M$, if there are no prescribed nodal loads, i.e. if $\mathbf{p}_0 = 0$, and finally if there are no support settlements, (i.e.) if $\mathbf{u}_0 = 0$. Therefore, for such a *no-load* case, as might be expected, $\mathbf{u}_{(1)} = 0$, from Eq. (15.44). From our discussions in Art. 11.3, we also know that Eq. (15.44) represents the force equilibrium equations in the directions of the components of $\mathbf{u}_{(1)}$. For example, the first scalar equation in Eq. (15.44) is the force-equilibrium equation, the left-hand side representing the internal forces and the right-hand side representing the external forces in the direction of the first component of $\mathbf{u}_{(1)}$.

15.3.2 Method utilizing the theorem of virtual deformations. This method requires that the deflection boundary conditions be available in the form of Eq. (15.21) and the elemental information in the form of Eq. (15.1). The complete

---

† Please note the use of boldface **δ** to denote the variation symbol. This usage, started in Chap. 11, is continued in Arts. 15.3.2, 15.3.3, 15.6.3, 16.2.2, 16.2.3, 16.4.2, 16.4.3, and 16.4.4. This boldface symbol **δ** is not to be confused with the boldface matrix symbols, which are usually upper- or lower-case Roman letters. In a very few cases, the Greek letters ε, σ, λ, Γ, and θ are printed boldface and used to designate matrices.

list of nodal concentrated loads $\mathbf{p}_0$ and the mesh-topology information $\mathbf{j}$ are also assumed available from the input data. As discussed in Art. 15.2.1, the matrix $\mathbf{N}$ of Eq. (15.10) is derivable from $\mathbf{j}$.

In this method, we first express the list of vertex forces $\mathbf{q}$ in terms of independent deflection components $\mathbf{u}_{(1)}$ and prescribed deflection components $\mathbf{u}_{(20)}$. Then, using internal virtual deflections $\delta\mathbf{w}$ and the corresponding external virtual deflections $\delta\mathbf{u}$ (both corresponding to arbitrary variations $\delta\mathbf{u}_{(1)}$), we obtain the force-equilibrium requirements between internal forces $\mathbf{q}$ and the external forces $\mathbf{p}_0$ in the directions of the components of $\mathbf{u}_{(1)}$.

We can rewrite Eq. (15.1) for $m = 1, \ldots, M$ compactly as

$$\mathbf{q} = \text{diag}\,(\mathscr{K}^m)\mathbf{w} + \mathbf{q}_0 \tag{15.45}$$

where column matrices $\mathbf{q}$, $\mathbf{w}$, and $\mathbf{q}_0$ are such that their mth partitions are $\mathbf{q}^m$, $\mathbf{w}^m$, and $\mathbf{q}_0^m$, respectively. We can substitute $\mathbf{w}$ from Eq. (15.25) into Eq. (15.45) and obtain

$$\mathbf{q} = \text{diag}\,(\mathscr{K}^m)\mathbf{A}_{(1)}\mathbf{u}_{(1)} + \text{diag}\,(\mathscr{K}^m)\mathbf{A}_{(0)}\mathbf{u}_{(20)} + \mathbf{q}_0 \tag{15.46}$$

which expresses internal forces $\mathbf{q}$ (the complete list of vertex forces) in terms of independent deflection components and prescribed deflections listed in $\mathbf{u}_{(1)}$ and $\mathbf{u}_{(20)}$. We know that the negatives of the vertex forces $\mathbf{q}$ and the prescribed nodal forces $\mathbf{p}_0$ are in equilibrium as a result of the fact that every node should be in force equilibrium (see Fig. 14.2c). We can obtain the equilibrium equations of the nodes by means of the theorem of virtual deformations (see Art. 11.4). We observe that we can assign arbitrary variations to the independent deflection components listed in $\mathbf{u}_{(1)}$. From Eq. (15.21), the variations $\delta\mathbf{u}$ corresponding to the arbitrary variations of $\delta\mathbf{u}_{(1)}$ can be obtained by taking the variations of both sides of Eq. (15.21) as

$$\delta\mathbf{u} = \mathbf{H}_{(1)}\,\delta\mathbf{u}_{(1)} \tag{15.47}$$

Similarly, from Eq. (15.25), $\delta\mathbf{w}$ variations corresponding to $\delta\mathbf{u}_{(1)}$ variations can be obtained as

$$\delta\mathbf{w} = \mathbf{A}_{(1)}\,\delta\mathbf{u}_{(1)} \tag{15.48}$$

Note that $\delta\mathbf{u}$ and $\delta\mathbf{w}$, as defined by Eqs. (15.47), and (15.48), are virtual external and virtual internal deflections, respectively, and both correspond to the arbitrary variations $\delta\mathbf{u}_{(1)}$ of $\mathbf{u}_{(1)}$. We can use virtual deflections $\delta\mathbf{u}$ and $\delta\mathbf{w}$ to establish the equilibrium conditions between $\mathbf{q}$ and $\mathbf{p}_0$. By Eq. (11.4) we can write

$$\delta\mathbf{w}^T\mathbf{q} = \delta\mathbf{u}^T\mathbf{p}_0 \tag{15.49}$$

Substituting $\delta\mathbf{u}$ and $\delta\mathbf{w}$ from Eqs. (15.47) and (15.48) into Eq. (15.49), and noting that Eq. (15.49) is an identity with respect to the arbitrary variations $\delta\mathbf{u}_{(1)}$, we obtain

$$\mathbf{A}_{(1)}^T\mathbf{q} = \mathbf{H}_{(1)}^T\mathbf{p}_0 \tag{15.50}$$

Now we can substitute $\mathbf{q}$ from Eq. (15.46) into the equilibrium equations in Eq. (15.50) and obtain

$$\bar{\mathbf{K}}\mathbf{u}_{(1)} = \bar{\mathbf{p}}_0 \tag{15.51}$$

where

$$\bar{\mathbf{K}} = \mathbf{A}_{(1)}^T \operatorname{diag}(\mathscr{K}^m)\mathbf{A}_{(1)} \tag{15.52}$$

and

$$\bar{\mathbf{p}}_0 = \mathbf{H}_{(1)}^T \mathbf{p}_0 - \mathbf{A}_{(1)}^T \mathbf{q}_0 - \mathbf{A}_{(1)}^T \operatorname{diag}(\mathscr{K}^m)\mathbf{A}_{(0)}\mathbf{u}_{(20)} \tag{15.53}$$

Recalling that $\mathbf{A}_{(1)}^m$ and $\mathbf{A}_{(0)}^m$ are the mth partitions of $\mathbf{A}_{(1)}$ and $\mathbf{A}_{(0)}$, we observe that Eqs. (15.52) and (15.53) are identical to Eqs. (15.41) and (15.42), respectively. We conclude that the virtual-work method described above produces the same governing equations for $\mathbf{u}_{(1)}$ as the stationary-energy method described in Art. 15.3.1.

15.3.3 Method utilizing the principle of stationary total potential energy (with Lagrange multipliers). This method can be used when the deflection boundary conditions are in the form of Eq. (15.16). We can write the total potential energy of the structure as in Eq. (15.31), or, substituting π^m from Eq. (15.2), we can write

$$\pi = \tfrac{1}{2}\mathbf{w}^T \operatorname{diag}(\mathscr{K}^m)\mathbf{w} + \mathbf{w}^T \mathbf{q}_0 - \mathbf{u}^T \mathbf{p}_0 \tag{15.54}$$

which is an alternate form of Eq. (15.32), if we recall that $\mathbf{w}^m$ and $\mathbf{q}_0^m$ are the mth partitions of $\mathbf{w}$ and $\mathbf{q}_0$, respectively. As a result of the mesh topology $\mathbf{j}$, we have Eq. (15.15), which relates $\mathbf{w}$ to $\mathbf{u}$. Substituting $\mathbf{w}$ from Eq. (15.15) into Eq. (15.54), we can write

$$\pi = \tfrac{1}{2}\mathbf{u}^T (\mathbf{N}^T \operatorname{diag}(\mathscr{K}^m)\mathbf{N})\mathbf{u} - \mathbf{u}^T(\mathbf{p}_0 - \mathbf{N}^T \mathbf{q}_0) \tag{15.55}$$

We would like to find the stationary point of π, subject to the constraint given in Eq. (15.16), reproduced below:

$$\mathbf{H}\mathbf{u} - \mathbf{h}_0 = 0 \tag{15.56}$$

In such constrained problems, the method of Lagrange multipliers may be used to advantage. Let λ denote the column of Lagrange multipliers which is of the same order as the row order of $\mathbf{H}$, that is, of order b. We premultiply both sides of Eq. (15.56) by λ^T and add the resulting equation to Eq. (15.55) such that

$$\pi = \tfrac{1}{2}\mathbf{u}^T (\mathbf{N}^T \operatorname{diag}(\mathscr{K}^m)\mathbf{N})\mathbf{u} - \mathbf{u}^T(\mathbf{p}_0 - \mathbf{N}^T \mathbf{q}_0) + \lambda^T(\mathbf{H}\mathbf{u} - \mathbf{h}_0) \tag{15.57}$$

Noting that $\mathbf{u}^T \mathbf{H} \lambda = \lambda^T \mathbf{H}^T \mathbf{u}$, we can rewrite Eq. (15.57) as

$$\pi = \tfrac{1}{2}[\mathbf{u}^T \quad \lambda^T]\begin{bmatrix} \mathbf{N}^T \operatorname{diag}(\mathscr{K}^m)\mathbf{N} & \mathbf{H}^T \\ \mathbf{H} & 0 \end{bmatrix}\begin{Bmatrix} \mathbf{u} \\ \lambda \end{Bmatrix} - [\mathbf{u}^T \quad \lambda^T]\begin{Bmatrix} \mathbf{p}_0 - \mathbf{N}^T \mathbf{q}_0 \\ \mathbf{h}_0 \end{Bmatrix} \tag{15.58}$$

If we consider λ as part of the quantities to be varied, it can be shown that the unconstrained stationary point of π in Eq. (15.58) corresponds to the constrained stationary point of π in Eq. (15.55). If we set $\delta\pi$ corresponding to variations $\delta\mathbf{u}$ and $\delta\lambda$ to zero, noting that the square matrix in Eq. (15.58) is symmetric, we obtain

$$\begin{bmatrix} \mathbf{N}^T \operatorname{diag}(\mathscr{K}^m)\mathbf{N} & \mathbf{H}^T \\ \mathbf{H} & 0 \end{bmatrix}\begin{Bmatrix} \mathbf{u} \\ \lambda \end{Bmatrix} = \begin{Bmatrix} \mathbf{p}_0 - \mathbf{N}^T \mathbf{q}_0 \\ \mathbf{h}_0 \end{Bmatrix} \tag{15.59}$$

which is the governing equation for $\mathbf{u}$.

Note that the order of the coefficient matrix in Eq. (15.59) is $Ne + b$, whereas the order of $\overline{\mathbf{K}}$ in Eqs. (15.44) or (15.51) is $Ne - b$. Since the number of arithmetic operations in solving linear equations in a digital computer is generally proportional to the order cubed, whenever possible we refrain from using the Lagrange-multipliers method, especially when b is large. Note that it is not possible to eliminate λ from Eq. (15.59) due to the zero diagonal submatrix. Therefore, during the solution of Eq. (15.59) for $\mathbf{u}$, as a by-product, we also obtain the numerical values of the Langrange multipliers λ. Unfortunately, these quantities do not carry any physical meaning which is useful to a structural engineer.

15.3.4   Problems for Solution

Problem 15.16   Consider the planar truss shown in Fig. 15.1. Show that Eq. (15.35) for this structures may be displayed as

$$\begin{Bmatrix} \mathbf{u}_1 \\ \mathbf{u}_2 \\ \mathbf{u}_3 \\ \mathbf{u}_4 \\ \mathbf{u}_5 \end{Bmatrix} = \begin{bmatrix} \mathbf{I} & \cdot & \cdot \\ \cdot & \mathbf{I} & \cdot \\ \cdot & \cdot & \mathbf{I} \\ \cdot & \cdot & \cdot \\ \cdot & \cdot & \cdot \end{bmatrix} \begin{Bmatrix} \mathbf{u}_1 \\ \mathbf{u}_2 \\ \mathbf{u}_3 \end{Bmatrix} + \begin{Bmatrix} \cdot \\ \cdot \\ \cdot \\ \cdot \\ \cdot \end{Bmatrix} \tag{a}$$

Problem 15.17   Considering the truss shown in Fig. 15.1, verify that

$$\mathbf{j}^T = [1\ 2 \quad 2\ 3 \quad 3\ 1 \quad 1\ 4 \quad 4\ 2 \quad 1\ 5 \quad 5\ 3] \tag{b}$$

is acceptable for the mesh topology.

Problem 15.18   Considering the truss shown in Fig. 15.1 and $\mathbf{j}$ given in Prob. 15.17, show that

$$\mathbf{N}^3 = \begin{bmatrix} \cdot & \cdot & \mathbf{I} & \cdot & \cdot \\ \mathbf{I} & \cdot & \cdot & \cdot & \cdot \end{bmatrix} \tag{c}$$

and obtain and display all $\mathbf{N}^m$, $m = 1, \ldots, 6$.

Problem 15.19   Using $\mathbf{H}_{(1)}$ from Eq. (a) of Prob. 15.16 and $\mathbf{N}^3$ from Eq. (c) of Prob. 15.18, show that $\mathbf{A}_{(1)}^3$ for the truss of Fig. 15.1 can be displayed as

$$\mathbf{A}_{(1)}^3 = \begin{bmatrix} \cdot & \cdot & \mathbf{I} \\ \mathbf{I} & \cdot & \cdot \end{bmatrix} \tag{d}$$

Obtain and display all $\mathbf{A}_{(1)}^m$, $m = 1, \ldots, 6$.

Problem 15.20   What are the $\mathbf{A}_{(0)}^m$, $m = 1, \ldots, 6$, matrices corresponding to $\mathbf{A}_{(1)}^m$, $m = 1, \ldots, 6$ matrices of Prob. 15.19? What are the components of $\mathbf{u}_{(20)}$ of the structure shown in Fig. 15.1?

Problem 15.21   Let free-free stiffness matrix of the third element of the structure shown in Fig. 15.1 be displayed as

$$\mathcal{K}^3 = \begin{bmatrix} \mathcal{K}_{11}^3 & \mathcal{K}_{12}^3 \\ \mathcal{K}_{21}^3 & \mathcal{K}_{22}^3 \end{bmatrix} \tag{e}$$

Show that

$$\mathbf{N}^{3T}\mathcal{K}^3\mathbf{N}^3 = \begin{bmatrix} \mathcal{K}_{22}^3 & \cdot & \mathcal{K}_{21}^3 & \cdot & \cdot \\ \cdot & \cdot & \cdot & \cdot & \cdot \\ \mathcal{K}_{12}^3 & \cdot & \mathcal{K}_{11}^3 & \cdot & \cdot \\ \cdot & \cdot & \cdot & \cdot & \cdot \end{bmatrix} \quad \text{and} \quad \mathbf{A}_{(1)}^{3T}\mathcal{K}^3\mathbf{A}_{(1)}^3 = \begin{bmatrix} \mathcal{K}_{22}^3 & \cdot & \mathcal{K}_{21}^3 \\ \cdot & \cdot & \cdot \\ \mathcal{K}_{12}^3 & \cdot & \mathcal{K}_{11}^3 \end{bmatrix}$$

Obtain all $\mathbf{A}_{(1)}^{mT}\mathcal{K}^m\mathbf{A}_{(1)}^m$, $m = 1, \ldots, 6$, triple products.

Problem 15.22   What is the matrix $\overline{\mathbf{K}}$ for the structure shown in Fig. 15.1 in terms of the triple matrix products obtained in Prob. 15.21? Display the submatrices of $\overline{\mathbf{K}}$. Check for symmetry.

*Problem 15.23   The only loading of the structure shown in Fig. 15.1 consists of $\mathbf{p}_{02}$.   What is the explicit expression for $\bar{\mathbf{p}}_0$?

Problem 15.24   If the structure in Fig. 15.1 is a planar frame, clamped at nodes 4 and 5, how should the answers of Probs. 15.16 to 15.23 be modified to fit this new structure?

*Problem 15.25   The three-bar structure shown in Fig. 15.3 is symmetrically loaded by the concentrated load P.   All bars have the same cross-sectional area A and Young's modulus E.   Express the total potential energy π in terms of the bar elongations, i.e., element deformations,

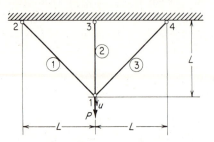

Figure 15.3   Problem 15.25

$v_1 = v_3$, $v_2 = u$.   Find the constraint involving v_1 and u, and compute u by the Lagrange-multipliers method.

*Problem 15.26   Consider the planar-truss problem sketched in Fig. 12.17 and discussed in Art. 12.4.7.   The deflection boundary conditions of the structure are given by Eqs. (n) of Art. 12.4. What is the order of $\bar{K}$ of this problem?   If you attempted to solve this analysis problem by the Lagrange-multipliers method, what would be the order of the coefficient matrix shown in Eq. (15.59)?

15.4   Properties and Method of Generation of $\bar{K}$ and $\bar{\mathbf{p}}_0$.   In Art. 12.4.3 we discussed how the deflection boundary conditions of structures can usually be expressed in the form of Eq. (12.118) without much difficulty.   In Art. 15.2.2, we saw that (again without much trouble) we can put these into the form of Eq. (15.21) or use them in the form of Eq. (15.35).   When the boundary conditions are expressed in such a form, we showed in Arts. 15.3.1 and 15.3.2 that the governing equations of the displacement method become as shown in Eqs. (15.44) or (15.51), with the coefficient matrix $\bar{K}$ and the right-hand column matrix $\bar{\mathbf{p}}_0$.   In this article we review the important properties of these matrices and discuss their systematic generation.

Various but equivalent definitions of the overall stiffness matrix $\bar{K}$ are given in Eqs. (15.39), (15.41), and (15.52).   From these definitions, we have already observed that $\bar{K}$ is *symmetrical* as stated in Eq. (15.43).   Matrix $\bar{K}$ is also *positive definite*.   We can see the presence of this property from Eq. (15.38), where the quadratic form $\frac{1}{2}\mathbf{u}_{(1)}^T \bar{K}\mathbf{u}_{(1)}$ is the total strain energy of the structure [see Eqs. (15.31) and (15.2)] in the supported state.   Since the strain energy is a positive quantity, the quadratic form is *positive*, and since the strain energy is zero only if all the components of $\mathbf{u}_{(1)}$ are zero, the quadratic form is also *definite* (see Art. 12.1.20).   Then, by definition, matrix $\bar{K}$ is also positive definite.   However, if we make an error in supporting the structure against the rigid-body movements, $\bar{K}$ will cease to be positive definite.   Likewise, if we use a material matrix which is not positive definite, e.g., an isotropic elastic material with Poisson's ratio larger than 0.5, $\overline{\mathbf{K}}$ will not be positive definite either.   Even if we support the structure properly, but fail to prevent the rigid-body movement of a node, as in the case of the node of collinear or coplanar bars of a framework subjected to a nodal force perpendicular to the bars, matrix $\overline{\mathbf{K}}$ will not be positive definite.

Any one of the definitions given in Eqs. (15.39), (15.41), and (15.52) can be used to generate the matrix $\overline{\mathbf{K}}$.   However, we can further manipulate these definitions to discover more facts about the structure of $\overline{\mathbf{K}}$ and to develop a method for its systematic generation.   In order to facilitate the operations which follow, let us rewrite Eq. (15.35), that is, the deflection boundary conditions, as

$$\mathbf{u} = \mathscr{H}^T \mathbf{u}_{(1)} + \mathbf{u}_0 \tag{15.60}$$

where
$$\mathscr{H}^T = \mathbf{H}_{(1)} \tag{15.61}$$

15.4.1 Matrix $\overline{\mathbf{K}}$ when there are no deflection constraints.

Let us first consider the form of matrix $\overline{\mathbf{K}}$ when there are no boundary conditions.   In such a case

$$\mathscr{H} = \mathbf{I} \quad \text{and} \quad \mathbf{u}_0 = \mathbf{0} \quad \text{when } b = 0 \tag{15.62}$$

implying that $\mathbf{u} = \mathbf{u}_{(1)}$, that is, all the deflection components are independent. Since the rigid-body movements are possible in this situation, matrix $\overline{\mathbf{K}}$ is only positive, and therefore it does not have an inverse and we cannot solve for $\mathbf{u}$ from Eq. (15.44).   However, in the present discussion, we are not interested in solving Eq. (15.44); we merely want to study the further properties of $\overline{\mathbf{K}}$.   With conditions in Eq. (15.62), the definition of $\overline{\mathbf{K}}$ in Eq. (15.39) becomes

$$\overline{\mathbf{K}} = \sum_{m=1}^{M} \mathbf{N}^{m^T} \mathscr{K}^m \mathbf{N}^m \tag{15.63}$$

Using the definition of $\mathbf{N}^m$ in Eq. (15.9) we can rewrite Eq. (15.63) as

$$\overline{\mathbf{K}} = \sum_{m=1}^{M} \left(\sum_{g=1}^{n^m} \mathbf{I}_{j_g^m} \mathbf{I}_g^{m^T} \right) \mathscr{K}^m \left(\sum_{h=1}^{n^m} \mathbf{I}_h^m \mathbf{I}_{j_h^m}^T \right) \tag{15.64}$$

Noting that

$$\mathbf{I}_g^{m^T} \mathscr{K}^m \mathbf{I}_h^m = \mathscr{K}_{gh}^m \tag{15.65}$$

we can rewrite Eq. (15.64) as

$$\overline{\mathbf{K}} = \sum_{m=1}^{M} \sum_{g=1}^{n^m} \sum_{h=1}^{n^m} \mathbf{I}_{j_g^m} \mathscr{K}_{gh}^m \mathbf{I}_{j_h^m}^T \tag{15.66}$$

which is a much better definition of $\overline{\mathbf{K}}$ than the one given by Eq. (15.63).   Equation (15.66) tells us that in the absence of deflection boundary conditions, in order to add the contribution of an element to the overall stiffness matrix, all we have to know are the free-free stiffness matrix of the element, in the overall coordinate system and in partitionable form with vertices, and its column matrix $\mathbf{j}^m$, that is, the column matrix listing the node labels of the vertices.

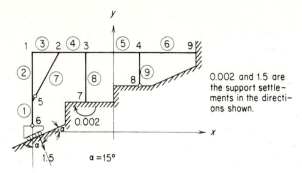

Figure 15.4   Key sketch of a planar frame

As an example, let us consider element 7 of the structure shown in Fig. 15.4. We note that for this structure $e = 3$, $N = 9$, $M = 9$, $n^m = 2$ for $m = 1, \ldots, 9$. The free-free stiffness matrix of element 7 in the overall coordinate system can be shown as

$$\mathcal{K}^7 = \begin{bmatrix} \mathcal{K}^7_{11} & \mathcal{K}^7_{12} \\ \mathcal{K}^7_{21} & \mathcal{K}^7_{22} \end{bmatrix} = \begin{bmatrix} \mathbf{A} & \mathbf{B} \\ \mathbf{C} & \mathbf{D} \end{bmatrix} \tag{a}$$

where each partition is of order $e = 3$.   Note that the hinged connection at node 5 does not change the orders of the partition, since we assume that the interelement-force boundary conditions are imposed as described in Art. 14.1.6.   Suppose we list the node labels of the vertices as

$$\mathbf{j}^7 = \begin{bmatrix} 5 & 2 \end{bmatrix} \tag{b}$$

According to Eq. (15.66), the contribution of $\mathcal{K}^7$ to $\overline{\mathbf{K}}$ can be expressed as the sum of four terms, since the number of vertices of the element is $n^7 = 2$. We can display these terms as follows, recalling that $\mathbf{I}_{j_1^7} = \mathbf{I}_5$ and $\mathbf{I}_{j_2^7} = \mathbf{I}_2$, where $\mathbf{I}_2$ is a 27 by 3 matrix representing the second partition column of $\mathbf{I}$, $\mathbf{I}_5$ is a 27 by 3 matrix representing the fifth partition column of $\mathbf{I}$, and $\mathbf{I}$ is a twenty-seventh-order identity matrix partitioned into 9 by 9 submatrices, each of order $e = 3$.   In the following we shall show $\mathbf{I}_5$ and $\mathbf{I}_2$ in partitioned form, where $\mathbf{0}$ submatrices will be shown by a boldface-dot and identity submatrices will be shown by $\mathbf{1}$.

$$\mathbf{I}_{j_1^7} \mathcal{K}^7_{11} \mathbf{I}^T_{j_1^7} = \begin{Bmatrix} \cdot \\ \cdot \\ \cdot \\ \cdot \\ \mathbf{1} \\ \cdot \\ \cdot \\ \cdot \\ \cdot \end{Bmatrix} \mathbf{A}[\,\cdot\ \cdot\ \cdot\ \cdot\ \mathbf{1}\ \cdot\ \cdot\ \cdot\ \cdot\,] = \begin{bmatrix} \cdot & \cdot & \cdot & \cdot & \cdot & \cdot & \cdot & \cdot & \cdot \\ \cdot & \cdot & \cdot & \cdot & \cdot & \cdot & \cdot & \cdot & \cdot \\ \cdot & \cdot & \cdot & \cdot & \cdot & \cdot & \cdot & \cdot & \cdot \\ \cdot & \cdot & \cdot & \cdot & \cdot & \cdot & \cdot & \cdot & \cdot \\ \cdot & \cdot & \cdot & \cdot & \mathbf{A} & \cdot & \cdot & \cdot & \cdot \\ \cdot & \cdot & \cdot & \cdot & \cdot & \cdot & \cdot & \cdot & \cdot \\ \cdot & \cdot & \cdot & \cdot & \cdot & \cdot & \cdot & \cdot & \cdot \\ \cdot & \cdot & \cdot & \cdot & \cdot & \cdot & \cdot & \cdot & \cdot \\ \cdot & \cdot & \cdot & \cdot & \cdot & \cdot & \cdot & \cdot & \cdot \end{bmatrix} \tag{c}$$

$$\mathbf{I}_{j_1}^7 \mathscr{K}_{12}^7 \mathbf{I}_{j_2}^{T_7} = \begin{Bmatrix} \cdot \\ \cdot \\ \cdot \\ 1 \\ \cdot \\ \cdot \\ \cdot \end{Bmatrix} \mathbf{B}[\,. \; 1 \; . \; . \; . \; . \; . \; .] = \begin{bmatrix} . & . & . & . & . & . & . & . \\ . & . & . & . & . & . & . & . \\ . & . & . & . & . & . & . & . \\ . & \mathbf{B} & . & . & . & . & . & . \\ . & . & . & . & . & . & . & . \\ . & . & . & . & . & . & . & . \\ . & . & . & . & . & . & . & . \\ . & . & . & . & . & . & . & . \end{bmatrix} \qquad (d)$$

$$\mathbf{I}_{j_2}^7 \mathscr{K}_{21}^7 \mathbf{I}_{j_1}^{T_7} = \begin{Bmatrix} \cdot \\ 1 \\ \cdot \\ \cdot \\ \cdot \\ \cdot \\ \cdot \end{Bmatrix} \mathbf{C}[\,. \; . \; . \; . \; 1 \; . \; . \; .] = \begin{bmatrix} . & . & . & . & . & . & . & . \\ . & . & . & . & \mathbf{C} & . & . & . \\ . & . & . & . & . & . & . & . \\ . & . & . & . & . & . & . & . \\ . & . & . & . & . & . & . & . \\ . & . & . & . & . & . & . & . \\ . & . & . & . & . & . & . & . \\ . & . & . & . & . & . & . & . \end{bmatrix} \qquad (e)$$

$$\mathbf{I}_{j_2}^7 \mathscr{K}_{22}^7 \mathbf{I}_{j_2}^{T_7} = \begin{Bmatrix} \cdot \\ 1 \\ \cdot \\ \cdot \\ \cdot \\ \cdot \\ \cdot \end{Bmatrix} \mathbf{D}[\,. \; 1 \; . \; . \; . \; . \; . \; .] = \begin{bmatrix} . & . & . & . & . & . & . & . \\ . & \mathbf{D} & . & . & . & . & . & . \\ . & . & . & . & . & . & . & . \\ . & . & . & . & . & . & . & . \\ . & . & . & . & . & . & . & . \\ . & . & . & . & . & . & . & . \\ . & . & . & . & . & . & . & . \\ . & . & . & . & . & . & . & . \end{bmatrix} \qquad (f)$$

and the sum of these four matrices can be displayed as

$$\sum_{g=1}^{2} \sum_{h=1}^{2} \mathbf{I}_{j_g}^7 \mathscr{K}_{gh}^7 \mathbf{I}_{j_h}^7 = \begin{bmatrix} . & . & . & . & . & . & . & . \\ . & \mathbf{D} & . & . & \mathbf{C} & . & . & . \\ . & . & . & . & . & . & . & . \\ . & \mathbf{B} & . & . & \mathbf{A} & . & . & . \\ . & . & . & . & . & . & . & . \\ . & . & . & . & . & . & . & . \\ . & . & . & . & . & . & . & . \\ . & . & . & . & . & . & . & . \end{bmatrix} \qquad (g)$$

We observe that the four submatrices of $\mathscr{K}^7$ are preserved as they are. However, they are placed at different positions in $\overline{\mathbf{K}}$ as dictated by the node labels of the vertices. We further observe that the diagonal submatrices of $\mathcal{K}^7$ are placed in the diagonal positions of $\overline{\mathbf{K}}$ and the off-diagonal submatrices of $\mathcal{K}^7$ are placed in the off-diagonal positions in $\overline{\mathbf{K}}$. We also note that the matrix in Eq. (g) is symmetrical, since $\mathbf{C} = \mathbf{B}^T$.

If we continue in this fashion, computing the contributions of each element and summing them according to Eq. (15.66), we obtain the overall stiffness matrix $\overline{\mathbf{K}}$. If we denote a nonzero submatrix by $\mathbf{1}$ and a zero submatrix by a boldface dot, we can show the sum as

$$\overline{\mathbf{K}} = \sum_{m=1}^{9} \sum_{g=1}^{2} \sum_{h=1}^{2} \mathbf{I}_{j_g^m} \mathcal{K}_{gh}^m \mathbf{I}_{j_h^m}^T = \begin{bmatrix} 1 & 1 & . & . & 1 & . & . & . & . \\ & 1 & 1 & . & 1 & . & . & . & . \\ & & 1 & 1 & . & . & 1 & . & . \\ & & & 1 & . & . & . & 1 & 1 \\ & & & & 1 & 1 & . & . & . \\ & & & & & 1 & . & . & . \\ & & & & & & 1 & . & . \\ & \text{Symmetric} & & & & & & 1 & . \\ & & & & & & & & 1 \end{bmatrix} \tag{h}$$

Note that the submatrices represented by $\mathbf{1}$ are nothing but appropriate sums of the submatrices $\mathcal{K}_{gh}^m$, $m = 1, \ldots, 9$; $g = 1, 2$; $h = 1, 2$. In fact, if we consider each of these as an identity matrix and therefore delete them from the sum in Eq. (h), for the general case we write

$$\mathcal{N} = \sum_{m=1}^{M} \sum_{g=1}^{n^m} \sum_{h=1}^{n^m} \mathbf{I}_{j_g^m} \mathbf{I}_{j_h^m}^T \tag{15.67}$$

which, when evaluated for the structure of Fig. 15.4, would produce the same pattern as Eq. (h). Note that matrix $\mathcal{N}$ is only a function of column matrices $\mathbf{j}^m$, $m = 1, \ldots, M$, that is, the list of node labels of the vertices. Matrix $\mathcal{N}$ is called the **connectivity matrix of the mesh** representing the structure. It consists of zero and nonzero submatrices of order e. In practice, we can represent a zero submatrix by a binary 0 and a nonzero submatrix by a binary 1. Given the mesh-topology column matrix $\mathbf{j}$, we can use the summation in Eq. (15.67) to obtain the connectivity matrix in a digital computer, using the OR operation instead of the addition (when we OR two binary bits, the result is 1 if either or both of the bits are 1 and it is 0 only if both of the bits are 0), and using the actual binary bits of the machine's memory. If a computer word consists of 32 binary bits, for a 1,024-node structure, i.e., when $N = 1,024$, we need only $(1,024)(1,025)/[(2)(32)] = 16,400$ computer words for the connectivity matrix if we consider only a full symmetric half.

Since the connectivity matrix shows the zero and nonzero submatrices of the overall stiffness matrix, it may be very useful to look into the connectivity matrix first, before the actual generation of the overall stiffness matrix. The connectivity matrix corresponding to the mesh shown in Fig. 15.4 can be reproduced from Eq. (h) as

$$\mathscr{N} = \begin{bmatrix}
1 & 1 & . & . & 1 & . & . & . & . \\
1 & 1 & 1 & . & 1 & . & . & . & . \\
. & 1 & 1 & 1 & . & . & 1 & . & . \\
. & . & 1 & 1 & . & . & . & 1 & 1 \\
1 & 1 & . & . & 1 & 1 & . & . & . \\
. & . & . & . & 1 & 1 & . & . & . \\
. & . & 1 & . & . & . & 1 & . & . \\
. & . & . & 1 & . & . & . & 1 & . \\
. & . & . & 1 & . & . & . & . & 1
\end{bmatrix} \qquad (i)$$

Note that if there are nine nodes in the structure, the connectivity matrix is of order 9, and for every node there is one row and one column in the matrix. We can also interpret the connectivity matrix in the following way. If node i is connected with node j, on the ith row at jth position we have a 1; otherwise we have 0. We always consider that a node is connected to itself, therefore the main diagonal of the connectivity matrix contains only 1s. With this interpretation of the connectivity matrix, we can generate it alternately, row by row, as follows. Suppose we come to the generation of the ith row. From the key sketch of the structure we obtain the **node set** of the ith node. Here the node set consists of the labels of the nodes which are connected to node i by means of actual structural elements. With this list we go to row i and place 1s at the columns corresponding to the node labels in the node set. If we repeat this process for all nodes, we end up with the connectivity matrix.

The truth of the second interpretation of the connectivity matrix can be seen directly from the equilibrium equations of Eq. (15.44). We can interpret $\bar{\mathbf{p}}_0$ as the forces at the nodes to keep the structure in the deformed configuration defined by deflections $\mathbf{u}_{(1)}$. Suppose we clamp all nodes and move only the ith node in the αth degree-of-freedom direction a unit amount. If we use the deflections corresponding to this case as $\mathbf{u}_{(1)}$, then the column matrix $\mathbf{u}_{(1)}$ will contain all zeros except for the ith partition, which will contain a 1 at αth position. If we premultiply this column matrix by $\bar{\mathbf{K}}$ as shown in Eq. (15.44), we obtain the αth column in the ith partition column of $\bar{\mathbf{K}}$, namely $\bar{\mathbf{k}}_{i,\alpha}$. The nonzero entries of $\bar{\mathbf{k}}_{i,\alpha}$ can now be interpreted as the holding forces at the nodes to keep the structure in the deformed configuration defined by our $\mathbf{u}_{(1)}$. We can see that if we clamp all nodes and move only one node, the holding forces will develop only at the node and its node set. Therefore, the ith column of $\mathscr{N}$ indeed contains 1s at the rows corresponding to the node set of the ith node. Since the connectivity matrix is symmetric, this is also true for the ith row.

As discussed in the next article, the number of arithmetic operations for the solution of Eq. (15.44) drastically decreases as the number of zeros at the upper right corner of $\bar{\mathbf{K}}$ (or lower left corner due to symmetry) increases. We have seen from the foregoing discussions that the structure of $\bar{\mathbf{K}}$ depends to a great extent on how we label the nodes. For example, when the vertex labels of element 7 in the structure shown in Fig. 15.4 are as in Eq. (b) above, that is, 5 and 2, we observe from Eq. (g) that positions (2,2), (2,5), (5,2) and (5,5) become nonzero. If the vertices of this element had the node labels 1 and 2, the positions (1,1), (1,2), (2,1), and (2,2) would be nonzero. The latter case is preferable, as far as increasing the number of zeros in the upper right corner of $\overline{\mathbf{K}}$ is concerned. We observe that if we decrease the node-label difference in all elements, the nonzero entries of $\overline{\mathbf{K}}$ will cluster around the main diagonal and thus create a very favorable situation for both the storage area and the number of arithmetic operations for the solution of Eq. (15.44). Matrices where the nonzero elements are clustered around the main diagonal are called **banded matrices.** We see that by proper labeling of the nodes we can make $\overline{\mathbf{K}}$ a banded matrix. The column-label difference of the last and the first nonzero entries in any one row of $\overline{\mathbf{K}}$ is called the **bandwidth** of that row. We observe that the bandwidth of the ith row is the maximum label difference in the node set of the ith node.

In order to increase the bandedness of $\overline{\mathbf{K}}$ we have to use a node-labeling system such that the label differences in all node sets are as small as possible. In Art. 12.2.1 we gave this rule without proof. We can materialize the objective of this rule in two-dimensional meshes if we start giving node labels at a corner node and continue doing so by moving in the direction of fewer nodes as much as possible. This rule is also applicable to three-dimensional meshes, where we again start labeling at a corner node and proceed to label the nodes by moving in the direction with the least number of nodes most often. When the structure is complicated, we can use the above rule to select a tentative labeling system. We find the connectivity matrix corresponding to this labeling system and then systematically apply a trial-and-error method to obtain a better labeling system.

To demonstrate the drastic improvement in the bandedness of $\overline{\mathbf{K}}$ let us consider the structure of Fig. 15.4 with the following node-labeling system: 6, 5, 1, 2, 3, 7, 4, 8, 9. In other words, the node which is labeled 6 in the figure should have the new label 1 (the position of 6 in the list), the node which is labeled 5 in the figure should have the new label 2 (the position of 5 in the list), etc. Keeping the element labels the same [actually the way we label the elements never enters in the connectivity matrix, as can be seen from Eq. (15.67)], if we first obtain the column matrix $\mathbf{j}$ and then the connectivity matrix, we find

$$
\mathscr{N} = \begin{bmatrix}
1 & 1 & . & . & . & . & . & . & . \\
 & 1 & 1 & 1 & . & . & . & . & . \\
 & & 1 & 1 & . & . & . & . & . \\
 & & & 1 & 1 & . & . & . & . \\
 & & & & 1 & 1 & 1 & . & . \\
\text{Symmetric} & & & & & 1 & . & . & . \\
 & & & & & & 1 & 1 & 1 \\
 & & & & & & & 1 & . \\
 & & & & & & & & 1
\end{bmatrix} \tag{j}
$$

where the upper right corner contains more zero submatrices than Eq. (i).

Let us conclude the discussion of $\overline{\mathbf{K}}$ in the absence of deflections boundary conditions by expressing Eqs. (15.66) and (15.67) in terms of unpartitioned elements. Using α and β for the degree-of-freedom directions at a node, and noting that the ranges of α and β are both e, we write

$$\overline{\mathbf{K}} = \sum_{m=1}^{M} \sum_{g=1}^{n^m} \sum_{\alpha=1}^{e} \sum_{h=1}^{n^m} \sum_{\beta=1}^{e} \mathbf{i}_{j_g^m \alpha} k_{g\alpha h\beta}^m \mathbf{i}_{j_h^m \beta}^T \qquad (15.68)$$

and

$$\mathcal{N} = \sum_{m=1}^{M} \sum_{g=1}^{n^m} \sum_{\alpha=1}^{e} \sum_{h=1}^{n^m} \sum_{\beta=1}^{e} \mathbf{i}_{j_g^m \alpha} \mathbf{i}_{j_h^m \beta}^T \qquad (15.69)$$

$\mathbf{i}_{j\alpha}$ is αth column in the jth partition of the identity matrix $\mathbf{1}$ of order Ne. Note that Eq. (15.68) implies a fivefold looping in order to generate $\overline{\mathbf{K}}$ from the free-free element matrices. The outer loop controls the elements, the next two loops scan the rows of the free-free element stiffness matrix, and the last two loops scan the columns of the free-free element stiffness matrix. In the case of $\mathcal{N}$, instead of using Eq. (15.69), we use Eq. (15.67) and work with the submatrices.

15.4.2 Matrix $\overline{\mathbf{K}}$ when deflection constraints exist. Now we consider the form of matrix $\overline{\mathbf{K}}$ when there are deflection boundary conditions. Suppose the boundary conditions are given as in Eq. (15.60). With the definition in Eq. (15.61) we can rewrite Eq. (15.39) as

$$\overline{\mathbf{K}} = \sum_{m=1}^{M} \mathcal{H} \mathbf{N}^{m^T} \mathcal{K}^m \mathbf{N}^m \mathcal{H}^T \qquad (15.70)$$

Noting that the $\mathcal{H}$'s do not depend on m, we may move them outside the summation sign to obtain

$$\overline{\mathbf{K}} = \mathcal{H} \left(\sum_{m=1}^{M} \mathbf{N}^{m^T} \mathcal{K}^m \mathbf{N}^m \right) \mathcal{H}^T \qquad (15.71)$$

where we recognize the parenthetical quantity on the right-hand side as the overall stiffness matrix of the structure without any deflection constraints [see Eq. (15.63)]. Let $\overline{\mathbf{K}}_0$ denote the overall stiffness matrix of the unsupported structure, namely,

$$\overline{\mathbf{K}}_0 = \sum_{m=1}^{M} \mathbf{N}^{m^T} \mathcal{K}^m \mathbf{N}^m \qquad (15.72)$$

Then we can rewrite Eq. (15.71) as

$$\overline{\mathbf{K}} = \mathcal{H} \overline{\mathbf{K}}_0 \mathcal{H}^T \qquad (15.73)$$

In some applications the generation of the overall stiffness matrix by means of Eq. (15.73) is very appropriate, especially when we would like to study the same structure with different support conditions. However, in other cases, the generation of $\overline{\mathbf{K}}$ from Eq. (15.73) is not appropriate for economical reasons. Since the order of $\overline{\mathbf{K}}$ is b smaller than that of $\overline{\mathbf{K}}_0$, we may need more storage than actually needed by $\overline{\mathbf{K}}$. We shall not discuss the generation of $\overline{\mathbf{K}}$ by Eq. (15.73) further.

Using the definition of $\mathbf{N}^m$ from Eq. (15.9) and taking advantage of the identity in Eq. (15.65), we can rewrite Eq. (15.70) as

$$\bar{\mathbf{K}} = \sum_{m=1}^{M} \sum_{g=1}^{n^m} \sum_{h=1}^{n^m} \mathscr{H} \mathbf{I}_{j_g^m} \mathscr{K}_{gh}^m \mathbf{I}_{j_h^m}^T \mathscr{H}^T \tag{15.74}$$

From Eq. (15.74) we observe that $\mathscr{H}$ is a $Ne - b$ by Ne matrix and partitionable by columns into N partitions, each of which contains e columns. Let $\mathscr{H}_j$ denote the jth column partition. Then we can write

$$\mathscr{H} \mathbf{I}_{j_g^m} = \mathscr{H}_{j_g^m} \tag{15.75}$$

Using this in Eq. (15.74), we obtain

$$\bar{\mathbf{K}} = \sum_{m=1}^{M} \sum_{g=1}^{n^m} \sum_{h=1}^{n^m} \mathscr{H}_{j_g^m} \mathscr{K}_{gh}^m \mathscr{H}_{j_h^m}^T \tag{15.76}$$

Note that $\mathscr{H}_{j_g^m}$ is the j_g^mth column partition of $\mathscr{H}$ or j_g^mth row partition of $\mathbf{H}_{(1)}$. The scalar j_g^m is the node label of the gth vertex of the mth element. According to Eq. (15.60), the row numbers of nonzero entries of $\mathscr{H}_{j_g^m}$ are the component numbers of those components of the independent deflections column $\mathbf{u}_{(1)}$ which are used in describing the deflections of node j_g^m. This information is almost directly provided by the DBC input units (see Art. 12.4.4). If the deflections of node j_g^m are not dependent on any other deflection, $\mathscr{H}_{j_g^m}$ is identical with $\mathbf{I}_{j_g^m}$; that is, it consists of $N - 1$ zero submatrices of order e and one identity matrix of order e as the j_g^mth submatrix. On the other hand, if all the deflections of node j_g^m are prescribed, then $\mathscr{H}_{j_g^m}$ consists of N zero submatrices of order e. If the deflections of node j_g^m depend on the deflections of nodes, say p and q, then all submatrices of $\mathscr{H}_{j_g^m}$ will be zero submatrices of order e, excluding those corresponding to the pth and the qth nodes. Given m, g, and h, if $\mathscr{H}_{j_g^m}$ contains s nonzero submatrices and $\mathscr{H}_{j_g^m}$ contains t nonzero submatrices, then submatrix $\mathscr{K}_{gh}^m$ will contribute st submatrices of $\bar{\mathbf{K}}$. Note that when s or t or both s and t are zero, there will not be any contribution to $\bar{\mathbf{K}}$ submatrix $\mathscr{K}_{gh}^m$.

As an example, let us sketch the contribution of the seventh element of the structure shown in Fig. 15.4 to the overall stiffness matrix. For this structure, $N = 9$, $e = 3$, $b = 9$, and therefore the order of the independent-deflections column matrix $\mathbf{u}_{(1)}$ is $Ne - b = 18$. Let us assume that the independent deflection components are ordered with their node labels; i.e., the independent deflection component of a node with smaller label comes before that of a node with larger label in the list of $\mathbf{u}_{(1)}$. Suppose that the vertices of element 7 are ordered as shown in Eq. (b) of this article. Denoting the zero submatrices by boldface points and the nonzero ones by $\mathbf{1}$, we can verify that

$$\mathscr{H}_{j_1^7}^T = \mathscr{H}_5^T = [.\ \ .\ \ .\ \ .\ \ \mathbf{1}\ \ .\ \ .] \tag{k}$$

and

$$\mathscr{H}_{j_2^7}^T = \mathscr{H}_2^T = [.\ \ \mathbf{1}\ \ .\ \ .\ \ .\ \ .\ \ .] \tag{l}$$

Note that when $\mathbf{u}_{(1)}$ is partitioned with nodes, there will be seven partitions. This is why $\mathcal{H}_5$ and $\mathcal{H}_2$ contain only seven partitions. If we denote the submatrices of $\mathcal{K}^7$ as in Eq. (a), we can display the contributions of $\mathcal{K}^7$ to $\overline{\mathbf{K}}$ as follows:

$$\sum_{g=1}^{2}\sum_{h=1}^{2}\mathcal{H}_{j_g^7}\mathcal{K}_{gh}^7\mathcal{H}_{j_h}^{T\,7} = \begin{bmatrix} \cdot & \cdot & \cdot & \cdot & \cdot & \cdot & \cdot \\ \cdot & \mathbf{D} & \cdot & \cdot & \mathbf{C} & \cdot & \cdot \\ \cdot & \cdot & \cdot & \cdot & \cdot & \cdot & \cdot \\ \cdot & \cdot & \cdot & \cdot & \cdot & \cdot & \cdot \\ \cdot & \mathbf{B} & \cdot & \cdot & \mathbf{A} & \cdot & \cdot \\ \cdot & \cdot & \cdot & \cdot & \cdot & \cdot & \cdot \\ \cdot & \cdot & \cdot & \cdot & \cdot & \cdot & \cdot \end{bmatrix} \qquad (m)$$

If we obtain the contributions of each of the elements in this way and sum them according to Eq. (15.76), we obtain

$$\sum_{m=1}^{9}\sum_{g=1}^{2}\sum_{h=1}^{2}\mathcal{H}_{j_g^m}\mathcal{K}_{gh}^m\mathcal{H}_{j_h}^{T\,m} = \begin{bmatrix} 1 & 1 & \cdot & \cdot & 1 & \cdot & \cdot \\ & 1 & 1 & \cdot & 1 & \cdot & \cdot \\ & & 1 & 1 & \cdot & \cdot & \cdot \\ & & & 1 & \cdot & \cdot & 1 \\ & & & & 1 & 1 & \cdot \\ & & \text{Symmetric} & & & 1 & \cdot \\ & & & & & & 1 \end{bmatrix} = \overline{\mathbf{K}} \qquad (n)$$

where $\mathbf{1}$ represents nonzero submatrices, and boldface dots represent zero submatrices. Note that in Eq. (n) the partitioning is not uniform and $\overline{\mathbf{K}}$ corresponds to the following ordering of $\mathbf{u}_{(1)}$:

$$\mathbf{u}_{(1)}^T = [d_{1x} \quad d_{1y} \quad \theta_{1z} \mid d_{2x} \quad d_{2y} \quad \theta_{2z} \mid d_{3x} \quad d_{3y} \quad \theta_{3z} \mid d_{4x} \quad d_{4y} \quad \theta_{4z} \mid$$
$$d_{5x} \quad d_{5y} \quad \theta_{5z} \mid d_{6x} \quad \theta_{6z} \mid \theta_{8z}] \qquad (o)$$

where the partitioning lines corresponding to those in Eq. (n) are indicated. So long as we are not interested in the numerical values of submatrices $\mathbf{1}$ in Eq. (n), we can replace the submatrices $\mathcal{K}_{gh}^m$ by the identity matrices of the same order and write from Eq. (15.76)

$$\mathcal{N} = \sum_{m=1}^{M}\sum_{g=1}^{n^m}\sum_{h=1}^{n^m}\mathcal{H}_{j_g^m}\mathcal{H}_{j_h}^{T\,m} \qquad (15.77)$$

If we consider the nonzero submatrices of $\mathcal{H}_{j_g^m}$ as binary 1s and zero submatrices as binary 0s and replace the addition operations by OR operations, matrix $\mathcal{N}$ generated in this manner is called the **connectivity matrix of the supported structure**. For the structure of Fig. 15.4, we can display the connectivity matrix as

$$\mathcal{N} = \begin{bmatrix} 1 & 1 & \cdot & \cdot & 1 & \cdot & \cdot \\ & 1 & 1 & \cdot & 1 & \cdot & \cdot \\ & & 1 & 1 & \cdot & \cdot & \cdot \\ & & & 1 & \cdot & \cdot & 1 \\ & & & & 1 & 1 & \cdot \\ & & & & & 1 & \cdot \\ & \text{Symmetric} & & & & & 1 \end{bmatrix} \qquad (p)$$

where binary zeros are shown by boldface points. Comparing this matrix with the one in Eq. (n), we observe that the connectivity matrix of the supported structure is of the same format as matrix $\overline{\mathbf{K}}$.

The arguments discussed for the connectivity matrix of the structure without the deflection boundary conditions are applicable also to the connectivity matrix of the supported structure. However, it should always be remembered that here the partitioning may not be uniform and the node set of a node contains not only the node labels of the nodes which are connected to the node by means of actual structural elements but also through the deflection boundary conditions. We observe that to use boundary conditions which relate the deflections of one node to those of other nodes may increase the bandwidth. To decrease the bandwidth in $\overline{\mathbf{K}}$, we can try to reorder the components of $\mathbf{u}_{(1)}$. For example, for the structure of Fig. 15.4, if we order the deflection components of $\mathbf{u}_{(1)}$ as

$$\mathbf{u}_{(1)}^{*T} = [d_{6x} \quad \theta_{6z} \mid d_{5x} \quad d_{5y} \quad \theta_{5z} \mid d_{1x} \quad d_{1y} \quad \theta_{1z} \mid d_{2x} \quad d_{2y} \quad \theta_{2z} \mid$$
$$d_{3x} \quad d_{3y} \quad \theta_{3z} \mid d_{4x} \quad d_{4y} \quad \theta_{4z} \mid \theta_{8z}] \qquad (q)$$

the corresponding connectivity matrix of the supported structure with the partitioning lines marked in Eq. (q) will be

$$\mathcal{N}^* = \begin{bmatrix} 1 & 1 & . & . & . & . & . \\ & 1 & 1 & 1 & . & . & . \\ & & 1 & 1 & . & . & . \\ & & & 1 & 1 & . & . \\ & & & & 1 & 1 & . \\ & \text{Symmetric} & & & & 1 & 1 \\ & & & & & & 1 \end{bmatrix} \qquad (r)$$

which is much better than the one corresponding to the ordering of $\mathbf{u}_{(1)}$ given in Eq. (o).

We can express $\overline{\mathbf{K}}$ and $\mathcal{N}$ of Eqs. (15.76) and (15.77) in terms of unpartitioned elements. Using α and β for the degree-of-freedom directions at a node and noting that the ranges of α and β are both e, we write

$$\overline{\mathbf{K}} = \sum_{m=1}^{M} \sum_{g=1}^{n^m} \sum_{\alpha=1}^{e} \sum_{h=1}^{n^m} \sum_{\beta=1}^{e} \hbar_{j_g^m \alpha} k_{g\alpha h\beta}^m \hbar_{j_h^m \beta}^{T_m} \qquad (15.78)$$

and

$$\mathcal{N} = \sum_{m=1}^{M} \sum_{g=1}^{n^m} \sum_{\alpha=1}^{e} \sum_{h=1}^{n^m} \sum_{\beta=1}^{e} \hbar_{j_g^m \alpha} \hbar_{j_h^m \beta}^{T_m} \qquad (15.79)$$

where $\hbar_{j_g^m \alpha}$ is the αth column in the j_g^mth partition of $\mathcal{H}$ (the order being $Ne - b$). Note that in order to generate $\overline{\mathbf{K}}$ or $\mathcal{N}$, we must go through a fivefold looping. In the case of $\mathcal{N}$, however, instead of Eq. (15.79), we use Eq. (15.77) and work with the submatrices rather than scalar entries.

15.4.3 The overall-load column matrix $\overline{\mathbf{p}}_0$. Various but equivalent definitions are given in Eqs. (15.40), (15.42), and (15.53) for the overall-load column matrix $\overline{\mathbf{p}}_0$. From these equations we have already established that if the structure is not subjected to concentrated nodal loads, element loads, and support settlements, there will be no loading on the structure, since in such a case $\bar{\mathbf{p}}_0$ becomes zero. We observe that $\bar{\mathbf{p}}_0$ is the only quantity where the contributions of various loading are accumulated. Note that the loading can in no way affect the overall-stiffness matrix $\bar{\mathbf{K}}$. That is why we refer to $\bar{\mathbf{K}}$ as representing the intrinsic properties of the structure, whereas $\bar{\mathbf{p}}_0$ represents the excitations. We give below expressions which facilitate the generation of $\bar{\mathbf{p}}_0$.

We can use the definitions given in Eqs. (15.61) and (15.9) in Eq. (15.40) and rewrite the latter as

$$\bar{\mathbf{p}}_0 = \mathscr{H}\mathbf{p}_0 - \sum_{m=1}^{M} \mathscr{H}\left(\sum_{g=1}^{n^m} \mathbf{I}_{j_g^m} \mathbf{I}_g^{m^T}\right)\mathbf{q}_0^m - \sum_{m=1}^{M} \mathscr{H}\left(\sum_{g=1}^{n^m} \mathbf{I}_{j_g^m} \mathbf{I}_g^{m^T}\right)\mathscr{K}^m\left(\sum_{h=1}^{n^m} \mathbf{I}_h^m \mathbf{I}_{j_h^m}^T\right)\mathbf{u}_0 \tag{15.80}$$

Noting that in addition to the identities given by Eqs. (15.75) and (15.65) we also have

$$\mathbf{I}_g^{m^T}\mathbf{q}_0^m = \mathbf{q}_{0g}^m \tag{15.81}$$

and

$$\mathbf{I}_{j_h^m}^T\mathbf{u}_0 = \mathbf{u}_{0j_h^m} \tag{15.82}$$

we can rewrite Eq. (15.80) as

$$\bar{\mathbf{p}}_0 = \sum_{i=1}^{N} \mathscr{H}_i\mathbf{p}_{0i} - \sum_{m=1}^{M}\sum_{g=1}^{n^m} \mathscr{H}_{j_g^m}\mathbf{q}_{0g}^m - \sum_{m=1}^{M}\sum_{g=1}^{n^m}\sum_{h=1}^{n^m} \mathscr{H}_{j_g^m}\mathscr{K}_{gh}^m\mathbf{u}_{0j_h^m} \tag{15.83}$$

Note that the first term involves a loop on the nodes and the second and the third terms involve looping on the elements. The first term is the contribution of prescribed nodal concentrated loads, the second term is the contribution of element loads, and the last term is the contribution of the support settlements.

15.4.4 Node relabeling. We have already discussed the effect of ordering the components of $\mathbf{u}_{(1)}$ on the bandedness of $\bar{\mathbf{K}}$. In practice, we usually order the independent deflection components in the following way. Following the node-labeling system used in the key sketch of the structure, we list the complete list of nodal deflections $\mathbf{u}$ in a manner partitionable with the nodes. For example, if the structure has N nodes and the degree of freedom per node is e, then $\mathbf{u}$ can be displayed as

$$\mathbf{u}^T = [u_{1,1} \quad \cdots \quad u_{1,e} \quad u_{2,1} \quad \cdots \quad u_{N,e}] \tag{15.84}$$

Then by examining the DBC input units we identify the dependent deflection components (those defined by the first index pairs in the DBC input units) and cross them out of $\mathbf{u}$. The remaining components are independent deflection components; they constitute the column matrix $\mathbf{u}_{(1)}$. This process carries the node labeling used in the key sketch of the structure directly to $\mathbf{u}_{(1)}$ and therefore to the connectivity and overall-stiffness matrices. When we apply this process for the structure shown in Fig. 15.4, we obtain $\mathbf{u}_{(1)}$ as displayed in Eq. (o). The connectivity matrix of the supported structure for this $\mathbf{u}_{(1)}$ is given in Eq. (p).

Suppose we change the labeling used in Fig. 15.4 into the order of

$$\jmath^{*T} = [6 \quad 5 \quad 1 \quad 2 \quad 3 \quad 7 \quad 4 \quad 8 \quad 9] \qquad (s)$$

from the order of

$$\jmath^{T} = [1 \quad 2 \quad 3 \quad 4 \quad 5 \quad 6 \quad 7 \quad 8 \quad 9] \qquad (t)$$

used in the figure. The new order given in Eq. (s) tells us that the node which carries label 6 with the ordering of Eq. (t) now carries label 1, the node with the old label 5 should carry label 2, etc. If we list all the deflection components with the ordering of Eq. (s) and then cross out the dependent components, we obtain $\mathbf{u}^{*}_{(1)}$ given by Eq. (q) and the corresponding connectivity matrix as shown in Eq. (r). Obviously the ordering system in Eq. (s) is better than the one used in Fig. 15.4. We write the relationship between $\jmath$ and $\jmath^{*}$ as

$$\jmath = \mathbf{P}\jmath^{*} \qquad (15.85)$$

where $\mathbf{P}$ is the permutation matrix, which can be displayed for the present case as

$$\mathbf{P} = [\mathbf{i}_6 \quad \mathbf{i}_5 \quad \mathbf{i}_1 \quad \mathbf{i}_2 \quad \mathbf{i}_3 \quad \mathbf{i}_7 \quad \mathbf{i}_4 \quad \mathbf{i}_8 \quad \mathbf{i}_9] \qquad (u)$$

where $\mathbf{i}_j$ is the jth column of the ninth-order identity matrix. Note that the indices of $\mathbf{i}$ in Eq. (u) are the same as those given in Eq. (s). If we consider the entries of the binary connectivity matrix $\mathcal{N}$ as scalars, we can write the relationship between $\mathcal{N}$ of Eq. (p) and $\mathcal{N}^{*}$ of Eq. (r) by means of a suitably selected permutation matrix. If we list the partition node labels in Eq. (o) as column $\jmath$, we have

$$\jmath^{T} = [1 \quad 2 \quad 3 \quad 4 \quad 5 \quad 6 \quad 8] \qquad (v)$$

Listing the partition node labels in Eq. (q) as column $\jmath^{*}$, we write

$$\jmath^{*T} = [6 \quad 5 \quad 1 \quad 2 \quad 3 \quad 4 \quad 8] \qquad (w)$$

The permutation matrix relating $\jmath$ to $\jmath^{*}$ as in Eq. (15.85) can be displayed as

$$\mathbf{P} = [\mathbf{i}_6 \quad \mathbf{i}_5 \quad \mathbf{i}_1 \quad \mathbf{i}_2 \quad \mathbf{i}_3 \quad \mathbf{i}_4 \quad \mathbf{i}_7] \qquad (x)$$

where $\mathbf{i}_j$ is the jth column of the seventh-order identity matrix $\mathbf{I}$. Now if we use this $\mathbf{P}$ and $\mathcal{N}$ of Eq. (p) and $\mathcal{N}^{*}$ of Eq. (r) in

$$\mathcal{N}^{*} = \mathbf{P}^{T}\mathcal{N}\mathbf{P} \qquad (15.86)$$

we observe that the equation holds. If we consider $\mathcal{N}$ the stiffness matrix, subject to the coordinate transformation as in Eq. (15.85), and call the transformed stiffness matrix $\mathcal{N}^{*}$, we obtain Eq. (15.86) as in Art. 14.5.3.

In many problems it may be beneficial to use the labeling system in the key sketch to generate the connectivity matrix of the supported structure and then look for a better labeling system which would produce a more compact matrix. Let $\jmath_{(0)}$ denote the node-labeling system used in the key sketch of the structure, and let $\mathcal{N}_{(0)}$ denote the associated connectivity matrix. Let a_0 denote the number of zero entries at the upper right corner of $\mathcal{N}_{(0)}$. Suppose we interchange any two successive (or otherwise) entries of $\jmath_{(0)}$ to obtain a new labeling system $\jmath_{(1)}$ and thus the permutation matrix $\mathbf{P}_{(1)}$ such that $\jmath_{(0)} = \mathbf{P}_{(1)}\jmath_{(1)}$, as in Eq. (15.85).

Using this permutation matrix, we can find the connectivity matrix corresponding to $\mathcal{S}_{(1)}$ as $\mathcal{N}_{(1)} = \mathbf{P}_{(1)}^T \mathcal{N}_{(0)} \mathbf{P}_{(1)}$ by using Eq. (15.86). We write the general recurrence formula at the ith iteration step as

$$\mathcal{N}_{(i)} = \mathbf{P}_{(i)}^T \mathcal{N}_{(i-1)} \mathbf{P}_{(i)} \tag{15.87}$$

where $\mathcal{N}_{(i)}$ corresponds to labeling system $\mathcal{S}_{(i)}$ and $\mathcal{N}_{(i-1)}$ corresponds to labeling system $\mathcal{S}_{(i-1)}$. Let the number of upper-right-corner zeros of $\mathcal{N}_{(i)}$ and $\mathcal{N}_{(i-1)}$ be denoted by a_i and a_{i-1}, respectively. If

$$a_i > a_{i-1} \tag{15.88}$$

for all i, this means that the effort spent in finding a better labeling system is paying off. During the iteration, if at any time we observe that $a_i \le a_{i-1}$, we may disregard $\mathcal{S}_{(i)}$ and try to find another one for which Eq. (15.88) would be true. Note that if $\mathcal{S}_{(i)}$ is obtained from $\mathcal{S}_{(i-1)}$ by interchanging only two entries, say, the pth and qth, the permutation matrix would look like

$$\mathbf{P}_{(i)} = \begin{bmatrix} 1 & & & \vdots & & & \vdots & & & \\ & 1 & & \vdots & & & \vdots & & & \\ & & \ddots & \vdots & & & \vdots & & & \\ & & & 1 & \vdots & & \vdots & & & \\ \cdots & & & 0 & \cdots & & 1 & \cdots & & \\ & & & & 1 & & \vdots & & & \\ & & & & & \ddots & \vdots & & & \\ & & & & & 1 & \vdots & & & \\ \cdots & & & 1 & \cdots & & 0 & \cdots & & \\ & & & & & & 1 & & & \\ & & & & & & & \ddots & & \\ & & & & & & & & 1 & \\ & & & & & & & & & 1 \end{bmatrix} \begin{matrix} \\ \\ p \\ \\ \\ \\ q \\ \\ \\ \\ \end{matrix} \tag{15.89}$$

which is obtained by interchanging the pth and the qth columns of the identity matrix of the same order. If we use this $\mathbf{P}_{(i)}$ in Eq. (15.87), the triple product on the right-hand side amounts to interchanging the pth and the qth rows and also interchanging the pth and qth columns of $\mathcal{N}_{(i-1)}$. Therefore the iteration mentioned above is not an expensive one. In determining the sequence of labeling systems for trial, that is, $\mathcal{S}_{(i)}$, $i = 1, 2, \ldots$, one may use various stratagems. One popular one is the sequence obtained by interchanges in successive pairs, one from top and one from bottom of $\mathcal{S}_{(0)}$, proceeding in this manner to the middle of $\mathcal{S}_{(0)}$ repeating this many times. For example, if $\mathcal{S}_{(0)}$ is given as in Eq. (*t*), one may try (1,2), (9,8), (2,3), (8,7), ... etc. interchanges and repeat this many times. If any labeling system in the standard sequence violates the rule in Eq. (15.88), we should discard it. We may cut off the iteration when improvement ceases in a predetermined number of successive trials. One should bear in mind that in this type of iterative methods, there is no mathematical guarantee that the final system is the best possible. The best system can be obtained by trying all $N!$ possibilities if the order of $\mathcal{S}_{(0)}$ is N. For large N, $N!$ is staggeringly large. We can spend some effort in finding a better labeling system provided the effort is acceptable in view of the expected gains.

15.4.5 How to store overall-stiffness matrix $\overline{\mathbf{K}}$. The connectivity matrix of the supported structure not only is useful in obtaining a better labeling system which would cluster the nonzero entries of the overall-stiffness matrix $\overline{\mathbf{K}}$ around the main diagonal but it also provides information about the necessary storage area for the entries of $\overline{\mathbf{K}}$. Since $\overline{\mathbf{K}}$ is symmetrical, we need keep only one symmetrical half of the matrix. Furthermore, as discussed in Art. 15.5, there is no need to allocate storage space for the zeros in the upper right corner of $\overline{\mathbf{K}}$ when we keep only the upper half of $\overline{\mathbf{K}}$. Let us assume that the connectivity matrix of a supported structure corresponding to the final node-labeling system is as displayed below

$$\mathscr{N} = \quad \text{Symmetrical} \qquad\qquad\qquad (15.90)$$

We can observe, first, two distinct zones in the upper half of the matrix: the bordered zone and the upper right zero zone. We should never save storage for the upper right zero zone. In the bordered zone we can identify three different subzones: unshaded zero pockets, shaded nonzero pockets, and doubly shaded zero pockets. Depending upon the solution method used to solve the independent deflection components $\mathbf{u}_{(1)}$ from Eq. (15.44), we may assign storage for only nonzero pockets, or for only nonzero pockets and doubly shaded zero pockets, or for the whole bordered zone. Whatever the case may be, the connectivity matrix provides all the necessary information for the locations and sizes of these various zones. Sometimes, the core memory of the computer may be too small to accommodate these pockets all at once. In such cases, the connectivity matrix provides information about the partitioning of the bordered zone which is best suited to that structure and the available rapid-access memory. Perhaps this is the single most important characteristic of the connectivity matrix, which provides the automatic partitioning as an alternative to partitioning by the engineer, i.e., substructuring.

15.4.6 Problems for Solution

Problem 15.27 Going through the algebra, obtain Eq. (15.66) from Eq. (15.64).

Problem 15.28 Express explicitly the nonzero matrices of Eq. (*h*) in Art. 15.4, in terms of $\mathscr{K}_{gh}^{m}$, $g = 1, 2$; $h = 1, 2$; $m = 1, \ldots, 9$.

Problem 15.29 Solve Prob. 12.13 using the structure shown in Fig. 12.12 as a model.

Problem 15.30 Prove that matrix $\bar{\mathbf{K}}$ of a properly supported structure is not positive definite if the material matrix $\mathscr{D}$ is not positive definite.

Problem 15.31 In a planar truss, if there are only two collinear bars meeting at a node, prove that $\bar{\mathbf{K}}$ is only positive.

Problem 15.32 Obtain the connectivity matrix of the structure shown in Fig. 15.1 using Eq. (15.77). Then obtain it from the key sketch directly.

Problem 15.33 Express the nonzero submatrices of matrix $\bar{\mathbf{K}}$ of the planar truss shown in Fig. 15.1 in terms of $\mathscr{K}_{gh}^{m}$, $g = 1, 2$; $h = 1\ 2$; $m = 1, \ldots, 6$.

Problem 15.34 Display the connectivity matrix of the supported structure of Fig. 15.2.

Problem 15.35 Display the connectivity matrix of the supported structure of Fig. 12.17.

Problem 15.36 Display explicitly the contribution of element 56 to matrix $\bar{\mathbf{K}}$ of the structure shown in Fig. 15.2 in terms of submatrices $\mathscr{K}_{gh}^{56}$, $g = 1, 2, 3$; $h = 1, 2, 3$, using Eq. (15.76) (one term only in the outer sum).

Problem 15.37 Prove that when a structure is supported in a statically determinate fashion, there is no contribution to the overall load matrix $\bar{\mathbf{p}}_0$ from the support settlements $\mathbf{u}_0$.

Problem 15.38 Display explicitly the contribution of element 56 to matrix $\bar{\mathbf{p}}_0$ of the structure shown in Fig. 15.2 in terms of subcolumns $\mathbf{q}_{0g}^{56}$, $g = 1, 2, 3$, using Eq. (15.83).

Problem 15.39 Apply the relabeling stratagem explained in Art. 15.4.4 to obtain the connectivity matrix and the corresponding labeling system given in Eqs. (*r*) and (*w*), respectively, of Art. 15.4, starting from the connectivity matrix corresponding to $\jmath^T = [5\ 6\ 2\ 1\ 3\ 8\ 4]$.

Problem 15.40 Write a computer procedure which will generate the connectivity matrix in binary form from the column matrix $\mathbf{j}$ representing the mesh topology. Write another computer procedure which will generate the connectivity matrix of the supported structure in binary form from column matrix $\mathbf{j}$ and i_b number of DBC input units. Assume that $e = i_d$ is a known variable.

15.5 Methods for Solving the Governing Equations.

Various methods can be applied to Eq. (15.44) to compute the numerical values of $\mathbf{u}_{(1)}$, that is, the independent deflection components. Since the coefficient matrix $\bar{\mathbf{K}}$ is symmetric, positive definite, and usually banded, the solution technique should take advantage of these desirable properties. In choosing the solution technique, we must take into account the number of loading conditions and the characteristics of the computing facility. With these in mind, we can narrow down the preferable solution techniques to iteration with single steps (Gauss-Seidel iteration) and the Cholesky method. We shall not consider relaxation methods here since they are not compatible with the objectives of the systematic analysis. A relaxation method is basically an iterative method in which human judgment is needed at each iterative step. In systematic analysis, however, our objective is to automate the whole analysis to the point of eliminating (or at least minimizing) human intervention.

In order to simplify the notation in this article, let us redefine $\bar{\mathbf{K}}$, $\bar{\mathbf{p}}_0$, and $\mathbf{u}_{(1)}$ as

$$\mathbf{A} = \bar{\mathbf{K}} \tag{15.91}$$

$$\mathbf{c} = \bar{\mathbf{p}}_0 \tag{15.92}$$

and

$$\mathbf{x} = \mathbf{u}_{(1)} \tag{15.93}$$

With these definitions, we can rewrite Eq. (15.44) as

$$\mathbf{Ax} = \mathbf{c} \tag{15.94}$$

where $\mathbf{A}$ is symmetric, positive definite, and banded; it will be assumed to be of order n.

15.5.1 Iteration with single steps. In this method, as far as matrix $\mathbf{A} = \overline{\mathbf{K}}$ is concerned, we need store only the nonzero coefficients, corresponding to the nonzero pockets within the border zone of $\mathcal{N}$ shown in Eq. (15.90). Also needed are two columns: one for the right-hand column matrix $\mathbf{c} = \overline{\mathbf{p}}_0$ and the other for storing the computed values of components $\mathbf{x} = \mathbf{u}_{(1)}$. We start the iteration by setting all components of $\mathbf{x}$ to zero. We call this the initial estimate of the solution and denote it by $\mathbf{x}^{(0)}$. Then, from the first scalar equation, using all the available information about the unknowns $\mathbf{x}$ but the first, that is, $x_i^{(0)}$, $i = 2, \ldots, n$, we compute a better value of the first component and store it over the old value. This new value is named $x_1^{(1)}$. Note that the superscripts in parentheses denote the iteration count. We proceed to compute the new value of the second component of $\mathbf{x}$, that is, $x_2^{(1)}$, from the second scalar equation, using the latest available values of all components of $\mathbf{x}$, that is, $x_1^{(1)}$, $x_i^{(0)}$, $i = 3, \ldots, n$, excluding the second component, and store it over the old value. In this manner we continue until the last component's new value, $x_n^{(1)}$, is computed and stored over the old value. This constitutes the first iteration cycle, called the **first sweep**. We apply many sweeps, until the difference between the magnitudes of two successive estimates for all components is acceptably small.

We can express the above procedure mathematically as follows. We first scale Eq. (15.94) by dividing both sides of each equation by its own diagonal element to obtain

$$\overline{\mathbf{A}}\mathbf{x} = \overline{\mathbf{c}} \tag{15.94'}$$

where

$$\overline{\mathbf{A}} = \text{diag}\,(a_{ii}^{-1})\mathbf{A} \tag{15.95}$$

and

$$\overline{\mathbf{c}} = \text{diag}\,(a_{ii}^{-1})\mathbf{c} \tag{15.96}$$

We observe that the main-diagonal entries of $\overline{\mathbf{A}}$ are all 1s. With this in mind we assume that $\overline{\mathbf{A}}$ consists of three components, $\mathbf{I}$, $\overline{\mathbf{L}}$, and $\overline{\mathbf{U}}$, such that

$$\overline{\mathbf{A}} = \mathbf{I} - \overline{\mathbf{L}} - \overline{\mathbf{U}} \tag{15.97}$$

Substituting $\overline{\mathbf{A}}$ from Eq. (15.97) into Eq. (15.94'), we obtain

$$\mathbf{x} = \overline{\mathbf{L}}\mathbf{x} + \overline{\mathbf{U}}\mathbf{x} + \overline{\mathbf{c}} \tag{15.98}$$

The iteration with single steps explained in the previous paragraph can now be expressed by the recurrence formula

$$\mathbf{x}^{(k+1)} = \overline{\mathbf{L}}\mathbf{x}^{(k+1)} + \overline{\mathbf{U}}\mathbf{x}^{(k)} + \overline{\mathbf{c}} \qquad \text{for } k = 0, 1, \ldots \tag{15.99}$$

with

$$\mathbf{x}^{(0)} = \mathbf{0} \tag{15.100}$$

The convergence is guaranteed as long as $\mathbf{A}$ is positive definite.[1]

The solution technique described in this subsection is also called Gauss-Seidel iteration. Since $\overline{\mathbf{K}} = \mathbf{A}$ is positive definite, it can be used in solving Eq. (15.44). The method has the following advantages:

1. Rapid-access memory requirements from the computing facility are minimal since we are interested only in the nonzero entries of $\overline{\mathbf{K}} = \mathbf{A}$. As discussed in Art. 15.4, $\overline{\mathbf{K}}$ is usually a very sparse matrix, i.e., it probably has more zero elements than the nonzero ones.

2. Relabeling the nodes is not necessary since relabeling merely moves the entries of $\overline{\mathbf{K}}$ around without changing either the values or the total number of nonzero entries.

3. The computer procedure related to the algorithm described by Eqs. (15.99) and (15.100) is very simple.

However, the method has the following disadvantages:

1. Although the convergence to the true solution, i.e., the statement in Eq. (15.105), is guaranteed, the convergence may be very slow; thus we may need many iteration sweeps in order to arrive at an acceptable solution.

2. Efforts spent to obtain the solution corresponding to one right-hand side cannot be utilized for another right-hand side.

Because of the above disadvantages, the Guass-Seidel iteration is usually confined to small problems and hand computations. For example, the **moment-distribution method** is nothing but an engineered Gauss-Seidel iteration.

15.5.2　Cholesky method. In this method, as far as matrix $\overline{\mathbf{K}} = \mathbf{A}$ is concerned, we must store the matrix elements falling in the bordered zone of Eq. (15.90), with or without the unshaded zero pockets marked in this equation. Storage area for the right-hand column matrix $\overline{\mathbf{p}}_0 = \mathbf{c}$ is also required. Other than these, basically no additional storage is necessary. In this method, we first systematically

---

[1] The error in $\mathbf{x}^{(k)}$ is defined as

$$\mathbf{e}^{(k)} = \mathbf{x} - \mathbf{x}^{(k)} \tag{15.101}$$

Since $\mathbf{x}$ is unknown, instead of $\mathbf{e}^{(k)}$, one may use the **residual $\mathbf{r}^{(k)}$**, defined as

$$\mathbf{r}^{(k)} = \bar{\mathbf{c}} - \bar{\mathbf{A}}\mathbf{x}^{(k)} \tag{15.102}$$

as the measure of the error. Using Eqs. (15.99) and (15.98) in Eq. (15.101) gives

$$\mathbf{e}^{(k+1)} = (\mathbf{I} - \bar{\mathbf{L}})^{-1}\bar{\mathbf{U}}\mathbf{e}^{(k)} \tag{15.103}$$

or using this equation recursively leads to

$$\mathfrak{z}^{(k+1)} = [(\mathbf{I} - \bar{\mathbf{L}})^{-1}\bar{\mathbf{U}}]^{k+1}\mathbf{e}^{(0)} \tag{15.104}$$

where the exponent of the bracketed factor shows power. From Eq. (15.101) we see that

$$\lim_{k \to \infty} \mathbf{x}^{(k)} \to \mathbf{x} \tag{15.105}$$

if $\mathbf{e}^{(k)}$ vanishes for large k. From Eq. (15.104), since $\mathbf{e}^{(0)}$ is a constant column matrix, its coefficient must vanish for large k, so that $\mathbf{e}^{(k)}$ vanishes. Luckily, this is ensured so long as $\mathbf{A}$ is positive definite.

modify the matrix elements within the bordered zone, leaving the unshaded zero pockets intact. This operation is called **decomposition.** Then using the decomposed matrix and the right-hand column matrix **c**, we modify the elements of **c**. This is called the **forward pass.** Finally, using the decomposed matrix and the modified **c**, we compute the numerical values of $u_{(1)} = x$ and store them over the modified **c** values. This last step is called the **backward pass.**

The method is based on the fact that any real symmetric positive definite matrix can be factored so that it is equal to the product of an upper triangular matrix premultiplied by its transpose. In what follows we shall not initially consider the coefficient matrix **A** to be banded, but later we shall see how to take advantage of this property.

For positive definite and symmetric real matrix **A** we write

$$A = U^T U \tag{15.106}$$

where **U** is an upper triangular matrix. We can display Eq. (15.106) as

$$
\begin{bmatrix} a_{11} & a_{12} & \cdots & a_{1n} \\ & a_{22} & \cdots & a_{2n} \\ & & \cdots\cdots\cdots \\ \text{Symmetric} & & & a_{nn} \end{bmatrix}
=
\begin{bmatrix} u_{11} & & & \\ u_{12} & u_{22} & & \\ \cdots\cdots\cdots \\ u_{1n} & u_{2n} & \cdots & u_{nn} \end{bmatrix}
\begin{bmatrix} u_{11} & u_{12} & \cdots & u_{1n} \\ & u_{22} & \cdots & u_{2n} \\ & & \cdots\cdots\cdots \\ & & & u_{nn} \end{bmatrix}
\tag{15.107}
$$

where zero entries are not written and it is assumed that the order of **A** is n. By actual multiplication of the matrices on the right-hand side, we obtain explicit expressions for u_{ij} entries of **U** in terms of the a_{ij} entries of **A**. For example, $a_{11} = u_{11}u_{11}$, which gives $u_{11} = \sqrt{a_{11}}$, and $a_{12} = u_{11}u_{12}$, which gives $u_{12} = a_{12}/u_{11}$, etc. If we continue in this fashion, proceeding in row major order in the upper portion of **A**, we obtain the following expressions, which define u_{ij} elements in terms of a_{ij}:

$$u_{11} = \sqrt{a_{11}} \tag{15.108}$$

$$u_{1j} = a_{1j}/u_{11} \qquad j = 2, \ldots, n \tag{15.109}$$

and

$$u_{ii} = \sqrt{a_{ii} - \sum_{k=1}^{i-1} u_{ki}u_{ki}} \tag{15.110}$$

$$u_{ij} = \frac{a_{ij} - \sum_{k=1}^{i-1} u_{ki}u_{kj}}{u_{ii}} \qquad j = i+1, \ldots, n \tag{15.111}$$

$$\left.\right\} \quad i = 2, \ldots, n$$

Note that the equations require that the u_{ij} elements be generated in row major order, i.e., first the first-row elements, then the second-row elements, etc. During the generation of u_{ij} entries in this order, when a reference is made to any a_{ij} element, that element is never referred again. This means that we can store the u_{ij} quantities in the same locations as the corresponding a_{ij} elements. Thus, at the end of the process, we complete the modification of a_{ij} numbers into u_{ij} numbers. As mentioned earlier, this is called the decomposition step.

When the matrix is positive definite, the quantities under the radical sign in Eqs. (15.108) and (15.110) are always larger than zero. We can prove this statement as follows. Suppose one of the radical signs contains a negative argument; then that diagonal element and the row of that diagonal will become *pure imaginary*. When we finish the decomposition, certain elements of U will be pure imaginary. However, when A is positive definite, this is not possible, since for any real column matrix v

$$\mathbf{U v = t} \tag{15.112}$$

and

$$\mathbf{v}^T \mathbf{A v} = \mathbf{t}^T \mathbf{t} = s \tag{15.113}$$

which follows from Eq. (15.106). Suppose U contains a pure imaginary row. Then we can select v such that t is nonnegative pure imaginary to produce a negative s, thus contradicting the positive definiteness of A. If A is only positive, at least one of the u_{ii} elements will become zero, and the decomposition will fail according to Eq. (15.111) due to division by u_{ii}.

In Art. 15.4, we discussed the circumstances which may produce a nonpositive definite coefficient matrix. We can use the sign and the magnitude of the quantity under the radical sign in u_{ii}, $i = 1, \ldots, n$, expressions to diagnose whether anything is wrong with the structure as far as the material, deflection boundary conditions, and geometry are concerned. From Eqs. (15.109) and (15.111) we also observe that the portion of A corresponding to the unshaded zero pockets within the bordered zone in Eq. (15.90) will remain zero after decomposition; however, all doubly shaded zero pockets will be filled in by nonzero elements. We also observe that the zeros in the upper right corner of A will remain zero after decomposition.

The real cost of decomposition comes from the computations involved in Eq. (15.111). Suppose the form of A is

From Eq. (15.111) we observe that the number of multiplications involved for the decomposition of A is basically proportional to T

$$T = n \bar{w}^2 \tag{15.115}$$

where n is the order of matrix $\mathbf{A}$ and $\bar{w}$ is the **half bandwidth,** as shown in Eq. (15.114).   If $\bar{w} = n$, the matrix is full and the number of multiplications for the decomposition is proportional to n^3 as seen from Eq. (15.115).   On the other hand, if $\bar{w} = n/5$, the number of multiplications is 25 times less than when the matrix is full.   The necessary **storage area** S for a full matrix can be expressed as

$$S = \tfrac{1}{2}n(n + 1) \approx n^2 \tag{15.116}$$

and in the case of the banded matrix of Eq. (15.114)

$$S = n\bar{w} \tag{15.117}$$

In summary, the storage necessary for $\mathbf{A}$ is proportional to $n\bar{w}$, and the number of multiplications for the decomposition is proportional to $n\bar{w}^2$.   This is why we would like to use a node-labeling system which will cluster the nonzero elements of the overall-stiffness matrix around its main diagonal as much as possible.

Now that we know how to decompose a given positive definite real symmetric matrix, we can proceed to obtain expressions for obtaining the solution of Eq. (15.94).   Substituting $\mathbf{A}$ from Eq. (15.106) into Eq. (15.94), we write

$$\mathbf{U}^T\mathbf{y} = \mathbf{c} \tag{15.118}$$

where
$$\mathbf{U}\mathbf{x} = \mathbf{y} \tag{15.119}$$

From Eq. (15.118) we can solve $\mathbf{y}$ easily and explicitly.   Noting the form of $\mathbf{U}^T$ in Eq. (15.107), we observe that we can easily solve y_1 as

$$y_1 = \frac{c_1}{u_{11}} \tag{15.120}$$

Having computed the value of y_1, we obtain the value of y_2 from the second scalar equation as $y_2 = (c_2 - u_{12}y_1)/u_{22}$.   Continuing in this fashion, we can obtain all of the components of $\mathbf{y}$.   We express this procedure mathematically as

$$y_i = \frac{c_i - \sum_{k=1}^{i-1} u_{ki}\,y_k}{u_{ii}} \qquad i = 2, \ldots, n \tag{15.121}$$

The process defined by Eqs. (15.120) and (15.121) constitutes the forward pass of the Cholesky method.   We observe that if we calculate y_i quantities in the increasing order of their subscripts, i.e., in the order of $y_1, y_2, \ldots, y_n$, we never refer to a c_i element after it is referred once.   Therefore, we can use the storage locations of $\mathbf{c}$ to store the values of $\mathbf{y}$.

Finally we can use Eq. (15.119) to obtain the values of the unknowns $\mathbf{x}$.   Noting the form of $\mathbf{U}$ in Eq. (15.107), we can solve x_n from the last scalar equation of Eq. (15.119) as

$$x_n = \frac{y_n}{u_{nn}} \tag{15.122}$$

Having computed x_n, we calculate the value of x_{n-1} from the $(n-1)$st scalar equation as $x_{n-1} = (c_{n-1} - u_{n-1,n}x_n)/u_{n-1,n-1}$.   Continuing in this fashion, we can obtain all the unknowns.   This process is expressed as

$$x_i = \frac{y_i - \sum\limits_{k=i+1}^{n} u_{ik} x_k}{u_{ii}} \qquad i = n-1, \ldots, 1 \qquad (15.123)$$

Equations (15.122) and (15.123) constitute the backward pass of the Cholesky method.   Note that when we compute the unknowns in the order $x_n, x_{n-1}, \ldots, x_1$, a y_i element is never required again once it is referred to.   Therefore, we can use the storage locations of $\mathbf{y}$ for the storage of $\mathbf{x}$.

15.5.3 An example for the Cholesky method.   As an example, let us solve the following set of linear simultaneous equations by the Cholesky method, assuming that the coefficient matrix is positive definite:

$$\begin{bmatrix} 9 & 6 & . & 3 & . & . & . \\ & 8 & . & . & 4 & . & . \\ & & 16 & 4 & -4 & -20 & . \\ & & & 28 & 2 & . & . \\ & \text{Symmetric} & & & 15 & 9 & 3 \\ & & & & & 63 & 13 \\ & & & & & & 14 \end{bmatrix} \begin{Bmatrix} x_1 \\ x_2 \\ x_3 \\ x_4 \\ x_5 \\ x_6 \\ x_7 \end{Bmatrix} = \begin{Bmatrix} 33 \\ 42 \\ -76 \\ 137 \\ 154 \\ 454 \\ 191 \end{Bmatrix} \qquad (a)$$

where the zero entries are shown by boldface dots.   We first apply Eqs. (15.108) and (15.109) to obtain the first row of the upper decomposed matrix and write the results on the first row of the coefficient matrix, erasing the original a_{1i}, $i = 1, \ldots, n$ numbers.   Then we apply Eqs. (15.110) and (15.111) to obtain the remainder of the decomposed matrix $\mathbf{U}$.   In applying Eq. (15.110), the following should be remembered.   To obtain u_{ii} corresponding to a_{ii}, we multiply the column of numbers above a_{ii} by itself in the inner-product sense, subtract the resulting scalar from a_{ii}, and obtain u_{ii} as the square root of this number.   This is sketched below:

$$u_{ii} = \sqrt{a_{ii} - \underline{\qquad\qquad}}$$

$$(b)$$

where $*$ are used for u_{ij} elements, and $\times$ are used for a_{ij} elements.   To obtain the u_{ij} corresponding to a given a_{ij}, we multiply the column of numbers above a_{ij} with the column of numbers above u_{ii} in the inner-product sense, subtract the resulting scalar from a_{ij}, and obtain u_{ij} as this number divided by u_{ii}.   This is sketched below:

$$u_{ij} = (a_{ij} - \overleftarrow{\quad\quad} \uparrow\,)/u_{ii}$$

(c)

With the help of these sketches we can obtain the decomposed matrix $\mathbf{U}$ corresponding to the coefficient matrix given in Eq. (a) as

$$\mathbf{U} = \begin{bmatrix} 3 & 2 & . & 1 & . & . & . \\ & 2 & . & -1 & 2 & . & . \\ & & 4 & 1 & -1 & -5 & . \\ & & & 5 & 1 & 1 & . \\ & & & & 3 & 1 & 1 \\ & & & & & 6 & 2 \\ & & & & & & 3 \end{bmatrix} \qquad (d)$$

Note that the upper right zeros of $\mathbf{U}$ and $\mathbf{A}$ are the same. With the terminology used in conjunction with Eq. (15.90), the unshaded zero pockets within the bordered zone of $\mathbf{A}$ remained zero, and the doubly shaded zero pockets within the bordered zone of $\mathbf{A}$ filled with nonzero entries. To continue the solution, after we obtain y_1 from Eq. (15.120) as 11, we obtain y_i corresponding to a given c_i as follows. We multiply the column of numbers above c_i by the column of numbers above u_{ii} in the inner-product sense, subtract this scalar from c_i, and write y_i as this number divided by u_{ii}. This is the direct interpretation of Eq. (15.121), sketched below:

$$y_i = (c_i - \overleftarrow{\quad\quad} \uparrow\,)/u_{ii}$$

(e)

With the help of the sketch in Eq. (e), we can obtain the column matrix $\mathbf{y}$ corresponding to Eq. (a) from Eqs. (d) and (a) as

$$\mathbf{y}^T = [11 \quad 10 \quad -19 \quad 31 \quad 28 \quad 50 \quad 21] \tag{f}$$

Using Eq. (15.122) we readily obtain $x_7 = 7$. For the rest of unknowns we can use Eq. (15.123). We sketch the operation involved as follows:

$$x_i = (y_i - \overset{\longrightarrow}{\longleftarrow})/u_{ii} \tag{g}$$

With the help of this sketch, using $\mathbf{U}$ from Eq. (d) and $\mathbf{y}$ from Eq. (f), we obtain the values of the unknowns as

$$\mathbf{x}^T = [1 \quad 2 \quad 3 \quad 4 \quad 5 \quad 6 \quad 7] \tag{h}$$

which can easily be verified by substituting into Eq. (a).

15.5.4 Advantages and disadvantages of Cholesky-type methods. There are many other methods similar to the Cholesky method, but their treatment is outside of the scope of this book. The advantages of these methods are:

1. They fully exploit the symmetry, the positive definite property, and the bandedness of the coefficient matrices.

2. They are very fast and arrive at the solution with a known number of operations.

3. By saving the decomposed matrix, most of the effort in solving the equations can be saved for another right-hand side. Note that if the coefficient matrix is of the type of Eq. (15.114), the number of multiplications in the forward pass can be expressed as

$$T' \approx n\bar{w} \tag{15.124}$$

from Eqs. (15.121) or (15.123). Note that T' is $\bar{w}$ times less than T given in Eq. (15.115).

The disadvantages of these methods are:

1. Referring to Eq. (15.90), they require storage for the whole bordered zone, with possible exclusion of the unshaded zero pockets. In many cases, relabeling helps to decrease the area of the bordered zone.

2. The computer procedure related to these algorithms is a little more involved than that of the iteration.

15.5.5 Problems for Solution

*Problem 15.41 Solve the following linear simultaneous equations by Gauss-Seidel iteration. First test whether the coefficient matrix is positive definite.

$$\begin{bmatrix} 5 & -1 & 1 \\ & 7 & 2 \\ \text{Symmetric} & & 6 \end{bmatrix} \begin{Bmatrix} x_1 \\ x_2 \\ x_3 \end{Bmatrix} = \begin{Bmatrix} 6 \\ 19 \\ 23 \end{Bmatrix}$$

Problem 15.42 Write a computer procedure to solve Eq. (15.44) by Gauss-Seidel iteration. Assume that each of the nonzero entries of $\bar{K}$ is given by an index pair and a value. For example, if $\bar{k}_{ij}$ is a nonzero entry at the (i, j) position, it is identified by the index pair (i, j) and the value of $\bar{k}_{ij}$. It may be economical to pack i and j into one computer word and use another word for the value.

*Problem 15.43 Solve the following equations by the Cholesky method:

$$\begin{bmatrix} 16 & 12 & 4 & 0 \\ & 13 & 5 & -2 \\ & & 11 & 2 \\ \text{Symmetric} & & & 3 \end{bmatrix} \begin{Bmatrix} x_1 \\ x_2 \\ x_3 \\ x_4 \end{Bmatrix} = \begin{Bmatrix} 52 \\ 45 \\ 55 \\ 14 \end{Bmatrix}$$

Problem 15.44 Write a computer procedure to decompose a positive definite symmetric real matrix which is of the form of that given in Eq. (15.114).

Problem 15.45 Write a computer procedure to perform the forward pass of the Cholesky method which can be used in conjunction with the procedure of Prob. 15.44.

Problem 15.46 Write a computer procedure to perform the backward pass of the Cholesky method which can be used in conjunction with the procedures of Probs. 15.44 and 15.45.

Problem 15.47 Prove that if u_{ii} is pure imaginary, the ith row of the decomposed matrix in Cholesky decomposition is also pure imaginary but the contributions from this row to the succeeding rows are always real.

*Problem 15.48 Suppose that the ith row of the decomposed matrix U is pure imaginary. Find a column matrix v such that the Uv product column matrix is nonnegative pure imaginary.

15.6 Computation of the Complete List of Deflections, Reactions, and Stresses.

So far in this chapter, we have discussed the governing equations of equilibrium in terms of deflections and the solution of these equations for the unknown deflections. If the governing equations are obtained by means of Lagrange multipliers, the solution will yield the complete list of deflections u, in addition to the Lagrange multipliers [see Eq. (15.59)]. Otherwise, the governing equations are for the independent components of u, that is, for $u_{(1)}$, and their solution will yield only the components of $u_{(1)}$ [see Eqs. (15.44) or (15.51)]. In the latter case we have yet to obtain the complete list of nodal deflections u. After the nodal deflections are found, we can proceed to obtain the vertex forces and the reactions. For structures of a two- and three- dimensional continuum, we should also compute the stresses in the elements.

15.6.1 Computation of the complete list of nodal deflections.

If the method for the governing equations is the Lagrange-multipliers method, the solution of Eq. (15.59) will directly yield the complete list of nodal deflections u in the overall coordinate system.   Because of large size of equations, we seldom use the Lagrange-multipliers method.   Generally we obtain the governing equations in the form of Eq. (15.44) and obtain from such equations the independent components listed in $\mathbf{u}_{(1)}$.   After obtaining the numerical values of these deflection components, we use them in Eq. (15.60) to obtain the numerical values of $\mathbf{u}$.   Note that Eq. (15.60) is an alternate form of Eqs. (15.35) and (15.21).   Note also that $\mathbf{u}_i$, $i = 1, \ldots, N$, are expressed in the overall coordinate system.

15.6.2   Computation of the vertex forces of elements.   When the complete list of deflections is available, we can use them in Eq. (15.8) to obtain the vertex deflections of the elements.   Changing the dummy indices, we rewrite Eq. (15.8) as

$$\mathbf{w}^m = \sum_{h=1}^{n^m} \mathbf{I}_h^m \mathbf{u}_{j_h^m} \tag{15.125}$$

where we also used the identity of

$$\mathbf{I}_{j_h^m}^T \mathbf{u} = \mathbf{u}_{j_h^m} \tag{15.126}$$

Now using $\mathbf{w}^m$ from Eq. (15.125) in Eq. (15.1), for the mth-element vertex forces in the overall coordinate system we obtain

$$\mathbf{q}^m = \mathcal{K}^m \sum_{h=1}^{n^m} \mathbf{I}_h^m \mathbf{u}_{j_h^m} + \mathbf{q}_0^m \qquad m = 1, \ldots, M \tag{15.127}$$

Alternately, noting that $\mathcal{K}^m \mathbf{I}_h^m$ product is the hth partition column of $\mathcal{K}^m$, that is, $\mathcal{K}_h^m$, we rewrite Eq. (15.127) as

$$\mathbf{q}_g^m = \sum_{h=1}^{n^m} \mathcal{K}_{gh}^m \mathbf{u}_{j_h^m} + \mathbf{q}_{0_g}^m \qquad \begin{array}{l} g = 1, \ldots, n^m \\ m = 1, \ldots, M \end{array} \tag{15.128}$$

which is the expression for the vertex forces in the overall cordinate system.   All quantities on the right-hand side of these expressions are known.   Note that in order to compute the numerical values of the vertex forces, we must have the free-free element-stiffness matrices $\mathcal{K}^m$ and the free-free element-load matrices $-\mathbf{q}_0^m$.   We used these quantities in obtaining the governing equations for the deflections.   When we generate them, we may choose to save them in an auxiliary storage for multiple use.   Instead of storing, we may also choose to regenerate them from the basic input data whenever the need arises.   The choice between these two options depends on the computer system used.

15.6.3   Equilibrium check of nodes and computation of reactions.   Once we compute the numerical values of the vertex forces of each element, using Eqs. (15.127) or (15.128), we can properly sum them to see whether the nodes are in static equilibrium.   This process will also yield the numerical values of the reactions if some or all of the deflection components are prescribed at a node.   In Art. 15.1 we defined column matrix $\mathbf{p}_0$ as denoting the list of prescribed nodal forces.   Matrix column $\mathbf{p}_0$ is of order Ne and contains nonzero and known values at the positions corresponding to those nodes and degree-of-freedom directions where there are prescribed concentrated loads.   All other entries of $\mathbf{p}_0$ are zeros.

Let us define a column matrix $\mathbf{p}^*$ of order Ne as listing the reactions components so that if a nodal degree-of-freedom direction is not constrained, the corresponding position in $\mathbf{p}^*$ will contain a zero; otherwise it will contain the symbolic name of the reaction at that node and in that direction. Let column matrix $\mathbf{p}$ denote the sum of $\mathbf{p}_0$ and $\mathbf{p}^*$:

$$\mathbf{p} = \mathbf{p}_0 + \mathbf{p}^* \tag{15.129}$$

Column matrix $\mathbf{p}$ lists all the external forces acting at the nodes; some of the components are prescribed forces, and some of the components are the reactions. On the other hand, column matrix $\mathbf{q}$, that is, $\mathbf{q}^T = [\mathbf{q}^{1^T} \cdots \mathbf{q}^{M^T}]$, list the vertex forces of all elements and represents the internal forces.

We can apply the theorem of virtual deformations to establish the equilibrium relationship between $\mathbf{p}$ and $\mathbf{q}$. Let $\delta\mathbf{u}$ denote a virtual-deflection column matrix applied to all nodes, and let $\delta\mathbf{w}^T = [\delta\mathbf{w}^{1^T} \cdots \delta\mathbf{w}^{M^T}]$ denote the corresponding vertex deflections. With the statement of the virtual-deformation theorem in Eq. (11.6) we can write

$$\delta\mathbf{u}^T \mathbf{p} = \delta\mathbf{w}^T \mathbf{q} \tag{15.130}$$

or, using the components on the right-hand side,

$$\delta\mathbf{u}^T \mathbf{p} = \sum_{m=1}^{M} \delta\mathbf{w}^{m^T}\mathbf{q}^m \tag{15.131}$$

From Eq. (15.8) we can write

$$\delta\mathbf{w}^m = \sum_{g=1}^{n^m} \mathbf{I}_g^m \mathbf{I}_{j_g}^T \delta\mathbf{u} \tag{15.132}$$

Substituting $\delta\mathbf{w}^m$ from Eq. (15.132) into Eq. (15.131) and noting that the latter equation is an identity with respect to $\delta\mathbf{u}$ variations, we obtain

$$\mathbf{p} = \sum_{m=1}^{M} \sum_{g=1}^{n^m} \mathbf{I}_{j_g^m} \mathbf{q}_g^m \tag{15.133}$$

where we also used the identity

$$\mathbf{q}_g^m = \mathbf{I}_g^T \mathbf{q}^m \tag{15.134}$$

Finally substituting $\mathbf{p}$ from Eq. (15.129) into Eq. (15.133), we obtain

$$\mathbf{p}^* = -\mathbf{p}_0 + \sum_{m=1}^{M} \sum_{g=1}^{n^m} \mathbf{I}_{j_g^m} \mathbf{q}_g^m \tag{15.135}$$

which is the explicit expression for the reactions.

If a certain entry of $\mathbf{p}^*$ does not contain any reaction, the corresponding quantity of the right-hand column in Eq. (15.135) should be zero. Therefore, this equation may also be used as an *equilibrium check*. In many problems, such a check is important. For example, if the Gauss-Seidel iteration is used to obtain the deflections from the governing equations, we can use the equilibrium check to determine the success of the iteration [see, for example, Eq. (15.102)]. In large problems, the solution of linear simultaneous equations necessarily involves a certain amount of accumulation of round-off errors. The equilibrium check can sometimes be used as an economical way of determining whether numerical results suffered extensively by the round-off-error accumulations. If an entry of $\mathbf{p}^*$ is computed as a nonzero quantity whereas it should have been a zero quantity, the computed value is called the **residual force,** and it is, in general, a good measure of the numerical errors.

The computation of reactions by Eq. (15.135) is applicable whether or not the governing equations are obtained by means of the Lagrange multipliers method. If we are not using the Lagrange multipliers method, there is an alternative way of computing the reactions without the equilibrium check. This we obtain by applying a virtual-deflections set where the external component $\delta\mathbf{u}$ is such that it has nonzero entries only in the directions of the reactions. From Eq. (15.21), fixing all directions associated with $\mathbf{u}_{(1)}$ and varying only $\mathbf{u}_{(20)}$ components, we write

$$\delta\mathbf{u} = \mathbf{H}_{(0)}\,\delta\mathbf{u}_{(20)} \tag{15.136}$$

where $\delta\mathbf{u}_{(20)}$ variations are in the directions of the prescribed deflections, namely, in the directions of the reactions. Note that the order of $\delta\mathbf{u}_{(20)}$ is b. Let $\mathbf{p}_r$ denote the list of reactions only. This column matrix is of order b, and its components are ordered so that the work done by the reactions under virtual deflections $\delta\mathbf{u}_{(20)}$ can be expressed as $\delta\mathbf{u}_{(20)}^T\,\mathbf{p}_r$. Let $\delta\mathbf{w}$ denote the internal virtual deflections corresponding to the external virtual deflections $\delta\mathbf{u}_{(20)}$. Using the virtual-deformation theorem in Eq. (11.6), we write

$$\delta\mathbf{u}_{(20)}^T\,\mathbf{p}_r + \delta\mathbf{u}^T\,\mathbf{p}_0 = \delta\mathbf{w}^T\,\mathbf{q} \tag{15.137}$$

or using the components on the right-hand side and Eq. (15.136), we have

$$\delta\mathbf{u}_{(20)}^T\,(\mathbf{p}_r + \mathbf{H}_{(0)}^T\mathbf{p}_0) = \sum_{m=1}^{M} \delta\mathbf{w}^{mT}\,\mathbf{q}^m \tag{15.138}$$

We can express $\delta\mathbf{w}^m$ corresponding to $\delta\mathbf{u}_{(20)}$ by substituting $\delta\mathbf{u}$ from Eq. (15.136) into Eq. (15.132)

$$\delta\mathbf{w}^m = \sum_{g=1}^{n^m} \mathbf{I}_g^m \mathbf{I}_{j_g^m}^T \mathbf{H}_{(0)}\,\delta\mathbf{u}_{(20)} \tag{15.139}$$

Substituting $\delta\mathbf{w}^m$ from Eq. (15.139) into Eq. (15.138) and using the identity in Eq. (15.134), we write

$$\mathbf{p}_r = \sum_{m=1}^{M} \sum_{g=1}^{n^m} \mathbf{H}_{(0)}^T \mathbf{I}_{j_g^m} \mathbf{q}_g^m - \mathbf{H}_{(0)}^T \mathbf{p}_0 \tag{15.140}$$

since the equation is an identity with respect to $\delta\mathbf{u}_{(20)}$. If we define

$$\mathscr{H}_0 = \mathbf{H}_{(0)}^T \tag{15.141}$$

and note the identity

$$\mathscr{H}_0\,\mathbf{I}_{j_g^m} = \mathscr{H}_{0\,j_g^m} \tag{15.142}$$

we can rewrite Eq. (15.140) as

$$\mathbf{p}_r = \sum_{m=1}^{M} \sum_{g=1}^{n^m} \mathcal{H}_{0j_g^m} \mathbf{q}_g^m - \mathcal{H}_0 \mathbf{p}_0 \tag{15.143}$$

which expresses the reactions explicitly. Note that during the generation of the governing equations we never referred to the matrix $\mathbf{H}_{(0)}$ individually. As indicated by Eq. (15.36), instead of referring to $\mathbf{H}_{(0)}$ and $\mathbf{u}_{(20)}$ individually for the generation of the governing equations, it was sufficient to refer to the product of the two, that is, $\mathbf{u}_0$. It is seen from the above treatment that if we want to compute reactions alone, without the equilibrium check, we must keep track of $\mathbf{H}_{(0)}$ also.

15.6.4 Computation of stresses in discrete structures. In these structures we can compute the vertex forces in the local coordinates. Recall from Eq. (12.119) and (12.120) that the vertex forces are nothing but the stress resultants and stress couples in the discrete structures. Since it is a very simple transformation to go from stress resultants and stress couples to actual stresses, we shall not continue the discussion beyond the point of computing the vertex forces in the local coordinates.

Using Eq. (15.128), we systematically obtain the description of forces in the overall coordinate system at any vertex of any element. Let $\mathbf{q}_g^m$ denote the vertex force at the gth vertex of the mth element computed in the overall coordinate system by means of Eq. (15.128). If we denote the actual vertex point of the mth element corresponding to g by $\mathcal{g}$, for the description of the vertex force at point $\mathcal{g}$ in the local coordinate system we can write, with the help of Eq. (13.75),

$$\mathbf{q}_{\mathcal{g}}^{\prime m} = \mathcal{T}_{\mathcal{g}g}^{\prime m - T} \mathbf{q}_g^m \tag{15.144}$$

where $\mathcal{T}_{\mathcal{g}g}^{\prime m - T}$ is the general force transfer matrix from point g to point $\mathcal{g}$ in the local coordinate system located at point $\mathcal{g}$. Here we considered the possibility that points g and $\mathcal{g}$ may be different but connected by a rigid bar and that each of the actual vertex points $\mathcal{g}$ of the mth element may have a different coordinate system. If points g and $\mathcal{g}$ are coincident, from Eqs. (13.55) and (13.14) we see that $\mathcal{T}_{gg}^{\prime m - T} = \mathbf{R}_g^{mT}$, and if there is only one coordinate system in the element, we have $\mathbf{R}_g^m = \mathbf{R}^m$. With these assumptions we rewrite Eq. (15.144) as

$$\mathbf{q}_g^{\prime m} = \mathbf{R}^{mT} \mathbf{q}_g^m \tag{15.145}$$

The detailed description of $\mathbf{R}^m$ matrices is given in Arts. 13.2.2 and 13.2.3. It is sufficient to remind the reader that the columns of $\mathbf{R}^m$, not the rows, contain the direction cosines of the local axes in the overall coordinate system.

15.6.5 Computation of stresses in continuum structures. A detailed treatment of this subject is outside the scope of this book. We shall discuss here only the basics related to the material covered earlier. In order to compute the stresses for an element, first we have to express its vertex deflections in the local coordinate system of that element. From Eq. (15.8), and using the identity of $\mathbf{I}_{j_g^m}^T \mathbf{u} = \mathbf{u}_{j_g^m}$, we write

$$\mathbf{w}^m = \sum_{g=1}^{n^m} \mathbf{I}_g^m \mathbf{u}_{j_g^m} \tag{15.146}$$

where $\mathbf{w}^m$ is the list of vertex deflections of the mth element in the overall coordinate system. The vertex deflection at the gth vertex of mth element is, then, simply

$$\mathbf{w}_g^m = \mathbf{u}_{j_g^m} \tag{15.147}$$

Suppose that the actual vertex point is $\mathscr{g}$ and that it is connected to g by means of a rigid bar. Then, with the help of Eq. (13.57) we write

$$\mathbf{w}_{\mathscr{g}}^{'m} = \mathscr{T}_{\mathscr{g}g}^{'m} \mathbf{w}_g^m \tag{15.148}$$

where $\mathbf{w}_{\mathscr{g}}^{'m}$ is the list of deflections at point $\mathscr{g}$ of the mth element in the local coordinate system of that point and $\mathscr{T}_{\mathscr{g}g}^{'m}$ is the general deflection transfer matrix from point g to point $\mathscr{g}$. Substituting $\mathbf{w}_g^m$ from Eq. (15.147) into Eq. (15.148) we obtain

$$\mathbf{w}_{\mathscr{g}}^{'m} = \mathscr{T}_{\mathscr{g}g}^{'m} \mathbf{u}_{j_g^m} \tag{15.149}$$

or for the complete list of vertex deflections of the element we write

$$\mathbf{w}^{'m} = \sum_{g=1}^{n^m} \mathbf{I}_{\mathscr{g}}^m \mathscr{T}_{\mathscr{g}g}^{'m} \mathbf{u}_{j_g^m} \tag{15.150}$$

If points g and $\mathscr{g}$ are coincident, from Eq. (13.55) and (13.14) we observe that $\mathscr{T}_{\mathscr{g}g}^{'m} = \mathbf{R}_g^{m^T}$, and if there is only one local coordinate system in the element, $\mathbf{R}_g^m = \mathbf{R}^m$, $g = 1, \ldots, n^m$. With these assumptions we can rewrite Eq. (15.150) as

$$\mathbf{w}^{'m} = \sum_{g=1}^{n^m} \mathbf{I}_g^m \mathbf{R}^{m^T} \mathbf{u}_{j_g^m} \tag{15.151}$$

which gives the list of vertex deflections of the mth element in the local coordinate system.

Then using **numerical-differentiation** techniques in the strain-displacement equations of the type of Eq. (12.104), we can first approximate strains and subsequently compute stresses in the element from Eq. (12.111).

15.6.6   Problems for Solution

Problem 15.49   Consider the structure given in Fig. 15.4. The DBC input units of this structure are as those given in Eq. (h) of Art. 12.4 except that at node 6 the rotation is not prescribed and the independent deflection components and their ordering are as shown in Eq. (o) of Art. 15.4. Suppose $\mathbf{u}_{(1)}^T = 10^{-3}[1 \quad 2 \quad 0.01 \quad -1 \quad -2 \quad 0.03 \quad 3 \quad 1 \quad -0.02 \quad 1 \quad 3 \quad 0.01 \quad 2 \quad 2 \quad 0.07 \quad -1$ $0.02 \quad 0.04]$. What is the complete list of deflections $\mathbf{u}$? First obtain $\mathscr{H}$ and $\mathbf{u}_0$ to satisfy Eq. (15.60).

*Problem 15.50   The equilibrium check and the reactions matrix $\mathbf{p}^*$ of a certain analysis problem is computed as

$$\mathbf{p}^{*T} = [-100.25 \quad 40.10 \quad 0.5 \quad 0.01 \quad 35.12 \quad 0.05 \quad 28.13]$$

Which entries may be reactions and which residual forces? Would you accept the related numerical solution of the problem if you knew that only the first, second, fifth, and seventh entries were indeed reactions? Why?

*Problem 15.51   Show that $\mathscr{H}_0$ matrix as defined by Eq. (15.141) is a b by Ne matrix containing all zeros but b 1's such that a column of $\mathscr{H}_0$ is either all 0's or all 0's with one 1. How would you store this matrix?

Problem 15.52   In local coordinate system the vertex forces at the first vertex of the fifteenth element of a space-frame structure are given as

$$\mathbf{q_1'^{15T}} = [100 \quad 40 \quad 15 \quad -10 \quad 500 \quad -750]$$

The cross section is circular with a radius of 2 units.   What are the stresses at the points with local coordinates (0,2,0) and (0,0,2)?

15.7   Important Parameters of the Method.   In this chapter, devoted to the displacement method, we have discussed ways of obtaining the governing equations, their solutions for the deflections, and computation of reactions and stresses.   In this article we discuss the important parameters of the method which directly determine the cost of analysis by this method.   In the discussion we shall assume that we have a structure with N nodes, e degrees of freedom per node, b constraints on the nodal deflections, and a constant bandwidth of w in the binary connectivity matrix of the supported structure.

15.7.1   Storage requirements.   From the discussions in Art. 15.5 we observe that the maximum storage required for the problem is proportional to $(Ne - b)\,we$, which is the area necessary for storing the upper half of the overall-stiffness matrix (assuming $w \approx \bar{w}/e$).   Let S denote the necessary storage area.   Ignoring the lower-order terms (such as those required for the storage of the right-hand sides), we write

$$S \approx Nwe^2 \tag{15.152}$$

which indicates that the *storage requirement is proportional to the square of the degree of freedom per node*.   Referring to Table 12.4, we see that the storage necessary for a space frame is $(6/2)^2 = 9$ times more than that of a planar truss if both structures have the same number of nodes N and the same half bandwidth w. Half bandwidth w is a function of mesh topology.   It is always less than or equal to N.   In three-dimensional structures it is closer to N than in two-dimensional structures.   Let d denote the dimension of the space in which the mesh is placed. As a very crude approximation we can write

$$N \geq w = N^{(d-1)/d} \tag{15.153}$$

From the discussions in Art. 15.4 we observe that w is closer to its upper bound if we increase the number of vertices in the elements.   Using Eq. (15.153) in Eq. (15.152), we write

$$N^2e^2 \geq S = N^{(2d-1)/d}e^2 \tag{15.154}$$

which tells us that the necessary storage area is at worst proportional to the square of the total number of nodes, or crudely it is proportional to the total number of nodes raised to a power which is less than 2.   Using the approximation on the right-hand side of Eq. (15.154), we can say that in two-dimensional meshes S is, on the average, proportional to $N^{1.5}$ and in three-dimensional meshes S is roughly proportional to $N^{1.67}$.

15.7.2   Number of arithmetic operations.   The total number of arithmetic operations in the displacement method is dominated by those arithmetic operations performed in solving the governing equations.   Let T denote a quantity proportional to the number of arithmetic operations required for solving the governing equations. Assuming that the Cholesky method is used for solving the governing equations, from our discussions in Art. 15.5.2, and using Eq. (15.115) with $w \approx \bar{w}/e$, we can write

$$T = Nw^2e^3 \tag{15.155}$$

and using Eq. (15.153) in Eq. (15.155), we obtain

$$N^3e^3 \geq T = N^{(3d-2)/d}e^3 \tag{15.156}$$

which indicates that the *number of arithmetic operations is always proportional to the cube of the number of degrees of freedom at a point* and *at worst it is also proportional to the cube of the total number of the nodes in the structure*. Using the approximation on the right-hand side of Eq. (15.156), we can say that for two-dimensional meshes T is roughly proportional to N^2 and for three-dimensional meshes it is approximately proportional to $N^{2.33}$.

As an example, consider again a planar truss and a space frame. If the two have the same number of nodes N and the same bandwidth w, the number of arithmetic operations for the space-frame problem will be $(6/2)^3 = 27$ times more than that of planar truss. As another example, consider two structures with the same N and e. Suppose the half bandwidth of one structure is twice that of the other. The number of arithmetic operations for the structure with the larger half bandwidth will be $2^2 = 4$ times as much.

15.7.3 Conclusions. From the foregoing arguments we can draw the following conclusions. Since the cost of a numerical solution in a digital computer is usually a linear function of S and T, one should model the structure with N and e as small as possible and prefer elements with fewer vertices since they will decrease the bandwidth. In discrete structures we usually do not have much choice in determining these parameters. However, in the continuum structures there will be a variety of choices open to the engineer. The parameters N, e, the number of vertices per element, and the accuracy of the results are closely interrelated in these structures.

15.7.4 Problems for Solution

*Problem 15.53 A certain continuum structure is idealized such that $N = 500$, $b = 20$, and $e = 6$. The results of the analysis indicate that for acceptably accurate results the number of nodes should be doubled. In such a refined mesh, the half bandwidth w is 50 per cent greater than that of the previous mesh. Estimate the ratios of the storage areas and solution times.

Problem 15.54 Consider two planar meshes for the same continuum structure which are rectangular. In the first the mesh elements are equilateral triangles; in the second the mesh elements are regular hexagons. Both meshes have the same number of nodal points. What are the ratios of half bandwidths, approximate storage areas, and approximate number of arithmetic operations?

*Problem 15.55 In thin shells a **membrane analysis** usually provides a good estimate of the stresses in most of the structure. However, near the boundaries, around the concentrated loads, and at the vicinity of other similar discontinuities, membrane analysis is not satisfactory. Referring to Table 12.4, indicate the increases in the approximate storage area and the number of arithmetic operations when one uses **membrane and bending analysis** instead of the membrane analysis in general shells. Repeat the exercise for shells of revolution undergoing axisymmetrical deformations.

*Problem 15.56 A certain solid foundation block is to be analyzed using two different three-dimensional meshes. In one, the nodal points establish a 5 by 10 by 20 grid, and in the other they establish a 10 by 10 by 10 grid. What are the values of N and w for these meshes? Find the approximate ratios of storage requirements and the number of arithmetic operations.

Problem 15.57 In order to increase the convergence characteristics, one may also take the middle points of the sides of a triangular continuum element as vertices, thus effectively doubling the number of vertices in the element. Such elements are referred to as **higher-order elements** because of the improvements in the interpolation rule. With the help of Prob. 15.54, find the increases in the storage and the number of arithmetic operations if one uses the higher-order triangular element instead of the ordinary ones in a given structure and a given mesh.

*Problem 15.58 In order to increase the convergence characteristics, one may increase the number of degrees of freedom at a point in a triangular continuum element. Because of the improvements in the interpolation rules, such elements are also called higher-order elements. Find the increases in storage requirements and the number of arithmetic operations if one doubles e, that is, the number of degrees of freedom at a point, for the same structure and for the same mesh. What are the increases if one triples e?

15.8 References

1. Utku, Ş.: "ELAS: A General Purpose Computer Program for the Equilibrium Problems of Linear Structures," vol. II, "Documentation of the Program," *Calif. Inst. Technol Jet Propul. Lab. Tech. Rep.* 32–1240, Pasadena, Calif., 1969.

2. Hildebrand, F. B.: "Methods of Applied Mathematics," 2d ed. chap. 2, Prentice-Hall, Inc., Englewood Cliffs, N.J., 1965.

3. Akyuz, F. A., and Ş. Utku: An Automatic Node-Relabelling Scheme for Bandwidth Minimization of Stiffness Matrices, *AIAA Journal*, vol. 6, no 4, pp. 728–730, April 1968.

4. Fox, L., "An Introduction to Numerical Linear Algebra," chap. 4, Oxford University Press, New York, 1965.

5. Schwarz, H. R., H. Rutishauser, and S. Stiefel: "Numerical Analysis of Symmetric Matrices," chap. 1, Prentice-Hall Inc., Englewood Cliffs, N.J., 1973.

16

Systematic Analysis by the Force Method

16.1 General Remarks. In this chapter we shall study systematic analysis by the force method. No distinction will be made between discrete structures and structures of two- and three-dimensional continua. We assume that a key sketch of the structure to be analyzed has already been prepared, as detailed in Art. 12.2.1. The number of elements in the structure will be denoted by M, and the number of nodes will be denoted by N, throughout. The number of degrees of freedom at a node e will be assumed constant for a given structure. Unless otherwise stated, the degree-of-freedom directions are those implied by the overall coordinate system. Table 12.4 may be consulted for the values of e and the associated degree-of-freedom directions in various structures.

16.1.1 Definitions. We shall use superscripts exclusively for the element labels. Thus, the number of vertices in the mth element will be shown by n^m. Given an element, say the mth one, we shall assume that $\mathbf{F}^m$, $\mathbf{v}_0^m$, $\mathbf{j}^m$ and also $\mathbf{B}^m$ and $\mathbf{q}^{0m}$ quantities are available. Here $\mathbf{F}^m$ is the mth **element-flexibility matrix** in the local coordinate system of the mth element, and $\mathbf{v}_0^m$ is the mth **element-deformation column matrix** corresponding to the element loads and expressed in the local coordinate system. Column matrix $\mathbf{j}^m$ is the list of node labels of the vertices of the mth element. Quantities $\mathbf{F}^m$ and $\mathbf{v}_0^m$ were studied in Art. 14.1.2, and $\mathbf{j}^m$ was introduced in Art. 12.2.1 and used extensively in Chap. 15. Column matrix $\mathbf{q}^{0m}$ [the n^mth partition of which is as given in Eq. (14.7)] and matrix $\mathbf{B}^m$ will be discussed in detail in Art. 16.2. $\mathbf{B}^m$ and $\mathbf{q}^{0m}$ relate the element forces $\mathbf{s}^m$ (those associated with the definition of $\mathbf{F}^m$) to the complete list of vertex forces of the element in the overall coordinate system, that is, $\mathbf{q}^m$, in the form

$$\mathbf{q}^m = \mathbf{B}^m \mathbf{s}^m + \mathbf{q}^{0m} \tag{16.1}$$

which represents the equilibrium of all forces acting on the element in the free state. We shall derive this in Art. 16.2

We can see from Eq. (16.1) that $\mathbf{q}^{0^m}$ is the list of vertex forces when the element forces, i.e., the components of $\mathbf{s}^m$, are zero. We use this definition to compute the numerical values of the components of $\mathbf{q}^{0^m}$. We place the element on statically determinate supports, apply the element loads, compute the resulting reactions, and list them, together with the zero vertex forces, with components in the overall coordinate system.

Column matrix $\mathbf{j}^m$ lists the node labels of the actual vertices of the mth element in the strict finite-element sense. However, for the elements of continuum structures, if the interpolation rule uses information outside the element, $\mathbf{j}^m$ contains the labels of all nodes which are involved in the interpolation rule. Considering such possibilities, we can interpret n^m as the order of $\mathbf{j}^m$.

As discussed in Art. 14.2.1, the orders of $\mathbf{F}^m$ and $\mathbf{v}_0^m$ (and also that of $\mathbf{s}^m$) are normally equal to $n'^m e$, where $n'^m = n^m - 1$. In such a case, we can observe from Eq. (16.1) that $\mathbf{B}^m$ is a $n^m e$ by $n'^m e$ matrix, since the orders of $\mathbf{q}^m$ and $\mathbf{q}^{0^m}$ are equal to $n^m e$. Let e' denote the number of independent reaction components to hold an element in the statically determinate state (see footnote on page 477). Usually $e = e'$; however, if $e \neq e'$, the orders of $\mathbf{F}^m$, $\mathbf{v}_0^m$, and $\mathbf{s}^m$ are equal to $n^m e - e'$. Since the orders of $\mathbf{q}^m$ and $\mathbf{q}^{0^m}$ are always equal to $n^m e$, $\mathbf{B}^m$ is a $n^m e$ by $n^m e - e'$ matrix. We assume that if there is one or more interelement-force boundary conditions related to the mth element, $\mathbf{F}^m$, $\mathbf{v}_0^m$, and $\mathbf{B}^m$ are already modified to reflect them. Suppose there is only one interelement-force boundary condition in the mth element. Since the element force related to the condition is zero, we exclude that component from $\mathbf{s}^m$ and also delete the related row from $\mathbf{F}^m$ and $\mathbf{v}_0^m$ and related column from $\mathbf{F}^m$ and $\mathbf{B}^m$ (see Art. 14.1.6 for the handling of interelement-force boundary conditions in $\mathbf{F}^m$ and $\mathbf{v}_0^m$). In such a case, the orders of $\mathbf{F}^m$, $\mathbf{v}_0^m$, and $\mathbf{s}^m$ are equal to $n^m e - e' - 1$, and those of $\mathbf{q}^m$ and $\mathbf{q}^{0^m}$ are $n^m e$; $\mathbf{B}^m$ is a $n^m e$ by $n^m e - e' - 1$ matrix.

In addition to the elemental information consisting of $\mathbf{F}^m$, $\mathbf{v}_0^m$, $\mathbf{B}^m$, and $\mathbf{q}^{0^m}$, we assume that the deflection boundary conditions of the structure are available as in either Eq. (12.118) or (12.113). We also assume that the column matrix listing the nonzero components of the concentrated loads acting at the nodes is also available. From such a list we can create a column matrix $\mathbf{p}_0$ such that its ith partition $\mathbf{p}_{0_i}$ will list the components in the overall coordinate system of the prescribed loads of node i. Note that the order of $\mathbf{p}_{0_i}$ is e and that of $\mathbf{p}_0$ is Ne.

The list of nodal deflections in the overall coordinate system will be denoted by $\mathbf{u}$, so that its ith partition $\mathbf{u}_i$ will list the deflection components of the ith node. Note that the order of $\mathbf{u}_i$ is e and that of $\mathbf{u}$ is Ne. The deflections and the forces at the gth vertex of the mth element in the overall coordinate system will be shown by $\mathbf{w}_g^m$ and $\mathbf{q}_g^m$, respectively. Note that $\mathbf{q}_g^m$ is the gth partition of $\mathbf{q}^m$. If we denote the list of vertex deflections of the mth element by $\mathbf{w}^m$, $\mathbf{w}_g^m$ is its gth partition. Column matrices $\mathbf{w}^m$ and $\mathbf{q}^m$ are of order $n^m e$, and their gth partitions $\mathbf{w}_g^m$ and $\mathbf{q}_g^m$ are of order e. When we place the mth element on statically determinate supports, the list of vertex deflections, excluding those on supports, in the local coordinate system will be shown by $\mathbf{v}^m$ (which is the list of element deformations) and the vertex forces in the directions associated with $\mathbf{v}^m$ will be shown by $\mathbf{s}^m$ (which is the list of element forces). Note that the orders of $\mathbf{v}^m$ and $\mathbf{s}^m$ are equal to $n^m e - e' - f^m$, where f^m is the number of interelement force-boundary conditions of the mth element.

As discussed in Arts. 14.1.4 and Art. 14.4, the flexibility relation of the mth element can be written

$$\mathbf{v}^m = \mathbf{F}^m \mathbf{s}^m + \mathbf{v}_0^m \tag{16.2}$$

Note that this relation is only approximate if the element is part of a two- or three-dimensional continuum. In place of Eq. (16.2) we may use the total complementary potential energy of the element to convey the same information. Let π^{*m} denote the total complementary potential energy of the mth element. For π^{*m} we write

$$\pi^{*m} = \tfrac{1}{2}\mathbf{s}^{m^T}\mathbf{F}^m\mathbf{s}^m + \mathbf{s}^{m^T}\mathbf{v}_0^m \tag{16.3}$$

as described in Arts. 14.1.5 and 14.4. Here, too, the expression for the total complementary potential energy is only approximate if the element is part of a two- or three-dimensional continuum. Although we know $\mathbf{F}^m$ and $\mathbf{v}_0^m$ numerically at the onset of the analysis problem, the numerical value of $\mathbf{s}^m$ is not known at the beginning. The following articles of this chapter will develop procedures for computing the unknown element forces $\mathbf{s}^m$, $m = 1, \ldots M$, first and then the nodal deflections $\mathbf{u}$ from the element forces.

16.1.2 Problems for Solution

*Problem 16.1 Consider a tetrahedral solid element. Suppose the element label is m. What are the values of n^m, e, and e'? What are the orders of $\mathbf{F}^m$, $\mathbf{v}_0^m$, $\mathbf{B}^m$, and $\mathbf{q}^{0m}$? How many different statically determinate support states are possible for the tetrahedral element? (*Hint:* See Art. 14.2.6.)

*Problem 16.2 Consider a space-frame element. Suppose the element label is m. What are the values of n^m, e, and e'? What are the orders of $\mathbf{F}^m$, $\mathbf{v}_0^m$, $\mathbf{B}^m$, and $\mathbf{q}^{0m}$? How many different statically determinate support states are possible for the element? (*Hint:* See Art. 14.2.6.)

Problem 16.3 Consider a space-truss element. Suppose its element label is m. What are the values of n^m, e, and e'? What are the orders of $\mathbf{F}^m$, $\mathbf{v}_0^m$, $\mathbf{B}^m$, and $\mathbf{q}^{0m}$? How many different statically determinate support states are possible for the element? (*Hint:* See Art. 14.2.6.)

Problem 16.4 Consider Eqs. (14.5) and (14.8), which are given for the elements of discrete structures where $e = e'$. With the terminology of these equations, show that $\mathbf{B}^m$ and $\mathbf{q}^{0m}$ can be displayed as

$$\mathbf{q}^{0m^T} = [\mathbf{0}^T \quad \cdots \quad \mathbf{0}^T \quad \mathbf{q}_n^{0m^T}] \tag{16.4}$$

and

$$\mathbf{B}^m = \begin{bmatrix} \mathbf{I} & & & \\ & \mathbf{I} & & \\ & & \ddots & \\ & & & \mathbf{I} \\ -\mathbf{T}_{n1}^{-T} & \cdots & & -\mathbf{T}_{nn'}^{-T} \end{bmatrix} \tag{16.5}$$

assuming that the local coordinate system is parallel to the overall coordinate system. Show that $\mathbf{B}^m$ of Eq. (16.5) is the transpose of $\boldsymbol{\Gamma}$ given by Eq. (14.75). Why?

Problem 16.5 If the local coordinate system is not parallel to the overall coordinate system, how should the expressions given for $\mathbf{q}^{0m}$ and $\mathbf{B}^m$ in Eqs. (16.4) and (16.5) be modified within the context of Prob. 16.4?

16.2   Relations between Forces; Equilibrium.   In this article we study the relationship which exist between various forces.   These relationships are all based on the conditions of equilibrium.   Actually we may classify them into two categories: those related to the equilibrium of elements and those related to the equilibrium of nodes.   In the following subsections we shall study these categories individually.

16.2.1   Equilibrium equations related to elements.   If a structure is in static equilibrium under the effect of some loading, every particle of it should be in static equilibrium when considered as a free body.   Consequently every element of a structure must be in static equilibrium when the elements are considered individually as free bodies.   When we consider the mth element of a structure as a free body, the forces acting on this element are the vertex forces $\mathbf{q}'^m$ and the element loads. We used a prime in $\mathbf{q}'^m$ to indicate that at this stage we prefer to express the vertex forces in the local coordinate system of the element.   Let $-\mathbf{q}_1^{0'm}$ denote the statical equivalent of the element loads at the first vertex of the mth element, i.e., at vertex 1.   Since the vertex forces $\mathbf{q}'^m$ and the element loads are in static equilibrium, when we transfer all these forces to the same point, say to vertex 1, we should obtain a null force as the sum.

Let us transfer all the forces acting on the mth element in the free state to vertex 1. Let $\mathbf{T}_{1i}$ denote the deflection transfer matrix from point i to point 1.   According to Eq. (13.68), the force transfer matrix from point i to point 1 is $\mathbf{T}_{1i}^{-T}$.   Let the vertex forces at the ith vertex be denoted by $\mathbf{q}_i'^m$.   With these we can write

$$-\mathbf{q}_1^{0'm} + \sum_{i=1}^{n^m} \mathbf{T}_{1i}^{-T}\mathbf{q}_i'^m = 0 \qquad (16.6)$$

which is the force-equilibrium equation of the mth element as expressed in the local coordinate system.   By defining

$$\mathbf{T}^m = [\mathbf{T}_{11}^{-T} \quad \cdots \quad \mathbf{T}_{1n^m}^{-T}] \qquad (16.7)$$

we can rewrite Eq. (16.6) as

$$-\mathbf{q}_1^{0'm} + \mathbf{T}^m\mathbf{q}'^m = 0 \qquad (16.8)$$

In Eq. (16.8) the numerical values of $\mathbf{q}_1^{0'm}$ and $\mathbf{T}^m$ are known, and only those entries of $\mathbf{q}'^m$ which are involved in the interelement-force boundary conditions of the mth element are also known as zero quantities.   Let e' denote the row order of $\mathbf{T}^m$ and the order of $\mathbf{q}_1^{0'm}$.   Obviously, we can solve e' number of components of $\mathbf{q}'^m$ in terms of the remaining components.   Let $\mathbf{q}_{(2)}^m$ denote the e' number of components, and let $\mathbf{s}^m$ denote the remaining components.   Denoting the matrix consisting of those columns of $\mathbf{T}^m$ associated with the entries of $\mathbf{q}_{(2)}'^m$ by $\mathbf{T}_{(2)}^m$ and the matrix consisting of the remaining columns of $\mathbf{T}^m$ by $\mathbf{T}_{(1)}^m$, we can rewrite Eq. (16.8) as

$$-\mathbf{q}_1^{0'm} + \mathbf{T}_{(1)}^m\mathbf{s}^m + \mathbf{T}_{(2)}^m\mathbf{q}_{(2)}'^m = 0 \qquad (16.9)$$

where $\mathbf{T}_{(2)}^m$ is a square matrix of order e'.   Actually, since matrix $\mathbf{T}^m$ possesses $n^m e - f^m$ number of columns (note that those involving with interelement-force boundary conditions are assumed deleted), we can establish the $\mathbf{T}_{(2)}^m$ matrix in

$\begin{pmatrix} n^m e - f^m \\ e' \end{pmatrix}$ different ways (see Art. 14.2.6). Let us assume that we have established $\mathbf{T}^m_{(2)}$ so that it has an inverse. Then we can solve $\mathbf{q}'^m_{(2)}$ from Eq. (16.9) as

$$\mathbf{q}'^m_{(2)} = -\mathbf{T}^{m^{-1}}_{(2)}\mathbf{T}^m_{(1)}\mathbf{s}^m + \mathbf{T}^{m^{-1}}_{(2)}\mathbf{q}'^{0'm}_1 \tag{16.10}$$

Considering the identity of

$$\mathbf{s}^m = \mathbf{I}\mathbf{s}^m + \mathbf{0}\mathbf{q}'^{0'm}_1 \tag{16.11}$$

we can combine the identity with Eq. (16.10) to write

$$\begin{Bmatrix} \mathbf{s}^m \\ \overline{\mathbf{q}'^m_{(2)}} \end{Bmatrix} = \left[\begin{array}{c|c} \mathbf{I} \\ \hline -\mathbf{T}^{m^{-1}}_{(2)} & \mathbf{T}^m_{(1)} \end{array} \right] \mathbf{s}_m + \left[\begin{array}{c} \mathbf{0} \\ \hline \mathbf{T}^{m^{-1}}_{(2)} \end{array} \right] \mathbf{q}'^{0'm}_1 \tag{16.12}$$

Let $\mathbf{P}$ denote a permutation matrix such that

$$\mathbf{q}'^m = \mathbf{P} \begin{Bmatrix} \mathbf{s}^m \\ \overline{\mathbf{q}'^m_{(2)}} \end{Bmatrix} \tag{16.13}$$

We can easily find the permutation matrix $\mathbf{P}$, noting that the column matrix on the left-hand side of Eq. (16.12) is, at worst, a reordered form of $\mathbf{q}'^m$. Premultiplying both sides of Eq. (16.12) by $\mathbf{P}$ and using Eq. (16.13), we obtain

$$\mathbf{q}'^m = \mathbf{B}'^m\mathbf{s}^m + \mathbf{q}^{0'm} \tag{16.14}$$

where

$$\mathbf{B}'^m = \mathbf{P} \left[\begin{array}{c|c} \mathbf{I} \\ \hline -\mathbf{T}^{m^{-1}}_{(2)} & \mathbf{T}^m_{(1)} \end{array} \right] \tag{16.15}$$

and

$$\mathbf{q}^{0'm} = \mathbf{P} \left[\begin{array}{c} \mathbf{0} \\ \hline \mathbf{T}^{m^{-1}}_{(2)} \end{array} \right] \mathbf{q}'^{0'm}_1 \tag{16.16}$$

Note that Eq. (16.14) is of the type of Eq. (16.1), the only difference being that $\mathbf{q}^{0'm}$ and $\mathbf{q}'^m$ are expressed in the local coordinate system and related to the actual vertex points, which may differ from those attached to the nodes directly. Let $\acute{\imath}$ denote an actual vertex point, and let i denote the corresponding point which is attached to the node directly. We assume that $\acute{\imath}i$ is a rigid link. Of course points $\acute{\imath}$ and i may be coincident. Let $\acute{\imath}$th partition of $\mathbf{q}'^m$ be denoted by $\mathbf{q}'^m_{\acute{\imath}}$, and let the general force transfer matrix from vertex $\acute{\imath}$ to point i be denoted by $\mathscr{T}^{m^{-T}}_{i\acute{\imath}}$. Let the statically equivalent counterpart of $\mathbf{q}'^m_{\acute{\imath}}$ at point i be denoted by $\mathbf{q}^m_i$, which is described in the overall coordinate system. Then, by Eq. (13.74), we have

$$\mathbf{q}^m_i = \mathscr{T}^{m^{-T}}_{i\acute{\imath}}\mathbf{q}'^m_{\acute{\imath}} \tag{16.17}$$

or, considering all the vertices, we write

$$\mathbf{q}^m = \mathrm{diag}\,(\mathscr{T}^{m^{-T}}_{i\acute{\imath}})\mathbf{q}'^m \tag{16.18}$$

By premultiplying both sides of Eq. (16.14) by diag $(\mathcal{T}_{i\acute{e}}^{m-T})$ and using Eq. (16.18) we obtain

$$\mathbf{q}^m = \mathbf{B}^m \mathbf{s}^m + \mathbf{q}^{0m} \tag{16.19}$$

where
$$\mathbf{B}^m = \text{diag}\,(\mathcal{T}_{i\acute{e}}^{m-T})\mathbf{P}\left[\begin{array}{c} \mathbf{I} \\ \hline -\mathbf{T}_{(2)}^{m-1} \quad \mathbf{T}_{(1)}^m \end{array}\right] \tag{16.20}$$

and
$$\mathbf{q}^{0m} = \text{diag}\,(\mathcal{T}_{i\acute{e}}^{m-T})\mathbf{P}\left[\begin{array}{c} \mathbf{0} \\ \hline \mathbf{T}_{(2)}^{m-1} \end{array}\right]\mathbf{q}_1^{0'm} \tag{16.21}$$

where we used Eqs. (16.15) and (16.16).

We observe that Eq. (16.19) is exactly the same as Eq. (16.1). Equation (16.19) is a modified version of the element force-equilibrium equations given in Eq. (16.6). When $e = e'$, the process is extremely simplified, as worked out in Probs. 16.4 and 16.5. However, the expressions given by Eqs. (16.20) and (16.21) represent the general case. In order to prevent the ambiguity in selecting the $\mathbf{T}_{(2)}^m$ matrix, out of $\binom{n^m e - \ell^m}{e'}$ possibilities, as explained in Art. 16.2.2, we usually choose the one associated with the largest determinant. In practice, the procedure which generates $\mathbf{F}^m$ and $\mathbf{v}_0^m$ from the input data also generates $\mathbf{B}^m$ and $\mathbf{q}^{0m}$ in accordance with Eqs. (16.20) and (16.21). Note that $\mathbf{F}^m$ and $\mathbf{v}_0^m$ should be those associated with $\mathbf{s}^m$ of Eq. (16.19). In other words, we should obtain $\mathbf{F}^m$ and $\mathbf{v}_0^m$ by placing the mth element on statically determinate supports, where the reactions should be in the degree-of-freedom directions associated with the components of $\mathbf{q}_{(2)}^{'m}$.

16.2.2   Equilibrium equations related to the nodes in directions associated with u. In order to have the structure in static equilibrium, not only the elements but also the nodes, when considered as free bodies, should be in static equilibrium. In Art. 15.6.3 we studied the equilibrium check for the nodes. Using the theorem of virtual deformations, we obtained the relationships between vertex forces $\mathbf{q}^m$ and the nodal forces $\mathbf{p}$ which, according to Eq. (15.129), lists the prescribed nodal concentrated forces and the reactions. We reproduce here Eq. (15.135), interchanging the places of $\mathbf{p}_0$ and $\mathbf{p}^*$:

$$\mathbf{p}_0 = -\mathbf{p}^* + \sum_{m=1}^{M}\sum_{g=1}^{n^m} \mathbf{I}_{j_g^m}\mathbf{q}_g^m \tag{16.22}$$

where $\mathbf{p}_0$ = list of prescribed nodal forces
$\quad\quad \mathbf{p}^*$ = list of reactions
$\quad\quad \mathbf{q}_g^m$ = vertex forces at gth vertex of mth element
$\quad\quad j_g^m$ = node label of gth vertex of mth element
$\quad\quad \mathbf{I}_p$ = pth partition of Ne-order identity matrix
Column matrices $\mathbf{p}_0$ and $\mathbf{p}^*$ are of order Ne, $\mathbf{I}_{j_g^m}$ is a Ne by e matrix, and $\mathbf{q}_g^m$ is a column matrix of order e. Note that $\mathbf{p}^*$ contains only b nonzero entries (see Art. 15.6.3). First using the identity in Eq. (15.134) for $\mathbf{q}_g^m$ and then using Eq. (16.19) for the vertex forces $\mathbf{q}^m$, we can rewrite Eq. (16.22) as

$$\mathbf{p}_0 = -\mathbf{p}^* + \sum_{m=1}^{M}\sum_{g=1}^{n^m} \mathbf{I}_{j_g^m}\mathbf{I}_g^{mT}(\mathbf{B}^m\mathbf{s}^m + \mathbf{q}^{0m}) \tag{16.23}$$

Above Eq. (15.137) we have defined the list of the nonzero components of $\mathbf{p}^*$ as $\mathbf{p}_r$, which is a column matrix of order b.   Note that $\mathbf{p}_r$ reactions are in the directions of $\mathbf{u}_{(20)}$ prescribed deflections [see, for example, Eq. (15.21)], whereas $\mathbf{p}^*$ components are in the directions of $\mathbf{u}$.

We may consider that $\mathbf{p}_r$ and $-\mathbf{p}^*$ constitute a force system in equilibrium.   In fact, we may consider $\mathbf{p}_r$ forces as internal and $\mathbf{p}^*$ forces as external forces of a hypothetical system.   Using the theorem of virtual deformations, as in Eq. (15.137), with virtual deflections $\boldsymbol{\delta}\mathbf{u}$ and $\boldsymbol{\delta}\mathbf{u}_{(20)}$ of Eq. (15.136), we write

$$\mathbf{p}_r = \mathscr{H}_0\,\mathbf{p}^* \tag{16.24}$$

where Eq. (15.141) is also used.   The same equation can be obtained directly from Eqs. (15.135) and (15.143) using Eq. (15.142).   By premultiplying both sides of Eq. (16.24) by $\mathscr{H}_0^T$ and noting that $\mathscr{H}_0^T\mathscr{H}_0$ is a diagonal matrix with 1's on diagonal locations corresponding to nonzero entries of $\mathbf{p}^*$ and 0's on diagonal locations corresponding to zero entries of $\mathbf{p}^*$ (see Prob. 15.51) we obtain

$$\mathbf{p}^* = \mathscr{H}_0^T\,\mathbf{p}_r \tag{16.25}$$

With this, we can rewrite Eq. (16.23) as

$$\mathbf{p}_0 = -\mathscr{H}_0^T\mathbf{p}_r + \sum_{m=1}^{M}\sum_{g=1}^{n^m}\mathbf{I}_{j_g}^{m}\mathbf{I}_{g}^{m^T}(\mathbf{B}^m\mathbf{s}^m + \mathbf{q}^{0m}) \tag{16.26}$$

Let us define a column matrix listing the element forces of all elements such that if we call this matrix $\mathbf{s}$, its mth partition is $\mathbf{s}^m$:

$$\mathbf{s}^T = [\mathbf{s}^{1^T} \quad \cdots \quad \mathbf{s}^{M^T}] \tag{16.27}$$

Now let us consider an identity matrix $\mathbf{I}_*$ such that its order is equal to that of $\mathbf{s}$ and it is partitioned in the same manner as $\mathbf{s}$.   If we denote the order of $\mathbf{s}$ by $\jmath$, we can write

$$\jmath = \sum_{m=1}^{M}(n^m e - e' - \jmath^m) \tag{16.28}$$

where, as defined in Art. 16.1.1, $\jmath^m$ is the number of interelement-force boundary conditions, n^m is the number of vertices, e is the number of degrees of freedom at a vertex, and e' is the number of independent reaction components to keep the element in the statically determinate state.   Let $\mathbf{I}_*^m$ denote the mth column partition of $\mathbf{I}_*$. Then we can write the identity

$$\mathbf{s}^m = \mathbf{I}_*^{m^T}\mathbf{s} \tag{16.29}$$

Using Eq. (16.29) in Eq. (16.26), we can rewrite the latter as

$$\bar{\mathbf{B}}\mathbf{z} = \bar{\mathbf{p}} \tag{16.30}$$

where

$$\bar{\mathbf{B}} = \left[\; \sum_{m=1}^{M}\sum_{g=1}^{n^m}\mathbf{I}_{j_g}^{m}\mathbf{I}_{g}^{m^T}\mathbf{B}^m\mathbf{I}_*^{m^T} \;\bigg|\; -\mathscr{H}_0^T \right] \tag{16.31}$$

$$\bar{\mathbf{p}} = \mathbf{p}_0 - \sum_{m=1}^{M}\sum_{g=1}^{n^m}\mathbf{I}_{j_g}^{m}\mathbf{I}_{g}^{m^T}\mathbf{q}^{0m} \tag{16.32}$$

and

$$\mathbf{z}^T = [\mathbf{s}^T \;\big|\; \mathbf{p}_r^T] \tag{16.33}$$

Equation (16.30) represents the equilibrium equations of the nodes in the directions associated with **u**. From the descriptive data of the analysis problem, matrices $\overline{\mathbf{B}}$ and $\overline{\mathbf{p}}$ are known. Matrix $\overline{\mathbf{B}}$ is of order Ne by $s + b$, and $\mathbf{p}_0$ is a column matrix of order Ne. Column matrix $\mathbf{z}$ is of order $s + b$ and lists all the unknown forces. Let $\bar{r}$ denote the rank of matrix $\overline{\mathbf{B}}$. If

$$\bar{r} = Ne = s + b \tag{16.34}$$

we call the structure **statically determinate** and **stable**. In this case the solution of $\mathbf{z}$ from Eq. (16.30) provides the numerical values of all unknown forces, without the use of $\mathbf{F}^m$ and $\mathbf{v}_0^m$, $m = 1, \ldots, M$, which are the quantities related with the material properties of the structure. Then we say that *in statically determinate structures, the stress distribution in the structure is independent of the material properties.* If

$$\bar{r} < Ne \le s + b \tag{16.35}$$

or

$$\bar{r} < s + b \le Ne \tag{16.36}$$

we call the structure **geometrically unstable**. Such systems are called **mechanisms** or **semimechanisms**, respectively, and fall outside the scope of the theory of structures, since they cannot carry loads. If

$$\bar{r} = Ne < s + b \tag{16.37}$$

we call the structure **statically indeterminate** or **hyperstatic** to the $(s + b - Ne)$th degree and **stable**. In this case, we note that the nodal force-equilibrium equations (16.30) provide fewer scalar equations than the number of unknown forces. In order to find the numerical values of the forces, we must refer to $\mathbf{F}^m$ and $\mathbf{v}_0^m$, $m = 1, \ldots, M$. *In statically indeterminate structures, the stress distribution in the structure is dependent on the material properties.* In Art. 16.3 to 16.5 we shall study the systematic solution of statically indeterminate but stable structures.

In order to get an idea of the dimensions of matrix $\overline{\mathbf{B}}$, let us consider the following example. Suppose the total number of nodes in a structure is $N = 100$; the number of elements $M = 60$; the number of degrees of freedom at a point $e = 3$; the number of vertices in an element $n^m = 8$, $m = 1, \ldots, 60$; the number of reaction components in an element to keep it in the statically determinate state is $e' = 6$; and the number of independent reaction components of the whole structure is $b = 25$. We assume that there are no interelement-force boundary conditions. We can then see that

$$\begin{array}{ll} Ne = (100)(3) = 300 & s = (60)((8)(3) - 6)) = 1{,}080 \\ b = 25 & s + b = 1{,}105 \end{array} \tag{a}$$

Then $\overline{\mathbf{B}}$ is a 300 by 1,105 matrix. Note that $\overline{\mathbf{B}}$ is a rectangular matrix, which cannot be symmetrical. Although it is sparse, during manipulation its sparsity disappears, and we have to allocate a full $(Ne)(s + b)$ words of memory space for $\overline{\mathbf{B}}$. In the above example this space is

$$(300)(1{,}105) = 331{,}500 \tag{b}$$

computer words. We immediately recognize that matrix $\overline{\mathbf{B}}$ is one of the stumbling blocks of the systematic force method of analysis. Any effort to decrease its size should be welcome. Next, we shall study how the row and the column orders of $\overline{\mathbf{B}}$ can indeed be reduced, thanks to the fact that since $\mathbf{s}$ and $\mathbf{p}_r$ are not independent, $\mathbf{p}_r$ can be eliminated from the equilibrium equations of the nodes, i.e., from Eq. (16.30)

16.2.3 Equilibrium equations related to nodes in directions associated with $\mathbf{u}_{(1)}$.

We may write the equilibrium equations of the nodes only in the directions associated with the components of the independent deflections $\mathbf{u}_{(1)}$. Since the order of $\mathbf{u}_{(1)}$ is $Ne - b$, we shall have $Ne - b$ scalar equations, involving only the components of $\mathbf{s}$ and not involving the components of $\mathbf{p}_r$. This means that the coefficient matrix in the equilibrium equations of the nodes will be an $Ne - b$ by s matrix, instead of an Ne by $s + b$ matrix. Such equilibrium equations are obtained in the next paragraph.

We can obtain the force-equilibrium equations of the nodes in the directions of the components of $\mathbf{u}_{(1)}$ by application of the theorem of virtual deformations, as already done in Art. 15.3.2. Excluding the reactions $\mathbf{p}_r$, we see that the forces acting on the nodes are the prescribed forces $\mathbf{p}_0$ and the internal forces $\mathbf{q}^m$, $m = 1$, ..., M, that is, the vertex forces. From Eq. (15.60) we can write

$$\delta\mathbf{u} = \mathcal{H}^T \, \delta\mathbf{u}_{(1)} \tag{16.38}$$

We can substitute $\delta\mathbf{u}$ from Eq. (16.38) into Eq. (15.132) and obtain

$$\delta\mathbf{w}^m = \sum_{g=1}^{n^m} \mathbf{I}_g^m \mathbf{I}_{j_g^m}^T \mathcal{H}^T \, \delta\mathbf{u}_{(1)} \tag{16.39}$$

The virtual-deformation theorem can be used to write

$$\delta\mathbf{u}^T \mathbf{p} = \sum_{m=1}^{M} \delta\mathbf{w}^{m^T} \mathbf{q}^m \tag{16.40}$$

where $\mathbf{p}$ is as defined in Eq. (15.129). Substituting $\delta\mathbf{u}$ and $\delta\mathbf{w}^m$ from Eqs. (16.38) and (16.39) and noting that the variations in the directions of the nonzero components of $\mathbf{p}^*$ are all zero, we obtain

$$\delta\mathbf{u}_{(1)}^T \mathcal{H} \mathbf{p}_0 = \delta\mathbf{u}_{(1)}^T \sum_{m=1}^{M} \sum_{g=1}^{n^m} \mathcal{H} \mathbf{I}_{j_g^m} \mathbf{I}_g^{m^T} \mathbf{q}^m \tag{16.41}$$

Since this is an identity with respect to the variations $\delta\mathbf{u}_{(1)}$,

$$\mathcal{H} \mathbf{p}_0 = \sum_{m=1}^{M} \sum_{g=1}^{n^m} \mathcal{H}_{j_g^m} \mathbf{I}_g^{m^T} \mathbf{q}^m \tag{16.42}$$

where the identity in Eq. (15.142) is also used. Now we can substitute $\mathbf{q}^m$ from Eq. (16.19) into Eq. (16.42) to obtain

$$\mathcal{H} \mathbf{p}_0 = \sum_{m=1}^{M} \sum_{g=1}^{n^m} \mathcal{H}_{j_g^m} \mathbf{I}_g^{m^T} (\mathbf{B}^m \mathbf{s}^m + \mathbf{q}^{0m}) \tag{16.43}$$

which, with the use of Eq. (16.29), can be rewritten

$$\mathbf{B}^0\mathbf{s} = \mathbf{p}^0 \tag{16.44}$$

where

$$\mathbf{B}^0 = \sum_{m=1}^{M} \sum_{g=1}^{n^m} \mathscr{H}_{jg}^m \ \mathbf{I}_g^{m^T}\mathbf{B}^m\mathbf{I}_*^{m^T} \tag{16.45}$$

and

$$\mathbf{p}^0 = \mathscr{H}\mathbf{p}_0 - \sum_{m=1}^{M} \sum_{g=1}^{n^m} \mathscr{H}_{jg}^m \ \mathbf{q}_g^{0m} \tag{16.46}$$

Equation (16.44) represents the force-equilibrium equations of the nodes in the directions of the components of $\mathbf{u}_{(1)}$.

The coefficient matrix $\mathbf{B}^0$ is of order $Ne - b$ by $\jmath$, which is smaller than $\bar{\mathbf{B}}$ of Eq. (16.30). Let r^0 denote the rank of $\mathbf{B}^0$. We can also classify structures with the rank and row and column orders of $\mathbf{B}^0$. If

$$r^0 = Ne - b = \jmath \tag{16.47}$$

we call the structure statically determinate and stable. In this case we can solve the element forces $\mathbf{s}$ directly from Eq. (16.44), and then, using Eqs. (16.19) and (15.143), we can obtain the reactions without ever referring to $\mathbf{F}^m$ and $\mathbf{v}_0^m$, $m = 1$, $\ldots, M$. If

$$r^0 < Ne - b \le \jmath \tag{16.48}$$

or

$$r^0 < \jmath \le Ne - b \tag{16.49}$$

the structure is called geometrically unstable and it constitutes either a mechanism or a semimechanism. If

$$r^0 = Ne - b < \jmath \tag{16.50}$$

the structure is called statically indeterminate or hyperstatic to the $(\jmath - Ne + b)$th degree and stable.

We observe that writing the nodal force-equilibrium equations in the form of Eq. (16.44) is a perfect equivalent of writing them in the form of Eq. (16.30). However, the former requires smaller storage area than the latter. For the example given in Art. 16.2.2, we found that $\bar{\mathbf{B}}$ was a 300 by 1,105 matrix. Using the quantities given in the example, we have

$$Ne - b = (100)(3) - 25 = 275 \tag{c}$$

and

$$\jmath = 1,080$$

therefore $\mathbf{B}^0$ is a 275 by 1,080 matrix, and the necessary storage for this matrix is

$$(275)(1,080) = 297,000 \tag{d}$$

which is about 10 per cent less than that required for $\bar{\mathbf{B}}$. As b increases, the storage area required by $\mathbf{B}^0$ is reduced even more sharply compared with $\bar{\mathbf{B}}$.

Whether we write the force-equilibrium equations of the nodes in the form of Eq. (16.30) or (16.44), for stable but statically indeterminate structures we are confronted with a set of linear simultaneous equations where there are more unknowns than available equations. In other words, the column order of the co-efficient matrix is greater than the row order. Actually the difference between the column order and the row order is the degree of statical indeterminancy of the structure. In such cases, we may take the unknowns in two groups and consider one group as independent force components and the remaining group as the dependent force components.

In statically indeterminate structures, the independent components are called **redundants**. The number of redundants is equal to the degree of indeterminacy of the structure, since the number of dependent force components is equal to the row order of the coefficient matrix, namely, the number of scalar equations. Which force components are to be considered as redundants and which as dependent is a major task in the systematic analysis of a given structure. Suppose the coefficient matrix is an i by $ɟ$ matrix, i being less than $ɟ$. Theoretically we have

$$\binom{ɟ}{i} = \frac{ɟ!}{(ɟ - i)\,!\,i!}$$

possibilities in selecting the dependent force components and therefore redundants. In practice, we usually prefer the selection associated with the largest-magnitude determinant among the ith-order minors which can be formed by the columns selected from the coefficient matrix. In Art. 16.3 the algorithm which selects the redundants in this fashion is detailed.

16.2.4 Vertex forces as a function of redundants. Suppose that we have the force-equilibrium equations of the nodes in the form of Eq. (16.44). Suppose also that we have already decided that certain $Ne - b$ components of s are to be the dependent force components and the remaining $ɟ - Ne + b$ components are to be the independent, or redundant, force components. Let $s_{(2)}$ denote the list of dependent components, and let x denote the independent components, or redundants. Let the matrix consisting of those columns of B^0 associated with the components of $s_{(2)}$ be denoted by $B^0_{(2)}$ and the matrix consisting the columns of B^0 associated with the components of x be denoted by B^0_x. With these definitions we can rewrite Eq. (16.44) as

$$B^0_{(2)} s_{(2)} + B^0_x x = p^0 \tag{16.51}$$

where $B^0_{(2)}$ is a square matrix of order $Ne - b$. Assuming that we selected the components of $s_{(2)}$ so that $B^0_{(2)}$ is not singular, we can solve Eq. (16.51) for $s_{(2)}$:

$$s_{(2)} = -B^{0^{-1}}_{(2)} B^0_x x + B^{0^{-1}}_{(2)} p^0 \tag{16.52}$$

Considering the identity

$$x = Ix + 0p^0 \tag{16.53}$$

we can combine Eqs. (16.52) and (16.53) as

$$\left\{ \frac{s_{(2)}}{x} \right\} = \left[\frac{-B^{0^{-1}}_{(2)} B^0_x}{I} \right] x + \left[\frac{B^{0^{-1}}_{(2)}}{0} \right] p_0 \tag{16.54}$$

Let $\mathbf{P}$ denote a permutation matrix of order $Ne - b$ such that

$$\mathbf{s} = \mathbf{P} \left\{ \frac{\mathbf{s}_{(2)}}{\mathbf{x}} \right\} \tag{16.55}$$

Premultiplying both sides of Eq. (16.54) by $\mathbf{P}$, we obtain

$$\mathbf{s} = \mathscr{B}\mathbf{x} + \mathscr{B}_0\,\mathbf{p}^0 \tag{16.56}$$

where

$$\mathscr{B} = \mathbf{P} \left[\frac{-\mathbf{B}_{(2)}^{0-1}\,\mathbf{B}_x^0}{\mathbf{I}} \right] \tag{16.57}$$

and

$$\mathscr{B}_0 = \mathbf{P} \left[\frac{\mathbf{B}_{(2)}^{0-1}}{\mathbf{0}} \right] \tag{16.58}$$

Note that Eq. (16.56) is a form of Eq. (16.44) modified so that for all possible values of $\mathbf{x}$ the equilibrium of nodes is always satisfied.

Consider the *actual structure*. The forces listed in $\mathbf{x}$ are some of the forces acting between nodes and the element vertices. To assign zero values to all the components of $\mathbf{x}$ is equivalent to introducing **physical cuts** in the associated locations of the actual structure. Such a cut structure is called the **statically determinate base system.** From Eq. (16.56) we observe that the ith column of $\mathscr{B}$, $1 < i < s -$ $Ne + b$, lists the values of the components of $\mathbf{s}$ when we apply $x_i = 1$ loading alone on the statically determinate base system. Similarly, from Eq. (16.56), we can see that the ith column of $\mathscr{B}_0$ lists the values of the components of $\mathbf{s}$ when we apply a unit load in the direction of the ith component of $\mathbf{u}_{(1)}$, $1 < i < Ne - b$, on the statically determinate base system. As discussed in Part I, in ordinary structural analysis we calculate the entries of matrices $\mathscr{B}$, and $\mathscr{B}_0$ from the sketch of the statically determinate base system, as explained above. In systematic analysis, we generate the equilibrium equations of the nodes as in Eq. (16.44) automatically and obtain matrices $\mathscr{B}$ and $\mathscr{B}_0$, again automatically, by the algorithm detailed in the next article.

16.2.5 Problems for Solution

Problem 16.6 Consider the following equation as Eq. (16.8):

$$\begin{Bmatrix} 1 \\ 2 \\ 3 \end{Bmatrix} + \begin{bmatrix} 1 & 3 & 1 & 4 & 2 & 2 \\ 0 & 0 & 2 & 2 & 1 & 1 \\ -1 & -3 & 3 & 0 & 3 & -3 \end{bmatrix} \begin{Bmatrix} x_1 \\ x_2 \\ x_3 \\ x_4 \\ x_5 \\ x_6 \end{Bmatrix} = \begin{Bmatrix} 0 \\ 0 \\ 0 \end{Bmatrix}$$

where $\mathbf{x}$ stands for $\mathbf{q}'^m$ of Eq. (16.6). Recast this equation into the format of Eq. (16.14). For this case, what is $\mathbf{s}^m$, and what are $\mathbf{B}'^m$ and $\mathbf{q}^{0'm}$? If you choose $q_{(2)}'^{mT} = [x_1 \ \ x_3 \ \ x_5]$, what is $\mathbf{s}^m$, and what is the permutation matrix $\mathbf{P}$ which can be used in Eq. (16.13)? If you have to store this $\mathbf{P}$ in the rapid-access memory, how would you store it?

Problem 16.7 Consider a gridwork-frame element where the element loads consist of the uniformly distributed load f_{tz}. In the overall coordinate system, the vertices of the element are located at $\mathbf{c}_1^T = [L \ \ 0 \ \ 0]$ and $\mathbf{c}_2^T = [L \ \ L \ \ 0]$, and they are attached to the nodes located at

$\mathbf{c}_1^{*T} = [0 \quad 0 \quad 0]$ and $\mathbf{c}_2^{*T} = [0 \quad 2L \quad 0]$, respectively, by rigid bars. What are $\mathbf{B}^m$ and $\mathbf{q}^{0m}$ matrices of this element if we take the vertex forces at vertex 1 in the local coordinate system as the element forces $\mathbf{s}^m$?

Problem 16.8   Suppose we want to obtain $\mathbf{B}^m$ and $\mathbf{q}^{0m}$ matrices for a triangular plane-stress element, the element load of which consists of a single concentrated force of magnitude P_y acting at the area center of the element in the overall y-axis direction. Let the position vectors of the vertices in the overall coordinate system be $\mathbf{c}_1^T = [4 \quad 4 \quad 0]$, $\mathbf{c}_2^T = [8 \quad 4 \quad 0]$, and $\mathbf{c}_3^T = [6 \quad 6 \quad 0]$. For the triangle in plane stress, we have $n^m = 3$, $e = 2$, $e' = 3$, $f^m = 0$. Show that the order of the column matrix of element forces is 3. Let the element forces be selected such that the first component acts at vertex 1 in the direction vertex 3 to vertex 1, the second acts at vertex 2 in the direction vertex 1 to vertex 2, and the third acts at vertex 3 in the direction vertex 2 to vertex 3. Give explicit expressions for $\mathbf{B}^m$ and $\mathbf{q}^{0m}$.

*Problem 16.9   For a space-truss structure $N = 400$, $M = 1,000$, $b = 350$. Noting that $n^m = 2$, $f^m = 0$, $e = 3$, $e' = 5$ for all bars and assuming that the structure is geometrically stable, compute the orders of the $\overline{\mathbf{B}}$ and $\mathbf{B}^0$ matrices. What is the degree of indeterminacy? If one stores $\mathbf{B}^0$ instead of $\overline{\mathbf{B}}$, what is the saving in storage?

*Problem 16.10   For a planar-frame structure $N = 200$, $M = 180$, $b = 100$. Noting that $n^m = 2$, $f^m = 0$, $e = 3$, $e' = 3$ for all elements and assuming that the structure is geometrically stable, compute the orders of the $\overline{\mathbf{B}}$ and $\mathbf{B}^0$ matrices. What is the degree of indeterminacy? If one stores $\mathbf{B}^0$ instead of $\overline{\mathbf{B}}$, what is the saving in storage?

*Problem 16.11   A structure consists of quadrilateral plane-stress elements. It is known that $N = 600$, $M = 551$, $b = 100$. Noting that $n^m = 4$, $f^m = 0$, $e = 2$, $e' = 3$ for all elements and assuming that the structure is geometrically stable, compute the orders of $\overline{\mathbf{B}}$ and $\mathbf{B}^0$ matrices. What is the degree of indeterminacy? If one stores $\mathbf{B}^0$ instead of $\overline{\mathbf{B}}$, what is the saving in storage?

Problem 16.12   Prove that $\mathscr{H}\mathbf{p}^* = 0$, where $\mathbf{p}^*$ and $\mathscr{H}$ are as defined by Eqs. (15.129) and (15.61), respectively.

16.3   Method of Identifying the Redundants.

On various occasions, we tried to change the form of a given set of linear equations with more unknowns than equations so that in the new form when we assign arbitrary values to those unknowns called independent variables, the equations are automatically satisfied. This we did for Eqs. (12.113), (16.8), and (16.44) so that we could obtain the forms in Eqs. (12.118), (16.14), and (16.56), respectively. In each case, we did not elaborate upon how to choose the independent components or how to find the permutation matrix $\mathbf{P}$ in order to arrive at the original ordering of the variables. In this article we detail a procedure which is applicable in all these cases. The procedure automatically selects the dependent components and also keeps track of the permutation matrix. It selects the dependent variables so that the magnitude of the determinant of the coefficient matrix related to the dependent variables is the largest possible. The name of the procedure is Gauss-Jordan elimination with pivot search.

16.3.1   Gauss-Jordan elimination with partial pivot search.

Although we describe the method for Eq. (16.44), it is applicable for all the other cases cited. Let us rewrite Eq. (16.44) in the form

$$\mathbf{B}^0\mathbf{s} = \mathbf{I}\mathbf{p}^0 \tag{16.59}$$

where $\mathbf{B}^0$ is a $Ne - b$ by $\jmath$ matrix and $\mathbf{I}$ is an identity matrix of order $Ne - b$. We would like to modify the equation into the form of Eq. (16.56). In order to keep track of the permutation matrix of the operations, we need a column matrix $\jmath$

of order s listing the column numbers of matrix $\mathbf{B}^0$. Initially, the column matrix contains sequential positive integers 1 through s. We can display s with its initial values as

$$s^T = [1 \quad 2 \quad \ldots \quad s] \tag{16.60}$$

We start the procedure by noting the column number q of the largest entry in magnitude in the first row of $\mathbf{B}^0$. This is called **partial pivot search.** (If the search is done to cover the complete matrix, it is called **complete pivot search.**) Then we interchange the qth column with the first. Since when we perform the interchanges, we must also interchange the corresponding entries in $\mathbf{s}$, we indicate this by interchanging the qth entry of s with the first entry. This brings the largest-magnitude entry of the first row to the $(1, 1)$ position, and we have the qth force component as the first dependent force component. Then by subtracting the proper multiples of the first equation from the other equations, we eliminate the first dependent force component from the other equations. Next we focus our attention to the equations without the first dependent force component. We repeat the operation to find the second dependent force component and eliminate it from the equations. We continue in this fashion until all the dependent force components are found and eliminated. When the last dependent force component is found, it certainly is contained in all of the equations preceding the last equation. This is also true for the other dependent force components. This phase of the procedure is called the forward pass.

We start the backward pass of the procedure by first scaling the equations such that the coefficient of the ith dependent force component in the ith equation is 1. Then we subtract the proper multiple of the second equation from the preceding one in order to eliminate the second dependent force component. Then we subtract the proper multiples of the third equation from the preceding equations in order to eliminate the third dependent force component from the preceding equations. Continuing in this manner, we successively eliminate the dependent force components from the equations such that the ith equation contains only the ith dependent force component.

Schematically, we can display the configurations of $\mathbf{B}^0$, $\mathbf{I}$, and s at three important stages of the procedure as follows. Denoting nonzero entries by x and zero entries by a boldface point, display these matrices at the beginning as

$$\mathbf{B}^0 \leftarrow \begin{bmatrix} x & x & x & x & x & x & x & x & x & x & x \\ x & x & x & x & x & x & x & x & x & x & x \\ x & x & x & x & x & x & x & x & x & x & x \\ x & x & x & x & x & x & x & x & x & x & x \\ x & x & x & x & x & x & x & x & x & x & x \end{bmatrix} \qquad \mathbf{I} \leftarrow \begin{bmatrix} 1 & . & . & . & . \\ . & 1 & . & . & . \\ . & . & 1 & . & . \\ . & . & . & 1 & . \\ . & . & . & . & 1 \end{bmatrix} \tag{16.61}$$

$$s^T \leftarrow [1 \quad 2 \quad 3 \quad 4 \quad 5 \quad 6 \quad 7 \quad 8 \quad 9 \quad 10 \quad 11]$$

where a left arrow $\leftarrow$ means that the quantities on the right will replace the quantities in the corresponding positions on the left. Assuming that the dependent force components are found such that $\mathbf{s}_{(2)}^T = [s_7 \quad s_3 \quad s_9 \quad s_1 \quad s_5]$, we display the same matrices at the end of the forward pass as

$$\mathbf{B}^0 \leftarrow \begin{bmatrix} X & X & X & X & X & X & X & X & X & X & X \\ . & X & X & X & X & X & X & X & X & X & X \\ . & . & X & X & X & X & X & X & X & X & X \\ . & . & . & X & X & X & X & X & X & X & X \\ . & . & . & . & X & X & X & X & X & X & X \end{bmatrix} \qquad \mathbf{I} \leftarrow \begin{bmatrix} 1 & . & . & . & . \\ x & 1 & . & . & . \\ x & x & 1 & . & . \\ x & x & x & 1 & . \\ x & x & x & x & 1 \end{bmatrix} \qquad (16.62)$$

$$\mathbf{\mathit{s}}^T \leftarrow [7 \quad 3 \quad 9 \quad 1 \quad 5 \quad 6 \quad 4 \quad 8 \quad 2 \quad 10 \quad 11]$$

Note that the nonzero entries developed in $\mathbf{I}$ are equal in number to the zero elements developed in $\mathbf{B}^0$. When we complete the backward pass, the same matrices will look like

$$\mathbf{B}^0 \leftarrow \begin{bmatrix} 1 & . & . & . & . & X & X & X & X & X & X \\ . & 1 & . & . & . & X & X & X & X & X & X \\ . & . & 1 & . & . & X & X & X & X & X & X \\ . & . & . & 1 & . & X & X & X & X & X & X \\ . & . & . & . & 1 & X & X & X & X & X & X \end{bmatrix} \qquad \mathbf{I} \leftarrow \begin{bmatrix} X & X & X & X & X \\ X & X & X & X & X \\ X & X & X & X & X \\ X & X & X & X & X \\ X & X & X & X & X \end{bmatrix} \qquad (16.63)$$

$$\qquad\qquad \mathbf{I} \qquad\qquad\quad \mathbf{B}^{0-1}_{(2)}\mathbf{B}^0_x \qquad\qquad\qquad \mathbf{B}^{0-1}_{(2)}$$

$$\mathbf{\mathit{s}}^T \leftarrow [7 \quad 3 \quad 9 \quad 1 \quad 5 \mid 6 \quad 4 \quad 8 \quad 2 \quad 10 \quad 11]$$

$$\qquad\quad \mathbf{s}_{(2)} \qquad\qquad\qquad \mathbf{x}$$

where we indicate the meaning of each group with the notation of Eq. (16.52). The labels below the display for $\mathbf{\mathit{s}}^T$ imply that

$$\mathbf{s}^T_{(2)} = [s_7 \quad s_3 \quad s_9 \quad s_1 \quad s_5] \qquad \text{and} \qquad \mathbf{x}^T = [s_6 \quad s_4 \quad s_8 \quad s_2 \quad s_{10} \quad s_{11}]$$

We observe that the order of the identity which matrix developed in the $\mathbf{B}^0$ area is the same as the order of $\mathbf{B}^{0-1}_{(2)}$ which developed in the $\mathbf{I}$ area, as seen from Eq. (16.63). Therefore, for the elimination procedure, we may not require additional storage area for matrix $\mathbf{I}$. On this basis, the configurations at the three stages of Eqs. (16.61) to (16.63) can be displayed as follows. At the beginning

$$\mathbf{B}^0 \leftarrow \begin{bmatrix} X & X & X & X & X & X & X & X & X & X & X & X \\ X & X & X & X & X & X & X & X & X & X & X & X \\ X & X & X & X & X & X & X & X & X & X & X & X \\ X & X & X & X & X & X & X & X & X & X & X & X \\ X & X & X & X & X & X & X & X & X & X & X & X \end{bmatrix} \qquad (16.64)$$

$$\mathbf{\mathit{s}}^T \leftarrow [1 \quad 2 \quad 3 \quad 4 \quad 5 \quad 6 \quad 7 \quad 8 \quad 9 \quad 10 \quad 11]$$

At the end of forward pass

$$\mathbf{B}^0 \leftarrow \begin{bmatrix} X & X & X & X & X & X & X & X & X & X & X \\ X & X & X & X & X & X & X & X & X & X & X \\ X & X & X & X & X & X & X & X & X & X & X \\ X & X & X & X & X & X & X & X & X & X & X \\ X & X & X & X & X & X & X & X & X & X & X \end{bmatrix} \qquad (16.65)$$

$$\mathbf{\mathit{s}}^T \leftarrow [7 \quad 3 \quad 9 \quad 1 \quad 5 \mid 6 \quad 4 \quad 8 \quad 2 \quad 10 \quad 11]$$

At the end of the backward pass

$$\mathbf{B}^0 \leftarrow \begin{bmatrix} x & x & x & x & x & x & x & x & x & x & x \\ x & x & x & x & x & x & x & x & x & x & x \\ x & x & x & x & x & x & x & x & x & x & x \\ x & x & x & x & x & x & x & x & x & x & x \\ x & x & x & x & x & x & x & x & x & x & x \end{bmatrix}$$

(16.66)

$$\underbrace{\qquad}_{\mathbf{B}_{(2)}^{0^{-1}}} \qquad \underbrace{\qquad}_{\mathbf{B}_{(2)}^{0^{-1}} \mathbf{B}_x^0}$$

$$\jmath^T \leftarrow [\underbrace{7 \quad 3 \quad 9 \quad 1 \quad 5}_{s_{(2)}} \mid \underbrace{6 \quad 4 \quad 8 \quad 2 \quad 10 \quad 11}_{x}]$$

where the meaning of each group is marked with the notation of Eq. (16.52). Matrix $\mathbf{B}^0$ is in general a very sparse matrix at the beginning; at the end, however, it is always a full matrix.

We can readily obtain the permutation matrix $\mathbf{P}$ from the final configuration of $\jmath$ so that Eq. (16.55) is valid. For the case in Eq. (16.66) we have

$$\mathbf{P} = [\mathbf{i}_7 \quad \mathbf{i}_3 \quad \mathbf{i}_9 \quad \mathbf{i}_1 \quad \mathbf{i}_5 \quad \mathbf{i}_6 \quad \mathbf{i}_4 \quad \mathbf{i}_8 \quad \mathbf{i}_2 \quad \mathbf{i}_{10} \quad \mathbf{i}_{11}] \qquad (16.67)$$

where $\mathbf{i}_j$ is the jth column of the identity matrix of order $N = 11$. Note that the subscripts of columns $\mathbf{i}$ are those listed in $\jmath$ of Eq. (16.66). The reader should verify the validity of $\mathbf{P}$ given above.

Having defined $\mathbf{P}$, $\mathbf{B}_{(2)}^{0^{-1}}$, and $\mathbf{B}_{(2)}^{0^{-1}} \mathbf{B}_x^0$ matrices in this way, it is a simple matter to obtain $\mathcal{B}$ and $\mathcal{B}_0$ matrices from Eqs. (16.57) and (16.58), respectively.

It can be shown that just after the forward pass, the product of the main-diagonal elements of the square matrix in the left partition of $\mathbf{B}^0$ is the determinant of $\mathbf{B}_{(2)}^0$. By searching and bringing the largest-magnitude elements to the main diagonal we are selecting the dependent force components such that the associated $\mathbf{B}_{(2)}^0$ matrix will give us largest possible determinant in the absolute value. We may, of course, change the criterion of selecting the dependent force components. For example, we may select the dependent force components such that they are associated with the structural elements which are most rigid in the whole structure. We may devise many other criteria to select the dependent force components. These criteria affect only how we interchange rows and columns of $\mathbf{B}^0$; otherwise, the basic Gauss-Jordan elimination process remains unaltered. Since the procedure automatically identifies the dependent and redundant force components and thus selects for us a good statically determinate base system, it is called a **structure cutter.**

The computational steps for transforming the arrays in Eq. (16.64) into the form of those of Eq. (16.66) are:

1. Set the equation counter $i = 1$.

2. Find the column number q of the element which is largest in magnitude in the ith row of matrix $\mathbf{B}^0$, where $i \leq q \leq \jmath$. If the largest-magnitude element in this row is zero, signal that the structure is geometrically unstable and terminate the procedure.

3. Interchange the ith column with the qth column; namely,

$$(c \leftarrow b_{ri}^0, \ b_{ri}^0 \leftarrow b_{rq}^0, \ b_{rq}^0 \leftarrow c, \ r = 1, \ldots, Ne - b).$$

4. Interchange the ith and qth entries of $\jmath$; namely, $(c \leftarrow \jmath_i, \jmath_i \leftarrow \jmath_q, \jmath_q \leftarrow c)$. If $i = Ne - b$, go to step 8; otherwise continue.

5. Subtract the proper multiples of the ith row from the succeeding rows such that the entries below b_{ii}^0 are all zeros; namely, $((b_{rt}^0 \leftarrow (b_{rt}^0 - b_{it}^0 b_{ri}^0/b_{ii}^0), t = 1, \ldots, \jmath;$ $t \neq i), r = i + 1, \ldots, Ne - b)$.

6. Modify the entries below b_{ii}^0 for $\mathbf{B}_{(2)}^{0-1}$; namely, $(b_{ri}^0 \leftarrow -b_{ri}^0/b_{ii}^0, r = i + 1, \ldots, Ne - b)$.

7. Set $i \leftarrow i + 1$ and go to step 2.

8. Signal the start of the backward pass. Scale all rows by their respective main-diagonal elements and save scaling factors in the main-diagonal positions; namely, $(c \leftarrow b_{rr}^0, (b_{rt}^0 \leftarrow b_{rt}^0/c, t = 1, \ldots, \jmath), b_{rr}^0 \leftarrow 1/c, r = 1, \ldots, Ne - b)$. Set $i = 1$; then go to step 11.

9. Subtract proper multiples of the ith row from the preceding rows such that the entries above b_{ii}^0 are all zeros; namely, $((b_{rt}^0 \leftarrow (b_{rt}^0 - b_{ri}^0 b_{it}^0), t = 1, \ldots, \jmath; t \neq i), r = i - 1, \ldots, 1)$.

10. Modify the entries above b_{ii}^0 for $\mathbf{B}_{(2)}^{0-1}$; namely, $(b_{ri}^0 \leftarrow -b_{ri}^0 b_{ii}^0, r = i - 1, \ldots, 1)$.

11. Set $i \leftarrow i + 1$. If new $i \leq Ne - b$, go to step 9; otherwise continue.

12. End of the procedure.

In the above procedure it is assumed that $\jmath > Ne - b > 1$.

16.3.2 Example for Gauss-Jordan Elimination. As an example, let us consider the following as the values of $\mathbf{B}^0$ and $\jmath$ at the beginning:

$$\mathbf{B}^0 \leftarrow \begin{bmatrix} 2 & 0 & -2 & 2 & 4 \\ -2 & 1 & 3 & -1 & -2 \\ 1 & -1 & 1 & 2 & 2 \end{bmatrix} \tag{a}$$

$$\jmath^T \leftarrow \begin{bmatrix} 1 & 2 & 3 & 4 & 5 \end{bmatrix} \tag{b}$$

At the end of the forward pass, the values with the procedure outlined in Art. 16.3.1 will be

$$\mathbf{B}^0 \leftarrow \begin{bmatrix} 4 & -2 & 0 & 2 & 2 \\ \frac{1}{2} & 2 & 1 & 0 & -1 \\ -1 & -1 & -2 & 1 & 1 \end{bmatrix} \tag{c}$$

$$\jmath^T \leftarrow \begin{bmatrix} 5 & 3 & 2 & 4 & 1 \end{bmatrix} \tag{d}$$

At the end of the backward pass we shall have

$$\mathbf{B}^0 \leftarrow \begin{bmatrix} \frac{1}{4} & \frac{1}{8} & \frac{1}{8} & \frac{5}{8} & \frac{3}{8} \\ 0 & \frac{1}{4} & \frac{1}{4} & \frac{1}{4} & -\frac{1}{4} \\ \frac{1}{2} & \frac{1}{2} & -\frac{1}{2} & -\frac{1}{2} & -\frac{1}{2} \end{bmatrix} \tag{e}$$

$$\jmath^T \leftarrow \begin{bmatrix} 5 & 3 & 2 & 4 & 1 \end{bmatrix} \tag{f}$$

Note that in Eq. (e) the partition on the left is $\mathbf{B}_{(2)}^{0-1}$ and the partition on the right is $\mathbf{B}_{(2)}^{0-1}\mathbf{B}_x^0$. From Eq. (f) we can write that $s_{(2)}^T = \begin{bmatrix} s_5 & s_3 & s_2 \end{bmatrix}$ and $\mathbf{x}^T = \begin{bmatrix} s_4 & s_1 \end{bmatrix}$. With the help of Eq. (16.67), we can obtain the permutation matrix $\mathbf{P}$ from Eq. (f) as $\mathbf{P} = [\mathbf{i}_5 \quad \mathbf{i}_3 \quad \mathbf{i}_2 \quad \mathbf{i}_4 \quad \mathbf{i}_1]$, where $\mathbf{i}_j$ is the jth column of the fifth-order identity matrix. Using these values in Eqs. (16.57) and (16.58), we obtain

$$\mathscr{B} = \begin{bmatrix} 0 & 1 \\ \tfrac{1}{2} & \tfrac{1}{2} \\ -\tfrac{1}{4} & \tfrac{1}{4} \\ 1 & 0 \\ -\tfrac{5}{8} & -\tfrac{3}{8} \end{bmatrix} \qquad \mathscr{B}_0 = \begin{bmatrix} 0 & 0 & 0 \\ \tfrac{1}{2} & \tfrac{1}{2} & -\tfrac{1}{2} \\ 0 & \tfrac{1}{4} & \tfrac{1}{4} \\ 0 & 0 & 0 \\ \tfrac{1}{4} & \tfrac{1}{8} & \tfrac{1}{8} \end{bmatrix} \qquad (g)$$

When matrix $\mathbf{B}^0$ of Eq. (16.44) is given as in Eq. (a), the matrices $\mathscr{B}$ and $\mathscr{B}_0$ of the equivalent form in Eq. (16.56) are as in Eq. (g).

16.3.3   Problems for Solution

Problem 16.13   Matrix $\mathbf{B}^0$ of Eq. (16.44) is given as

$$\mathbf{B}^0 = \begin{bmatrix} -1 & 2 & 4 & 3 & 2 \\ -2 & 4 & 4 & 6 & 7 \\ -1 & 6 & -4 & 2 & 3 \end{bmatrix}$$

Compute matrices $\mathscr{B}$ and $\mathscr{B}_0$ of the equivalent form in Eq. (16.56) by the method of Gauss-Jordan elimination with partial pivot search.

Problem 16.14   Show that the number of multiplications to obtain $\mathbf{B}_{(2)}^{0-1}$ from $\mathbf{B}^0$ is proportional to $(Ne - b)^2 \partial$ in the Gauss-Jordan elimination method.

Problem 16.15   During the Gauss-Jordan elimination with partial pivot search, if the largest-magnitude element of a row is zero, explain why the structure is geometrically unstable.

Problem 16.16   See if the itemized computation procedure given in Art. 16.3.1 will work when $\partial > Ne - b = 1$.

Problem 16.17   Using Gauss-Jordan elimination with partial pivot search, write a computer procedure to transform matrix $\mathbf{B}^0$ into the one where the left and the right partitions will contain $\mathbf{B}_{(2)}^{0-1}$ and $\mathbf{B}_{(2)}^{0-1}\mathbf{B}_x^0$, respectively, and to transform the initial column-label column matrix ∂ into one where column labels corresponding to $\mathbf{s}_{(2)}$ and $\mathbf{x}$ will be contained in the left and the right partitions, respectively.

Problem 16.18   See if the itemized procedure given in Art. 16.3.1 will work when $\partial = Ne - b$. Note that this corresponds to the case of statically determinate structures, where there are no redundant forces.

Problem 16.19   Modify the computational procedure given in Art. 16.3.1 so that it works with complete rather than partial pivot search. What are the advantages and disadvantages of working with complete pivot search?

16.4   Governing Equations for Forces; Completion of Analysis.

Once the elemental matrices $\mathbf{F}^m$, $\mathbf{v}_0^m$, $m = 1, \ldots, M$, and the prescribed deflections $\mathbf{u}_0$ as in Eq. (15.36) are known in the overall coordinate system, we can obtain the governing equations for the redundant forces in various ways, depending upon the form of the nodal force-equilibrium equations. The alternate forms of these equilibrium equations are given in Eqs. (16.30), (16.44), and (16.56). If the nodal force-equilibrium equations are in the form of Eq. (16.56), we can use the principle of stationary total complementary potential energy with substitution or the theorem of virtual forces to obtain the governing equations for the redundants $\mathbf{x}$. After computing $\mathbf{x}$ from these equations, we can use Eq. (16.56) to compute the complete list of element forces $\mathbf{s}$ and Eq. (16.23) to compute the reactions. Once $\mathbf{s}$ is known, we can use Eq. (16.14) to obtain, in the local coordinate systems, the complete list of vertex forces for each element. If the nodal force-equilibrium equations are given in the form of Eq. (16.30) or (16.44), we can use the method based on the principle of stationary total complementary potential energy with Lagrange multipliers, to obtain the governing equations for the complete list of element forces and reactions $\mathbf{z}$ or the element forces $\mathbf{s}$ alone. In order to minimize the number of equations to be solved simultaneously, we always prefer the form of the nodal force-equilibrium equations in Eq. (16.56). Because of simplicity, the substitution method using the principle of stationary total complementary potential energy may be preferred over the virtual-work approach, although these two methods yield the same set of equations for the redundants. The following articles detail these three methods, in the order of preference.

16.4.1 Method using principle of stationary total complementary potential energy with substitution. This method requires that the nodal force-equilibrium equations be available in the form of Eq. (16.56). We first express the total complementary potential energy in terms of the element forces $\mathbf{s}$ and reactions $\mathbf{p}^*$. Next, using Eq. (16.23), we obtain the expression for the total complementary potential energy in terms of $\mathbf{s}$ only. Then, by means of Eq. (16.56) the energy expression is obtained as a quadratic form plus a linear form of $\mathbf{x}$. The application of the principle of stationary energy yields the governing equations for $\mathbf{x}$. From these equations we obtain the numerical values of $\mathbf{x}$; using them in Eq. (16.56) gives the complete list of element forces, and Eq. (16.23) yields the values of the reactions.

Let π^* denote the total complementary potential energy of the structure. Noting that the potential of the prescribed nodal deflections $\mathbf{u}_0$ [see Eq. (15.36)] is $-\mathbf{p}^{*T}\mathbf{u}_0$ [see Eq. (11.10)], we write

$$\pi^* = \left(\sum_{m=1}^{M} \pi^{*m}\right) - \mathbf{p}^{*T}\mathbf{u}_0 \tag{16.68}$$

or, substituting π^{*m} from Eq. (16.3), we have

$$\pi^* = \left(\frac{1}{2}\sum_{m=1}^{M} \mathbf{s}^{m^T}\mathbf{F}^m\mathbf{s}^m + \mathbf{s}^{m^T}\mathbf{v}_0^m\right) - \mathbf{p}^{*T}\mathbf{u}_0 \tag{16.69}$$

Using $\mathbf{p}^*$ from Eq. (16.23), we can express π^* in terms of $\mathbf{s}$ alone

$$\pi^* = \frac{1}{2}\mathbf{s}^T \operatorname{diag}(\mathbf{F}^m)\mathbf{s} + \mathbf{s}^T\mathbf{v}_0 - \mathbf{s}^T\left(\sum_{m=1}^{M}\sum_{g=1}^{n^m} \mathbf{I}_*^m \mathbf{B}^{m^T}\mathbf{I}_g^m\mathbf{I}^T j_g^m\mathbf{u}_0\right) + \text{const} \tag{16.70}$$

where $\mathbf{v}_0$ is the column matrix listing $\mathbf{v}_0^m$, $m = 1, \ldots, M$, and const represents the terms without $\mathbf{s}$. Substituting $\mathbf{s}$ from Eq. (16.56), we rewrite Eq. (16.70) as

$$\pi^* = \frac{1}{2}\mathbf{x}^T\overline{\mathbf{F}}\mathbf{x} - \mathbf{x}^T\overline{\mathbf{v}}_0 + \text{const} \tag{16.71}$$

where
$$\overline{\mathbf{F}} = \mathscr{B}^T \operatorname{diag}(\mathbf{F}^m)\mathscr{B} \tag{16.72}$$

and

$$\overline{\mathbf{v}}_0 = -\mathscr{B}^T\mathbf{v}_0 + \mathscr{B}^T\left(\sum_{m=1}^{M}\sum_{g=1}^{n^m} \mathbf{I}_*^m \mathbf{B}^{m^T}\mathbf{I}_g^m\mathbf{I}^T j_{g}^m\mathbf{u}_0\right) - \mathscr{B}^T \operatorname{diag}(\mathbf{F}^m)\mathscr{B}_0\mathbf{p}^0 \tag{16.73}$$

where we have used the fact that diag $(\mathbf{F}^m)$ is symmetrical. Noting that

$$\text{diag} (\mathbf{F}^m) = \sum_{m=1}^{M} \mathbf{I}_*^m \mathbf{F}^m \mathbf{I}_*^{m^T} \tag{16.74}$$

and defining

$$\mathcal{B}^T \mathbf{I}_*^m = \mathcal{B}^{m^T} \tag{16.75}$$

and

$$\mathcal{B}_0^T \mathbf{I}_*^m = \mathcal{B}_0^{m^T} \tag{16.76}$$

we can rewrite Eqs. (16.72) and (16.73) as

$$\bar{\mathbf{F}} = \sum_{m=1}^{M} \mathcal{B}^{m^T} \mathbf{F}^m \mathcal{B}^m \tag{16.77}$$

and

$$\bar{\mathbf{v}}_0 = -\left(\sum_{m=1}^{M} \mathcal{B}^{m^T} \mathbf{v}_0^m \right) + \left(\sum_{m=1}^{M} \sum_{g=1}^{n^m} \mathcal{B}^{m^T} \mathbf{B}^m \mathbf{I}_g^m \mathbf{u}_{0 j_g^m} \right) - \left(\sum_{m=1}^{M} \mathcal{B}^{m^T} \mathbf{F}^m \mathcal{B}_0^m \mathbf{p}^0 \right) \tag{16.78}$$

where the identity

$$\mathbf{v}_0 = \sum_{m=1}^{M} \mathbf{I}_*^m \mathbf{v}_0^m \tag{16.79}$$

and the one in Eq. (15.82) are also used. Note that $\mathcal{B}^m$ and $\mathcal{B}_0^m$ are the mth row partitions of $\mathcal{B}$ and $\mathcal{B}_0$, which are defined by Eqs. (16.57) and (16.58).

Matrix $\bar{\mathbf{F}}$ is called the **flexibility matrix associated with the directions of the redundants**, and the column matrix $\bar{\mathbf{v}}_0$ is called the **deformations associated with the directions of the redundants**. Since $\mathbf{F}^m$, $m = 1, \ldots, M$, are all symmetric, from Eq. (16.77) we observe that $\bar{\mathbf{F}}$ is also symmetric, namely,

$$\bar{\mathbf{F}} = \bar{\mathbf{F}}^T \tag{16.80}$$

Moreover, since the quadratic form at the right of Eq. (16.71) is the total complementary strain energy of the structure, it is always larger than or equal to zero, being zero only if $\mathbf{x}$ is a null matrix; we therefore conclude that $\bar{\mathbf{F}}$ is also positive definite. With Eq. (16.80) in mind, and according to the principle of stationary total complementary potential energy (see Art. 11.5), which states that for true $\mathbf{x}$ values $\delta\pi^* = 0$, from Eq. (16.71) by differentiation we obtain

$$\bar{\mathbf{F}}\mathbf{x} = \bar{\mathbf{v}}_0 \tag{16.81}$$

From our discussions in Art. 11.5 we know that Eq. (16.81) represents geometric compatibility in the directions associated with the redundant forces listed in $\mathbf{x}$.

Since the coefficient matrix is symmetric and positive definite, we can use the Cholesky method to solve the numerical values of $\mathbf{x}$. Having computed the values of $\mathbf{x}$, we obtain the element forces from Eq. (16.56), which can be written alternately as

$$\mathbf{s}^m = \mathcal{B}^m \mathbf{x} + \mathcal{B}_0^m \mathbf{p}^0 \qquad m = 1, \ldots, M \tag{16.82}$$

From Eq. (16.19) we obtain the vertex forces in the overall coordinates, i.e.,

$$\mathbf{q}^m = \mathbf{B}^m \mathbf{s}^m + \mathbf{q}^{om} \qquad m = 1, \ldots, M \tag{16.83}$$

Using the numerical values of $\mathbf{q}^m$ in Eq. (16.22), we obtain the reactions as

$$\mathbf{p}^* = -\mathbf{p}_0 + \sum_{m=1}^{M} \sum_{g=1}^{n^m} \mathbf{I}_{j_g^m} \mathbf{q}_g^m \tag{16.84}$$

The values of the vertex forces in the local coordinate systems can be obtained by means of Eq. (13.75), namely,

$$\mathbf{q}_i'^m = \mathcal{T}_{\varepsilon i}^{m'-T} \mathbf{q}_i^m \qquad \begin{aligned} i &= 1, \ldots, n^m \\ m &= 1, \ldots, M \end{aligned} \tag{16.85}$$

If the structure is part of a two- or three-dimensional continuum, from the vertex forces in local coordinates we can calculate the actual stresses in the element by some approximation satisfying only the force-equilibrium requirements.

16.4.2　Method using the theorem of virtual forces.　This method requires that the nodal force-equilibrium equations be available in the form of Eq. (16.56). In this method, we first express the deflections in the directions of the element forces in terms of the element forces, as in Eq. (16.2).　Then we seek the geometrical relationship between these deflections and the prescribed nodal deflections $\mathbf{u}_0$, namely, the support settlements, by means of the theorem of virtual forces.　We give an arbitrary variation to the redundant forces $\delta\mathbf{x}$ and express the corresponding virtual variations in the element forces $\delta\mathbf{s}$ and in the reactions $\delta\mathbf{p}^*$ with the help of Eq. (16.23).　Then we obtain the geometric relationship from the fact that the work done by the internal virtual forces $\delta\mathbf{x}$ and $\delta\mathbf{s}$ are equal to the work done by the external virtual forces listed in $\delta\mathbf{p}^*$ when these virtual forces are subjected to the actual internal and external deflections.

The deflections in the directions of the element forces can be written, from Eq. (16.2), as

$$\mathbf{v}^m = \mathbf{F}^m \mathbf{s}^m + \mathbf{v}_0^m \qquad m = 1, \ldots, M \tag{16.86}$$

We can write them compactly as

$$\mathbf{v} = \text{diag} \, (\mathbf{F}^m)\mathbf{s} + \mathbf{v}_0 \tag{16.87}$$

where $\mathbf{s}$ and $\mathbf{v}_0$ are as defined in Eqs. (16.27) and (16.79) and

$$\mathbf{v}^T = [\mathbf{v}^{1^T} \quad \cdots \quad \mathbf{v}^{M^T}] \tag{16.88}$$

From Eqs. (16.56) we can write the relationship between $\delta\mathbf{x}$ and $\delta\mathbf{s}$ as

$$\delta\mathbf{s} = \mathcal{B} \, \delta\mathbf{x} \tag{16.89}$$

by taking the variations of both sides of Eq. (16.56).　Similarly, we obtain the relationship between $\delta\mathbf{x}$ and $\delta\mathbf{p}^*$ by taking the variations of both sides of Eq. (16.23):

$$\delta\mathbf{p}^* = \left(\sum_{m=1}^{M} \sum_{g=1}^{n^m} \mathbf{I}_{j_g^m} \mathbf{I}_g^{m^T} \mathbf{B}^m \mathbf{I}_*^{m^T} \right) \delta\mathbf{s} \tag{16.90}$$

where we also used Eq. (16.29). Now, we can use the theorem of virtual forces [see Eq. (11.5)] to write

$$\delta\mathbf{s}^T\mathbf{v} + \delta\mathbf{x}^T\mathbf{0} = \delta\mathbf{p}^{*T}\mathbf{u}_0 \tag{16.91}$$

Note that the integrity of the structure requires that there be no deflections in the directions of the redundant forces listed in $\mathbf{x}$, that is, no gaps and overlaps in these directions. Substituting $\delta\mathbf{s}$ and $\delta\mathbf{p}^*$ from Eqs. (16.89) and (16.90) into Eq. (16.91), we obtain

$$\delta\mathbf{x}^T\left(\mathscr{B}^T\mathbf{v} - \mathscr{B}^T\sum_{m=1}^{M}\sum_{g=1}^{n^m}\mathbf{I}_*^m\mathbf{B}^{mT}\mathbf{I}_g^m\mathbf{u}_{0j_g^m}\right) = 0 \tag{16.92}$$

where we made use of Eq. (15.82). Since Eq. (16.92) is an identity with respect to the arbitrary variation $\delta\mathbf{x}$, its coefficient must be a null column matrix; therefore

$$\mathscr{B}^T\mathbf{v} - \mathscr{B}^T\sum_{m=1}^{M}\sum_{g=1}^{n^m}\mathbf{I}_*^m\mathbf{B}^{mT}\mathbf{I}_g^m\mathbf{u}_{0j_g^m} = 0 \tag{16.93}$$

Substituting $\mathbf{v}$ from Eq. (16.87) and $\mathbf{s}$ from Eq. (16.56) into Eq. (16.93), we obtain

$$\bar{\mathbf{F}}\mathbf{x} = \bar{\mathbf{v}}_0 \tag{16.94}$$

where $\bar{\mathbf{F}}$ and $\bar{\mathbf{v}}_0$ are as in Eqs. (16.72) and (16.73) or (16.77) and (16.78). Note that Eq. (16.94) is identical to Eq. (16.81). After we solve the numerical values of the redundants from Eq. (16.94), we obtain the numerical values of other forces from Eqs. (16.82) to (16.85).

16.4.3 Method using the principle of stationary total complementary potential energy with Lagrange multipliers. This method can be used when the nodal force-equilibrium equations are available in the form of Eq. (16.30) or (16.44). Let us suppose that the equilibrium equations of the nodes are available in the form of Eq. (16.30). From our discussion in Art. 15.6.3 we note that

$$\mathbf{p}^{*T}\mathbf{u}_0 = \mathbf{p}_r^T\mathbf{u}_{(20)} \tag{16.95}$$

Using this in Eq. (16.69), we write the total complementary potential energy as

$$\pi^* = \tfrac{1}{2}\mathbf{s}^T\,\mathrm{diag}\,(\mathbf{F}^m)\mathbf{s} + \mathbf{s}^T\mathbf{v}_0 - \mathbf{p}_r^T\mathbf{u}_{(20)} \tag{16.96}$$

where $\mathbf{s}$ and $\mathbf{v}_0$ are as defined in Eqs. (16.27) and (16.79), $\mathbf{p}_r$ lists the b reaction components, i.e., the nonzero entries of $\mathbf{p}^*$, and $\mathbf{u}_{(20)}$ lists the prescribed deflections in the directions of the components of $\mathbf{p}_r$. In the right-hand side of Eq. (16.96) only the quantities $\mathbf{s}$ and $\mathbf{p}_r$ are not known. If $\mathbf{s}$ and $\mathbf{p}_r$ are **admissible**, i.e., if they satisfy the force-equilibrium requirements, we can apply the stationary-total-complementary-energy principle to obtain the governing equations for $\mathbf{s}$ and $\mathbf{p}_r$. Forces $\mathbf{s}$ and $\mathbf{p}_r$ are admissible only if they satisfy the equilibrium requirements given by Eq. (16.30). In other words, we are looking for the stationary point of π^* in Eq. (16.96), subject to the constraints given by Eq. (16.30). We can handle this problem by the Lagrange-multipliers method. Using the definition given in Eq. (16.33), we first express π^* of Eq. (16.96) in terms of $\mathbf{z}$:

$$\pi^* = \tfrac{1}{2}\mathbf{z}^T \mathcal{F} \mathbf{z} - \mathbf{z}^T v \tag{16.97}$$

where

$$\mathcal{F} = \begin{bmatrix} \operatorname{diag} \mathbf{F}^m & 0 \\ \hline 0 & 0 \end{bmatrix} \tag{16.98}$$

and

$$v^T = [-\bar{\mathbf{v}}_0^T \mid \mathbf{u}_{(20)}^T] \tag{16.99}$$

We reproduce the constraint equation (16.30) as

$$\bar{\mathbf{B}}\mathbf{z} - \bar{\mathbf{p}} = 0 \tag{16.100}$$

Let λ denote the column matrix of Lagrange multipliers, which is of the same order of $\bar{\mathbf{B}}$, that is, of order Ne. Premultiplying both sides of Eq. (16.100) by λ^T and adding the resulting equation to Eq. (16.97), we write

$$\pi^* = \tfrac{1}{2}\mathbf{z}^T \mathcal{F} \mathbf{z} - \mathbf{z}^T v + \lambda^T(\bar{\mathbf{B}}\mathbf{z} - \bar{\mathbf{p}}) \tag{16.101}$$

Noting that $\mathbf{z}^T \bar{\mathbf{B}}^T \lambda = \lambda^T \bar{\mathbf{B}}\mathbf{z}$, we can rearrange Eq. (16.101) as

$$\pi^* = \tfrac{1}{2}[\mathbf{z}^T \mid \lambda^T]\begin{bmatrix} \mathcal{F} & \bar{\mathbf{B}}^T \\ \hline \bar{\mathbf{B}} & 0 \end{bmatrix}\begin{Bmatrix} \mathbf{z} \\ \lambda \end{Bmatrix} - [\mathbf{z}^T \mid \lambda^T]\begin{Bmatrix} v \\ \bar{\mathbf{p}} \end{Bmatrix} \tag{16.102}$$

If we consider λ as part of the quantities to be varied, it can be shown that the unconstrained stationary point of π^* in Eq. (16.102) corresponds to the constrained stationary point of π^* in Eq. (16.97). If we set the variation of π^* corresponding to the variations $\delta\mathbf{z}$ and $\delta\lambda$ to zero, noting that the square matrix in Eq. (16.102) is symmetric, after cancellations we obtain

$$\begin{bmatrix} \mathcal{F} & \bar{\mathbf{B}}^T \\ \hline \bar{\mathbf{B}} & 0 \end{bmatrix}\begin{Bmatrix} \mathbf{z} \\ \lambda \end{Bmatrix} = \begin{Bmatrix} v \\ \bar{\mathbf{p}} \end{Bmatrix} \tag{16.103}$$

Note that the order of the linear equations in Eq. (16.103) is the sum of orders $\mathbf{z}$ and λ, namely, $\sigma + b + Ne$. For the 60-element structure summarized in Eqs. (a) of Art. 16.2, the order of the linear equations of the type of Eq. (16.103) is $1{,}080 + 25 + 300 = 1{,}405$. The coefficient matrix in Eq. (16.103) is symmetric but not positive definite. If the same problem is handled with the substitution method of Art. 16.4.1, we must first apply the Gauss-Jordan elimination method to $\bar{\mathbf{B}}$, which is a 300 by 1,080 matrix, and then solve a linear-equation set in the form of Eq. (16.81), which is of order $1{,}080 - 300 + 25 = 805$. (Note that solving the same problem by the displacement method requires the solution of only $300 - 25 = 275$ equations.) Since the number of arithmetic operations in solving linear equations is generally proportional to the cube of the order of equations, we refrain from using the Lagrange-multipliers method as much as possible. Note that it is not possible to eliminate λ from Eq. (16.103) due to the zero diagonal submatrix. Therefore, during the solution of Eq. (16.103) for $\mathbf{z}$, as a by-product, we also obtain the numerical values of Lagrange multipliers λ. Unfortunately, these quantities do not carry any physical meaning which is useful to a structural engineer.

By a small amount of substitution, we can decrease the order of the linear equations in the Lagrange-multipliers method somewhat. First of all, instead of using the constraint equations in the form of Eq. (16.30), we can use the form in Eq. (16.44). We may write the total complementary potential energy of the structure from Eq. (16.70), where the reactions $\mathbf{p}^*$ are substituted from Eq. (16.23),

$$\pi^* = \tfrac{1}{2}\mathbf{s}^T \operatorname{diag}(\mathbf{F}^m)\mathbf{s} - \mathbf{s}^T \boldsymbol{v}_0 + \text{const} \tag{16.104}$$

where

$$\boldsymbol{v}_0 = -\mathbf{v}_0 + \sum_{m=1}^{M} \sum_{g=1}^{n^m} \mathbf{I}_*^m \mathbf{B}^{m^T} \mathbf{I}_g^m \mathbf{u}_{0 j_g^m} \tag{16.105}$$

and we have made use of Eq. (15.82). In Eq. (16.104), only $\mathbf{s}$ is not known. In order to be admissible, $\mathbf{s}$ has to satisfy the constraints in Eq. (16.44), which can be reproduced as

$$\mathbf{B}^0 \mathbf{s} - \mathbf{p}^0 = 0 \tag{16.106}$$

where the row order of $\mathbf{B}^0$ is $Ne - b$. Let the Lagrange-multiplier column matrix $\boldsymbol{\lambda}$ be of order $Ne - b$. Then, as before, we can combine π^* of Eq. (16.104) with the constraints of Eq. (16.106) and write

$$\pi^* = \tfrac{1}{2}[\mathbf{s}^T \,|\, \boldsymbol{\lambda}^T]\left[\begin{array}{c|c} \operatorname{diag}\mathbf{F}^m & \mathbf{B}^{0^T} \\ \hline \mathbf{B}^0 & 0 \end{array}\right]\left\{\begin{array}{c} \mathbf{s} \\ \hline \boldsymbol{\lambda} \end{array}\right\} - [\mathbf{s}^T \,|\, \boldsymbol{\lambda}^T]\left\{\begin{array}{c} \boldsymbol{v}_0 \\ \hline \mathbf{p}^0 \end{array}\right\} \tag{16.107}$$

which leads to

$$\left[\begin{array}{c|c} \operatorname{diag}\mathbf{F}^m & \mathbf{B}^{0^T} \\ \hline \mathbf{B}^0 & 0 \end{array}\right]\left\{\begin{array}{c} \mathbf{s} \\ \hline \boldsymbol{\lambda} \end{array}\right\} = \left\{\begin{array}{c} \boldsymbol{v}_0 \\ \hline \mathbf{p}^0 \end{array}\right\} \tag{16.108}$$

The order of this linear set is $\mathit{s} + Ne - b$, that is, $2b$ less than the linear set in Eq. (16.103). For the example in Eq. (a) of Art. 16.2, the order of this set would be $1,080 + 300 - 25 = 1,355$. Since the method will not provide the numerical values of the reactions, we can use Eqs. (16.83) and (16.84) to calculate the reactions after the element forces listed in $\mathbf{s}$ are obtained.

Having computed the element forces and the reactions, we can proceed to compute other forces as in Art. 16.4.1.

16.4.4 Computation of nodal deflections. After the computation of element forces $\mathbf{s}$, it is a relatively easy matter to compute the deflections. However, the computation of deflections requires that the nodal force-equilibrium equations be expressed in the form of Eq. (16.56). This is an especially severe drawback for methods where the Lagrange multipliers are used. From our discussions in Art. 16.4.3 we know that in the Lagrange-multipliers methods we can use the nodal force-equilibrium equations in the form of Eqs. (16.30) or (16.44). In these methods, as far as the computation of forces are concerned, there is no need to apply the expensive Gauss-Jordan elimination to transform Eqs. (16.30) or (16.44) into the form of Eq. (16.56). However, when it comes to the computation of nodal deflections, we must have the nodal force-equilibrium equations in the form of Eq. (16.56). Note that we compute nodal deflections last in the force method of analysis.

In computing the nodal deflections, we first compute the independent deflection components as listed in $\mathbf{u}_{(1)}$ by the theorem of virtual forces. Let $\delta\mathbf{p}^0$ denote virtual forces in the directions of external deflections $\mathbf{u}_{(1)}$, and let the corresponding internal virtual forces be $\delta\mathbf{s}$, which are in the directions of internal deflections $\mathbf{v}$ [see Eq. (16.88)]. We obtain the relationship between the internal virtual forces $\delta\mathbf{s}$ and the external virtual forces $\delta\mathbf{p}^0$ from Eq. (16.56) by taking the variations of both sides and noting that during this process we keep the redundants fixed:

$$\delta\mathbf{s} = \mathscr{B}_0 \, \delta\mathbf{p}^0 \tag{16.109}$$

According to the theorem of virtual forces, work done by internal virtual forces should be equal to the work done by the external virtual forces when these are subjected to the actual internal and external deflections (see Art. 11.4). With this we can write

$$\delta\mathbf{s}^T \mathbf{v} = \delta\mathbf{p}^{0T} \, \mathbf{u}_{(1)} \tag{16.110}$$

Substituting $\delta\mathbf{s}$ from Eq. (16.109) into this equation and noting that the resulting equation is an identity with respect to $\delta\mathbf{p}^0$, we obtain

$$\mathbf{u}_{(1)} = \mathscr{B}_0^T \mathbf{v} \tag{16.111}$$

Since

$$\mathbf{v} = \sum_{m=1}^{M} \mathbf{I}_*^m \mathbf{v}^m \tag{16.112}$$

we can rewrite Eq. (16.111) as

$$\mathbf{u}_{(1)} = \sum_{m=1}^{M} \mathscr{B}_0^T \mathbf{I}_*^m \mathbf{v}^m \tag{16.113}$$

or, using the identity in Eq. (16.75),

$$\mathbf{u}_{(1)} = \sum_{m=1}^{M} \mathscr{B}_0^{mT} \mathbf{v}^m \tag{16.114}$$

Now, substituting $\mathbf{v}^m$ from Eq. (16.86), we obtain

$$\mathbf{u}_{(1)} = \sum_{m=1}^{M} \mathscr{B}_0^{mT} (\mathbf{F}^m \mathbf{s}^m + \mathbf{v}_0^m) \tag{16.115}$$

which expresses the independent deflection components $\mathbf{u}_{(1)}$ in terms of element forces $\mathbf{s}^m$ and element deformations $\mathbf{v}_0^m$ which are due to element loads. We can obtain the complete list of nodal deflections by using the numerical values of $\mathbf{u}_{(1)}$ above in Eq. (15.60).

Note that matrix $\mathscr{B}_0$ explicitly referred to only for the computation of deflections. During the computation of forces, instead of $\mathscr{B}_0$ we need the $\mathscr{B}_0 \mathbf{p}^0$ product [see Eq. (16.73)]. Considering the possibility of deflection computation in Art. 16.3.1, we made sure that the Gauss-Jordan elimination process explicitly yielded the $\mathbf{B}_{(2)}^{0-1}$ matrix, which is the main component of $\mathscr{B}_0$ [see Eq. (16.58)]. If we were interested in only the product of $\mathscr{B}_0 \mathbf{p}^0$, the Gauss-Jordan process would be somewhat simpler.

16.4.5   Problems for Solution

*Problem 16.20   Consider a truss structure consisting of five bars with equal stiffness, that is, AE/L is the same for all bars.   Suppose the external forces in the directions of independent deflections $\mathbf{u}_{(1)}$ are $\mathbf{p}^{0T} = P\,[1\quad 2\quad 3]$ and the force-equilibrium equations of the nodes in these directions are as in Eq. (16.44) with $\mathbf{B}^0$ as in Eq. (*a*) of Art. 16.3.   Solve for the bar forces **s** by the method of stationary total complementary potential energy with substitution.

Problem 16.21   Solve Prob. 16.20 by the method of stationary total complementary potential energy with Lagrange multipliers.

*Problem 16.22   Suppose that the structure in Prob. 16.20 is subjected to a support settlement of u_0 and the reaction in the direction of the settlement can be expressed as

$$p_r = [1\quad -1\quad 2\quad 0\quad 0]\mathbf{s}.$$

Compute the bar forces due to the support settlement by the substitution method of Art. 16.4.1.

Problem 16.23   Solve Prob. 16.22 by the Lagrange-multipliers method.

*Problem 16.24   Suppose the structure in Prob. 16.20 is given such that its matrix $\mathbf{B}^0$ is as in Prob. 16.13.   Solve for the bar forces in shortest possible way.

*Problem 16.25   Suppose in Prob. 16.24 there is a support settlement u_0 in the direction of reaction $p_r = [0\quad 1\quad 0\quad -1\quad 2]\mathbf{s}$.   Solve for the bar forces due to the settlement in the shortest possible way.

*Problem 16.26   Consider the space-framework structure described in Prob. 16.9.   Assuming that the solution of a linear set of equations and the Gauss-Jordan elimination process described in Art. 16.3.1 both require amounts of arithmetic operations proportional to the cube of the row orders, express quantitatively the relative amounts of arithmetic operations for the computation of element forces **s** in substitution methods using Eq. (16.56) and the two Lagrange-multipliers methods using Eqs. (16.30) and (16.44).   Assume that the nodal force-equilibrium equations are given initially as in Eq. (16.30) or (16.44) but not in the form of Eq. (16.56).

Problem 16.27   Repeat Prob. 16.26 for the structure of Prob. 16.10.

Problem 16.28   Repeat Prob. 16.26 for the structure of Prob. 16.11.

*Problem 16.29   What are the rapid-access memory requirements for solving in a computer without auxiliary storage the structure of Prob. 16.9 by the substitution methods for which the nodal force-equilibrium equations are given as in Eq. (16.44) and the two Langrange-multipliers methods.

Problem 16.30   Solve Prob. 16.29 assuming that the structure is the one given in Prob. 16.10.

Problem 16.31   Solve Prob. 16.29 assuming that the structure is the one given in Prob. 16.11.

Problem 16.32   Show that the linear equations (15.59), (16.103), and (16.108) cannot be solved by the Cholesky method without pivot search as described in Art. 15.5.2 if they are arranged by row and column interchanges such that $\boldsymbol{\lambda}$ would be the first partition.   Since the linear equations associated with the Langrange-multiplier methods are not banded and positive definite, we can use a generalized Cholesky algorithm for nonbanded, nonpositive definite, symmetric, real systems with pivot search without destroying the symmetry.   Write down explicitly the computational steps of such a procedure.

*Problem 16.33   Compute the independent deflections $\mathbf{u}_{(1)}$ in Prob. 16.20.   As implied in the statement of the problem, assume that $v_0^m = 0$, $m = 1, \ldots, 5$.

Problem 16.34   Compute the independent deflections $\mathbf{u}_{(1)}$ in Prob. 16.22.

Problem 16.35   Compute the independent deflections $\mathbf{u}_{(1)}$ in Prob. 16.24.

Problem 16.36   Compute the independent deflections $\mathbf{u}_{(1)}$ in Prob. 16.25.

Problem 16.37   Modify the itemized computational steps in Art. 16.3.1 so that instead of $\mathbf{B}_{(2)}^{0-1}$ we obtain only $\mathbf{B}_{(2)}^{0-1}\mathbf{p}^0$.   Let $\mathbf{p}^{0\prime}$ denote the column matrix $\mathbf{B}_{(2)}^{0-1}\mathbf{p}^0$.   Show that

$$\mathscr{B}_0\,\mathbf{p}^0 = \mathbf{P}\!\left(\!\frac{\mathbf{p}^{0\prime}}{\mathbf{0}}\!\right) \tag{16.116}$$

where **P** is the same permutation matrix as that used in Eq. (16.58).

16.5 Discussion of Systematic Analysis by the Force Method. In previous articles of this chapter we detailed the procedures of systematic analysis by the force method. Since the objective of systematic analysis is to obtain accurate and economical solutions by means of digital computers, in retrospect, we may favor certain procedures of the method over other available alternatives. In order to minimize the cost of analysis and maximize its accuracy, we must confine ourselves to smaller-order matrices. Such a criterion dictates our preference for generating the nodal force-equilibrium equations in the form of Eq. (16.44) rather than Eq. (16.30). Therefore we use Eqs. (16.45) and (16.46) to generate $\mathbf{B}^0$ and $\mathbf{p}^0$ matrices systematically from the elemental matrices $\mathbf{j}^m$, $\mathbf{B}^m$, and $\mathbf{q}^{0m}$, $m = 1, \ldots, M$, and matrix $\mathscr{H}$ of the deflection boundary conditions given by Eq. (15.60). With the same criterion and considering that we also have to compute the nodal deflections) we should not use the methods associated with the Lagrange multipliers in obtaining the governing equations for the forces. We should use the substitution methods, which require that we transform the nodal force-equilibrium equations in Eq. (16.44) into the form of Eq. (16.56) by means of the Gauss-Jordan process, which is detailed in Art. 16.3.1 and Eqs. (16.57) and (16.58). In using Eqs. (16.57) and (16.58), we must not waste the rapid-access memory locations by storing $\mathbf{0}$ and $\mathbf{I}$ submatrices; by increasing the programming logic slightly, we can handle $\mathscr{B}$ and $\mathscr{B}_0$ as in Eq. (16.66). We can obtain the governing equations for the redundants as in Eqs. (16.81), where the coefficient matrix $\bar{\mathbf{F}}$ and the right-hand column matrix $\bar{\mathbf{v}}_0$ can be systematically generated by Eqs. (16.77) and (16.78).

16.5.1 Storage requirements. From the outline above, we see that the two main matrices of the force method are $\mathbf{B}^0$ of Eq. (16.45) and $\bar{\mathbf{F}}$ of Eq. (16.77). Matrix $\mathbf{B}^0$ is a rectangular matrix of row order $Ne - b$ and column order $\mathscr{J}$. Initially it is a sparse matrix; however, it is a full matrix at the end of the Gauss-Jordan elimination process. Matrix $\bar{\mathbf{F}}$ is a symmetric positive definite matrix of order $\mathscr{J} - Ne + b$, that is, the difference between the column and the row orders of $\mathbf{B}^0$. The largest single storage required by the force method of analysis is that of matrix $\mathbf{B}^0$. Obtaining the value of $\mathscr{J}$, that is, the column order of $\mathbf{B}^0$, from Eq. (16.28) and denoting the necessary storage area for $\mathbf{B}^0$ by $\mathscr{S}$, we write

$$\mathscr{S} = (Ne - b)\left(\sum_{m=1}^{M} (n^m e - e' - f^m) \right) \tag{16.117}$$

Assuming that b and f^m are negligible and $e = e'$, $n^m = n$ for $m = 1, \ldots, M$, we obtain from Eq. (16.117)

$$\mathscr{S} = NM(n - 1)e^2 \tag{16.118}$$

which states that the *storage area in the force method is proportional to the square of the degrees of freedom per node and increases linearly with the number of nodes, the number of elements, and the number of vertices per element.* To compare the storage requirements of the force and displacement methods, by dividing Eqs. (16.118) and (15.152) side by side, we write

$$\frac{\mathscr{S}}{S} = \frac{M(n-1)}{w} >> 1 \tag{16.119}$$

which states that the storage requirement in the force method is much more than that of the displacement method. As discussed in Art. 15.7.1, w appearing in Eq. (16.119) is usually a fraction of N.

16.5.2   Number of arithmetic operations.   It can be shown that the number of arithmetic operations in the Gauss-Jordan elimination process increases with the square of the row order of $\mathbf{B}^0$ and with the first power of the column order. Since this is much larger than any other step in the force method, we can approximate the number of arithmetic operations in the force method $\mathcal{T}$ by that of the Gauss-Jordan elimination process. Assuming that b and ℓ^m are negligible and $e = e'$ and $n^m = n$, $m = 1, \ldots, M$, we write

$$\mathcal{T} = N^2 M(n - 1)e^3 \tag{16.120}$$

which states that the *number of arithmetic operations in the force method increases with the cube of the number of degrees of freedom per node, with the square of the number of nodes, and linearly with the number of vertices per element.* To compare the total number of arithmetic operations in the force and displacement methods, by dividing Eqs. (16.120) and (15.155) side by side, we write

$$\frac{\mathcal{T}}{T} = \frac{NM(n - 1)}{w^2} > > > 1 \tag{16.121}$$

which states that the number of arithmetic operations in the force method is much more than that of the displacement method. As discussed in Art. 15.7.1, w appearing in Eq. (16.121) is usually a fraction of N.

16.5.3   Other considerations.   In the force method, the matrices related to the structural elements are $\mathbf{F}^m$, $\mathbf{v}_0^m$, $\mathbf{j}^m$, $\mathbf{B}^m$, $\mathbf{q}^{0m}$, and relevant transformation and transfer matrices. In the displacement method, instead of these, we need $\mathcal{K}^m$, $\mathbf{q}_0^m$, $\mathbf{j}^m$, and relevant transformation and transfer matrices. We observe that the force method of analysis requires two more basic elemental matrices than those of the displacement method.

The summation operations involved in defining the matrices $\mathbf{B}^0$, $\mathbf{p}^0$, $\overline{\mathbf{F}}$, $\overline{\mathbf{v}}_0$, $\mathbf{p}^*$, or $\mathbf{p}_r$ in the force method and the matrices $\overline{\mathbf{K}}$, $\overline{\mathbf{p}}_0$, $\mathbf{p}^*$, or $\mathbf{p}_r$ in the displacement method are called **assembly procedures.** We observe that the force method requires two more assembly procedures than the displacement method, i.e., the assembly procedures defined by Eqs. (16.45) and (16.46) for $\mathbf{B}^0$ and $\mathbf{p}^0$.

Considering that the Cholesky method is nothing but an engineered Gauss-elimination process, in the force method we have two Gauss eliminations, one for transforming Eq. (16.44) into the form of Eq. (16.56) and one for obtaining the numerical values of the redundants from Eq. (16.81), whereas in the displacement methods we have only one to obtain the numerical values of the independent deflection components from Eq. (15.44).

From the foregoing discussions we can easily see that from the systematic-analysis standpoint the force method is not as favorable as the displacement method. A quick review of the available structural-analysis software will readily verify this. It should be remembered, however, that during the analysis the force method produces much information pertaining to the redundants which has been, and still is,

important to the design engineer.   For example, given a structure, a good structure
cutter will tell the designer which structural elements are essential and which ones
are redundant in carrying the loads [see matrix $\jmath$ in Eq. (16.66)].   This kind of
information is not available in the displacement method unless one is willing to
repeat the analysis many times.   Considering the relative costs of the two methods
of systematic analysis, it appears that the displacement method of analysis should
be preferred by the majority of designers.

16.5.4 Problems for Solution

*Problem 16.38   For the structure described by Eq. (*a*) in Art. 16.2, compute the first approximation of the necessary storage area in the force method of analysis as multiples of that required
by the displacement method.   Assume $w = 20$.

*Problem 16.39   For the structure of Prob. 16.38, compute the first approximation of the
number of arithmetic operations required in the force method of analysis as multiples of that of
the displacement method.

*Problem 16.40   Consider the following planar-frame structure: $M = 135$, $N = 100$, $e = 3$,
$b = 30$, $e' = 3$, and $f^m = 1$, $n^m = 2$, $m = 1, \ldots, 135$.   First show that this structure is statically determinate if not geometrically unstable.   Next show that if the deflections are also required, the
displacement method of analysis is still preferable over the force method of analysis.   Assume
$w = 25$.

Problem 16.41   Consider the following planar-truss structure: $M = 200$, $N = 120$, $e = 2$,
$e' = 3$, $f^m = 0$, $n^m = 2$, $m = 1, \ldots, 200$.   First show that the structure is statically determinate
if not geometrically unstable.   Then show that if the deflections are also required, the displacement method of analysis is still preferable to the force method of analysis.   Assume $b = 40$ and
$w = 40$.

Problem 16.42   Suppose one has the option of using either triangular or quadrilateral elements
in the force method of analysis of a plane-stress problem with a given number of nodes.   Which
element type is preferable? Why? Give approximate expressions for the ratios of arithmetic
operations and the storage requirements of the two choices.

16.6   References

1. Robinson, John: "Integrated Theory of Finite Element Methods," John Wiley & Sons,
London, 1973.

2. Carnahan, Brice, H. A. Luther, and James O. Wilkes: "Applied Numerical Methods,"
sec. 5.4, John Wiley & Sons Inc., New York, 1969.

Part III

Introduction to Advanced Structural Analysis

Basic principles involved in the analysis of buckling, dynamic response, and the stress distribution in complex elements. Consideration of the analysis of plates and shells and of the plastic behavior of structures.

17

Introduction to Advanced
Structural Mechanics

17.1 Introduction. In general, structures may be built up out of one or more members or panels. A **member** is an element which is relatively long compared with its cross-sectional dimensions, such as a beam or strut. **Panels** may be defined as elements the length and the breadth of which are of the same order of magnitude, but the thickness of which is considerably smaller, such as plates, slabs, and shells.

Elementary theories for the stress distribution in structural members are developed in introductory courses in mechanics and strength of materials. Such theories are adequate to handle many of the conventional structural problems concerning member-type structures. Little consideration is given in most introductory courses to the analysis of plates and shells or to the analysis of buckling or vibrational characteristics of either member or panel elements.

These more advanced aspects are discussed in advanced structural mechanics courses. Here, a very brief introduction to the following areas of advanced structural mechanics is presented.

Theory of elasticity. This subject denotes the area of structural mechanics in which are developed methods of determining the stresses, strains, and displacements in *linearly elastic bodies*. Actually, the fundamentals regarding force equilibrium and the geometry of deformation are applicable to nonlinear elastic or to plastic bodies. The restriction of linear elasticity enters only when the stress-strain relations are introduced. The fundamental relations of the theory of elasticity are applicable to both member- and panel-type elements. They also pertain both to the determination of stress distributions and to the analysis of buckling and dynamic behavior.

Theory of plates and shells. The elastic theory of plates and shells is developed by introducing certain simplifications into the fundamental relations of the theory of elasticity and by adapting these relations to the solution of typical plate and shell problems. The nonlinear elastic theory and the plastic theory of plates and shells are beyond the scope of the present discussion.

Buckling of structural elements. Usually the static equilibrium position of an actual structure or structural element is a *stable* one to which the system returns after being slightly disturbed. Sometimes, however, the equilibrium position is an *unstable* one from which the system continues to move or *buckle* after being slightly disturbed. The stability of the equilibrium position is dependent on the stiffness or rigidity of the system rather than on the strength. Often the buckling of a system can be studied by using the stiffness obtained from approximate load-deflection characteristics defined by elementary theory.

Dynamic response of structures.   The static deflection of a structure is a function only of the applied static loads and the stiffness of the system.   For a dynamic loading, however, the response is also a function of the mass of the system since accelerations are involved.

For detailed discussion of these more advanced subjects, the reader is referred to the books listed on page 6 of the Foreword.

17.2   Basic Definitions and Relationships Involved in the Theory of Elasticity.   The methods of the theory of elasticity are developed for application to a body the material of which is uniformly distributed over a volume of a specific undeformed shape and which is supported and then subsequently loaded and/or deformed in some specific manner.   It is assumed that the material is *homogeneous, isotropic,* and *linearly elastic*—all of which are properties that can be approached but never realized exactly in actuality.   For such a specified problem, it is desired to compute the state of stress, strain, and displacement throughout the body.

For purposes of the mathematical solution of such problems, it is necessary to locate points and portions of the body with respect to some suitable system of coordinate axes.   Although special coordinate systems are more convenient for certain problems, the conventional rectangular coordinate system of x, y, and z axes is the most general and will be used here.

Notation and convention.   The **external forces** acting on a body are of two types, **surface forces** and **body forces.**   Isolation of a portion of a body and consideration of its equilibrium demonstrate the existence of **internal forces** on the exposed internal faces of such a portion.   The notation and convention used for describing external and internal forces are shown in Fig. 17.1.   In all cases, forces are shown in their positive senses.   For isolated differential portions such as shown, the variation of intensity over a differential area (or volume) is so small that the total force on an area (or volume) may be represented as the intensity at the center of the face (or volume) multiplied by the area of the face (or by the volume of the portion).

The surface forces are distributed over the surface of the body and can be resolved into three coordinate components, the intensities (force per unit area) of which are denoted by $\overline{X}$, $\overline{Y}$, and $\overline{Z}$. The body forces are distributed over the volume of the body and may be caused by gravity, magnetism, or acceleration (inertia forces); they can be represented by three coordinate components, the intensities (force per unit volume) of which are denoted by X, Y, Z.   Note that the positive sense of the surface or body-force components is the same as the positive sense of the corresponding coordinate axis.

The resultant internal force distributed over an internal face may be resolved into two components, a **normal** component perpendicular to the face and a **shear** component parallel to the face. Usually this shear component is further resolved into two components acting in the plane of the face. The intensities (force per unit area) of the normal and shear components are called, respectively, the **normal stress** σ and the **shear stress** τ.   Usually the stress condition at a point is defined by specifying the normal and shear stresses on the three coordinate planes at a point, the x, y, or z plane being perpendicular to the x, y, or z coordinate axis, respectively.

One subscript is used with σ to denote the plane on which the normal stress acts; thus σ_y denotes the normal stress on a plane normal to the y axis.   Normal stresses will be considered plus when tensile.

Two subscripts are used with τ, the first to denote the plane on which it acts and the second to denote the direction in which it acts in this plane; thus τ_{xy} denotes the shear stress on the x plane acting parallel to the y axis.   Shear stresses will be considered plus when acting in plus coordinate directions on the most positive face of a portion.

As a result of deformation, points on a body are displaced.   The resultant displacement of a point may be represented by its three coordinate components u, v, and w, which represent components in the x, y, and z directions, respectively.   Displacement components are considered plus when in the positive sense of the corresponding coordinate axes.

Once the displacements of all points are defined, it is easy to compute the strains.   The **lineal strain** of a line element is denoted by ϵ and considered plus when an elongation.   One subscript is used with ϵ to denote the original direction of the element involved.   The change in the right angle between two originally perpendicular line elements is called the **shear strain** between these elements and is denoted by γ and is considered plus when the original right angle between the

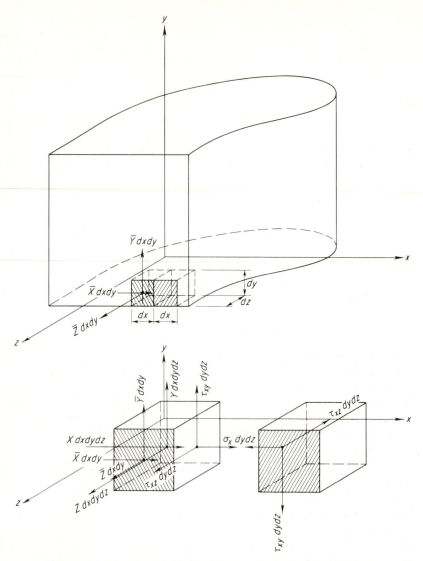

Figure 17.1 Notation and convention for force and stress intensities

positive extension of these line elements is reduced. Two subscripts are used with γ to denote the original direction of these two line elements.

Independent components of stress. It is easy to show that the normal and shear stresses on any plane through a point can be computed if the normal stress and two shear-stress components are known on each of three mutually perpendicular planes (i.e., three orthogonal planes) at a point. It would appear therefore that nine stress components (three on each of three planes) must be known at each point in order to define completely the state of stress at each point. It is easy to show, however, that there are only six independent components of stress at each point.

Usually for convenience the independent components of stress at a point are considered to be those acting on the three coordinate plates at a point. Consider the stresses on three such planes

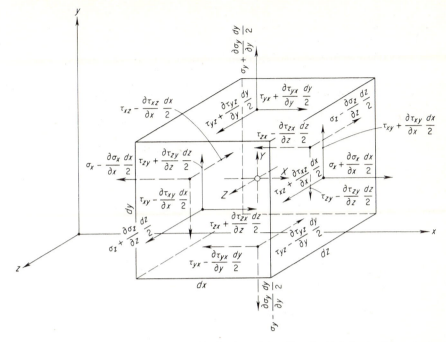

Note: Shown are stresses and body-force intensities. To obtain forces on faces multiply stresses shown by face area; to obtain body forces multiply body-force intensities by particle volume

Figure 17.2 Equilibrium considerations of differential particle

at the center of the differential particle shown in Fig. 17.2 to be σ_x, τ_{xy}, and τ_{xz}; σ_y, τ_{yx}, and τ_{yz}; and σ_z, τ_{zx}, and τ_{zy}. Then, considering the change of stress which occurs between the center of the particle and the center of one of the faces, the stresses at the center of each face are as shown. Suppose now that the equation for moment equilibrium about the x axis, $\sum M_x = 0$, is written. If only the lowest-order differential terms (third order) are retained, this equation becomes

$$\tau_{yz}\, dx\, dy\, dz - \tau_{zy}\, dx\, dy\, dz = 0 \qquad \text{or} \qquad \tau_{yz} = \tau_{zy}$$

Similarly, moment-equilibrium equations about the y axis and about the z axis lead to the conclusions that

$$\tau_{xz} = \tau_{zx} \qquad \text{and} \qquad \tau_{xy} = \tau_{yx}$$

Thus, it is apparent that *there are only six independent components of stress at each point:* three normal stresses, σ_x, σ_y, and σ_z; and three shear stresses, $\tau_{xy}\ (=\tau_{yx})$, $\tau_{xz}\ (=\tau_{zx})$, and $\tau_{yz}\ (=\tau_{zy})$. If these six stress components are known for the coordinate planes at any particular point, it is easy to compute the stresses on any other plane through this point by considering the force equilibrium of a differential tetrahedron at this location formed by the intersection of this plane with three coordinate planes. In this manner, so-called **stress-transformation equations** are obtained. Such equations may be found in any standard textbook on elasticity.

Such considerations can be extended to prove that through any point there are three mutually perpendicular planes on which the resultant stress is normal to the plane; i.e., on which there are no shear stresses. Such planes are called the **principal planes,** and the normal stresses on these planes are called **principal stresses.** One of these principal stresses is the algebraically largest normal stress on any plane through this point and is called the **major principal stress;** one is the algebraically smallest and is called the **minor principal stress;** and the third is an intermediate value.

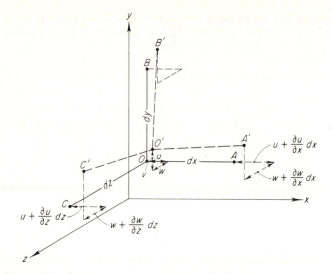

Figure 17.3 Geometrical relations between displacements and strains

It is important to note that all these conclusions regarding stress components are based only on equilibrium considerations and therefore are valid for any material whether it exhibits linear or nonlinear stress-strain properties.

Definition of components of strain. Once the displacements u, v, and w are known at each and every point in a body, it is easy to compute by simple geometrical considerations any aspect of the deformation of the body. Such computations are particularly easy if the displacements are small so that the small-angle assumptions ($\alpha \approx \sin \alpha \approx \tan \alpha$ and $\cos \alpha \approx 1$) are valid and so that squares of the first derivatives of the displacement can be neglected in comparison with the first derivative itself.

For example, as shown in Fig. 17.3, consider three line elements at point O: OA, OB, and OC. After deformation, these elements have moved to the positions $O'A'$, $O'B'$, and $O'C'$, the displacements having been exaggerated tremendously to clarify the geometry involved. Note that, if u, v, and w are the displacements of point O, the displacements of A, B, and C are

Point A:
$$u + \frac{\partial u}{\partial x}\,dx \qquad v + \frac{\partial v}{\partial x}\,dx \qquad \text{and} \qquad w + \frac{\partial w}{\partial x}\,dx$$

Point B:
$$u + \frac{\partial u}{\partial y}\,dy \qquad v + \frac{\partial v}{\partial y}\,dy \qquad \text{and} \qquad w + \frac{\partial w}{\partial y}\,dy$$

Point C:
$$u + \frac{\partial u}{\partial z}\,dz \qquad v + \frac{\partial v}{\partial z}\,dz \qquad \text{and} \qquad w + \frac{\partial w}{\partial z}\,dz$$

In view of the small displacements, substantially the true lengths of and the true angle between elements such as $O'A'$ and $O'C'$ will be seen in the projection of these lines on the xz plane. In such a projection, for example, it will be apparent that $\Delta(OA)$, the change in the length of OA, is substantially

$$\Delta(OA) = O'A' - OA = \frac{\partial u}{\partial x}\,dx$$

Hence, the **unit elongation,** or **lineal strain,** of this element which was originally parallel to the x axis is called ϵ_x and is equal to $\partial u/\partial x$. In a similar manner, $\Delta(\angle AOC)$, the change in the angle between the originally perpendicular elements OA and OC, is substantially

$$\Delta(\angle AOC) = \angle AOC - \angle A'O'C' = \frac{\partial u}{\partial z} + \frac{\partial w}{\partial x}$$

This reduction in the original right angle between these two elements which were originally parallel to the x and z axis is called **shear strain** γ_{xz}. Thus, the six strain components at point O (three lineal strains and three shear strains) are found in this manner to be related to the three displacements u, v, and w by the following expressions:

$$\epsilon_x = \frac{\partial u}{\partial x} \qquad \gamma_{xy} = \frac{\partial u}{\partial y} + \frac{\partial v}{\partial x}$$

$$\epsilon_y = \frac{\partial v}{\partial y} \qquad \gamma_{xz} = \frac{\partial u}{\partial z} + \frac{\partial w}{\partial x} \qquad (17.1)$$

$$\epsilon_z = \frac{\partial w}{\partial z} \qquad \gamma_{yz} = \frac{\partial v}{\partial z} + \frac{\partial w}{\partial y}$$

Note that these six components of strain are not independent since they are expressed in terms of only three quantities, u, v, and w.

This interdependency of the six strain components is very important. In the terminology of elasticity literature, it is said that *the strain must be compatible.* The test for compatibility is that the six strain components must satisfy six compatibility equations, three of the type

$$\frac{\partial^2 \epsilon_x}{\partial y^2} + \frac{\partial^2 \epsilon_y}{\partial x^2} = \frac{\partial^2 \gamma_{xy}}{\partial x \, \partial y} \qquad (17.2a)$$

and three of the type $\qquad 2\dfrac{\partial^2 \epsilon_x}{\partial y \, \partial z} = \dfrac{\partial}{\partial x}\left(-\dfrac{\partial \gamma_{yz}}{\partial x} + \dfrac{\partial \gamma_{xz}}{\partial y} + \dfrac{\partial \gamma_{xy}}{\partial z}\right) \qquad (17.2b)$

The validity of these indentities can be verified by substitution from Eq. (17.1). Actually, the satisfaction of these six compatibility equations is the unique characteristic of an exact elasticity solution for a given problem. *There are an infinite number of stress distributions that satisfy the statical requirements and the boundary-force requirements, but there is only one that also satisfies the strain compatibility equations and the boundary-displacement requirements.*

Once the six strain components have been computed at a point for three elements originally parallel to x, y, and z axes, it is easy to compute the six strain components with respect to any other three orthogonal directions at the point. This may be done using the **strain-transformation equations** found in standard elasticity textbooks.

As in the case of stresses, it can be shown that there are at a point three directions which were mutually perpendicular before deformation and which remain so after deformation; i.e., there are three mutually perpendicular axes between which there is no shear strain. These three directions are called the **axes of principal strain.** The algebraically largest of these three lineal strains is called the **major principal strain;** and the algebraically smallest, the **minor principal strain.**

Note that these considerations regarding strain are limited only by the geometrical approximations dependent on small displacements and they are therefore valid for a material of any stress-strain characteristics. Also, note that if the material is homogeneous and isotropic, the directions of the axes of principal strain and the directions of the principal stresses are coincident.

Differential equations of equilibrium. In the previous discussion of stress, only the relations of stress at a point were considered. The variation of stress between adjacent points must now be considered to establish the equilibrium requirements between the rates of change of the six stress components.   Consider the particle shown in Fig. 17.2, and first write the equation for force equilibrium in the x direction, $\sum F_x = 0$.   This equation simplifies to

$$\frac{\partial \sigma_x}{\partial x} + \frac{\partial \tau_{yx}}{\partial y} + \frac{\partial \tau_{zx}}{\partial z} + X = 0$$

and $\sum F_y = 0$ to

$$\frac{\partial \tau_{xy}}{\partial x} + \frac{\partial \sigma_y}{\partial y} + \frac{\partial \tau_{zy}}{\partial z} + Y = 0 \qquad (17.3)$$

and $\sum F_z = 0$ to

$$\frac{\partial \tau_{xz}}{\partial x} + \frac{\partial \tau_{yz}}{\partial y} + \frac{\partial \sigma_z}{\partial z} + Z = 0$$

These three equations, called the **differential equations of equilibrium,** must be satisfied at all points throughout the volume of a body.   The solution for the stresses which satisfies Eq. (17.3) is also required to be in equilibrium with the specified forces at all points on the boundary of the body.

Linearly elastic stress-displacement equations.   So far no specific stress-strain relations have been introduced.   Such relations can be discussed in a rigorous mathematical manner.   The following, however, will be presented simply on the basis of experimental evidence regarding *linearly elastic, homogeneous, and isotropic materials*.   Such evidence shows that a normal stress such as σ_x produces a lineal strain of σ_x/E in the x direction, lineal strains of $-v\sigma_x/E$ in both the y and z directions, and no shear strains between x and y, x and z, or y and z axes.   Likewise, a shear stress such as τ_{xy} produces only a shear strain γ_{xy} but no other strains regarding the x, y, and z elements.   Thus, the following six stress-strain relations can be written by superimposing the contributions of all six strain components:

$$\epsilon_x = \frac{1}{E}[\sigma_x - v(\sigma_y + \sigma_z)] \qquad \gamma_{xy} = \frac{\tau_{xy}}{G}$$

$$\epsilon_y = \frac{1}{E}[\sigma_y - v(\sigma_x + \sigma_z)] \qquad \gamma_{xz} = \frac{\tau_{xz}}{G} \qquad (17.4)$$

$$\epsilon_z = \frac{1}{E}[\sigma_z - v(\sigma_x + \sigma_y)] \qquad \gamma_{yz} = \frac{\tau_{yz}}{G}$$

where E = Young's modulus of elasticity in tension and compression
v = Poisson's ratio
$G = E/2(1 + v)$ = modulus of elasticity in shear

The above stress-strain relations can be converted into the desired stress-displacement equations by substitution from Eqs. (17.1):

$$\frac{\partial u}{\partial x} = \frac{1}{E}[\sigma_x - v(\sigma_y + \sigma_z)] \qquad \frac{\partial u}{\partial y} + \frac{\partial v}{\partial x} = \frac{\tau_{xy}}{G}$$

$$\frac{\partial v}{\partial y} = \frac{1}{E}[\sigma_y - v(\sigma_x + \sigma_z)] \qquad \frac{\partial u}{\partial z} + \frac{\partial w}{\partial x} = \frac{\tau_{xz}}{G} \qquad (17.5)$$

$$\frac{\partial w}{\partial z} = \frac{1}{E}[\sigma_z - v(\sigma_x + \sigma_y)] \qquad \frac{\partial v}{\partial z} + \frac{\partial w}{\partial y} = \frac{\tau_{yz}}{G}$$

17.3   Formulation of the Solution of the Elasticity Problem.   Now that the basic definitions and relations of the theory of elasticity are available, it is possible to formulate the method of solution of typical problems.   Given in such problems are the geometry and material characteristics of the body involved, together with the definition of the forces on, and/or displacements of, the boundary surface of this body.   Required are the stresses and displacements throughout the volume of the body.   *As unknowns, therefore, there are nine unknown functions of x, y, and z.   Six of these* functions (σ_x, σ_y, σ_z, τ_{xy}, τ_{xz}, and τ_{yz}) *define the six independent stress components throughout the body, and three of these* (u, v, and w) *define the three displacement components.*

Available to determine these nine unknown functions are *nine partial differential equations, three of which are the different equations of equilibrium* (17.3) *and six of which are the stress-displacement equations* (17.5). *The nine unknown functions must satisfy not only these nine equations but also the force and/or displacement conditions on the boundary in order to be the true elasticity solution to a given problem.*

Note that, in general, the elasticity problem is a *statically indeterminate* one. There are only three equations of static equilibrium, which alone are insufficient to determine the six stress components involved in these equations. Only when the six stress-displacement equations are also considered can a unique solution be obtained.

Elasticity problems may be categorized as being one of three types, depending on the nature of the boundary conditions:

Type I   To determine the stress distribution and displacements when the body forces and boundary forces are prescribed and when there are no prescribed boundary displacements. *Example:* Simple end-supported beam.

Type II   To determine the stress distribution and displacements when the body forces are prescribed and when the boundary forces are prescribed except at discrete points where, instead, displacements are prescribed.   *Example:* Two-span continuous beam.

Type III   To determine the stress distribution and displacements when the body forces are prescribed and when the displacements are prescribed over the entire boundary.

Civil engineering problems usually belong to the type I or II.   Further, by using the familiar primary-structure technique described in Chap. 9, type II problems can be converted into several type I problems.

To solve type I problems, the nine partial differential equations are condensed to six by eliminating u, v, and w from the stress-displacement equations. This procedure results in six equations which are exactly the same as those which would be obtained by substituting for the strains in the compatibility equations (17.2) from the stress-strain relations (17.4). These six equations will involve only the six independent stress components as the unknowns. When the stress components have been determined so as to satisfy these six equations and the boundary-force conditions, these values may be substituted into Eqs. (17.5) and the displacements u, v, and w determined from the resulting equations.

The reader is referred to standard elasticity textbooks for solutions to typical problems. Actually, relatively few problems have been solved, and those which have are relatively simple compared with many actual practical problems. Nevertheless, these solutions have been particularly useful as a means of improving our understanding and knowledge of the behavior of structures and structural members.

17.4 Small-Deflection Theory of Flat Plates.   In the Foreword, the behavior of a plate is visualized as being analogous to that of two layers of beam strips. One approximate method of computing plate behavior is based on this analogy and involves approximating a plate by a rectangular grid of beams. The analysis involves dividing the applied load between the two systems of intersecting beams so that the deflection and slope of each system are identical at all intersection points of the grid. In the following discussion, however, the classical theory for analyzing the behavior of the continuum of a flat plate is summarized.

Plates are subjected primarily to loadings that act transversely to the original middle plane of the plate.   Most commonly the resulting transverse bending deflections of the plate are small and can be evaluated by a so-called **small-deflection theory.**   According to this theory, even when the *transverse loading* is accompanied by *middle-plane loads* (which act parallel to the original position of the middle plane of the plate), only the transverse loading contributes to the transverse bending of the plate.   This is comparable, in cases of beam struts that are subjected simultaneously to transverse and axial loading, to assuming that the transverse bending of the beam struts is caused only by the transverse loading and is independent of the axial loading.

For more flexible plates, bending deflections increase and the modification of the transverse bending by the middle-plane loading becomes relatively more important. A more refined theory that includes this modification in the computation of the transverse bending of a plate is called **approximate large-deflection theory.** This is comparable to computations for slender beam struts that include the bending moment contributed by the axial forces acting with lever arms which depend on the transverse bending deflections of the member.

The most accurate form of plate theory is called **large-deflection theory.** In this theory, not only is the effect of the middle-plane loading on the transverse bending considered, but in addition the following refinements are evaluated. As a plate bends, the edges of the plate tend to move parallel to its original plane. If the edges are restrained against such movement, middle-plane reactions (usually called **membrane reactions**) are developed. Moreover, plates usually are bent into nondevelopable surfaces, and as a result further modification of the middle-plane (or membrane) stresses are induced. In large-deflection theory, therefore, the complete interaction of membrane and transverse bending behavior is considered. It is a nonlinear theory and difficult to apply, but it must be considered for the thin plates undergoing large transverse bending deflection in the presence of large membrane forces. Such plates deflect so much that they are behaving much more like shallow shells than flat plates; and, in the latter stages of such behavior, the loading is primarily carried by membrane forces.

Simplifying assumptions and approximate elasticity relations. The following limitations are imposed on the ensuing discussion: (1) the thickness of the plate h is very small compared with the length and breadth; (2) if the thickness is not constant, the rate of change of thickness is very small; and (3) the transverse bending deflections are small compared with the thickness of the plate.

Actually, the problem of finding the stresses and deflections of a flat plate is a three-dimensional elasticity problem. By introducing certain valid approximations, it is possible in the case of **small-deflection theory** to separate the three-dimensional problem into two two-dimensional elasticity problems, each of which gives a reasonably good solution for the unknowns involved. These approximations may be justified by considering the order of magnitude of the six independent stress components involved.

Consider a flat plate such as shown in Fig. 17.4a. The six stresses may be grouped into three classifications: the transverse normal stress σ_z; the transverse shear stresses τ_{xz} and τ_{yz}; and the stresses parallel to the middle plane of the plate σ_x, σ_y, and τ_{xy}.

Note that σ_z is of the order of magnitude of q, where q is usually in the range of 1 to 10 psi, occasionally 10 to 100 psi, and rarely in the range of hundreds of pounds per square inch. The total transverse load on the plate is of the order of qL^2. For transverse equilibrium, this load is balanced by transverse shearing forces of the order of $\tau_{xz}Lh$ or $\tau_{yz}Lh$. Therefore, τ_{xz} and τ_{yz} are

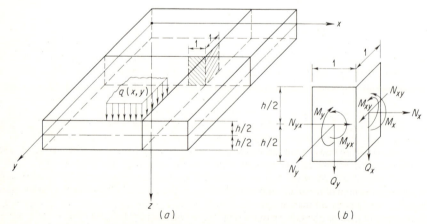

Figure 17.4 Thin plate notation and convention

of the order of $(q)(L/h)$. If a bending of a strip of the plate of unit width is considered, the bending moment is of the order of qL^2, whereas the resisting moment is of the order of $\sigma_x h^2$ or $\sigma_y h^2$. Therefore, σ_x or σ_y (and, it is assumed, τ_{xy} also) is of the order of $(q)(L/h)^2$. Thus, since L/h is relatively large for thin plates, σ_x, σ_y, and τ_{xy} are greater than τ_{xz} and τ_{yz} and much greater than σ_z.

Since τ_{xz}, τ_{yz}, and σ_z are relatively small, their effect on the deflections of the plate will be neglected. As a result, the exact stress-displacement equations, Eqs. (17.5), will be approximated by the following approximate stress-displacement equations for thin plates:

$$\frac{\partial u}{\partial x} = \frac{1}{E}(\sigma_x - \nu\sigma_y) \qquad \frac{\partial u}{\partial y} + \frac{\partial v}{\partial x} = \frac{\tau_{xy}}{G}$$

$$\frac{\partial v}{\partial y} = \frac{1}{E}(\sigma_y - \nu\sigma_x) \qquad \frac{\partial u}{\partial z} + \frac{\partial w}{\partial x} = 0 \qquad (17.6)$$

$$\frac{\partial w}{\partial z} = 0 \qquad \frac{\partial v}{\partial z} + \frac{\partial w}{\partial y} = 0$$

Whereas τ_{xz}, τ_{yz}, and σ_z are being neglected as far as deformations are concerned, *all six stress components will be retained for any equilibrium considerations.* Note that this is comparable to elementary beam theory wherein transverse shearing stresses are considered in equilibrium considerations but are usually neglected in the computation of bending deflections.

Notation and convention. When considering the equilibrium of a plate, it is convenient to isolate portions by passing transverse cutting planes or surfaces. The stresses on such surfaces, σ_x, σ_y, τ_{xy}, τ_{xz}, and τ_{yz}, vary over the thickness of the plate, of course. To avoid having to deal with these distributed stresses, it is convenient to represent them by their resultant forces and couples. For this purpose, it is convenient to introduce the definitions of the **stress resultants** N_x, N_y, N_{xy} $(=N_{yx})$, Q_x, and Q_y and the **stress couples** M_x, M_y, and M_{xy} $(=-M_{yx})$, as shown in their positive sense in Fig. 17.4b and as defined mathematically in Eqs. (17.7):

$$N_x = \int_{-h/2}^{+h/2} \sigma_x\,dz \qquad\qquad N_y = \int_{-h/2}^{+h/2} \sigma_y\,dz$$

$$N_{xy} = \int_{-h/2}^{+h/2} \tau_{xy}\,dz \quad = \quad N_{yx} = \int_{-h/2}^{+h/2} \tau_{yx}\,dz$$

$$Q_x = \int_{-h/2}^{+h/2} \tau_{xz}\,dz \qquad\qquad Q_y = \int_{-h/2}^{+h/2} \tau_{yz}\,dz \qquad (17.7)$$

$$M_x = \int_{-h/2}^{+h/2} \sigma_x z\,dz \qquad\qquad M_y = \int_{-h/2}^{+h/2} \sigma_y z\,dz$$

$$M_{xy} = -\int_{-h/2}^{+h/2} \tau_{xy} z\,dz = -M_{yx} = -\int_{-h/2}^{+h/2} \tau_{yx} z\,dz$$

In other words, these stress resultants and stress couples are, respectively, the force per unit length (pounds per inch) and the couple per unit length (inch-pounds per inch) supplied by the stresses involved. Note $N_{xy} = N_{yx}$ and $M_{xy} = -M_{yx}$ in view of the fact that $\tau_{yx} = \tau_{xy}$ and because of the sign convention adopted for all of these quantities. Note that each of the stress couples and stress resultants is defined by a function only of x and y.

The displacements of points lying in the middle surface of the plate in the x, y, or z direction are denoted by u_o, v_o, and w_o, respectively, all of which are functions only of x and y.

Formulation of small-deflection theory. If the stress condition in a plate is represented by the stress resultants and stress couples, there will be a total of 11 independent unknowns required in order to define completely the stress and displacement condition of a plate:

Five stress resultants: N_x, N_y, N_{xy} $(=N_{yx})$, Q_x, and Q_y
Three stress couples: M_x, M_y, and M_{xy} $(=-M_{yx})$
Three displacements: u_o, v_o, and w_o

In order to solve for these unknowns, eleven equations are required. As shown below, there are six independent stress-displacement equations and five independent equilibrium equations available.

According to the third of Eqs. (17.6), w is independent of z. Therefore, at any location (x, y) the value of w is the same regardless of the value of z and equal to w_o, the value of w at the middle surface. Then, by integrating the last two of Eqs. (17.6),

$$u(x, y, z) = u_o(x, y) - z \frac{\partial w_o}{\partial x} (x, y)$$

(17.8)

and

$$v(x, y, z) = v_o(x, y) - z \frac{\partial w_o}{\partial y} (x, y)$$

As shown by Fig. 17.5a, these equations are equivalent to stating that lines perpendicular to the middle surface before bending of the plate will remain perpendicular to it after bending (this is called the Bernoulli-Navier hypothesis). Equations (17.8) may be substituted into the first, second, and fourth of Eqs. (17.6) to obtain

$$\sigma_x = \frac{E}{1 - v^2} \left[\left(\frac{\partial u_o}{\partial x} + v \frac{\partial v_o}{\partial y} \right) - z \left(\frac{\partial^2 w_o}{\partial x^2} + v \frac{\partial^2 w_o}{\partial y^2} \right) \right]$$

$$\sigma_y = \frac{E}{1 - v^2} \left[\left(\frac{\partial v_o}{\partial y} + v \frac{\partial u_o}{\partial x} \right) - z \left(\frac{\partial^2 w_o}{\partial y^2} + v \frac{\partial^2 w_o}{\partial x^2} \right) \right]$$

(17.9)

$$\tau_{xy} = G \left(\frac{\partial u_o}{\partial y} + \frac{\partial v_o}{\partial x} - 2z \frac{\partial^2 w_o}{\partial x \, \partial y} \right)$$

The six stress-displacement equations can then be obtained by substituting from Eqs. (17.9) into the expressions of Eqs. (17.7) for the N's and the M's:

$$N_x = \frac{Eh}{1 - v^2} \left(\frac{\partial u_o}{\partial x} + v \frac{\partial v_o}{\partial y} \right)$$

$$N_y = \frac{Eh}{1 - v^2} \left(\frac{\partial v_o}{\partial y} + v \frac{\partial u_o}{\partial x} \right)$$

(17.10)

$$N_{xy} = N_{yx} = Gh \left(\frac{\partial u_o}{\partial y} + \frac{\partial v_o}{\partial x} \right)$$

and

$$M_x = - \frac{Eh^3}{12(1 - v^2)} \left(\frac{\partial^2 w_o}{\partial x^2} + v \frac{\partial^2 w_o}{\partial y^2} \right)$$

$$M_y = - \frac{Eh^3}{12(1 - v^2)} \left(\frac{\partial^2 w_o}{\partial y^2} + v \frac{\partial^2 w_o}{\partial x^2} \right)$$

(17.11)

$$M_{xy} = - M_{yx} = \frac{2Gh^3}{12} \frac{\partial^2 w_o}{\partial x \, \partial y}$$

The required differential equations of equilibrium can be obtained by considering the equilibrium of the differential portion of the plate shown in Fig. 17.5b. Considering that N_x, N_y, M_x, etc., represent the stress couples and stress resultants on x and y planes at the center of the element, the forces and couples acting on the positive faces of the element will be as shown. In addition, but not shown, are similar forces and couples on the negative faces, such as $\left(N_x - \dfrac{\partial N_x}{\partial x}\dfrac{dx}{2}\right) dy$, etc. For equilibrium of this portion, the following equations must be satisfied:

$$\sum F_x = 0 \qquad \frac{\partial N_x}{\partial x} + \frac{\partial N_{yx}}{\partial y} = 0$$

$$\sum F_y = 0 \qquad \frac{\partial N_{xy}}{\partial x} + \frac{\partial N_y}{\partial y} = 0 \tag{17.12}$$

$$\sum F_z = 0 \qquad \frac{\partial Q_x}{\partial x} + \frac{\partial Q_y}{\partial y} + q = 0$$

$$\sum M_y = 0 \qquad \frac{\partial M_x}{\partial x} + \frac{\partial M_{yx}}{\partial y} - Q_x = 0 \tag{17.13}$$

$$\sum M_x = 0 \qquad -\frac{\partial M_{xy}}{\partial x} + \frac{\partial M_y}{\partial y} - Q_y = 0$$

Note that $\sum M_z = 0$ results simply in $0 = 0$ since $N_{xy} = N_{yx}$ and does not yield a usable equation.

Consideration of the 11 equations from Eqs. (17.10) to (17.13) shows that they separate into two groups: (1) Eqs. (17.10) and (17.12) containing N_x, N_y, N_{xy} ($= N_{yx}$), u_o, and v_o; (2) Eqs. (17.11) and (17.13) containing M_x, M_y, M_{xy} ($= -M_{yx}$), Q_x, Q_y, and w_o. Thus, the small-deflection theory for thin plates separates the solution into two parts: the first of which is called the **membrane problem** involving the **membrane stress resultants** N_x, N_y, and N_{xy} and the **membrane displacements** u_o and v_o, which will not be considered further; and the second of which is called the **transverse-bending problem**, which is discussed below.

Transverse-bending problem. In the following discussion of the transverse-bending portion of small-deflection plate theory, it is assumed that the thickness h of the plate is constant. The symbol D is used to designate the flexural rigidity of the plate:

$$D = \frac{Eh^3}{12(1 - \nu^2)} \tag{17.14}$$

The six unknowns involved in the transverse-bending problem will be designated as

$$w_o = \textbf{transverse-bending deflection}$$
$$M_x, M_y = \textbf{bending stress-couples}$$
$$M_{xy} (= -M_{yx}) = \textbf{twisting stress-couples}$$
$$Q_x, Q_y = \textbf{transverse shear stress-resultants}$$

Available for the solution of these unknowns are the three differential equations of equilibrium, Eqs. (17.13), and the three stress-displacement equations, Eqs. (17.11).

These six equations can be consolidated into one partial differential equation involving only w_o as the unknown. Substitute for Q_x and Q_y in the first of Eqs. (17.13) from the last two of these equations. Then substitute in the resulting equation from Eqs. (17.11) and obtain

$$D\left(\frac{\partial^4 w_o}{\partial x^4} + 2\frac{\partial^4 w_o}{\partial x^2\,\partial y^2} + \frac{\partial^4 w_o}{\partial y^4}\right) = q(x, y) \tag{17.15}$$

This is the equation which governs the transverse-bending deflection of the plate. If w_o can be determined so as to satisfy Eq. (17.15) and the boundary conditions concerning w_o, then w_o can be

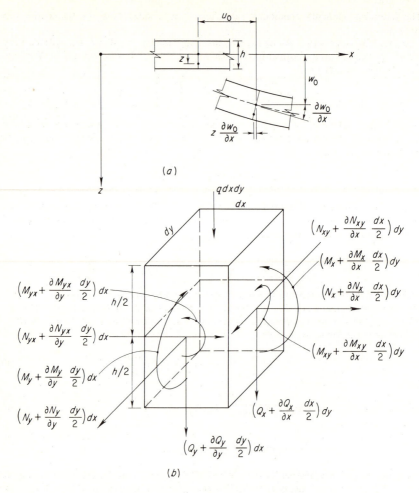

Figure 17.5 Small deflection theory for thin plates

substituted into Eqs. (17.11) to obtain M_x, M_y, and M_{xy} ($= -M_{yx}$). Subsequently, these stress-couples can be substituted into the last two of Eqs. (17.13) to obtain Q_x and Q_y.†

Once the stress-couples and stress-resultants are known, it is easy to compute the stress intensities associated with the transverse bending of the plate, using the expressions

$$\sigma_x = \frac{12z}{h^3} M_x \qquad \sigma_y = \frac{12z}{h^3} M_y \qquad \tau_{xy} = -\frac{12z}{h^3} M_{xy}$$

$$\tau_{xz} = \frac{3}{2} \frac{Q_x}{h} \left(1 - 4\frac{z^2}{h^2}\right) \qquad \text{and} \qquad \tau_{yz} = \frac{3}{2} \frac{Q_y}{h} \left(1 - 4\frac{z^2}{h^2}\right)$$

† Note that this proposed method of analysis is one of the type described in Chap. 15 as displacement methods of analysis. Note also the similarity between Eq. (17.15) and the differential equation for the elastic curve of an ordinary beam.

The following boundary conditions regarding w_o on various types of boundaries can be developed.

Fixed (or Clamped) Edge. Along such an edge, both the transverse bending deflection and the slope of the middle surface normal to the edge must be zero. For example, if the fixed edge is parallel to the x axis at $y = b$,

$$[w_o]_{y=b} = 0 \quad \text{and} \quad \left[\frac{\partial w_o}{\partial y} \right]_{y=b} = 0$$

Simply Supported Edge. Along such an edge, both the transverse bending deflection and the bending stress-couple normal to the edge are zero, which results in the following conditions for an edge parallel to the x axis at $y = b$:

$$[w_o]_{y=b} = 0 \quad \text{and} \quad \left[\frac{\partial^2 w_o}{\partial y^2} \right]_{y=b} = 0$$

Free Edge. Such an edge must be entirely free of bending or twisting stress-couples and of transverse shear stress-resultants. It can be shown that such conditions are substantially obtained if, along an edge $x = a$, for example,

$$\left[\frac{\partial^3 w_o}{\partial x^3} + (2 - \nu) \frac{\partial^3 w_o}{\partial x\, \partial y^2} \right]_{x=a} = 0 \quad \text{and} \quad \left[\frac{\partial^2 w_o}{\partial x^2} + \nu \frac{\partial^2 w_o}{\partial y^2} \right]_{x=a} = 0$$

Solution of transverse-bending problems. Application of the above theory to various plate problems will be found in standard textbooks on plates and shells and will not be discussed here. For example, the solution for a rectangular plate simply supported on all four edges as shown in Fig. 17.6 would be found to be given by the following Fourier series:

$$w_o = \sum_{m=1}^{\infty} \sum_{n=1}^{\infty} A_{mn} \sin \frac{m\pi x}{a} \sin \frac{n\pi y}{b} \tag{17.16}$$

where

$$A_{mn} = \frac{P_{mn}}{\pi^4 D[(m/a)^2 + (n/b)^2]^2} \tag{17.17}$$

and

$$P_{mn} = \frac{4}{ab} \int_0^a \int_0^b q(x, y) \sin \frac{m\pi x}{a} \sin \frac{n\pi y}{b} \, dx \, dy \tag{17.18}$$

from which P_{mn} can be evaluated for any specified transverse loading $q(x, y)$. Although this solution is not derived here, it is easy to verify that the expression for w_o given by Eq. (17.16) satisfies Eq. (17.15) and the boundary conditions for the simply supported edges along all four sides of this rectangular plate.

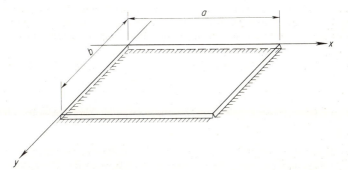

Figure 17.6 Rectangular plate simply supported on all four edges

17.5 Large-Deflection Theory of Plates. As the transverse bending deflections become larger and larger, the contribution of the middle-plane forces to the transverse bending becomes more and more significant.   This effect may be accounted for by considering the equilibrium of the element in Fig. 17.5b in its deflected, rather than its undeflected, position.   As a result of the displacement of this particle, the side faces and the stress-resultants and stress-couples acting on them are rotated through small angles.   If these slight changes in direction are considered in rewriting the equilibrium equations, it is found that only the first of Eqs. (17.13) is revised significantly and becomes

$$\frac{\partial Q_x}{\partial x} + \frac{\partial Q_y}{\partial y} + q + N_x \frac{\partial^2 w_o}{\partial x^2} + N_y \frac{\partial^2 w_o}{\partial y^2} + 2N_{xy} \frac{\partial^2 w_o}{\partial x \, \partial y} = 0 \qquad (17.19)$$

The last two of Eqs. (17.13) and Eqs. (17.12) remain unchanged.

In addition to this modification, large transverse-bending deflections have an effect on the strains of the middle surface of the plate.   To evaluate this effect refer to Fig. 17.3, and suppose that in this figure the displacements v in the y direction were much larger than the displacements in the xz plane.   Such large displacements would now modify the strain of fibers OA and OC, whereas previously, when these displacements were small, it was found that they had only a second-order effect which was legitimately neglected.   If such effects were included in the plate theory derived above, the stress-displacement equations, Eqs. (17.10), would become

$$N_x = \frac{Eh}{1 - \nu^2} \left[\frac{\partial u_o}{\partial x} + \frac{1}{2}\left(\frac{\partial w_o}{\partial x}\right)^2 + \nu \frac{\partial v_o}{\partial y} + \frac{\nu}{2}\left(\frac{\partial w_o}{\partial y}\right)^2 \right]$$

$$N_y = \frac{Eh}{1 - \nu^2} \left[\frac{\partial v_o}{\partial y} + \frac{1}{2}\left(\frac{\partial w_o}{\partial y}\right)^2 + \nu \frac{\partial u_o}{\partial x} + \frac{\nu}{2}\left(\frac{\partial w_o}{\partial x}\right)^2 \right] \qquad (17.20)$$

$$N_{xy} = N_{yx} = Gh\left(\frac{\partial u_o}{\partial y} + \frac{\partial v_o}{\partial x} + \frac{\partial w_o}{\partial x} \frac{\partial w_o}{\partial y} \right)$$

The remaining three stress-displacement equations, *Eqs.* (17.11) would remain unchanged, however.

When these refinements are introduced to obtain large-deflection theory, it becomes apparent that the solution no longer separates into the two independent parts, the membrane problem and the transverse-bending problem, as it does in small-deflection plate theory.   There are still 11 equations involving 11 unknowns, but the solution of these equations is very difficult.   A few simple problems have been solved by large-deflection theory.   The reader is referred to the literature for further discussion of this subject.[1]

17.6 Analysis of Thin Shells. In the Foreword, the structural behavior of shells was described in a qualitative manner.   In the following discussion, fundamentals of plate theory are extended to include shells.   This brief presentation covers little more than an introduction to the classical theory of shells and is intended to assist the reader in starting his studies of the standard textbooks and the extensive periodical literature devoted to the analysis and design of shell structures.

Limitations, assumptions, notation, and convention. In the following discussion, it will be assumed that h, the thickness of the shell, is small compared with the radii of curvature and the span of the shell.   Essentially the same assumptions are used for thin shells as are used for thin plates, namely:

1. Normal stresses perpendicular to the middle surface of the shell are disregarded.

2. Contributions of the radial shear stresses to the deformation and deflection of the shell are neglected.

3. Points on lines normal to the middle surface of the shell before deformation remain on lines normal to the middle surface after deformation.

---

[1] See, for example, S. Timoshenko and S. Woinowsky-Krieger, "Theory of Plates and Shells," 2nd ed., McGraw-Hill Book Company, New York, 1959.

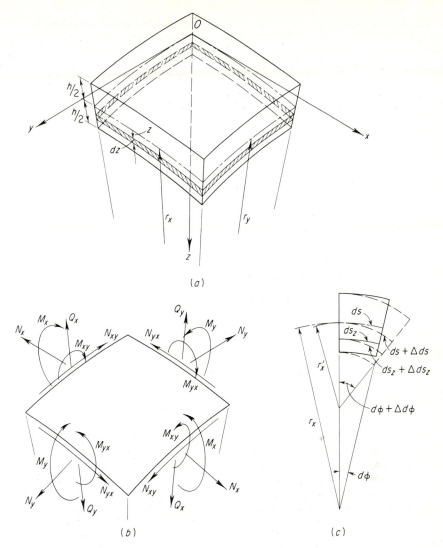

Figure 17.7 Thin shell notation

Basically the same definitions and notation as used for plates are used for thin shells.

Strains, stresses, and displacements at a given location in the shell are referenced to orthogonal coordinate axes with an origin O at a point lying on the middle surface. The x and y axes are taken coincident with the tangents to the lines of curvature of the middle surface at point O, and the z axis is taken coincident with the normal to the middle surface. The principal radii of curvature of the surface in the xoz and yoz planes are called r_x and r_y, respectively. Refer to Fig. 17.7a.

The internal stresses acting on a differential shell element such as shown in Fig. 17.7a can be represented by a system of equivalent stress-resultants and stress-couples which is similar to that used for plates. For a small element of a shell, the exposed side faces, because of the initial curvature of the shell, are of a trapezoidal form rather than rectangular as for plates. For thin shells, however, *where h is very small compared with r_x or r_y*, the effect of this trapezoidal form may be neglected and the stress-couples and stress-resultants may be defined by Eqs. (17.7), the same definition as used for plates.   These definitions are in terms of per unit length of the side faces *at the middle surface of the shell*.   The stress-couples and stress-resultants are shown in their positive senses in Fig. 17.7*b*.

The definition of strains in a *thin shell* can be developed by considering the deformation of a small element of the shell such as shown in Fig. 17.7*c*, where lines normal to the middle surface are considered to be so both before and after deformation.   Without considering the details of the derivation, the strains ϵ_x, ϵ_y, and γ_{xy} for x and y elements lying at a distance z from the middle surface may be defined as

$$\epsilon_x = \epsilon_{xo} - z\chi_x$$
$$\epsilon_y = \epsilon_{yo} - z\chi_y \tag{17.21}$$
$$\gamma_{xy} = \gamma_{xyo} - 2z\chi_{xy}$$

where χ_x = change of curvature of middle surface in *xoz* plane
χ_y = change of curvature of middle surface in *yoz* plane
χ_{xy} = change of twist of middle surface

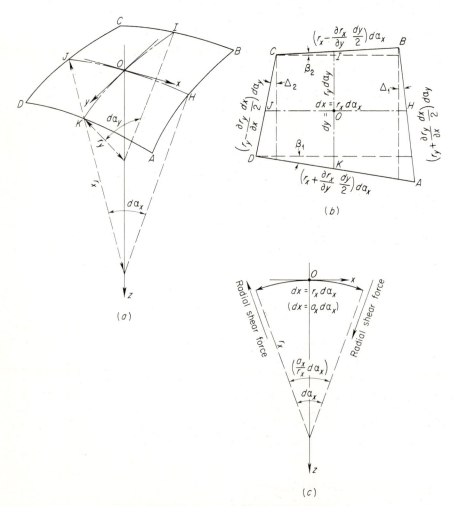

Figure 17.8   Geometry of differential shell element

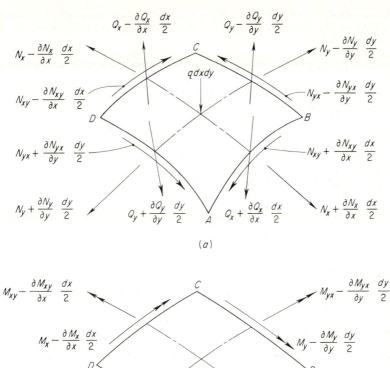

Figure 17.9   Stress-resultants and stress-couples

and where ϵ_{xo}, ϵ_{yo}, and γ_{xyo} refer to the strain of x and y elements lying in the middle surface of the shell.

The strain expressions (17.21) may be substituted into approximate stress-strain equations, comparable to Eqs. (17.6). From the resulting equations, expressions for the stresses in terms of the strains may be substituted into Eqs. (17.7), the definitions of the stress-couples and stress-resultants, to obtain the following relationships.

Stress-Strain Relations for Membrane Stress-Resultants:

$$N_x = \frac{Eh}{1-\nu^2}\,(\epsilon_{xo}+\nu\epsilon_{yo}) \qquad N_y = \frac{Eh}{1-\nu^2}\,(\epsilon_{yo}+\nu\epsilon_{xo}) \qquad N_{xy}=N_{yx}=\frac{Eh}{2(1+\nu)}\,\gamma_{xyo} \qquad (17.22)$$

Stress-Strain Relations for Bending and Twisting Stress-Couples:

$$M_x = -D(\chi_x+\nu\chi_y) \qquad M_y=-D(\chi_y+\nu\chi_x) \qquad M_{xy}=-M_{yx}=D(1-\nu)\chi_{xy} \qquad (17.23)$$

In any given problem, these equations can be converted into stress-displacement equations by replacing the strains and changes in curvature by suitable expressions in terms of the displacements of the shell, u_o, v_o, and w_o.

Formulation of analysis of shells. As in plate analysis, the analysis of a shell involves a total of eleven unknown functions of x and y, which describe the three membrane stress-resultants N_x, N_y, and N_{xy} ($=N_{yx}$); the five bending stress-couples and stress-resultants, two of which are the bending stress-couples M_x and M_y, one of which is the twisting stress-couple M_{xy} ($=-M_{yx}$), and two of which are the radial shear stress-resultants Q_x and Q_y; and the three displacements u_o, v_o, and w_o. Available to find these unknowns are eleven partial differential equations: the six stress-displacement equations which can be obtained in any given problem by converting the six stress-strain equations, Eqs. (17.22) and (17.23); and the five differential equations of equilibrium which can be derived as described below.

For purposes of the following derivation, a different element such as shown in Fig. 17.7a is used. *This element is isolated by pairs of adjacent planes which are normal to the middle surface and are assumed to contain lines of curvature of the middle surface.* Only the middle surface of this element is shown in Figs. 17.8 and 17.9. The origin is selected at the center of the element, and the x and y axes lie in the tangent plane and coincide with the tangents to the lines of curvature. Note that none of the four sides of this element is exactly normal or parallel to *any* of the three coordinate axes. Further, none of the middle surface lengths of the four faces will be the same. For purposes of writing the three force-equilibrium equations, all five stress-resultants are shown in Fig. 17.9a, while for purposes of deriving the moment-equilibrium equations, the double-arrowhead moment vectors representing the stress-couples are shown in Fig. 17.9b. The intensity of the resultant load is designated by q, the x, y, and z components of which are designated by q_x, q_y, and q_z.

Five equations of equilibrium may be written for the shell element: $\sum F_x = 0$, $\sum F_y = 0$, $\sum F_z = 0$, $\sum M_x = 0$, and $\sum M_y = 0$. Of course, the equation $\sum M_z = 0$ is identically satisfied and as in the case of thin plates does not yield a usable equation. In order to save space, the detailed derivation of these equations is omitted. These derivations are straight-forward but somewhat tedious to perform because of the inclinations of the four faces. These five equations become as follows when the central angles α_x and α_y are used as the independent variables in the differentiation:

$$\frac{\partial}{\partial \alpha_x}(N_x r_y) + \frac{\partial}{\partial \alpha_y}(N_{yx} r_x) + N_{xy}\frac{\partial r_x}{\partial \alpha_y} - N_y\frac{\partial r_y}{\partial \alpha_x} - Q_x r_y + q_x r_x r_y = 0$$

$$\frac{\partial}{\partial \alpha_y}(N_y r_x) + \frac{\partial}{\partial \alpha_x}(N_{xy} r_y) + N_{yx}\frac{\partial r_y}{\partial \alpha_x} - N_x\frac{\partial r_x}{\partial \alpha_y} - Q_y r_x + q_y r_x r_y = 0$$

$$\frac{\partial}{\partial \alpha_x}(Q_x r_y) + \frac{\partial}{\partial \alpha_y}(Q_y r_x) + N_x r_y + N_y r_x + q_z r_x r_y = 0 \qquad (17.24)$$

$$\frac{\partial}{\partial \alpha_x}(M_{xy} r_y) - \frac{\partial}{\partial \alpha_y}(M_y r_x) - M_{yx}\frac{\partial r_y}{\partial \alpha_x} + M_x\frac{\partial r_x}{\partial \alpha_y} + Q_y r_x r_y = 0$$

$$-\frac{\partial}{\partial \alpha_y}(M_{yx} r_x) - \frac{\partial}{\partial \alpha_x}(M_x r_y) + M_{xy}\frac{\partial r_x}{\partial \alpha_y} + M_y\frac{\partial r_y}{\partial \alpha_x} + Q_x r_x r_y = 0$$

In the above derivation, it is presumed that the shell surface is defined by the lines of curvature which are two orthogonal families of curves along any one of which the normals to the shell surface at consecutive points intersect. *It is also presumed that the osculating planes,* i.e., the planes containing three consecutive points, or two consecutive tangents, on a space curve, *of these lines of curvature are normal to the surface.* However, *in general, the osculating planes of the lines of curvature are not normal to the surface,* e.g., in the case of a surface of revolution the lines of curvature of which are the meridians and the parallels of latitude. The meridional plane is normal to the surface, and the radius of curvature of the meridian at any point is identical to one of the principal radii of curvature of the surface at that point. However, the plane of the parallel is not normal to the surface, and the radius of curvature of the parallel is not identical to the other principal radius of curvature of the surface at that point.

In such cases, the differential element shown in Figs. 17.8 and 17.9 would be isolated by pairs of curves from the two families of lines of curvature. The lengths of the sides would be described by the radii of curvature of these lines of curvature, which would be called a_x and a_y, but these might not be principal radii of curvature of the surface. However, the z axis would still be taken radially and normal to the tangent plane of the surface at O. The radial shears Q_x and Q_y would also act normal to the surface and in the direction of the principal radii of curvature. Therefore, in computing the component of Q_x in the x direction, for example, the angle involved would not be $d\alpha_x/2$ but (as shown in parentheses in Fig. 17.8c) would be $a_x\,d\alpha_x/2r_x$. Likewise, in setting up the equation $\sum F_z = 0$, components of the normal stress resultants N_x and N_y would be computed in the z direction, using this same modified angle.

Thus, when the surface is defined by lines of curvature the osculating planes of one or both of which are not perpendicular to the surface, the following version of Eqs. (17.24) must be used:

$$\frac{\partial}{\partial \alpha_x}(N_x\,a_y) + \frac{\partial}{\partial \alpha_y}(N_{yx}\,a_x) + N_{xy}\frac{\partial a_x}{\partial \alpha_y} - N_y\frac{\partial a_y}{\partial \alpha_x} - Q_x\frac{a_x\,a_y}{r_x} + q_x a_x a_y = 0$$

$$\frac{\partial}{\partial \alpha_y}(N_y\,a_x) + \frac{\partial}{\partial \alpha_x}(N_{xy}\,a_y) + N_{yx}\frac{\partial a_y}{\partial \alpha_x} - N_x\frac{\partial a_x}{\partial \alpha_y} - Q_y\frac{a_x\,a_y}{r_y} + q_y a_x a_y = 0$$

$$\frac{\partial}{\partial \alpha_x}(Q_x\,a_y) + \frac{\partial}{\partial \alpha_y}(Q_y\,a_x) + N_x\frac{a_x\,a_y}{r_x} + N_y\frac{a_x\,a_y}{r_y} + q_z a_x a_y = 0 \qquad (17.25)$$

$$\frac{\partial}{\partial \alpha_x}(M_{xy}\,a_y) - \frac{\partial}{\partial \alpha_y}(M_y\,a_x) - M_{yx}\frac{\partial a_y}{\partial \alpha_x} + M_x\frac{\partial a_x}{\partial \alpha_y} + Q_y a_x a_y = 0$$

$$-\frac{\partial}{\partial \alpha_y}(M_{yx}\,a_x) - \frac{\partial}{\partial \alpha_x}(M_x\,a_y) + M_{xy}\frac{\partial a_x}{\partial \alpha_y} + M_y\frac{\partial a_y}{\partial \alpha_x} + Q_x a_x a_y = 0$$

Membrane theory of shells. Over the major portion of most civil engineering shells, there are substantially no bending and twisting stress-couples and no radial shear resultants. Therefore, the load is carried substantially by the membrane stress-resultants N_x, N_y, and $N_{xy}(=N_{yx})$. Bending and twisting couples and radial shears are significant only in the vicinity of the boundaries or near discontinuities in the loading or in the geometry of the shell surface.

If there are no bending and twisting couples or radial shears, i.e., if $M_x = M_y = M_{xy} = Q_x = Q_y = 0$, the *five* equilibrium equations are reduced to *three* equations involving the membrane stress-resultants: N_x, N_y, and $N_{xy}(=N_{yx})$. Or, from Eqs. (17.25),

$$\frac{\partial}{\partial \alpha_x}(N'_x\,a_y) + \frac{\partial}{\partial \alpha_y}(N'_{yx}\,a_x) + N'_{xy}\frac{\partial a_x}{\partial \alpha_y} - N'_y\frac{\partial a_y}{\partial \alpha_x} + q_x a_x a_y = 0$$

$$\frac{\partial}{\partial \alpha_y}(N'_y\,a_x) + \frac{\partial}{\partial \alpha_x}(N'_{xy}\,a_y) + N'_{yx}\frac{\partial a_y}{\partial \alpha_x} - N'_x\frac{\partial a_x}{\partial \alpha_y} + q_y a_x a_y = 0 \qquad (17.26)$$

$$N'_x\frac{a_x\,a_y}{r_x} + N'_y\frac{a_x\,a_y}{r_y} + q_z a_x a_y = 0$$

The values of the membrane stresses defined by these equations constitute the analysis of a shell by so-called **membrane theory.** These values are designated by the prime marks N'_x, N'_y, and $N'_{xy}(=N'_{yx})$ to indicate that they are first approximations of the membrane stress-resultants that would be determined as part of the general solution of the problem, that included the effect of the bending and twisting couples and the radial shears.

It is important to note that *the membrane-theory solution is a statically determinate one defined simply by the above three equilibrium equations and the boundary conditions of the structure.*[1] Such a structure, because of its initial curvature, could maintain a *continuous distributed loading* in equilibrium even if it had no bending stiffness and therefore simply utilized its membrane strength. In making this statement, the possibility of buckling of the shell is ignored. Note that a flat plate is incapable of doing this, as is evident from the first of the equilibrium equations (17.13). If the bending and twisting couples and radial shears are zero, there is no way of satisfying the equilibrium equation $\sum F_z = 0$. Only after the plate has undergone considerable transverse deflection and curvature can the membrane forces be brought to bear and help to stabilize the plate. (This is one of the factors considered in the large-deflection theory for plates.)

It is interesting to note that the structural action of a shell is very similar to that of a truss. A truss does not *need* rigid joints or bending stiffness in its members (if the possibility of buckling is ignored) in order to carry a system of *joint* loads and reactions. It can carry these loads simply by using axial force, i.e., "membrane forces" in its members. If the truss *does have* rigid joints and the bending capacity of its members *is recognized*, it *does* mobilize some of this strength in carrying joint loads, but most of such loads are carried by the axial forces in the members. In other words, just as for a shell carrying *distributed loadings*, a "membrane solution" for a truss computed on the basis that it carries its joint loads as a pin-connected truss is a very good first approximation for the axial forces in the members of the actual truss with rigid joints.

Application of membrane theory to shells of revolution. Shells of revolution such as cones, spheres, etc., are generated by rotation of a plane curve about an axis lying in the plane of the curve, as shown in Fig. 17.10. Various positions of the generating curve are called meridians of the surface. At any point O on the surface, the meridional plane through that point is one of the planes of principal curvature, and the meridian is one of the lines of curvature, the radius of curvature of which is denoted by r_1. The other plane of principal curvature is normal to the meridional plane, being also normal to the tangent plane, contains the normal oz. The intercept of this normal plane on the surface is the curve JOH, the radius of curvature of which is the distance OL, which will be denoted by r_2.

For purposes of analysis, it is convenient to define the middle surface of the shell by the lines of curvature, one family of which is the meridians and the other of which is the parallels of latitude intercepted on the surface by planes which are normal to the axis of revolution. A differential element of the middle surface isolated by pairs of these parametric curves is shown in Fig. 17.10b and positioned by the angles θ and ϕ. Note further that the radius of curvature of the parallel is r_0.

Suppose that the distributed loading on this shell is symmetrical with respect to the axis of revolution. Thus, q_x, the circumferential component of the load, is zero, and q_y and q_z are functions only of ϕ. In such a case, because of the symmetry of the loading and of the shell, the stress condition of the shell is independent of θ and varies only with ϕ. Thus, there are no shearing forces $N_{\theta\phi}$ or $N_{\phi\theta}$. The normal stress-resultants in the circumferential and meridional directions are denoted by N_θ and N_ϕ as shown.

The membrane-theory solution can be obtained by applying Eqs. (17.26) and noting that

$$\alpha_x = \theta \qquad a_x = r_0 \qquad r_x = r_2 \qquad r_0 = r_2 \sin\phi \qquad N'_{xy} = N'_{\theta\phi} = 0$$

$$\alpha_y = \phi \qquad a_y = r_1 \qquad r_y = r_1 \qquad N'_x = N'_\theta \qquad N'_y = N'_\phi$$

Hence,

$$\frac{d}{d\phi}(N'_\phi r_0) - N'_\theta \frac{dr_0}{d\phi} + q_y r_0 r_1 = 0$$

$$\hfill (17.27)$$

$$N'_\theta \frac{\sin\phi}{r_0} + \frac{N'_\phi}{r_1} + q_z = 0$$

---

[1] In the membrane-theory solution, it is assumed that the boundary supports supply only the membrane-type boundary forces as required by the theory and that the boundaries are free to displace as required by the membrane-theory strains.

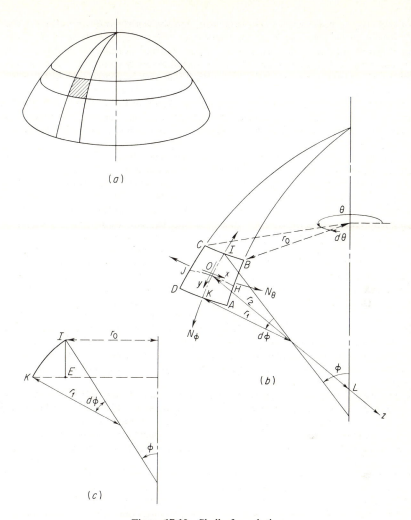

Figure 17.10   Shell of revolution

From Fig. 17.10c, note that

$$\frac{dr_0}{d\phi}\, d\phi = KE \approx KI \cos \phi \approx r_1 \cos \phi \, d\phi \qquad \therefore \; \frac{dr_0}{d\phi} \approx r_1 \cos \phi \qquad (a)$$

The second of Eqs. (17.27) becomes

$$N_\theta' = -\frac{r_0}{\sin \phi}\left(\frac{N_\phi'}{r_1} + q_z\right) \qquad (17.28)$$

By substituting from (a) and Eq. (17.28) in the first of Eqs. (17.27) multiplied by $\sin \phi$ and then by integrating from O to ϕ, the following expression is obtained:

$$N_\phi' = -\frac{1}{2\pi r_0 \sin \phi} \int_0^\phi (q_y \sin \phi + q_z \cos \phi)(2\pi r_0)r_1 \, d\phi$$

The integral in this expression may be interpreted as the total vertical load F applied above a certain latitude ϕ. Hence,

$$N'_\phi = -\frac{F}{2\pi r_0 \sin \phi}$$

and from Eq. (17.28),   (17.29)

$$N'_\theta = +\frac{F}{2\pi r_1 \sin^2 \phi} - q_z \frac{r_o}{\sin \phi}$$

The membrane-theory solution for a symmetrically loaded shell of revolution can easily be computed by substituting into Eqs. (17.29) the value F.

Application of general theory to shell analysis.   The general theory of shell analysis is formulated above.   The application of this theory is difficult and has been done successfully only for a limited number of cases.[1]

Usually a successive-approximation technique is used to obtain the solution.   First, a membrane-theory solution is obtained, to give a first approximation of the membrane stress-resultants, and the corresponding displacements are then computed.   These displacements imply certain changes in curvature which, in turn, imply certain bending couples and radial shears.   The radial shears would imply certain changes in the membrane stress-resultants, and thus establish a second approximation of the membrane stress-resultants.   Such calculations can be continued through as many cycles as are necessary to obtain a satisfactory convergence on the exact answers.   Such final answers would be realized in the shell only if the supports supplied the boundary forces implied by these answers.   In general the supports do not do so; therefore further interacting forces and couples have to be introduced between the shell and its support to reconcile the discrepancies.   Similar forces and couples would have to be introduced on internal sections at locations of load or geometry discontinuities to reconcile internal discrepancies in the continuity of the strains.   It is these corrective forces and couples which must be applied at the boundaries or at internal sections that are the most difficult to compute.   They also cause the major difference between the stresses obtained by the general theory and those obtained by simple membrane theory.

17.7 Analysis of Buckling Behavior.   In the Foreword, various types of structural failure were discussed, and it was emphasized that a buckling failure is potentially very dangerous and may trigger the collapse of many types of engineering structures.   It is very important, therefore, for a structural engineer to develop the capability to analyze buckling behavior.

The possibility of buckling exists in any compressed portion.   It may take the form of buckling of the structure as a whole or the localized buckling of an individual member or component part thereof, which may or may not precipitate the failure of the entire structure.   *The load at which buckling occurs depends upon the stiffness of a structure or portion thereof, rather than upon the strength of the material involved.*   As a result, stress intensities, if involved, are involved only to the extent of identifying the point on the stress-strain curve at which the modulus (or the slope) of the stress-strain curve is desired.   Hence, for a hypothetical linearly elastic material with an infinite elastic limit, stress intensity would never enter the buckling considerations.   Of course, for a nonlinearly elastic material or an elastoplastic material, stress intensity is involved in defining stress-strain characteristics.

It is also interesting to note that buckling can occur without the material involved being subjected to compressive stresses.   For example, consider a tubular column of circular cross section.   The axial force at which the column buckled would be the same whether the tube were compressed by uniformly distributed axial loads applied directly to its end cross sections or whether the tube were filled with water and the loads applied to the fluid through pistons inserted in each end so that the only stresses induced in the tube would be hoop tensions.

---

[1] For a discussion of such solutions see, for example, Timoshenko and Woinowsky-Krieger, op. cit., or K. Girkmann, "Flächentragwerke," 4th ed., Springer-Verlag, Berlin, 1956.

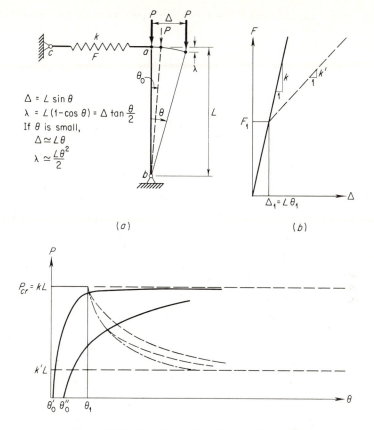

Figure 17.11   Buckling of idealized elastic system

Fundamentally, the analysis of buckling behavior is simply an investigation of whether or not the equilibrium configuration of a system is stable. *The stability or instability of the system depends upon the answer to the following question: If a system is displaced slightly from its equilibrium position, does it tend to return to its original position or does it tend to displace further when the disturbance is removed?   If it returns, the system is stable; if it displaces further, it is unstable.*

Methods of evaluating elastic buckling behavior.   Several approaches may be used to obtain the answer to the fundamental question stated above regarding the stability of an equilibrium position. For purposes of simplifying the discussion and the computations involved, consider the simple idealized system shown in Fig. 17.11a.   Assume that the compressed bar *ab* is weightless but perfectly rigid and nondeformable and that the only way the system can deflect is for the spring *ac* to elongate.   Initially it is assumed that the spring possesses the linearly elastic force-deflection characteristics[1] shown in Fig. 17.11b (though later in the discussion the effect of the nonlinear characteristics shown by the dashed line is considered).

The buckling behavior of this system may be studied by one of the following three methods.

Method I: Equilibrium Method.   Assume that the member is initially perfectly vertical and concentrically loaded by the force *P*, so that there is no tendency for the bar to rotate about point *b* and nominally it is in equilibrium in its original vertical position.   However, is there some value of

---

[1] The spring stiffness factor *k* is equal to the force required to stretch the spring a unit distance and has units of pounds per inch.

P (called the **critical value** and denoted by P_{cr}) for which the bar would also be in equilibrium in an alternative inclined position to which it is moved by a disturbance? If so, for equilibrium in the inclined position, $\sum M_b = 0$. Therefore, $P\Delta = FL \cos \theta \approx FL$, if θ is small. But $F = k\Delta$; therefore, $P\Delta = (k\Delta)L$, or $P = kL$. Hence, if $P = kL$, the inclined position is also a possible alternative equilibrium position. This value of P is the critical value, or $P_{cr} = kL$. Note that the value of θ is indeterminate provided that it is small. In other words, the value of P_{cr} is independent of Δ (that is, of θ) and remains the same for any slightly inclined position of bar ab. This solution is represented by the horizontal line shown in Fig. 17.11c. Physically, when P reaches the value P_{cr}, the nominal vertical position is an unstable one because at this load the bar could also be in equilibrium at any one of an infinite number of slightly inclined positions. That is, at P_{cr}, the system is unstable, and buckling is possible.

Method II: Energy Method. This method involves a consideration of the energy changes that occur as a system is displaced from its nominal equilibrium position into a nearby possible alternative position. Consider again the vertical member shown in Fig. 17.11a, which nominally is in equilibrium in this vertical position when acted upon by a concentric load P. Suppose that the bar is rotated slightly to the alternative position defined by θ. During this rotation, the load P does work denoted by $\Delta \mathscr{W}_E$ and equal to $P\lambda$ or $PL\theta^2/2$. Also, strain energy denoted by $\Delta \mathscr{U}$ is stored in the spring of an amount equal to $F\Delta/2$ or $k\Delta^2/2$ or $kL^2\theta^2/2$. If $\Delta \mathscr{U}$ is greater than $\Delta \mathscr{W}_E$, the disturbance has to do work *against* the spring to get the system into this position; therefore, when the disturbance is removed, the energy stored in the spring will pull the bar back to its original vertical position. If $\Delta \mathscr{U}$ is less than $\Delta \mathscr{W}_E$, the disturbance has to work *with* the spring to prevent the bar from rotating more and more; and if it were removed, the system would collapse. If $\Delta \mathscr{U}$ is exactly equal to $\Delta \mathscr{W}_E$, the system is in a *neutral* and *indifferent* condition, in which the system can be displaced from its nominal vertical position without the disturbance doing any work. This is the buckling condition which delineates the end of stability of the system and the start of instability. Therefore, when P is equal to P_{cr}, $\Delta \mathscr{W}_E$ is equal to $\Delta \mathscr{U}$. Hence, $P_{cr}L\theta^2/2 = kL^2\theta^2/2$ and therefore $P_{cr} = kL$, thereby checking the value found by method I.

Method III: Load-Deflection Curve Method. The critical value of P corresponding to the buckling condition can be identified simply by studying the characteristics of the load-deflection curve for the system when the system is assumed to have some slight imperfection. For example, in this present case assume that the bar initially is not perfectly vertical but has some slight inclination θ_o. In this initial position, however, the spring is unstressed. Now, as P is applied, the member rotates about the base. For equilibrium at any particular load, $P\Delta = FL$, where now $F = k(\theta - \theta_o)L$. Therefore, $P = kL(1 - \theta_o/\theta)$. If kL is replaced by P_{cr}, then $P = P_{cr}(1 - \theta_o/\theta)$. When this equation is used, the solid-line load-deflection curve shown in Fig. 17.11c can be drawn. Note that as P approaches P_{cr}, θ approaches infinity. In other words, as P approaches P_{cr}, the system begins to lose its capability to resist rotation and very large rotation (or buckling) is imminent.

In the above considerations, the spring is assumed to be linearly elastic, with an indefinite elastic limit. It is also interesting to consider the effect of nonlinear elastic spring characteristics such as indicated by the dashed line in Fig. 17.11b, where the stiffness is reduced to k' when the force in the spring exceeds F_1 or $k\Delta_1$. For such a spring, the above solutions are valid only up to the point where θ equals $\theta_1 + \theta_o$, in which expression θ_1 is equal to Δ_1/L. Application of method III to this case leads to the following results. If $\Delta_o \leq \Delta \leq \Delta_o + \Delta_1$ or $0 \leq F \leq F_1$, or $\theta_o \leq \theta \leq \theta_o + \theta_1$,

$$P = kL\left(1 - \frac{\theta_o}{\theta}\right) \tag{a}$$

from which, when $\theta = \theta_o + \theta_1$,

$$[P]_{\theta_o + \theta_1} = kL\frac{\theta_1}{\theta_o + \theta_1}$$

However, if $\Delta \geq \Delta_o + \Delta_1$,

$$F = kL\theta_1 + k'L(\theta - \theta_1 - \theta_o)$$

but, for equilibrium,

$$P\Delta = FL \quad \text{or} \quad P = \frac{F}{\theta}$$

Therefore, if $\theta \geq \theta_o + \theta_1$,

$$P = \frac{kL\theta_1}{\theta} + k'L\left(1 - \frac{\theta_1 + \theta_o}{\theta}\right) \qquad (b)$$

Curves plotted in accordance with Eqs. (a) and (b) are plotted in Fig. 17.11c for two values of θ_o smaller than θ_1: θ_o' and θ_o''. When the spring force reaches F_1, such curves break down sharply, as shown by the dashed-line portions, and become asymptotic to the horizontal line corresponding to $P = k'L$.

The nonlinear spring characteristics used in this hypothetical problem are actually quite representative of the stiffness characteristics of actual structural members, which are characterized by an initial stiff portion up to a certain elastic limit, above which the stiffness decreases markedly. The above results indicate that the critical load kL associated with the initial stiffness k can be approached only if the initial imperfection θ_o is small compared with θ_1. For large imperfections, it is dangerous to count on the system to carry more than $k'L$.[†]

Buckling of columns. Elastic buckling of structures and structural elements can be evaluated by any of the three approaches used above. The mathematical details become more cumbersome to handle, but the fundamental concepts still are valid. To illustrate, method I is used below to study the buckling characteristics of the simple column shown in Fig. 17.12.

This column, hinged at both ends, is assumed to be a *perfect* compressed member which is homogeneous, linearly elastic, without initial curvature or eccentricity, and not subjected to any lateral loads. It is assumed to be constrained to deflect in the plane of the paper. Nominally, such a member simply undergoes only axial strain and remains unbent. Under certain conditions, the straight form, however, may not be the only deformed position which satisfies the equilibrium requirements for the member. Consider the possibility of the laterally bent position shown in Fig. 17.12. If the column is to be in equilibrium in this position, the differential equation of the elastic curve must be satisfied, or

$$\frac{d^2v}{dx^2} = -\frac{M}{EI} = -\frac{Pv}{EI}$$

Therefore,

$$\frac{d^2v}{dx^2} + \frac{P}{EI}v = 0$$

the solution of which is

$$v = A \cos \sqrt{\frac{P}{EI}}\,x + B \sin \sqrt{\frac{P}{EI}}\,x$$

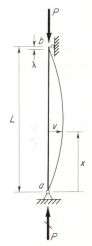

Figure 17.12
Buckling of
simple column

[†] In the periodical and textbook literature, a number of authors have reported some very valuable studies of buckling phenomena using simplified mathematical models such as that in Fig. 17.11. For example, D. C. Drucker and E. T. Onat, On the Concept of Stability of Inelastic Systems, *J. Aeronaut. Sci.*, pp. 543–548, August 1954; and F. R. Shanley, "Strength of Materials," McGraw-Hill Book Company, New York, 1957.

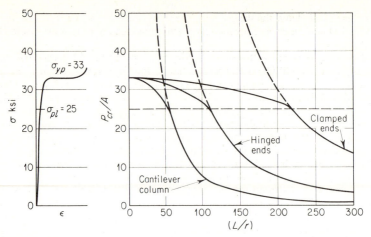

Figure 17.13   Buckling curves of typical columns (no eccentricity, initial curvature, or other imperfections)

If such a deflection occurred, the boundary conditions at a and b would have to be satisfied:

When $x = 0$: $[v]_{x=0} = 0 = A(1) + 0$ $\therefore$ $A = 0$

When $x = L$: $[v]_{x=L} = 0 = 0 + B \sin \sqrt{\dfrac{P}{EI}} L$

To satisfy the latter condition, either B or $\sin \sqrt{P/EI}\, L$ has to be zero. If B equals zero, then no lateral deflection would be possible. However, if $\sin \sqrt{P/EI}\, L = 0$, then B does not have to be zero and could, in fact, have any value, in which case

$$v = B \sin \frac{\pi}{L} x$$

In other words, if the axial load P has such a value that $\sqrt{P/EI}\, L = \pi$, or if $P = (\pi^2/L^2)EI$, the straight form of the column is no longer the only equilibrium position, since at this load a laterally bent (or buckled) position is also possible. This value of the load is called the critical load P_{cr}. Sometimes it is called the **Euler load** after Leonhard Euler, who first developed the theory for the elastic buckling of flexible columns over two hundred years ago.

If the exact buckled shape were used in applying method II, the energy method, to this problem, exactly the same value would be obtained for the critical buckling load. If the column were assumed to have initial curvature, an eccentricity, or a lateral loading, method III could be used to obtain a plot of the axial load vs. the lateral deflection of the column. Such a plot would show that as P approached P_{cr}, the lateral deflection would approach infinity again indicating instability or buckling when $P = (\pi^2/L^2)EI$.

Actually these solutions are based on the assumption of linearly elastic behavior and are not valid if the proportional limit of the material is exceeded. If the stresses exceed this limit, the buckling load P_{cr} of a column which is hinged at both ends, concentrically loaded, and without initial curvature is found to be $(\pi^2/L^2)E_t\, I$, where E_t is the **tangent modulus** (or slope of the stress-strain curve at the stress corresponding to P_{cr}/A). The buckling curve (P_{cr}/A vs. L/r)† for such a column is shown in Fig. 17.13. Also shown for comparison are curves for a cantilever column and for a column which is fixed against rotation at both ends. If the material were linearly elastic

---

† In this sense, r is the radius of gyration of the cross-sectional area and is equal to $(I/A)^{1/2}$.

with an infinite elastic limit, the dashed-line extensions would be valid and are shown for comparison with the buckling curves corresponding to the typical stress-strain curve for structural steel shown.

17.8 Evaluation of Dynamic Response. In the past, the design of most civil engineering structures has been controlled by static (or substantially static) loading conditions. Such dynamic loadings as were involved made minor contributions to the total effect, so that the dynamic contribution was handled approximately by use of an equivalent static load, or by an impact factor, or by a modification of the factor of safety.

In recent years, a number of developments have led to a growing interest in a more precise evaluation of the contribution of dynamic loadings. Among these have been the imposition on structures of heavier machinery and vehicles operating at higher speeds, the construction of higher towers and buildings and longer bridges involving more severe and important wind-loading conditions, the necessity of developing blast-resistant construction, and the desire to improve earthquake-resistant construction. The growing interest in this area has stimulated the publication of many papers in the periodical literature and a number of books.[1]

The response of a structure to dynamic loading depends upon the definition of the loading, the resistance of the structure to deflection, and (since acceleration is involved) the mass of the structure. Of course, if the supports of the structure are not immovable, their motion must also be defined in terms of time.

One of the best ways to develop a feeling and appreciation for dynamic response is to use a step-by-step numerical procedure to integrate the differential eqations of motion. Such procedures are generally applicable for any type of loading and for elastic or elastoplastic structural resistance, in which cases more formal mathematical methods of solution become impractical or impossible to apply. For multimass systems step-by-step numerical procedures become laborious but are adaptable to modern computational techniques and machines.

Consider the simple idealized system shown at rest by the solid lines in Fig. 17.14. This system is allowed to move only in a horizontal translatory motion in the plane of the paper. This rigid body of weight W is restrained by a linearly elastic spring of stiffness k. Possible damping or frictional effects are ignored. Consider the dynamic response of this system to some known dynamic force P. The spring force kx is designated as R, as shown. If, at any time t_o the displacement x_o were known, thereby defining the value of the spring force R_o, the net accelerating force acting on the mass would be $P - R_o$ and the acceleration $\ddot{x}_o$ at this instant would therefore be

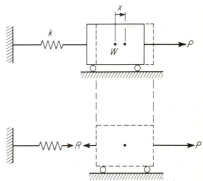

Figure 17.14   Notation for step-by-step procedure

$$\ddot{x}_o = \frac{P - R_o}{W/g}$$

[1] Books dealing with structural dynamics include S. Timoshenko, "Vibration Problems in Engineering," 3d ed., D. Van Nostrand Company, Inc., Princeton, N.J., 1955; G. L. Rogers, "Dynamics of Frame Structures," John Wiley & Sons, Inc., New York, 1959; C. H. Norris, R. J. Hansen, M. J. Holley, Jr., J. M. Biggs, S. Namyet, and J. K. Minami, "Structural Design for Dynamic Loads," McGraw-Hill Book Company, New York, 1959; J. M. Biggs, "Introduction to Structural Dynamics," McGraw-Hill Book Company, New York, 1964; W. C. Hurty and M. F. Rubinstein, "Dynamics of Structures," Prentice-Hall, Inc., Englewood Cliffs, N.J., 1964; R. W. Clough and J. Penzien, "Dynamics of Structures," McGraw-Hill Book Company, New York, 1975.

If the velocity $\dot{x}_o$ at the time t_o were also known and if, during the following short interval of time Δt, the *acceleration were assumed to remain constant* at the value $\ddot{x}_o$, the velocity $\dot{x}_a$ at the end of this interval, i.e., at time $t_o + \Delta$, would be

$$\dot{x}_a = \dot{x}_o + \ddot{x}_o \, \Delta t$$

In addition, the displacement x_a at the end of the interval would be

$$x_a = x_0 + \dot{x}_0 \, \Delta t + \ddot{x}_0 \, \frac{(\Delta t)^2}{2}$$

Knowing this displacement x_a defines a new value of the spring force R_a, and the accelerating force $P - R_a$ at the end of this interval. The above process could then be repeated for the next short time interval Δt to predict the velocity $\dot{x}_b$ and the displacement x_b at the time $t_o + 2\Delta t$. By this step-by-step numerical process the motion of the mass could be tracked indefinitely. Obviously the accuracy of the above procedure depends on the length of the time interval and the degree of the approximation of the acceleration within it.

Figure 17.15 summarizes in a qualitative way the step-by-step development of the response of this simple system to a suddenly applied load. Suppose that at $t = 0$ the system has no displace-

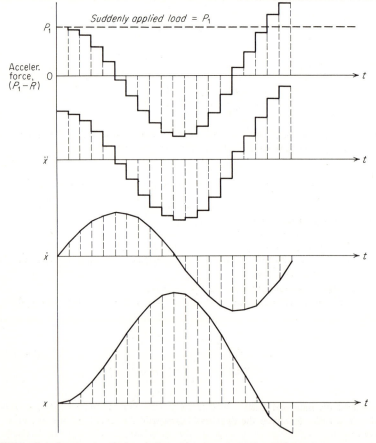

Figure 17.15 Step-by-step analysis of dynamic response of one-degree system to suddenly applied load

ment or velocity. Consider the response for a series of intervals Δt, which are short compared with the natural period of the system. There being no displacement at $t = 0$, the spring force R is 0, and the accelerating force during the first interval is P_1. As the displacement increases, R gradually builds up and the accelerating force, and hence the acceleration, decreases. When R exceeds P_1, the system starts decelerating. Subsequently the velocity is reduced to zero when the displacement reaches its maximum value of $2P_1/k$. Then the mass starts rebounding and eventually reaches zero velocity and minimum deflection at its original starting position, thereby completing one complete cycle of this periodic response. Neglecting damping, the response continues indefinitely, repeating the first cycle in a periodic fashion with a period equal to the natural period of the system.

Of course, for a simple elastic system such as this, it is easy to solve the differential equation of motion involved for a general dynamic load P,

$$P = P_1 f(t)$$

where P_1 is the maximum value attained by the dynamic load. To derive the differential equation of motion for the undamped simple system shown in Fig. 17.14, apply d'Alembert's principle and add the inertia force $\ddot{x}W/g$ to the free-body sketch, showing it acting in the negative x direction. The condition for dynamic equilibrium $\sum H = 0$ for the isolated body is

$$P - kx - \frac{W}{g}\ddot{x} = 0$$

which can be rewritten as

$$\frac{d^2x}{dt^2} + \frac{kg}{W}x = \frac{P_1 g}{W}f(t)$$

The derivation is not included here, but the general solution of this equation to obtain the displacement x at any time t can be expressed in the form

$$x = x_o \cos \omega t + \frac{\dot{x}_o}{\omega}\sin \omega t + x_{st}\left[\omega \int_0^t f(t')\sin \omega(t - t')\, dt'\right] \qquad (17.30)$$

where $t' =$ any intermediate time between 0 and t

   $x_o =$ initial displacement at $t = 0$

   $\dot{x}_o =$ initial velocity at $t = 0$

   $x_{st} = P_1/k =$ static deflection of spring if P_1, the maximum value of dynamic load, were applied as a static load

   $\omega = \sqrt{kg/W} =$ natural (circular) frequency of system

Note that if the system were at rest at $t = 0$, then x_o and $\dot{x}_o$ would both be zero and Eq. (17.30) would become

$$x = x_{st}(\text{DLF}) \qquad (17.31)$$

where

$$\text{DLF} = \omega \int_0^t f(t')\sin \omega(t - t')\, dt' \qquad (17.32)$$

and is called the **dynamic-load factor.** This factor is the value by which x_{st} should be multiplied at any time t in order to obtain the dynamic displacement at that time of a system which was initially at rest.

The dynamic-load factor for any given loading may be computed from Eq. (17.32). This factor is a function of time, i.e., varies with time. Usually, the designer is most interested in the maximum value of the dynamic-load factor $\text{DLF}_{\max}$, since this value defines the maximum displacement of the system. The results for three typical dynamic loadings are summarized below:

Suddenly Applied Load (Fig. 17.16):

$$0 \leq t \qquad P = P_1 \qquad \therefore f(t) = 1$$

$$\text{DLF} = 1 - \cos \omega t \qquad \text{DLF}_{\max} = 2$$

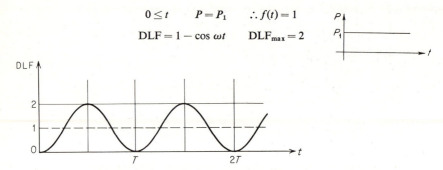

Figure 17.16 Dynamic load factor for suddenly applied load

Gradually Applied Load (Fig. 17.17):

$$0 \leq t \leq t_1 \qquad P = P_1 \frac{t}{t_1} \qquad \therefore f(t) = \frac{t}{t_1}$$

$$t_1 \leq t \qquad P = P_1 \qquad \therefore f(t) = 1$$

$$0 \leq t \leq t_1 \qquad \text{DLF} = \frac{t}{t_1} - \frac{\sin \omega t}{\omega t_1} = \frac{t}{t_1} - \frac{T}{2\pi t_1} \sin \omega t$$

$$t_1 \leq t \qquad \text{DLF} = 1 + \frac{T}{2\pi t_1} [\sin \omega(t - t_1) - \sin \omega t]$$

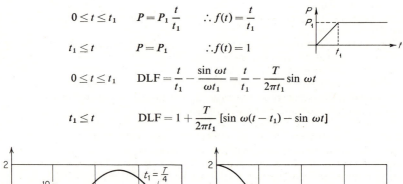

Figure 17.17 Dynamic load factor for gradually applied load

Rectangular Pulse Load (Fig. 17.18):

$$0 \leq t \leq t_1 \qquad P = P_1 \qquad \therefore f(t) = 1$$

$$t_1 \leq t \qquad P = 0 \qquad \therefore f(t) = 0$$

$$0 \leq t \leq t_1 \qquad \text{DLF} = 1 - \cos \omega t$$

$$t_1 \leq t \qquad \text{DLF} = 2 \sin \frac{\omega t_1}{2} \sin \omega \left(t - \frac{t_1}{2}\right)$$

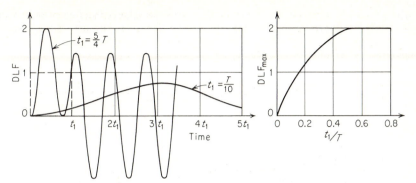

Figure 17.18   Dynamic-load factor for rectangular-pulse load

To study the dynamic behavior of a structure, one must be able to define its position at any instant.   If at any instant the position of a structure can be defined by *one* number, or *one* coordinate, the structure is said to have *one degree of freedom*.   The simple system considered above illustrates such a case.   The two-mass system shown in Fig. 17.19b represents a structure with *two degrees of freedom*, since *two* distances x_1 and x_2 are required to define the position of the system at any instant.   The beam shown in Fig. 17.19c has an *infinite number of degrees of freedom* because the definition of its positions requires enumerating an infinite number of vertical ordinates of its deflection curve.

Consider the rigid frame shown in Fig. 17.19d, which is constrained to deflect in the plane of the paper.   Assume that the weight of the members is small compared with the concentrated masses shown, which are assumed to be concentrated at a point at mid-lengths of the members.   Also assume that the rotation of each mass is small and therefore can be neglected.   Consider only the effect of bending deformation of the structure.   On this basis *five* deflections (horizontal deflection of masses 1, 2, and 4 and vertical deflection of masses 1 and 3) completely define the configuration of the frame, and therefore the structure is said to have *five degrees of freedom*.

The present discussion will not be extended to include the multi-degree-of-freedom system which is representative of actual structures.   The reader should refer to the references noted for such a discussion.   It will be found that a general method can be developed for evaluating the dynamic response of linearly elastic multi-degree-of-freedom systems, which utilizes many of the

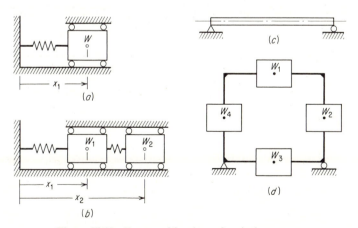

Figure 17.19   Degree of freedom of typical systems

basic ideas discussed above for the simple one-degree system.   This method involves first determining the natural frequency and deflected shape of each of the *normal modes of vibration* of the system, there being as many such modes as there are degrees of freedom of the system.   The dynamic response of the system to a given dynamic load is then evaluated by superimposing proper contributions of each of the various normal modes of vibration.   The contribution of each mode depends on the *participation factor* and the *dynamic-load factor* for that mode.   The participation factor depends upon the characteristic shape of the mode in comparison with the distribution of the dynamic load over the structure.   The dynamic-load factor depends upon the natural frequency of the mode and the time variation of the dynamic load.

18

Plastic Behavior of Structures

18.1 Introduction. In the previous discussions of the computation of deflections and the analysis of statically indeterminate structures, attention was focused almost exclusively on linearly elastic behavior. However, structural engineers are becoming increasingly interested in basing some of their designs on the ultimate strength of the structure. It is therefore important to consider the plastic behavior of some typical structural members as background for the plastic-design procedures which are being developed.

Many structural materials are ductile. Of these, structural steel exhibits the most suitable postelastic properties from the standpoint of applying the principles of plastic design. Of course, as soon as yield stress intensities are developed at the most highly stressed section, *statically determinate* steel structures can carry very little more load without undergoing excessive deflections. In the case of a *statically indeterminate* structure, however, plastic deformation can occur at such a section with no substantial loss of resistance while additional load is applied to the structure. During this additional loading, the structure is somewhat more flexible but not excessively so. The load may be increased still further until yield stresses are developed at enough locations for the structure to be on the point of deflecting as a mechanism and undergoing excessive deflections. In certain cases, this reserve load-carrying capacity of a statically indeterminate steel structure, over and beyond the load which first produced yield stresses, can be utilized in design.

This reserve capacity beyond the yield load P_{yp} (that is, the load which produced yield stresses at the first location) is most impressive in structures where material is used inefficiently. Note that this does not necessarily mean structures which are uneconomically designed. For example, it is often economical to use a rolled beam the capacity of which is utilized only at the one most highly stressed section. On the other hand, if it is economical to vary the size of the members and thereby design the "one-horse-shay" type of structure with substantially constant stresses throughout, excessive deflections impend at loads only slightly greater than the yield load.

Plastic design is most appropriate when the dominant portion of the design loading is an immovable and essentially nonreversing loading. Under severe reversing loadings, **plastic fatigue** could occur, as exemplified by breaking a nail or wire by bending it back and forth into the plastic range. If heavy moving loads were involved, it is possible to produce additional plastic deformation with each passage of the load and eventually accumulate excessive permanent deflections. In many building design situations, the most important design loadings may be considered as static loadings, and it is to such situations that plastic design is intended to apply.

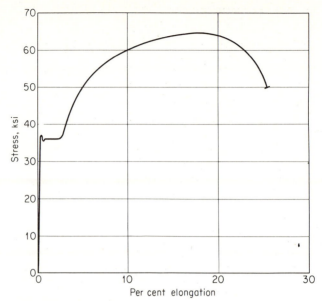

Figure 18.1   Typical stress-strain curve for A36 structural steel

Here the plastic behavior of certain important cases will be discussed.   This is not intended, however, to be a comprehensive treatment of either the plastic analysis or the plastic design of structures.[1]

18.2 Mechanical Properties of Steel.   The important properties of structural steel from a design standpoint are depicted in Figs. 18.1 and 18.2.   In Fig. 18.1 is shown the complete (engineering) stress-strain curve of a typical A36 structural steel, illustrating the exceptional ductility of such a steel before actual tensile rupture occurs.   From a practical design viewpoint, major interest is focused on the initial portion of this curve, say up to an elongation of 2 per cent.   This initial portion is idealized and shown with an expanded scale for strain in Fig. 18.2.

In the idealized stress-strain curve, structural steel is assumed to be linearly elastic up to the **yield-point stress** σ_{yp} of 36 kips per sq in.   Note that the **yield-point strain** ϵ_{yp} is a little more than 0.001, or 0.1 per cent.   Actually, steel is linearly elastic up to a **proportional limit** of about two-thirds of the yield-point stress.   Strains are still recoverable upon unloading up to the **elastic limit,** which is about three-quarters of the yield-point stress.   When steel is unloaded after being loaded beyond the elastic limit, some permanent set or residual strain will be apparent.   In other words, plastic behavior actually starts at the elastic limit.   Such behavior is not marked, however, until the yield-point stress level is reached.   Actually at this level, the steel goes through an upper yield and subsequently a lower yield before settling down to the yield stress level commonly referred to.   The magnitude of the wiggle associated with this upper and lower yield-point phenomenon is accentuated with the speed of loading and is depicted at the start of the yield plateau

---

[1] For such information the reader is referred to such books as J. F. Baker, M. R. Horne, and J. Heyman, "The Steel Skeleton," vol. II, "Plastic Behavior and Design," Cambridge University Press, London, 1956; L. S. Beedle, "Plastic Design of Steel Frames," John Wiley & Sons, Inc., New York, 1958; "Plastic Design in Steel," American Institute of Steel Construction, New York, 1958; P. G. Hodge, Jr., "Plastic Analysis of Structures," McGraw-Hill Book Company, New York, 1959; L. S. Beedle et al., "Structural Steel Design," The Ronald Press Company, New York, 1964; B. Bresler, T. Y. Lin, and J. B. Scalzi, "Design of Steel Structures," 2nd. ed., John Wiley & Sons, Inc., New York, 1968.

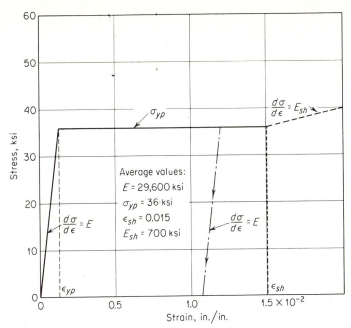

Figure 18.2   Idealized stress-strain curve for A36 stuctural steel (idealization of initial portion of complete stress-strain curve shown in Fig. 18.1).

in Fig. 18.1. For static or quasi-static loading, this phenomenon has no importance and is therefore ignored in the idealized curve shown in Fig. 18.2. For all practical purposes, when the yield-point stress level is reached, structural steel will deform plastically at no increase in stress up to a strain of 1 or 1.5 per cent. At this point, so-called **strain hardening** begins, and again an increase in stress is required to produce an increase of strain. In the early stages of strain hardening the slope of the stress-strain curve corresponds to the modulus E_{sh} of about 700 kips per sq in. compared with Young's modulus of elasticity E of 29,000 kips per sq in. in the elastic range.

In the subsequent discussion, structural steel will be assumed to be linearly elastic up to the (tensile) yield point and then to yield *indefinitely* at the constant yield-point stress (in other words, strain hardening will be ignored).[1] If, after yielding a certain amount, an element is unloaded, it will be assumed to unload (as shown by the dot-dash line) along a path which is parallel to the initial elastic line. If the element is then loaded in the opposite sense, it will be assumed to continue straining, as shown by the dot-dash line, until a yield stress level of -36 kips per sq in. is reached. At this point, if required, it would be assumed to be capable of yielding indefinitely in compression at this constant stress level.

For simplicity in the following discussion, the assumed behavior of structural steel is depicted by the sketches shown in Fig. 18.3. Note that the yield-point stress in both tension and compression is rounded off to a value of 30 kips per sq in.

18.3   Plastic Behavior of Simple Truss Structure. For purposes of this discussion, the very simple three-bar truss shown in Fig. 18.4 is used. It will be assumed that buckling is prevented. Therefore, these bars with a cross-sectional area of 1 sq in. will yield indefinitely at a bar force of $+$ or -30 kips for a yield stress of 30 kips per sq in., as shown in Fig. 18.3. In actuality, slender

---

[1] In plastic design, the maximum strains in the structure corresponding to the ultimate load at which excessive deflections impend will seldom exceed 0.015. Neglect of the strain-hardening range therefore seldom has any important effect on the computation of the ultimate load. Further, neglect of this effect leads to a conservative estimate of the ultimate load.

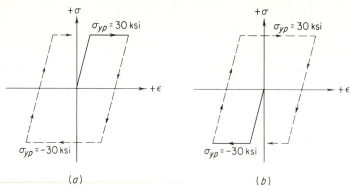

Figure 18.3 Behavior of steel assumed for discussion of plastic behavior (*a*) starting with tensile strains, (*b*) starting with compressive strains

compression members have an F-vs.-ΔL relationship as shown by the dotted line in Fig. 18.4*b*, where in this case ΔL is interpreted as the relative movement of the two ends of a buckled bar rather than the axial change in length of the bar.[1] Whereas this hypothetical example is not representative of actual trussed structures, it does provide an excellent vehicle for developing the important concepts involved in the plastic behavior of structures.

Before starting the discussion of the truss shown in Fig. 18.4, it is interesting to note the behavior of several statically determinate modifications of this structure. Suppose that the structure consisted simply of two bars, say *ab* and *ac* (or *ab* and *ad*). For such a structure, when the vertical load P reached a value of 30 kips, the force in bar *ab* would reach its yield value of 30 kips and excessive vertical deflections of joint *a* would impend. Therefore, the value of both P_{yp} and P_{ult} is 30 kips.[2] On the other hand, if the structure consisted of simply the two bars *ac* and *ad*, a vertical load P of 48 kips would produce yield stresses of 30 kips per sq in. in both bars simultaneously. Therefore, the value of both P_{yp} and P_{ult} is again identical but equal to 48 kips.

Considering now the actual indeterminate structure in Fig. 18.4, it is easy to analyze by any of the standard methods its behavior during the initial linearly elastic range. In this simple case, it is even easier to recognize that, for consistent deformation of the three bars,

$$\Delta L_{ac} = \Delta L_{ad} = 0.8 L_{ab}$$

as shown by the dotted lines in Fig. 18.4*a*. Therefore,

$$\frac{(F_{ac})(25)}{(1)(E)} = \frac{(0.8)(F_{ab})(20)}{(1)(E)}$$

or $F_{ac} = F_{ad} = 0.64 F_{ab}$. However, for vertical equilibrium of joint *a*, $0.8F_{ac} + 0.8F_{ad} + F_{ab} = P$. Hence, $2.024F_{ab} = P$. Since bar *ab* starts to yield when $F_{ab} = 30$ kips, $P_{yp} = 60.72$ kips. For this load, note that bars *ac* and *ad* are still well within their elastic range since

$$F_{ac} = F_{ad} = (0.64)(30) = 19.2 \text{ kips}$$

Since bars *ac* and *ad* are available to resist further loading of the structure even though bar *ab* is yielding at a constant resisting force of 30 kips, load P may be increased an additional 17.28 kips

---

[1] This characteristic inability of compression members to hold their maximum compressive load indefinitely over the same yield range as a tension member is the main reason why the plastic behavior of truss structures cannot be relied on in design.

[2] In this discussion, the subscripts *yp* and *ult* attached to the load symbol have the following significance: P_{yp} is the load at which yield stress intensities are reached in the most highly stressed element, and P_{ult} is the load at which excessive deflection of the structure impends.

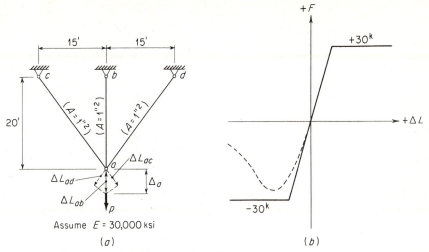

Figure 18.4   Simple truss structure

to a total value of 78 kips before the force in members ac and ad is built up an additional 10.8 kips to a total of 30 kips. Since all three bars are now on the point of yielding, excessive deflections impend and P_{ult} therefore is equal to 78 kips. Note that this value could also be computed directly by noting that excessive deflections impend when all three bars are at their yield value of 30 kips. Hence, simply from $\sum V = 0$,

$$P_{ult} = 30 + (2)(0.8)(30) = 78 \text{ kips}$$

During the initial elastic stage, Δ_a, the vertical deflection of joint a, is equal to the change in length of bar ab, or

$$\Delta_a = \frac{(F_{ab})(20)}{(1)(E)} = \frac{(P)(20)}{(2.024)(E)}$$

If E is 30,000 kips per sq in., $\Delta_a = P/3{,}036$. Thus at P_{yp}, $\Delta_a = 0.02$ ft. Of course, during the increment of loading between P_{yp} and P_{ult}, only the bars ac and ad are resisting the additional load ΔP, resulting in a structure about twice as flexible. Thus, $\Delta(\Delta_a) = (1.25)(\Delta F_{ac})(25)/(1)(E)$, or $\Delta(\Delta_a) = \Delta P/1{,}536$. Hence, for the last 17.28 kips of load, the additional deflection is $17.28/1{,}536$, or 0.01125 ft. Thus, at P_{ult} the total deflection of a is $0.02 + 0.0112 = 0.0312$ ft.

The behavior of this structure through this stage is depicted graphically by the initial-loading portions of the graphs shown in Fig. 18.5. Plotted are both the bar forces vs. P and the vertical deflection of joint a vs. P.

Now suppose that the structure is gradually unloaded. During the removal of the load, the stress and strain decrease, as indicated by the initial portion of the dashed lines in Fig. 18.3a. In other words, the structure unloads in a linearly elastic fashion parallel to the initial elastic loading stage. As a result, as shown graphically in Fig. 18.5, after the load has been entirely removed, there is a residual deflection of joint a equal to 0.0054 ft downward, and there are the following residual bar forces: $F_{ab} = -8.6$ kips and $F_{ac} = F_{ad} = +5.4$ kips.

If the structure were now reloaded, it would behave in a linearly elastic manner. In fact, over the entire reapplication of the 78 kips, the structure would behave elastically and come back to the identical stresses, strains, and deflections which existed when P_{ult} of 78 kips was first acting on the structure. Further, the structure can be unloaded and reloaded an indefinite number of times without any further plastic deformation beyond that which occurred during the initial loading up to 78 kips.

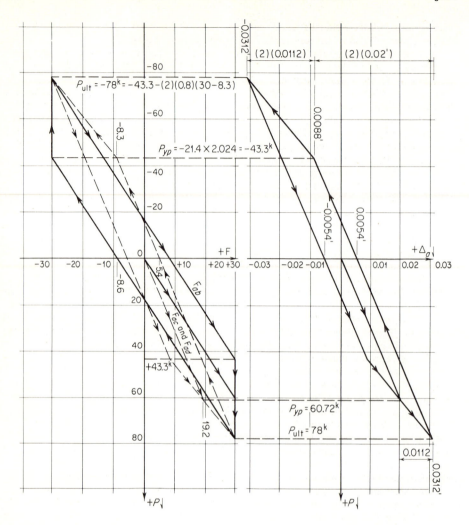

Figure 18.5   Behavior of three-bar truss structure shown in Fig. 18.4

The behavior of this simple structure illustrates very effectively the elastic-plastic behavior of indeterminate structures when subjected to a load system the points of application, distribution, and directions of which are unchanging.   During the initial application of such a load, plastic deformation occurs only during the stage in which the structure mobilizes its reserve strength between P_{yp} and P_{ult}.   As a result of this permanent set (which, in the illustrative example, is an elongation of 0.0112 ft in bar *ab*), *the structure has in effect "prestressed itself" so that this particular loading can be removed and reapplied an indefinite number of times in an elastic manner and with no further plastic deformation.*

Note further in this structure the situation if the members are disconnected from one another at joint *a* after the first cycle of loading and unloading.   It would be found that members *ac* and *ad* are still 25 ft long, but member *ab* would be 0.0112 ft longer than its original length of 20 ft.   To reassemble the members at *a* would require that bars *ac* and *ad* be stretched with tensions of 5.4 kips and that bar *ab* be compressed 0.0057 ft with a compressive force of 8.6 kips.   This of course corresponds with the residual stresses which are noted above to exist in the unloaded structure after the load of 78 kips had been applied and removed the first time.

Now consider what would happen if the direction of the load were reversed and an upward load were applied to the structure. (Remember that it is assumed that the members can carry compressive forces without buckling.) During the initial stages of this upward loading, the structure would behave elastically. However, when an upward load has been applied sufficient to increase the force in ab from its residual value of -8.6 kips to its compressive yield value of -30 kips, the structure is on the point of undergoing plastic deformation in the upward sense. This upward load P_{yp} will be $P_{yp} = (2.024)(-21.4) = -43.3$ kips. During the application of this load, the force in bars ac and ad will have changed from $+5.4$ kips to a value of $+5.4 + (0.64)(-21.4) = -8.3$ kips. Whereas bar ab is now on the point of being compressed plastically, bars ac and ad still have considerable reserve capacity before they too are on the point of being compressed plastically. An additional load of $(2)(0.8)(30 - 8.3) = 34.7$ kips can be applied before bars ac and ad reach their compressive yield force of 30 kips. Thus, the upward $P_{ult} = -43.3 - 34.7 = -78$ kips. *This is of course the same as the downward P_{ult} and likewise can be computed simply from statics and the yield capacity of the three bars.*

Note that during the application of the last 34.7 kips of upward load, bar ab is compressed plastically $(2)(0.0112) = 0.0224$ ft. If it were disconnected from the other bars at joint a, it would be 0.0112 ft shorter than its original length of 20 ft.

If the upward load of 78 kips is removed, the unloaded structure will have residual bar forces and deflections numerically the same but opposite in sense to those remaining after the removal of the downward load of 78 kips.

If the structure were now reloaded with a downward load, bar forces and deflections would change in a linearly elastic manner for the first 43.3 kips of loading. Bar ab would then have a force of $+30$ kips and would stretch plastically during the application of the additional 34.7 kips of loading which is required to bring bars ac and ad to their tension yield value of 30 kips. *Thus, a reversing load of 78 kips applied first down and then up can be applied without producing excessive deflection. With each cycle of downward and upward loading, bar ab is subjected to the plastic-fatigue condition of first being stretched plastically 0.0224 ft and then compressed plastically the same amount.* A steel member could not stand too many cycles of such abuse.

To avoid alternating plastic deformation with each cycle of downward and upward loading, note that the absolute sum of the downward and upward load cannot exceed $2P_{yp}$, or $(2)(60.72) = 121.4$ kips. If 60.7 kips is applied downward, 60.7 kips can be applied upward; if 78 kips is applied downward, only 43.4 kips can be applied upward; if 65 kips is applied downward, 56.4 kips can be applied upward; etc. This behavior is typical of the effect of reversing loads in situations involving plastic deformation.

Now consider the behavior of this structure in its original unstressed condition but acted upon by an entirely different loading, say a horizontal load H acting to the right at joint a. Under linear elastic conditions, such a loading would produce the following bar forces: $F_{ab} = 0$, $F_{ac} = -F_{ad} = 5H/6$. In this case, $H_{yp} = 1.2F_{ac} = 36$ kips. Since both ac and ad are on the point of yielding, one in tension and one in compression, no reserve load-carrying capacity is available. Therefore, $H_{ult} = H_{yp} = 36$ kips. *This illustrates the property that the reserve strength above the yield load depends both on the properties of the structure and on the nature of the loading.*

This simple problem has served to illustrate all the most important characteristics of plastic behavior except the effect of alternating loadings. Such loadings can cause an accumulation of plastic deformation first at one location then at another location in the structure. With each such cycle of loading, the structure deflects more and more out of its original shape. Such behavior is illustrated in Art. 18.6.

18.4 Moment-Curvature Relations for Flexural Members. It is convenient to start the discussion of the plastic behavior of flexural members by considering the case of pure bending of a simple beam of rectangular cross section acted upon by equal end couples. Such a case involves no shear or axial force and no transverse loads. Hence, plane cross sections before bending remain plane after bending, and it is easy to develop, for both the elastic and the elastic-plastic ranges, relations between σ_1 (the stress in the extreme fiber) and M (the bending moment) and between ϕ (the curvature) and M. Note that the curvature ϕ is equal to $d\theta/ds$, the rate of change of slope of the cross sections of the beam. In other words, the curvature of ϕ is equal to the relative rotation of two adjacent cross sections a unit distance apart.

In Fig. 18.6 are shown several stages in the bending of a differential element of this beam, starting with the end of the elastic range where σ_1 has just reached the value σ_{yp} and concluding with the fully plasticized cross section where fibers close to the center of the cross section have reached the yield strain ϵ_{yp}. It is assumed that the material is following the idealized stress-strain curve shown in Fig. 18.2. The width of the rectangular cross section is b and the depth d.

Throughout the elastic range, the bending moment M is equal to $\sigma_1 \mathscr{S}$, where $\mathscr{S}$ is the section modulus of the cross section and the curvature ϕ is equal to M/EI. At the end of the elastic range, the stress in the extreme fiber σ_1 reaches the yield-point stress σ_{yp}. Therefore,

$$M_{yp} = \sigma_{yp}\mathscr{S} \tag{18.1}$$

$$\phi_{yp} = \frac{M_{yp}}{EI} \tag{18.2}$$

A beam is an assemblage of an infinite number of elements or fibers, and the internal stress distribution among these elements must result in consistent deformation of them as well as satisfying equilibrium requirements. In other words, a beam is a structure which is statically indeterminate to an infinite degree. Just as in the case of the simple indeterminate truss discussed in Art. 18.3, the fact that the outer fibers of the beam are on the point of yielding does not mean that excessive deflections impend immediately, because the entire inner core of the beam is still elastic and capable of resisting additional bending moment.

The postelastic behavior of a beam can be studied by considering successive stages of deformation, as shown in Fig. 18.6 for a beam of rectangular cross section. In these considerations it is assumed that cross sections continue to remain plane and therefore that the bending strains vary linearly across the beam. Since the material is being assumed to follow the idealized stress-strain curve, the stress in any fiber which is strained more than ϵ_{yp} is constant and equal to σ_{yp}. Thus, in any postelastic stage, there is an inner elastic core in which the bending strains are equal to or less than ϵ_{yp} and over which the normal stresses σ vary in a linear manner. This core is flanked above and below by plasticized layers in which the bending strains are equal to or more than ϵ_{yp} and in which the normal stresses are constant and equal to σ_{yp}. At any stage it is easy to compute the resisting moment furnished by the bending stresses and thereby define the equivalent bending moment involved. Note that the curvature at any stage is defined by the depth of the elastic core of the beam.

By considering various stages of plastification and computing the bending moment and curvature associated with each, it is possible to obtain the data for plotting the nondimensional moment-curvature relationship shown in Fig. 18.7. In the limit, the elastic core shrinks to a differential depth, the upper half of the beam is plasticized compressively, and the lower half is plasticized in tension. This is called the **fully plasticized stage** and represents the ultimate bending resistance M_{pl} the cross section can provide:

$$M_{pl} = \sigma_{yp}\mathscr{Z} \tag{18.3}$$

where $\mathscr{Z}$ is called the **plastic modulus** of the cross section. For the rectangular cross section, $\mathscr{Z} = bd^2/4$. Note that

$$\frac{M_{pl}}{M_{yp}} = \frac{\mathscr{Z}}{\mathscr{S}} = \mathscr{F} \tag{18.4}$$

where $\mathscr{F}$ is called the **form factor** of the cross section. For the rectangular cross section, $\mathscr{F} = 1.5$.

Beams of other types of cross sections could be studied in exactly the same manner as described above for the rectangular cross section. If this is done, results such as shown in Fig. 18.7 will be obtained for typical cross sections. Of these, the most important is the idealized W^F-beam cross

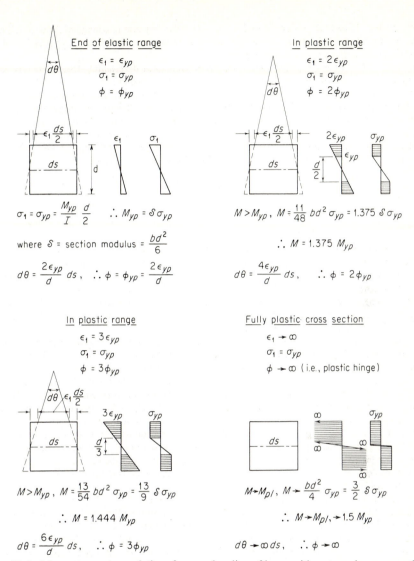

Figure 18.6   Moment-curvature relations for pure bending of beam with rectangular cross section

section shown in Fig. 18.8.   For the W^F cross section, the following values would be obtained for the section modulus $\mathscr{S}$ and the plastic modulus $\mathscr{Z}$:

$$\mathscr{S} = \frac{b_1 h_1^3}{6h} + \frac{b(h^2 - h_1^2)(h + h_1)}{8h} \tag{18.5}$$

$$\mathscr{Z} = \frac{b_1 h_1^2}{4} + \frac{b(h^2 - h_1^2)}{4} \tag{18.6}$$

If these formulas are applied to typical W^F beams, the form factor $\mathscr{F}$ will be found to range from 1.10 to 1.18 and to be about 1.12 for the most commonly used sizes.   Note that in the limit, if half

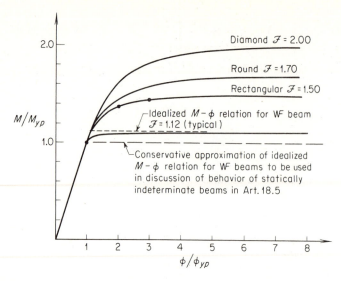

Figure 18.7 Nondimensional moment-curvature relationships for typical beam cross sections

the area of the beam were concentrated on a line at the top edge and the other half on a line at the bottom edge with an infinitesimally thin web in between, the form factor $\mathscr{F}$ of such an I beam would be 1.0.

The moment-curvature relation of the W^F beam is such a flat curve for the portion between M_{yp} and M_{pl} that often the idealized version shown by the short-dash line in Fig. 18.7 is used. In fact, sometimes the more conservative version shown by the long-dash line is used, as is done in the discussion of the statically indeterminate beams in Art. 18.5. When either of these idealizations is used, the concept is that when the moment reaches the level defined by the horizontal lines,

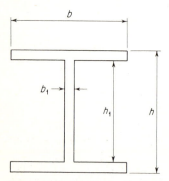

Figure 18.8 Idealized wide-flange (W^F) beam cross section

an indefinite amount of flexing or curvature can occur without increase in resisting moment. In other words, the beam acts from then on as if a friction hinge had been installed, which allowed flexing to occur while it resisted at a constant moment. In the subsequent discussions, this concept will be referred to by stating that "a **plastic hinge** with a resisting moment of ――― kip-ft forms ―――."

The above considerations are based on the assumption of pure bending. This study of the moment-curvature relations could be extended to consider the effect of using the actual stress-strain curve, including the effect of strain hardening, the effect of residual stresses, the effect of shear and axial force in combination with bending moment, etc. The reader is referred to books such as noted in Art. 18.1 for a more detailed discussion. The introduction of such refinements would not alter in any major way the subsequent discussion of the plastic behavior of beams, and these details are therefore not included in this presentation.

18.5 Plastic Behavior of Typical Beams. The following discussion of the plastic behavior of beams parallels that given in Art. 18.3 regarding trusses and is prefaced by the consideration of a beam supported in a statically determinate manner.

Simple end-supported beam. Consider the simple end-supported beam acted upon by a single concentrated load P at mid-span, as shown in Fig. 18.9. A beam of rectangular cross section and a

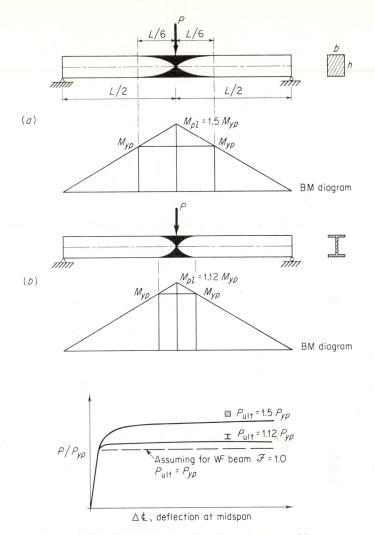

Figure 18.9 Plastic behavior of simple end-supported beam

W^F beam are considered. Either of these beams behaves in a linearly elastic manner up to the load P_{yp}, which develops a moment M_{yp} at mid-span.

Consider first that the true $M - \phi$ relations shown by the solid lines in Fig. 18.7 pertain. Whereas either of these beams starts to undergo plastic deformation at the yield load, either will accept more load before excessive deflections occur. The beam of rectangular cross section can be loaded to a P_{ult} of $1.5P_{yp}$ before the mid-span cross section is fully plasticized. At this ultimate load, the plasticized zone (shown in black) will have spread out one-sixth of the span on either side of the load on the top and bottom edges of the beam. In the case of the W^F beam if $\mathscr{F} = 1.12$, the plasticized zone is much more localized and will have spread out only $0.0536L$ on either side of the ultimate load of $1.12P_{yp}$.

Note also how flat the load-deflection curve of the W^F beam is for loads greater than P_{yp}. In fact, if the conservative version of the idealized $M - \phi$ relation in Fig. 18.7 were used, it could be

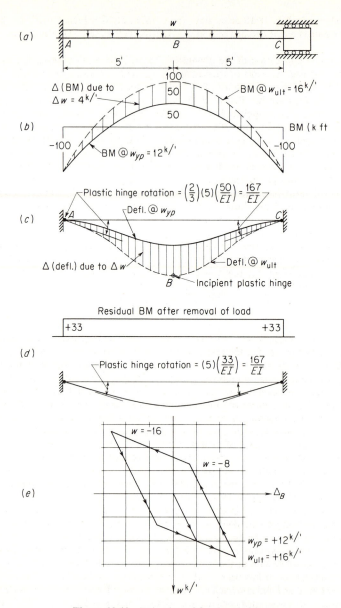

Figure 18.10   Behavior of fixed-ended beam

said that a plastic hinge forms when the load reaches P_{yp} and excessive deflections impend immediately.

Fixed-end beam.   In the following discussion of the behavior of a W^F beam fixed at both ends, it is assumed that the conservative idealization of the $M - \phi$ curve shown in Fig. 18.7 pertains.   For illustrative purposes, it will be assumed that for this W^F cross section

$$M_{yp} = M_{pl} = 100 \text{ kip-ft}$$

The beam is shown in Fig. 18-10 acted upon by a uniformly distributed load of w kips per ft. Note that the clamped end is detailed so that it is impossible to develop even secondary horizontal reactions.

During the initial application of the downward loading, the beam behaves in a linear elastic manner, developing end moments of $wL^2/12$ and a moment at mid-span of $wL^2/24$. A load of 12 kips per ft will develop end moments of -100 kip-ft, the plastic moment capacity of the beam. Therefore $w_{yp} = 12$ kips per ft. Plastic hinges are now on the point of forming at A and C if additional downward load is applied, but the member is capable of carrying an additional increment Δw as an end-supported beam. The load Δw may be increased to a value such that the simple beam moments caused by it increase the moment at B by 50 kip-ft, thereby bringing the total moment at B up to $+100$ kip-ft. A plastic hinge is now on the point of forming at B, and the beam will collapse as a two-bar mechanism if any further loading is attempted. Therefore, $\Delta w = (50)(8)/100 = 4$ kips per ft, and $w_{ult} = 12 + 4 = 16$ kips per ft.

The fixed-ended beam deflections under the first 12 kips per ft and the simple beam deflections under the last 4 kips per ft are shown in Fig. 18.10c. Also noted is the plastic rotation which occurs at the plastic hinges A and C under this last 4 kips per ft of loading, with the beam acting as a simple end-supported one.

If the load were now removed, the beam would act in the linear elastic manner of a fixed-ended beam. The change in bending moments during the removal of the load would be identical to the fixed-ended beam bending moments caused by an upward load of 16 kips per ft, or

$$\frac{+(16)(100)}{12} = +133 \text{ kip-ft}$$

at A and C and $-(16)(100)/(24) = -67$ kip-ft at B. When these values are superimposed on the dashed-line curve in Fig. 18.10b, the residual moments of $+33$ kip-ft shown in Fig. 18.10d are obtained. These residual moments and the residual deflections are locked in the beam as a result of the permanent plastic rotation which occurred at the plastic hinges at A and C.

Note that in effect the beam is now "prestressed" so that w_{ult} of 16 kips per ft may be applied and removed an indefinite number of times without further plastic deformation taking place.

However, if an upward loading were now applied to the unloaded structure, only 8 kips per ft could be applied elastically, since this load would bring the moment at A and C up to a value of $+100$ kip-ft. An additional upward load of 8 kips per ft could be applied with plastic hinges acting at A and C, before the collapse as a two-bar mechanism is imminent due to the formation of a third plastic hinge at point B. During the application of this last 8 kips per ft of loading, plastic rotation of $333/EI$ (in an upward sense) would take place at the plastic hinges A and C. When this total upward loading of 16 kips per ft is removed, residual moments and deflections numerically the same but opposite in sense to those shown in Fig. 18.10d would now be locked in the beam. This reversing plastic deformation which would have then taken place at A and C would be exactly the same as the phenomenon used to break a wire or nail.

As in the case of the truss example of Art. 18.3, the only way to avoid reversing plastic deformation is to keep the absolute sum of the upward and downward load less than twice w_{yp}. Figure 18.10e shows the behavior of this beam in the same manner as is used in Fig. 18.5.

Comparison of yield and ultimate loadings for several typical beams. For a uniform downward loading, the fixed-ended beam just considered exhibits a reserve load capacity of $4\!/_{12}$, or $33\frac{1}{3}$ per cent, between w_{yp}, the load at which plastic deformation starts, and w_{ult}, the load at which excessive deflections or collapse impends. *For a nonreversing and immovable loading on any indeterminate structure, the magnitude of this reserve load capacity depends both on the distribution of the strength characteristics of the structure and on the distribution of the loading on the structure.* As a result, a particular structure may exhibit considerable reserve capacity under one loading but very little or no reserve under another loading.

To illustrate this principle, consider the same fixed-ended beam as used above but loaded by a single concentrated load P at mid-span, as shown in Fig. 18.11. For this loading, in the elastic range the bending moments at both ends and at mid-span are all numerically equal to $PL/8$.

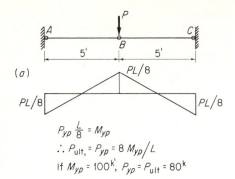

Legend regarding formation of plastic hinges:

1st hinge:

2nd hinge: etc.

Incipient final hinge:

In all cases, $M_{yp} = M_{pl}$

(a)

$$P_{yp} \frac{L}{8} = M_{yp}$$

$$\therefore P_{ult,} = P_{yp} = 8\, M_{yp}/L$$

If $M_{yp} = 100^{k'}$, $P_{yp} = P_{ult} = 80^{k}$

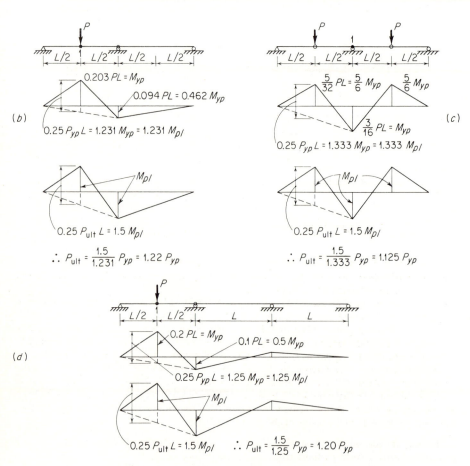

Figure 18.11   Yield and ultimate loadings for typical continuous beams

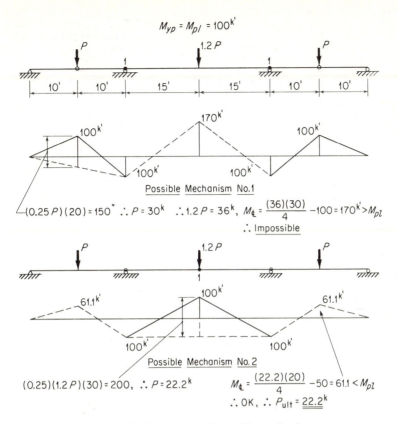

Figure 18.12   Comparison of possible mechanisms

Therefore, M_{yp} (and in this case M_{pl}) is developed simultaneously at both ends and at mid-span under a load of $(100)(8)/10 = 80$ kips. Therefore, both P_{yp} and P_{ult} equal 80 kips, and there is *no* reserve capacity exhibited under this type of loading.

In Fig. 18.11b, c, and d, both the yield and ultimate loads for several typical continuous beams and loadings are illustrated in order to define the reserve capacity which is developed during the plastification of the structure. Note that it is possible to compute the ultimate loading without first computing the yield loading, as is illustrated in Fig. 18.12. *To compute the yield load, it is necessary to perform an elastic indeterminate analysis of the structure. To compute the ultimate load, it is necessary to perform only a plastic analysis of the structure.*

The reader is referred to "Plastic Design of Steel Frames" by L. S. Beedle for a full discussion of various methods of plastic analysis. Here, it will simply be noted that there are primarily two basic methods, the mechanism method and the statical method. The mechanism method is illustrated in Fig. 18.12 and consists of the following steps:

1. Identify the location of potential plastic hinges (supports, joints, concentrated load points, and points of zero shear in portions subjected to distributed loadings).
2. Consider various combinations of these hinges to form possible collapse mechanisms.[1]

---

[1] Note that the ultimate loading may be associated with a collapse mechanism which *may* or *may not* involve the entire structure. For example, in Fig. 18.11d, only the left span is on the point of collapsing; the other two spans are well within the elastic range.

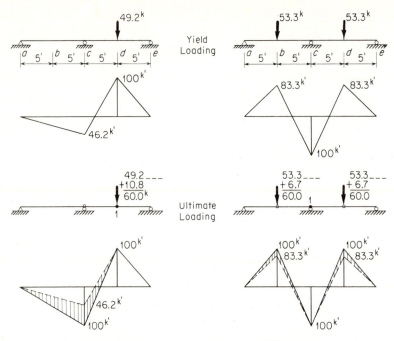

Figure 18.13   Two-span continuous W beam

3. Compute the loads associated with each of these possible mechanisms.   The correct mechanism of failure will be the one associated with the lowest load.   This will be the ultimate loading, provided that the moment at any section does not exceed M_{pl} at that section.

18.6   Plastic Behavior of Continuous Beam under Alternating Loading.   One more aspect of plastic behavior will be discussed using as an illustration the two-span continuous beam shown in Fig. 18.13.   Assume that for this W beam $M_{yp} = M_{pl} = 100$ kip-ft.   Shown in Fig. 18.13 are the yield and the ultimate values of two loading conditions: (1) a single concentrated load applied at the center of one span and (2) equal concentrated loads applied simultaneously at the center of both spans.   These ultimate loads are computed on the assumption that each of the loadings is nonreversible and immovable.

During the application of the ultimate value of the single concentrated load, plastic rotation occurs at a plastic hinge which forms in the beam at the load point d.   During the application of the ultimate value of the two concentrated loads, plastic rotation occurs at a plastic hinge located over the center support.   In the case of the single load, during the application of the loading between the yield and ultimate condition, rotation occurs at the plastic hinge at d and the last 10.8 kips of load is carried by portion $abcd$, which is supported at a and c and cantilevered from c to d.   The bending moments change by the shaded amount between the dashed and solid lines. Similarly, in the case of the two equal loads, the last 6.7 kips of each load is carried by ac and ce acting as end-supported beams while rotation occurs at the plastic hinge located at point c.   Of course, in the case of either of these loading conditions, if it were the only loading condition involved, it could be removed and reapplied an indefinite number of times without any further plastic deformation beyond that which occurred during the first application of the ultimate value of the loading.

Consider now the behavior of the beam if it is subjected alternately to the ultimate loading of each of these two loading systems.   First the single 60-kip load is applied at d and then removed; then the two 60-kip loads are applied one at b and one at d and later removed; then the single

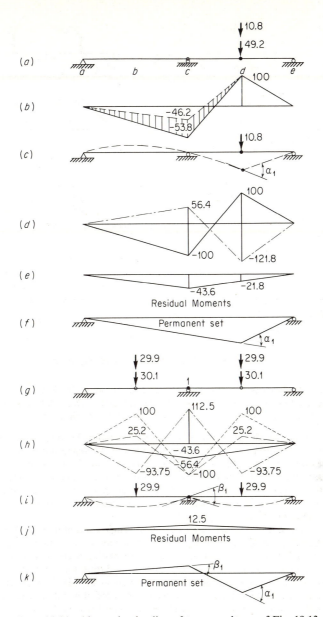

Figure 18.14 Alternating loading of two-span beam of Fig. 18.13

60-kip load is again applied at d and then removed; then the two 60-kip loads are applied and removed; etc.

The following sequence of events is recorded in Fig. 18.14. As described above, the last 10.8 kips of the 60-kip load at d causes the plastic rotation α_1 of the plastic hinge at point d. When this 60-kip load is removed, the beam behaves elastically and causes changes in the bending moments, as noted on the dot-dash diagram in (d). These changes in moment are $-60/49.2$ times those shown in Fig. 18.13 for the elastic application of the 49.2-kip load at d. The superposition of the dot-dash diagram on the solid-line diagram in (d) gives the residual moments shown in (e). Note that, if the support at c were removed, the member would be kinked as shown in (f) because of the plastic rotation α_1 at d. The upward force required to push c up to its proper position would be found to be equal to 8.72 kips and therefore would produce the residual moments noted in (e).

Suppose that two equal loads are now applied at b and d to the beam with these residual moments, noted again by the solid line in (h). During the first 30.1 kips of such loads, the member would behave elastically and the change in moments thereby produced would be 30.1/53.3 times those shown in Fig. 18.13 for the 53.3-kip loads. These changes in moment would bring the moment c to a value of -100, and when superimposed on the residual moments give the dashed-line curve in (h). An additional 29.9 kips can be applied at b and d, with the structure behaving as the two end-supported beams shown in (i) and with a plastic rotation β_1 occurring at the plastic hinge at c. As the two 60-kip loads are now removed, the member behaves elastically and the change in moments developed during the removal is $-60/53.3$ times those shown in Fig. 18.13 and is indicated by the dot-dash line in (h). The residual moments left in the beam after the removal of the two 60-kip loads are shown in (j). Note that if the support at c were again removed, the member would be kinked as shown in (k) because of the plastic deformations α_1 at d and β_1 at c.

If the single 60-kip loads were applied again at d, the member would behave elastically for the first 46.2 kips, thereby bringing the moment at d up to 100 kip-ft. During the last 13.8 kips of loading, the load would be carried similarly to the manner shown in (c), but the plastic rotation at d would be α_2 and equal to 13.8/10.8 times α_1. Removal of this 60-kip load would leave the same residual moments as shown in (e), but the permanent set would involve two kinks: β_1 at c and $\alpha_1 + \alpha_2$ at d.

If the two 60-kip loads were applied at b and d and then removed, an additional plastic rotation of β_1 would be produced at c. If the single 60-kip load were applied again at d and removed, an additional plastic rotation of α_2 would be produced at d. Subsequently, alternating the two 60-kip loads at b and d with the single 60-kip load at d would cause with every cycle an additional plastic rotation β_1 at c and α_2 at d. As a result, with each cycle the member is kinked more and more at c and d, and before long the member will be so badly deformed that it is no longer service-able.

Computations show that plastic deformations will not accumulate if no more than a 50.5-kip load at d is alternated with two 50.5-kip loads, one at b and one at d. During the first 49.2 kips of the single load at d, the member behaves elastically. During the next 1.3 kips, plastic deforma-tion occurs at the plastic hinge at d. As a result, when the 50.5-kip load is removed, triangularly varying residual moments are left in the beam with peak value of 5.3 kip-ft at c. When the two 50.5-kip loads are now applied at b and d, the member behaves elastically, bringing the total moment at c just up to 100 kip-ft but producing no plastic deformation at this point. Such loadings can subsequently be alternated an indefinite number of times with no further plastic deformation. In other words, the first application of the 50.5-kip load at d "prestressed" the beam so that subsequent alternations of this load with the two 50.5-kip loads could take place elastically with no further plastic deformation. Thus, 50.5 kips is called the value of the **shakedown loading**. Note that it is very little greater than the yield value of a single load at d.

18.7 Stress Control of Indeterminate Structures. It may not seem appropriate to discuss stress control and the normal and hybrid action of statically indeterminate structures in this chapter on plastic behavior. Actually, however, it will be shown that all these subjects are very closely related.

First note that a statically determinate structure can be designed to operate with any stress desired, since the structure is free to change its shape to correspond. For example, the members of a statically determinate truss can be designed to operate at any suitable stress level, since the truss is free to adjust its configuration as may be required. *In general, however, the stress level (and hence the strain level, also) cannot be set arbitrarily in the various portions of a statically indeterminate structure since the various portions must fit together in a consistent manner, so that simultaneously both the equilibrium and the deformation requirements for the structure are satisfied.*

It is possible, however, to control the erection of an indeterminate structure in such a manner that a *desired stress (and strain) pattern is realized for one particular static load condition.* For

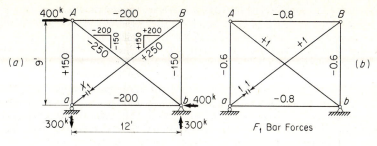

$$(1)(\Delta_1) = \sum F_1 \frac{F}{A} \frac{L}{E} = \frac{1}{E} \sum F_1 \frac{F}{A} L = \frac{3684}{30 \times 10^3}, \quad \therefore \Delta_1 = 0.1228 \text{ in. (together)}$$

$$\therefore \text{ Make bar } aB \text{ 0.1228 in. short}$$

Bar	L (in.)	$\frac{F}{A}$ (ksi)	F_1	$F_1 \frac{F}{A} L$
AB	144	−14	− 0.8	+ 1,612
ab	144	−14	− 0.8	+ 1,612
Aa	108	+18	− 0.6	− 1,167
Bb	108	−14	− 0.6	+ 907
Ab	180	−14	+ 1.0	− 2,520
aB	180	+18	+ 1.0	+ 3,240

$$\Sigma = + 3,684$$

Figure 18.15   Example of stress control

example, the two-span continuous beam shown in Fig. 18.14 (and assumed as weightless for this discussion) could be erected as follows: (1) the beam could be connected to the supports at a and e, both of which are capable of providing either an upward or a downward reaction; (2) then a jack could be placed under the beam at c and an upward force of 8.72 kips applied to the beam; (3) the support at c could then be shimmed to fit the jacked position of the beam; and (4) if the jack were then removed, the reactions of the beam would be 8.72 kips upward at c and 4.36 kips downward at a and e and the initial moments would be as shown in Fig. 18.14e.   Now the beam has been prestressed so that a 60-kip downward load can be applied at d with the structure behaving completely elastically and no plastic deformation would be involved.   This process is called **stress control** of a statically indeterminate structure.

The truss structure shown in Fig. 18.4 furnishes another convenient example of stress control. Refer to the previous discussion in Art. 18.3.   Suppose that members ac and ad of this structure were fabricated exactly 25 ft long but member ab was deliberately made 0.0112 ft too long.   Then during the erection the member could be forced into place, thereby creating a compressive force of 8.6 kips in ab and tensile forces of 5.4 kips in ac and ad.   This structure has now been prestressed so that a vertical load of 78 kips at a will bring the force in all three bars up to their tensile yield capacity of 30 kips simultaneously.

A third example of stress control is given for the simple truss structure shown in Fig. 18.15. Suppose that it is desired to operate the compression members at 14 kips per sq in. and the tension members at 18 kips per sq in.   Further, suppose that it is desired to divide the horizontal load equally between the two diagonals.   In order to satisfy static equilibrium, this choice results in the bar forces shown in Fig. 18.15a.   Suppose that the areas of the bars were now selected so as to obtain exactly the desired stress intensities.   Select aB as the redundant bar, cut it, and apply the tensile redundant forces of 250 kips acting toward each other on each side of the cut.   If the desired conditions are to be obtained, there must be no relative deflection of the two sides of the cut.   The computations in Fig. 18.15 show, however, that the two sides of the cut have come together, i.e tried to telescope one another, 0.1228 in.   If, however, member aB had been made 0.1228 in. too short, the two sides of the cut would fit together perfectly. Therefore, if in the fabrication of this truss member aB were made 0.1228 in. too short and all other members were made perfectly, then during the erection aB could be forced into place. This would create a prestressed condition in the truss such that when the 400-kp load was applied, the desired stress conditions would be obtained exactly.

18.8 Normal and Hybrid Action of Statically Indeterminate Trusses. In the previous article, it was shown that statically indeterminate structures are susceptible to stress control. By this technique, *any* statically indeterminate structure can be made to operate at any prescribed stress condition (consistent with equilibrium requirements) for *any particular static loading condition.* To accomplish this, it is necessary to control the displacements of the point of application of the redundants in a suitable way, as illustrated above.

If stress control is to be used, a designer usually wants to use it to achieve maximum economy (which for illustrative purposes will be assumed to be coincident with minimum volume of material). He therefore starts investigating what values he should use for the redundants to obtain minimum volume of material. Such a study leads him to recognize two distinct types of action of indeterminate structures, **normal action** and **hybrid action.**[1] These ideas apparently are applicable to any type of statically indeterminate structure and to any degree of indeterminacy. However, here only a simple indeterminate truss, indeterminate to the first degree, is considered.

Consider the structure shown in Fig. 18.16. It is recognized that by using stress control, X, the bar force in the redundant bar aB, may have any desired value and the tension bars can be made to operate at 18 kips per sq in. and the compression bars at 14 kips per sq in. Assuming that the required bar areas are exactly obtained, the volume in cubic inches for each bar could be computed easily, as shown for various values of X. The volumes so obtained are plotted in Fig. 18.16 for each bar, and the volumes of all bars are added to obtain the total volume of material in the entire truss, as shown by the dot-dash line. Note that the vertical scale used for the individual bars is twice that of the total volume of the truss.

The interesting fact here is that the minimum volume for the truss is obtained when $X = 0$. In other words, the minimum volume is obtained when the truss is reduced to the two-bar statically determinate truss formed by aA and Ab. This illustrates **hybrid action** of a statically indeterminate structure.

Now consider the same truss but this time loaded at B as shown in Fig. 18.17. By proceeding in the same fashion as above, it is easy to obtain the volume curves noted. The interesting feature of this case is the minimum range of the volume curve for the truss, between $X = 0$ and $X = 500$ kips. The volume of material is the same for any value selected for the redundant between these limits. This range is called the **normal range,** and the action illustrated is called the **normal action** of a statically indeterminate structure.

For values of redundants selected within the normal range, direct design is possible without using stress control. In other words, if X is selected within this range and the areas are determined to satisfy the desired stresses, it will be found that the two sides of the cut in the redundant match perfectly without making any adjustment in the length of this bar. Also within this range, the deflected configurations of the structure are identical.

Note the structures defined by the values of the redundants at the ends of the normal range. When $X = 0$, the structure reduces to the statically determinate truss composed of AB, Aa, and Ab. When $X = +500$, the structure reduces to the statically determinate truss composed of Ba, Bb, and ab.

18.9 Application of Principles of Normal and Hybrid Action to Structural Design. Whereas the considerations of the previous article are restricted to a very simple indeterminate truss, more extensive studies indicate that the following conclusions regarding normal and hybrid action are

[1] This general subject has interested a number of structural engineers over the years. Apparently one of the first was F. H. Cilley about 1900; see F. H. Cilley, The Exact Design of Statically Indeterminate Frameworks: An Exposition of Its Possibility, but Futility, *Trans. ASCE*, June 1900. See also L. M. Laushey, Direct Design of Optimum Indeterminate Trusses, *J. Struct. Div. ASCE*, vol. 84, December 1958.

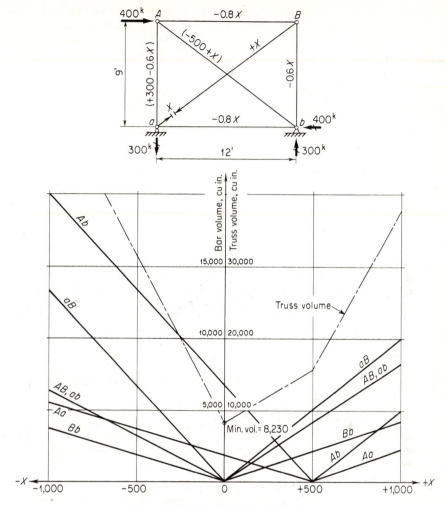

Figure 18.16 Example of hybrid action

valid for any type of indeterminate structure which is to be designed for *a system of immovable static loads:*

1. For every statically indeterminate structure, there is a statically determinate version which, if designed by itself, will have the minimum volume of material, and for the volume used, minimum deflections.

2. For some structures under some loadings, a normal range exists through which identical minimum volumes of material and identical deflections are obtained.

3. Through the normal range, direct design is possible without resorting to stress control.

4. For any statically indeterminate structure for which a direct design is not obtainable, successive approximation leads to the most economical design which will be statically determinate if the original layout and loading involve hybrid action.

5. Apparently from the standpoint of strength and stiffness, no theoretical economy is achieved by using statically indeterminate structures for static load systems.

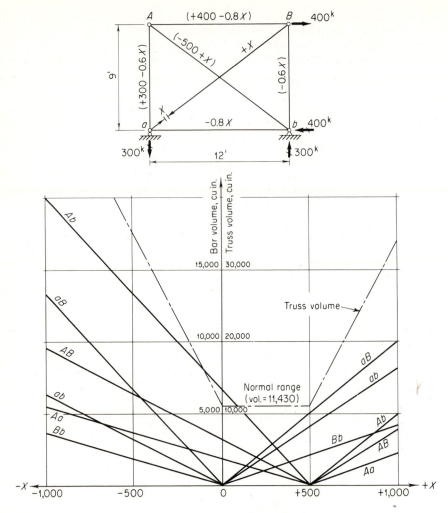

Figure 18.17 Example of normal action

These conclusions do not pertain to situations where significant moving or movable loads must be provided for in the design.   Under such loadings, maximum stresses at various locations in the structure occur for different positions of the moving loads.

It is interesting to note that standard practice is consistent with the above conclusions.   For the very-long-span structures where the design is controlled largely by the dead weight of the structure, i.e., a static load, the most commonly used structures are statically determinate, such as cantilever trusses or girders, three-hinged arches, etc.

18.10 Plastic Design of Steel Structures. The above considerations are not presented as a complete and comprehensive treatment of the subject of plastic analysis and design.   Rather this chapter is intended as an introduction to some of the fine textbooks and reference books and to the extensive periodical literature which is now available.

The concept of basing design on the ultimate load capacity of a structure is over fifty years old, though most of the present knowledge in this area has been developed in the last thirty-five years. At the present time, there is a great deal of interest in incorporating these ideas into standard practices wherever they are appropriate. Of course, there are many examples of the tacit accep-tance of plastic behavior: joint designs based on uniform distribution of load to rivets, bolts, and welds; neglect of stress concentrations under static load conditions where brittle fracture is unlikely; neglect of residual stresses due to straightening or other fabrication and manufacturing procedures; etc.

Plastic-design ideas are most appropriate for substantially static load conditions. When severe reversing or alternating loads are involved, the possibility of plastic fatigue or the accumula-tion of plastic deformation must be recognized. Modern construction and design usually involve statically indeterminate structures even though the above considerations of hybrid and normal action suggest that on the basis of strength and stiffness requirements there is no economy in statically indeterminate construction for substantially immovable and static loadings. It has also been shown that for static loads stress-control techniques lead to the same design as plastic-design procedures. The plastic-design philosophy, however, offers the structure the opportunity to exercise its own stress control by introducing plastic deformations at its plastic hinges.

From many points of view, it is important for the modern structural engineer to understand both the elastic and the plastic behavior of his structures and to blend this knowledge into his analytical and design practices.

18.11 Problems for Solution

Problem 18.1 Compute the key points of and draw the load-deflection curve (P vs. Δ_a) for the structure shown in Fig. 18.18 for each of the following assumptions for the stress-strain curve for

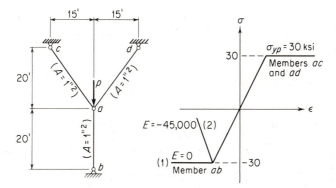

Figure 18.18   Problem 18.1

member ab: (1) that the member yields indefinitely at a constant yield stress of 30 kips per sq in. in compression; in other words, this member yields indefinitely with an effective E of zero; (2) that the member yields, as shown by line 2, where the effective E is $-45,000$ kips per sq in.

Problem 18.2 In the following discussion of plastic behavior and plastic design vs. elastic design of beams, assume that I-beam cross sections are involved and, in such cases, that the form factor $\mathscr{F}$ of the cross sections is equal essentially to 1.0; or in other words the section modulus and the plastic modulus of the cross section are essentially equal. Consider the design of a two-span beam of constant EI to carry the two 40-kip loads shown in Fig. 18.19. Assume that σ_{yp} of the structural steel is 33 kips per sq in.

1. On an *elastic-design* basis, using a factor of safety of 1.65, the allowable bending stress is $33/1.65 = 20$ kips per sq in. Therefore, the section modulus required to carry the maximum moment of 75 kip-ft is $\mathscr{S} = (75)(12)/20 = 45$ in.³

2. On a *plastic-design* basis, using the factor of safety of 1.65, the beam should be designed for ultimate loads of $(40)(1.65) = 66$ kips. Therefore, $1.5M_{pl} = P_{ult}L/4$, or $M_{pl} = (66)(10)/(1.5)(4) = 110$ kip-ft. The plastic modulus (and also the section modulus if $\mathscr{F} = 1.0$) required to provide this

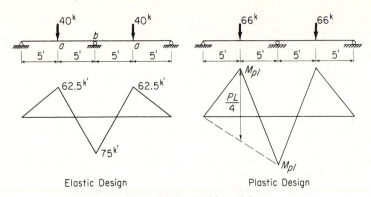

Elastic Design   Plastic Design

Figure 18.19   Problem 18.2

M_{pl} at σ_{yp} of 33 kips per sq in. is $\mathscr{Z} = (110)(12)/33 = 40$ in.³, thereby showing a saving of 40 vs. 45 over the elastic design above.

3. Stress control can be used to equalize the moments at points a and b for the elastic design. For example, during construction the end supports can be jacked up enough to create the initial moments shown in Fig. 18.20. When these are superimposed on the elastic moments produced by

Figure 18.20   Initial moments

the 40-kip loads, the totals shown in Fig. 18.21 are obtained.

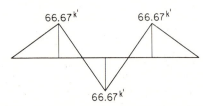

Figure 18.21   Superposition of initial and elastic moments

The section modulus required to carry the maximum moments of 66.67 kip-ft is $\mathscr{S} = (66.67)(12)/20 = 40$ cu in. This demonstrates that the elastic-design approach in combination with stress control leads to the same cross section as plastic design.

Suppose that this beam is designed on an elastic basis, with the intention that stress control will be used to produce the initial moments noted above. Suppose in fact, however, that because of an error the beam was jacked 12 times the intended amount.

(a) Describe the behavior of the beam under these circumstances when the 40-kip loads are applied.

(b) What would be the ultimate value of the loads and what would be the factor of safety against excessive deflections?

Problem 18.3 Consider the two-span continuous beam shown in Fig. 18.22. Assume that it is an I beam and that the form factor $\mathscr{F}$ is essentially equal to 1.0, and therefore that $M_{pl} = M_{yp} = 100,000$ ft-lb.

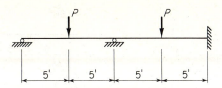

Figure 18.22   Problem 18.3

(*a*) Compute the ultimate value of equal concentrated loads *P* applied at the center of each span.

(*b*) Compute in terms of *EI* of the beam the ductile deformation (i.e., the relative rotation of the cross sections on each side of a plastic hinge) which would occur at the plastic hinges developed in the beam during the first application of these ultimate loads.

(*c*) In the example shown in Fig. 18.13, which had a roller support at the right end instead of being fixed, the ultimate value of the concentrated loads was 60 kips.   What would be the ultimate value if the middle support in this example settled ¼ in. ?

Problem Answers

Following are the answers (or partial answers) to the problems preceded by an asterisk.

Chapter 2

2.1 (*e*) unstable; (*f*) unstable; (*i*) stable, indeterminate; (*l*) stable, determinate; (*m*) unstable; (*n*) stable, determinate; (*o*) unstable; (*t*) stable, externally indeterminate; (*u*) stable, determinate; if β is such that EF does not pass through D; (*v*) stable, determinate

2.2 (*a*) $R_{ax} = 0$, $R_{ay} = 26.0$ kips up; $R_{by} = 46$ kips up
   (*b*) $R_{ax} = 1.48$ kips to right, $R_{ay} = 71.03$ kips up
   $R_{bx} = 25.48$ kips to left, $R_{by} = 50.97$ kips up
   (*c*) $R_{ax} = 30.0$ kips to left, $R_{ay} = 22.92$ kips up
   $R_{by} = 90.42$ kips up; $R_{ey} = 54.26$ kips up
   $R_{fx} = 22.19$ kips to right; $R_{fy} = 42.32$ kips down
   (*d*) $R_{ax} = 10.00$ kips to left, $R_{ay} = 16.36$ kips up
   $R_{by} = 13.64$ kips up
   (*e*) $R_{cx} = 26.\dot{6}$ kips to left, $R_{cy} = 11.85$ kips down
   $R_{dy} = 11.85$ kips up
   (*f*) $R_{ax} = 5.5$ kips to left, $R_{ay} = 4.6$ kips up
   $M_a = 30.0$ kip-ft counterclockwise; $R_{dx} = 1.0$ kips to right
   $R_{dy} = 7.\dot{3}$ kips up
   (*g*) $R_{ax} = 56.0$ kips to left, $R_{ay} = 42.0$ kips up
   $R_{bx} = 24.0$ kips to left, $R_{by} = 12.0$ kips down
   $R_{ex} = 80.0$ kips to right, $R_{ey} = 30.0$ kips up

Chapter 3

3.2 Shear in pounds: left of $A = -1,350$; right of $A = +5,400$; left of $B = -9,000$; right of $B = +5,850$
   Moment in foot-pounds at: $A = -1,800$; midway between A and $B = +16,200$; $B = -9,000$

3.3 Shear in kips at: $a = +15$; $b = -15$; right of $c = +5.417$; $d = +5.417$
   Moment in kip-feet at: midway between a and $b = +50$; $c = -100$; 10 ft right of $c = -75.83$

3.5 Shear in kips at: $A = -20.0$; right of $C = +50.0$; left of $D = +100.0$; $B = +20.0$
   Moment in kip-feet at: $C = -180, -292$; $D = +608, -120$; $B = +40$

3.6 Shear in kips at: $a = +26.46$; right of $b = -24.57$; right of $c = +26.46$; $d = -15.53$
Moment in kip-feet at: $a = -16.0$; $b = +195.73$, $+291.97$; $c = -2.87$, $+93.40$

3.9 Shear in kips at: $a = +45.3$; 20 ft right of $a = +15.3$; 20 ft left of $b = -18.7$; $b = -52.7$
Moment in kip-feet at: 9 ft right of $a = +407.7$; 24 ft right of $a = +637.2$; 6 ft left of $b = 316.2$

Chapter 4

4.1 (a) determinate, stable; (b) indeterminate to first-degree reactions, stable; (c) determinate, stable; (d) indeterminate to first-degree reactions, stable; (e) indeterminate to first-degree bar forces, stable; (f) unstable; (g) unstable; (h) determinate, stable; (k) determinate, stable; (l) indeterminate to first-degree bar forces, stable; (m) unstable

4.2 (A) $a = +18.03$, $b = -8.3$, $c = +25.0$, $d = -18.03$
(B) $a = -43.75$, $b = +24.04$, $c = +2.083$
(C) $a = +3.535$, $b = +27.95$, $c = +17.50$, $d = -15.00$
(D) $a = -1.41$, $b = -4.87$, $c = -5.17$, $d = +28.18$
(E) $a = -20.00$, $b = +24.74$, $c = +32.50$, $d = +38.89$
(F) $a = -45.44$, $b = -42.76$, $c = -5.26$, $d = +60.57$, $e = +26.32$
(G) $a = -22.36$, $b = -56.57$, $c = 0.0$, $d = +60.00$, $e = -20.00$

4.5 (a) Member: $HI = -65.086$; $DE = +65.086$; $Fg = -13.017$; $cD = +45.56$
(b) Geometrically unstable
(c) Geometrically unstable
(d) Member: $Bc = -16.406$; $cd = -24.745$; $De = +4.841$; $fg = -5.156$

4.6 (a) Member: $BD = -28.125$; $Bc = +1.563$; $ab = +27.188$
Member BD: Shear at: $B = +5.0$, $D = -5.0$
Moment at: $B = 0$, $C = +150$, $D = 0$
(b) Member: $CE = +5.625$; $bC = +8.125$; $Ab = -5.0$
Member AC: Shear at: $A = +4.00$; $C = -4.00$
Moment at: $A = 0$, under load $= +120.0$; $C = 0$
Member CE: Shear at: $C = +2.5$; $E = -2.5$
Moment at: $C = 0$; under load $= +75$; $E = 0$

4.8 $R_{ay} = 1,000$ lb down; $R_{az} = 866$ lb up; $R_{by} = 1,000$ lb up; $R_{bz} = 866$ lb down; $R_{cy} = 0$; $R_{cx} = 1,000$ lb to left

4.9 Bar: $ab = -500$; $bc = +1,000$; $ca = -1,000$; $ed = -1,000$; $eb = -1,000$; $ea = +1,414$ lb; all others $= 0$

4.10 $R_{ay} = 1,154$ down; $R_{az} = 250$ up; $R_{by} = 577$ up; $R_{bz} = 750$ down; $R_{cy} = 577$ up; $R_{cx} = 866$ lb to left
Bar: $ab = -433$; $bc = +866$; $ca = -289$; $ed = eb = -577$; $da = +577$; $dc = -815$; $ea = +815$ lb; all others $= 0$

4.11 (a) 11 bars, 8 reactions, 6 joints, therefore indeterminate to first degree
(b) $R_{ay} = 5,000$ down; $R_{az} = 5,000$ to left; $R_{by} = 5,000$ up; $R_{bx} = 0$; $R_{cy} = 5,000$ down; $R_{cx} = 0$; $R_{dy} = 5,000$ up; $R_{dz} = 5,000$ lb to left
Bar: $ef = -5,000$; $ed = -7,500$; $af = +7,500$
$ec = +5,600$; $bf = -5,600$; $bd = +2,500$
$ca = -2,500$ lb; all others $= 0$

Chapter 5

5.1 (a) Ordinate at: left end $= +0.25$; left of $a = -0.5$; right of $a = +0.5$
(b) Ordinate at: left end $= -2.5$; $b = 0$; $a = +5.0$; right end $= -5.0$
(c) Influence line varies linearly from $+1.25$ at left end to -0.5 at right end

5.2 (a) 14,062.5 lb; (b) $+50,000$ lb-ft, -40.625 lb-ft; (c) $+10,312.5$ lb, $-3,750$ lb; (d) $+68,750$ lb-ft

5.5   (a) Influence line varies linearly from -0.75 at C to 0.0 at E to -0.75 at G
         (b) Influence line varies linearly from 0.0 at E to -15.0 at G
5.9   Bar $a = -112,500$ lb; bar $b = +30,300$ lb, $-17,045$ lb
5.10 (a) Influence line varies linearly from 0.0 at left to -1.25 at panel point 3 to 0.0 at right
         end, panel point 7
         (b) Influence line varies linearly from 0.0 at left to $+0.833$ at panel point 3 to -0.125 at
         panel point 4 to 0.0 at right end
         (c) Influence line varies linearly from 0.0 at left to $+0.300$ at panel point 5 to -0.750 at
         panel point 6 to 0.0 at right end
         (d) Influence line varies linearly from 0.0 at left to $+0.333$ at panel point 4 to 0.0 at right
         end

Chapter 6

6.1   (a) Member: $L_0 L_1 = -105.0$; $U_1 L_2 = +84.85$; $U_2 U_3 = +45.0$ kips
         (b) Member: $L_2 L_3 = -31.62$; $U_1 U_2 = +45.0$, -10.0; $L_2 U_3 = -49.50$ kips
         (c) No.
6.4   (a) Influence line varies linearly from 0.0 midway between i and b to 1.0 at b to 0.0 at a
         (b) Influence line varies linearly from 0.0 at h to 1.5 at b to 0.0 at a
         (c) Influence line varies linearly from 0.0 at b to 1.8 at a to 0.0 at f
         (d) 747,000 lb
6.5   Influence line varies linearly from 0.0 at a to $+0.220$ at b to -1.781 at e to 0.0 at i
6.6 $Bc = +19,110$ lb, $-16,930$ lb

Chapter 7

7.1   (a) $+47,812.8$ lb; (b) $\mp 8,437.5$ lb, $\pm 3,750$ lb
7.5   (a) Shear in pounds: from base to knee $= +2,500$; above knee $= -3,750$
         Moment in pound-feet at: base $= -18,750$; knee $= +18,750$; top $= 0.0$
         (b) Axial force: in left post, $-2,846$ lb; in right post, $+2,846$ lb
         Shear: in left post, 3,750 lb to right; in right post, 3,750 lb to right; force in left knee
         brace $= +8,520$; in right knee brace $= -8,520$ lb
7.7   Bending moment at supports, from left to right: 15.0, 30.0, 30.0, 15.0 kip-ft
7.8   Bending moment at supports, from left to right: 13.5, 31.5, 31.5, 13.5 kip-ft
7.9   Bending moment at supports, from left to right: 23.80, 25.45, 25.45, 23.80 kip-ft

Chapter 8

8.2   0.0390 ft to right
8.3   0.375 in. to right
8.5   0.00642 ft down
8.7   (a) 0.0774 ft down; (b) 0.75 in. up, 0 in. horizontal
8.8   1.905 in. up
8.9   0.01695 ft, down and to right at 27.7° to vertical
8.12 (a) 0.00585 ft down; (b) 0.003635 radian counterclockwise
8.14 0.000877 ft down
8.17 (a) At a, $115/EI$ down; at c, $17.5/EI$ up; at d, $10/EI$ down; at b, $23.33/EI$ counterclockwise;
         and at c, $6.67/EI$ clockwise. (b) At a, 0.02204 in.; at b, 0.000373 radian
8.20 0.021 ft down
8.22 0.00698 ft up at 8.56 ft from a
8.25 0.01182 ft down at 5.1 ft from a

Chapter 9

9.1   At A, 83^k up and 44^k to left; at B, 44^k to right; and at C, 17^k up

9.3 $F_{ad} = +3.06^k$

9.4 $F_{ab} = +14.4^k$; $F_{ed} = -15.6^k$; $F_{bc} = +16.1^k$; etc.

9.5   (a) 58.65^k to right; (b) 222.5^k to left; (c) 391^k to right

9.9 M at base of column, $270^{k'}$ causing tension in left-hand fibers

9.10 $R_{ax} = 0$; $R_{ay} = 31.1^k$ up; $M_a = 195.4^{k'}$ counterclockwise; $R_{dy} = 28.9^k$ up

9.11 $R_{ax} = 18.3^k$ to right; $R_{ay} = 30^k$ up; $R_{cx} = 18.3^k$ to left; $R_{cy} = 10^k$ up

9.12 Shear constant and equal to 3^k. $M_A = 15^{k'}$ tension in outer fibers; $M_B = 15^{k'}$ tension in inner fibers

9.13 Shear A to load $= +5.31^k$; shear load to $B = -4.69^k$; $M_A = 17.81^{k'}$; M at load $= +14.07^{k'}$

9.15 $F_{L1b} = +47.0^k$

9.19 Slope-deflection signs. $M_{AB} = -275^{k'}$; $M_{BA} = +170^{k'}$; $M_{BD} = -70^{k'}$; $M_{DB} = -35^{k'}$

9.20 Slope-deflection signs. $M_{ab} = -160^{k'}$; $M_{ba} = -20^{k'}$; $M_{bc} = -90^{k'}$; $M_{be} = -80^{k'}$; $M_{eb} = -40^{k'}$

9.22 Slope-deflection signs. $M_{AB} = -164.6^{k'}$; $M_{BA} = -150.2^{k'}$; $M_{BC} = +150.2^{k'}$; $M_{CB} = +147.4^{k'}$; $M_{CD} = -147.4^{k'}$; $M_{DC} = -163.5^{k'}$

9.25 Slope-deflection signs. $M_{AB} = -180.8^{k'}$; $M_{BA} = +88.3^{k'}$; $M_{BD} = -28.3^{k'}$; $M_{DB} = -10.3^{k'}$; $M_{DE} = +5.1^{k'}$; $M_{DF} = +5.1^{k'}$

9.27 Slope-deflection signs. $M_{AB} = -338^{k'}$; $M_{BA} = -340^{k'}$; $M_{BC} = +294^{k'}$; $M_{BD} = +46^{k'}$; $M_{DB} = +98.6^{k'}$; $M_{DE} = -98.6^{k'}$

9.28 Slope-deflection signs. $M_{AB} = +26.8^{k'}$; $M_{AD} = -26.8^{k'}$; $M_{DA} = -33.2^{k'}$; $M_{DC} = +10.4^{k'}$; $M_{DE} = +15.2^{k'}$; $M_{DG} = +7.7^{k'}$

9.30 Slope-deflection signs. $M_{AB} = -158.2^{k'}$; $M_{BA} = -123.6^{k'}$; $M_{CD} = -223.6^{k'}$; $M_{DC} = -208.2^{k'}$

Chapter 12

12.3 $\det \mathbf{A} = -32$

12.5 $(\text{diag } d_i)^{-1} = \text{diag}(1/d_i)$

12.7 $\mathbf{j}^T = [1 \quad 4 \quad 2 \quad 5 \quad 3 \quad 6 \quad 3 \quad 2 \quad 1 \quad 3 \quad 1 \quad 2 \quad 3 \quad 5 \quad 1 \quad 5]$
            $\mathbf{c}^T = [8 \quad 5 \quad 4 \quad -6 \quad 0 \quad 8 \quad 8 \quad -5 \quad 4 \quad 8 \quad 5 \quad 0 \quad -8 \quad 0 \quad 0 \quad 8 \quad -5 \quad 0]$

12.8 $L = 17.23$

12.12 $A = 43.25$

12.20 $K = 8 \times 10^6$

12.21 $(15,1)(15,1)$ 2; $(15,1)(15,2)$ -0.5; $(15,1)(15,3)$ -0.5

12.24 $(1,2)(1,2)$ 0; $(1,3)(1,3)$ 0; $(1,4)(1,4)$ 0; $(1,6)(1,6)$ 0; $(2,1)(2,1)$ 0; $(2,2)(2,2)$ 0; $(2,3)(2,3)$ 0; $(2,4)(2,4)$ 0; $(2,5)(2,5)$ 0; $(2,6)(2,6)$ 0

12.29 $\mathbf{p}^T = [0 \quad 0 \quad \pi/100]$

12.30 $\Delta t = 20°C$; $\Delta t_{,y} = 20°C/m$; $\Delta t_{,z} = 50°C/m$

12.31 36; 96

12.32 90; 270; 45

12.33 48

12.34 48; 144; 72

12.35 8; 4

12.36 Ne; $M(n-1)e$; $Ne + M(n-1)e$

12.37 Let $a = N + M(n-1)$; then $(a/N)^3$; $(a/N)^2$

Chapter 13

13.1   Local: [258.55   10   20]; overall: [66.04   264.62   96.11]

13.2   Local: [25.47   11.30   0]; overall: [17.38 -4.22 9.08]

13.9 $(15,5)(15,4)$ -0.640; $(15,5)(15,6)$ -0.444

13.10 $\mathbf{d}_{25}^T = 10^{-2}[-2.598 \quad 7.000 \quad -1.500]$; $(25,1)(10,1)$ 0.866; $(25,2)(10,2)$ 1; $(25,3)(10,1)$ 0.5

13.13 Normal force: 7.830; shear y: -101.207; shear z: -98.779; torsion: 3.066; bending y: -72.064; bending z: 2.704

13.15 Normal force: 66.635; shear: -5.367; moment: 30.0

13.24 $[v_{2x} \quad v_{3x} \quad v_{3y}] = 10^{-2}[-5.0656 \quad -7.2817 \quad -2.4941]$

13.29 $[-12 \quad 5 \quad 180]$

Chapter 14

14.6 $\mathbf{v}_0^T = [0 \quad L^4(4\rho_2 + 11\rho_1)/120EI_{zz} \quad -L^3(\rho_2 + 3\rho_1)/24EI_{zz}]$

14.7 $\mathbf{v}_0^T = [0 \quad 7\rho_0 L^4/90EI_{zz} \quad -\rho_0 L^3/10EI_{zz}]$

14.14 $-\mathbf{q}_0^T = [0 \quad 9\rho_2 + 21\rho_1 \quad (2\rho_2 + 3\rho_1)L \quad 0 \quad 21\rho_2 + 9\rho_1 \quad -(2\rho_1 + 3\rho_2)L\,]L/60$

14.15 $-\mathbf{q}_0^T = [0 \quad \rho_0 L/3 \quad \rho_0 L^2/15 \quad 0 \quad \rho_0 L/3 \quad -\rho_0 L^2/15]$

14.17 $\mathbf{q}_0'^T = [0 \quad 0 \quad -(3\rho_1 + \rho_2)L^2/40 \quad 0 \quad -(\rho_1 + \rho_2)L/2 \quad -(5\rho_1 + 3\rho_2)L^2/24]$

14.18 $\mathbf{q}_0'^T = [0 \quad -7\rho_0 L/30 \quad 0 \quad 0 \quad -13\rho_0 L/30 \quad \rho_0 L^2/10]$

14.25 $\mathscr{T}_{ii} = \mathscr{T}_{jj} = \begin{bmatrix} 0.9578 & 0.2873 & 0.6705 \\ -0.2873 & 0.9578 & -1.2451 \\ 0 & 0 & 1 \end{bmatrix}$

Chapter 15

15.1 $e = 3$; $N = 12$; $M = 15$; $n^m = 2$, $m = 1, \ldots, 15$; assuming that the nodes and the elements are labeled from left to right and proceeding from bottom to top, $\mathbf{j}^T = [1,4 \quad 2,5 \quad 3,6 \quad 4,5 \quad 5,6 \quad 4,7 \quad 5,8 \quad 6,9 \quad 7,8 \quad 8,9 \quad 7,10 \quad 8,11 \quad 9,12 \quad 10,11 \quad 11,12]$; $(1,1)(1,1)\ 0$, $(1,2)(1,2)\ 0$, $(1,3)(1,3)\ 0$, $(2,1)(2,1)\ 0$, $(2,2)(2,2)\ 0$, $(2,3)(2,3)\ 0$, $(3,1)(3,1)\ 0$, $(3,2)(3,2)\ 0$, $(3,3)(3,3)\ 0$

15.10 $e = 3$; $N = 3$; $M = 2$; $n^m = 2$, $m = 1,2$; $\mathbf{u}_{(1)}^T = [d_{2x} \quad d_{2y} \quad \theta_{2z} \quad d_{3x} \quad \theta_{3z}]$; $\mathbf{u}_{(2)}^T = [d_{1x} \quad d_{1y} \quad \theta_{1z} \quad d_{3y}]$; $\mathbf{H}_{(21)} = 0$; $\mathbf{u}_{(20)} = 0$; $\mathbf{P} = [\mathbf{i}_4 \quad \mathbf{i}_5 \quad \mathbf{i}_6 \quad \mathbf{i}_7 \quad \mathbf{i}_9 \quad \mathbf{i}_1 \quad \mathbf{i}_2 \quad \mathbf{i}_3 \quad \mathbf{i}_8]$; $\mathbf{H}_{(1)} = [\mathbf{i}_4 \quad \mathbf{i}_5 \quad \mathbf{i}_6 \quad \mathbf{i}_7 \quad \mathbf{i}_9]$; $\mathbf{H}_{(0)} = [\mathbf{i}_1 \quad \mathbf{i}_2 \quad \mathbf{i}_3 \quad \mathbf{i}_8]$; $\mathbf{n}_0^T = [6 \quad 6 \quad 6 \quad 1 \quad 2 \quad 3 \quad 4 \quad 6 \quad 5]$; $\mathbf{a}_0^T = [0 \quad 0 \quad 0 \quad 1 \quad 1 \quad 1 \quad 1 \quad 0 \quad 1]$

15.23 $\bar{\mathbf{p}}_0^T = [0 \quad 0 \quad p_{02x} \quad p_{02y} \quad 0 \quad 0]$

15.25 $\pi = v_1 AEv_1/(\sqrt{2}L) + uAEu/2L - uP$; $v_1 - u/\sqrt{2} = 0$; $\pi' = v_1 AEv_1/(\sqrt{2}L) + uAEu/2L - uP + \lambda(v_1 - u/\sqrt{2})$; solving $\pi'_{,u} = 0$, $\pi'_{,v_j} = 0$, and $\pi'_{,\lambda} = 0$, one finds

$$u = \frac{PL/AE}{1 + 1/\sqrt{2}}$$

15.26 $Ne - b = 8$; $Ne + b = 52$

15.41 The coefficient matrix is positive definite since all the leading principal minors are larger than zero (see Art. 12.1.20). If $\mathbf{x}^{(0)} = 0$, then

$$\mathbf{x}^{(1)T} = [1.200 \quad 2.886 \quad 2.671]$$
$$\mathbf{x}^{(2)T} = [1.243 \quad 2.129 \quad 2.916]$$
$$\mathbf{x}^{(3)T} = [1.043 \quad 2.030 \quad 2.983]$$
$$\mathbf{x}^{(4)T} = [1.009 \quad 2.006 \quad 2.997]$$
$$\mathbf{x}^{(5)T} = [1.002 \quad 2.001 \quad 2.999]$$
$$\mathbf{x}^{(6)T} = [1.000 \quad 2.000 \quad 3.000]$$

15.43 $\mathbf{U} = [\mathbf{u}_1 \quad \mathbf{u}_2 \quad \mathbf{u}_3 \quad \mathbf{u}_4]$, $\mathbf{u}_1^T = [4 \quad 3 \quad 1 \quad 0]$, $\mathbf{u}_2^T = [0 \quad 2 \quad 1 \quad -1]$, $\mathbf{u}_3^T = [0 \quad 0 \quad 3 \quad 1]$, $\mathbf{u}_4^T = [0 \quad 0 \quad 0 \quad 1]$; $\mathbf{y}^T = [13 \quad 3 \quad 13 \quad 4]$; $\mathbf{x}^T = [1 \quad 2 \quad 3 \quad 4]$

15.48 $\mathbf{v} = \mathbf{B}^{-1}\mathbf{i}_i$, where $\mathbf{B} = \text{diag}(1, \ldots, 1, 1/\sqrt{-1}, 1, \ldots, 1)\mathbf{U}$

15.50 Possible reactions: -100.25, 40.10, 35.12, 28.13. The largest of the other entries, that is, 0.5, is at least 50 times smaller in magnitude. Plausible solution if $\bar{\mathbf{K}}$ is not ill-conditioned.

15.51 For each of the Ne columns store one integer number (0 if the column is all zeros, the row number of the nonzero entry if the column is not all zeros).

15.53 $\dfrac{S_2}{S_1} = \dfrac{N_2}{N_1}\dfrac{w_2}{w_1} = (2)(1.5) = 3;$

$$\dfrac{T_2}{T_1} = \dfrac{N_2}{N_1}\left(\dfrac{w_2}{w_1}\right)^2 = 4.5$$

15.55 $\dfrac{S_2}{S_1} = \left(\dfrac{e_2}{e_1}\right)^2 = 4;\ \dfrac{T_2}{T_1} = \left(\dfrac{e_2}{e_1}\right)^3 = 8$

15.56 $N_1 = N_2 = 1{,}000;\ w_1 = 51,\ w_2 = 101;\ S_2/S_1 = w_2/w_1 \approx 2;\ T_2/T_1 = (w_2/w_1)^2 \approx 4$

15.58 If e is doubled, $S_2/S_1 = 4;\ T_2/T_1 = 8$; if e is tripled $S_2/S_1 = 9,\ T_2/T_1 = 27$

Chapter 16

16.1 $n^m = 4;\ e = 3;\ e' = 6;$ 6 by 6; 6 by 1; 12 by 6; 12 by 1; at most 924

16.2 $n^m = 2;\ e = 6;\ e' = 6;$ 6 by 6; 6 by 1; 12 by 6; 12 by 1; at most 924

16.9 1,200 by 1,350; 850 by 1,000; 150; 770,000 words

16.10 600 by 640; 500 by 540; 114,000 words

16.11 1,200 by 2,855; 1,100 by 2,755; 1,655; 395,500 words

16.20 $\mathbf{s}^T = [-1{,}376 \quad 1{,}856 \quad 10{,}144 \quad 5{,}088 \quad 5{,}568]P/9{,}408$

16.22 $\mathbf{s}^T = [136 \quad 8 \quad 64 \quad -120 \quad 24]AEu_0/147L$

16.24 $\mathbf{s}^T = [-0.063 \quad 0.453 \quad -0.044 \quad 0.108 \quad -0.059]P$

16.25 $\mathbf{s}^T = [0.187 \quad -0.483 \quad 0.558 \quad -1.542 \quad 0.745]AEu_0/L$

16.26 Let $\eta = (s + b - Ne)/s$, $\gamma = Ne/(Ne - b)$, $\beta =$ order of the flexibility matrix of a typical element and $\mathcal{T}_s$, $\mathcal{T}_{L1}$, $\mathcal{T}_{L2}$ denote the total number of multiplications for the computation of element forces in the substitution method and the first and the second Lagrange-multipliers methods. Then, one obtains

$$\mathcal{T}_s = \tfrac{1}{8}s^3[2 - 4\eta + (\beta + 2)\eta^2 - (\beta - \tfrac{2}{3})\eta^3] + \text{lower-order terms}$$
$$\mathcal{T}_{L1} = \tfrac{1}{8}s^3[1 + (2\gamma - 1)(1 - \eta)]^3 + \text{lower-order terms}$$
$$\mathcal{T}_{L2} = \tfrac{1}{8}s^3(2 - \eta)^3 + \text{lower-order terms}$$

For the structure of Prob. 16.9: $\mathcal{T}_{L1}/\mathcal{T}_s \approx 7.5$ and $\mathcal{T}_{L2}/\mathcal{T}_s \approx 2.9$

16.29 Let $\mathcal{S}_s$, $\mathcal{S}_{L1}$, and $\mathcal{S}_{L2}$ denote the storage area required by the substitution method and the first and the second Lagrange-multiplier methods. With the notation of the answer to Prob. 16.26: $\mathcal{S}_s = s^2(1 - \eta)^2$; $\mathcal{S}_{L1} = \tfrac{1}{2}s^2[1 + (2\gamma - 1)(1 - \eta)]^2$; $\mathcal{S}_{L2} = \tfrac{1}{2}s^2(2 - \eta)^2$. For the structure of Prob. 16.9: $\mathcal{S}_{L1}/\mathcal{S}_s \approx 4.5$ and $\mathcal{S}_{L2}/\mathcal{S}_s \approx 2.4$

16.33 $\mathbf{u}_{(1)}^T = [145 \quad 260 \quad 144]PL/588AE$

16.38 $\mathcal{S}/S = 21$

16.39 $\mathcal{T}/T = 105$

16.40 $s + b - Ne = (2)(135) + 30 - (100)(3) = 0;\ \mathcal{T}/T = 15$

Name Index

Akyuz, F. A., 564
Andrée, W. L., 351
Andrews, E. S., 268
Archimedes, 3
Argyris, J. H., 269, 380
Aristotle, 3

Baker, J. F., 630
Beedle, L. S., 630, 643
Bernoulli, John, 5, 215
Betti, E., 276
Biggs, J. M., 623
Bleich, F., 7
Borg, S. F., 343
Bresler, B., 630

Carnahan, B., 593
Castigliano, A., 266, 268, 304
Cilley, F. H., 648
Clapeyron, B. P. E., 309
Clough, R. W., 623
Coulomb, C. A., 5
Crandall, S. H., 73, 380
Cross, Hardy, 7, 328, 352

Dahl, N. C., 73
Darling, A. R., 2
da Vinci, L., 4
Drucker, D. C., 621
Duchemin, N. V., 167n.

Engesser, F., 379
Euler, L., 622

Feld, J., 16
Fife, W. M., 59, 116, 351
Flügge, W., 7
Föppl, A., 6
Föppl, L., 6
Fox, L., 564
Freudenthal, A. M., 12

Galilei, G., 4
Garrelts, J. M., 12
Garvey, G., 31
Gennaro, J. J., 343
Girkmann, K., 7, 618
Goodier, J. N., 6
Grinter, L. E., 343

Hall, A. S., 472
Hansen, R. J., 623
Henneberg, L., 121
Heyman, J., 630
Hildebrand, F. B., 447, 564
Hodge, P. G., Jr., 630
Holley, M. J., Jr., 623
Horne, M. R., 630
Hovey, O. E., 175
Hurty, W. C., 623

Subject Index